Lecture Notes in Applied and Computational Mechanics

Volume 32

Series Editors

Prof. Dr.-Ing. Friedrich Pfeiffer
Prof. Dr.-Ing. Peter Wriggers

Lecture Notes in Applied and Computational Mechanics

Edited by F. Pfeiffer and P. Wriggers

Further volumes of this series found on our homepage: springer.com

Vol. 32: Bardzokas, D.I.; Filshtinsky, M.L.; Filshtinsky, L.A. (Eds.)
Mathematical Methods
in Electro-Magneto-Elasticity
530 p. 2007 [978-3-540-71030-1]

Vol. 31: Lehmann, L. (Ed.)
Wave Propagation in Infinite Domains
186 p. 2007 [978-3-540-71108-7]

Vol. 30: Stupkiewicz, S. (Ed.)
Micromechanics of Contact and Interphase Layers
206 p. 2006 [978-3-540-49716-5]

Vol. 29: Schanz, M.; Steinbach, O. (Eds.)
Boundary Element Analysis
571 p. 2006 [978-3-540-47465-4]

Vol. 28: Helmig, R.; Mielke, A.; Wohlmuth, B.I. (Eds.)
Multifield Problems in Solid and Fluid Mechanics
571 p. 2006 [978-3-540-34959-4]

Vol. 27: Wriggers P., Nackenhorst U. (Eds.)
Analysis and Simulation of Contact Problems
395 p. 2006 [978-3-540-31760-9]

Vol. 26: Nowacki, J.P.
Static and Dynamic Coupled Fields in Bodies
with Piezoeffects or Polarization Gradient
209 p. 2006 [978-3-540-31668-8]

Vol. 25: Chen C.-N.
Discrete Element Analysis Methods
of Generic Differential Quadratures
282 p. 2006 [978-3-540-28947-0]

Vol. 24: Schenk, C.A., Schu°ller. G
Uncertainty Assessment of Large
Finite Element Systems
165 p. 2006 [978-3-540-25343-3]

Vol. 23: Frmond M., Maceri F. (Eds.)
Mechanical Modelling and Computational Issues
in Civil Engineering
400 p. 2005 [978-3-540-25567-3]

Vol. 22: Chang C.H.
Mechanics of Elastic Structures with Inclined Members:
Analysis of Vibration, Buckling and Bending of X-Braced
Frames and Conical Shells
190 p. 2004 [978-3-540-24384-7]

Vol. 21: Hinkelmann R.
Efficient Numerical Methods and Information-Processing
Techniques for Modeling Hydro- and Environmental
Systems
305 p. 2005 [978-3-540-24146-1]

Vol. 20: Zohdi T.I., Wriggers P.
Introduction to Computational Micromechanics
196 p. 2005 [978-3-540-22820-2]

Vol. 19: McCallen R., Browand F., Ross J. (Eds.)
The Aerodynamics of Heavy Vehicles:
Trucks, Buses, and Trains
567 p. 2004 [978-3-540-22088-6]

Vol. 18: Leine, R.I., Nijmeijer, H.
Dynamics and Bifurcations
of Non-Smooth Mechanical Systems
236 p. 2004 [978-3-540-21987-3]

Vol. 17: Hurtado, J.E.
Structural Reliability: Statistical Learning Perspectives
257 p. 2004 [978-3-540-21963-7]

Vol. 16: Kienzler R., Altenbach H., Ott I. (Eds.)
Theories of Plates and Shells:
Critical Review and New Applications
238 p. 2004 [978-3-540-20997-3]

Vol. 15: Dyszlewicz, J.
Micropolar Theory of Elasticity
356 p. 2004 [978-3-540-41835-1]

Vol. 14: Frmond M., Maceri F. (Eds.)
Novel Approaches in Civil Engineering
400 p. 2003 [978-3-540-41836-8]

Vol. 13: Kolymbas D. (Eds.)
Advanced Mathematical and Computational
Geomechanics
315 p. 2003 [978-3-540-40547-4]

Vol. 12: Wendland W., Efendiev M. (Eds.)
Analysis and Simulation of Multifield Problems
381 p. 2003 [978-3-540-00696-1]

Vol. 11: Hutter K., Kirchner N. (Eds.)
Dynamic Response of Granular and Porous Materials
under Large and Catastrophic Deformations
426 p. 2003 [978-3-540-00849-1]

Vol. 10: Hutter K., Baaser H. (Eds.)
Deformation and Failure in Metallic Materials
409 p. 2003 [978-3-540-00848-4]

Vol. 9: Skrzypek J., Ganczarski A.W. (Eds.)
Anisotropic Behaviour of Damaged Materials
366 p. 2003 [978-3-540-00437-0]

Vol. 8: Kowalski, S.J.
Thermomechanics of Drying Processes
365 p. 2003 [978-3-540-00412-7]

Vol. 7: Shlyannikov, V.N.
Elastic-Plastic Mixed-Mode Fracture Criteria
and Parameters
246 p. 2002 [978-3-540-44316-2]

Vol. 6: Popp K., Schiehlen W. (Eds.)
System Dynamics and Long-Term Behaviour
of Railway Vehicles, Track and Subgrade
488 p. 2002 [978-3-540-43892-2]

Mathematical Methods in Electro-Magneto-Elasticity

Demosthenis I. Bardzokas · Michael L. Filshtinsky ·
Leonid A. Filshtinsky

Springer

Authors
D.I. Bardzokas
National Technical University of Athens
School of Applied Mathematics
and Physical Sciences
Department of Mechanics Laboratory
of Testing and Materials
Athens, Hellas, Greece

L.A. Filshtinsky
Dept. of Mathematical Physics Sumy
State University
Sumy, Ukraina

M.L. Filshtinsky†

Library of Congress Control Number: 2007923590

ISSN 1613-7736

ISBN 978-3-540-71030-1 Springer Berlin Heidelberg New York

Springer is a part of Springer Science+Business Media
springer.com

Cover design: Künkel & Lopka, Heidelberg
Typesetting by the authors
Production: Integra Software Services Pvt. Ltd., Pondicherry, India

Printed on acid-free paper SPIN: 11944416 45/3100/Integra 5 4 3 2 1 0

This book is dedicated to Michael Filshtinsky who had always a vision to create. This creation was expressed both in science and music.

Preface

The science of mechanics of coupled fields is a dynamic research area, studying in a unified way continuum mechanics, heat transfer phenomena and electromagnetism, i.e. branches of science that are usually studied separately.

For a more rigorous and accurate description of the influence of static and dynamic loadings, high temperatures and strong electromagnetic fields on elastic media and constructions we need a single approach, i.e. one that has the potential to integrate all the above mentioned physical fields. The investigation of the mechanics of stressed-deformed media subject to coupled fields is one of the most important tasks in today's material science with major technological importance. Our scope is the rigorous qualitative and quantitative analysis of the stressed-deformed states of materials and constructions and their physical and mechanical properties within the phenomenological theory of electro-magneto-thermo elasticity. For that reason we derive the constitutive equations that govern the behavior of the problem under study, we pose the appropriate boundary conditions and finally we present and apply the appropriate mathematical methods for their solution.

General relations of the mechanics of deformed bodies interchanging with electromagnetic field are used and special attention is paid to the construction of the theory of shells and plates made from piezoelectric materials. We present in detail art of the state mathematical methods for the solution of a broad class of two-dimensional problem of electroelasticity for multiconnected bodies. We also examine several static and dynamic problems for piecewise-homogeneous piezoelectric plates, weakened by cracks and openings. We mostly consider questions connected with the application of the method of boundary integral equations for the investigation of the problem of deformation of electroelastic waves on heterogeneities of various types. Special emphasis is given in the definition of strength and breakdown characteristics of bodies with defects. We state and solve some inverse problems of electroelasticity about optimal, in some sense, equations by parameters of strength and breakdown.

The interplay of coupled physical fields in anisotropic media introduces additional difficulties into the analysis of boundary problems of electro-magneto-elasticity. The boundary conditions are not usually separated, which leads to the necessity of consideration of compound boundary problems of mathematical physics.

We believe that this monograph provides the theoretical background and the tools for the systematic analysis of electromegneto-elastic media with a broad spectrum of applications.

This book is the condensation of a long scientific collaboration of the authors despite the fact that Dr. M. L. Filshtinsky, passed away unexpectedly 3 years ago. Michael L. Filshitinsky was born in 1961 in the town of Novosibirsk. In 1989 he received his PhD in Mathematical Physics at the Mechanic-Mathematical Faculty of the State University of Moscow (Lomonosov). His Thesis was entitled: "Solution of two-dimensional dynamical problems of elasticity and electroelasticity for cracked bodies". The most characteristic aspect of the work of M.L. Filshtinsky was the combination of his natural talent and his hard work, for solving a vast number of novel and real-world problems. Another interesting component of his personality was his interest in music composition. M.L. Filshtinsky and us have published together many articles in international scientific journals and congress proceedings as well as 5 monographs in Russian language [[190, 254]–[256]]

The present work is a small component of the knowledge of the great Soviet School in the area of Coupled Fields Mechanics and is part of an effort to bring out the rich Russian literature in the area.

At this point Prof. D.I. Bardzokas would also like to thank, with all his heart, his associate, friend and student George I. Sfyris as well as his colleague and good friend Constantinos I. Siettos for the help they provided in writing this monograph. Without their valuable help the publication of this work would be impossible.

This book is for the specialists in Continuous Mechanics, Acoustics and Defectoscopy, and also for advanced undergraduate and graduate –level students in Applied Mathematics, Physics, Engineering Mechanics and Physical Sciences.

D.I. Bardzokas
Professor NTUA, Section of Mechanics, Greece

L.A. Filshtinsky
Professor of Sumy State Univerisity, Ukraine

Contents

Introduction

For the last ten years the section of the mechanics of the deformed solid body, that is usually called electroelasticity, is intensely developed due to rapid development of piezoceramic transformers in ultrasonics, radioelectronics, measuring and computer engineering. This direction of science, based on the use of definite physical properties of natural crystals as well as artificial ceramics (that have undergo certain treatment), studies the mechanics of coupled electric and mechanical fields in the corresponding elements of constructions.

After the discovery of a piezoeffect by Jack and Pier Curie and the classical treatise by W. Voigt [250] the theory of electroelasticity was systematically developed in works [25, 75, 122, 125, 135, 161], and [228]. Further development of the mechanics of coupled electrical and elastic fields in piezoelectric media is connected with the statement of boundary problems (of electroelasticity) and elaboration of the methods for their solution that are contained in works [6, 11, 12, 21, 23, 26, 27, 29, 39, 55, 59, 83, 85, 86, 92, 93, 94, 95, 96, 101, 102, 110, 115, 139, 142, 149, 152, 156, 163, 169, 170, 171, 190, 192, 199, 204, 207, 209, 222, 229, 230, 231, 232, 234, 235, 236, 241, 242, 254, 255, 256, 257, 258, 259, 260]. The fact that the linear equations of state for preliminarily polarized piezoceramics similar form with the corresponding relations for crystals with hexagonal symmetry allows us to formalize the statement of fundamental boundary problems in the same framework. Over the past years a large body of work, in the field of statistic and dynamic problems of electroelasticity are gathered in the literature. The results are the outcome of the work of scientists of various disciplines.

The coupling of mechanical with electrical fields and anisotropy causes additional difficulties into the analysis of the boundary problems of electroelasticity. The order of the differential equations for the field quantities increases, and the fact that the mechanical and electric boundary conditions are not usually separable forces us to consider the complex boundary problems of the mathematical physics.

The two-dimensional static problems of electroelasticity for plane and antiplane deformation of bodies are much easier to analyze, as well as the

problem of bending a plate. The use of complex variables for the two-dimensional
problems of electroelasticity are reduces them to the corresponding boundary problems of the theory of analytical functions. The mechanical and electric field quantities are expressed for piezoceramics, for example, by three analytical functions with complex variables. Such variants of the complex representations are suggested by various authors in works [40, 59, 86]. Based on their results we considered many statical problems for piecewise-homogeneous piezoceramic bodies. So, in [85] on the basis of methods, developed in [86] the stress concentration was studied on the contour of an elliptic hole in a halfplane under the influence of a point electric charge on its boundary. The boundary problem of fracture mechanics for a linear crack located on the boundary between a piezoelectric and a conductive medium is solved in [93]. In monograph [59] fundamental static problems of electroelasticity for a piezoceramic plane and halfplane are examined, and also Green's functions for an infinite plate weakened by a linear crack or a rigid linear inclusion are constructed; the problem of averaging the electroelastic properties of piecewise-homogeneous piezoceramic structures is also considered using the method of integral equations.

While considering the boundary problems of electroelasticity for bodies with cracks, which in the non-deformed state are associated with mathematical cuts, the most important factor is the correct statement of the electric boundary conditions on the surfaces of the cracks. This problem is discussed in reports [152, 242].

The study of coupled mechanical and electric fields in composite plates consisting of two heterogeneous piezoceramic halfplanes, when on the common boundary a linear crack exists is of major importance. Green's function of the corresponding two-dimensional problems of electroelasticity for the case, when an crack exists on the interphase boundary, is constructed in [178]. It was observed that, similarly with the classical elasticity theory, in the vicinity of the crack tip appears a power singularity [157, 200]. Review of the investigations of interphase cracks are given in [67], a set of results in this field is obtained in [76, 90, 93, 157].

An energy criterion of fracture for a piezoelectric body with a crack is constructed in [93, 94]. The criterion of fracture for an electroelastic body initiated by the concentration of the electric field on the edges of the electrodes is suggested in [14]. On the basis of electromechanic analogy, relations of fracture mechanics with respect to the electric breakdown of a dielectric are used. They obtained the relation from which it follows that in case of electromechanic fracture of an electroelastic dielectric the surface energy consists of mechanical and electric components, each of which contains a criterion of fracture of the elastic medium and a criterion of an electric breakdown for a dielectric.

Devices using the Rayleigh surface acoustic waves (SAW) have practical application in various fields of acoustoelectronics; in order to excite these waves we must place a finite number of electrodes on the crystal surface, the location of which and their potential may vary widely. In [12] a general method of calculations for the devices is presented, based on the use of Green's function, the construction of which, for coupled problems of electroelasticity, is a separate and complicated task. Similar methods were used in works [26, 27, 54, 55, 56], and some additional assumptions were also introduced, which allowed for simplification of the obtained integral equations. In [68] the problem of excitation of Rayleigh SAW by the finite system of electrodes is solved based on the assumption that on the electrodes the charge densities are prescribed, which are determined by the solution of the electrostatic problem for the case of a periodical system of electrodes. Review of the solution methods for the problems of SAW excitation in piezoelectrics is given in [220].

Assuming that electrodes are weightless and have negligible rigidity, we considered many static and dynamic boundary problems of electroelasticity for piezoelectrics with surface electrodes. The cases when the position of the electrodes on the surface of the body has either periodical (infinite number of electrodes) or symmetric character were studied, which allowed us to use the effective, in such cases, methods of series and integral transformation to solve the problem. If we need to investigate an electroelastic field in a body, excited by an arbitrary system of electrodes, the usage of the above mentioned methods has difficulties of mathematical character. However, such investigations are of importance for practical applications. The approach to the investigation of two-dimensional physical fields in a piezoceramic cylinder excited by an arbitrary system of electrodes (the approach is based on the method of integral equations) is given in [175].

Many actual scientific and technical problems are connected with the investigation of the process of wave propagation of distribution of the waves in piezoelectric media and the determination of dynamic stress intensity factor near the heterogeneities of various types. Solution of the appearing in this case complex problems requires the use of modern mathematical means and, in particular, the methods and approaches of the dynamic theory of elasticity. The development of these methods is reflected in monographs reviews and reports, which appeared the last decades [3, 4, 5, 6, 8, 9, 10, 30, 34, 38, 44, 45, 46, 49, 61, 62, 63, 64, 65, 68, 79, 80, 81, 82, 87, 97, 98, 99, 100, 105, 112, 113, 114, 116, 117, 119, 123, 134, 136, 137, 138, 143, 144, 147, 150, 151, 153, 154, 158, 159, 160, 162, 164, 173, 194, 195, 196, 201, 237, 238, 241, 247, 251, 252].

It is necessary to describe the work of a piezotransducer, which exists in an elastic medium, when analyzing the dynamic stress intensity factor of a piecewise-homogeneous piezoceramic body. Consequently, it is of great importance to study the problems of diffraction of electroelastic waves on heterogeneities such as cracks, openings or heterogeneous inclusions. In the vicinity of such heterogeneities large gradients of mechanical and electric field quantities exist, which usually cause concentration of mechanical stresses and an increased value of the electric field. The effect of coupling of electroacoustical fields appears more clearly in zones near the vicinity of the peak of excitation frequency.

From the theoretical point of view the crack may be modeled as a mathematical cut on the sides of which a displacement jump appears. If the displacements jumps are not known beforehand, an accurate mathematical description of the wave field, appearing because of the existence of crack, is extremely difficult.

Most investigations which can be found in the scientific literature refer to problems of diffraction of elastic waves on linear finite or half-infinite cuts. However, in reality a crack may differ from a linear one, and as the investigations show, the curvature of the defect may substantially influence the behavior of the stress intensity factor.

The characteristics of the fracture also depend on the proximity of the body boundary to the fracture source, because besides the running wave, the source is affected by a reflected, from the body boundary, wave. The wave process in the bounded medium is characterized by reciprocal transformation of the waves due to the boundary type. Therefore, it is necessary to have algorithms allowing evaluation of the influence not only of the curvature of the defect but also of the body boundary on the stress intensity factors.

A larger group of diffraction problems of the classical elasticity and electroelasticity is considered for steady and harmonic waves. Here, owing to the separation of exponential multiplier $e^{-i\omega t}$ the time is excluded and this allows us to solve the problem only with respect to the unknown complex amplitudes. But even in this case the solution of the dynamic problems requires the usage of coarse mathematics and the known analytical solutions in the literature are rather few. Therefore it is actual, while investigating the diffraction and propagation of waves in piecewise-homogeneous piezoelectric bodies not only to obtain the analytical mathematical solution but also to develop methods permitting investigation of the dynamic stress intensity factor of the body in the vicinity of a heterogeneity, as well as influence of the coupling of physical fields, and the characteristics of the field in a distance from the singularity.

The investigation of the diffraction of elastic waves on the cuts was initiated by the work of Sommerfeld [69] where the optical problem of wave diffraction on a screen was studied (mathematically equivalent to the stationary dynamic problem of the theory of elasticity of a half-infinite crack of a longitudinal shear). Maue [226, 227] gave the solution of this problem for a case of a plane deformation. The oscillations of the plane with a half-infinite cut were also investigated in [261, 262]. The diffraction problems of electromagnetic waves were studied by many authors, for example, in [31, 72, 127, 128, 129, 130, 131, 197, 198].

The influence of dynamic effects on the increase of hazard of brittle fracture may be evaluated by an example of the problem of interaction of elastic waves with cracks-cuts of finite sizes. In this case, as the investigations show, the angular distribution of the stress field in the vicinity of the tip coincides with the corresponding distribution at static loading. This stress field is singular, and the main term in the vicinity of the tip has the form $Kf(\theta)/\sqrt{2\pi r}$, where r is the distance to the tip of the crack, and K is the corresponding *stress intensity factor* (SIF). Quantity K under the dynamic loading depends on the time. If the body is under the conditions of a stationary wave process (harmonic oscillations), the amplitude of factor K depends on circular frequency ω, and there is a large frequency interval where the amplitude may considerably exceed its static analogue [196, 238].

When solving the diffraction problems for canonical regions we usually use the method of separation of variables. References concerning these results may be found in [30, 62, 63, 115]. The development of this method, connected with the passing to multicoupled regions, with variation of the surface form, is given in works [64, 65].

In reports [1, 2, 110] the problems of diffraction of elastic waves of shear and compression are considered on a circular piezoelectric cylinder, and also the diffraction of a shear wave on a piezo-semiconductive cylinder with hexagonal class of symmetry $6mm$. The problem of diffraction of a shear wave on a linear cut in an infinite piezoelectric medium with hexagonal symmetry is studied in [139]. As well as in the analogous problem of the theory of elasticity [32], the solution is constructed in an elliptic system of coordinates using Mathieu functions. As the result of the solution of the problem formulas for the stress intensity factors at the tips of the cuts are obtained.

In monograph [115] the applied theories of deformation of piezoceramic plates and shells are constructed, the correct statement of the boundary problems of electroelasticity is discussed, the energetic theory of electromechanical transformation is developed, and the solutions of a series of boundary space problems for canonical bodies are obtained. A lot of works of this cycle are devoted to the investigation of wave processes in piezoceramic bodies, plates and shells at different directions of the preliminary polarization. Various results in the field of the problem of oscillation of piezoceramic bodies are presented in [199].

The appearance of powerful computers allowed passing from formal mathematical solutions of boundary problems to constructive procedures of bringing the solution to final results. As a rule, these methods represent a combination of analytical approaches (integral transformations, separation of variables, integral representations, etc.), which give possibility to bring the problem to integral equations and numerical methods for their approximate solution. In [238] the problem of dispersion of P – and SV – elastic waves is considered, using Fourier integral transform over the transverse to the front crack coordinate. Similar procedures are used in work [241]. The stationary dynamic problems for cuts crossing the boundary of the half-space or approaching it are considered in reports [202, 203, 220, 241].

Solving the direct problems of diffraction of elastic waves on curvilinear cuts there specific difficulties appear. At the same time, the investigation of such problems are also actual because the characteristics of fracture considerably depends on the configuration of the defects, and also due to the necessity of solution of inverse problems, permitting us to determine the changes of the geometric parameters of the defect as a function of the characteristics of the remote field. The problem of identification of the defects was considered, for example, in [203].

Development of the methods of the potentials theory in the problems of the mathematical physics, electrodynamics and elasticity brought to the design of a rather powerful and efficient approach usually called the method of *boundary integral equations* (MBI). It is based on the representation of the inner field quantities by integrals over the boundary areas which are the convolutions of some "densities" with definite differential operators that are functions of the fundamental solutions of the corresponding equations. The densities in the integral representations of the solutions make sense only in the boundary area, and therefore the problem of their determination has one less dimension than the initial boundary problem. Substitution of the limiting values of the solutions, into the boundary conditions brings to the boundary integral equations with respect to the sought-for densities. The latter are solved numerically.

The important fact, when considering the problems of diffraction of elastic waves on the linear heterogenities of crack or rigid inclusion types, is the construction of the integral representations for the solutions, in a way that they should provide the existence of the jump of the displacement vector (or stress) on the front of the defect. Solving the diffraction problem under more complex boundary conditions these representations should have the necessary completeness.

On the interphase (or on the line of the crack) certain components of the coupled fields have discontinuities, while others cross it continuously. Therefore the integral representations of the solutions of the boundary problem should be correct, i.e. they can satisfy these conditions of coupling and also definite behaviour of the solutions at infinity.

There are several similar approaches to the construction of integral representations of discontinuous solutions of the equations of elasticity and of equations and systems of elliptic type in general. They are: the procedure of the distortion method, potential representations, representations of solutions in the form of generalized integrals of Cauchy type, etc. More usual in these types of problems is the method of the theory of generalized functions.

The mathematical problems of the boundary integral equations method for the solution of the stationary dynamic problem of the theory of elasticity is developed in the works of V.D. Kupradze and his colleagues [97, 98, 99, 100]. Some useful modifications of the method of the boundary integral equations are given in the works [33, 88].

Recently there has been rapid development in the field of magnetoelasticity; its connected with the fact that the theory is applicable in various fields such as geophysics (propagation of elastic waves) under the influence of magnetic field, acoustics (fast damping of acoustic waves when being inside a strong magnetic field), technology and embiomechanics. We see that the problem of the interaction of the electromagnetic fields on conductive media that are deformed due to the action of electromagnetic forces has been under systematic research for the last 30 years.

Up to date we may point out some evolvement in the field of magnetoelasticity such as:

a) The deformable body is under the influence of a strong and static magnetic field; the interaction of the stress field with the electromagnetic field is done through Lorentz forces. These forces are included in the motion equations and the generalized Ohm's Law for the particle that moves within the magnetic field. The state equations are described through a non-linear system of coupled equations of electrodynamics and the movement of the elastic medium. This system may be linearized under the assumption that the disturbed electromagnetic field that is produced due to the mechanical action can be ignored. This model of magnetoelasticity does not consider the magnetic or electrical polarization. The before mentioned process is effective for the formulation of the magnetoelasticity problems for thin shells and plates under the influence of steady magnetic field.
b) The magnetoelastic interaction of ferromagnetic media is studied taking into account the magnetization that appears due to the presence of an external magnetic field. We have to comment here that in the study of the interaction between elastic and magnetic field on ferromagnetic media results to ana extremely difficult problem of linearization of the state equations, for soft ferromagnetics. These materials have broad application in modern technology, thus the study of such phenomena is of great theoretical and practical importance.
c) Solving problems of magnetoelasticity for the body with cracks that are under the influence of an electromagnetic field. The study of the interaction between the electromagnetic field with the elastic cracked body depends on the properties of the medium. For a conductive elastic body, that can be magnetized and is under the influence of a stationary magnetic field, the study of stress-strain state at the tip of the crack is possible through the linear equations of magnetoelasticity.

One axis of this monograph is the treatment of a series of actual problems of magnetoelasticity for multiply connected bodies.

The problems of optimal (in any sense) control of the characteristics of strength and fracture of piecewise-homogeneous bodies, where physical fields of various type are coupled with each other, are a new class of problems appearing in the process of development of modern technologies. In many cases the parameters regulating the fracture of a piezoelement are in the form of functionals that are determined by the solutions of integral equations for the corresponding primal problem, fact that considerably simplifies the solution of inverse (optimizing) problems. The methods of the theory of optimal control of the systems with concentrated and distributed parameters are considered, for example, in [223, 36, and 107]. The mathematical aspects of the problem of moments was considered in [239], and in particular the ℓ-problem of moments was developed in [89], while solutions of optimal control problems were studied in [107].

In [60] some optimal problems of the theory of elasticity are stated as well as the methods of bringing the given problems to the general problems of moments in the space of continuous functions. Using the approximation of non-standard moment functions by polynomials the general problems of moments are reduced to the classical exponential problems of moments or ℓ-problems of moments [7, 89]. Some problems connected with the control of SIF on the boundary of a piezoceramic halfspace with cracks (and also with their movement) in elastic plates, are considered in [174, 179]. According to the character of the optimal problem, the control should be chosen from a certain functional class.

The literature on diffraction of electroelastic waves on heterogeneous types of cracks, openings, inclusions in piezoelectric media is rather scanty. This is explained by the complexity of the dynamic problem of electroelasticity and mathematical difficulties arising during the process for their solution.

The present monograph, which to some extent fills in the indicated gaps in the scientific literature, presents an approach, based on to the solution of two-dimensional static and dynamic problems of electroelasticity and to the construction on that bases structural procedures for investigation of coupled electric fields in piecewise-homogeneous piezoceramic bodies.

The developed approach contains the following principal parts: the construction of the fundamental solutions of the corresponding equations of electroelasticity, the derivation of the integral representations of the boundary problems solutions; the reduction of the latter to the system of singular or regular integral equations; and an approximate numerical solution of the obtained equations.

In some particular cases, from the constructed analytical algorithms follow the solutions of the corresponding boundary problems of the theory of elasticity for piecewise-homogeneous bodies.

1

Physical Fields in Solid Bodies

1.1 Heat Field. Heat Conduction Equation in Solid Bodies

1.1.1 Heat Field

One of the central properties of physical bodies is temperature, a concept closely related to that of heat equilibrium. Our ability to quantitatively measure the temperature using a thermometer is based on the establishement of an equilibrium state in the bodies of reference. If such a equilibrium has not been established then the temperature cannot be defined for a body of finite size. However, the mathematical simulation of real physical processes with the aid of modern thermodynamics, is based on the concept of finite body V; the later is defined as a collection of material particles (assempled by material points $P = (x_1, x_2, x_3)$) of negligible representive volume ΔV), that during a small interval of time Δt their state is considered to be in equilibrium having a definite value of temperature given by

$$T = T(P, t) = T(x_1, x_2, x_3, t) \tag{1.1}$$

The above scalar function is used to determine the temperature field in the volume V.

1.1.2 Equilibrium Heat Equation

The transfer of heat within an irregularly heated solid body from one part of the body to another is carried out by heat conduction. For its quantitative description we use the *vector of heat flow* $\vec{q} = \vec{q}(P, t)$, by which we may express, heat quantity ΔQ_1 that flows during time Δt through an element $\Delta\Sigma$ lying on the body surface (Fig. 1.1) as

$$\Delta Q_1 = q^{(n)} \Delta\Sigma\Delta t = \vec{q} \cdot \vec{n} \Delta\Sigma\Delta t \tag{1.2}$$

where $q^{(n)}$ is the normal component of the heat flow along the normal to the surface $\overrightarrow{n}$ (Fig. 1.1).

Consider now an arbitrary volume V within the body bounded by the closed surface Σ. With the help of relation (1.2) we may express Q_1-the heat flowing into volume V through surface Σ during time Δt as

$$Q_1 = -\oint\!\!\!\oint_{\Sigma} \overrightarrow{q} \cdot \overrightarrow{n} \Delta t d\Sigma \tag{1.3}$$

The minus sign reflects the fact that $\overrightarrow{n}$ is pointing outwards the surface.

If there is a heat sourse inside volume V (e.g. due to the interaction of an electromagnetic field with the medium), Q_2 -the heat quantity released in volume V per time Δt, is given by

$$Q_2 = \iiint_V W \Delta t dV \tag{1.4}$$

$W = W(P, t)$ denotes the heat quantity released per unit time and volume.

The sum of Q_1, Q_2 i.e., $Q_1 + Q_2$ in V actually causes a change in the temperature:

Accuming the specific heat capacity and density being functions of pressure, i.e. $c = c(P)$ and $\rho = \rho(P)$, the heat ΔQ_3 which changes the temperature of an element of ΔV volume for a infitesimely $\Delta T \approx \frac{\partial T}{\partial t} \Delta t$ reads:

$$\Delta Q_3 = c\rho \frac{\partial T}{\partial t} \Delta t,$$

The above expression can be also written in the integral form as

$$Q_3 = \iiint_V c\rho \frac{\partial T}{\partial t} \Delta t dV \tag{1.5}$$

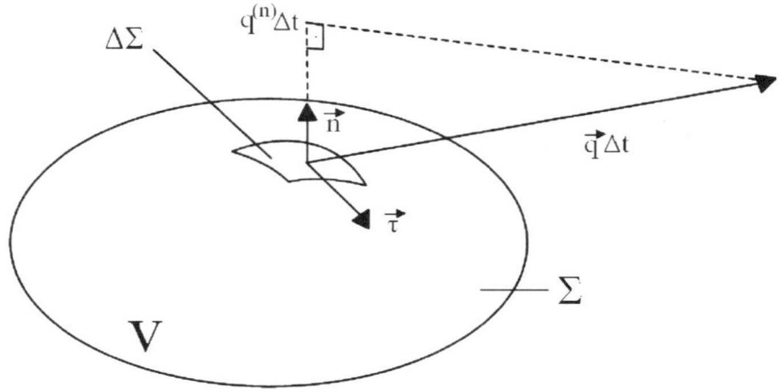

Fig. 1.1. Calculation of the heat flow passing through surface Σ of an unevenly heated body

Applying the Gauss-Ostogradsky theorem to (1.3) we get

$$Q_1 = -\iiint\limits_V div\vec{q}\,\Delta t dV \tag{1.6}$$

Using (1.4)–(1.6) the heat balance $Q_3 = Q_1 + Q_2$, now reads

$$\iiint\limits_V \left(c\rho\frac{\partial T}{\partial t} + div\vec{q} - W \right)\Delta t dV = 0 \tag{1.7}$$

But, since V is an arbitrary chosen volume (1.7) is satisfied when

$$c\rho\frac{\partial T}{\partial t} = -div\vec{q} + W, \tag{1.8}$$

The above equation is the differential form of the heat balance.

1.1.3 Heat Conduction Equation

For an isotropic body the heat conduction is determined by

$$\vec{q} = -\lambda gradT \tag{1.9}$$

(known as *Fourier's law of heat conduction*)

In the above equation, $\lambda = \lambda(P) > 0$ is the *heat conduction coefficient* and the minus sign reflects the fact, that at a point P the heat flows in the direction of decreasing temperature (what exactly the vector grad T indicates).

Combining relations (1.8) and (1.9) we obtain the *heat conduction equation for a heterogeneous isotropic body*

$$c\rho\frac{\partial T}{\partial t} = div\left(\lambda gradT\right) + W \tag{1.10}$$

If the body is homogeneous, then c, ρ and λ are constants and (1.10) reads

$$\frac{\partial T}{\partial t} = a^2\Delta T + w\left(P, t\right) \tag{1.11}$$

where $a = \sqrt{\dfrac{\lambda}{c\rho}}$, $w\left(P, t\right) = \frac{1}{c\rho}W\left(P, t\right)$, and Δ is the Laplase operator $\left(\Delta = \partial^2/(\partial x_i \partial x_i)\right)$.

The parameter

$$k = \frac{\lambda}{c\rho} \tag{1.12}$$

is the so-called *temperature conductivity coefficient.*

For anisotropic bodies, such as crystals, the flow direction of heat $\vec{q}$, in thegeneral case, does not coincide with the temperature gradient. Hence, instead of formula (1.9) a more general relation is the:

$$\overrightarrow{q} = -\boldsymbol{\lambda} gradT \tag{1.13}$$

or in a suffix notation:

$$q_i = -\lambda_{ij} T_{,j} \tag{1.14}$$

$\boldsymbol{\lambda}$ or λ_{ij} is called the *thermal conductivity tensor* (a second rank tensor) *heat conduction tensor* (or *tensor of the heat conduction*). Each of tensor components λ_{ij} has a definite physical meaning. For example, if the unit gradient of temperature points in the direction of Ox_2 axis ($gradT = \{0, 1, 0\}$), the heat flow $\overrightarrow{q} = \{-\lambda_{12}, -\lambda_{22}, -\lambda_{32}\}$ has a component $-\lambda_{22}$ along that axis, and also two transverse components, namely $-\lambda_{12}$ along Ox_1 axis and λ_{32} along Ox_3 axis.

In most of the cases, due to properties associated with statistical physics the *conductivity tensor* is symmetric, i.e.

$$\lambda_{ij} = \lambda_{ji}. \tag{1.15}$$

An asymmetry of tensor λ_{ij} would signify that the heat from a point source would flow on a plate along untwisting spirals. During last century scientists intensively tried to discover the spiral propagation of the heat, but all these attempts gave negative results.

Equally with tensor λ_{ij} we often use tensor

$$k_{ij} = \frac{\lambda_{ij}}{\rho c}, \tag{1.16}$$

this is called the *temperature conduction tensor.*

In the principal axes coordinate system the only non-zero components are the diagonal $\lambda_i = \lambda_{ii}$ $(i = 1, 2, 3)$ (no summation over i here). In this coordinate system the law of heat conduction (1.14) is given by

$$\overrightarrow{q} = \{-\lambda_1 T_{,1}; -\lambda_2 T_{,2}; -\lambda_3 T_{,3}\} \tag{1.17}$$

and the heat flows in the direction of the principal axes. The directions of the principal axes of a crystal are associated with its symmetry.

By substituting (1.14) into (1.10) the equation of heat conduction for homogeneous crystals reads

$$\frac{\partial T}{\partial t} = k_{ij} T_{,ij} + w\,(P, t) \tag{1.18}$$

1.1.4 Boundary Conditions

To solve the above set of (1.10), (1.11) or (1.18) we need to specify the temperature field inside the body at an initial time $t = t_0$

$$T = T_0\,(P, t_0) = T_0(x_1, x_2, x_3, t_0) \tag{1.19}$$

(the *initial condition*) and the heat flowing at the boundary Σ (*boundary or edge conditions*).

Depending on the problem formulation one can impose different kind of boundary conditions: if, for example, over the surface of the body flows a liquid of known temperature T(P,t) then the BC may read:

$$T(P,t) = T_*(P,t) \qquad (P \in \Sigma) \tag{1.20}$$

(these are boundary conditions of the first kind).

Another type of BC are the so-called *boundary conditions of the second kind* where, for example, the value of the normal component of outer main flow $\overrightarrow{q}_*$ is given (case $q_*^{(n)} = 0$ is called the *condition of heat insulation*). In that case the B.C. read:

$$q^{(n)}(P,t) = -q_*^{(n)}(P,t) \qquad (P \in \Sigma, n \text{ is the external normal}) \tag{1.21}$$

Other conditions of heat exchange could also be imposed. For example, when there is convective or radiant heat exchange on the surface, in the balance of heat flows on the boundary of the body (1.21) we will have terms, depending on the temperature values of the body and environment. Here we should note that in the general case, where the edge conditions of different kind are prescribed on various parts of the boundary, we have to solve the mixed problem of heat conduction.

Equations of heat conduction are of parabolic type, the characteristic properties of which are the facts, that they predict an infinite speed of propagation of the disturbances. Theoretically that means that at an abrupt change in the temperature at any point will cause the immediate change in the temperature at every point. Immediate changes in temperature at a distance from the heating point (sourse) are practically insignificant a fact which is in agreement with the established theory.

1.2 Electromagnetic Fields. Maxwell's Equations

1.2.1 The Laws of Electrodynamics in Integral Forms

Maxwell's equations describe the behaviour of electromagnetic fields as they interact with matter. They represent the principal laws of electromagnetism and one can say that they play a similar role as Newton's laws in mechanics.

The first pair of Maxwell's equations contains the equation of Faraday's law of electromagnetic induction and the equation of Gauss law for closed force lines of an electromagnetic field.

Faraday's Law

Faraday's Law reads:

$$\oint_L \overrightarrow{E} \cdot d\overrightarrow{r} = -\iint_S \frac{\partial \overrightarrow{B}}{\partial t} \cdot \overrightarrow{n}\, dS \tag{1.22}$$

The above equation links the velocity of changes of the magnetic induction flow $\vec{B}$ through surface S with an electromotive force, which is expressed as the circulation of electric field $\vec{E}$ along a closed loop contour $L, +$ and the boundary surface S (Fig. 1.2). The direction of tracing the contour L in formula (1.22) is positive as described in Fig. 1.2.

Gauss's law
Gauss's Law reads:

$$\oiint_{\Sigma} \vec{B} \cdot \vec{n}\, d\Sigma = 0 \tag{1.23}$$

The above equation expresses the equality to zero of the vector of magnetic induction $\vec{B}$ in a volume V bounded by closed surface Σ (Fig. 1.3). This law demonstates the absence of a magnetic field when magnetic charges or currents that act as sources of a field don't exist. The lines of forces of the magnetic field due to (1.23) are closed or infinite (i.e. are closed across an infinitely distant point of space).

The second pair of Maxwell's equations contains the equation suggested by Maxwell, given by:

$$\oint_L \vec{H} d\vec{r} = \iint_S \left(\vec{j} + \frac{\partial \vec{D}}{\partial t} \right) \cdot \vec{n}\, dS, \tag{1.24}$$

The above equation connects the flow of conduction current density vector $\vec{j}$ and the displacement current $\frac{\partial \vec{D}}{\partial t}$ across surface S with the integral of vector $\vec{H}$ on the closed contour L (Fig. 1.2).

Besides (1.24) to the second pair of Maxwell's equations we refer the equation of the theorem of Gauss for an electric field inside a material

$$\oiint_{\Sigma} \vec{D} \cdot \vec{n}\, d\Sigma = \iiint_V \rho_e dV \tag{1.25}$$

This equation connects the flow of the electric displacement vector $\vec{D}$ from closed surface Σ with the enclosed (by the surface) free electric charge (Fig. 1.3), while ρ_e is the volume density of free electric charges. Equation (1.25) shows that the field lines of vector $\vec{D}$ may start and end on electric charges.

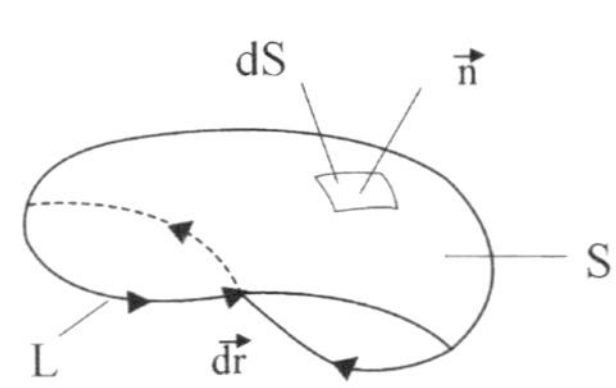

Fig. 1.2. About choice of a positive direction on the contour of integration L

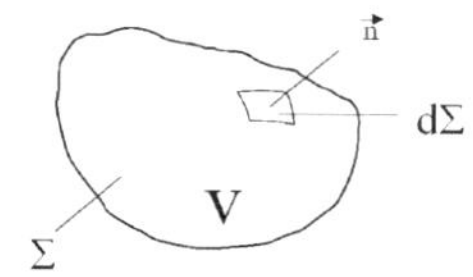

Fig. 1.3. Calculation of the vector flow from arean V bounded by closed surface Σ

1.2.2 Maxwell's Equations in Differential Forms

Using Gauss-Ostrogradsky theorem, Maxwell's (1.22)–(1.25) can be also expressed in *in differential form*, connecting the vectors $\overrightarrow{E}$, $\overrightarrow{D}$, $\overrightarrow{H}$, $\overrightarrow{B}$ and their time derivatives at any point in space.

Applying Stock's theorem to (1.22) we obtain, for arbitrary surface S with unity normal $\overrightarrow{n}$, that

$$\iint\limits_S rot\overrightarrow{E} \cdot \overrightarrow{n} dS = -\iint\limits_S \frac{\partial \overrightarrow{B}}{\partial t} \cdot \overrightarrow{n} dS$$

or,

$$\iint\limits_S \left(rot\overrightarrow{E} + \frac{\partial \overrightarrow{B}}{\partial t} \right) \cdot \overrightarrow{n} dS = 0$$

The equality is valid only when the integrand function is equal to zero at any point of space. The second Maxwell's equation of Maxwell in differential form reads:

$$rot\overrightarrow{E} = -\frac{\partial \overrightarrow{B}}{\partial t}$$

Applying Stocks theorem to (1.24) we obtain by analogy that

$$rot\overrightarrow{H} = \overrightarrow{j} + \frac{\partial \overrightarrow{D}}{\partial t}$$

Using Gauss-Ostrogradsky theorem in (1.25) we transform the surface integral into a volumic one:

$$\iiint\limits_V div\overrightarrow{D} dV = \iiint\limits_V p_e dV.$$

For an arbitrary volume V, the above equality is satisfied only if the integrand functions at every point in space are equal:

$$div\overrightarrow{D} = \rho_e$$

Making use of the Gauss-Ostrogradsky theorem in (1.23) we get:

$$div\vec{B} = 0$$

Hence, the differential form of Maxwell's equations is given by the following set of equations:

First pair of equations:

$$rot\vec{E} = -\frac{\partial\vec{B}}{\partial t}, \tag{1.26}$$

$$div\vec{B} = 0, \tag{1.27}$$

Second pair of equations:

$$rot\vec{H} = \vec{j} + \frac{\partial\vec{D}}{\partial t}, \tag{1.28}$$

$$div\vec{D} = \rho_e, \tag{1.29}$$

If we multiply with del (1.28) and take into account (1.29) together with identity

$$div[rot\vec{H}] = 0$$

as a consequence of the second pair of Maxwell's equation we will come to the known equation of continuity of the electric charge

$$div\vec{j} + \frac{\partial\rho_e}{\partial t} = 0 \tag{1.30}$$

The system of Maxwell's equations is completed by relations determining the *vector of electric polarization* $\vec{P}$ and *vector of magnetization* $\vec{M}$. For a solid body we have

$$\vec{D} = \ni_0 \vec{E} + \vec{P} \tag{1.31}$$

$$\vec{B} = \mu_0\left(\vec{H} + \vec{M}\right) \tag{1.32}$$

here $\ni_0$ is the electric constant $\left(\ni_0 = 8,854 \cdot 10^{-12} F/m\right)$, and μ_0 is the magnetic constant $\left(\mu_0 = 12,566 \cdot 10^{-7} H/m\right)$.

In physics it is acceptable to consider vector $\vec{H}$ as an axial vector, determining its direction to the direction of the polar vector of current $\vec{j}$ according to the right-hand screw rule. With this choice Maxwell's equations are satisfied if we get the vectors $\vec{H}$, $\vec{B}$, $\vec{M}$ to be axial, and the vectors $\vec{E}$, $\vec{D}$, $\vec{P}$, $\vec{j}$ to be polar. (At this point we should mention that Maxwell's equations could be satisfied by the opposite method of determination of the vectors.

1.2.3 Magnetization

In many isotropic (not ferromagnetic) media, magnetization $\overrightarrow{M}$ is directly proportional to the field strength $\overrightarrow{H}$:

$$\overrightarrow{M} = \mu_0 \psi \overrightarrow{H} \tag{1.33}$$

where ψ is a dimensionless constant called the *magnetic susceptibility.* If $\psi > 0$, the medium is called *paramagnetic*; if $\psi < 0$ the medium is *diamagnetic.*

Combining (1.32) and (1.33) we obtain

$$\overrightarrow{B} = \mu_0 (1 + \psi) \overrightarrow{H} = \mu_0 \mu \overrightarrow{H} = \mu_a \overrightarrow{H} \tag{1.34}$$

where, the dimensional constant μ_a is the *absolute magnetic permeability* of the medium, and the dimensionless constant $\mu = 1 + \psi$ is the *relative magnetic permeability.*

In the general case of anisotropic crystals, the vectors $\overrightarrow{B}$, $\overrightarrow{H}$ and $\overrightarrow{M}$ are not parallel, and are defined by the followin relations that are analogous with (1.33) and (1.34),

$$M_i = \mu_0 \psi_{ij} H_j \tag{1.35}$$

$$B_i = \mu_{ij} H_i \tag{1.36}$$

In the above equations ψ_{ij} is the *tensor of magnetic susceptibility* and $\mu_{ij} = \mu_0(\delta_{ij} + \psi_{ij})$ is the *tensor of magnetic permeability.* The tensor $M_{ij} = \delta_{ij} + \psi_{ij}$, which is the *tensor of relative magnetic permeability of a crystal* is also commonly used.

From the expression giving the *energy density of a magnetic field*:

$$w_m = \frac{1}{2} \overrightarrow{B} \cdot \overrightarrow{H} = \frac{1}{2} \mu_{ij} H_i H_j \tag{1.37}$$

it follows that the tensors of the magnetic permeability μ_{ij} and M_{ij} respectively, as well that of magnetic susceptibility are symmetric. These may be transformed to main axes, and if the field $\overrightarrow{H}$ is directed along one of them, then the vectors $\overrightarrow{M}$ and $\overrightarrow{B}$ are also directed along the same axis. The magnetic susceptibility of the crystal is completely defined by the principal values of the susceptibilities ψ_1, ψ_2, ψ_3 along the main axes.

A crystal is called *paramagnetic* or *diamagnetic* along a given main axis, if the principal value of susceptibility along it is positive or negative, respectively. Here we should also note that some crystals are paramagnetic along one axis and diamagnetic along another.

As the principal susceptibilities of paramagnetic and diamagnetic crystals are usually less than a unity $\left(\sim 10^{-5}\right)$, the field created by magnetized crystals is smaller in comparison with the outer field, and it may be neglected. The situation is quite different in the case of *ferromagnetics*, where the quantity of magnetic permeability is very high (e.g. $\mu = 5000$ for iron). But in ferromagnetics the magnetic permeability is not constant and strongly dependent on the amplitude and the direction of the magnetic field $\overrightarrow{H}$.

1.2.4 Electric Polarization

For isotropic media the connection between vectors of electric field $\overrightarrow{E}$, polarization $\overrightarrow{P}$ and electric displacement $\overrightarrow{D}$ are analogous to the connection between vectors $\overrightarrow{H}$, $\overrightarrow{M}$ and $\overrightarrow{B}$ as follows:

$$\overrightarrow{P} =\ni_0 \chi \overrightarrow{E} \tag{1.38}$$

$$\overrightarrow{D} =\ni_0 (1+\chi) \overrightarrow{E} = \kappa \overrightarrow{E} =\ni_0 K \overrightarrow{E} \tag{1.39}$$

where χ is the *electric susceptibility*, κ is the *dielectric permeability* and $K = 1+\chi$ is the *dielectric constant*. For anisotropic crystals instead of (1.38) and (1.39) the following set is used:

$$P_i =\ni_0 \chi_{ij} E_j, \tag{1.40}$$

where χ_{ij} is the *tensor of dielectric susceptibility*, and

$$D_i = \kappa_{ij} E_j, \tag{1.41}$$

where $\kappa_{ij} =\ni_0 \ (\delta_{ij} + \chi_{ij})$ is the tensor of *dielectric permeability*.

In terms of the expression of the *energy density of the electric field*

$$w_e = \frac{1}{2} \overrightarrow{D} \cdot \overrightarrow{E} = \frac{1}{2} k_{ij} E_i E_j \tag{1.42}$$

we derive the symmetry of tensors k_{ij} and χ_{ij} and thus, we can map them to the main axes.

Polarization and magnetization phenomena have an obvious analogy in their description. The electric field is created by the availability of electric charges, and the magnetic field by their movement. We will not examine practically basic effects such as the depolarization of a dielectric body. Here we should note that unlike the magnetic susceptibilities, the principal dielectric susceptibilities are always positive (the above are true under the assumption that we don't have *depolarization* of the electric field in a dielectric, and leakage of charges due to the presence of conduction). Besides, their values are larger than unity (for example, for chloride of sodium they are quantities of order 5,6.

Dielectric and magnetic constants of the crystal are subjected to several constraints imposed by the symmetry of the crystal. This problem will be considered in the next chapter which is dedicated to crystal physics.

1.2.5 The Balance Equation of the Electromagnetic Field's Energy. Umov-Poynting Vector

Consider a region V bounded by a closed surface Σ. Let the electromagnetic field described by Maxwell's (1.26)–(1.29). In the general case electric currents flow in the body, and due to Joule's law, the heat energy with specific power that is being released is given by:

$$w_J = \vec{j} \cdot \vec{E} \tag{1.43}$$

Let us now multiply (1.28) by $\vec{E}$, and (1.26) by $\vec{H}$, subtract the second equation from the first and integrate the result over volume V, taking as result:

$$\iiint\limits_V \left(\vec{E} \cdot rot\vec{H} - \vec{H} \cdot rot\vec{E}\right) dV = \iiint\limits_V \left(\vec{E} \cdot \frac{\partial \vec{D}}{\partial t} + \vec{H} \cdot \frac{\partial \vec{B}}{\partial t} + \vec{j} \cdot \vec{E}\right) dV$$

In order to transform the obtained expressions we use the known identity from the vector analysis

$$\vec{E} \cdot rot\vec{H} - \vec{H} \cdot rot\vec{E} = -div\left(\vec{E} \times \vec{H}\right) \tag{1.44}$$

(the reader should try to derive the above expression as a homework), and then apply the Gauss-Ostrogradsky theorem:

$$\iiint\limits_V div\left(\vec{E} \times \vec{H}\right) dV = \oiint\limits_\Sigma \left(\vec{E} \times \vec{H}\right) \cdot \vec{n}\, d\Sigma \tag{1.45}$$

where $\vec{n}$ is the unit normal vector to the surface.

By applying the above transformations, we get:

$$-\oiint\limits_\Sigma \vec{S} \cdot \vec{n}\, d\Sigma = \iiint\limits_V \left(\vec{E} \cdot \frac{\partial \vec{D}}{\partial t} + \vec{H} \cdot \frac{\partial \vec{B}}{\partial t} + \vec{j} \cdot \vec{E}\right) dV, \tag{1.46}$$

where we have introduced the vector:

$$\vec{S} = \vec{E} \times \vec{H} \tag{1.47}$$

which is called *Umov-Poynting* vector.

Equation (1.46), which is direct mathematical consequence of Maxwell's equations, may be interpreted as a balance equation of the electromagnetic energy. Indeed, the first two summands of the first side give the velocity of changes of the energy density of the magnetic field at a given point within volume V:

$$\vec{E} \cdot \frac{\partial \vec{D}}{\partial t} + \vec{H} \cdot \frac{\partial \vec{B}}{\partial t} = \frac{\partial}{\partial t}\left(w_e + w_M\right) \tag{1.48}$$

(which is not difficult to derive directly with the help of expressions (1.37) and (1.42)). The last term is the density of Joule heat (1.43), i.e. the rate at which heat energy is supplied to a particle. The expression on the left side of (1.46) is the flow of Umov-Pointing vector, describing the inflow of the electromagnetic energy across surface Σ of the body from the surrounding. A more complete equation of energy balance which takes into account the deformation of the body will be obtained in the later sections of the chapter.

1.2.6 Vector and Scalar Potentials of Electromagnetic Field

For the solution of problems concerning the determination of electromagnetic fields it is convenient to introduce potentials into the system of Maxwell's equations. From (1.27) it follows that the solenoid vector of the magnetic induction may be represented as a rotor of some vector $\overrightarrow{C}$

$$\overrightarrow{B} = rot\overrightarrow{C} \tag{1.49}$$

However, relation (1.49) does not simply determine vector $\overrightarrow{C}$ over vector $\overrightarrow{B}$. As for any scalar function ψ we have the identity: rot grad $\psi = 0$, (may be easily checked), we may determine the *vector potential* $\overrightarrow{A}$ as

$$\overrightarrow{B} = rot\overrightarrow{A}, \qquad \overrightarrow{A} = \overrightarrow{C} - grad\psi \tag{1.50}$$

By substituting (1.49) and (1.50) into (1.26) we get:

$$rot\left(\overrightarrow{E} + \frac{\partial\overrightarrow{A}}{\partial t}\right) = 0. \tag{1.51}$$

Any irrotational vector field (1.51) can be described as a function of a potential, i.e. there is available *scalar potential* φ, where

$$\overrightarrow{E} + \frac{\partial\overrightarrow{A}}{\partial t} = -grad\varphi. \tag{1.52}$$

With the help of (1.50) and (1.52) the system of Maxwell's equations can be writte as follows:

$$rot\overrightarrow{H} = \overrightarrow{j} + \frac{\partial\overrightarrow{D}}{\partial t} \tag{1.53}$$

$$div\overrightarrow{D} = \rho_e \tag{1.54}$$

$$\overrightarrow{B} = rot\overrightarrow{A} \tag{1.55}$$

$$\overrightarrow{E} = grad\varphi - \frac{\partial\overrightarrow{A}}{\partial t} \tag{1.56}$$

For the calculation of the vector potential $\overrightarrow{A}$ and the scalar potential φ we need additional conditions which are called *calibration conditions*. It is assumed that:

$$div\overrightarrow{A} = 0, \tag{1.57}$$

A condition called *Coulomb calibration*. It is also assumed that the potentials at infinity become zero.

For piezoelectrics-non-conductive non-magnetized and electrically neutral bodies the following relations hold:

$$\overrightarrow{j} = 0, \quad \overrightarrow{M} = 0, \quad \rho_e = 0$$

1.2.7 Boundary Conditions

To complete the system of Maxwell's equations and the defining relations it is necessary to formulate the conditions on the boundary of two different media. We will deduce these conditions using Maxwell's equations in the integral formulation given by (1.22)–(1.25).

Consider the elements just below and above the boundary at a distance of $\pm\frac{1}{2}h$ at the normal (to the boundary) direction (Fig. 1.4). We will assume that normal $\overrightarrow{n}$ is directed from medium "–" to medium "+", and that means

$$\overrightarrow{n}^{+} = \overrightarrow{n} = -\overrightarrow{n}^{-} \tag{1.58}$$

In (1.23) let us consider the limiting transition at $h \to 0$ while making the assumption that $\overrightarrow{B}$ is infinite. The integrals along the side surfaces of the layer disappear and (1.23) turns into

$$\iint\limits_{S} \overrightarrow{B^{+}} \cdot \overrightarrow{n^{+}} dS + \iint\limits_{S} \overrightarrow{B^{-}} \cdot \overrightarrow{n^{-}} dS = 0 \tag{1.59}$$

where B^{+} and B^{-} designate the limiting values of field $\overrightarrow{B}$ at approximation to the interphase from the corresponding side. Due to the arbitrariness of S we will obtain the sought for boundary condition from (1.59) taking into account relation (1.58)

$$\left(\overrightarrow{B^{+}} - \overrightarrow{B^{-}}\right) \cdot \overrightarrow{n} = 0 \quad \text{or} \quad B_n^{+} = B_n^{-}, \tag{1.60}$$

designating the continuation of the normal component of vector of magnetic induction $\overrightarrow{B}$ on the interphase. The limiting transition condition in (1.25) is analogous for the electric displacement $\overrightarrow{D}$:

$$\left(\overrightarrow{D}^{+} - \overrightarrow{D}^{-}\right) \cdot \overrightarrow{n} = 0 \quad \text{or} \quad D_n^{+} = D_n^{-}, \tag{1.61}$$

under the assumptions that the volumic integral from ρ_e at limiting transition disappears, the density of free charges ρ_e is bounded (and the body is electrically neutral, $\rho_e = 0$).

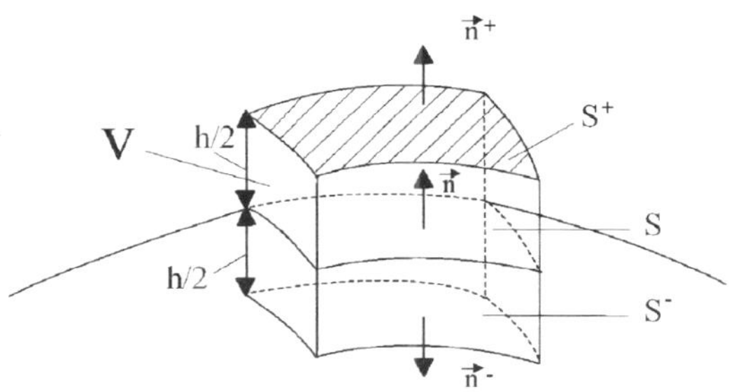

Fig. 1.4. The scheme of integration during the deducing the boundary conditions with the help of the theorem Gauss-Ostrogradsky

In the case when additional (*outside*) charges are applied to the material electric neutrality is then disrupted. For example, if a thin surface electrode exists on the interphase, we introduce the concept of surface density of charges determined with the help of relation

$$\lim_{h\to 0}\iiint_V \rho dV = \iint_S \sigma dS \qquad (1.62)$$

When $h \to 0$ (1.25) gives the surface integral of relation (1.62). If we use the plus and minus notation (1.25) will result to

$$\iint_S \overrightarrow{D}^+ \cdot \overrightarrow{n}^+ dS + \iint_S \overrightarrow{D}^- \cdot \overrightarrow{n}^- dS = \iint_S \sigma dS$$

For an electroded surface, due to the arbitraty choise in$S$, the following boundary condition applies

$$D_n^+ - D_n^- = \sigma \qquad (1.63)$$

Now lets condider a surface, S crossing the interphase of media (see Fig. 1.5).

The other boundary conditions can be derived as follows from Maxwell's (1.22) and (1.24), if we consider the surface that may be obtained by parallel transition of an element of arc AB lying on the interphase on $\pm\frac{1}{2}h$ along the normal to this surface.

For $h \to 0$ the surface integrals on the right-hand side of (1.22) and (1.24), disappears if the integrand functions are bounded and if they are not connected to specific surface currents. Besides, the linear integrals disappear on the regions which are perpendicular to the surface interphase and as a result we get the following relations:

$$\int_{AB} \overrightarrow{E}^+ \cdot d\overrightarrow{r} + \int_{BA} \overrightarrow{E}^- \cdot d\overrightarrow{r} = 0$$

$$\int_{AB} \overrightarrow{H}^+ \cdot d\overrightarrow{r} + \int_{BA} \overrightarrow{H}^- \cdot d\overrightarrow{r} = 0$$

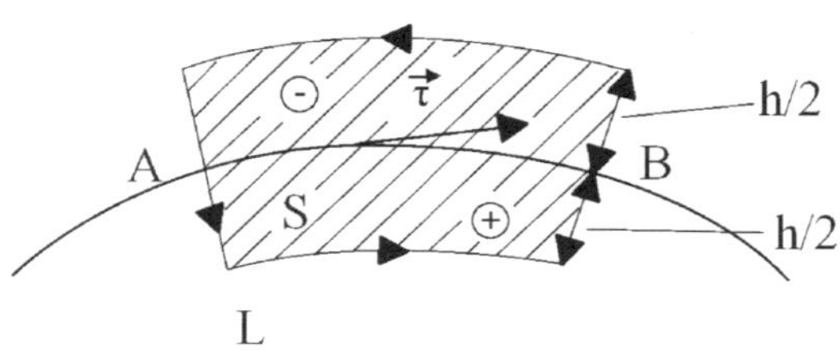

Fig. 1.5. The scheme of integration during the deducing of the boundary conditions with the help of the theorem of Stokes

Changing the direction of integration in the integrals over BA and taking into account the arbitrary choice of AB, we conclude that at every point on the surface, the tangential components of vectors $\overrightarrow{E}$ and $\overrightarrow{H}$ (along unit vector $\overrightarrow{\tau}$ which is tangential to interphase) remain continuous, i.e.:

$$\left(\overrightarrow{E}^{+} - \overrightarrow{E}^{-}\right) \cdot \overrightarrow{\tau} = 0 \quad \text{or} \quad E_{\tau}^{+} = E_{\tau}^{-} \tag{1.64}$$

$$\left(\overrightarrow{H}^{+} - \overrightarrow{H}^{-}\right) \cdot \overrightarrow{\tau} = 0 \quad \text{or} \quad H_{\tau}^{+} = H_{\tau}^{-} \tag{1.65}$$

On the outer boundary of the body bounded by vacuum or air we require the same conditions (1.60), 1.61), (1.61), (1.64), (1.65) and in the general case we construct solutions of the problems of electrodynamics for the body and for surrounding connected with these conditions. In the particular case of piezoelectric media, the dielectric permeability of which is of order $O(10^2)$ larger than that of the air, we may use the approximate condition

$$\overrightarrow{D} \cdot \overrightarrow{n} = 0 \quad \text{or} \quad D_n = 0 \tag{1.66}$$

on the body surface.

1.2.8 Stationary Electromagnetic Field. Electrostatic Approximation

In stationary electromagnetic problems Maxwell's equations are time independent (no time derivatives) and the equations reduce to the following system of equations (equations of electrostatics)

$$rot\,\overrightarrow{E} = 0 \tag{1.67}$$

$$div\,\overrightarrow{D} = \rho_e \tag{1.68}$$

while the second pair now reads

$$rot\,\overrightarrow{H} = \overrightarrow{j} \tag{1.69}$$

$$div\,\overrightarrow{B} = 0 \tag{1.70}$$

The above equations are used to calculate the stationary magnetic field of stationary currents, if any.

From (1.56) it follows, that the electrostatic field is determined by a scalar potential φ by

$$\overrightarrow{E} = -grad\varphi, \quad E_i = -\varphi_{,i} \tag{1.71}$$

Here (1.71) replaces (1.67) in the set of the electrostatic equations. Potential φ should be sought for from the equation obtained by substitution (1.71) and determinant (1.41) into (1.68)

$$\left(\kappa_{ij}\varphi_{,j}\right)_{,i} + \rho_e = 0. \tag{1.72}$$

This equation is solved together with boundary conditions (1.61), (1.63), (1.64). The last condition, taking into account (1.71) is reduced to the equality

$$\frac{\partial \varphi^+}{\partial \tau} = \frac{\partial \varphi^-}{\partial \tau}$$

which after integration we get the continuation condition of the potential at the interphase (1.73):

$$\varphi^+ = \varphi^- \tag{1.73}$$

(here, the constant of the integration is omitted as its value does not influence the values of electric field $\overrightarrow{E}$).

In conductors the strength of the electric field is equal to zero, which means that on the boundary between a dielectric and a conductor the next relations hold:

$$E_\tau = 0, \quad D_n = \sigma \tag{1.74}$$

With the help of (1.71), (1.73) the first of these conditions is reduced to the condition of constant potential φ

$$\varphi = V = const, \tag{1.75}$$

while the second condition can be used to find the distribution of charges on the electrode.

Concluding, we should mention that during the design of piezoelectronics devices we have to model the coupled elastic electromagnetic waves, i.e. the elastic waves during their interaction with electric fields and electromagnetic waves accompanied by the deformation of media. Due to the fact that the value of the velocity of the elastic waves is smaller, by a factor of $O(10^5)$, than the propagation velocity of electromagnetic waves, it is possible to neglect the changes in the magnetic field during the investigation of the elastic waves. Hence, in most problems in the design of piezotransducers and the propagation of electroacoustic waves magnetic effects are neglected, while we use the quasistatic approximation for an electric field, i.e. the electrostatic (1.67), (1.68) and (1.71).

1.3 Stresses and Deformations. Hooke's Law

1.3.1 State of Stress

The basic forces acting on body are the volumetric (body) forces and the surface ones. To *volumetric forces* belong, for example, the gravitational, electromagnetic, and inertial forces. These forces are distributed throughout the volume and to assign them we use vector $\overrightarrow{F}$ relating to the volume unit so that the force affecting volume dV is equal to $\overrightarrow{F} = \overrightarrow{F}dV$.

Surface forces occur when two or more bodies are in contact. There two kinds of surface forces: the outer surface forces resulting from the surrounding bodies and the inner surface forces resulting from the interaction of various parts within the body.

Lets us consider an infinitesimal tetrahedron having its vertex at a given point P and lateral basis located along the axes of coordinates (Fig. 1.6). Let the influence of the surrounding portions of the body on the inclined faces Q_1, Q_2, Q_3 with area dS and unity outer normal $\overrightarrow{n}$ is equivalent to surface force $\overrightarrow{T}_n dS$, vector $\overrightarrow{T}_n$ is called *inner stress*. The surface forces acting on the lateral faces of the tetrahedron may be designated as $\overrightarrow{T}_1 dS_1$, $\overrightarrow{T}_2 dS_2$, $\overrightarrow{T} dS_2$, $\overrightarrow{T}_3 dS_3$. Here index i denotes the axis perpendicular to a face, and dS_i denote the face areas given by the projection of basis dS, i.e. $dS_i = n_i dS_i$.

Employing force balance on the the tetrahedron we get the following equation:

$$\overrightarrow{T}_n dS - \overrightarrow{T}_i dS_i - \overrightarrow{F} dV = \left(\overrightarrow{T}_n - \overrightarrow{T}_i n_i\right) dS - \overrightarrow{F} dV = 0 \tag{1.76}$$

During the contraction of the tetrahedron to point P the resultant of mass forces will tend to zero faster than the resultant of the surface forces, and therefore from (1.76) it follows that the value in brackets should be equal to zero and therefore

$$\overrightarrow{T}_n = \overrightarrow{T}_i n_i \tag{1.77}$$

The coordinates of vectors T_i are called stress components and are given by

$$\overrightarrow{T}_i = \sigma_{ij} \overrightarrow{e}_j \quad (e_j, j = 1, 2, 3 \text{ are the unit vectors of the reference system}) \tag{1.78}$$

The stresses σ_{11}, σ_{22}, σ_{33} are called normal stresses and they act along the coordinate axes perpendicular to the faces of an elementary tetrahedron.

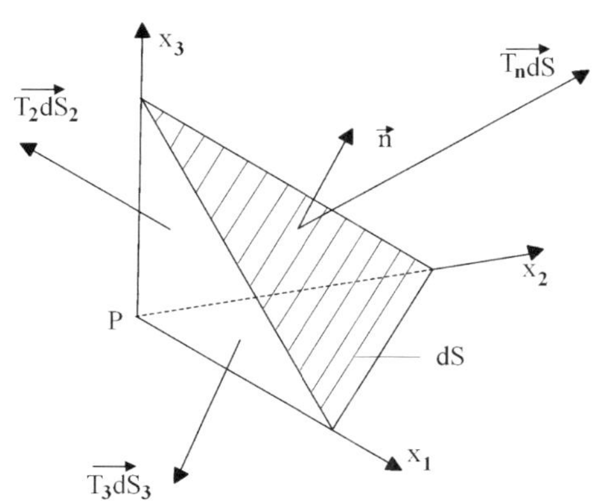

Fig. 1.6. Equilibrium of an elementary tetrahedron

The stress components T_{12}, T_{21}, T_{13}, T_{31}, T_{23}, T_{32} are directed along the tangentials and are called *tangential stresses* (shear stresses). The signs of the stress components shown in Fig. 1.7 are taken to be positive according to definition (1.78). From condition (1.78) it also follows that σ_{ij} forms a second rank tensor (as its convolution with vectors $\vec{e}$ is a vector).

1.3.2 Equations of Equilibrium and Motion. Symmetry of Stress Tensors

Let us set up the equations of equilibrium for a finite volume V bounded by a closed surface Σ. Every volume element dV is affected by the body force $F_j dV \vec{e}_j$, and every element, dS, on its surface is affected by the surface force $\sigma_{ij} n_i \vec{e}_j dS$. Equlibrium requires that the sum of all the acting forces to be zero, i.e.:

$$\oiint_{\Sigma} \sigma_{ij} n_i d\Sigma + \iiint_{V} F_j dV = 0 \tag{1.79}$$

Using the theorem of Ostrogradsky-Gauss we obtain

$$\iiint_{V} (\sigma_{ij,i} + F_j) dV = 0. \tag{1.80}$$

Now let us consider the moments of forces acting on the body. The body force $\vec{F} dV$ creates a moment $\vec{r} \times \vec{F} dV = \varepsilon_{ikj} x_k F_j dV \vec{e}_i$ which is applied at the point with radius-vector $\vec{r} = x_k \vec{e}_k$. The surface force acting on an element dS creates a moment $\vec{r} \times \vec{T}_n dS = \varepsilon_{ikj} x_k \sigma_{\ell j} n_l \vec{e}_i$. Again, equilibrium requires the sum of moments of all the forces acting on the the body to be zero:

$$\oiint_{\Sigma} \varepsilon_{ikj} x_k \sigma_{\ell j} n_l d\Sigma + \iiint_{V} \varepsilon_{ikj} x_k F_j dV = 0 \tag{1.81}$$

Applying Gauss- Ostogradsky theorem we get

$$\begin{aligned} \oiint_{\Sigma} \varepsilon_{ikj} x_k \sigma_{\ell j} n_l d\Sigma &= \iiint_{V} \left(\varepsilon_{ikj} x_k \sigma_{\ell j}\right)_{,l} dV \\ &= \iiint_{V} \left(\varepsilon_{ikj} x_{k,l} \sigma_{\ell j} + \varepsilon_{ikj} x_k \sigma_{lj,l}\right) dV \end{aligned}$$

Using equalities $x_{k,\ell} = \delta_{k\ell}$ and $\varepsilon_{ikj} \delta_{k\ell} \sigma_{\ell j} = \varepsilon_{ikj} \sigma_{kj}$, (1.81) now reads:

$$\iiint_{V} \varepsilon_{ikj} \left[\sigma_{kj} + x_k \left(\sigma_{\ell j,\ell} + F_j\right)\right] dV = 0 \tag{1.82}$$

Assuming continuity of the the integrated functions in (1.80) and (1.82) ar it can be easily shown that, due to the arbitray choice of V:

$$\sigma_{ij,i} + F_i = 0 \tag{1.83}$$

and

$$\varepsilon_{ikj}\sigma_{kj} = 0 \tag{1.84}$$

Equations (1.83) represent the *equations of equilibrium* of a body which is under the influence of body forces. Conditions (1.84) are reduced to the requirement of the symmetry of stress tensor (the law of *twoness of tangential stresses*)

$$\sigma_{ij} = \sigma_{ji} \tag{1.85}$$

Relations (1.85) are deduced by assuming that the interactions of the portions of the body are only due to surface forces. In the case where surface moments exist, the stress tensor becomed asymmetric. Such situation appears when we consider the interaction of a material medium with an electromagnetic field. In such interaction they form a single object and during the set up of the mathematical model we have to determine what should be related to the medium and what to the field. In practice, although there are known models with asymmetric stress tensor, approximation (1.85) is quite common and we will also use it in what follows. The symmetrical stress tensor may be transformed to the main axes, i.e. three axes perpendicular to each other, where the stress tensor contains only three non-zero components $\sigma_1 = \sigma_{11}$, $\sigma_2 = \sigma_{22}$, $\sigma_3 = \sigma_{33}$, (*the main stresses*) and the stress state is reduced to the deduction of these three components, which are perpendicular to each other, while the tangential stresses on the faces of the cube (Fig. 1.7) will disappear. Here we should note that the ledt-hand side of equilibrium (1.83) expresses the resultant of outer forces applied to the material particle of the medium. Therefore, according to Newton's second law the *equation of motion* reads:

$$\sigma_{ij,i} + F_j = \rho\ddot{u}_i \tag{1.86}$$

$\ddot{u}_i = \frac{\partial^2 u_i}{\partial t^2}(\overrightarrow{u} = u_i\overrightarrow{e}_i$ displacement) is the particle acceleration and ρ its density.

Summarizing we should underline that the tensor, discussed here, differs from the tensor of the second rank which we considered earlier, such as the

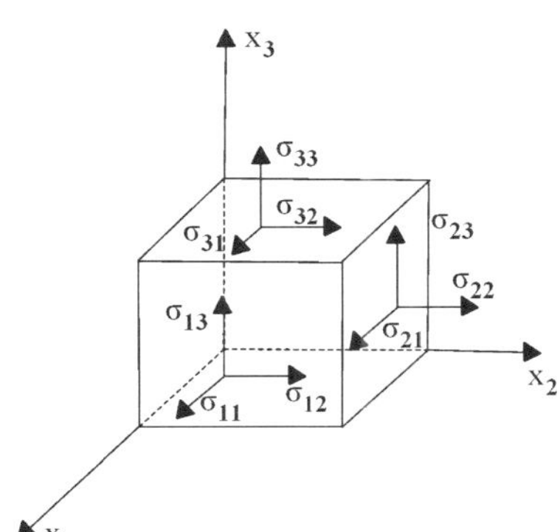

Fig. 1.7. About the determination of the components of a stress tensor

heat conduction tensor or the tensor of dielectric permeability. These tensors express the properties of a crystal, have a definite orientation in the crystal and do not match with the symmetry of the crystal (this will be discussed later). These tensors are called *material tensors.* On the contrary, the stress tensor like the deformation tensor, which will be studied in the next sections, may have any orientation in a crystal. In a sense it is similar to an electric field, which may have any direction. Such tensors are called field tensors.

1.3.3 Deformed State

Let us consider the deformation of a body, where a point P in a Cartesian system of coordinates (x_1, x_2, x_3) undergoes an displacement getting to point P' with coordinates (x'_1, x'_2, x'_3). $\overrightarrow{u} = \overrightarrow{PP'}$ denotes the *displacement vector*,

$$u_i = x'_i - x_i. \tag{1.87}$$

We define the *relative deformation* of an element ds, which after deformation equals to ds', as

$$e = \frac{ds' - ds}{ds} \tag{1.88}$$

Let us express it by the displacement vector u_i:

$$\begin{aligned} ds^2 &= dx_i dx_i, \\ ds'^2 &= dx'_i dx'_i = (dx_i + du_i)(dx_i + du_i) = ds^2 + 2dx_i du_i + du_i du_i, \\ du_i &= u_{i,j} dx_i, \\ dx_i du_i &= u_{i,j} dx_i dx_j, \\ du_i du_i &= u_{k,i} u_{k,j} dx_i dx_j, \end{aligned} \tag{1.89}$$

where

$$ds'^2 - ds^2 = (2u_{i,j} + u_{k,i}u_{k,j})\, dx_i dx_j$$

The expression in the brackets represents the asymmetric-in the general case-tensor, which, as it was shown in the preceeding chapter, may be given in the form of the sum of the *symmetric tensor of deformation* ε_{ij}:

$$\varepsilon_{ij} = \frac{1}{2}(u_{i,j} + u_{j,i} + u_{k,i}u_{k,j}) \tag{1.90}$$

And the *asymmetric tensor of rotation*:

$$\omega_{ij} = \frac{1}{2}(u_{i,j} - u_{j,i}) \tag{1.91}$$

Substituting $\varepsilon_{ij} + \omega_{ij}$ into (1.89) due to the convolution with $dx_i dx_j$ the tensor of rotation disappears:

$$ds'^2 - ds^2 = 2\varepsilon_{ij} dx_i dx_j \tag{1.92}$$

and the deformation of the medium (changes of the distances between the points) is determined only by the deformation tensor ε_{ij}. The rotation tensor ω_{ij} characterizes the turn of the particle medium as a rigid unit.

Using (1.88) we may express $ds'^2 - ds^2$ in terms of the deformation e as follows

$$ds' = (e+1)\,ds$$
$$ds'^2 - ds^2 = \left[(e+1)^2 - 1\right] ds^2 = \left(2e + e^2\right) ds^2 \tag{1.93}$$

In what follows we shall consider only finite deformations $e << 1$. Under this assumption (from (1.92), (1.93)) we obtain:

$$e = \frac{\varepsilon_{ij} dx_i dx_j}{ds^2} \tag{1.94}$$

Considering the cosines direction s_i^0 of section a $ds : dx_i = s_i^0 ds$, (1.94) reads

$$e = \varepsilon_{ij} s_i^0 s_j^0 \tag{1.95}$$

In (1.95) the tensor of deformations needs to be infinitesimal, that is $|\varepsilon_{ij}| << 1$, as well as the rotation tensor $|\omega_{ij}| << 1$. In that case all derivatives are infinitesimal too, that is $|u_{ij}| << 1$. Under these assumptions the expression for the deformation tensor is simplified to

$$\varepsilon_{ij} = \frac{1}{2}\left(u_{i,j} + u_{j,i}\right) \tag{1.96}$$

The components of the tensor ε_{ij} have a particular mechanical meaning: the diagonal components ε_{11}, ε_{22}, ε_{33} represent tension per unit length along the axes Ox_1, Ox_2 and Ox_3, respectively, while the off-diagonal components, e.g. component $\varepsilon_{12} = \varepsilon_{21}$, reflect the change of the angle between the initially orthogonal elements of the media, taken along axes Ox_1 and Ox_2 etc. The off-diagonal components of deformation ε_{ij} $(i \neq j)$ are called *shear deformations.*

Figure 1.8 shows the orthogonal elements $\overrightarrow{PQ_1} = \{dx_1, 0\}$ and $\overrightarrow{PQ_2} = \{0, dx_2\}$. After the deformation moves them into $\overrightarrow{P'Q'_1} = \{dx_1 + du_1, du_2\}$ and $\overrightarrow{P'Q'_2} = \{du_1, dx_2 + du_2\}$ respectively. The turn of element $\overrightarrow{PQ_1}$ is described by the small angle $\alpha_1 \approx tg\alpha_1 = \frac{du_2}{dx_1 + du_1} \approx \frac{\partial u_2}{\partial x_1}$, the turn of element $\overrightarrow{PQ_2}$ is described by the small angle $\alpha_2 \approx tg\alpha_2 = \frac{du_1}{dx_2 + du_2} \approx \frac{\partial u_1}{\partial x_2}$. Hence the initial angle between $\overrightarrow{PQ_1}$ and $\overrightarrow{PQ_2}$, which is equal to $\frac{\pi}{2}$, changes, during the deformation to $\alpha_1 + \alpha_2 \approx u_{2,1} + u_{1,2} = 2\varepsilon_{12}$.

The symmetrical tensor of deformation may be transformed to main axes, where all shear components disappear, and only non-zero ones are the main deformations $\varepsilon_1 = \varepsilon_{11}$, $\varepsilon_2 = \varepsilon_{22}$ and $\varepsilon_3 = \varepsilon_{33}$. The main deformations can be geometrically explained as follows:

Let us consider a cube, the side of which is equal to unity and parallel to the main axes. After the deformation it becomes a rectangular parallelepiped

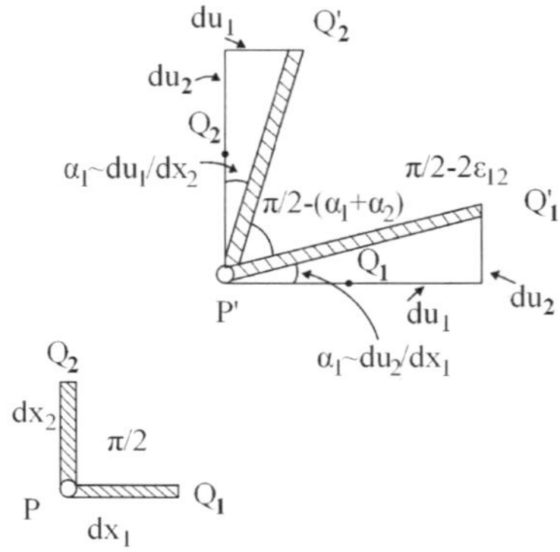

Fig. 1.8. About the determination of the geometric meaning of shear deformation ε_{12}

with sides $(1+\varepsilon_1)$, $(1+\varepsilon_2)$ and $(1+\varepsilon_3)$. During the deformation the cube turns as a unit and this turn is described by the rotation tensor ω_{ij}.

The relative change of the volume is called the *volumetric expansion*:

$$\varepsilon = (1+\varepsilon_1)(1+\varepsilon_2)(1+\varepsilon_3) - 1 \approx \varepsilon_1 + \varepsilon_2 + \varepsilon_3 \tag{1.97}$$

due to the fact that ε_i is infinitesimal. The given quantity is the first invariant tensor of deformation and in the arbitrary system of coordinates is expressed as

$$\varepsilon = \varepsilon_{ii} \tag{1.98}$$

1.3.4 Equations of Compatibility

The body deformation ε_{ij} $(i, j = 1, 2, 3)$ is completely determined by the displacement field u_i $(i = 1, 2, 3)$. However the displacement field u_i does not necessarily corresponds to any field of deformation tensor ε_{ij}. Imagine that we cut the body into small cubes. After the deformation, assigned by the tensor ε_{ij}, each one of the blocks turns into an oblique-angled parallelepiped. It is unlikely to expect that the resulting parallelepipeds can form a solid body again. Formally, relations (1.96) may be considered as a system of six equations for the derivatives of the three unknown, u_1, u_2, u_3, functions. The additional conditions required for ε_{ij} in order to provide the solvability of the given problem are called compatibility equations. These are obtained by the following relation:

$$\omega_{ij,k} = \frac{1}{2}\left(u_{i,jk} - u_{j,ik}\right) = \frac{1}{2}\left(u_{i,kj} + u_{k,ij} - u_{k,ji} - u_{j,ki}\right) = \varepsilon_{ik,j} - \varepsilon_{kj,i} \tag{1.99}$$

where ω_{ij} statisfies the next equation

$$\omega_{ij,kl} - \omega_{ij,\ell k} = 0 \tag{1.100}$$

Substituting (1.99) into (1.100) *equations of compatibility read*

$$\varepsilon_{ik,jl} - \varepsilon_{kj,i\ell} - \varepsilon_{il,jk} + \varepsilon_{lj,ik} = 0 \tag{1.101}$$

From the 81 relations of (1.101), only six of them are independent. These are the following:

$$\begin{aligned} \varepsilon_{11,22} + \varepsilon_{22,11} &= 2\varepsilon_{12,12} \\ \varepsilon_{22,33} + \varepsilon_{33,22} &= 2\varepsilon_{23,23} \\ \varepsilon_{11,33} + \varepsilon_{33,11} &= 2\varepsilon_{13,13} \end{aligned} \tag{1.102}$$

$$\begin{aligned} \left(\varepsilon_{12,3} + \varepsilon_{31,2} - \varepsilon_{23,1}\right)_{,1} &= \varepsilon_{11,23} \\ \left(\varepsilon_{23,1} + \varepsilon_{12,3} - \varepsilon_{31,2}\right)_{,2} &= \varepsilon_{22,31} \\ \left(\varepsilon_{31,2} + \varepsilon_{32,1} - \varepsilon_{12,3}\right)_{,3} &= \varepsilon_{33,12} \end{aligned} \tag{1.103}$$

1.3.5 Elastic Body

A body is called *elastic* when the stress at each point is a single-valued function of the deformation at that point. This can be represented by

$$\sigma_{ij} = \varphi_{ij}\left(\varepsilon_{k\ell}\right) \tag{1.104}$$

In the one-dimensional case the law of elasticity is given by

$$\sigma = \varphi\left(\varepsilon\right) \tag{1.105}$$

Let us now consider the homogeneous deformation $\varepsilon = u/\ell$ of an elastic rod (u is the displacement of its end, ℓ is the initial length) under the action of a force $F = \sigma \cdot S$ (S is the rod section), applied to the end of the rod (Fig. 1.9). The work of the outer force is given by

$$A = \int_0^u F du = S\ell \int_0^\varepsilon \sigma\, d\varepsilon = V \int_0^\varepsilon \varphi\left(\varepsilon\right) d\varepsilon \tag{1.106}$$

($V = S\ell$ is the volume of the rod) passes into the elastic energy.
The quantity

$$U = \int_0^\varepsilon \sigma\, d\varepsilon = \int_0^\varepsilon \varphi\left(\varepsilon\right) d\varepsilon \tag{1.107}$$

is called the *volumetric density of the elastic energy.*

The stress in the rod is expressed in terms of elastic energy as

$$\sigma = \frac{\partial U}{\partial \varepsilon}, \qquad (dU = \sigma d\varepsilon) \tag{1.108}$$

i.e. the elastic energy is the *stress potential.*

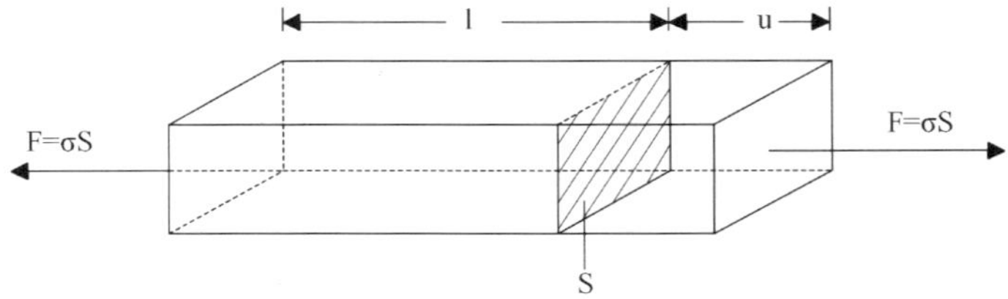

Fig. 1.9. Homogeneous deformation of the rod

In the three-dimensional case, an analogous relation holds:

$$dU = \sigma_{ij} d\varepsilon_{ij} \tag{1.109}$$

To the requirement of a single-valued connection between the stresses and deformations (1.104) in an elastic body the requirement of absence of non-mechanical losses is usually added (e.g. absence of temperature effects). Mathematically, this is expressed as the condition of zero work over an arbitrary chosen closed cycle:

$$\oint_L dU = \oint_L \sigma_{ij} d\varepsilon_{ij} = 0 \tag{1.110}$$

THe above is fulfilled when the integrand can be represeneted as a full differential, hence

$$\sigma_{ij} = \frac{\partial U}{\partial \varepsilon_{ij}} \tag{1.111}$$

i.e. the elastic energy $U = U(\varepsilon_{ij})$ is the *stress potential.* In the modern literature the bodies for which conditions (1.110) and (1.111) are satisfied, are called hyperelastic, but more often the term *elastic bodies* is used.

1.3.6 Hooke's Law. Stress Potential

For infinitesimal deformations of elastic bodies, relations (1.104), (1.105) are linear, obeying to Hooke's law:

$$\sigma = E\varepsilon \tag{1.112}$$

or in the general case

$$\sigma_{ij} = c_{ijk\ell}\varepsilon_{k\ell} \tag{1.113}$$

where $c_{ijk\ell}$ are the *constants of rigidity* for a crystal. The number of the constants, depending on the case under study, may reach a maximum of 81.

The single-valued linear (1.113) assumes the possibility of obtaining the reverse linear relation

$$\varepsilon_{ij} = s_{ijk\ell}\sigma_{k\ell} \tag{1.114}$$

where $s_{ijk\ell}$ are the *constants of compliance* of a crystal, in all there are 81 constants.

Relations (1.113), (1.114) express Hooke's law, and they describe an arbitrary finite deformation of the crystal. Physically these relations reflect the fact that under the action of a simple tension, a crystal may bend and curl. From the mathematical point of view in (1.113) and (1.114) $c_{ijk\ell}$ and $s_{ijk\ell}$ form the fourth rank *tensor of rigidity*. In these tensors the number of independent constants is much less than 81. This is due to: the symmetry of the tensors of stresses and deformations, the condition of potentiality (1.111), and the symmetry of the crystal. The latter is the subject of crystalophysics, which will is studied e.g. in the monograph [ ]. The first two situations are investigated later in the current chapter.

If we apply a shear stress σ_{12} to a crystal, according to the law of twoness (1.85) σ_{21} (1.114) reads

$$\varepsilon_{ij} = s_{ij12}\sigma_{12} + s_{ij21}\sigma_{21} = (s_{ij12} + s_{ij21})\,\sigma_{12}$$

Therefore, it is unlikely to design an experiment where we can independenly set values in s_{ij12} or s_{ij21}, since they are connected through Hooke's law. Therefore, in order to avoid the appearance of arbitrary constants we assume that

$$s_{ijk\ell} = s_{ij\ell k} \tag{1.115}$$

The symmetry of the tensor of deformation (1.96) is expressed as

$$s_{ijk\ell} = s_{jik\ell} \tag{1.116}$$

Due to (1.115) and (1.116) the number of independent components of the tensor of compliance is decreased from 81 to 36. The relations which are analogous to (1.115) and (1.116) are also valid for the tensor of rigidity $c_{ijk\ell}$

$$c_{ijk\ell} = c_{ij\ell k} \tag{1.117}$$

$$c_{ijk\ell} = c_{jik\ell}. \tag{1.118}$$

Due to relations (1.117) and (1.118) the number of independent components in the tensor of rigidity is also decreased from 81 to 36.

Further decrease in the number of components of tensors $c_{ijk\ell}$ and $s_{ijk\ell}$ is related with (1.111), from which we get

$$\frac{\partial\sigma_{ij}}{\partial\varepsilon_{k\ell}} = \frac{\partial^2 U}{\partial\varepsilon_{ij}\partial\varepsilon_{k\ell}} \text{ and } \frac{\partial\sigma_{k\ell}}{\partial\varepsilon_{ij}} = \frac{\partial^2 U}{\partial\varepsilon_{k\ell}\partial\varepsilon_{ij}},$$

and due to the equality of mixed derivatives we obtain:

$$\frac{\partial\sigma_{ij}}{\partial\varepsilon_{k\ell}} = \frac{\partial\sigma_{k\ell}}{\partial\varepsilon_{ij}} \tag{1.119}$$

Substituting (1.113) into (1.119) we get the condition of symmetry with respect to the first and second pair of indexes:

$$c_{ijk\ell} = c_{k\ell ij} \tag{1.120}$$

Also, relations (1.119) may be written as

$$\frac{\partial \varepsilon_{k\ell}}{\partial \sigma_{ij}} = \frac{\partial \varepsilon_{ij}}{\partial \sigma_{k\ell}} \tag{1.121}$$

due to one-to one connection of the tensors of stresses and deformations. Substituting (1.114) into (1.121), the condition of symmetry for the tensor of compliance reads

$$s_{ijk\ell} = s_{k\ell ij} \tag{1.122}$$

Relations (1.120) and (1.122) indicate that the independent components in tensors $c_{ijk\ell}$ and $s_{ijk\ell}$ are not more than 21 out of 81.

1.3.7 Matrix Designations

Because of the symmetry of tensors $s_{ijk\ell}$ and $c_{ijk\ell}$ (1.120), (1.222) we may use the matrix designations. Let us assign the indexes 1 to 6 to the components of the symmetric tensors of stresses and deformations as follows:

$$\begin{bmatrix} \sigma_{11} & \sigma_{12} & \sigma_{13} \\ \sigma_{12} & \sigma_{22} & \sigma_{23} \\ \sigma_{13} & \sigma_{23} & \sigma_{33} \end{bmatrix} \to \begin{bmatrix} \sigma_1 & \sigma_6 & \sigma_5 \\ \sigma_6 & \sigma_2 & \sigma_4 \\ \sigma_5 & \sigma_4 & \sigma_3 \end{bmatrix} \tag{1.123}$$

$$\begin{bmatrix} \varepsilon_{11} & \varepsilon_{12} & \varepsilon_{31} \\ \varepsilon_{12} & \varepsilon_{22} & \varepsilon_{23} \\ \varepsilon_{31} & \varepsilon_{23} & \varepsilon_{33} \end{bmatrix} \to \begin{bmatrix} \varepsilon_1 & \frac{1}{2}\varepsilon_6 & \frac{1}{2}\varepsilon_5 \\ \frac{1}{2}\varepsilon_6 & \varepsilon_2 & \frac{1}{2}\varepsilon_4 \\ \frac{1}{2}\varepsilon_5 & \frac{1}{2}\varepsilon_4 & \varepsilon_3 \end{bmatrix}$$

It is possible to unite the first and second indexes of tensors s_{ijkl} and c_{ijkl}, as well as the third and the fourth according to the next scheme

$$\begin{gathered} 11 \to 1; \quad 22 \to 2; \quad 33 \to 3; \\ 23, 32 \to 4; \quad 31, 13 \to 5; \quad 12, 21 \to 6. \end{gathered}$$

And in this case we have to introduce the multipliers 2 and 4 as follows:

$s_{ijk\ell} = s_{mn}$, when m and n are equal to 1,2 or 3
$2s_{ijk\ell} = s_{mn}$, when either m or n are equal to 4, 5 or 6
$4s_{ijk\ell} = s_{mn}$, when m as well as n are equal to 4, 5 or 6

Under this representation Hooke's law ((114)) can be written in a more compact form as

$$\varepsilon_i = s_{ij}\sigma_j \quad (i, j = 1, 2, \ldots, 6) \tag{1.124}$$

Multiplies 2 and 4 are introduced in order to avoid their appearance in (1.124).

For $c_{ijk\ell}$ the multipliers should not be introduced, therefore if we will simply write

$$c_{ijk\ell} = c_{mn} \quad (i, j, k, \ell = 1, 2, 3; \quad m, n = 1, 2, \ldots 6), \tag{1.125}$$

the (1.113) may be represented in the form

$$\sigma_i = c_{ij}\varepsilon_j \quad (i, j = 1, 2, \ldots 6) \tag{1.126}$$

Matrixes c_{ij} and s_{ij} are symmetric, but they are not the tensors of the second rank. In order to makes transformations of other axes the use of the tensor form is imperative.

1.4 Singular Physical Fields

In scientific and engineering calculations we often confront problems where scalar or tensor physical quantities or components of vectors tend to infinity to some *special points* in the body. Such physical fields are called *singular*. It is important to note that the observed singularities do not refer to a real object, but to its mathematical model. Therefore it is evident that their appearance does not indicate that the situation is catastrophic, it is just the result of the simplifications assumed for the construction of the model.

As an example we may refer to models describing point electric charges or concentrated heat sources. Here the singularity of the electric and temperature field respectively, is the result of a complete neglection of the charged body or area, where the heat release takes place. Besides, during the construction of these models we use electropolation and linear laws of electrostatics or heat conduction on the area of unbounded force fields or unbounded high temperatures. At this point we should note that the linearized models describe in a successful way the electrostatic and heat processes only at the points of the space which lie at a large distance from the charged or heated body (at least at a distance several times as large as its size). If we need a detailed description of the processes occurring in the vicinity of the charge or heat source, we should use a more complex non-linear mode. The solutions in both areas should be in agreement with each other - the asymptotics of the inner solution at infinity should match with the asymptotics of the outer solution at a special point.

Analogous problems are caused when we use wedgewise areas for the modeling of electric fields at the sharp edge of a flat electrode, or mechanic fields near a sharp notch, and also in the case of modeling of electric fields in the vicinity of a pointed end of a conductor in conical areas

In cases where the sizes of the zone of non-linearity are small compared to the characteristic sizes of the body, singular models have full right to exist.

We may accept the singularity of the physical fields at special points, and use the analysis of the asymptotic fields at special points to construct effective mathematical models of such complex phenomena as mechanical fracture or electric break-down.

Main importance, in the process of mathematical modeling, has the analysis of the energy flow directed into special points or special lines, where the energy is discharged from (or flows into) the system by mechanisms that cannot be represented by closed-form equations of a physical field. The energy flow coming into the end zone determines the intensity of all these processes of fracture (mechanical or electric), and hence its analysis permits us to estimate the condition under which the global fracture is being initiated, and also to predict its behaviour (resistant or non-resistant, quasistatic or dynamic).

1.4.1 Analysis of the Singularities of Physical Fields

The analysis of physical fields in the vicinity of special points is a rather complex problem, and its solution should be mathematically correct (the solution should be unique and robust with respect to perturbations of boundary conditions and shapes of the body) and physical coherence (continuation and boundness of the displacements in the end area in the absence of concentrated effects, boundness of the energy).

Let us consider, as a simple example, a plane problem for the determination of an electric field in the vicinity of a wedgelike edge of a conductor bounded with vacuum (Fig. 1.10). According to (1.72) the electrostatics electric potential $\varphi = \varphi(r,\theta)$ outside the onductor is a harmonic function satisfying the Laplace equation in cylindrical coordinates r, θ, z (axis z tends along the rib of a wedge):

$$\frac{\partial^2\varphi}{\partial r^2} + \frac{1}{r}\frac{\partial\varphi}{\partial r} + \frac{1}{r^2}\frac{\partial^2\varphi}{\partial\theta^2} = 0 \tag{1.127}$$

It is assumed that the wedge is surrounded by vacuum occupying area $-\alpha \leq \theta \leq \alpha$, on the faces of wedge $\theta = \pm\alpha$ (equipotential surfaces), while the the potential is equal to zero

$$\varphi(r, \pm\alpha) = 0 \tag{1.128}$$

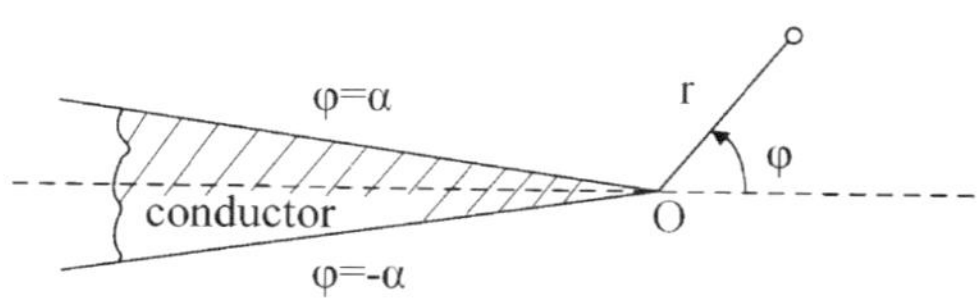

Fig. 1.10. Analysis of an electric field in the vicinity of a wedgelike edge of a conductor bounded with vacuum

We look for a solution to the Dirichlet problem, as given by (1.27)–(1.28), of the form

$$\varphi(r,\theta) = r^{\lambda} f(\theta) \tag{1.129}$$

Using (1.129) to (1.127) we get the following ordinary differential equation, with respect to function $f(\theta)$:

$$f'' + \lambda^2 f = 0,$$

Based on the above form, the general solution of (1.127) reads

$$\phi(r,\theta) = r^{\lambda}(A\cos\lambda\theta + B\sin\lambda\theta) \tag{1.130}$$

Substituting (1.130) into boundary conditions repsresented by (1.128) the following set of homogeneous algebraic equations is obtained:

$$\begin{aligned} A\cos\lambda\,\alpha + B\sin\lambda\alpha &= 0 \\ A\cos\lambda\alpha - B\sin\lambda\alpha &= 0 \end{aligned} \tag{1.131}$$

where the constants A and B are the unknowns.

The non-trivial solution of system (1.131) may be found by setting its determinant to be equal to zero

$$\begin{vmatrix} \cos\lambda\alpha & \sin\lambda\alpha \\ \cos\lambda\alpha & -\sin\lambda\alpha \end{vmatrix} = -\sin 2\lambda\alpha = 0 \tag{1.132}$$

which means that:

$$\lambda = \frac{\pi}{2\alpha}n \qquad (n = 0, \pm1, \pm2, \ldots) \tag{1.133}$$

When n is even, $A = 0$, and when n is odd, $B = 0$. Thus, the solution of the homogeneous problem of Dirihlet for wedgewise $-\alpha \leq \theta \leq -\alpha$ is represented in the form of an infinite series

$$\phi(r,\theta) = \sum_{n=\pm1,\pm3,\ldots} A_n r^{\frac{\pi n}{2\alpha}} \cos\left(\frac{\pi}{2\alpha}n\theta\right) + \sum_{n=\pm2,\pm4,\ldots} B_n r^{\frac{\pi n}{2\alpha}} \sin\left(\frac{\pi}{2\alpha}n\theta\right) \tag{1.134}$$

In the above equation we should omit the summands corresponding to negative values of n, as they describe an electric field with unbounded energy. Considering the wedgwide area as a limited case of the areas bounded by even surface, we can show that only those solutions, where the energy is bounded, have a physical meaning. Now, take a look at the temr

$$A_1 r\frac{\pi}{2\alpha}\cos\left(\frac{\pi}{2\alpha}\theta\right). \tag{1.135}$$

This term has an unbounded derivative over r at the top, only at $\alpha > \pi/2$. In the case of a wedgewise edge conductor, this describes the electric field, continuously increasing at approximation to the top of the wedge. The sharp increase in the electric field near the point on the surface of the conductor is a well-know physical phenomenon, providing various non-linear processes, including electric break-down.

In the limiting case of an electrode of zero thickness $\alpha = \pi$, we have

$$\begin{aligned}
\varphi(r,\theta) &= A_1\sqrt{r}\cos\frac{\theta}{2} + \dots \\
E_r(r,\theta) &= -\frac{\partial\varphi}{\partial r} = -\frac{A_1}{2\sqrt{r}}\cos\frac{\theta}{2} + \dots \\
E_\theta(r,\theta) &= \frac{1}{r}\frac{\partial\varphi}{\partial\theta} = -\frac{A_1}{2\sqrt{r}}\sin\frac{\theta}{2} + \dots
\end{aligned}$$

The continuation notation in the last formulas represents the terms of a series which remain bounded or tend to zero as $r \to 0$. The above relations are usually written in the following form:

$$\begin{aligned}
\varphi(r,\theta) &= -K_E\sqrt{\frac{2r}{\pi}}\cos\frac{\theta}{2} + \dots \\
E_r(r,\theta) &= \frac{K_E}{\sqrt{2\pi r}}\cos\frac{\theta}{2} + \dots \\
E_\theta(r,\theta) &= \frac{K_E}{\sqrt{2\pi r}}\sin\frac{\theta}{2} + \dots
\end{aligned} \tag{1.136}$$

Here instead of A_1 we use the the proportional to it constant K_E, which is called the *electric field intensity factor.*

The value of the intensity factor is extremely important, because it determines the field in the vicinity of the sharp edge of the electrode, hence the intensity of the process of preliminarily fracture in the end zone, and the condition of the electric break-down. In order to find the dependence of the intensity factor on the outer external action (for example, on the potential of the electrode) we have to solve the problem for the whole conductive body and the field which surrounds it.

Using (1.136) for the electric potential we can calculate the distribution of the charges on the surface of an electrode near its edge. Taking into account the fact that inside the conductor no field exists, we obtain from (1.63) that the density of the charges on the surface of the electrodes (at $\theta = \pi$)

$$\sigma = D_\theta = \ni_0 E_\theta = -\ni_0 \frac{1}{r}\frac{\partial\varphi}{\partial\theta}\bigg|_{\theta=\pi} = \ni_0 \frac{K_E}{\sqrt{2\pi r}}\sin\frac{\theta}{2}\bigg|_{\theta=\pi} + \dots = \frac{\ni_0 K_E}{\sqrt{2\pi r}} + \dots \tag{1.137}$$

At the edge of the electrode ($r \to 0$) the above tends to infinity.

Relation (1.137) is usually represented as

$$\sigma = \frac{K_\sigma}{\sqrt{2\pi r}} + \dots \tag{1.138}$$

The above repsrentation introduces the *intensity factor of the density charges* at the edges of the electrode. This factor can be also determined from the solution of the electric problem for the whole body. The question of choice between the introduced intensity factor K_E or K_σ (and K_D for *electric induction*) is not trivial and may be solved only on the basis of the analysis of the experimental data. Depending on the analysis of experiments (also including experiments on the break-down between the surface electrodes), our opinion is that is better to use the concept of the intensity factor of charges K_σ for the calculation of the break down stresses.

1.4.2 The Electric Field of a Charged Conductive Disk

Let us carry out the analysis of the singularities of the electric field which is produced by an infinitely thin electrode in the form of a round disk in vacuum. In order to get a physically correct solution of this singular problem, we must obtain its solution by solving the classical problem of a conductive ellipsoid in case when its thickness tends to zero.

Consider an oblated conductive ellipsoid created by rotation

$$\frac{x^2}{R^2} + \frac{y^2}{R^2} + \frac{z^2}{c^2} = 1 \qquad (c < R)\,, \tag{1.139}$$

with a full charge, Q, (Fig. 1.11). The surface of the ellipsoid (1.139) is equipotential (denoting its potential with V). In order to analyze the electric field in the surrounding space we use the cylindrical system of coordinates $\rho,\ \phi,\ z$

$$\left(x = \rho\cos\phi, y = \rho\sin\phi, \rho = \sqrt{x^2 + y^2}\right)$$

and also the system of coordinates of the oblated ellipsoid created by ellipsoid $\sigma,\ \tau,\ \phi$. Their relation with the cylindrical coordinates is prescribed by the equations

$$\rho = a\sqrt{(1+\sigma^2)(1-\tau^2)}, \quad z = a\sigma\tau \tag{1.140}$$

As a curvilinear coordinate we can consider the oblated ellipsoid created by rotation $\sigma = const$ $(0 \leq \sigma < +\infty)$ (Fig. 1.12):

$$\frac{\rho^2}{a^2(1+\sigma^2)} + \frac{z^2}{a^2\sigma^2} = 1, \tag{1.141}$$

and also one-pole rotation hyperboloides for which $\tau = const$ $(-1 \leq \tau \leq 1)$

$$\frac{\rho^2}{a^2(1-\tau^2)} - \frac{z^2}{a^2\tau^2} = 1$$

and planes $\phi = const$ $(0 \leq \phi < 2\pi)$

$$y = xtg\phi$$

crossing across Oz axis.

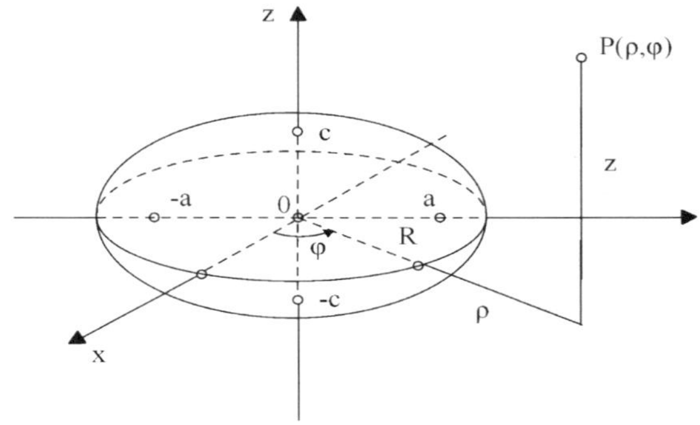

Fig. 1.11. Use of the cylindrical system of coordinates for investigation of a field of a charged conductive and oblated ellipsoid of rotation (the half-axes of the ellipsoid are R, R an $c\,(c < R)$)

The surface of the charged ellipsoid is coordinated surface $\sigma = \sigma_0$. Constants a and σ_0 are connected with the sizes of ellipsoids R and c by the relations

$$R = a\sqrt{1 + \sigma_0^2}, \qquad c = a\sigma_0 \tag{1.142}$$

The z-axis is described by the relations $\tau = \pm 1$, while the xOy plane is described by $\tau = 0$. The focuses of the systems of ellipsoids and hyperboloids lie in this plane on circle $\rho = a$.

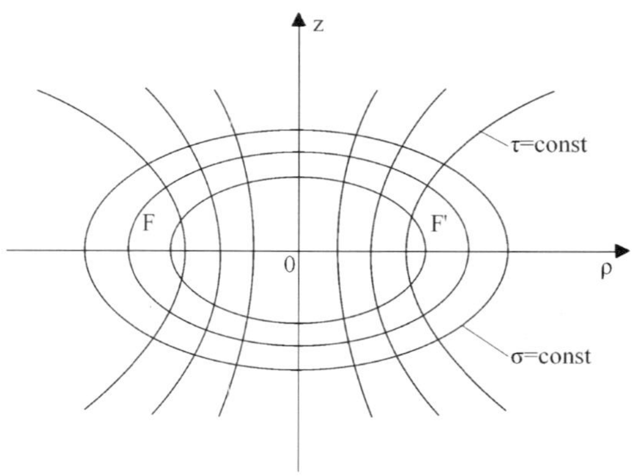

Fig. 1.12. The orthogonal system of co-focused ellipse ($\sigma = const$) and hyperbola ($\tau = const$) with focuses F and F′ (OF = OF′ = α). This system determines the system of coordinates of an oblated ellipsoid of rotation, if it is rotated with respect to axis OZ

The potential of the asymmetric electric field outside the ellipsoid $\Phi(\sigma, \tau)$ is a harmonic function, satisfying Laplase equation

$$\Delta\Phi = \frac{1}{a^2(\sigma^2+\tau^2)}\left\{\frac{\partial}{\partial\sigma}\left[\left(1+\sigma^2\right)\frac{\partial\Phi}{\partial\sigma}\right]+\frac{\partial}{\partial\tau}\left[\left(1-\tau^2\right)\frac{\partial\Phi}{\partial\tau}\right]\right\}=0 \quad (1.143)$$

On the surface of charged ellipsoid $\sigma = \sigma_0$ the condition of equipotentiality must be satisfied

$$\Phi(\sigma, \tau) = V = const$$

This condition is satisfied automatically if the field potential is a function only of σ. In this case, Laplace (1.143) is can be rewritten as

$$\frac{\partial}{\partial\sigma}\left[\left(1+\sigma^2\right)\frac{\partial\Phi}{\partial\sigma}\right]=0,$$

The general solution of the above equation reads

$$\Phi(\sigma) = C_1 arctg\sigma + C_2 \quad (1.144)$$

At infinity (when $\tau = \sqrt{\rho^2+z^2} \to \infty$) the electric field tends to zero, based on the same law, as the Coulomb field of a point charge Q:

$$\Phi_E = \frac{Q}{4\pi \ \ni_0 r} \to 0 \quad (1.145)$$

From (1.140) at $r \to \infty$ it follows that $r \approx a\sigma$. If we use this result in (1.145) we will get

$$\Phi(\sigma) = \frac{Q}{4\pi \ \ni_0 a}\left(\frac{\pi}{2} - arctg\sigma\right). \quad (1.146)$$

Because on the surface of the ellipsoid the potential has constant value V, from (1.146) it follows that

$$V = \frac{Q}{4\pi \ni_0 a}\left(\frac{\pi}{2} - arctg\sigma_0\right) \quad (1.147)$$

with the help of (1.147) we can express (1.146) in the form

$$\Phi(\sigma) = V\left(\frac{\pi}{2} - arctg\sigma\right)\Big/\left(\frac{\pi}{2} - arctg\sigma_0\right) \quad (1.148)$$

Here we can use both relations (1.146) and (1.148) according to the prescribed values of full charge Q or potential V.

Using (1.146) or (1.148) we can calculate the density of distribution of electric charges σ^* on the surface of the ellipsoid $\sigma = \sigma_0$. Using (1.63) we find that

$$\sigma^* = \ni_0 E_n = - \ni_0 \left.\frac{\partial\Phi}{\partial n}\right|_{\sigma=\sigma_0} = -\frac{\ni_0}{\sqrt{g_{\sigma\sigma}}}\left.\frac{\partial\Phi}{\partial\sigma}\right|_{\sigma=\sigma_0} \quad (1.149)$$

where $g_{\sigma\sigma} = a^2\frac{\sigma^2+\tau^2}{1+\sigma^2}$.

Substitution of (1.146) into (1.149) gives

$$\sigma * = \frac{q}{4\pi a^2} \frac{1}{\sqrt{(\sigma_0^2 + \tau^2)(1 + \sigma_0^2)}} \tag{1.150}$$

With the help of (1.141) and (1.142) we can work with the cylindric system of coordinates in (1.150) and obtain the density of distribution of the electric charges on the surface of the ellipsoid as:

$$\sigma * = \frac{Q}{4\pi \sqrt{R^4 - (R^2 - c^2)\rho^2}} \tag{1.151}$$

From tha above relation it is seen that at a distance from the axis of symmetry the density of charges increases from the minimal value $\sigma^* = \frac{Q}{4\pi R^2}$ at $\rho = 0$ on the axis of symmetry to the maximal value $\sigma^* = \frac{Q}{4\pi Rc}$ on the edge of the ellipsoid at $\rho = R$.

At the limiting transition $c/R \to 0$ to the infinitely thin circular conductive disk $(0 \leq \rho \leq R,\ z = 0)$ we obtain, that the density of charges tends to infinity approaching the edge at $\rho \to R$

$$\sigma * (\rho) = \frac{Q}{4\pi R \sqrt{R^2 - \rho^2}} \tag{1.152}$$

The assympotic analytical expansion into power series of $R - \rho$ order brings to the separation of the main term, which increases at $\rho \to R$

$$\sigma * = \frac{Q}{4\sqrt{\pi R^3}\sqrt{2\pi (R - \rho)}} + \ldots \tag{1.153}$$

(other terms omitted in expansion (1.153) tend to zero at $\rho \to R$).

Comparing (1.153) and (1.138) we can get the factor of the density intensity of charges, which in the general case is determined by

$$K_\sigma = \lim_{\rho \to R} \left[\sigma * (\rho) \sqrt{2\pi (R - \rho)} \right]. \tag{1.154}$$

Taking into account (1.152) or (1.151), we also have

$$K_\sigma = \frac{Q}{4\sqrt{\pi R^3}} \tag{1.155}$$

Now consider in a more detail the electric field created by a charged conductive circular disk of radius R. In that case it is necessary to assume $\sigma_0 = 0$ in (1.142), (1.146)–(1.148), which means that $a = R$

$$\Phi(\sigma) = \frac{Q}{4\pi \ni_0 R} \left(\frac{\pi}{2} - arctg\sigma \right) \tag{1.156}$$

Here

$$Q = 8 \ni_0 RV,$$

(hence, the capacitance of a circular disk in vacuum is given by $C = Q/V = 8 \ni_0 R$).

Let us now work in the cylindrical system of coordinates. Expressing σ by ρ and z and proceeding from (1.141) we obtain

$$\sigma = \left(z^2 + \rho^2 - R^2 + \left(\left(z^2 + \rho^2 - R^2 \right)^2 + 4R^2 z^2 \right)^{1/2} \right) \Big/ \left(a\sqrt{2} \right) \tag{1.157}$$

Equations (1.156) and (1.157) describe the distribution of the potential of a circular disk in the whole space. We find more important the asymptotic analysis of the field near the edge of the disk. For this purpose we introduce the local system of polar coordinates in ϕ-const plane passing through Oz-axis (Fig. 1.13)

$$\begin{aligned} \rho &= R + \Delta \cos\theta \\ z &= \Delta \sin\theta \\ (0 &\le \Delta < +\infty, -\pi \le \theta \le \pi) \end{aligned} \tag{1.158}$$

Substituting (1.158) into (1.157) we obtain at $\Delta << R$ that

$$\sigma = \sqrt{\frac{2\Delta}{R}} \cos\frac{\theta}{2} + \dots \tag{1.159}$$

The subsequent substitution of (1.159) into (1.156) gives rise to

$$\Phi(\Delta, \theta) = \frac{Q}{4\pi \ni_0 R} \left(\frac{\pi}{2} - \sqrt{\frac{2\Delta}{R}} \cos\frac{\theta}{2} \right) + \dots, \tag{1.160}$$

which describes an electric field in the vicinity of the edge of electrode ($\Delta << R$).

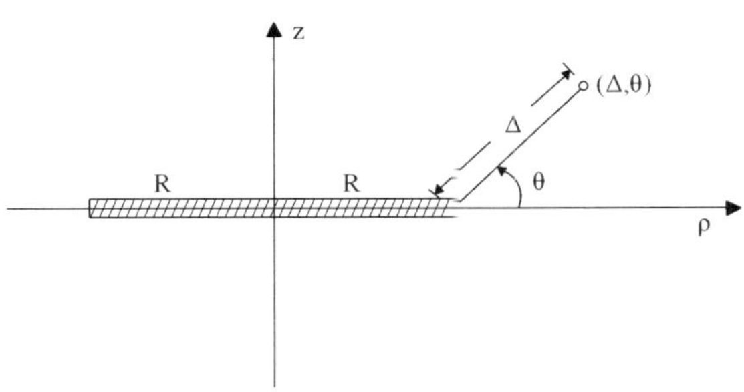

Fig. 1.13. The local polar system of coordinates (Δ, Θ) for the analyzing an electric field at the edge of a charged thin conductive disc with radius R

Here we should mention that (1.160) differs from (1.136) by a constant since (1.136) is derived with the assumption that on the surface of the electrode $\theta = \pm\pi$ the potential is equal to zero; hence (1.160) considers the potential, V, on the surface to be non-zero.

Calculating the strength of the electric field in the Δ, θ system of coordinates as

$$E_\Delta = -\frac{\partial \Phi}{\partial \Delta}, \quad E_\theta = -\frac{1}{\Delta}\frac{\partial \Phi}{\partial \theta}$$

we obtain

$$\begin{aligned} E_\Delta &= \frac{Q}{4 \ni_0 \sqrt{\pi R^3}} \frac{1}{\sqrt{2\pi \Delta}} \cos\frac{\theta}{2} + \ldots \\ E_\Theta &= \frac{Q}{4 \ni_0 \sqrt{\pi R^3}} \frac{1}{\sqrt{2\pi \Delta}} \sin\frac{\theta}{2} + \ldots \end{aligned} \tag{1.161}$$

Equation (1.161) are in agreement with (1.136), and their comparison allows us to determine the intensity factor of the electric field for a circular disk, writting:

$$K_E = \frac{Q}{4 \ni_0 \sqrt{\pi R^3}} \tag{1.162}$$

or, taking into account (1.157)

$$K_E = \frac{2V}{\sqrt{\pi R}} \tag{1.163}$$

The intensity factor of the density of charges K_σ (1.155) differs from K_E by the multiplier $\ni_0$

$$K_\sigma = \frac{2 \ni_0 V}{\sqrt{\pi R}} \tag{1.164}$$

Concluding, the above analysis indicates that near the edges of the electrode there is a high concentration of the strength of the electric field (in the considered ideal model this concentration is infinite, and the electric field is singular on the sharp edges of the electrode). The intense heterogeneity of the electric field in the vicinity of the electrode initiates complex physical phenomena such as catastrophs- the electric breakdown - the macroscopic electric fracture of the body. The mathematical modeling of this phenomenon is extremely important and interesting and it will be discussed extensively in the next sections. To solve it we use the methods which were developed in preceding sections for the analysis of the problem of mechanical fracture of rigid bodies. In the next chapters we will try to explain briefly the fundamental ideas of the mechanics of fracture.

1.4.3 Mathematical Idealization of Cracks in an Elastic Body

Stress Concentration

In the mechanics of fracture the real cracks are considered as strongly extended cavities (voids) which already exist in the body and/or are formed under the influence of outer loads. The precence of a cavity in the body substantially changes its stress state. For example, at uniform tension of a thin rectangular plate with a small circular hole with radius r ($r/\ell << 1, r/c << 1$) (Fig. 1.14) normal stresses σ_{12} on line $|x_1| > r,\ x_2 = 0$ are determined by

$$\sigma_{22}(x_1, 0) = p\left(1 + \frac{r^2}{2x_1^2} + \frac{3r^2}{2x_1^4}\right) \tag{1.165}$$

From (1.165) it is seen that the existence of holes causes the local increase of stresses: if at distances $r,\ 2r,\ 3r$ from the edge of a void the stresses σ_{12} exceed the nominal stress by 22, 7 and 4 percent, respectively, at the very edge we have that $\sigma_{22} = \rho$, i.e. the stress increases three times as large, although the opening itself does not practically decreases the cross-section area. In such a case we agree for the creation of the *concentration of stresses.*

A more dense concentration of stresses may be observed at a smaller opening of an elliptic form with half-axes a and b in a uniformly stretched plate (Fig. 1.15). Here the highest value of σ_{22} is observed at the tips of the ellipse

$$x_1 = \pm a, \quad x_2 = 0: \tag{1.166}$$
$$\sigma_{22} = p\left(1 + 2\sqrt{\frac{a}{\rho}}\right),$$

where $\rho = b^2/a$ is the radius of the curvature at the tip of the ellipse.

For an infinitely thin elliptic cut-out ($\rho \to 0$) the concentration of stresses at the indicated tips gets the largest possible value. The example, shown here, visualizes a substantial dependence of the concentration of stresses on the curvature radius at the given point of the opening contour.

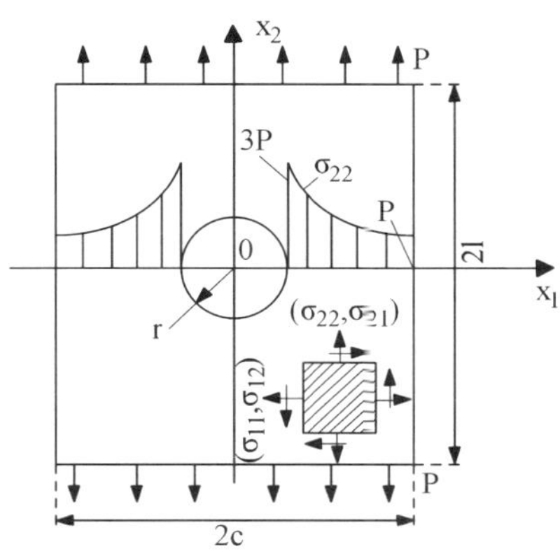

Fig. 1.14. Concentration of stresses at uniform tension of a thin rectangular plate with a small circular opening

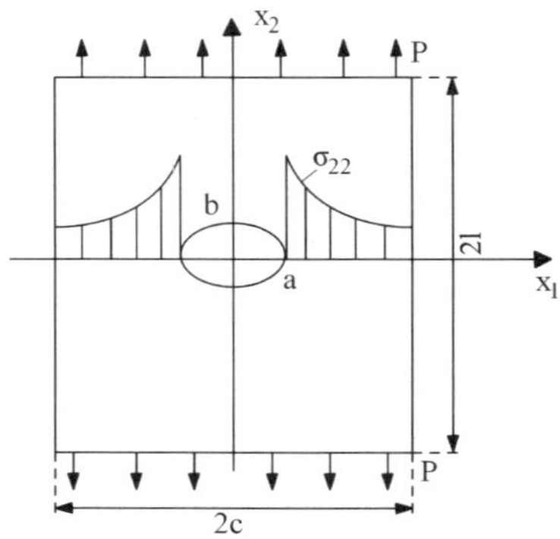

Fig. 1.15. Concentration of stresses at uniform tension of a thin rectangular plate with a small elliptic opening

A Mathematical Model of a Crack

Considering a real crack (Fig. 1.16) in a deformed solid body we may always isolate a line I on its boundary – the *front of cra*cks on which the surfaces of cavity 2 are joint – the tips of *cracks*. It is evident that in the vicinity of the front there is a high concentration of stresses and exactly there, local fractures of the material take place.

From the point of view of the statement and solution of the problem of the theory of elasticity the crack edges play the role of an additional boundary of the body. Taking into account the small distances between the boundaries, the real crack may be considered as a *mathematical cut*, i.e. the cavity of zero volume bounded by two geometrically coinciding surfaces – the edges of the cut. For example, transition from a cavity to a mathematical cut, may be obtained as follows:

Consider an unbounded body with a cavity in the form of an elliptic cylinder the element of which is parallel to Ox_3 – axis (Fig. 1.17a).

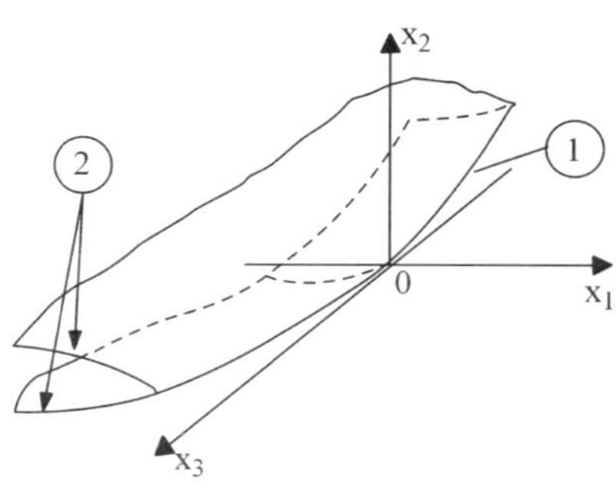

Fig. 1.16. A real crack in a deformed solid body

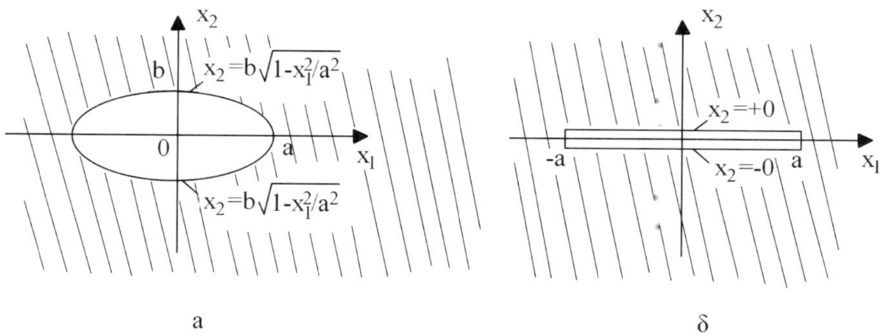

Fig. 1.17. Transition from the cavity to the mathematic cut in an unbounded body: a cavity in the form of an elliptic cylinder (a), a mathematical tunnel of a rectangular cut (b)

At the limiting transition $b \to 0$, the ellipse becomes a linear tunnel cut in section $(-a, a)$ of Ox_1- axis. The upper half-ellipse $x_2 = b\sqrt{1 - x_1^2/a^2}$ moves to the upper edge of the cut $|x_4| < a$, $x_2 = +0$, and the lower half-ellipse $x_2 = -b\sqrt{1 - x_1^2/a^2}$ to the lower edge of the cut $|x| < a$, $x_2 = -0$ (Fig. 1.17b).

Here, we should mention that the linear tunnel cut in an unbounded body or linear through cut in a thin plate is a basic idealized form of a real crack: in an arbitrary small vicinity of the point, O, of the front (Fig. 1.16) the crack may be investigated as a flat crack with a linear front. Hence, in the vicinity of any point of the front of a crack, the study of the stressed deformed state may be carried out in the framework of a plane problem of the theory of elasticity.

A Plane Problem of the Theory of Elasticity

Before we proceed, we should note that a problem is called *plane* when the displacements u_1 and u_2 depend only on x_1 and x_2 coordinates, and the displacement u_3 within the whole body is constant. In the problem under sudy only the following deformations

$$\varepsilon_{11} = \frac{\partial u_1}{\partial x_1}, \quad \varepsilon_{22} = \frac{\partial u_2}{\partial x_2}, \quad \varepsilon_{12} = \frac{1}{2}\left(\frac{\partial u_1}{\partial x_2} + \frac{\partial u_2}{\partial x_1}\right) \tag{1.167}$$

and stresses

$$\begin{aligned} \sigma_{11} &= (\lambda + 2\mu)\,\varepsilon_{11} + \lambda\varepsilon_{22}, \quad \sigma_{12} = 2\mu\varepsilon_{12} \\ \sigma_{22} &= \lambda\varepsilon_{11} + (\lambda + 2\mu)\,\varepsilon_{22}, \end{aligned} \tag{1.168}$$

differ from zero. λ and μ are the elastic constants of the material.

Such a state approximately appears in a long cylinder (Fig. 1.18) on the surface of which normal loads act, which are uniformly distributed along its elements (*state of plane deformation*). The same appear in a thin plate

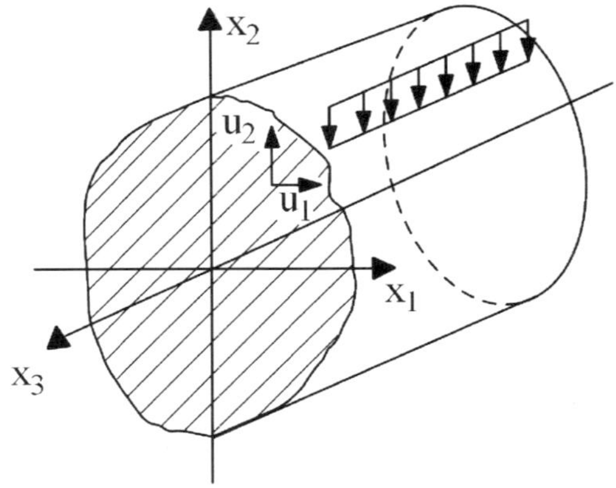

Fig. 1.18. Realization of a plane stressed state in a thin plate

(Fig. 1.19) under the influence of the system of outer load, which are parallel and symmetrically distributed with respect to its middle surface (*plane stressed state*).

For a plane deformation the elastic factors λ and μ are related with Young's modulus E and Poisson's ration as follows

$$\lambda = \frac{Ev}{(1+\nu)(1-2\nu)}, \quad \mu = \frac{E}{2(1+\nu)},$$

and for the plane stressed state the constant λ in (1.168) should be substituted by $\lambda(1-2\nu)/(1-\nu)$ and by stresses, deformations and displacements everywhere it is necessary to assume the values averaged according to the thickness of the plate. For example the displacement will be

$$\bar{u}(x_1, x_2) = \frac{1}{2h} \int_{-h}^{+h} u(x_1, x_2, x_3)\, dx_3.$$

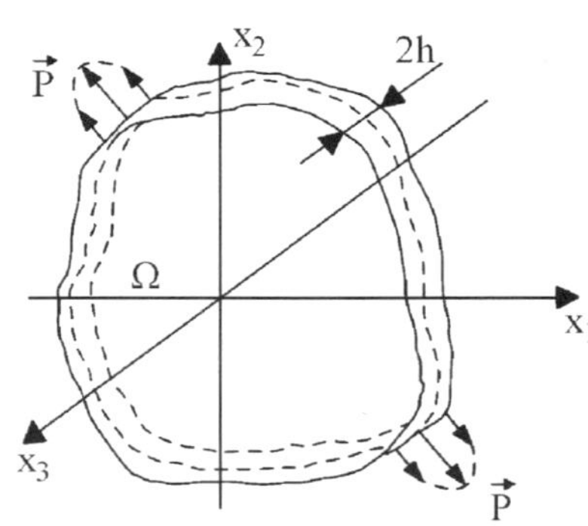

Fig. 1.19. Realization of a plane stressed state in a thin plate

Stressed State in the Vicinity of Linear Cracks

Cracks of Normal Fracture

Consider a plane problem in the framework of elasticity for an unbounded body with a linear cut on the part of the body $|x_1| < a,\ x_2 = 0$ (Fig. 1.20a).

The cut is considered to be an unbounded tunnel in the direction of x_3- axis if we solve the problem of plane deformation. The cut runs through the body, if we study the plane stressed state of a thin plate. Let us assume that the edges of the cut be free from outer loading, and that far from the cut act constant stretched stresses p_0 (see Fig. 1.20a). In that case the linear crack is called the *crack of normal break*, as under the influence of the load p_0 its tips are separated in opposite directions symmetrically to the plane of the crack.

Due to the linearity of the equations in the framewrok of elasticity theory, the solution of the problem may be represented in the form of a sum of solutions of the two problems (Fig. 1.20b,c). In the first problem (see, Fig. 1.20b) we consider a uniformly stretched plane without a break, while in the second problem a stressed state of the plane with a cut on the edges of which normal stresses $\sigma_{22} = -p_0$ act, is determined. The normal stresses with opposite signs which originate on the place of the cut in the first problem (Fig. 1.20b). Besides, taking into account the symmetry of the load and the geometry of the area we set $\sigma_{12} = 0$ at $x_2 = 0,\ |x| < \infty$.

The solution of the first problem is elementary ($\sigma_{11} = \sigma_{12} = 0, \sigma_{22} = p_0$), and the solution of the second problem may be limited to the investigation

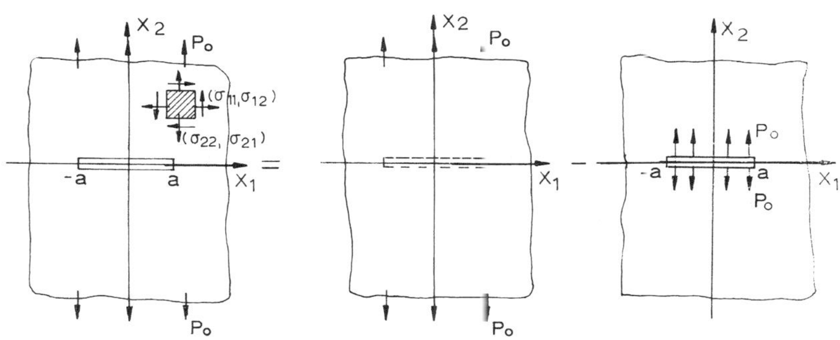

Fig. 1.20. Representation of the solution of the problem of tension of an unbounded body with a transversal cut (a), in the form of a superposition of the solutions of the problem of uniform tension of a body without a cut (b) and the problem of a body with a cut, on the edges of which there act normal loads (c)

of the stressed deformed state of the upper half-plane ($x_2 > 0$) only, on the boundary of which the following conditions should be satisfied

$$\sigma_{12}(x_1,0) = 0, \quad |x_1| < \infty, \tag{1.169}$$

$$\sigma_{22}(x_1,0) = -p_0, \quad |x_1| < a, \tag{1.170}$$

$$u_2(x_1,0) = 0, \quad |x_1| > a. \tag{1.171}$$

It is obvious that the stressed state of the half-plane $x_2 > 0$, satisfying conditions (1.169)–(1.171) is localized in the area of the cut, $|x| < a,\ x_2 = 0$ and at $x_1^2 + x_2^2 \to \infty$ all the stresses and displacements tend to zero.

The analysis of stresses and displacements in the vicinity of the tips of the cut (point $x_1 = a,\ x_2 = 0$) requires the introduction of polar coordinates with the center at the point $x_1 = a,\ x_2 = 0$; $x_1 = a + r\cos\varphi,\ x_2 = r\sin\varphi$. It is possible to show that in a small vicinity around the tip of the cut, i.e. at $r \to 0$ the components of the stresses and displacements are determined by the following asymptotic relations

$$\sigma_{11} = \frac{K_I}{\sqrt{2\pi r}}\left(\cos\frac{\varphi}{2} - \frac{1}{2}\sin\varphi\sin\frac{3\varphi}{2}\right), \tag{1.172}$$

$$\sigma_{22} = \frac{K_I}{\sqrt{2\pi r}}\left(\cos\frac{\varphi}{2} + \frac{1}{2}\sin\varphi\sin\frac{3\varphi}{2}\right),$$

$$\sigma_{12} = \frac{K_I}{\sqrt{2\pi r}}\frac{1}{2}\sin\varphi\cos\frac{3\varphi}{2},$$

$$u_1 = \frac{K_I\sqrt{r}}{\mu\sqrt{2\pi}}\left(\frac{\kappa+1}{2} - \cos^2\frac{\varphi}{2}\right)\cos\frac{\varphi}{2}, \tag{1.173}$$

$$u_2 = \frac{K_I\sqrt{r}}{\mu\sqrt{2\pi}}\left(\frac{\kappa+1}{2} - \cos^2\frac{\varphi}{2}\right)\sin\frac{\varphi}{2}.$$

Here $\kappa = (3-\nu)/(1+\nu)$ for the plane stressed state and $\kappa = 3 - 4\nu$ for the plane deformation, while K_I is calculated by,

$$K_I = \lim_{x_1 \to a}\left[\sigma_{22}(x_1,0)\sqrt{2\pi(x_1 - a)}\right], \tag{1.174}$$

The latter is called the *stress intensity factor of a normal break.* We will show that the behaviour of the crack and fracture of the body depends on K_I.

In the considered case

$$\sigma_{22}(x_1,0) = \frac{p_0 x_1}{\sqrt{x_1^2 - a^2}}, \quad (x_1 > a). \tag{1.175}$$

If we substitute (1.175) into (1.174) and calculate the limit we will get

$$K_I = p_0\sqrt{\pi a} \tag{1.176}$$

At this point, we should note that (1.172) and (1.173) are valid for other cracks of normal break in finite and infinite bodies, and depending on the values of loads, the size of the cracks and the sizes of the body, the corresponding intensity factor is given by (1.174).

For example, for a crack, the tips of which break by arbitrarily distributed normal loads with intensity $p_0(x_1)$, we have that

$$K_I = \frac{1}{\sqrt{\pi a}} \int_{-a}^{a} p_0(\xi) \sqrt{\frac{a+\xi}{a-\xi}} d\xi, \tag{1.177}$$

$$K_I = \frac{1}{\sqrt{\pi a}} \int_{-a}^{a} p_0(\xi) \sqrt{\frac{a-\xi}{a+\xi}} d\xi. \tag{1.178}$$

Here, (1.177) refers to the first tip of the crack, and (1 178) to the left tip. It is obvious that for a symmetrically loaded crack ($p_0(-x_1) = p_0(x_1)$) both equations give the same results.

Consider the cracks loaded by a uniformly distributed normal pressure p_0 on the section of length 2ε in the center of the crack:

$$p_0(\xi) = p_0 \text{ at } |\xi| < \varepsilon \text{ and } p_0(\xi) = 0 \text{ at } \varepsilon < |\xi| < a.$$

Substitution of these expressions into (1.174) gives

$$K_I = \frac{p_0}{\sqrt{\pi a}} \int_{-\varepsilon}^{\varepsilon} \sqrt{\frac{a+\xi}{a-\xi}} d\xi = \frac{2p_0 a}{\sqrt{\pi a}} \arcsin \frac{\varepsilon}{a}. \tag{1.179}$$

In the case of loading of the crack by two concentrated forces P (Fig. 1.21) we may obtain the factor K_I, if we consider $\varepsilon \to 0$ and $p_0 \to \infty$, thus that $2p_0\varepsilon \to P$. This limiting transition in (1.179) leads to

$$K_I = P/\sqrt{\pi a}. \tag{1.180}$$

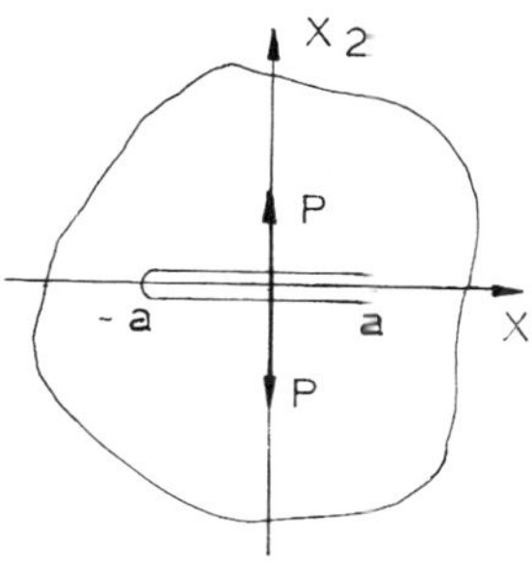

Fig. 1.21. A crack of a normal break loaded by a pair of concentrated forces applied in the middle of the edges of the crack

A Crack of a Transversal Shear

Let us assume that an unbounded plane is under the influence of known and permanent tangential stresses $\sigma_{12} = \tau$, at infinity. The plane contains a linear crack located on x_1 axis ($x_2 = 0$) described by equation $|x_1| < a$, the tips of which have no outer loads (Fig. 1.22a).

In this case the edges of the cut are displaced with respect to each other along the line perpendicular to the front of thickness (called the thickness of a *transversal shear*). The solution of the plane problem for cracks of transversal shear, in the framework of elasticity theory, may also be represented in the form of a sum of solutions of two problems: in the first problem (Fig. 1.22b) we consider an unbounded plane without a cut that has known load at infinity, that is tangential stresses $\sigma_{12} = \tau_0$. The second problem is an unbounded plate loaded by shear forces $\sigma_{12} = -\tau_0$.

The solution of the first problem is the trivial

$$\sigma_{11} = 0, \qquad \sigma_{12} = \tau_0, \qquad \sigma_{22} = 0, \tag{1.181}$$

and for the solution of the second problem (Fig. 1.22c) it is enough to consider the half-plane $x_2 > 0$ under the following conditions on the boundary $x_2 = 0$

$$\sigma_{22}(x_1, 0) = 0, \quad |x_1| < \infty, \tag{1.182}$$

$$\sigma_{12}(x_1, 0) = -\tau_0, \quad |x_1| < a, \tag{1.183}$$

$$u_1(x_1, 0) = 0, \quad |x_1| > a. \tag{1.184}$$

To describe the distribution of stresses and displacements in the vicinity of the tips of the crack of a transversal shear we introduce the polar coordinates

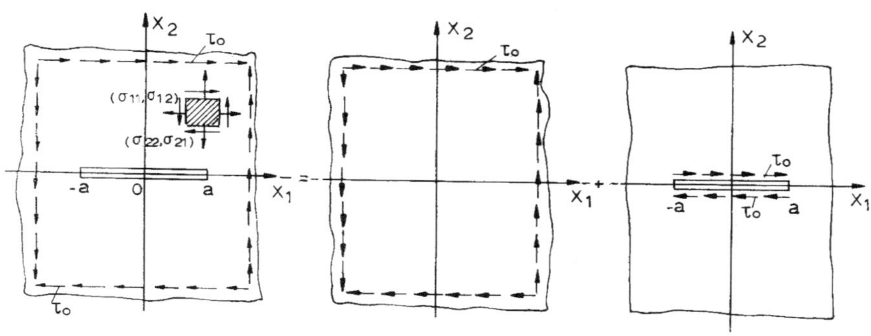

Fig. 1.22. Representation of the solution of the problem of a transversal shear of a body with a cut (a), in the form of a superposition of the solution of the problem of a transversal shear of solid body (b) and the problem of a body with a cut, to the edges of which there are applied transversal tangential loads (c)

r, φ with a pole at the tips $x_1 = a,\ x_2 = 0$. Then at $r \to 0$ the following asymptotic equations are obtained

$$\sigma_{11} = -\frac{K_{II}}{\sqrt{2\pi r}} \sin\frac{\varphi}{2}\left(2 + \cos\frac{\varphi}{2}\cos\frac{3\varphi}{2}\right), \tag{1.185}$$

$$\sigma_{22} = \frac{K_{II}}{\sqrt{2\pi r}} \sin\frac{\varphi}{2}\cos\frac{\varphi}{2}\cos\frac{3\varphi}{2},$$

$$\sigma_{12} = \frac{K_{II}}{\sqrt{2\pi r}} \cos\frac{\varphi}{2}\left(1 - \sin\frac{\varphi}{2}\sin\frac{3\varphi}{2}\right),$$

$$u_1 = \frac{K_{II}\sqrt{r}}{\mu\sqrt{2\pi}} \sin\frac{\varphi}{2}\left(\frac{1+\kappa}{2} + \cos^2\frac{\varphi}{2}\right), \tag{1.186}$$

$$u_2 = \frac{K_{II}\sqrt{r}}{\mu\sqrt{2\pi}} \cos\frac{\varphi}{2}\left(\frac{\kappa-1}{2} + \sin^2\frac{\varphi}{2}\right).$$

here $K_{II} = \tau_0\sqrt{\pi a}$ is the *stress intensity factor of a transversal shear.*

The above equations are valid at an arbitrary distribution of tangential stresses along the cracks $\tau_0 = \tau_0(x_1)$, but the value of the stress intensity factor is determined by the equations

$$K_{II} = \lim_{x_1 \to a}\left[\sigma_{12}(x_1, 0)\sqrt{2\pi(x_1 - a)}\right] = \frac{1}{\pi a}\int_{-a}^{a} \tau_0(\xi)\sqrt{\frac{a+\xi}{a-\xi}}d\xi, \tag{1.187}$$

$$K_{II} = \lim_{x_1 \to -a}\left[\sigma_{12}(x_1, 0)\sqrt{2\pi(x_1 + a)}\right] = \frac{1}{\pi a}\int_{-a}^{a} \tau_0(\xi)\sqrt{\frac{a-\xi}{a+\xi}}d\xi.$$

The above relations hold for the left and right tips of the cracks, respectively.

A Crack of a Longitudinal Shear

Let us now consider one more variant of the stressed deformed state appearing in some 2-dimensional elements of real constructions, where all main characteristics, that is stress, deformations and displacements are functions of the two variables $x_1,\ x_2$

$$u_1 = 0, \qquad u_2 = 0, \qquad u_3 = u_3(x_1, x_2) \tag{1.188}$$

In this case only the following deformations differ from zero

$$\varepsilon_{23} = \frac{1}{2}\frac{\partial u_3}{\partial x_2}, \qquad \varepsilon_{13} = \frac{1}{2}\frac{\partial u_3}{\partial x_1} \tag{1.189}$$

As well as the following stresses

$$\sigma_{23} = \mu\frac{\partial u_3}{\partial x_2}, \qquad \sigma_{13} = \mu\frac{\partial u_3}{\partial x_1}, \tag{1.190}$$

and function $u_3(x_1, x_2)$ is harmonic

$$\frac{\partial^2 u_3}{\partial x_1^2} + \frac{\partial^2 u_3}{\partial x_2^2} = 0.$$

Such a state is called the state of *antiplane* or *longitudinal shear*. It is approximately realized in a long cylindrical body whose surface is under the influence of only tangential forces directed along the element of the cylinder (Ox_3-axis).

Under the condition of the anti-plane stressed deformed state we consider an unbounded elastic body with a tunnel crack located on the Ox_1x_3 - plane (Fig. 1.23) and loaded along the edges of the tangential stresses

$$\sigma_{23}(x_1, \pm 0) = -\tau_1(x_1), \qquad |x_1| < a, \qquad |x_3| < \infty.$$

From the solution of that problem we may find the distribution of stresses σ_{23}, σ_{13} and displacement u_3 in the vicinity of a tunnel crack. Here the asymptotic equations are valid for a small vicinity of the point $x_1 = a, x_2 = 0$ and they are given by

$$\sigma_{23} = \frac{K_{III}}{\sqrt{2\pi r}} \cos\frac{\varphi}{2}, \quad \sigma_{13} = \frac{K_{III}}{\sqrt{2\pi r}} \sin\frac{\varphi}{2}, \tag{1.191}$$

$$u_1 = u_2 = 0, \quad u_3 = \frac{K_{III}}{\mu}\frac{\sqrt{2r}}{\sqrt{\pi}} \sin\frac{\varphi}{2}, \tag{1.192}$$

where $K_{III} = \frac{1}{\sqrt{\pi a}} \int\limits_{-a}^{a} \tau_1(\xi) \sqrt{\frac{a+\xi}{a-\xi}} d\xi$ is the *stress intensity factor for a crack with a longitudinal shear*, determined as

$$K_{III} = \lim_{x_1 \to a} \left[\sigma_{23}(x_1, 0)\sqrt{2\pi(x_1 - a)}\right].$$

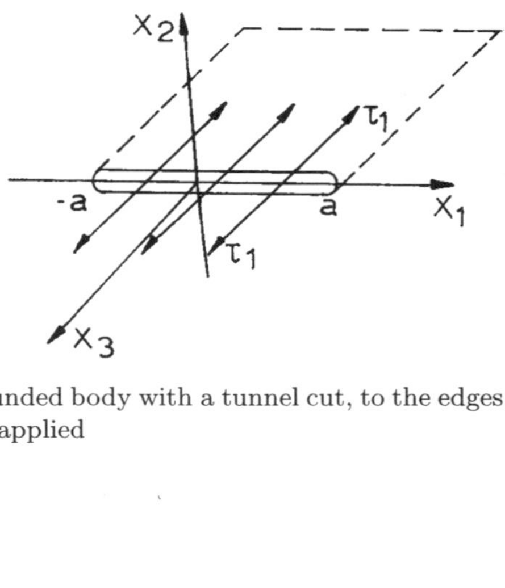

Fig. 1.23. An unbounded body with a tunnel cut, to the edges of which longitudinal tangential loads are applied

Classification of a Crack

According to the character of the deformed state in the vicinity of the front of cracks we may distinguish three main types of deformation for the local elements of the medium, containing the front of cracks, and, hence, three types of cracks: a) *the crack of a normal break*, b) *the crack of a transversal shear* and c) *the crack of a longitudinal shear* (Fig. 1.24).

In all three cases the stressed-deformed state, in a small vicinity of the origin of the coordinates, is determined by the asymptotic (1.172), (1.173), (1.185), (1.186), (1.191) and (1.192). These equations contain three parameters, namely the *stress intensity factors* K_I, K_{II}, K_{III}. The values of these parameters are obtained from the solution of the corresponding problems of the mathematical theory of elasticity and are important linear characteristics of the fields of stresses and displacements in the front of the cracks. The stress intensity factors depend on the outer load, geometry of the body with a crack and the coordinates of the considered point of the front of the cracks.

With the solutions of the problems for cracks of normal breaks for transversal and longitudinal shear at hand, we can investigate the stressed state in the vicinity of the point of the front of the cracks with arbitrary normal and tangential loads on its edges. It is obvious, that in the general case of loading, the fields of stresses and displacements are determined as a sum of the corresponding fields for the indicated three states of deformation of the local element.

It is important to note that the corresponding asymptotic equations are obtained as a result of the solution for the problem of the cracks which in the framework of the linear theory of elasticity are considered as mathematical cuts. The consequences of such a strong idealization are, for example, infinite stresses at points of the front of the cracks, which cannot be observed in reality as well as infinite density of a substance at a material point. In the front of the real crack the stresses are bounded by the limits of the strength or flow of the material. The use of the linear theory of elasticity gives us an almost-true picture for the behavior of the body with cracks in those practically important

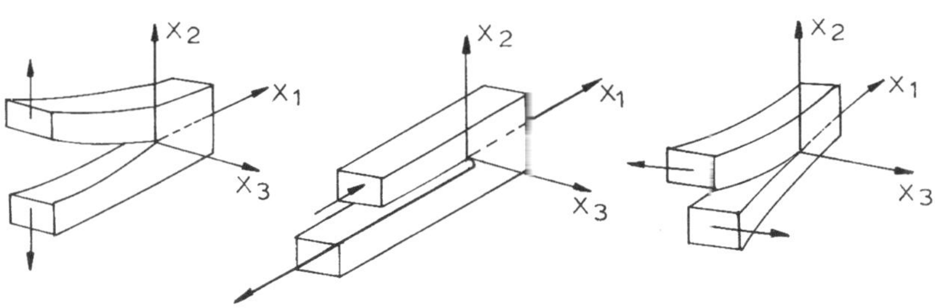

Fig. 1.24. Classification of main types of displacement of the surface of a crack: a crack of a normal break (a), a crack of a transversal shear (b) and a crack of a longitudinal shear (c)

cases where the deviation from the linear theory of elasticity takes place only in the small vicinity at the front of the cracks.

1.4.4 Criterion of Fracture of a Body with a Crack

The Theory of Griffits

It was shown that in the vicinity of the crack tip there is a considerable concentration of stresses, which can cause the break of the bounded atoms at the crack tip, growth of the cracks and eventually the complete failure of the body. Thus, the main problem of the mechanics of fracture of a deformed body is the determination of the value of the critical load. If we apply the critical load to the crack, the crack will enlarge catastrophically.

If we want the crack to propagate and its surface to enlarge by ΔS, we need to offer work $\Delta\Gamma = 2\gamma\Delta S$ to overcome the coalescence of the material particles (γ is the specific *work of fracture* with respect to the unit area of the crack surface). The energy that it is required for fracture appears at the expense of the reduction of the energy of deformation of ΔU and at the expense of the work of outer forces $\Delta \mathrm{A}$

$$\Delta\Gamma = \Delta U - \Delta \mathrm{A} = -\Delta \mathrm{E}.$$

The variable $E = U - A$ represents the potential energy of the body. The crack stayes motionless if

$$2\gamma\Delta S > -\Delta \mathrm{E} \tag{1.193}$$

and may start its growth when the sign in inequality (1.193) changes

$$-\Delta \mathrm{E} \geq 2\gamma\Delta S. \tag{1.194}$$

The equality sign in (1.194) determines the critical state, at which the crack loses its stability. Moving to the limit at $\Delta S \to 0$, we can write down the limiting condition in the form of

$$-\frac{d\mathrm{E}}{dS} = 2\gamma \tag{1.195}$$

The quantity γ is assumed to be a constant related to the material which may be measured experimentally. Griffiths, who derived condition (1.195), assumed that γ is the surface energy of the body; however such a representation appeared to be admissible only for very brittle materials of glass type. However, Griffiths' condition (1.195) may successfully describe the fracture of many constructive metallic and non-metallic materials, if by γ we imply the specific energy used for plastic deformation and fracture in the thin layers of the material adjoining to the crack.

From condition (1.195) we can either determine the *critical load* for the body with a crack of a given size, or the *critical size* of the crack in the body, on which given load acts.

For a crack of length $2a$ in an unbounded plate of unit thickness ($S = 2a \cdot 1$) stretched by stresses $\sigma_{22} = p$ in the direction perpendicular to the crack, the potential energy is calculated by the formula

$$\mathrm{E} = \mathrm{E}(a) = const - \frac{\pi p^2 a^2}{E} \tag{1.196}$$

Using condition (1.195) we obtain $\pi p^2 a = 2\gamma E$. Hence

$$p_{cr} = \sqrt{\frac{2\gamma E}{\pi a}} \tag{1.197}$$

is the *critical load* causing the fracture of the plate with a crack of the given length of $2a$ and

$$2a_{cr} = \frac{4\gamma E}{\pi p^2} \tag{1.198}$$

is the *critical size* of the crack under the given loading p.

The direct calculation of the flow of energy, $G = -\frac{d\mathrm{E}}{dS}$, through the tip of the crack becomes a very complex mathematical problem even in the simple case of (1.196). In order to calculate the growth of the crack surface on ΔS we may firstly use an imaginary cut which begins at the front of the given crack, at the same direction. The size of the imaginary crack is to be found, while the crack is loaded as described in Fig. 1.25a. Then if we slowly decrease these stresses to zero, we can obtain a crack of area increased by a factor of ΔS (Fig. 1.25b)

The estimation of the flow of energy $G = -d\mathrm{E}/dS$ in that way is valid only close to the vicinity of the front of the crack, where we may divide the stressed deformated state into three components: normal break, transversal shear and longitudinal shear, and we can also use the corresponding asymptotic equations.

Hence, for the crack of normal break (see Fig. 1.24) we obtain

$$G = -\lim_{\Delta a \to 0} \frac{1}{2\Delta a} \int\limits_0^{\Delta a} \sigma_{22} 2u_2 dx_1 \tag{1.199}$$

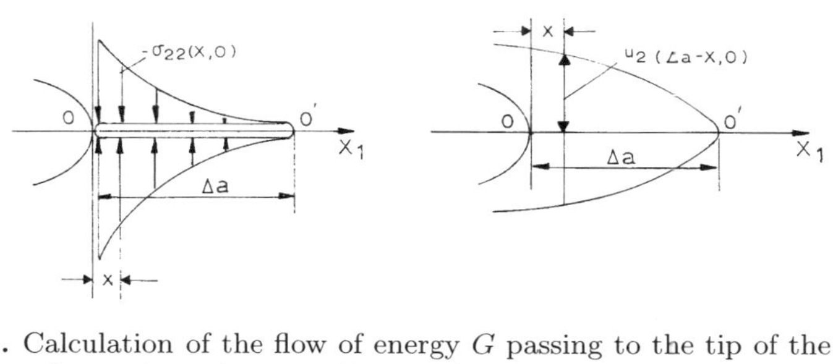

Fig. 1.25. Calculation of the flow of energy G passing to the tip of the crack; the initial state - the crack on the area of length Δa is closed by compressing stresses (a), the final state - the edges of the crack are free from loads, the crack is completely open

The asymptotic (1.172) and (1.173)

$$\sigma_{22} = K_I/\sqrt{2\pi x_1}, \quad (r = x_1, \phi = 0) \tag{1.200}$$
$$u_2 = \frac{4\left(1-\nu^2\right)}{E} K_I \sqrt{\frac{\Delta a - x_1}{2\pi}}, \quad (r = \Delta a - x_1, \varphi = \pi)$$

together with the condition, given by (1.199) give

$$G = \frac{4\left(1-\nu^2\right) K_I^2}{2\pi E} \lim_{\Delta a \to 0} \frac{1}{\Delta a} \int\limits_0^{\Delta a} \sqrt{\frac{\Delta a - x_1}{x_1}} dx_1. \tag{1.201}$$

This integral, with the help of the variable

$$x_1 = \Delta \alpha \sin^2 \theta$$

reads:

$$G = \frac{4\left(1-\nu^2\right) K_I^2}{2\pi E} \int\limits_0^{\pi/2} 2\cos^2\theta d\theta = \frac{\left(1-\nu^2\right) K_I^2}{E}. \tag{1.202}$$

In the same way we can determine the flow of energy for cracks of transversal and longitudinal shear and for cracks of the general form

$$G = -\frac{d\mathrm{E}}{da} = \frac{1+\nu}{E}\left[(1-\nu)\left(K_I^2 + K_{II}^2\right) + K_{III}^2\right] \tag{1.203}$$

Substituting G into (1.195) we obtain the condition of local fracture at a point of the front of the crack of the general type

$$\frac{1+\nu}{E}\left[(1-\nu)\left(K_I^2 + K_{II}^2\right) + K_{III}^2\right] = 2\gamma \tag{1.204}$$

Force Criterion of Fracture

The energetic condition of fracture (1.204) may be written in the following form:

- *for a crack of normal break* $(K_{II} = K_{III} = 0)$

$$K_I = K_{I_c}, \tag{1.205}$$

where $K_{I_c} = \sqrt{E\gamma/\left(1-\nu^2\right)}$ is the constant material which is called the *critical stress intensity factor* or the *viscosity of fracture.*

- *for a crack of transversal shear* $(K_I = K_{III} = 0)$

$$K_{II} = K_{I_c} \tag{1.206}$$

- *for a crack of longitudinal shear* $(K_I = K_{II} = 0)$

$$K_{III} = K_{III_c}. \tag{1.207}$$

where $K_{III_c} = \sqrt{2E\gamma/\left(1+\nu\right)}$ is the constant material.

According to (1.205)–(1.207), the local fracture for the given material takes place when the constant of the critical values of the stress intensity factor is reached. This is the force criterion of fracture equivalent to Griffiths' criterion.

Let us now consider some examples of application of the force criterion of fracture for a crack of normal break.

Example 1

Let an unbounded thin plate have a thorough crack of length $2a$ and the plate stretches at infinity of uniform load $\sigma_{22} = p$ (see Fig. 1.20). In this case the stress intensity factor is determined by (1.176). Using the criterion (1.205) we find the ultimate load, for which the crack will start developing:

$$p_{cr}\sqrt{\pi a} = K_c \tag{1.208}$$

Here the quantity K_c (which is different from K_{I_C}) appears, and as for thin plates the experiment gives the critical value of the stress intensity factor that exceeds K_{I_c} and depends on the thickness of the plate. Thus, the thickness of length $2a$ starts to grow if the load p reaches the ultimate value of

$$p_{cr} = K_c / \sqrt{\pi a} \tag{1.209}$$

If the applied methods of nondestructive testing cannot discover cracks of length less than $2a$, (1.209) allows us to determine the limiting level of load, at which the "defectless" plate may be used.

Example 2

Let the load of a through crack of a normal break be carried out by a couple of concentrated forces, applied to its edges in the middle of the crack (Fig. 1.21).

In the given case, according to (1.180) and the criterion of fracture (1.205) the limiting values of force P which causes the increase of the crack, is determined by the expression

$$P_{cr} = K_c\sqrt{\pi a} \tag{1.210}$$

Example 3

By applying a compressing force $-P$ we may decrease or even remove the stress concentration at the tip of the crack on the plate stretched by stress p_0. The stress intensity for such a loading may be calculated by (1.180) and (1.176) (see Figs. 1.20, 1.21 and 1.26)

$$K_I = p_0\sqrt{\pi a} - \frac{P}{\sqrt{\pi a}}. \tag{1.211}$$

In engineering manufacturing is widely used to imposing reinforcing constraints in order to prevent brittle fractures.

Stability of a Crack Growth

In the example of a crack of normal break we can investigate the problem of stability of a limiting state of a body with respect to small disturbances of the load p and the size of crack α. The growth of the crack by small quantity $dp > 0$ may be accompanied by the increase of the load parameter $dp > 0$ or its decrease by $dp < 0$. The crack is considered to be *stable* if $dp/da > 0$, i.e. if for the further growth of the crack the applied load is increased. On the other hand when $dp/da < 0$, the crack will grow spontaneously even under permanent or decreasing load and it is called *unstable*.

According to the local criterion of fracture (1.205) we have

$$K_{I_c} = K_I\,(p, a) \tag{1.212}$$

Then

$$\frac{dp}{da} = -\left(\frac{\partial K_I}{\partial a}\right) \Big/ \left(\frac{\partial K_I}{\partial p}\right) \tag{1.213}$$

As $\partial K_I/\partial p > 0$, the crack of normal break is stable, if the stress intensity factor is a decreasing function of the length of crack $\partial K_I/\partial a < 0$. The instability is observed in those cases, where the stress intensity factor is an increasing function of the length of crack $\partial K_I/\partial a > 0$.

In particular, for cracks on a stretched plate according to (1.176) we get

$$K_I = p\sqrt{\pi\, a}, \qquad \frac{\partial K_I}{\partial a} = \frac{p\sqrt{\pi}}{2\sqrt{a}} > 0,$$

i.e. such a crack in the limited state becomes unstable. The opposite result is obtained for cracks of normal break, loaded by two concentrated forces. The stress intensity factor at the tip is determined by (1.180) and for such a crack the following equality is satisfied

$$\frac{\partial K_I}{\partial a} = -\frac{P}{2a\sqrt{\pi\, a}} < 0.$$

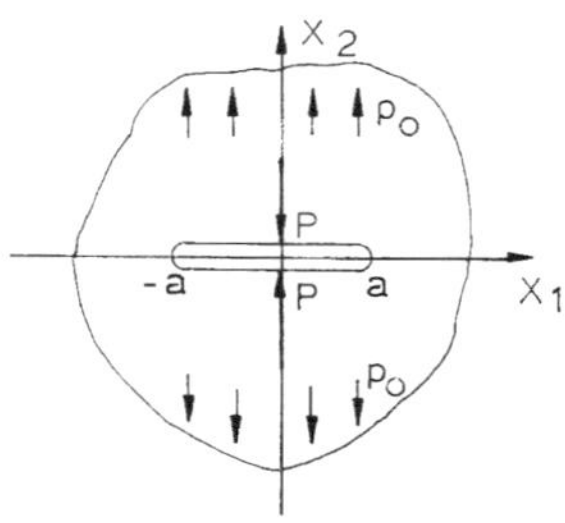

Fig. 1.26. Modeling of the influence of reinforced connections on the concentration of stresses near the tips of the crack: a pair of oblated concentrated forces applied to the edges of a transversal crack in a stretched plate

Thus, a crack of normal break loaded by two concentrated forces is stable. If we apply a force P to a crack of length $2a_0$, the quantity of which exceeds the critical ones (1.210)

$$P > K_c\sqrt{\pi\, a_0}$$

Beyond this value the crack will start to grow; it will stop when its length reaches $2a_*$ which is defined by the critical condition $P = K_C\sqrt{\pi a_*}$.

The unstable growth of a crack of a arbitrary form is accompanied by an increase of the velocity of the elastic energy release, upon increasing the length of crack $dG/da > 0$, i.e. in order to analyze the stability of the crack of general form by formula (1.203) it is enough to know the dependence of the stress intensity factor on the length of the crack. The condition of instability is the inequality

$$\frac{d}{da}\left[(1-\nu)\left(K_I^2 + K_{II}^2\right) + K_{III}^2\right] > 0.$$

2

Basic Equations of the Linear Electroelasticity

2.1 The Linear Theory of the Piezoelectricity

A piezoelectric material is subject to the coupled action of mechanical and electrical fields. This is caused by the asymmetry of the atomic lattice of the piezoelectric material (ferroelectric). These materials pose a very interesting property: under the action of an electric field the material undergoes a deformation and vice versa, a deformation of a piezoelectric material gives birth to an electric field. Based on that, the piezoeffect may be classified as a reverse and direct one.

For describing in a complete and accurate way the behavior of such materials, it is vital to consider not only the relation between the stressed and deformed state of the body, but also the equations of the electromagnetic field in the medium.

Contrary to the case of vacuum, in a material medium, which is under the influence of an external electomagnetic field both electric polarization and magnetization effects appear. Maxwell's equations for that case are

$$div \overrightarrow{D} = \rho_*, \qquad rot \overrightarrow{E} = -\partial \overrightarrow{B} \big/ \partial t, \qquad (2.1)$$
$$rot \overrightarrow{H} = \partial \overrightarrow{D} \big/ \partial t + \overrightarrow{j}, \qquad div \overrightarrow{B} = 0.$$

where $\overrightarrow{D}$ is the vector of the electric displacement, $\overrightarrow{E}$ is the vector of the electric field strength, $\overrightarrow{H}$ and $\overrightarrow{B}$ are the vectors of the magnetic field strength and magnetic induction, respectively, ρ_* is the free charge density, $\overrightarrow{j}$ is the free charge current density. The components of the vectors $\overrightarrow{D}$ and $\overrightarrow{H}$ are defined by the equalities

$$D_i = \ni_0 E_i + P_i, \quad H_i = \frac{B_i}{\mu_0} - M_i. \qquad (2.2)$$

where the vectors $\overrightarrow{P}$ and $\overrightarrow{M}$ characterize the properties of polarization and magnetization of the medium under study and the constants $\ni_0 = 8.85 \cdot$

$10^{-12}\,F/m$ and $\mu_0 = 12.56 \cdot 10^{-7}\,\mathrm{H}/m$ are the dielectric and magnetic permittivity of the vacuum respectively.

If we consider vacuum as the surrounding medium, it is necessary to introduce $\overrightarrow{P} = \overrightarrow{M} = 0$ into (2.2); in that case, (2.1) may be written in terms of two vectors, e.g. the vectors $\overrightarrow{E}$ and $\overrightarrow{H}$. Under this formulation the system (2.1) becomes solvable.

If medium under consideration is a material (2.1), (2.2) can be solved assuming availability of the functional dependences $\overrightarrow{\boldsymbol{P}} = \overrightarrow{\boldsymbol{P}}\left(\overrightarrow{\boldsymbol{E}}\right)$, $\overrightarrow{\boldsymbol{M}} = \overrightarrow{\boldsymbol{M}}\left(\overrightarrow{\boldsymbol{B}}\right)$ and $\overrightarrow{\boldsymbol{J}} = \overrightarrow{\boldsymbol{J}}\left(\overrightarrow{\boldsymbol{E}}\right)$; for a given material, these are usually estimated experimentally and take the form of linear relations. In particular, for an anisotropic medium the linear relations connecting the components of the vectors $\overrightarrow{D}$ and $\overrightarrow{E}$ can be represented in an index notation as (f course, the summation is over repeated indexes)

$$D_i = \ni_{ij} E_j. \tag{2.3}$$

where $\ni_{ij}$ is the symmetrical tensor of dielectric constants.

In reality (2.3) reflect only the physical processes in the definite interval of E_i and correspond to zero polarization for $\overrightarrow{E} = 0$. However the last condition is not true for ferroelectric materials where the polarization occurs when $E_j = 0$.

The experiments show that in the case of a direct piezoeffect and in the absence of an electric field, the vector of polarization of the piezoelectric is related in a linear way with the components of the mechanical stress tensor $\sigma_{k\ell}$ $(k, \ell = 1, 2, 3)$ as

$$P_i = d_{ik\ell}\sigma_{k\ell}, \tag{2.4}$$

where $d_{ik\ell}$ are the piezoelectric components of a third-rank tensor.

The linear connection between the electric field strength E_i and the deformation tensor ε_{ij} of a mechanically free crystal $(\sigma_{ij} = 0)$ is a mathematical interpretation of the reverse piezoeffect:

$$\varepsilon_{ij} = d_{kij} E_k. \tag{2.5}$$

Since $\varepsilon_{ij} = \varepsilon_{ji}$, tensor d_{kij} is symmetrical, according to (2.5) regarding the last two indexes. Hence the number of the independent components of the tensor decreases from $3^3 = 27$ to 18. Depending on the crystal symmetry it is possible to obtain a further decrease in the number of independent piezoelectric modulus. By using elements of tensor analysis it is possible to show that in the presence of a centre of symmetry, the crystal cannot poses piezoelectric properties.

The crystal structure of ferroelectrics, such as titanite barium, niobate lithium, and ceramics PZT, reveals that they have a non-zero dipole moment. Above a temperature (which is called Curie point) the directions of the dipoles are random. It is possible to polarize (i.e. to orientate the dipoles in the chosen direction, of the external field) the ferroelectric material, by applying an intensive electric field at a temperature close to the Curie point. In that case most of the domains within the material which are macroscopic areas of ordered directions of the dipoles have a similar orientation. The polarization process is used in practice for the creation of artificial piezoelectric materials with the desirable electroacoustic characteristics.

At this point, it is necessary to show the difference of the piezoeffect from the electrostriction, which occurs in any dielectric under the effect of an electric field and is proportional to the square of its strength In materials, except those that have high dielectric permittivity, the electrostriction- when compared to the piezoelectric effect- may be neglected. This is related with the fact that internal fields produced by the asymmetry of the lattice exceed the possible external fields by several orders.

It is interesting to mention that (2.4), (2.5) have similar piezoelectric tensor coefficient. This fact can be obtained from thermodynamic analysis, on the basis of the law of conservation of energy formulated for a piezoelectric media.

Under the action of external loads and a electromagnetic field, the macroscopic form of the energy balance for a piezoelectric with volume V reads

$$\frac{d}{dt}\int_V \left(\frac{1}{2}\rho\dot{u}_i\dot{u}_i + U\right)dV = \int_V \left(X_i\dot{u}_i + E_i\dot{D}_i\right)dV + \int_{\partial V} q_i\dot{u}_i dS. \tag{2.6}$$

where U is the internal energy density, u_i is the vector of the mechanical displacement, X_i are the mass forces, $q_i = \sigma_{ij}n_j$ are the surface forces, ρ is the density of the material, n_i is the vector of a single normal to the surface ∂V, while time derivatives are designated by a dot.

In the right-hand side of (2.6) the expression

$$U_{e\ell} = \int_V E_i\dot{D}_i dV$$

is for the flow of electromagnetic energy in the electrostatic approximation [137] and the quantity

$$L = \int_V X_i\dot{u}_i dV + \int_{\partial V} q_i\dot{u}_i dS$$

for the power of the external forces.

Applying the equations of motion

$$\partial_j\sigma_{ij} + X_i = \rho\ddot{u}_i, \qquad \partial_j = \partial/\partial x_j\ (i,j = 1,3) \tag{2.7}$$

and using the symmetry of the stress tensor σ_{ij}, the surface integral in (2.6) is transformed in

$$\int_{\partial V} q_i \dot{u}_i dS = \int_{\partial V} \sigma_{ij} \dot{u}_i n_j dS = \int_V (\dot{u}_i \partial_j \sigma_{ij} + \sigma_{ij} \partial_j \dot{u}_i)\, dV =$$
$$= \int_V (-X_i \dot{u}_i + \rho_i \ddot{u}_i \dot{u}_i + \sigma_{ij} \dot{\varepsilon}_{ij})\, dV, \tag{2.8}$$

where $\varepsilon_{ij} = \frac{1}{2}(\partial_j u_i + \partial_i u_j)$ is the deformation tensor. Taking into account (2.8) the equation of the energy balance (2.6) becomes

$$\int_V \dot{U} dV = \int_V \left(\sigma_{ij} \dot{\varepsilon}_{ij} + E_i \dot{D}_i\right) dV. \tag{2.9}$$

Since the equality (2.9) is valid for any arbitrary chosen volume, the local energy equation reads

$$\dot{U} = \sigma_{ij} \dot{\varepsilon}_{ij} + E_i \dot{D}_i. \tag{2.10}$$

The internal energy U is a function of the independent thermodynamic variables ε_{ij}, D_i and therefore

$$\dot{U} = \frac{\partial}{\partial \varepsilon_{ij}} \dot{\varepsilon}_{ij} + \frac{\partial U}{\partial D_i} \dot{D}_i. \tag{2.11}$$

From expressions (2.10), (2.11), because of the arbitrary time derivatives we can get the following relations

$$\sigma_{ij} = \frac{\partial U}{\partial \varepsilon_{ij}}, \qquad E_i = \frac{\partial U}{\partial D_i}. \tag{2.12}$$

Thus on the basis of (2.10), (2.12) the total differential of the internal energy is given by

$$dU = \sigma_{ij} d\varepsilon_{ij} + E_m dD_m. \tag{2.13}$$

where

$$\sigma_{ij} = \left(\frac{\partial U}{\partial \varepsilon_{ij}}\right)_D, \qquad E_m = \left(\frac{\partial U}{\partial D_m}\right)_\varepsilon, \tag{2.14}$$

and the lower indexes show that D and E are constant for the process of partial differentiation of function U.

The total derivatives that depend on variables σ_{ij} and E_m can be written as

$$d\sigma_{ij} = \left(\frac{\partial \sigma_{ij}}{\partial \varepsilon_{k\ell}}\right)_D d\varepsilon_{k\ell} + \left(\frac{\partial \sigma_{ij}}{\partial D_m}\right)_\varepsilon dD_m, \tag{2.15}$$
$$dE_m = \left(\frac{\partial E_m}{\partial \varepsilon_{ij}}\right)_D d\varepsilon_{ij} + \left(\frac{\partial E_m}{\partial D_k}\right)_\varepsilon dD_k.$$

The above equations can be written in a more compact form as

$$\begin{aligned} d\sigma_{ij} &= c^D_{ijk\ell} d\varepsilon_{k\ell} - h_{ijm} dD_m, \\ dE_m &= -h_{mij} d\varepsilon_{ij} + \beta^\varepsilon_{mk} dD_k. \end{aligned} \tag{2.16}$$

where $c^D_{ijk\ell}$ denotes the tensor of the modulus of electricity at constant electric induction; β^ε_{mk} denote the dielectric constants at fixed deformations and h_{mij} the tensor of piezoelectric constants.

When the range of the values of the variables D_k and ε_{ij} is small the equalities (2.16) can be integrated. As result we get the following linear equations describing the piezoelectric medium state:

$$\begin{aligned} \sigma_{ij} &= c^D_{ijk\ell} \varepsilon_{k\ell} - h_{ijm} D_m, \\ E_m &= -h_{mij} \varepsilon_{ij} + \beta^\varepsilon_{mk} D_k. \end{aligned} \tag{2.17}$$

Other forms of the equations of the state can be derived, in an analogous manner, with the help of other thermodynamic potentials. If we take the deformations ε_{ij} and the strength of the electric field E_i to be independent variables, then the stresses σ_{ij} and electric induction D_i are the dependent ones. Then, using the Gibbs electric function $G = U - E_m D_m$ [140] we find

$$\begin{aligned} \sigma_{ij} &= c^E_{ijk\ell} \varepsilon_{k\ell} - e_{ijm} E_m, \\ D_m &= \ni^\varepsilon_{mk} E_k + e_{mij} \varepsilon_{ij}. \end{aligned} \tag{2.18}$$

Where $\ni^\varepsilon_{mk}, e_{ijm}$ are tensors of dielectric permittivity (at the constant deformation) and of piezoelectric constants.

Using the expression for the complete Gibb's potential $G = U - \sigma_{ij}\varepsilon_{ij} - E_m D_m$ we can derive another form of the equations of state, reading

$$\begin{aligned} \varepsilon_{ij} &= s^E_{ijk\ell} \sigma_{k\ell} + d_{kij} E_k, \\ D_i &= d_{ijk} \sigma_{jk} + \ni^\sigma_{ij} E_j \end{aligned} \tag{2.19}$$

The optimal choice of any equation system out of (2.17)–(2.19) is defined by the specific range of problem of the theory of electroelasticity. When studying wave processes in piezoelectric media it is convenient to use (2.17), (2.18); in piezoelectric shells and plates, for the investigation of the associated fields it is recommended to consider (2.19).

Relations (2.17) as well as (2.18) and (2.19) should be written in matrix form, when solving problems.

Here we introduce the following rules

$$\Gamma_{11} = \Gamma_1, \Gamma_{22} = \Gamma_2, \Gamma_{33} = \Gamma_3, \Gamma_{23} = \Gamma_4, \Gamma_{13} = \Gamma_5, \Gamma_{12} = \Gamma_6,$$

where, Γ is for the stress tensor σ_{ij} or the deformation tensor ε_{ij}. The tensors $c_{ijk\ell}$, $s_{ijk\ell}$ are symmetrical by the first two (i and j) and the last two (k and l) indexes so we can write them as:

$$c_{ijk\ell} = c_{\alpha\beta}, \quad s_{ijk\ell} = s_{\alpha\beta}, \qquad (2.20)$$
$$(i, j, k, \ell = 1, 2, 3; \quad \alpha, \beta = 1, 2, ..., 6)\,.$$

Consequently, the coefficients $c_{\alpha\beta}$ $(\alpha, \beta = 1, 2, \ldots, 6)$ form a square matrix (6×6).

Taking into account the symmetry of the tensors e_{mij} and d_{mij} by the pair of the last two indexes we have

$$e_{mij} = e_{m\alpha}, \qquad d_{mij} = d_{m\alpha}, \qquad (2.21)$$
$$(m = 1, 2, 3; \qquad \alpha = 1, 2, ..., 6)\,,$$

Hence we get a 3x6 matrix having as elements the piezoelectric constants. As a result, the equations of state (2.18) read

$$\sigma_\alpha = c^{\mathrm{E}}_{\alpha\beta}\varepsilon_\beta - e_{\alpha m}E_m, \qquad (2.22)$$
$$D_m = \ni^{\varepsilon}_{mk} E_k + e_{m\alpha}\varepsilon_\alpha,$$
$$(m, k = 1, 2, 3; \qquad \alpha, \beta = 1, 2, ..., 6)\,.$$

In the general case of a crystal of the triclinic system (2.22) contain 21 modules of elasticity, 18 piezoelectric and 6 dielectric constants. The crystallographic symmetry (relating to chosen coordinate axes) in the deformed piezoelectric body decreases substantially the number of independent constants in the above equations.

In Table 2.1 are given the nonzero elements of the matrixes of elastic piezoelectric and dielectric constants for some crystal classes used in ultrasonics, such as Quartz (class 32), niobate lithium $LiNbO_3$ (class $3m$), sulphide cadmium CdS (class $6mm$) and polarized ceramics (class $6mm$).

In the absence of the conductance current $\overrightarrow{j}$ and free charges ρ_*, the solution to the equations of motion (2.7) and Maxwell's (2.1) determines the behavior of waves due to coupled electro-magneto-elastic effects, i.e. elastic waves accompanied by electric field, and electromagnetic waves accompanied by deformation of the medium. If we assume that the velocity of elastic waves propagation is $\overrightarrow{v}$, the corresponding velocity of an electromagnetic wave c will be of order $10v^5$ and therefore in the case of the elastic waves we can neglect the magnetic field generated by electricity. Hence, for important real-world problems related with the analysis of the dynamics of piezoelectric transducers and the propagation of electro-acoustic waves, the magnetic effects are neglected; application of the quasi-static approximation for an electric field, gives

$$rot\overrightarrow{E} = -\frac{\partial \overrightarrow{B}}{\partial t} \cong 0, \qquad div\overrightarrow{D} = 0. \qquad (2.23)$$

Based on the above assumptions the electric field strength E_i is expressed in terms of scalar potential ϕ as

$$E_i = -\partial_i \phi \qquad (2.24)$$

Table 2.1. The non-zero elements of the matrixes of elastic piezoelectric and dielectric constants for some crystal classes used in ultrasonics

Matrix of electric constants $c_{a\beta}$						Matrix of piezoelectric constants e_{ai}						Matrix of electric constants $\ni_{ik}$		
Crystals with triangular symmetry (Class 32, 3m)														
c_{11}	c_{12}	c_{13}	c_{14}	.	.	e_{11}	.	.	.	$-e_{22}$	e_{13}	$\ni_{11}$	.	.
c_{12}	c_{11}	c_{13}	c_{14}	.	.	$-e_{11}$	.	.	.	e_{22}	e_{13}	.	$\ni_{11}$	.
c_{13}	c_{13}	c_{33}	.	.	.	.	.	.	.	.	e_{33}	.	.	$\ni_{33}$
c_{14}	$-c_{14}$	.	c_{44}	.	.	e_{41}	.	.	.	e_{51}	.			
.	.	.	.	c_{44}	c_{14}	.	$-e_{41}$	.	e_{51}	.	.			
.	.	.	.	c_{14}	c_{66}	.	$-e_{41}$	.	$-e_{22}$	.	.			
$c_{66} = (c_{11} - c_{12})/2$														
Crystals with hexagonal symmetry (Class 6mm)														
c_{11}	c_{12}	c_{13}	.	.	.		.	.	e_{13}			$\ni_{11}$	.	.
c_{12}	c_{11}	c_{13}	.	.	.		.	.	e_{13}			.	$\ni_{11}$	.
c_{13}	c_{13}	c_{33}	.	.	.		.	.	e_{33}			.	.	$\ni_{33}$
.	.	.	c_{44}	.	.		.	e_{31}	.					
.	.	.	.	c_{44}	.		e_{51}	.	.					
.	.	.	.	.	c_{66}		.	.	.					
$c_{66} = (c_{11} - c_{12})/2$														
Crystals with cubic symmetry (Class $\bar{4}3m$)														
c_{11}	c_{12}	c_{12}	.	.	.		.	.	.			$\ni_{11}$	.	.
c_{12}	c_{11}	c_{12}	.	.	.		.	.	.			.	$\ni_{11}$	.
c_{12}	c_{12}	c_{11}	.	.	.		.	.	.			.	.	$\ni_{11}$
.	.	.	c_{44}	.	.		e_{41}	.	.					
.	.	.	.	c_{44}	.		.	e_{41}	.					
.	.	.	.	.	c_{44}		.	.	e_{41}					

The complete set of equations for describing a linear piezoelectric medium is acquired by inserting the equation of state (2.18) into the (2.7) and the equation of electrostatics $\partial_i D_i = 0$.

Expressions of mechanical stresses and electric inductions allowing for (2.24) will have the form

$$\begin{aligned} \sigma_{ij} &= c^E_{ijk\ell}\partial_\ell u_k + e_{kij}\partial_k \phi, \\ D_i &= e_{ik\ell}\partial_\ell u_k - \ni^{\varepsilon}_{ik}\, \partial_k \phi. \end{aligned} \tag{2.25}$$

Inserting (2.25) into the equations of motion (2.7) and the equation of electrostatics we obtain the coupled system of solvable equations in the framework of linear electroelasticity:

$$\begin{aligned} c^E_{ijk\ell}\partial_\ell\partial_j u_k + e_{kij}\partial_k\partial_j\phi &= \rho\frac{\partial^2 u_i}{\partial t^2}, \\ e_{ik\ell}\partial_\ell\partial_i u_k - \ni^{\varepsilon}_{ik}\, \partial_k\partial_i\phi &= 0. \end{aligned} \tag{2.26}$$

It should be noted that the above set of equations are sufficient to analyze the propagation of acoustic-electric waves in piezoelectric media of infinite size. When solving the basic problem of electroelasticity, e.g. in the case of bodies with stress concentrators such as cracks, cavities and inclusions it is necessary to consider the system (2.26) together with the corresponding boundary conditions referring to mechanical and electric quantities. The method of singular integral equations used for the determination of conjugate electric and elastic fields in the above mentioned problems assumes the knowledge of the fundamental solutions of the (2.26). Then it is possible to construct integral representations of the solutions that determine the connection between boundary values of the sought-for functions and their values in the body area. However, for piezocrystals possessing common classes of symmetry the determination of the the corresponding fundamental solutions of the system (2.26) becomes a rather difficult mathematical-task.

2.2 Equations of State for Piezoelectric Ceramics

The piezoelectric ceramics possess a series of remarkable properties and in some cases appear to be preferable to natural piezoelectric crystals. The essential advantage of the piezoelectric ceramics lies in the manufacturing process; depending on the transducer different geometric shapes, can be made.

In their initial state the piezoelectric ceramics are isotropic dielectrics which consist of polycrystal solid solution of titanite barium and zirconate lead containing as an adhesive polymeric glass phase. In that state polycrystal ceramics reveal a higher symmetry when compared with the symmetry of separate crystals. In order to make the ceramics piezoactive it is necessary to polarize them at a temperature below the Curie point [250]. After the removal of the electric field and the cooling of the material the polarization is preserved. This is due to the orientation of the polarization directions of the domains along the external field. Due to the preliminary polarization the polycrystal ceramics turn into piezoelectricones with mechanical and electric properties which are characteristic features of the materials of transversal and isotropic symmetry. A rigorous derivation of the defining relations for polarized piezoelectric ceramics is given as an example in [102, 122, 220]. These are obtained from general laws of motion for deforming polarizing media interacting with the electromagnetic fields. The linear equations of the state of piezoelectric ceramics polarized in the direction of the x_3 axis (in Cartesian coordinates) have the form of the corresponding equations for a piezocrystal of hexagonal symmetry in $6mm$ (the optical axis coincides with the x_3- axis):

$$\begin{aligned}
\sigma_{11} &= c_{11}^E \varepsilon_{11} + c_{12}^E \varepsilon_{22} + c_{13}^E \varepsilon_{33} - e_{31} E_3, \\
\sigma_{22} &= c_{12}^E \varepsilon_{11} + c_{11}^E \varepsilon_{22} + c_{13}^E \varepsilon_{33} - e_{31} E_3, \\
\sigma_{33} &= c_{13}^E (\varepsilon_{11} + \varepsilon_{22}) + c_{33}^E \varepsilon_{33} - e_{33} E_3, \\
\sigma_{23} &= 2c_{44}^E \varepsilon_{23} - e_{15} E_2, \\
\sigma_{13} &= 2c_{44}^E \varepsilon_{13} - e_{15} E_1, \\
\sigma_{12} &= \left(c_{11}^E - c_{12}^E\right) \varepsilon_{12}, \\
D_1 &= \ni_{11}^{\varepsilon} E_1 + 2e_{15}\varepsilon_{13}, \\
D_2 &= \ni_{11}^{\varepsilon} E_2 + 2e_{15}\varepsilon_{23}, \\
D_3 &= \ni_{33}^{E} E_3 + e_{31} (\varepsilon_{11} + \varepsilon_{22}) + e_{33}\varepsilon_{33}
\end{aligned} \tag{2.27}$$

Under this framework the mechanical, dielectric and piezoelectric properties of the ceramics are determined by a set of five moduli of elasticity c_{11}^E, c_{12}^E, c_{13}^E, c_{33}^E, c_{44}^E, two dielectric permittivities $\ni_{11}^{\varepsilon}$, $\ni_{33}^{\varepsilon}$ and three piezoelectric constants e_{31}, e_{33}, e_{15}.

We should note that the (2.27) describing the piezoeffect take into account additional electroelastic fields in preliminarily polarized ceramics which appear under the influence of the given mechanical and electric load.

If we consider a homogeneous distribution of the field of preliminary polarization inside of the body volume, then the above equations uniquely depict the functional associations. Otherwise the material factors in (2.27) should be considered as coordinate functions x_i $(i = 1, 2, 3)$.

The equations of the state analogous to (2.27) may also be represented by relations (2.17) as

$$\begin{aligned}
&\sigma_{11} = c_{11}^D \varepsilon_{11} + c_{12}^D \varepsilon_{22} + c_{13}^D \varepsilon_{33} - h_{31} D_3, \\
&\sigma_{22} = c_{12}^D \varepsilon_{11} + c_{11}^D \varepsilon_{22} + c_{13}^D \varepsilon_{33} - h_{31} D_3, \\
&\sigma_{33} = c_{13}^D (\varepsilon_{11} + \varepsilon_{22}) + c_{33}^D \varepsilon_{33} - h_{33} D_3, \\
&\sigma_{23} = 2c_{44}^D \varepsilon_{23} - h_{15} D_2, \quad \sigma_{13} = 2c_{44}^D \varepsilon_{13} - h_{15} D_1, \\
&\sigma_{12} = \left(c_{11}^D - c_{12}^D\right) \varepsilon_{12}, \qquad E_1 = -2h_{15}\varepsilon_{13} + \beta_{11}^{\varepsilon} D_1, \\
&E_2 = -2h_{15}\varepsilon_{23} + \beta_{11}^{\varepsilon} D_2, \quad E_3 = -h_{31} (\varepsilon_{11} + \varepsilon_{22}) - h_{33}\varepsilon_{33} + \beta_{33}^{\varepsilon} D_3.
\end{aligned} \tag{2.28}$$

Table 2.2. The values of material parameters for most ceramics, taken from [140]

	Piezoceramic PZT 65/35	
$c_{11}^E = 1,594 \cdot 10^{11}\ Pa$	$e_{31} = -6,127 cal/m^2$	$\ni_{11} = 5,66 \cdot 10^{-9}\ F/m$
$c_{12}^E = 7,385 \cdot 10^{10}\ Pa$	$e_{33} = 10,71 cal/m^2$	$\ni_{33} = 2,243 \cdot 10^{-9}\ F/m$
$c_{33}^E = 1,261 \cdot 10^{11}\ Pa$	$e_{15} = 8,387 cal/m^2$	
$c_{44}^E = 3,89 \cdot 10^{10}\ Pa$		
$c_{66}^E = 4,276 \cdot 10^{10}\ Pa$		
	Piezoceramic VTC-19	
$c_{11}^E = 11,22 \cdot 10^{10}\ N/m^2$	$e_{31} = -3,4 cal/m^2$	$\ni_{11} = 7,257 \cdot 10^{-9}\ F/m$
$c_{13}^E = 6,22 \cdot 10^{10}\ N/m^2$	$e_{33} = 15,1 cal/m^2$	$\ni_{33} = 8,274 \cdot 10^{-9}\ F/m$
$c_{33}^E = 10,6 \cdot 10^{10}\ N/m^2$	$e_{15} = 9,45 cal/m^2$	
$c_{44}^E = 2,49 \cdot 10^{10}\ N/m^2$		
	Piezoceramic PZT-4	
$c_{11}^E = 13,9 \cdot 10^{10}\ N/m^2$	$e_{31} = -5,2 cal/m^2$	$\ni_{11} = 6,45 \cdot 10^{-9}\ F/m$
$c_{12}^E = 7,78 \cdot 10^{10}\ N/m^2$	$e_{33} = 15,1 cal/m^2$	$\ni_{33} = 5,62 \cdot 10^{-9}\ F/m$
$c_{13}^E = 7,43 \cdot 10^{10}\ N/m^2$	$e_{15} = 12,7 cal/m^2$	
$c_{33}^E = 11,5 \cdot 10^{10}\ N/m^2$		
$c_{44}^E = 2,56 \cdot 10^{10}\ N/m^2$		
$c_{66}^E = 3,06 \cdot 10^{10}\ N/m^2$		
$S_{11}^E = 12,3 \cdot 10^{-12}\ m^2/N$	$d_{31} = 1,23 \cdot 10^{-10} cal/N$	
$S_{12}^E = -4,05 \cdot 10^{-12}\ m^2/N$	$d_{33} = 2,89 \cdot 10^{-10} cal/N$	
$S_{13}^E = -5,31 \cdot 10^{-12}\ m^2/N$	$d_{15} = 4,96 \cdot 10^{-10} cal/N$	
$S_{33}^E = 15,5 \cdot 10^{-12}\ m^2/N$		
$S_{44}^E = 39,5 \cdot 10^{-12}\ m^2/N$		
$S_{66}^E = 32,7 \cdot 10^{-12}\ m^2/N$		

Here $\varepsilon_{ij} = \frac{1}{2}(\partial_j u_i + \partial_i u_j)$ are the components of the tensors of small deformations; c_{ij}^D, β_{km}^ε and h_{lm} are the moduli of elasticity at a constant electric induction, the dielectric "impermittivities" at fixed deformations and constants of piezoelectric connectivity respectively. Table 2.2 gives the values of the material parameters for the most wide-spread piezoelectric ceramics, taken out of [140].

2.3 Two-Dimensional Problems of Electroelasticity

Let us consider the analysis of the differential (2.26) for different crystallographic classes in order to reveal the conditions under which it is possible to write down the plane problem of electroelasticity. Suppose that one of the components of the vector of elastic displacement is equal to zero, but all the other components of mechanical and electric fields do not depend on the corresponding coordinate, i.e.

$$u_\alpha = 0, \quad \frac{\partial}{\partial x_\alpha} = 0 \qquad (\alpha = 1, 2, 3).$$

Then the conditions of solution existence to the plane problem are zreduced to the following equalities

$$c_{\alpha\beta\beta\beta} = c_{\alpha\beta\beta\gamma} = c_{\alpha\beta\gamma\gamma} = c_{\alpha\gamma\beta\beta} = c_{\alpha\gamma\beta\gamma} = c_{\alpha\gamma\gamma\gamma} = 0, \tag{2.29}$$
$$e_{\beta\alpha\beta} = e_{\beta\alpha\gamma} = e_{\gamma\alpha\gamma} = 0, \qquad \gamma \neq \alpha, \quad \beta \neq \gamma \quad (\alpha, \beta, \gamma = 1, 2, 3).$$

Assuming that $u_2 = 0, \partial/\partial x_2 = 0$ conditions (2 29) will be applied to class $2mm$ of rhombic system, classes $4mm$ of square system and class $6mm, 6m2$ of hexagonal symmetry. For these classes the dynamic equations of the plane problem of electroelasticity allowing for matrix symbols are given by:

$$c_{11}^{E}\partial_1^2 u_1 + \left(c_{13}^{E} + c_{55}^{*}\right)\partial_1\partial_3 u_3 + c_{55}^{*}\partial_3^2 u_1 + \tag{2.30}$$
$$+ e_{11}^{*}\partial_1^2\phi + \left(e_{13}^{*} + e_{15}^{*}\right)\partial_1\partial_3\phi = \rho\frac{\partial^2 u_1}{\partial t^2},$$
$$c_{33}^{E}\partial_3^2 u_3 + \left(c_{13}^{E} + c_{55}^{*}\right)\partial_1\partial_3 u_1 + c_{55}^{*}\partial_1^2 u_3 +$$
$$+ e_{15}^{*}\partial_1^2\phi + e_{33}^{*}\partial_3^2\phi = \rho\frac{\partial^2 u_3}{\partial t^2},$$
$$e_{11}^{E}\partial_1^2 u_1 + \left(e_{13}^{*} + e_{15}^{*}\right)\partial_1\partial_3 u_1 + e_{33}^{*}\partial_3^2 u_3 +$$
$$+ e_{13}^{*}\partial_1^2 u_3 - \ni_{11}^{\varepsilon}\,\partial_1^2\phi - \ni_{33}^{\varepsilon}\,\partial_3^2\phi = 0.$$

Here $u_i = u_i(x_1, x_3)$ are the vector components of the mechanic displacement, $\phi = \phi(x_1, x_3)$ is the electric potential. Table 2.3 gives the values of the factors-marked with asterisks- for different classes of symmetry.

From (2.30) and Table 2.3 it follows that in the case of the deformation of a rhombic symmetry crystal of class $2mm$ the equations referring to elastic and electric fields are separated; however for class $4mm$ of the tetragonal system and for class $6mm$ of the hexagonal system they coincide.

Assuming that $u_3 = 0, \partial/\partial x_3 = 0$ the equations of the plane problem of electroelasticity are available for 422 class of tetragonal symmetry and for 6,

Table 2.3. The values of factors for different classes of symmetry

	Rhombic System 2mm	Square System 4mm	Hexagonal System	
			6mm	$\bar{6}m2$
c_{55}^{*}	c_{55}	c_{44}	c_{44}	c_{44}
e_{11}^{*}	0	0	0	e_{11}
e_{13}^{*}	0	e_{13}	e_{13}	0
e_{15}^{*}	0	e_{15}	e_{15}	0
e_{33}^{*}	0	e_{33}	e_{33}	0

622, $\bar{6}m2$ classes of hexagonal symmetry and they may be presented in the following form:

$$
\begin{aligned}
&c_{11}^E\partial_1^2u_1+\left(c_{12}^E+c_{66}^E\right)\partial_1\partial_2u_2+c_{66}^E\partial_2^2u_1+\\
&\qquad+e_{11}^*\partial_1^2\phi-3e_{22}^*\partial_1\partial_2\phi-2e_{11}^*\partial_2^2\phi=\rho\frac{\partial^2u_1}{\partial t^2},\\
&c_{11}^E\partial_2^2u_1+\left(c_{12}^E+c_{66}^E\right)\partial_1\partial_2u_1+c_{66}^E\partial_1^2u_2+\\
&\qquad+e_{22}^*\partial_2^2\phi-3e_{11}^*\partial_1\partial_2\phi-2e_{22}^*\partial_1^2\phi=\rho\frac{\partial^2u_2}{\partial t^2},\\
&e_{11}^*\partial_1^2u_1-3e_{22}^*\partial_1\partial_2u_1-2e_{11}^*\partial_2^2u_1+e_{22}^*\partial_2^2u_2-\\
&\qquad-3e_{11}^*\partial_1\partial_2u_2-2e_{22}^*\partial_1^2u_2-\ni_{11}^{\varepsilon}\left(\partial_1^2\phi+\partial_2^2\phi\right)=0,
\end{aligned}
\tag{2.31}
$$

The values of the coefficients e_{11}^*,e_{22}^* for the above-mentioned classes of symmetry are given in Table 2.4.

Furthermore let us consider the possibility of the problem statement of antiplane deformation for piezoelectric media. Conditions of solution existence in case of antiplane deformation lead to the following requirements:

$$
\begin{aligned}
&c_{\alpha\alpha\gamma\alpha}=c_{\alpha\alpha\gamma\beta}=c_{\beta\beta\gamma\alpha}=c_{\beta\beta\gamma\beta}=c_{\alpha\beta\gamma\alpha}=c_{\alpha\beta\gamma\beta}=0,\\
&e_{\alpha\alpha\alpha}=e_{\alpha\alpha\beta}=e_{\alpha\beta\beta}=e_{\beta\beta\alpha}=e_{\beta\beta\beta}=0,\qquad \alpha\neq\beta,\qquad \beta\neq\gamma.
\end{aligned}
\tag{2.32}
$$

Let the vector of displacement be $u=(0,0,u_3(x_1,x_2))$ and $\partial/\partial x_3=0$. Then the conditions given in (2.32) apply in the case of 222 and $2mm$ of classes of rhombic symmetry, for 4, $\bar{4}$, $\bar{4}2m$, 422 and $4mm$ classes of tetragonal symmetry, for classes 6,622,$6mm$ of hexagonal symmetry and for all classes of the cubic system. The corresponding equations of antiplane problem of electroelasticity for these classes have the following form:

$$
\begin{aligned}
&c_{55}^*\partial_1^2u_3+c_{44}^E\partial_2^2u_3+e_{15}^*\partial_1^2\phi+\\
&\qquad+e_{24}^*\partial_2^2\phi+\left(e_{14}^*+e_{25}^*\right)\partial_1\partial_2\phi=\rho\frac{\partial^2u_3}{\partial t^2},\\
&e_{15}^*\partial_1^2u_3+\left(e_{14}^*+e_{25}^*\right)\partial_1\partial_2u_3+e_{24}^*\partial_2^2u_3-\\
&-\ni_{11}^{\varepsilon}\partial_1^2\phi-\ni_{22}^*\partial_2^2\phi=0,
\end{aligned}
\tag{2.33}
$$

where the values of the factors marked with asterisks for the mentioned classes of symmetry are given in Table 2.5.

Table 2.4. The values of the coefficients e_{11}^* and e_{22}^* for some symmetry classes

	Square System, Class 422	Hexagonal System		
		Class 622	Class $\bar{6}$	Class 6m2
c_{11}^*	0	0	e_{11}	e_{11}
e_{22}^*	0	0	e_{33}	0

Table 2.5. The values of the factors marked with asterisks for the mentioned classes of symmetry

	Rhombic System		Square System				
	222	$2mm$	4	$\bar{4}$	$\bar{4}2mm$	$4mm$	422
c_{55}^*	c_{55}^E	c_{55}^E	c_{44}^E	c_{44}^E	c_{44}^E	c_{44}^E	c_{44}^E
e_{14}^*	e_{14}	0	0	e_{14}	e_{14}	0	0
e_{15}^*	*0*	e_{15}	e_{15}	e_{15}	0	e_{15}	0
e_{24}^*	0	e_{24}	e_{15}	$-e_{15}$	0	e_{15}	0
e_{25}^*	e_{25}	0	0	e_{14}	e_{14}	0	0
$\ni_{22}^*$	$\ni_{22}$	$\ni_{22}$	$\ni_{11}$	$\ni_{11}$	$\ni_{11}$	$\ni_{11}$	$\ni_{11}$

	Square System	Hexagonal System		
		6	622	$6mm$
c_{55}^E	c_{44}^E	c_{44}^E	c_{44}^E	c_{44}^E
e_{14}^*	e_{14}	0	0	0
e_{15}^*	0	e_{15}	0	e_{15}
e_{24}^*	0	e_{15}	0	e_{15}
e_{25}^*	e_{14}	0	0	0
$\ni_{22}^*$	$\ni_{11}$	$\ni_{11}$	$\ni_{11}$	$\ni_{11}$

Table 2.6. The values of constants c_{66}^*, e_{36}^* and $\ni_{33}^*$ for some symmetry classes

	222	422	$\bar{4}2m$	622	Cubic System
c_{66}^*	c_{66}^E	c_{66}^E	c_{66}^E	c_{66}^E	c_{44}^E
e_{36}^*	e_{36}	0	0	0	e_{14}^E
$\ni_{33}^*$	$\ni_{33}$	$\ni_{33}$	$\ni_{33}$	$\ni_{33}$	$\ni_{11}$

If the vector of the elastic field is $u = \{0, u_2(x_1, x_3, t), 0\}$ and $\partial/\partial x_2 = 0$, the antiplane problem statement is feasible only for classes 222 of rhombic symmetry, for classes $422, \bar{4}2m$ of tetragonal symmetry, for all cubic symmetries and for class 662 of hexagonal system. The corresponding equations for the given classes may be written as follows

$$c_{66}^* \partial_1^2 u_2 + c_{44}^E \partial_3^2 u_2 + (e_{14} + e_{36}^*)\, \partial_1 \partial_3 \phi = \rho \frac{\partial^2 u_2}{\partial t^2}, \qquad (2.34)$$
$$(e_{14} + e_{36}^*)\, \partial_1 \partial_3 u_2 - \ni_{11}^{\varepsilon}\, \partial_1^2 \phi - \ni_{33}^*\, \partial_3^2 \phi = 0.$$

In Table 2.6 are indicated values of the constants $c_{66}^*, e_{36}^*, \ni_{33}^*$.

2.4 Boundary Conditions

In the problems of electroelasticity, boundary conditions may be conventionally categorized into two groups: the mechanical and the electric ones. If the piezoelectric medium occupies the volume V surrounded by the surface S the mechanical boundary conditions are the ones of the classical elasticity

$$\sigma_{ij}n_j = q_i \text{ on } S_\sigma, \quad u_i = u_i^* \text{ on } S\backslash S_\sigma, \tag{2.35}$$

where n_j is a single normal to S; S_σ is a part of the surface S on which external load q_i is assigned.

Considering the electric boundary conditions in the general case when the surface of the piezoelectric body is contiguous with the external medium it is necessary to join the Maxwell's equation for the given medium to the (2.26). Also we should take into account the continuity of the tangential component of the vector of the electric field strength $\overrightarrow{E}$ and that of the normal component of the vector of the electric induction $\overrightarrow{D}$ on S (when these are no free electric charges at the interface of the two media).

In the particular case where the external medium is associated with vacuum (the dielectric permittivity $\ni_0$ of which is considerably smaller than the constants $\ni_{ij}$ for the majority of piezoelectric materials) the condition of continuity for the normal component of the vector D_i is approximated by

$$D_i n_i \approx 0 \text{ on } S. \tag{2.36}$$

For polarized piezoceramics we have that $\ni_{11}^{\varepsilon} \ / \ \ni_0 =$ 700–1000, $\ni_{33}^{\varepsilon} \ / \ \ni_0 =$ 700–850; therefore, when the surrounding medium is the vacuum (air) it is possible to use the approximate boundary condition given by (2.36).

It is evident that the form of the electric boundary conditions is related with the supplied electric energy to the piezoelement. If for example, the parts S_ϕ of the surface S are covered with electrodes which are supplied with an outlet potential $\pm V_0 e^{i\varpi t}$ from a voltage generator, the electric boundary conditions on S_ϕ read

$$\phi|_{S_\phi} = \pm V_0 e^{i\omega t}. \tag{2.37}$$

In a case where there is a crack in the piezoelectric body which is modelled by the mathematical section L it is natural to consider the conditions of continuity of the electric potential ϕ and of the normal component of the vector of electric induction $D_n = D_j n_j$ on L

$$\phi^+ = \phi^-, \quad D_n^+ = D_n^-. \tag{2.38}$$

Here the signs "plus" and "minus" correspond to the left and right lips of the crack L respectively, along the movement from the beginning a to end b.

It should be noted that in the presence real crack-like defects, which under the effect of mechanical and electric loads actually lead to failure of the piezoelectric body, the first condition in (2.38) is not always applicable.

To prove this statement let us consider the solution of the elementary plane problem for the distribution of an electric field in a isotropic dielectric with elliptic hole under the influence of a static electric field E_0 at infinity. Introducing elliptic coordinates ξ, η with the help of the ratio

$$x_1 + ix_2 = \ell ch\left(\xi + i\eta\right), \qquad 0 \le \xi \le \infty, \quad -\pi < \eta < \pi,$$

The solution to the two-dimensional equations of electrostatics for the dielectric region ($\xi_0 < \xi < \infty$) with dielectric permittivity $\ni_1$ and for the opening ($0 \le \xi \le \xi_0$) with dielectric constant of the vacuum $\ni_0$ is given by

$$\phi_1\left(\xi, \xi_0, \eta\right) = -E_0 \ell sh\xi \sin\eta - \frac{\left(1-\lambda\right) E_0 \ell e^{\xi_0} ch\xi_0}{\lambda cth\xi_0 + 1} e^{-\xi} \sin\eta \quad \left(\xi_0 < \xi < \infty\right), \tag{2.39}$$

$$\phi_0\left(\xi, \xi_0, \eta\right) = -\frac{E_0 \ell e^{\xi_0} sh\xi \sin\eta}{\lambda ch\xi_0 + sh\xi_0} \left(0 \le \xi \le \xi_0\right), \qquad \lambda = \ni_0/\ni_1.$$

The formulas are derived by applying the continuity conditions for the normal vector component of the electric induction and for the tangent component of the electric field strength on the dielectric boundary and the vacuum. Besides, for the opening region it is assumed that

$$\phi_0\left(0, \xi_0, \eta\right) = \phi_0\left(0, -\xi_0, \eta\right),$$

$$\left.\frac{\partial \phi_0\left(\xi, \xi_0, \eta\right)}{\partial \xi}\right|_{\xi=0} = -\left.\frac{\partial \phi_0\left(\xi, \xi_0, -\eta\right)}{\partial \xi}\right|_{\xi=0}.$$

Using (2.39), the potential jump on the edges of the open crack is given by

$$\phi_1\left(\xi_0, \xi_0, \eta\right) - \phi_1\left(\xi_0, \xi_0, -\eta\right) = -2E_0 \ell \sin\eta \left(sh\xi_0 + \frac{\left(1-\lambda\right) ch\xi_0}{\lambda cth\xi_0 + 1}\right). \tag{2.40}$$

From (2.40) it follows that the potential jump on the edges of the crack in the form of a narrow ellipse with semi-axis $\ell, w = \ell th\xi_0 \approx \ell\xi_0 << \ell$ ($2w$ is a maximum opening of the edges of the crack) is different from zero even for small values ξ_0 if the ratio $\ni_0 / \ni_1$ is of the order of ξ_0. In that case the value $\ni_0 / \ni_1\ cth\xi_0$ will be comparable to unity and then at $\xi_0 << 1$ the formula (2.40) becomes

$$\phi_1\left(\xi_0, \xi_0, \eta\right) - \phi_1\left(\xi_0, \xi_0, -\eta\right) \approx \frac{2E_0 \ell \sin\eta}{\lambda cth\xi_0 + 1}. \tag{2.41}$$

In materials with a bigger value of the dielectric constant (in comparison with the vacuum one) the continuity condition of the electric potential on the edges of the crack is not satisfied. In that case the crack is approximated with a cavity filled with vacuum (air) and for the sizes we may use the condition (2.36) on the surface of an opened crack. This condition is a consequence of the expression for the component of the electric field $E_\xi^{(1)}$ outside of the opening region

$$E_\xi^{(1)}(\xi,\xi_0,\eta)=\frac{E_0\sin\eta}{\sqrt{ch^2\xi-\cos^2\eta}}\left(ch\xi-\frac{(1-\lambda)\,ch\xi_0 e^{-(\xi-\xi_0)}}{\lambda\,(cth\xi_0+1)}\right).$$

When $\ni_0=0$ and $\xi=\xi_0$ it follows that

$$D_\xi^{(1)}(\xi_0,\xi_0,\eta)=\ni_1 E_\xi^{(1)}(\xi_0,\xi_0,\eta)=0.$$

We should also note that for $\ni_0=0$ the potential jump on the edges of the opened crack proceeding from (2.41) is determined by the expression

$$\phi_1(\xi_0,\xi_0,\eta)-\phi_1(\xi_0,\xi_0,-\eta)\cong-2E_0\ell\sin\eta.$$

2.5 Mechanics of Fracture of Piezoelectrics

The existence of defects like cracks, cuts, narrow cavities, etc. in real materials may lead in some cases to their rapid propagation and fracture of the whole body. The mathematical theory of the cracks based on the energetic approach of Griffith appeared to be very successful for the theoretic studies of different aspects of the process of fracture and for the development of practical methods of firmness analysis of the elements of the constructions. At present, a substantial progress is achieved on the investigation of materials fracture under the action of mechanical and thermal loads. However, the fracture of materials with coupled fields of different nature (e.g. elastic, thermal and electric) is not well-studied yet. Hence, the generalization of the method of fracture mechanics on piezoelectric materials is of great interest.

The most important issue in the theory of cracks is the formulation of the condition of the local fracture at a given point of the crack contour. For example, this condition can be obtained as an outcome from the determination of energy flow during the formation of a new breakage surface. Let us consider two deformed conditions of the piezoelectric body V having a crack of an arbitrary form. Suppose that there are no volume forces and free charges in the piezoelectric. Let $\sigma_{ij}^{(0)},\varepsilon_{ij}^{(0)},u_i^{(0)},\phi^{(0)}$ be components of stress and deformation tensors, displacement vector and also electric potential in some initial condition of the body "0", while $\sigma_{ij}^{(1)},\varepsilon_{ij}^{(1)},u_i^{(1)},\phi^{(1)}$ are the corresponding quantities in the condition "1" where the double-sided surface of the crack gets an increment $\Delta\Sigma$.

The expression for the internal energy of a piezoelectric body in the conditions "0" and "1" [14, 140, 141, 142, 251, 252, 254] is

$$u^{(k)}=\frac{1}{2}\int\limits_V\left(\sigma_{ij}^{(k)}\varepsilon_{ij}^{(k)}+E_j^{(k)}D_j^{(k)}\right)dV\quad(k=0,1)\tag{2.42}$$

The increase of the internal energy at transition from the condition "0" into the condition "1" reads

$$
\begin{aligned}
\Delta U &= \frac{1}{2}\int\limits_V \left(\sigma_{ij}^{(1)}\varepsilon_{ij}^{(1)} - \sigma_{ij}^{(0)}\varepsilon_{ij}^{(0)} + E_j^{(1)}D_j^{(1)} - E_j^{(0)}D_j^{(0)}\right) dV \\
&= \frac{1}{2}\int\limits_V \left[\left(\sigma_{ij}^{(1)} + \sigma_{ij}^{(0)}\right)\left(\varepsilon_{ij}^{(1)} - \varepsilon_{ij}^{(0)}\right) + \left(E_j^{(1)} + E_j^{(0)}\right)\left(D_j^{(1)} - D_j^{(0)}\right)\right] dV \\
&\quad -\frac{1}{2}\int\limits_V \left(\sigma_{ij}^{(0)}\varepsilon_{ij}^{(1)} + E_j^{(0)}D_j^{(1)} - \sigma_{ij}^{(1)}\varepsilon_{ij}^{(0)} - E_j^{(1)}D_j^{(0)}\right) dV. \qquad (2.43)
\end{aligned}
$$

Allowing for the reciprocity relation

$$
\sigma_{ij}^{(0)}\varepsilon_{ij}^{(1)} + E_j^{(0)}D_j^{(1)} = \sigma_{ij}^{(1)}\varepsilon_{ij}^{(0)} + E_j^{(1)}D_j^{(0)}, \qquad (2.44)
$$

we transform (2.43) as follows

$$
\begin{aligned}
\Delta U &= \frac{1}{2}\int\limits_V \left[\left(\sigma_{ij}^{(1)} + \sigma_{ij}^{(0)}\right)\left(\varepsilon_{ij}^{(1)} - \varepsilon_{ij}^{(0)}\right) + \left(E_j^{(1)} + E_j^{(0)}\right)\left(D_j^{(1)} - D_j^{(0)}\right)\right] dV \\
&= \frac{1}{2}\int\limits_V \left\{\partial_j\left[\left(\sigma_{ij}^{(1)} + \sigma_{ij}^{(0)}\right)\left(u_i^{(1)} - u_i^{(0)}\right)\right]\right. \\
&\quad \left. - \partial_j\left[\left(D_j^{(1)} - D_j^{(0)}\right)\left(\phi^{(1)} + \phi^{(0)}\right)\right]\right\} dV \\
&= \frac{1}{2}\int\limits_{\Sigma+\Delta\Sigma} \left[\left(\sigma_{ij}^{(1)} + \sigma_{ij}^{(0)}\right)\left(u_i^{(1)} - u_i^{(0)}\right)\right. \\
&\quad \left. - \left(D_j^{(1)} - D_j^{(0)}\right)\left(\phi^{(1)} + \phi^{(0)}\right)\right] n_j dS.
\end{aligned}
$$

Here it is taken into account that

$$
\partial_j \sigma_{ij}^{(k)} = 0, \quad \partial_i D_i^{(k)} = 0, \quad E_j^{(k)} = -\partial_j \phi^{(k)} \quad (k = 0, 1).
$$

The last equality may be written in the following form

$$
\begin{aligned}
\Delta U &= \Delta \mathrm{A}_\Sigma + \frac{1}{2}\int\limits_\Sigma \left[\sigma_{ij}^{(0)}u_1^{(1)} - \sigma_{ij}^{(1)}u_j^{(0)} - D_j^{(1)}\phi^{(0)} + D_j^{(0)}\phi^{(1)}\right] n_j dS + \\
&\quad + \frac{1}{2}\int\limits_{\Delta\Sigma} \left(\sigma_{ij}^{(1)} + \sigma_{ij}^{(0)}\left(u_1^{(1)} - u_i^{(0)}\right)\right) n_j dS \\
&\quad - \frac{1}{2}\int\limits_{\Delta\Sigma} \left(D_j^{(1)} - D_j^{(0)}\right)\left(\phi^{(1)} + \phi^{(0)}\right) n_j dS, \qquad (2.45)
\end{aligned}
$$

where

$$\Delta A_{\Sigma} = \frac{1}{2} \int\limits_{\Sigma} \left(\sigma_{ij}^{(1)} u_i^{(1)} - D_j^{(1)} \phi^{(1)} \right) n_j dS - \frac{1}{2} \int\limits_{\Sigma} \left(\sigma_{ij}^{(0)} u_i^{(0)} - D_j^{(0)} \phi^{(0)} \right) n_j dS$$

the increment of the internal force and electric field work on the surface Σ. Using the reciprocity relation (2.44) it is easy to show that

$$\int\limits_{\Sigma} \left[\sigma_{ij}^{(0)} u_i^{(1)} - \sigma_{ij}^{(1)} u_i^{(0)} - D_j^{(1)} \phi^{(0)} + D_j^{(0)} \phi^{(1)} \right] n_j dS = 0.$$

Besides, taking into account $u_i^{(0)}, \sigma_{ij}^{(0)}, \phi^{(0)}, D_i^{(0)}$ we have

$$\int\limits_{\Delta\Sigma} \sigma_{ij}^{(0)} u_i^{(0)} n_j dS = 0, \qquad \int\limits_{\Delta\Sigma} D_j^{(0)} \phi^{(0)} n_j dS = 0.$$

Thus provided that no mechanical loads are applied on the crack surface ($\sigma_{ij}^{(1)} = 0$ on $\Delta\Sigma$) we obtain

$$\Delta U = \Delta \mathrm{A}_{\Sigma} + \Delta \mathrm{A}_{\Delta\Sigma}, \tag{2.46}$$

where

$$\Delta A_{\Delta\Sigma} = \frac{1}{2} \int\limits_{\Delta\Sigma} \left(\sigma_{ij}^{(0)} u_i^{(1)} + D_j^{(0)} \phi^{(1)} \right) n_j dS - \frac{1}{2} \int\limits_{\Delta\Sigma} D_j^{(1)} \left(\phi^{(1)} + \phi^{(0)} \right) n_j dS. \tag{2.47}$$

The value $\Delta \mathrm{A}_{\Delta\Sigma}$, which appears in the expression (2.46) for the increment of the internal energy at transition of the piezoelectric body from condition "0" to condition "1", defines the energy flow during the formation of the breakage $\Delta\Sigma$, the integration in (2.47) taking place in two surfaces $\Delta\Sigma_1$, and $\Delta\Sigma_2$ of the additional breakage and the normal directions to $\Delta\Sigma_1$ and $\Delta\Sigma_2$ are directed inside the crack.

Using (2.47) the condition of propagation of the crack is given by

$$\gamma \left(\Delta\Sigma_1 + \Delta\Sigma_2 \right) = -\Delta \mathrm{A}_{\Delta\Sigma}, \tag{2.48}$$

where γ is the density of the surface energy.

Integrating along the double-sided surface $\Delta\Sigma$ in (2.47) we get-considering the continuity of $\phi^{(0)}, \sigma_{ij}^{(0)}, D_j^{(0)}$ on $\Delta\Sigma$-the following expression

$$\begin{aligned} \Delta A_{\Delta\Sigma} = & \frac{1}{2} \int\limits_{\Delta\Sigma_1} \sigma_{ij}^{(0)} n_j \left(u_j^{(1)+} - u_j^{(1)-} \right) dS + \frac{1}{2} \int\limits_{\Delta\Sigma_1} D_j^{(0)} n_j \left(\phi^{(1)+} - \phi^{(1)-} \right) dS \\ & - \frac{1}{2} \int\limits_{\Delta\Sigma_1} \left(D_j^{(1)+} \phi^{(1)+} - D_j^{(1)-} \phi^{(1)-} \right) n_j dS \\ & - \frac{1}{2} \int\limits_{\Delta\Sigma_1} \left(D_j^{(1)+} - D_j^{(1)-} \right) n_j \phi^{(0)} dS. \end{aligned} \tag{2.49}$$

where the signs "±" correspond to the limiting values of the quantities $u_i^{(1)}, \phi_i^{(1)}, D_j^{(1)}$ on the double-sided surface $\Delta\Sigma$, while the integration takes on one side $\Delta\Sigma$.

The formulation of the criterion for a crack spreading in a piezoelectric medium seems to depend on the type of electric boundary conditions on crack surfaces. When a crack is considered as a mathematical surface (line) allowing for the conditions of the continuity of the electric potential $\phi^{(1)}$ and the normal component of the electric induction vector $D_j^{(1)} n_j$ on the surface $\Delta\Sigma$ we get the following expression giving the energy flow during the breakage forming:

$$\Delta A_{\Delta\Sigma} = \frac{1}{2} \int_{\Delta\Sigma} \sigma_{ij}^{(0)} n_j \left(u_i^{(1)+} - u_i^{(1)-} \right) dS. \tag{2.50}$$

In this case the condition (2.48) takes the form

$$2\gamma = -\frac{1}{2} \lim_{\Delta\Sigma_1 \to 0} \frac{1}{\Delta\Sigma_1} \int_{\Delta\Sigma_1} \sigma_{ij}^{(0)} n_j \left(u_i^{(1)+} - u^{(1)-} \right) dS. \tag{2.51}$$

Therefore, allowing for the continuity of the electric potential and of the normal component of the electric induction vector on the crack surface the condition of the local fracture for the piezoelectric medium coincides with the known condition of the fracture in elastic bodies [140, 141].

For piezoelectric materials such as polarized piezoceramics on the edges of the opened crack an electric potential jump appears. Therefore for cracks in piezoceramics it is possible to use the approximate condition $D_j^{(1)} n_j = 0$ on $\Delta\Sigma_1$. In this case, using (2.48), (2.49) we obtain

$$2\gamma = -\frac{1}{2} \lim_{\Delta\Sigma_1 \to 0} \frac{1}{\Delta\Sigma_1} \int_{\Delta\Sigma_1} \left[\sigma_{ij}^{(0)} n_j \left(u_i^{(1)+} - u_i^{(1)-} \right) \right. \tag{2.52}$$
$$\left. + D_j^{(0)} n_j \left(\phi^{(1)+} - \phi^{(1)-} \right) \right] dS.$$

It is known that in the small vicinity of an arbitrary point of the crack contour the medium is in the conditions of a plane state; in addition it is possible to consider the crack as a semi-infinite and plane (linear). And corresponding asymptotic representations are valid for stresses, displacements and electric potentials.

Let a piezoelectric medium refer to an arbitrary orientated system of the Cartesian coordinate system x_k $(k = 1, 2, 3)$ and the linear crack 2ℓ $(|x_1| < \ell)$ in length is situated in the $x_2 = 0$.
Assuming that the electroelastic condition

(a) does not depend on the x_3 - coordinate, and that on the edges of the crack,
(b) no mechanical load is applied and

(c) the approximate equation $D_j n_j \cong 0$ is satisfied
the condition of a crack propagation (2.52) reads

$$2\gamma = \frac{1}{2} \lim_{\Delta\ell \to 0} \frac{1}{\Delta\ell} \int\limits_{\ell}^{\ell+\Delta\ell} \left\{ \sigma_{i2}^{+}(x_1, 0, \ell) \left[u_i (x_1, 0, \ell + \Delta\ell) \right] + D_2^{+}(x_1, 0, \ell) \left[\phi (x_1, 0, \ell + \Delta\ell) \right] \right\} dx_1. \tag{2.53}$$

Here $\sigma_{i2}(x_1, 0, \ell), D_2^{+}(x_1, 0, \ell)$ are stress components and a normal component of the electric induction vector on the extension of the crack 2ℓ in length; $[u_i], [\phi]$ are jumps of the displacement vector and of electric potential on the crack $2(\ell + \Delta\ell)$ in length. It should be noted that the second term of the right-hand side in (2.53) is equal to zero if we accept the condition of the continuity of the electric potential on the crack edges.

It is known that a crack edge is a peculiar discharge area for released energy that is distributed along the crack contour. Due to [196] we introduce the energy discharge density vector $\Gamma = (\Gamma_1, \Gamma_2)$ for piezoelectric medium considering a static and plane problem. The components Γ_k $(k = 1, 2)$ of this vector are determined by the following expressions

$$\Gamma_k = \oint\limits_{c} \left[H n_k - \sigma_{ij} n_i \partial_k u_j + D_i n_i E_k \right] dS, \tag{2.54}$$

where c is an arbitrary closed contour surrounding the crack tip, $H = U - E_i D_i$ is an electric enthalpy, n_k are the components of the unit normal vector to c.

We will now show that the integrals Γ_k do not depend on the choice of the contour c, if on the crack edges the conditions $\sigma_{ij} n_j = 0, D_j n_j = 0$ apply. Let us consider the integral Γ_1 of the closed contour consisting of arbitrary curves c, c_1 and linear regions on the crack edges (Fig. 2.1). Taking into account the fact that the sub-integral expression in (2.54) is equal to zero on the crack edges and using the equality

$$\partial_j \sigma_{ij} = 0, \quad \partial_i D_i = 0, \quad \partial_k H = \sigma_{ij} \partial_k \varepsilon_{ij} - D_i \partial_k E_i,$$

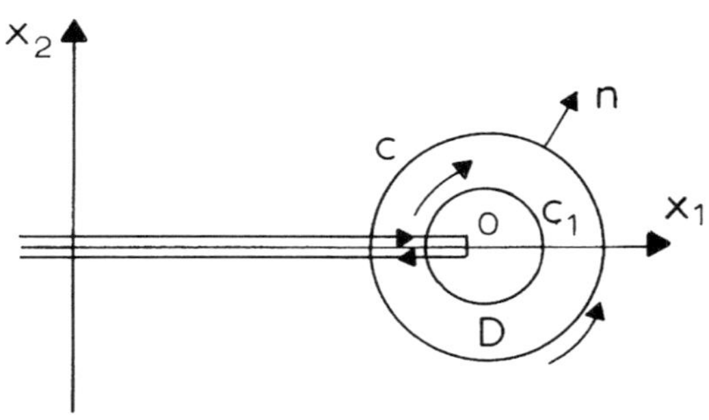

Fig. 2.1. Paths of integration for the Γ_k integral

we find that

$$
\begin{aligned}
\Gamma_1|_c - \Gamma_1|_{c_1} &= \oint\limits_{c_1+c} [Hn_1 - \sigma_{ij} n_i \partial_1 u_j + D_i n_i E_1] dS = \\
&= \iint\limits_D [\partial_1 H - \partial_i (\sigma_{ij} \partial_1 u_j) + \partial_i (D_i E_1)] dx_1 dx_2 = \\
&= \iint\limits_D [\sigma_{ij} \partial_1 \varepsilon_{ij} - D_i \partial_1 E_i - \partial_i \sigma_{ij} \partial_1 u_j - \sigma_{ij} \partial_1 \partial_i u_j - E_1 \partial_i D_i \\
&\qquad + D_i \partial_i E_1]\, dx_1 dx_2 = 0.
\end{aligned}
\tag{2.55}
$$

In the same way it is possible to show that the integral Γ_2 does not depend on the choice of the contour c.

If the development of a crack in a piezoelectric medium occurs in its initial plane $x_2 = 0$ the condition of fracture takes the form

$$\Gamma_1 = 2\gamma. \tag{2.56}$$

In the general case of an anisotropic material the quantity 2γ appears to depend on the position of the point O and on the orientation of the crack plane at that point. For a crack which deviates at an angle θ from its initial position the energy discharge is equal to the vector projection Γ on the direction of the crack growth. In this case the criterion determining the beginning of the crack development is given by

$$\Gamma_1 \cos\theta + \Gamma_2 \sin\theta = 2\gamma(\theta), \tag{2.57}$$

where $\gamma(\theta)$ is an experimentally determined quantity.

3

Static Problems of Electroelasticity for Bimorphs with Stress Concentrators

In this chapter we study some two-dimensional static problems of electroelasticity for piezoceramic bimorphs weakened by defects of crack and opening type.

The term bimorph implies a cylindrical body consisting of two continuously connected, along the common part of the side surface, heterogeneous cylindrical bodies. Such piecewise-uniform bodies are the compound plates consisting of two heterogeneous halfplates continuously connected along their plane boundary.

Construction of the analogous algorithms is fulfilled by application of complex solution representations allowing us to reveal the singularities of field quantities in the vicinity of the crack tips. By assuming the former we reduce the corresponding boundary problems of electroelasticity to systems of singular integral equations.

3.1 Complex Representations of Solutions in Two-Dimensions

Let us consider a piezoceramic medium in a Cartesian system of $Ox_1x_2x_3$ coordinates and, in that case, let x_3- axis be directed along the force lines of the electric field of the preliminarily polarization of the ceramics. Then, the linear equations of piezoeffect may be represented, in the form of [115, 140], as well as in other forms [25]

$$\begin{aligned}
&\varepsilon_{11} = s_{11}\sigma_{11} + s_{12}\sigma_{22} + s_{13}\sigma_{33} + d_{31}E_3,\\
&\varepsilon_{22} = s_{12}\sigma_{11} + s_{11}\sigma_{22} + s_{13}\sigma_{33} + d_{31}E_3,\\
&\varepsilon_{33} = s_{13}\left(\sigma_{11} + \sigma_{22}\right) + s_{13}\sigma_{33} + d_{33}E_3,\\
&2\varepsilon_{23} = s_{44}\sigma_{23} + d_{15}E_2, \quad 2\varepsilon_{13} = s_{44}\sigma_{13} + d_{15}E_1,\\
&2\varepsilon_{12} = s_{66}\sigma_{12} = 2\left(s_{11} - s_{12}\right)\sigma_{12}, \quad D_1 = \text{э}_{11}\, E_1 - d_{15}\sigma_{13},\\
&D_2 = \text{э}_{11}\, E_2 + d_{15}\sigma_{23}, \quad D_3 = e_{33}E_3 + d_{31}\left(\sigma_{11} + \sigma_{22}\right) + d_{33}\sigma_{33}.
\end{aligned} \tag{3.1}$$

Here $s_{11} = s_{11}^E, \ldots, s_{44} = s_{44}^E$ are the elastic compliances determined at a constant value of the electric field; d_{31}, d_{15}, d_{33} are the piezoelectric constants and $\ni_{11}=\ni_{11}^\sigma$, $\ni_{33}=\ni_{33}^\sigma$ are the dielectric permitivities defined at constant values of the mechanical stress; E_j and $D_j\,(j = 1, 2, 3)$ are the components of strength vector and electric field induction directed along x_j - axis, respectively. Here we should note that relations (3.1) coincide with the corresponding relations for the crystal in the hexagonal system $6mm$ [25] with optic.

Apart from the state (3.1), there are available the *equations of equilibrium* $\partial_j\sigma_{ij} = 0$ (in the absence of volume forces), the *Cauchy relations* $\varepsilon_{ij} = 0.5\,(\partial_i u_j + \partial_j u_i)$, the *conditions of deformation compatibility* [134] and the *electrostatic equations*, $\operatorname{rot}\vec{E} = 0$, $\operatorname{div}\vec{D} = 0$ [167].

Let us seek a solution of the electroelasticity equation that doesn't depend on coordinate x_2. Assuming in (3.1) that the deformation $\varepsilon_{22} = 0$ and introducing the *electrostatic potential* $\vec{E} = -\mathrm{grad}\phi$ and the *functions of stress* φ_1, φ_3 given by $(\partial_m = \partial/\partial x_m)$

$$\sigma_{11} = \partial_3^2\varphi_1, \quad \sigma_{33} = \partial_1^2\varphi_1, \quad \sigma_{13} = -\partial_1\partial_3\varphi_1,$$
$$\sigma_{12} = \partial_3\varphi_3, \quad \sigma_{23} = -\partial_1\varphi_3,$$

we write

$$\begin{aligned}
&\varepsilon_{11} = s_{11}^*\partial_3^2\varphi_1 + s_{13}^*\partial_1^2\varphi_1 - d_{31}^*\partial_3\phi, \quad \varepsilon_{12} = (s_{11} - s_{12})\,\partial_3\varphi_3,\\
&\varepsilon_{33} = s_{13}^*\partial_3^2\varphi_1 + s_{33}^*\partial_1^2\varphi_1 - d_{33}^*\partial_3\phi,\\
&2\varepsilon_{13} = -s_{44}\partial_1\partial_3\varphi_1 - d_{15}\partial_1\phi, \quad 2\varepsilon_{23} = -s_{44}\partial_1\varphi_3,\\
&D_1 = -d_{15}\partial_1\partial_3\varphi_1 - \ni_{11}\,\partial_1\phi, \quad D_2 = -d_{15}\partial_1\varphi_3,\\
&D_3 = d_{31}^*\partial_3^2\varphi_1 + d_{33}^*\partial_1^2\varphi_1 - \ni_{33}^*\,\partial_3\phi,
\end{aligned} \tag{3.2}$$

where

$$s_{11}^* = s_{11} - \frac{s_{12}^2}{s_{11}}, \quad s_{13}^* = s_{13}\left(1 - \frac{s_{12}}{s_{11}}\right), \quad s_{33}^* = s_{33} - \frac{s_{13}^2}{s_{11}},$$
$$d_{31}^* = d_{31}\left(1 - \frac{s_{12}}{s_{11}}\right), \quad d_{33}^* = d_{33} - d_{31}\frac{s_{13}}{s_{11}}, \quad \ni_{33}^* = \ni_{33} - \frac{d_{31}^2}{s_{11}}.$$

In that case the equations of equilibrium are satisfied identically. In conditions of deformation compatibility, only two of the independent relations remain, namely

$$\partial_3^2\varepsilon_{11} + \partial_1^2\varepsilon_{33} = 2\partial_1\partial_3\varepsilon_{13}, \quad \partial_3\varepsilon_{21} - \partial_1\varepsilon_{23} = 0. \tag{3.3}$$

Substituting into (3.3) (and also into electrostatic equation $\operatorname{div}\vec{D} = 0$) the expressions of deformations and the components of the electric induction vector from (3.2) we arrive at the following system of differential equations (summarizing over index j)

$$
\begin{aligned}
&l_{ij}\varphi_j = 0, \quad l_{33}\varphi_3 = 0 \quad (i,j = 1,2)\,,\\
&l_{11} = s_{33}^{*}\partial_1^4 + (s_{44} + 2s_{13}^{*})\,\partial_1^2\partial_3^2 + s_{11}^{*}\partial_3^4,\\
&l_{12} = l_{21} = (d_{15} - d_{33}^{*})\,\partial_1^2\partial_3 - d_{31}^{*}\partial_3^3, \quad \varphi_2 = \phi,\\
&l_{22} = \ni_{11}\,\partial_1^2 + \ni_{33}^{*}\,\partial_3^2, \quad l_{33} = s_{44}\partial_1^2 + 2\,(s_{11} - s_{12})\,\partial_3^2.
\end{aligned}
\tag{3.4}
$$

Thus, the state of plane and antiplane deformation in transverse isotropic piezoceramic medium is separated, and the mechanical and electric fields can be coupled only in the first case.

Let us now consider the problem of plane deformation in the Cartesian coordinates $Ox_1x_2x_3$ of a piezoceramic medium with the direction of preliminary polarization along Ox_3 -axis. The plane of deformation is the x_1Ox_3 -plane.

The density of inner energy of the piezoelectric medium is given by

$$
U = \frac{1}{2}\,(\sigma_{ij}\varepsilon_{ij} + D_iE_i) \quad (i,j = 1,2,3)
$$

Its physical meaning is related to the positively defined quadrature form of variables σ_{ij} and E_m. Therefore the system of equations $\ell_{ij}\varphi_j = 0$ in (3.4) is elliptic. Introducing the solvability function $V\,(x_1, x_3)$ by the formulas

$$
\varphi_1 = l_{22}V, \quad \varphi_2 = -l_{12}V, \tag{3.5}
$$

the system of (3.4) is given by the following 6-th order homogeneous differential equation

$$
\left(l_{11}l_{22} - l_{12}^2\right)V = 0. \tag{3.6}
$$

The general real solution of this equation reads

$$
V(x_1, x_3) = 2\mathrm{Re}\sum_{k=1}^{3} f_k\,(z_k)\,, \quad z_k = x_1 + \mu_k x_3, \tag{3.7}
$$

where μ_k are the roots of the characteristic equation

$$
\begin{aligned}
&\left\{s_{33}^{*} + (s_{44} + 2s_{13}^{*})\,\mu^2 + s_{11}^{*}\,\mu^4\right\}\left(\ni_{11} + \ni_{33}^{*}\,\mu^2\right) -\\
&- \mu^2\left(d_{15} - d_{33}^{*} - d_{31}^{*}\mu^2\right)^2 = 0.
\end{aligned}
\tag{3.8}
$$

In the case of plane stressed state and plane deformation the roots $\mu_k\,(k = 1,2,3)$ of (3.8) for some piezoceramics are given in Table A.2 (in Appendix A).

System (3.4) is elliptic so it follows that $\mathrm{Im}\mu_\kappa \neq 0$. For definiteness, we assume that $\mathrm{Im}\mu_\kappa > 0\,(k = 1,2,3)$. Formula (3.7) is valid in the case where the roots μ_k are simple. The functions $f_k\,(z_k)$ that appear in that equation are arbitrary analytical functions of z_k.

The stress function φ_1 and the electric potential $\phi = \varphi_2$ can be found using equalities (3.5)

$$\varphi_1 (x_1, x_3) = 2\mathrm{Re} \sum_{k=1}^{3} \gamma_k f''_k (z_k), \quad f''_k (z_k) = \frac{d^2 f_k}{dz_k^2}, \tag{3.9}$$

$$\varphi_2 (x_1, x_3) = -2\mathrm{Re} \sum_{k=1}^{3} \lambda_k f'''_k (z_k),$$

$$\gamma_k = \ni_{11} + \ni^*_{33} \mu_k^2, \quad \lambda_k = (d_{15} - d^*_{33}) \mu_k - d^*_{31} \mu_k^3.$$

Then using the electrostatic equations, formulas (3.2) and stress expressions by stress function φ_1 the components of mechanical and electric quantities [59] are given by

$$\sigma_{11} = 2\mathrm{Re} \sum_{k=1}^{3} \gamma_k \mu_k^2 \Phi_k (z_k), \quad \sigma_{33} = 2\mathrm{Re} \sum_{k=1}^{3} \gamma_k \Phi_k (z_k),$$

$$\sigma_{13} = -2\mathrm{Re} \sum_{k=1}^{3} \gamma_k \mu_k \Phi_k (z_k), \tag{3.10}$$

$$E_1 = 2\mathrm{Re} \sum_{k=1}^{3} \lambda_k \Phi_k (z_k), \quad E_3 = 2\mathrm{Re} \sum_{k=1}^{3} \lambda_k \mu_k \Phi_k (z_k),$$

$$D_1 = 2\mathrm{Re} \sum_{k=1}^{3} r_k \mu_k \Phi_k (z_k), \quad D_3 = -2\mathrm{Re} \sum_{k=1}^{3} r_k \Phi_k (z_k),$$

$$r_k = \ni_{11} \lambda_k \mu_k^{-1} - d_{15} \gamma_k, \quad \Phi_k (z) = f_k^{(4)} (z).$$

We obtain components of the displacement vector by integrating the expressions for deformations in (3.2) taking into account (3.9). We have

$$u_1 = 2\mathrm{Re} \sum_{k=1}^{3} p_k \varphi_k (z_k), \quad u_3 = 2\mathrm{Re} \sum_{k=1}^{3} q_k \varphi_k (z_k),$$

$$p_k = s^*_{11} \gamma_k \mu_k^2 + s^*_{13} \gamma_k + d^*_{31} \lambda_k \mu_k, \quad \Phi_k(z) = \varphi'_k (z), \tag{3.11}$$

$$q_k = s^*_{33} \gamma_k \mu_k^{-1} + s^*_{13} \gamma_k \mu_k + d^*_{33} \lambda_k.$$

The main vector and the main moment of the mechanical forces, which acts at the side of positive normal on arc AB in the area occupied by the medium, read

$$X_1 = \int\limits_{AB} X_{1n} ds = 2\mathrm{Re} \sum_{k=1}^{3} \gamma_k \mu_k \left[\varphi_k (z_k) \right]_A^B, \tag{3.12}$$

$$X_3 = \int\limits_{AB} X_{3n} ds = -2\mathrm{Re} \sum_{k=1}^{3} \gamma_k \left[\varphi_k (z_k) \right]_A^B,$$

$$M = \int\limits_{AB} (x_1 X_{3n} - x_3 X_{1n})\, ds = \left[\varphi_1 - (x_1 \partial_1 \varphi_1 + x_3 \partial_3 \varphi_1) \right]_A^B.$$

To formulate the electric boundary conditions it is necessary to write down the formula for the normal components of the electric induction vector D_n and the tangential of the component of electric field strength E_s vector on arc AB. Due to (3.10) we find

$$D_n = D_1 \cos\psi + D_3 \sin\psi = 2\mathrm{Re}\sum_{k=1}^{3} a_k(\psi)\, r_k \Phi_k(z_k), \tag{3.13}$$

$$E_s = E_3 \cos\psi - E_1 \sin\psi = 2\mathrm{Re}\sum_{k=1}^{3} a_k(\psi)\, \lambda_k \Phi_k(z_k),$$

$$a_k(\psi) = \mu_k \cos\psi - \sin\psi,$$

where ψ is an angle between the positive normal to line AB at point $(x_1,\ x_3) \in AB$ and arc Ox_1 (Fig. 3.1).

The flow of the electric induction vector AB and the circulation of the vector of the field strength follow directly from relation (3.13) and are given by

$$\int\limits_{AB} D_n ds = 2\mathrm{Re}\sum_{k=1}^{3} r_k \left[\varphi_k(z_k)\right]_A^B, \tag{3.14}$$

$$\int\limits_{AB} E_s ds = 2\mathrm{Re}\sum_{k=1}^{3} \lambda_k \left[\varphi_k(z_k)\right]_A^B.$$

According to formulas (3.12), (3.10) the mechanical conditions of the first basic problem on the full boundary of area L_j may be represented in the following form

$$2\mathrm{Re}\sum_{k=1}^{3} \gamma_k \mu_k \varphi_k(\zeta_k) = \int\limits_0^s X_{1n}(s)\, ds + c, \tag{3.15}$$

$$2\mathrm{Re}\sum_{k=1}^{3} \gamma_k \varphi_k(\zeta_k) = -\int\limits_0^s X_{3n}(s)\, ds + c^*,$$

$$\zeta_k = \mathrm{Re}\zeta + \mu_k \mathrm{Im}\zeta, \quad \zeta \in L.$$

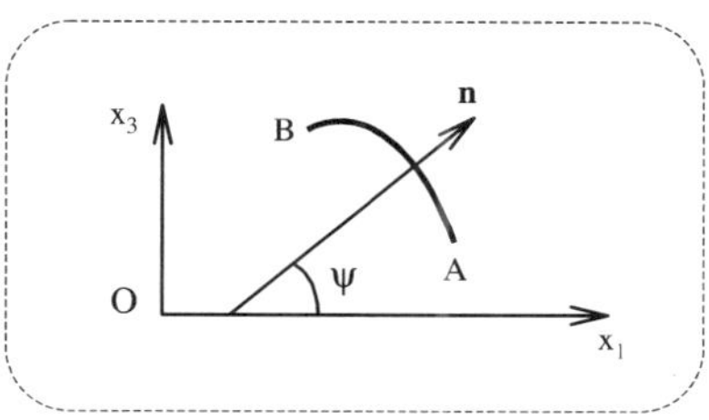

Fig. 3.1. For determination of angle Ψ

Here $c = \{c_j, \zeta \in L_j\}$, $c^* = \{c_j^*, \zeta \in L_j\}$ are the constants determined by the solution of the boundary problem; s is the arc abscissa of the boundary contour.

When the displacements $g_1(\zeta), g_3(\zeta)$ are given on the boundary of the body, the boundary conditions according to (3.11) are written as

$$2\mathrm{Re}\sum_{k=1}^{3} p_k\varphi_k(\zeta_k) = g_1(\zeta), \quad 2\mathrm{Re}\sum_{k=1}^{3} q_k\varphi_k(\zeta_k) = g_3(\zeta). \tag{3.16}$$

In some cases it is more convenient to represent the mechanical boundary conditions as derivatives of the function $\varphi_k(z_k)$. For that it is necessary to differentiate (3.15), (3.16) along the arc coordinate of contour L. For the first problem we have

$$2\mathrm{Re}\sum_{k=1}^{3} \gamma_k\mu_k^m a_k(\psi)\,\Phi_k(\zeta_k) = F_m(\zeta) \quad (m = 0, 1), \tag{3.17}$$

$$F_0(\zeta) = -X_{3n}(\zeta), \quad F_1(\zeta) = X_{1n}(\zeta), \quad \zeta \in L;$$

while for the second basic problem:

$$2\mathrm{Re}\sum_{k=1}^{3} p_k^{(m)} a_k(\psi)\,\Phi_k(\zeta_k) = F_m(\zeta) \quad (m = 0, 1), \tag{3.18}$$

$$p_k^{(0)} = p_k, \quad p_k^{(1)} = q_k, \quad F_0(\zeta) = \frac{dg_1}{ds}, \quad F_1(\zeta) = \frac{dg_3}{ds}.$$

The electric boundary conditions are formulated as follows [115, 140]: On the electroded area of boundary L^* the values of electric potential ϕ are prescribed. In that case due to (3.9) we have that

$$2\mathrm{Re}\sum_{k=1}^{3} \lambda_k\varphi_k(\zeta_k) = V(\zeta), \quad \zeta \in L^*, \tag{3.19}$$

or, that

$$2\mathrm{Re}\sum_{k=1}^{3} \lambda_k a_k(\psi)\,\Phi_k(\zeta_k) = \frac{dV(\zeta)}{ds}. \tag{3.20}$$

On the part of the boundary $L\backslash L^*$ that is not covered with electrodes (on the boundary with the vacuum) the approximate condition $D_n = 0$ holds [115], which allowing for (3.13), one obtains

$$2\mathrm{Re}\sum_{k=1}^{3} r_k a_k(\psi)\,\Phi_k(\zeta_k) = 0. \tag{3.21}$$

Thus, we have two mechanical and one electric boundary condition to define the three functions $\Phi_k(z_k)$. Here we should note that all the relations that have been obtained in this section for the field quantities are also valid in the case of a plane stress state (the thin-plate case). It is only necessary to drop the asterisks of the constants $s_{ik}^*, d_{ik}^*, \ni_{33}^*$.

3.2 A Bimorph with Cracks in One of the Pair Components

Consider a piecewise-homogeneous piezoceramic medium made up of two continuously connected, along x_1-axis, various piezoceramic halfplanes. To be more formalistic let us assume that the direction of the field of the preliminarily polarization of the piezoceramic is parallel to x_3 -axis. The upper halfplane contains crack $L_j (j = 1, 2, \ldots, k)$ that can be described as mathematical lines. In a homogeneous field of mechanical stresses the quantities $< \sigma_{11}^{(1)} >$, $< \sigma_{11}^{(2)} >$, $< \sigma_{13} >$ and $< \sigma_{33} >$ are prescribed at infinity while a stress vector (X_{1n}, X_{3n}) could possibly act on the cut edges.

It is assumed that the L_j's are smooth disconnected arcs, whose curves and functions X_{1n}, X_{3n} satisfy the Hölder condition (they are functions of class H [121]), contours L_j do not have common points and the given on them loading is continuous from one edge to the other. The mechanical and electric quantities in each halfplane are determined using three complex potentials $\Phi_\nu (z_\nu)$ given by formulas (3.10).

In addition, all quantities on the "r"-th halfplane will have the "r" index in the upper part. The condition of ideal mechanical contact $\left(\sigma_{33}^{(1)} = \sigma_{33}^{(2)}, \sigma_{13}^{(1)} = \sigma_{13}^{(2)}, u_1^{(1)} = u_1^{(2)}, u_3^{(1)} = u_3^{(2)}\right)$ must be fulfilled on the line of conjugation of $x_3 = 0$ where electric quantities $\overrightarrow{E_1}$ and $\overrightarrow{D_3}$ should also be continuous. The ideal mechanical contact conditions can be written as follows, similarly to (3.10)

$$
\begin{aligned}
&\operatorname{Re} \sum_{\nu=1}^{3} \left\{ c_{k\nu}^{(1)} \varPhi_\nu^{(1)} (x_1) - c_{k\nu}^{(2)} \varPhi_\nu^{(2)} (x_1) \right\} = 0 \quad (k = 1, 2, \ldots, 6), \\
&c_{1\nu}^{(r)} = \gamma_\nu^{(r)}, \quad c_{2\nu}^{(r)} = \gamma_\nu^{(r)} \mu_\nu^{(r)}, \quad c_{3\nu}^{(r)} = p_\nu^{(r)}, \\
&c_{4\nu}^{(r)} = q_\nu^{(r)}, \quad c_{5\nu}^{(r)} = \lambda_\nu^{(r)}, \quad c_{6\nu}^{(r)} = r_\nu^{(r)} \quad (r = 1, 2).
\end{aligned}
\tag{3.22}
$$

The mechanical boundary conditions on the edges L_j are formulated in the usual way. Electric conditions are presented as conditions of conjunction of the electric field on the interphase of two dielectrics [152]. In terms of the potentials we write

$$
\begin{aligned}
&2\operatorname{Re} \sum_{\nu=1}^{3} b_{p\nu} \left\{ \varPhi_\nu^{(1)} \left(\zeta_\nu^{(1)} \right) \right\}^{\pm} = f_p^{\pm} (\zeta) \quad (p = 1, 2), \\
&\operatorname{Re} \sum_{\nu=1}^{3} b_{n\nu} \left[\varPhi_\nu^{(1)} \left(\zeta_\nu^{(1)} \right) \right] = 0 \quad (n = 3, 4), \\
&b_{1\nu} = \gamma_\nu^{(1)} \mu_\nu^{(1)} a_\nu^{(1)} (\psi), \quad b_{2\nu} = \gamma_\nu^{(1)} a_\nu^{(1)} (\psi),
\end{aligned}
\tag{3.23}
$$

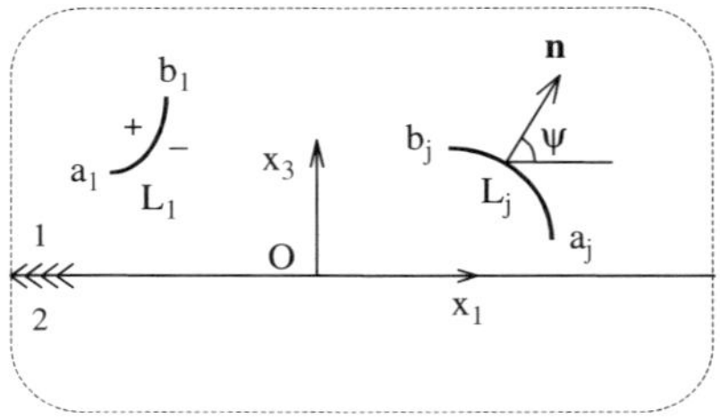

Fig. 3.2a. The bimorph scheme with cracks in one of the components of a pair

$$
\begin{aligned}
&b_{3\nu} = \lambda_\nu^{(1)} a_\nu^{(1)}(\psi), \quad b_{4\nu} = r_\nu^{(1)} a_\nu^{(1)}(\psi), \\
&a_\nu^{(1)}(\psi) = \mu_\nu^{(1)} \cos\psi - \sin\psi, \\
&f_1(\zeta) = X_{1n}(\zeta), \quad f_2(\zeta) = -X_{3n}(\zeta), \\
&\zeta_\nu^{(1)} = \mathrm{Re}\zeta + \mu_\nu^{(1)} \mathrm{Im}\zeta, \quad \zeta = \xi_1 + i\xi_2 \in L_j \quad (j = 1, 2, \ldots, k).
\end{aligned}
$$

Here ψ is the angle between the normal of the left edge L_j (during the movement from its beginning a_j to the end b_j) and Ox_1 - axis (Fig. 3.2a); the symbol [] signifies a jump of the corresponding quantity passing through the cut.

Thus, the problem is in defining of potentials $\Phi_\nu^{(r)}\left(z_\nu^{(r)}\right)$ by 1) conjugation conditions on the interphase of media (3.22), 2) the boundary equalities (3.23) and also 3) the conditions at infinity.

In order to obtain the integral equations of the boundary problem of electroelasticity let us first construct the fundamental solution for the compound piezoceramic plane in the no- crack case. Let us consider a concentrated force $P_1 + iP_3$ or a concentrated charge ρ (Fig. 3.2b) acting at a point (x_{10}, x_{30}) on the upper halfplane of the compound medium.

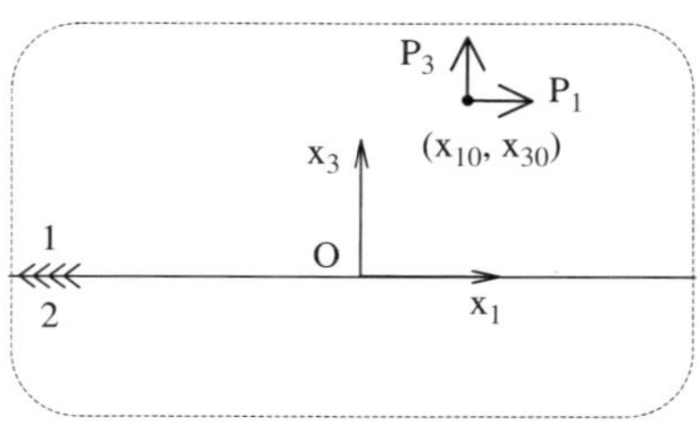

Fig. 3.2b. The bimorph scheme with concentrated force (charge) in the upper halfplane

The solution of this problem is called fundamental. Applying the concept of the *method of images* [120] it takes the following form [178]

$$
\begin{aligned}
&\Phi_\nu^{(1)}\left(z_\nu^{(1)}\right) = \frac{A_\nu^{(1)}}{z_\nu^{(1)} - z_{\nu 0}^{(1)}} - \sum_{m=1}^{3} \frac{\overline{\beta_{\nu m}^{(1)} A_m^{(1)}}}{z_\nu^{(1)} - \overline{z_{m0}^{(1)}}},\\
&\Phi_\nu^{(2)}\left(z_\nu^{(2)}\right) = \sum_{m=1}^{3} \frac{\beta_{\nu+3,m}^{(1)} A_m^{(1)}}{z_\nu^{(2)} - z_{m0}^{(1)}},\\
&z_\nu^{(r)} = \mathrm{Re}z + \mu_\nu^{(r)}\mathrm{Im}z, \quad z_{\nu 0}^{(r)} = \mathrm{Re}z_0 + \mu_\nu^{(r)}\mathrm{Im}z_0,\\
&z = x_1 + ix_3, \quad z_0 = x_{10} + ix_{30} \quad (x_{30} > 0; r = 1,2; \nu = 1,2,3)\,.
\end{aligned}
\tag{3.24}
$$

The first term in the first of the above equations corresponds to the fundamental solution for the homogeneous piezoceramic medium [59]. To calculate constants $A_\nu^{(1)}$ it is necessary to consider the three conditions of displacement singlevalueness, the condition referring to potential of the electric field and three conditions of balance and charge conservation

$$
\oint_c X_{1n} ds = -P_1, \quad \oint_c X_{3n} ds = -P_3, \quad \oint_c D_n ds = \rho,
$$

where $\boldsymbol{c}$ is an arbitrary closed contour containing the point (x_{10}, x_{30}). Realization of these conditions brings to a system of six algebraic equations w.r.t $A_\nu^{(1)}$

$$
\begin{aligned}
&2\mathrm{Im}\sum_{\nu=1}^{3} A_\nu^{(1)}\left(\mu_\nu^{(1)}\right)^{m-1} = \frac{B_m}{h} \quad (m = 0,1,\ldots,5)\,,\\
&B_0 = \frac{P_1}{2\pi s_{33}^{(1)}\backepsilon_{11}^{(1)}}\left(\frac{s_{33}^{(1)} d_{31}^{(1)}}{\Delta_1}\backepsilon_{33}^{(1)} - s_{13}^{(1)}\right),\\
&B_1 = \frac{1}{2\pi\Delta_1\backepsilon_{11}^{(1)}}\left[P_3\left(d_{15}^{(1)}\backepsilon_{33}^{(1)} + d_{31}^{(1)}\backepsilon_{11}^{(1)}\right) + \backepsilon_{33}^{(1)}\rho\right],\\
&B_2 = -\frac{P_1 d_{31}^{(1)}}{2\pi\Delta_1},\\
&B_3 = -\frac{1}{2\pi\Delta_1}\left(\rho + P_3 d_{33}^{(1)}\right),\\
&B_4 = \frac{P_1}{2\pi\Delta_1}\left(d_{33}^{(1)} - d_{15}^{(1)}\right),\\
&B_5 = \frac{\Delta_3}{2\pi\Delta_1\Delta_2}\left[P_3\left(s_{13}^{(1)}\frac{\Delta_1}{\Delta_3} - d_{33}^{(1)}\right) - \rho\right],\\
&\Delta_1 = \left(d_{33}^{(1)} - d_{15}^{(1)}\right)\backepsilon_{33}^{(1)} - \backepsilon_{11}^{(1)} d_{31}^{(1)}, \quad \Delta_2 = s_{11}^{(1)}\backepsilon_{33}^{(1)} - \left(d_{31}^{(1)}\right)^2,\\
&\Delta_3 = \left(d_{33}^{(1)} - d_{15}^{(1)}\right) d_{31}^{(1)} - s_{11}^{(1)}\backepsilon_{11}^{(1)},
\end{aligned}
\tag{3.25}
$$

where h is the thickness of the plate.

The determinant of the system (3.25) reads

$$\Delta = -8i\frac{\mathrm{Im}\mu_1\mathrm{Im}\mu_2\mathrm{Im}\mu_3}{\left|\mu_1\mu_2\mu_3\right|^2}\left|(\mu_2-\mu_1)(\mu_3-\mu_1)(\mu_3-\mu_2)\times(\mu_2-\overline{\mu}_1)\right.$$
$$\left.(\mu_3-\overline{\mu}_1)(\mu_3-\overline{\mu}_2)\right|^2 \neq 0$$

Demanding the determined in (3.24) solution to satisfy the conditions of conjugation (3.22) on line $x_3 = 0$ we come to the following system of linear algebraic equations w.r.t $\beta^{(1)}_{\nu m}$

$$\sum_{\nu=1}^{3}\left\{\overline{c^{(1)}_{n\nu}\beta^{(1)}_{\nu m}} + c^{(2)}_{n\nu}\beta^{(1)}_{\nu+3,m}\right\} = c^{(1)}_{nm} \tag{3.26}$$
$$(n = 1,2,\ldots,6; \quad m = 1,2,3)\,.$$

It may be shown that the determinants of these systems are different from zero. Hence, the fundamental solution is determined by (3.24)–(3.26).

If the concentrated force or the electric charge is applied at the point (x'_{10}, x'_{30}) on the lower halfplane, the corresponding fundamental solution has the form

$$\Phi^{(1)}_\nu\left(z^{(1)}_\nu\right) = \sum_{m=1}^{3}\frac{\beta^{(2)}_{\nu+3,m}A^{(2)}_m}{z^{(1)}_\nu - z^{(2)}_{m1}},$$
$$\Phi^{(2)}_\nu\left(z^{(2)}_\nu\right) = \frac{A^{(2)}_\nu}{z^{(2)}_\nu - z^{(2)}_{\nu1}} - \sum_{m=1}^{3}\frac{\overline{\beta^{(2)}_{\nu m}A^{(2)}_m}}{z^{(2)}_\nu - \overline{z^{(2)}_{m1}}},$$
$$z^{(2)}_{m1} = x'_{10} + \mu^{(2)}_m x'_{30} \quad (x'_{30} < 0, \nu = 1,2,3)\,, \tag{3.27}$$

where $A^{(2)}_\nu$, $\beta^{(2)}_{\nu m}$ are defined, from (3.25) and (3.26) respectively, while i the upper indexes 1 and 2 can be exchanged.

The complex potentials for the basic boundary problem are derived by substituting the constants $A^{(1)}_m$ $(m = 1,2,3)$ with distribution described in (3.24)

$$A^{(1)}_m = -\frac{1}{2\pi i}\int_L \omega_m(\zeta)\,a^{(1)}_m(\psi)\,ds, \quad \zeta \in L = \cup L_j. \tag{3.28}$$

Where ds is the element of the length of contour arc L_j

We have [177]:

$$\Phi^{(1)}_\nu\left(z^{(1)}_\nu\right) = B^{(1)}_\nu + \frac{1}{2\pi i}\int_L \frac{\omega_\nu(\zeta)\,d\zeta^{(1)}_\nu}{\zeta^{(1)}_\nu - z^{(1)}_\nu} + \sum_{m=1}^{3}\frac{\overline{\beta^{(1)}_{\nu m}}}{2\pi i}\int_L \frac{\overline{\omega_m(\zeta)}\overline{d\zeta^{(1)}_m}}{\overline{\zeta^{(1)}_m} - z^{(1)}_\nu},$$
$$\Phi^{(2)}_\nu\left(z^{(2)}_\nu\right) = B^{(2)}_\nu + \sum_{m=1}^{3}\frac{\beta^{(1)}_{\nu+3,m}}{2\pi i}\int_L \frac{\omega_m(\zeta)\,d\zeta^{(1)}_m}{\zeta^{(1)}_m - z^{(2)}_\nu}, \tag{3.29}$$
$$\omega_\nu(\zeta) = \{\omega_{\nu j}(\zeta), \zeta \in L_j\} \quad (j = 1,2,\ldots,k)\,,$$
$$\zeta^{(r)}_\nu = \mathrm{Re}\zeta + \mu^{(r)}_\nu \mathrm{Im}\zeta, \quad \zeta \in L \quad (r = 1,2; \nu = 1,2,3)\,.$$

The integration here takes place on the corresponding affine reflections of contour L_j on plane $z_\nu^{(1)}$ $(\nu = 1, 2, 3)$ and the constants $B_\nu^{(r)}$ must simultaneously support the existence of the given field of stresses and satisfy the conditions of media conjugation at infinity.

The integral representations in (3.29) satisfy the conditions of conjugation (3.22) irrespective of the choice of function $\omega_\nu(\zeta)$. The latter must be determined from the boundary equalities (3.23). Substituting the limiting values of functions (3.29) and using the *formulas Sohotsky-Plemmely* [143] for the generalized integrals of Cauchy type, we come to a real system of two singular integral and four algebraic equations w.r.t three complex "*densities*" $\omega_\nu(\zeta)$

$$2\mathrm{Re}\sum_{\nu=1}^{3} b_{p\nu}^0 \left\{ \frac{1}{\pi i} \int\limits_L \frac{\omega_\nu(\zeta)\, d\zeta_\nu^{(1)}}{\zeta_\nu^{(1)} - \zeta_{\nu 0}^{(1)}} + \sum_{m=1}^{3} \frac{\overline{\beta_{\nu m}^{(1)}}}{\pi i} \int\limits_L \frac{\overline{\omega_m(\zeta)} d\overline{\zeta_m^{(1)}}}{\overline{\zeta_m^{(1)}} - \zeta_{\nu 0}^{(1)}} \right\} = W_p(\zeta_0),$$

$$\mathrm{Re}\sum_{\nu=1}^{3} b_{n\nu}\omega_\nu(\zeta) = 0 \quad (p = 1, 2; n = 1, 2, \ldots, 4), \tag{3.30}$$

$$W_1(\zeta_0) = 2X_{1n} - 2\left(< \sigma_{11}^{(1)} > \cos\psi_0 + < \sigma_{13} > \sin\psi_0\right),$$

$$W_2(\zeta_0) = -2X_{3n} + 2\left(< \sigma_{13} > \cos\psi_0 + < \sigma_{33} > \sin\psi_0\right),$$

$$b_{n\nu} = b_{n\nu}(\zeta), \quad b_{n\nu}^0 = b_{n\nu}(\zeta_0), \quad \psi_0 = \psi(\zeta_0),$$

$$\zeta_{\nu 0}^{(1)} = \mathrm{Re}\zeta_0 + \mu_\nu^{(1)}\mathrm{Im}\zeta_0, \quad \zeta_0 \in L = \bigcup_{j=1}^{k} L_j.$$

The above system is augmented with the additional conditions

$$\mathrm{Re}\sum_{\nu=1}^{3} p_\nu^{(1)} \int\limits_{L_j} \omega_\nu(\zeta)\, d\zeta_\nu^{(1)} = 0, \tag{3.31}$$

$$\mathrm{Re}\sum_{\nu=1}^{3} q_\nu^{(1)} \int\limits_{L_j} \omega_\nu(\zeta)\, d\zeta_\nu^{(1)} = 0 \quad (j = 1, 2, \ldots, k),$$

which come out of the requirement of the displacement uniqueness in the compound medium.

Equations (3.30), (3.31) uniquely define the functions $\omega_\nu(\zeta)$ $(\nu = 1, 2, 3)$ in the class of functions not limited on the ends L_j [121]. The potentials $\Phi_\nu^{(r)}\left(z_\nu^{(r)}\right)$ are reproduced from those equations according to formulas (3.29). In addition, the mechanical and electric field quantities may be calculated, from (3.10) at any point of a compound medium.

The derivation of formulas for the stress intensity factors results from an asymptotic analysis of the corresponding stresses in the vicinity of the crack tip. In that case it is necessary to proceed from the determining expressions [138]

$$K_I = \ell im\sqrt{2\pi r^*}\sigma_n, \quad K_{II} = \ell im\sqrt{2\pi r^*}\tau_{ns} \quad (r^* \to 0), \tag{3.32}$$

where σ_n, τ_{ns} are the normal and tangential stresses, respectively, on the continuation of the crack over the tip, and r^* denotes the distance from the point to tip c.

The parameterization of the contour L_j (the index j is omitted) reads

$$\zeta = \zeta(\beta), \quad \zeta_0 = \zeta(\beta_0) \quad (-1 \leq \beta, \beta_0 \leq 1).$$

Due to this situation

$$\omega_m(\zeta) = \frac{\Omega_m(\beta)}{\sqrt{1-\beta^2}}, \tag{3.33}$$

where the functions $\Omega_m(\beta)$ $(m = 1, 2, 3)$ are continuous on section $[-1, 1]$ according to Hölder condition.

Calculating the stress asymptotic on the tip using (3.10), (3.24), (3.29) and (3.33), from (3.32) we find

$$\begin{aligned}
K_I^{\pm} &= \pm\sqrt{\pi s'(\pm 1)}\,\mathrm{Im}\sum_{m=1}^{3}\gamma_m^{(1)}\left\{a_m^{(1)}(\psi_c)\right\}^2\Omega_m(\pm 1),\\
K_{II}^{\pm} &= \pm\sqrt{\pi s'(\pm 1)}\,\mathrm{Im}\sum_{m=1}^{3}\gamma_m^{(1)}a_m^{(1)}(\psi_c)\,e_m^{(1)}(\psi_c)\,\Omega_m(\pm 1),\\
a_m^{(1)}(\psi_c) &= \mu_m^{(1)}\cos\psi_c - \sin\psi_c,\\
e_m^{(1)}(\psi_c) &= \mu_m^{(1)}\sin\psi_c + \cos\psi_c,
\end{aligned} \tag{3.34}$$

where the upper index corresponds to the tip $c = b$, the lower one is $-c = a$; ψ_c is the angle between the normal to L_j on the tip c and Ox_1 -axis.

Using the parameterization of the contour L_j on the section $-1 \leq \beta \leq 1$ the system (3.30) can be written in a more convenient (w.r.t. the calculations) way:

$$\sum_{\nu=1}^{3}\int_{-1}^{1}\left\{\omega_\nu^*(\beta)\,G_{\nu p}(\beta,\beta_0) + \overline{\omega_\nu^*(\beta)G_{\nu p}(\beta,\beta_0)}\right\}d\beta = W_p^*(\beta_0), \tag{3.35}$$

$$\sum_{\nu=1}^{3}\left\{\omega_\nu^*(\beta)\,b_{n\nu}^*(\beta) + \overline{\omega_\nu^*(\beta)b_{n\nu}^*(\beta)}\right\} = 0 \quad (p = 1, 2;\ n = 1, 2, \ldots, 4).$$

where

$$\begin{aligned}
G_{\nu p}(\beta,\beta_0) &= \frac{a_\nu^{(1)}(\psi)}{\pi i}\left\{\frac{b_{p\nu}^{*0}}{\zeta_\nu^{(1)} - \zeta_{\nu 0}^{(1)}} - \sum_{m=1}^{3}\frac{\overline{b_{pm}^{*0}}\beta_{m\nu}^{(1)}}{\zeta_\nu^{(1)} - \overline{\zeta_{m0}^{(1)}}}\right\}s'(\beta),\\
W_p^*(\beta_0) &= W_p(\zeta_0), \quad \omega_\nu^*(\beta) = \omega_\nu(\zeta), \quad \psi = \psi(\zeta),\\
b_{p\nu}^{*0} &= b_{p\nu}(\zeta_0), \quad \zeta = \zeta(\beta).
\end{aligned}$$

According to the scheme given in [210] we can calculate the approximate values of the unknown functions $\Omega_\nu(\beta)$ in (3.33) in the zero of Chebyshev polynomial of the first kind. The points of collocation are the roots of Chebyshev polynomial of the second kind (see: Appendix 12.8, par. B.1).

Applying the quadrature formulas of (B.5), (B.8) type on both the integral equations in (3.35) and the additional conditions (3.31), we come to a $6N$ set of linear algebraic equations w.r.t the values at n the nodes of interpolation $\beta_\nu\,(\nu = 1, 2, \ldots, N)$ of the functions, $\Omega_i(\beta)$ and $\overline{\Omega_i(\beta)}(i = 1, 2, 3)$ given by

$$
\begin{aligned}
&\frac{\pi}{N}\sum_{i=1}^{3}\sum_{\nu=1}^{N}\left\{\Omega_{\nu i}G_{ip}(\beta_\nu,\eta_m)+\overline{\Omega_{\nu i}}\;\overline{G_{ip}(\beta_\nu,\eta_m)}\right\}=W_p^*(\eta_m)\\
&(p=1,2;m=1,2,\ldots,N-1),\\
&\sum_{i=1}^{3}\sum_{\nu=1}^{N}\left\{b_{ni}^*(\beta_\nu)\,\Omega_{\nu i}+\overline{b_{ni}^*(\beta_\nu)\Omega_{\nu i}}\right\}=0\quad(n=1,2,\ldots,4),\\
&\mathrm{Re}\sum_{i=1}^{3}\sum_{\nu=1}^{N}p_i^{(1)}\Omega_{\nu i}a_{i\nu}^{(1)}s'(\beta_\nu)=0,\\
&\mathrm{Re}\sum_{i=1}^{3}\sum_{\nu=1}^{N}q_i^{(1)}\Omega_{\nu i}a_{i\nu}^{(1)}s'(\beta_\nu)=0,
\end{aligned}
\tag{3.36}
$$

where

$$
\begin{aligned}
&\Omega_{\nu i}=\Omega_i(\beta_\nu),\quad a_{i\nu}^{(1)}=a_i^{(1)}(\zeta_\nu),\quad s'(\beta_\nu)=\left.\frac{ds}{d\beta}\right|_{\beta=\beta_\nu},\\
&\zeta_\nu=\zeta(\beta_\nu),\quad \beta_\nu=\cos\frac{2\nu-1}{2N}\pi,\quad \eta_m=\cos\frac{\pi m}{N}.
\end{aligned}
$$

The last two equations in (3.36), corresponding to the additional conditions (3.31) are derived with the help of equality $d\zeta_\nu^{(1)} = a_\nu^{(1)}(\psi)\,ds$,

To calculate the quantities $\Omega_i(\pm 1)\,(i = 1, 2, 3,)$, appearing in expressions (3.34) for the stress intensity factors it is also necessary to use the formulas [126]

$$
\begin{aligned}
L_N[\Omega_i,-1]&=\frac{1}{N}\sum_{\nu=1}^{N}(-1)^{\nu+N}\,\Omega_{\nu i}\mathrm{tg}\frac{2\nu-1}{4N}\pi,\\
L_N[\Omega_i,1]&=-\frac{1}{N}\sum_{\nu=1}^{N}(-1)^{\nu}\,\Omega_{\nu i}\mathrm{ctg}\frac{2\nu-1}{4N}\pi.
\end{aligned}
\tag{3.37}
$$

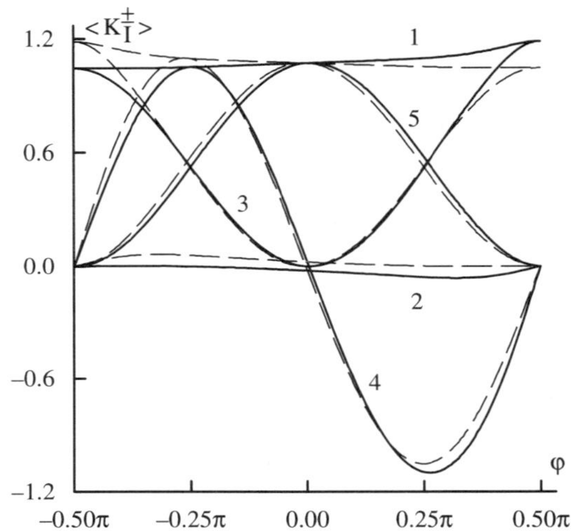

Fig. 3.3. Changes of quantity $< K_I >= (\Lambda\sqrt{\pi l})^{-1} K_I$ according to the orientation angle φ of the crack

By increasing the parameter N, the solution of the system of algebraic (3.36), which is a discrete analogue of (3.30), (3.31), converges uniformly to their exact solution.

As an example let us consider the bimorph (material of the upper halfplane is ceramics $BaTiO_3$, of the lower one is PZT-5), having a parabolic crack

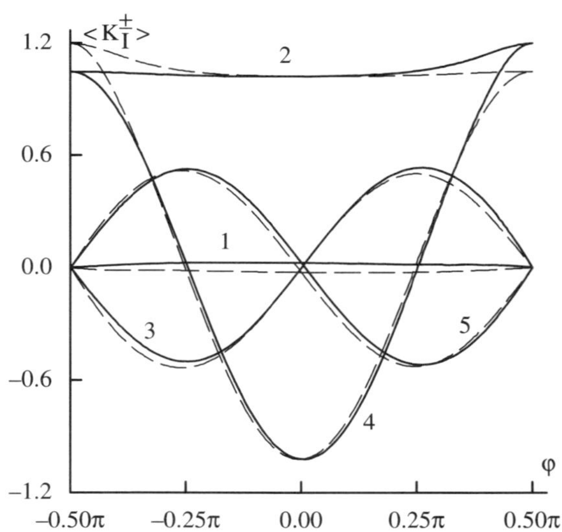

Fig. 3.4. Changes of quantity $< K_{II} >= (\Lambda\sqrt{\pi l})^{-1} K_{II}$ according to the orientation angle φ of the crack

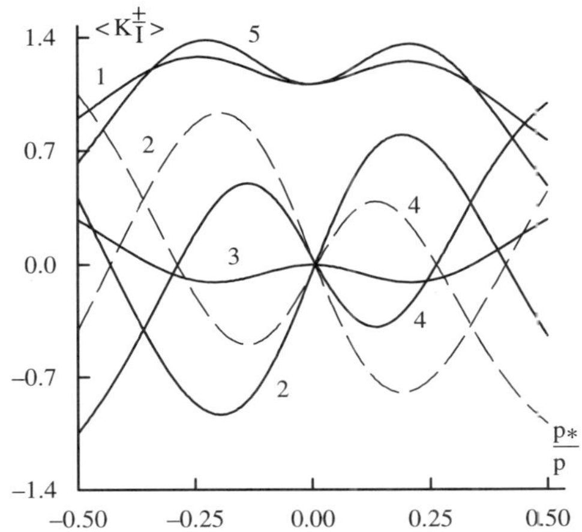

Fig. 3.5. Changes of quantity $< K_I >$ as a function of parameter p_*/p

above the interphase and the corresponding parametric equation having the form of

$$\xi_1 + i\xi_2 = \left(p\beta + ip_*\beta^2\right) e^{i\varphi} + ih \quad (-1 \le \beta \le 1)\,.$$

The number of Chebyshev nodes was taken to be equal to $N = 11, 15, 19$; further increase of N did not refine the solution.

Below are presented the results obtained for the relative stress intensity factors $< K_{I,II} >= \left(\Lambda\sqrt{\pi\ell}\right)^{-1} K_{I,II}$ (2ℓ is the crack length) w.r.t the angle of

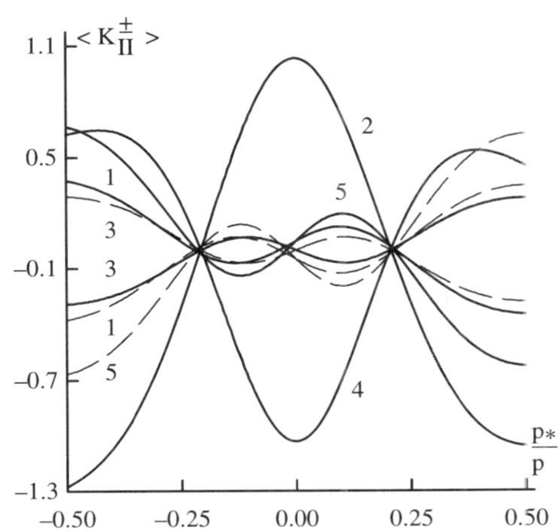

Fig. 3.6. Changes of quantity $< K_{II} >$ as a function of curvature parameter p_*/p

the crack orientation φ and curve parameter p_* for different types of stresses. The curves 1, 2 are constructed for uniform normal and tangential loading of intensity P on crack edges ($\Lambda = P$); curves 3, 4 and 5 are constructed for uniform tension along x_1 $\left(\Lambda =< \sigma_{11}^{(1)} >\right)$, pure shear ($\Lambda =< \sigma_{13} >$) and tension along axis x_3 ($\Lambda =< \sigma_{33} >$) at infinity. The solid- lines correspond to tip $c = a$, dashed- lines to $c = b$. Figs. 3.3 and 3.4 illustrate the quantities $< K_I >$ and $< K_{II} >$ for a linear crack ($p_*/p = 0,\ h/p = 1.05$) in Figs. 3.5 and 3.6 for parabolic crack ($\varphi = 0,\ h/p = 0.6$).

3.3 Bimorph with Openings in One of the Pair Components

Consider the analogous problem of bimorph weakened in the upper halfplane of opening Γ_j ($j = 1, 2, \ldots, n$). Let us assume that when on the edges of the openings bounded with vacuum, the stress vector (X_{1n}, X_{3n}) acts the homogeneous stress fields $< \sigma_{11}^{(1)} >$, $< \sigma_{11}^{(2)} >$, $< \sigma_{13} >$, $< \sigma_{33} >$ and the electric field with strengths $< E_1 >$, $< E_3^{(1)} >$, $< E_3^{(2)} >$ are applied at infinity. The prescribed at infinity fields are not quite arbitrary; conditions of joint deformation of the upper and lower half-planes cause connections which are described below. Let us suppose that the curves of contour Γ_j and function X_{1n}, X_{3n} belong to the class of functions, continuous by Hölder, and besides $\cap \Gamma_j = \emptyset$.

The above problem is reduced to the determination of six complex functions $\Phi_\nu^{(r)}\left(z_\nu^{(r)}\right)$ ($\nu = 1, 2, 3;\ r = 1, 2$) from conditions (3.22) on the interphase of media, conditions at infinity and also on boundary conditions on the contours of openings Γ_j. The latter may be presented as follows

$$2\mathrm{Re} \sum_{\nu=1}^{3} b_{p\nu} \left\{ \Phi_\nu^{(1)}\left(\zeta_\nu^{(1)}\right)\right\}^{-} = f_p(\zeta) \quad (p = 1, 2, 3), \tag{3.38}$$

$$b_{1\nu} = \gamma_\nu^{(1)} \mu_\nu^{(1)} a_\nu^{(1)}(\psi), \quad b_{2\nu} = \gamma_\nu^{(1)} a_\nu^{(1)}(\psi), \quad b_{3\nu} = r_\nu^{(1)} a_\nu^{(1)}(\psi),$$

$$a_\nu^{(1)}(\psi) = \mu_\nu^{(1)} \cos\psi - \sin\psi,$$

$$f_1(\zeta) = X_{1n}(\zeta), \quad f_2(\zeta) = -X_{3n}(\zeta), \quad f_3(\zeta) = 0,$$

$$\zeta_\nu^{(1)} = \mathrm{Re}\zeta + \mu_\nu^{(1)} \mathrm{Im}\zeta, \quad \zeta = \xi_1 + i\xi_2 \in \Gamma_j \quad (j = 1, 2, \ldots, n).$$

Here ψ is the angle between the normal to contour Γ_j and Ox_1 - axis (Fig. 3.7). As functions $\Phi_\nu^{(1)}\left(\zeta_\nu^{(1)}\right)$ we consider thei limiting values at $z \to \zeta \in \Gamma_j$ when this transition has direction from the body to the hole. The condition corresponding to $p = 3$ expresses the equality to zero of the normal component of the vector of electric induction on Γ_j.

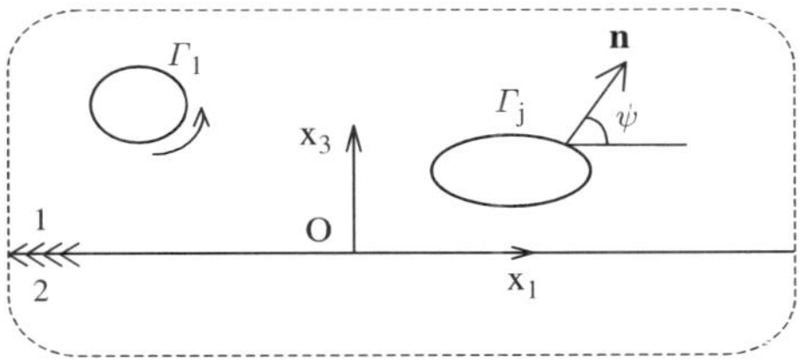

Fig. 3.7. A bimorph with an opening in one of the components of a pair

By applying the fundamental solution (3.24) we may present the analytical functions $\Phi_\nu^{(r)}\left(z_\nu^{(r)}\right)$ in the following form [188]

$$\Phi_\nu^{(1)}\left(z_\nu^{(1)}\right) = B_\nu^{(1)} + \sum_{k=1}^{3}\int_\Gamma q_k(\zeta)\left\{\frac{\omega_{k\nu}^{(1)}}{z_\nu^{(1)}-\zeta_\nu^{(1)}} - \sum_{m=1}^{3}\frac{\overline{\beta_{\nu m}^{(1)}\omega_{km}^{(1)}}}{z_\nu^{(1)}-\overline{\zeta_m^{(1)}}}\right\}ds, \quad (3.39)$$

$$\Phi_\nu^{(2)}\left(z_\nu^{(2)}\right) = B_\nu^{(2)} + \sum_{k=1}^{3}\int_\Gamma q_k(\zeta)\sum_{m=1}^{3}\frac{\beta_{\nu+3,m}^{(1)}\omega_{km}^{(1)}}{z_\nu^{(2)}-\zeta_m^{(1)}}ds \quad (\nu=1,2,3),$$

$$q_\nu(\zeta) = \{q_{\nu j}(\zeta), \zeta\in\Gamma_j\}, \quad \mathrm{Im}q_\nu(\zeta)=0, \quad \Gamma = \bigcup_{j=1}^{n}\Gamma_j.$$

Here ds is the element of the arc length of contour Γ, the constants $B_\nu^{(r)}$ reflect the, given at infinity, homogeneous fields of mechanical stresses and strengths of the electric field. The complex constants $\omega_{km}^{(1)}$ represent "standard" solutions of the system (3.25) which are connected by

$$A_m^{(1)} = P_1\omega_{1m}^{(1)} + \rho\omega_{2m}^{(1)} + P_3\omega_{3m}^{(1)} \quad (m=1,2,3).$$

The integral representations (3.39) satisfy the conditions of conjugation (3.22) no matter what the choice of "*densities*" $q_k(\zeta)$ is.

Substituting the limiting values of representations (3.39) at $z_\nu^{(1)} \to \zeta_{\nu 0}^{(1)}$ into the boundary conditions (3.38) we get a system of three real singular integral equations of the second kind:

$$\sum_{k=1}^{3}\left\{t_{pk}(\zeta_0)q_k(\zeta_0) + \int_\Gamma q_k(\zeta)G_{kp}(\zeta,\zeta_0)ds\right\} = N_p(\zeta_0) \quad (p=\overline{1,3}),$$

$$G_{kp}(\zeta,\zeta_0) = 2\mathrm{Re}\sum_{\nu=1}^{3}b_{p\nu}^0\left\{\sum_{m=1}^{3}\frac{\overline{\beta_{\nu m}^{(1)}\omega_{km}^{(1)}}}{\overline{\zeta_m^{(1)}}-\zeta_{\nu 0}^{(1)}} - \frac{\omega_{k\nu}^{(1)}}{\zeta_\nu^{(1)}-\zeta_{\nu 0}^{(1)}}\right\}, \quad (3.40)$$

$$t_{pk}(\zeta_0) = -2\pi\mathrm{Im}\sum_{\nu=1}^{3}\frac{b_{p\nu}^0\omega_{k\nu}^{(1)}}{a_\nu^{(1)}(\psi_0)},$$

$$N_p(\zeta_0) = f_p(\zeta_0) + f_p^*(\zeta_0), \quad b_{p\nu}^0 = b_{p\nu}(\psi_0), \quad \psi_0 = \psi(\zeta_0),$$
$$f_1^*(\zeta_0) = -<\sigma_{11}^{(1)}> \cos\psi_0 - <\sigma_{13}> \sin\psi_0,$$
$$f_2^*(\zeta_0) = <\sigma_{13}> \cos\psi_0 + <\sigma_{33}> \sin\psi_0,$$
$$f_3^*(\zeta_0) = -\left(d_{15}^{(1)} <\sigma_{13}> + \ni_{11}^{(1)} <E_1>\right)\cos\psi_0 -$$
$$-\left(d_{31}^{(1)} <\sigma_{11}^{(1)}> + d_{33} <\sigma_{33}> + \ni_{33}^{(1)} <E_3^{(1)}>\right)\sin\psi_0,$$
$$\zeta_{\nu 0}^{(1)} = \mathrm{Re}\zeta_0 + \mu_\nu^{(1)}\mathrm{Im}\zeta_0; \quad \zeta, \zeta_0 \in \Gamma.$$

The functions $b_{p\nu}(\zeta)$, $f_p(\zeta)$ are defined in (3.38).

To determine the stress concentration in a bimorph with openings, we first calculate the normal stress σ_ϑ on contour Γ. Allowing for representation (3.10) we have

$$\sigma_\vartheta(\zeta_0) = \sigma_{11}\sin^2\psi_0 + \sigma_{33}\cos^2\psi_0 - \sigma_{13}\sin 2\psi_0 = \tag{3.41}$$
$$= 2\mathrm{Re}\sum_{\nu=1}^{3}\gamma_\nu^{(1)}\left(\mu_\nu^{(1)}\sin\psi_0 + \cos\psi_0\right)^2 \Phi_\nu^{(1)}\left(\zeta_{\nu 0}^{(1)}\right).$$

Here by functions $\Phi_\nu^{(1)}\left(\zeta_{\nu 0}^{(1)}\right)$ we consider their limiting values at $z \to \zeta_0 \in \Gamma$ when this transition has direction from the body to the hole, ϑ is the angle between the positive tangential to the contour of the opening at point ζ_0 and axis Ox_1.

As an example, two variants of the compound plate were studied: in the first one the material of the upper halfplane is $BaTiO_3$, while that of the lower is PZT-5 (variant A); in the second one the material of the upper halfplane is PZT-5, while that of the lower one is $BaTiO_3$ (variant B). In these cases, compound plates with elliptic and "*square*" holes described by the following parametric equations (at a constant value, $C = 0.14036$),

$$\mathrm{Re}\zeta = R_1\cos\varphi, \quad \mathrm{Im}\zeta = h + R_2\sin\varphi, \quad \varphi \in [0, 2\pi];$$
$$\mathrm{Re}\zeta = R_0(\cos\varphi + C\cos 3\varphi), \tag{3.42}$$
$$\mathrm{Im}\zeta = R_0(\sin\varphi - C\sin 3\varphi) + h.$$

Using (3.42) the integral (3.40) can be reduced to a system of linear algebraic equations w.r.t. the relative values of $q_k(\zeta)$ calculated at the nodes of interpolation using the quadrature method [48, 84, and 137] (see: Appendix B 12.8, par. B.2).

Figure 3.8 demonstrates the behavior of relative normal stress $\lambda = \sigma_\vartheta/\Lambda$ (Λ is the loading intensity) on the contour of an elliptic opening when the stresses $<\sigma_{11}^{(1)}>$, $<\sigma_{13}>$ or $<\sigma_{33}>$ (curves 1,2 and 3, respectively) are acting at infinity for variant A. The values of the parameters were $h/R_1 = 4/5$, $R_1/R_2 = 5/3$. Dashed lines correspond to the changes of λ on the contour of the given opening in the homogeneous piezoceramic plate

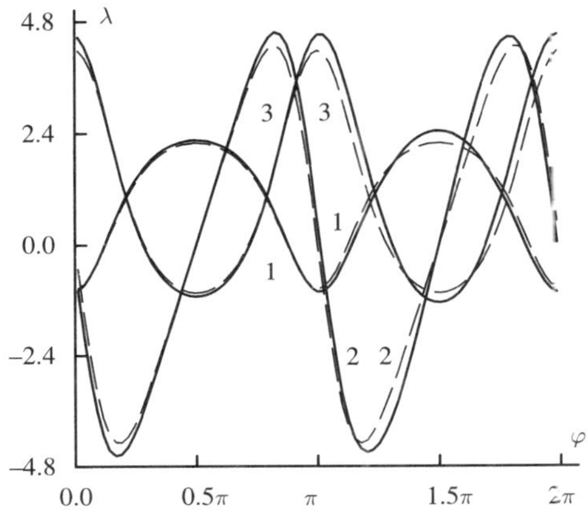

Fig. 3.8. The behaviour of relative stress $\lambda = \sigma_\theta/\Lambda$ on the contour of an elliptic opening under the influence of stresses $< \sigma_{11}^{(1)} >$ and $< \sigma_{13} >$, $< \sigma_{33} >$

(material $BaTiO_3$). We should note that the stress σ_ϑ was calculated by using (3.39) and (3.41).

The investigation of the stress concentration on a plate with an opening in the presence of an external electric field is of great interest. In Figs. 3.9 and 3.10 we present the stress σ_ϑ on the contour of a square opening under the action of fields $< E_1^{(1)} >$ (curves 1) and $< E_3^1 >$ (curves 2) by intensity $1V/m$ for variants A and B, respectively. For our illustrations we used $h/R_0 = 1.2,\ R_0 = 1m$.

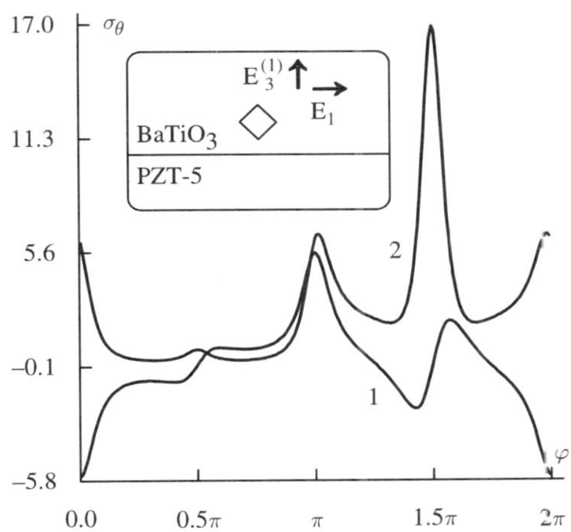

Fig. 3.9. Stressses σ_θ on the contour of a square opening under the influence of an electric field. $< E_1^{(1)} >$, and $< E_3^{(1)} >$ with intensity 1V/m (variant A)

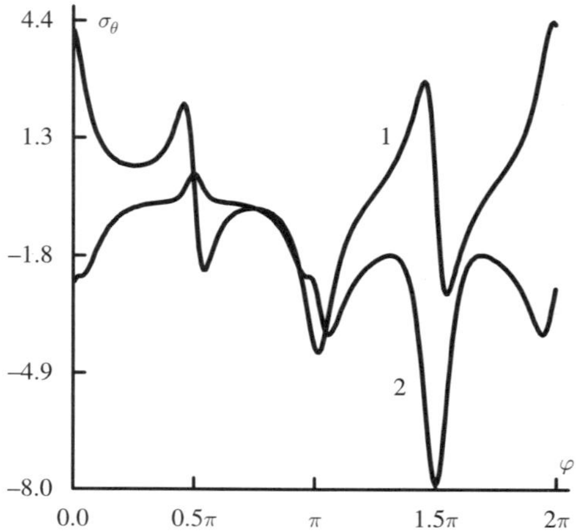

Fig. 3.10. The behaviour of relative stresses σ_θ on the contour of a square opening under the influence of electric field $< E_1^{(1)} >$ and $< E_3^{(1)} >$ with intensity 1V/m (variant B)

Figure 3.11 shows results in the case of a plate with a square opening the change of quantity λ for variants A. Curves 1, 2 and 3 correspond to the action of stresses $< \sigma_{11}^{(1)} >$, $< \sigma_{13} >$ and $< \sigma_{33} >$ $(h/R_0 = 1.2)$. The dashed lines have the same meaning as in Fig. 3.8.

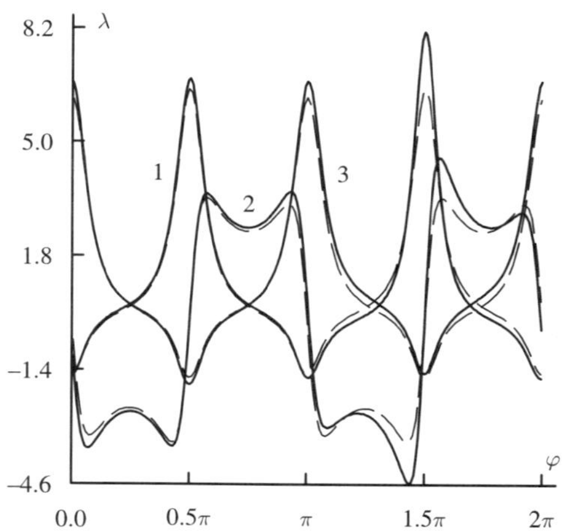

Fig. 3.11. Changes of quantity λ on the contour of an opening under the influence of stress $< \sigma_{11}^{(1)} >$ and $< \sigma_{13} >$, $< \sigma_{33} >$

Table 3.1. The approximate values of the stress concentration factors for various types of loading for the case of a plate with an elliptic opening

	$<\sigma_{11}^{(1)}>$	$<\sigma_{13}>$	$<\sigma_{33}>$	N	T	$<E_1^{(1)}>$	$<\mathrm{E}_3^{(1)}>$
			Variant A				
$\mathrm{R}_1/\mathrm{R}_2=1$	3.29	4.39	3.09	1.18	0.09	0.032	0.073
$\mathrm{R}_1/\mathrm{R}_2=2$	2.07	4.69	5.08	3.04	1.55	0.018	0.071
			Variant B				
$\mathrm{R}_1/\mathrm{R}_2=1$	3.01	3.87	2.69	1.04	0.11	0.073	0.114
$\mathrm{R}_1/\mathrm{R}_2=2$	2.01	4.35	4.44	2.64	1.48	0.033	0.109

In Table 3.1 we present, for various types of loadings, the approximate values of the stress concentration factors for the problem of a compound plate with elliptic ($h/R_1=1.5$) opening,. In the case where an electric field $<E_j^{(1)}>$ ($j=1,3$) acts at infinity the concentration factor is given by $k_j=d_{15}^{(1)}\left|\sigma_\vartheta\right|/\left(\text{э}_{11}^{(1)}<E_j^{(1)}>\right)$. Similar results for a homogeneous piezoceramic plate are presented in Table 3.2

The results presented in Figs. 3.9, 3.10 testify a strong influence of the pair material on the stress concentration under the action of external electric field.

It should be noted that the prescribed at infinity mechanical and electric quantities are not quite arbitrary. The stresses σ_{13} and σ_{33} and also the components of electric strength E_1 continue across the interphase of the media but σ_{11} and E_3 undergo jumps which may be determined from conditions of the joint deformation

$$
\begin{aligned}
& s_{11}^{(2)}<\sigma_{11}^{(2)}>+d_{31}^{(2)}<E_3^{(2)}>=s_{11}^{(1)}<\sigma_{11}^{(1)}>+ \\
& +\left(s_{13}^{(1)}-s_{13}^{(2)}\right)<\sigma_{33}>+d_{31}^{(1)}<E_3^{(1)}>. \\
& d_{31}^{(2)}<\sigma_{11}^{(2)}>+\text{э}_{33}^{(2)}<E_3^{(2)}>=d_{31}^{(1)}<\sigma_{11}^{(1)}>+ \\
& +\left(d_{33}^{(1)}-d_{33}^{(2)}\right)<\sigma_{33}>+\text{э}_{33}^{(1)}<E_3^{(1)}>.
\end{aligned}
\tag{3.43}
$$

Table 3.2. The approximate values of stress concentration factors for a homogeneous piezoceramic plate

	$<\sigma_{11}^{(1)}>$	$<\sigma_{13}>$	$<\sigma_{33}>$	N	T	$<E_1^{(1)}>$	$<E_3^{(1)}>$
			$BaTiO_3$				
$\mathrm{R}_1/\mathrm{R}_2=1$	3.0	4.05	2.91	1.05	0.05	0.0019	0.0023
$\mathrm{R}_1/\mathrm{R}_2=2$	2.0	4.52	4.83	2.87	1.52	0.0018	0.0043
			PZT-5				
$\mathrm{R}_1/\mathrm{R}_2=1$	3.08	4.06	2.83	1.06	0.06	0.022	0.022
$\mathrm{R}_1/\mathrm{R}_2=2$	2.04	4.48	4.67	2.76	1.5	0.019	0.038

Equalities (3.43) permit us to define $\left\langle \sigma_{11}^{(2)} \right\rangle$, $\left\langle E_3^{(2)} \right\rangle$ by the given quantities $\left\langle \sigma_{11}^{(1)} \right\rangle$, $\langle \sigma_{33} \rangle$, $\left\langle E_3^{(1)} \right\rangle$.

3.4 A Composite Plate with a Crack Crossing the Interphase

Let us investigate the behavior of the stress intensity factor at tension of an infinite composite piezoceramic plate with a crack crossing the line of media conjugation.

Let a curvilinear crack L cross the material interphase at node C, and at this point the crack is not smooth (Fig. 3.12). Let us suppose that at infinity, a homogeneous field of mechanical stresses $\left\langle \sigma_{11}^{(1)} \right\rangle, \langle \sigma_{13} \rangle, \langle \sigma_{33} \rangle$ is prescribed, and a force vector (X_{1n}, X_{3n}) acts on the crack edges. We assume that the direction of the electric field of the preliminary polarization of the piezoceramics is parallel to x_3-axis.

Let $L_r\ (r = 1, 2)$ be a part of the crack contour L lying on the r-th half-plane. Then, using (3.10) the mechanical and electric boundary conditions on the crack can be written as

$$2\mathrm{Re}\sum_{k=1}^{3} b_{pk}^{(r)} \left\{ \Phi_k^{(r)} \left(\zeta_k^{(r)} \right) \right\}^{\pm} = f_p^{(r)} \left(\zeta^{(r)} \right), \quad \zeta^{(r)} \in L_r, \tag{3.44}$$

$$\mathrm{Re}\sum_{k=1}^{3} b_{nk}^{(r)} \left[\Phi_k^{(r)} \left(\zeta_k^{(r)} \right) \right] = 0 \quad (p = 1, 2; n = 3, 4),$$

where

$$\begin{aligned} b_{1k}^{(r)} &= \gamma_k^{(r)} \mu_k^{(r)} a_k^{(r)} (\psi), \quad b_{2k}^{(r)} = \gamma_k^{(r)} a_k^{(r)} (\psi), \\ b_{3k}^{(r)} &= \lambda_k^{(r)} a_k^{(r)} (\psi), \quad b_{4k}^{(r)} = r_k^{(r)} a_k^{(r)} (\psi), \\ a_k^{(r)} (\psi) &= \mu_k^{(r)} \cos\psi - \sin\psi \quad (r = 1, 2), \\ f_1^{(r)} &= X_{1n}^{(r)}, \qquad f_2^{(r)} = -X_{3n}^{(r)}. \end{aligned}$$

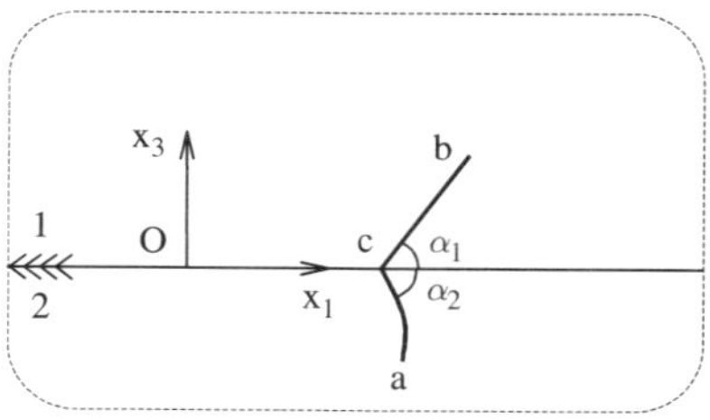

Fig. 3.12. The scheme of a composite plate with a crack

Here the symbol $[\cdot]$ denotes a jump of the corresponding quantity on L, Ψ is an angle between the normal to the left edge L (when it moves from its beginning to the end b) and Ox_2- axis. Here we should note that the last two conditions in (3.44) express the continuation of the normal component of the induction vector and the tangential component of the strength vector of the electric field crossing L_r.

Hence, the problem consists of defining six potentials $\Phi_\nu^{(r)}\left(z_\nu^{(r)}\right)$ $(\nu=\overline{1,3};\ r=1,2)$ by boundary conditions (3.44), conjugated conditions (3.22) and also conditions at infinity.

By applying the fundamental solution (3.24), (3.27) let us present the complex potentials as follows [16]

$$\Phi_\nu^{(r)}\left(z_\nu^{(r)}\right)=\frac{1}{2\pi i}\int\limits_{L_r}\frac{\omega_\nu^{(r)}\left(\zeta^{(r)}\right)d\zeta_\nu^{(r)}}{\zeta_\nu^{(r)}-z_\nu^{(r)}}++\frac{1}{2\pi i}\sum_{m=1}^{3}\overline{\beta_{\nu m}^{(r)}}\int\limits_{L_r}\frac{\overline{\omega_m^{(r)}\left(\zeta^{(r)}\right)d\zeta_m^{(r)}}}{\overline{\zeta_m^{(r)}}-z_\nu^{(r)}}+$$

$$+\frac{1}{2\pi i}\sum_{m=1}^{3}\beta_{\nu+3,m}^{(3-r)}\int\limits_{L_{3-r}}\frac{\omega_m^{(3-r)}\left(\zeta^{(3-r)}\right)d\zeta_m^{(3-r)}}{\zeta_m^{(3-r)}-z_\nu^{(r)}}+B_\nu^{(r)} \tag{3.45}$$

$$(r=1,2;\nu=\overline{1,3})\qquad z_\nu^{(r)}=x_1+\mu_\nu^{(r)}x_3\qquad \zeta_\nu^{(r)}=\xi_1^{(r)}+\mu_v^{(r)}\xi_3^{(r)}$$

$$\left(\xi_1^{(r)},\xi_3^{(r)}\right)\in L_r.$$

Here $x_3>0$ at $r=1$ and $x_3<0$ at $r=2$, and in this case the integration is carried out by the following affine reflection of contour L_r on $z_\nu^{(r)}$ -planes. The constants $B_\nu^{(r)}$ depend on the disturbance of the electric fields due to given at infinity stresses.

At this point we should also note that integral representations (3.45) satisfy the conjugated conditions (3.22) irrespective of the choice of the functions $\omega_\nu^{(r)}\left(\zeta^{(r)}\right)$. Substituting the limiting values of potentials $\Phi_\nu^{(r)}\left(z_\nu^{(r)}\right)$ into the boundary condition (3.44) at $z_\nu^{(r)}\to\zeta_{\nu o}^{(r)}$, we get a real system of four singular integral equations and eight algebraic equations w.r.t six complex "*densities*" $\omega^{(r)}\left(\zeta^{(r)}\right)$:

$$\mathrm{Re}\sum_{\nu=1}^{3}b_{p\nu0}^{(r)}\left\{\frac{1}{\pi i}\int\limits_{L_r}\frac{\omega_\nu^{(r)}\left(\zeta^{(r)}\right)d\zeta_\nu^{(r)}}{\zeta_\nu^{(r)}-\zeta_{\nu0}^{(r)}}+\frac{1}{\pi i}\sum_{m=1}^{3}\overline{\beta_{\nu m}^{(r)}}\int\limits_{L_r}\frac{\overline{\omega_m^{(r)}\left(\zeta^{(r)}\right)d\zeta_m^{(r)}}}{\overline{\zeta_m^{(r)}}-\zeta_{\nu0}^{(r)}}+\right.$$

$$\left.+\frac{1}{\pi i}\sum_{m=1}^{3}\beta_{\nu+3,m}^{(3-r)}\int\limits_{L_{3-r}}\frac{\omega_m^{(3-r)}\left(\zeta^{(3-r)}\right)d\zeta_m^{(3-r)}}{\zeta_m^{(3-r)}-\zeta_{\nu0}^{(r)}}\right\}=F_p^{(r)}\left(\zeta_0^{(r)}\right)\qquad(p=1,2), \tag{3.46}$$

$$\begin{aligned}
&\operatorname{Re}\sum_{\nu=1}^{3} b_{n\nu}^{(r)}\omega_{\nu}^{(r)}\left(\zeta^{(r)}\right)=0 \qquad (r=1,2;n=1,2,\ldots,4)\,,\\
&F_1^{(r)}\left(\zeta_0^{(r)}\right)=X_{1n}^{(r)}-\left\langle\sigma_{11}^{(r)}\right\rangle\cos\psi_0^{(r)}-\langle\sigma_{13}\rangle\sin\psi_0^{(r)},\\
&F_2^{(r)}\left(\zeta_0^{(r)}\right)=-X_{3n}^{(r)}+\langle\sigma_{13}\rangle\cos\psi_0^{(r)}+\langle\sigma_{33}\rangle\sin\psi_0^{(r)},\\
&\psi_0^{(r)}=\psi^{(r)}\left(\zeta_0^{(r)}\right), \qquad b_{n\nu 0}^{(r)}=b_{n\nu}^{(r)}\left(\zeta_0^{(r)}\right),\\
&\zeta^{(r)}=\xi_1^{(r)}+i\xi_3^{(r)}, \qquad \zeta_0^{(r)}\in L_r.
\end{aligned}$$

The system (3.46) is augmented by the additional conditions

$$\begin{aligned}
&\operatorname{Re}\sum_{r=1}^{2}\sum_{\nu=1}^{3}p_{\nu}^{(r)}\int_{L_r}\omega_{\nu}^{(r)}\left(\zeta^{(r)}\right)d\zeta_{\nu}^{(r)}=0,\\
&\operatorname{Re}\sum_{r=1}^{2}\sum_{\nu=1}^{3}q_{\nu}^{(r)}\int_{L_r}\omega_{\nu}^{(r)}\left(\zeta^{(r)}\right)d\zeta_{\nu}^{(r)}=0,
\end{aligned} \tag{3.47}$$

which come out of the requirement that the shift jumps at node c must be equal.

As it is presented in Sect. 3.3, the prescribed at infinity stresses are not quite arbitrary; the conditions of singlevalueness of deformations on the upper and lower halfplanes cause the connection expressed by the equalities (3.43). In the absence of electric field at infinity $\left(\left\langle E_3^{(1)}\right\rangle=0\right)$, the strength $\left\langle\sigma_{11}^{(2)}\right\rangle$ may be determined by

$$\begin{aligned}
\left\langle\sigma_{11}^{(2)}\right\rangle=&\frac{d_{31}^{(1)}d_{31}^{(2)}-\ni_{33}^{(2)}s_{11}^{(1)}}{\left(d_{31}^{(2)}\right)^2-\ni_{33}^{(2)}s_{11}^{(2)}}\left\langle\sigma_{11}^{(1)}\right\rangle\\
&+\frac{d_{31}^{(2)}\left(d_{33}^{(1)}-d_{33}^{(2)}\right)-\ni_{33}^{(2)}\left(s_{13}^{(1)}-s_{13}^{(2)}\right)}{\left(d_{31}^{(2)}\right)^2-\ni_{33}^{(2)}s_{11}^{(2)}}\langle\sigma_{33}\rangle\,.
\end{aligned} \tag{3.48}$$

Now, equality (3.48) can be used for the calculation of the function $F_1^{(2)}\left(\zeta_0^{(2)}\right)$ appearing in (3.46).

The kernels of the integral (3.46) besides the movable singularity of Cauchy type, thy also possess the immovable singularity at node c. Therefore, the approximate solution to the system (3.46) together with equality (3.47) should be sought as a function class not limited from the both ends of contour L_r.

After approximating the solution of system (3.46), (3.47) by using the functions $\omega_{\nu}^{(r)}\left(\zeta^{(r)}\right)$ according to formulas (3.45), the potentials $\Phi_{\nu}^{(r)}\left(z_{\nu}^{(r)}\right)$ are transformed, then considering complex representations (3.10) the coupled mechanical and electric quantities at any point of the compound media are defined.

Let us parametrize the contour L_r $(r = 1, 2)$ by making use of the formulas $\zeta^{(r)} = \zeta^{(r)}(\delta)$, $\delta \in [-1, 1]$. In the general case it is possible to write

$$\omega_\nu^{(1)}\left(\zeta^{(1)}\right) = \frac{\Omega_\nu^{(1)}(\delta)}{(1+\delta)^\gamma \sqrt{1-\delta}}, \qquad (3.49)$$

$$\omega_\nu^{(2)}\left(\zeta^{(2)}\right) = \frac{\Omega_\nu^{(2)}(\delta)}{(1-\delta)^\gamma \sqrt{1+\delta}},$$

where $\Omega_\nu^{(r)}(\pm 1) \neq 0$.

The order of power singularity γ in (3.46) may be determined from asymptotic analysis of singular integral equations in the vicinity of point c on arcs L_r. This rather complex investigation is not carried out here.

Assuming that the inequality $\operatorname{Re}\gamma < 1/2$ holds, the numerical realization of the constructed algorithm is accomplished writing

$$\omega_\nu^{(r)}\left(\zeta^{(r)}\right) = \frac{\Omega_\nu^{(r)}(\delta)}{\sqrt{1-\delta^2}} \qquad (\nu = \overline{1,3}; r = 1, 2), \qquad (3.50)$$

considering that

$$\Omega_\nu^{(1)}(-1) = 0, \qquad \Omega_\nu^{(2)}(1) = 0. \qquad (3.51)$$

Now we are ready to apply the quadrature formulas of Gauss-Chebyshev [210, 211] to the system of (3.46), (3.47) and by using (3.50) we come to a system of $12N - 2$ algebraic equations w.r.t $12N$ unknown values of the functions $\Omega_\nu^{(r)}(\delta)$ and $\overline{\Omega_\nu^{(r)}}(\delta)$ calculated at the nodes of the division of contour L_r (N is the number of the interpolation nodes). To complete the resulting system we need to add two algebraic equations which correspond to two equalities from (3.51). The calculations show that the arbitrary choice of these equalities does not practically influence the precision of the obtained results. The details of the numerical scheme in the case of broken and star cracks in isotropic medium can be found in [160].

By applying asymptotic analysis for the stresses on the tips of cracks a and using (3.10), (3.45) and (3.50) we find the following expressions for the stress intensity factor f

$$K_I^\pm = \pm\sqrt{\pi s'(\pm 1)}\operatorname{Im}\sum_{m=1}^{3}\gamma_m^{(r)}\left\{a_m^{(r)}(\psi_d)\right\}^2\Omega_m^{(r)}(\pm 1),$$

$$K_{II}^\pm = \pm\sqrt{\pi s'(\pm 1)}\operatorname{Im}\sum_{m=1}^{3}\gamma_m^{(r)}a_m^{(r)}(\psi_d)e_m^{(r)}(\psi_d)\Omega_m^{(r)}(\pm 1), \qquad (3.52)$$

$$s'(\pm 1) = \left.\frac{ds}{d\delta}\right|_{\delta=\pm 1}, \qquad a_m^{(r)}(\psi_d) = \mu_m^{(r)}\cos\psi_d - \sin\psi_d,$$

$$e_m^{(r)}(\psi_d) = \mu_m^{(r)}\sin\psi_d + \cos\psi_d.$$

Here the upper sign corresponds to the crack tip $d = p\,(r = 1)$, the lower one to $d = a, (r = 2)$, ψ_d is angle between the normal to L on tip d and axis Ox_1.

As an example let us consider a piezoceramic bimorph (the material of the upper halfplane is $BaTiO_3$, of the lower one is PZT-5), weakened by a broken crack with parametric equations

$$\xi_1^{(r)} = \frac{(1-(-1)^r\,\delta)\cos\alpha_r}{2} \qquad (-1 \le \delta \le 1)\,,$$
$$\xi_3^{(r)} = \frac{(\delta-(-1)^r\,\delta)\sin\alpha_r}{2} \qquad (r = 1, 2)\,.$$

In Figs. 3.13 and 3.14 we show the change of quantity $\langle K_I^{\pm}\rangle = K_I^{\pm}/\left(\Lambda\sqrt{\pi\ell}\right)$, as a function of crack orientation angle L_1 under the action of stress $\left\langle\sigma_{11}^{(1)}\right\rangle$ and $\langle\sigma_{13}\rangle$ respectively at infinity (edges L are free of forces). Here 2ℓ is the length of arc L_1, Λ is the intensity of the given stresses. The Curves 1 and 2 in Figs. 3.13, 3.14 are constructed for $a_2 = \pi/10$ and $\pi/2$. The solid lines correspond to tip b, the dashed ones to a. The graphs of $\langle K_{II}^{\pm}\rangle = K_{II}^{\pm}/\left(\Lambda\sqrt{\pi\ell}\right)$ at the same values of the parameters and in the same conditions are given in Figs. 3.15, 3.16.

To estimate the influence of the presence of interphase on the intensity factors $K_I^{\pm}$ and $K_{II}^{\pm}$, the system (3.46), (3.47) was numerically solved in the case of a homogeneous $BaTiO_3$ plate, weakened by broken crack. The results

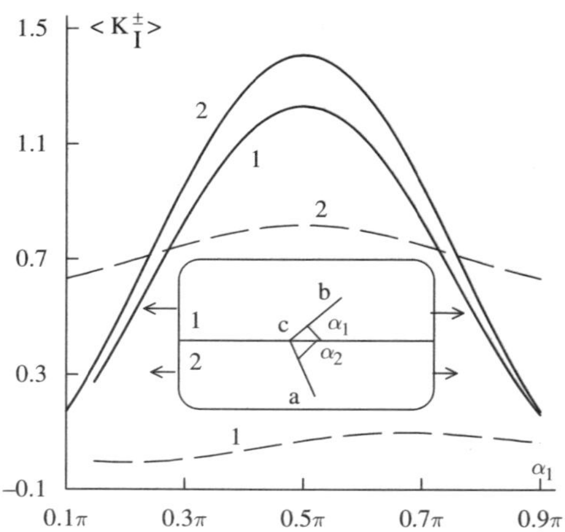

Fig. 3.13. Changes of quantity $< K_I^{\pm} >$ as a function angle α_1, under the influence of stress $< \sigma_{11}^{(1)} >$ at infinity

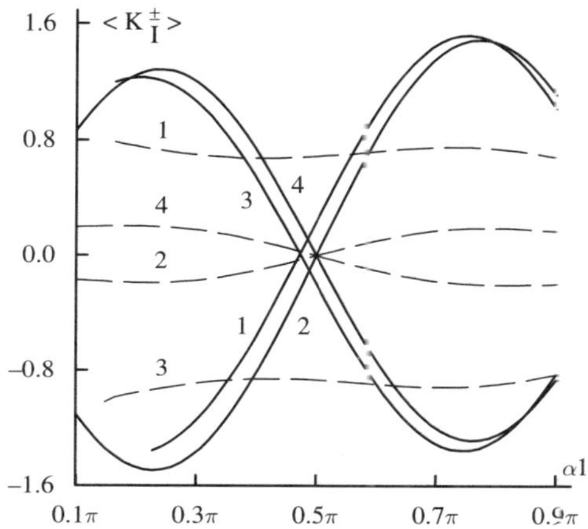

Fig. 3.14. Changes of quantity $< K_I^{\pm} >$ as a function of angle α_1, under the influence of stress $< \sigma_{13} >$ at infinity

are illustrated in Figs. 3.14, 3.15; curves 3 and 4, refer to $a_2 = \pi/10$ and $\pi/2$, respectively.

The graphs of $< K_I^{\pm} >$ at $\kappa = \ell/\ell^* = 0.1m$ (m is the number of the curve, $2\ell^*$ is the length of L_2) and $< \sigma_{13} > \neq 0$ is shown in Fig. 3.17.

From these results we conclude that under the action of tangential stresses $\langle \sigma_{13} \rangle$ at infinity, the presence of the interphase may substantially change the stress intensity factors.

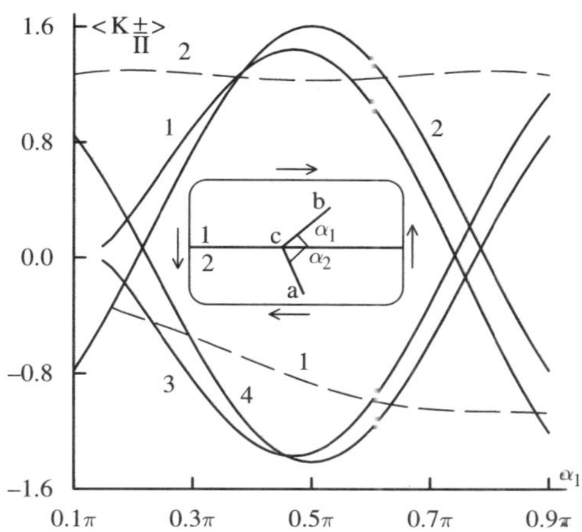

Fig. 3.15. Changes of quantity $< K_{II}^{\pm} >$ as a function of angle α_1, under the influence of stress $< \sigma_{13} >$ at infinity

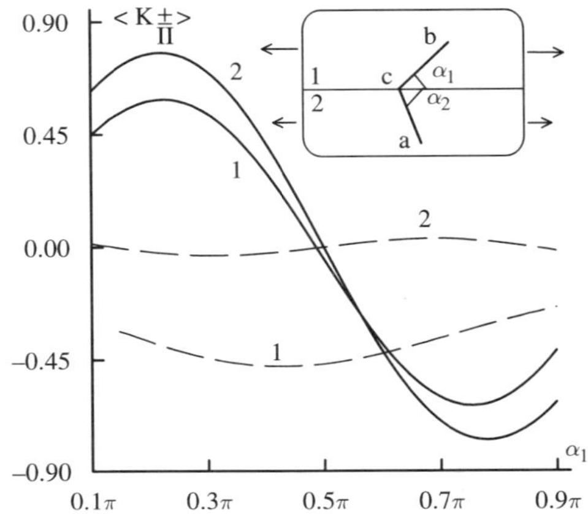

Fig. 3.16. Changes of quantity $< K^{\pm}_{II} >$ as a function of angle α_1, under the influence of stress $< \sigma^{(1)}_{11} >$ at infinity

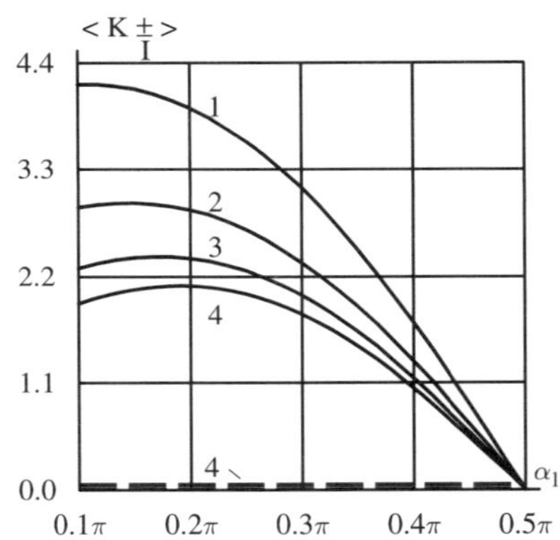

Fig. 3.17. The graphs of quantity $< K^{\pm}_{I} >$ for various values of $k = l/l^*$ under the influence of stress $< \sigma_{13} >$ at infinity

3.5 A Composite Plate with an Opening Crossing the Interphase

Here we investigate the stress concentration in a compound piezoceramic plate weakened by a hole Γ and crossing the boundary interphase of materials at points c_1 and c_2. At these points the contour may not be smooth (Fig. 3.18). A homogeneous field of mechanical stresses $\left\langle \sigma^{(1)}_{11} \right\rangle$, $\langle \sigma_{13} \rangle$, $\langle \sigma_{33} \rangle$ and electrical field $\langle E_1 \rangle$, $\left\langle E^{(1)}_3 \right\rangle$, $\left\langle E^{(2)}_3 \right\rangle$ is applied at infinity, the hole is free from mechanical forces, and bounded by vacuum (axis x_3 coincides with the direction of preliminarily polarized piezoceramics).

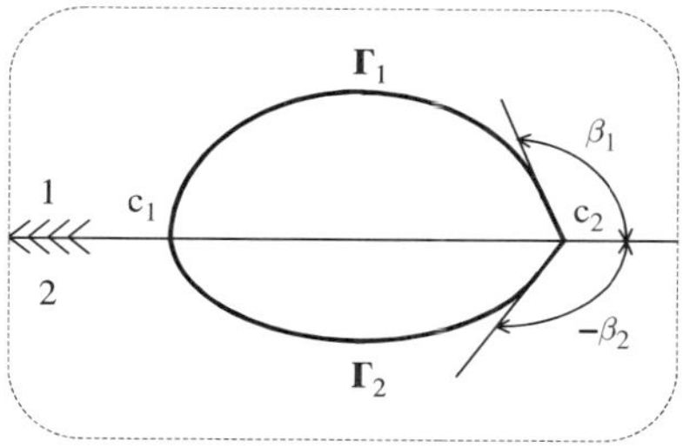

Fig. 3.18. The scheme of a composite plate with an opening crossing the interphase

Denoting by Γ_r $(r = 1, 2)$ the part of contour Γ lying on the r-th halfplate let us write down the mechanical and electric boundary conditions on the edges of the opening as follows

$$2\mathrm{Re}\sum_{k=1}^{3} b_{nk}^{(r)} \left\{\Phi_k^{(r)}\left(\zeta_k^{(r)}\right)\right\}^{-} = 0 \qquad (r = 1, 2; n = 1, 2, 3), \tag{3.53}$$

$$b_{1k}^{(r)} = \gamma_k^{(r)} \mu_k^{(r)} a_k^{(r)}\left(\psi^{(r)}\right), \qquad b_{2k}^{(r)} = \gamma_k^{(r)} a_k^{(r)}\left(\psi^{(r)}\right),$$
$$b_{3k}^{(r)} = r_k^{(r)} a_k^{(r)}\left(\psi^{(r)}\right),$$
$$a_k^{(r)}(\psi) = \mu_k^{(r)} \cos\psi^{(r)} - \sin\psi^{(r)},$$
$$\zeta_k^{(r)} = \xi_1^{(r)} + \mu_\nu^{(r)}\xi_3^{(r)}, \qquad \left(\xi_1^{(r)}, \xi_3^{(r)}\right) \in \Gamma_r.$$

Here $\psi^{(r)}$ is the angle between the normal to contour Γ_r and Ox_1- axis, while by functions $\Phi_k^{(r)}\left(\zeta_k^{(r)}\right)$ we consider their limiting values at $z_\nu^{(r)} \to \zeta_{\nu 0}^{(r)}$ moving from the body to the hole. The condition in (3.53) corresponding to $n = 3$ regulates the equality to zero the normal components of the electric induction vector on the boundary (air).

Using (3.24), (3.27) let us represent the analytical functions $\Phi_\nu^{(r)}\left(z_\nu^{(r)}\right)$ in the following way [17]

$$\Phi_\nu^{(r)}\left(z_\nu^{(r)}\right) = \sum_{k=1}^{3}\int_{\Gamma_r} q_k^{(r)}\left(\zeta^{(r)}\right)\left\{\frac{\omega_{k\nu}^{(r)}}{z_\nu^{(r)} - \zeta_\nu^{(r)}} + \sum_{m=1}^{3}\frac{\overline{\omega_{km}^{(r)}\beta_{\nu m}^{(r)}}}{\overline{\zeta_\nu^{(r)}} - z_\nu^{(r)}}\right\} ds +$$
$$+\sum_{k=1}^{3}\int_{\Gamma_{3-r}} q_k^{(3-r)}\left(\zeta^{(3-r)}\right)\sum_{m=1}^{3}\frac{\omega_{km}^{(3-r)}\beta_{\nu+3,m}^{(3-r)}}{z_\nu^{(r)} - \zeta_m^{(3-r)}} ds + B_\nu^{(r)}, \tag{3.54}$$
$$\mathrm{Im} q_k^{(r)} = 0, \quad z_\nu^{(r)} = x_1 + \mu_\nu^{(r)} x_3 \quad (r = 1, 2; \nu = 1, 2, 3).$$

where $x_3 = 0$ at $r = 1$ and $x_3 = 0$ at $r = 2$, ds is the element of the arc length of contour Γ; the constants $B_\nu^{(r)}$ reflect the conditions at infinity. The Integral representations (3.54) satisfy the conjugation conditions (3.22) irrespective of the choice of the function $q_k^{(r)}(\zeta)$.

Substituting the limiting values of the potentials $\Phi_\nu^{(r)}\left(z_\nu^{(r)}\right)$ at $z_\nu^{(r)} \to \zeta_{\nu 0}^{(r)}$ into the boundary conditions (3.53) we get a system of six singular integral equations of the second kind:

$$
\begin{aligned}
&\sum_{k=1}^{3} t_{nk}^{(r)} q_k^{(r)}\left(\zeta_0^{(r)}\right) + 2\mathrm{Re}\sum_{\nu=1}^{3} b_{n\nu 0}^{(r)} \left\{ \sum_{k=1}^{3} \int_{\Gamma_r} q_k^{(r)}\left(\zeta^{(r)}\right) \right. \\
&\left[\frac{\omega_{k\nu}^{(r)}}{\zeta_{\nu 0}^{(r)} - \zeta_\nu^{(r)}} + \sum_{m=1}^{3} \frac{\overline{\omega_{km}^{(r)} \beta_{\nu m}^{(r)}}}{\overline{\zeta_m^{(r)}} - \zeta_{\nu 0}^{(r)}} \right] ds + \\
&\left. + \sum_{k=1}^{3} \int_{\Gamma_{3-r}} q_k^{(3-r)}\left(\zeta^{(3-r)}\right) \sum_{m=1}^{3} \frac{\omega_{km}^{(3-r)} \beta_{\nu+3,m}^{(3-r)}}{\zeta_{\nu 0}^{(r)} - \zeta_m^{(3-r)}} ds \right\} = F_n^{(r)}\left(\zeta_0^{(r)}\right) \\
&\left(r = 1, 2; n = \overline{1,3}\right), \\
&t_{nk}^{(r)} = -2\pi\mathrm{Im}\sum_{\nu=1}^{3} \frac{b_{n\nu 0}^{(r)} \omega_{k\nu}^{(r)}}{a_\nu^{(r)}\left(\psi_0^{(r)}\right)}, \\
&F_1^{(r)}\left(\zeta_0^{(r)}\right) = -\left\langle \sigma_{11}^{(r)} \right\rangle \cos\psi_0^{(r)} - \langle \sigma_{13} \rangle \sin\psi_0^{(r)}, \\
&F_2^{(r)}\left(\zeta_0^{(r)}\right) = \langle \sigma_{13} \rangle \cos\psi_0^{(r)} + \langle \sigma_{33} \rangle \sin\psi_0^{(r)}, \\
&F_3^{(r)}\left(\zeta_0^{(r)}\right) = -\left(d_{15}^{(r)} \langle \sigma_{13} \rangle + \ni_{11}^{(r)} \langle E_1 \rangle\right) \cos\psi_0^{(r)} - \\
&\qquad - \left(d_{31}^{(r)} \left\langle \sigma_{11}^{(r)} \right\rangle + d_{33}^{(r)} \langle \sigma_{33} \rangle + \ni_{33}^{(r)} \left\langle E_3^{(r)} \right\rangle\right) \sin\psi_0^{(r)}, \\
&\psi_0^{(r)} = \psi^{(r)}\left(\zeta_0^{(r)}\right), \qquad b_{n\nu 0}^{(r)} = b_{n\nu}^{(r)}\left(\zeta_0^{(r)}\right), \\
&\zeta^{(r)} = \xi_1^{(r)} + i\xi_3^{(r)}; \qquad \zeta^{(r)}, \zeta_0^{(r)} \in \Gamma_r.
\end{aligned}
\tag{3.55}
$$

At nodes c_1 and c_2, the kernels of the above system have immobile singularities. Therefore, its solution should be sought in the class of functions undergoing breaks on the ends of arc Γ_r: the order of power singularities of the solution of system (3.55) is determined from asymptotic analysis of the singular integral equations in the vicinity of nodes on arcs Γ_r. From this analysis we get linear connections between left-side and right-side limits of the solution at nodes which should be considered as additional conditions for the single-value solution of system (3.55).

Let us now consider the approximate scheme for the numerical realization of integral (3.55) which allows us to avoid the exact determination of the order of singularities of their solution at the nodal points. Let us assume that the contour Γ crosses x_1 - axis at nodes c_1 and c_2 at an angle $(\beta_1 = -\beta_2 = \pi/2)$

and that its curvature is a continuous function of s arc-curvature. In this case we may assume that the order of power singularity γ of functions $q_k^{(r)}\left(\zeta^{(r)}\right)$ in (3.55) does not considerably differ from zero (Re $\gamma << 1/2$). To numerically solve the system of integral equations we use the quadrature formulas for the calculations of singular and regular integrals given on the closed contours [137]. Here, it should be mentioned that the method of quadratures is based on the approximation of the functions by continuous aggregates (Lagrange polynomial); hence the given approach is proved to be correct for the calculation of "*densities*" $q_k(\zeta)$ only at the points of contour Γ, which are in the vicinity of the nodes c_1 and c_2.

To investigate the concentration of stresses in a piecewise-uniform plate under the influence of the prescribed and electric fields we determine the normal stress σ_θ on contour Γ. Using the complex representations (3.10) we find

$$\sigma_\vartheta\left(\zeta_0^{(r)}\right) = 2\mathrm{Re}\sum_{\nu=1}^{3}\gamma_\nu^{(r)}\left(\mu_\nu^{(r)}\sin\psi_0^{(r)} + \cos\psi_0^{(r)}\right)^2 \Phi_\nu^{(r)}\left(\zeta_{\nu 0}^{(r)}\right). \qquad (3.56)$$

In (3.56), the function $\Phi_\nu^{(r)}\left(\zeta_{\nu 0}^{(r)}\right)$ $(r = 1, 2)$ is evaluated at its limiting values $z_\nu^{(r)} \to \zeta_{\nu 0}^{(r)}$.

As an example let us consider a composite plate (the material of the upper halfplane is $BaTiO_3$, and of the lower one is PZT-5) weakened by a circular opening with a center lying on x_1 -axis. In Fig. 3.19 the graphs display the change in the parameter $\lambda = \sigma_\theta/\Lambda$ (Λ is the intensity of loading) on the opening contour under the action of stresses $\left\langle\sigma_{11}^{(1)}\right\rangle$, $\langle\sigma_{13}\rangle$ or $\langle\sigma_{33}\rangle$ at infinity (curves 1, 2 and 3, respectively). Fig. 3.20 depicts the behavior of stresses σ_θ resulting from the applied, at infinity, electric field $\langle E_1\rangle$ or $\left\langle E_3^{(1)}\right\rangle$ with intensity $1V/m$, (curvatures 1 and 2, respectively). Fig. 3.21 illustrates the changes in E_s on the opening contour under the action of mechanical stresses $\left\langle\sigma_{11}^{(1)}\right\rangle$, $\langle\sigma_{13}\rangle$ or $\langle\sigma_{33}\rangle$ with intensity $1N/m^2$ (the curvatures are built in the same way as in Fig. 3.19).

In Figs. 3.19–3.21 dashed-lines depict the influence of the boundary interphase on the stress concentration on a homogeneous plate from $BaTiO_3$ with a circular opening. From the results, it follows that in the vicinity of the opening, a composite plate subjected to a homogeneous field of stresses, undergoes a considerable increase of stress concentration. This result is in analogy with the case of deformation of a homogeneous plate with an opening (see Fig. 3.19). Due to the reverse piezoelectric effect, under the influence of an external electric field, the presence of the boundary conjugation of halfplates brings to a considerable increase in stress concentration (Fig. 3.20).

The proposed simplified scheme for the numerical solution of the integral (3.55) is efficient when the knowledge of the distribution of electric fields in

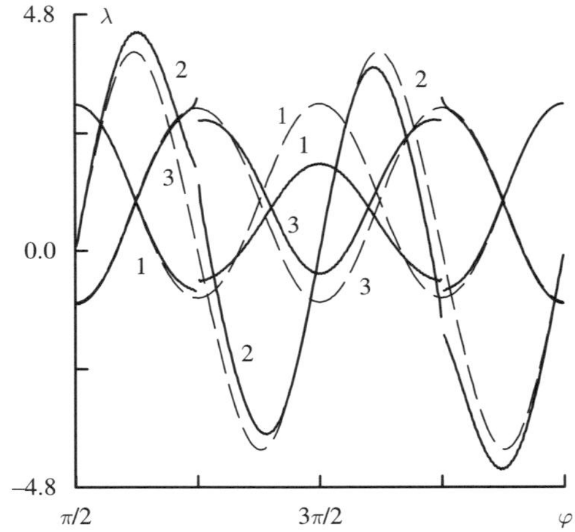

Fig. 3.19. Dependence of quantity λ on polar angle Φ under the influence of stress $< \sigma_{11}^{(1)} >$, $< \sigma_{13} >$ and $< \sigma_{33} >$ at infinity

the vicinity of points c_1 and c_2 is not required. If it is necessary to investigate the concentration of the quantities in the vicinity of these points, we must use more complex quadrature formulas (for example, formulas of Gauss-Jacobi [208, 243] which accurately reflect the preliminarily determined singularities of solutions at the nodes.

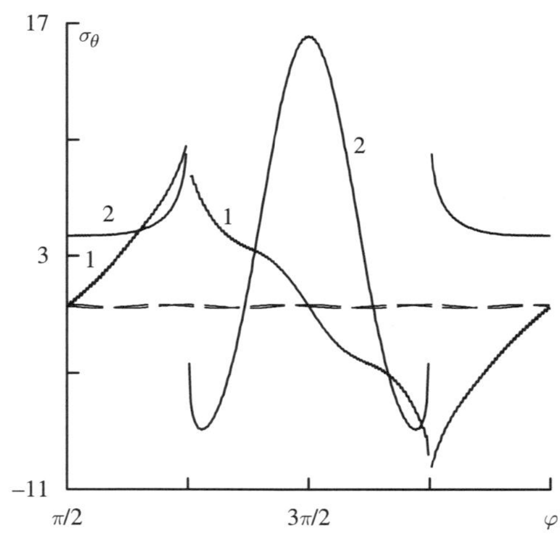

Fig. 3.20. Changes of stress σ_θ on the opening contour under the influence of an electric field of type $< E_1^{(1)} >$ and $< E_3 >$ with intensity 1V/m

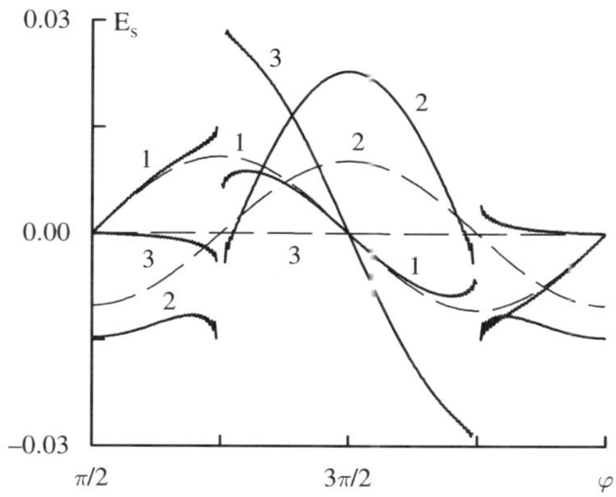

Fig. 3.21. Dependence of quantity E_S on polar angle Φ under the influence of stresses $<\sigma_{11}^{1}>$, $<\sigma_{13}>$ and $<\sigma_{33}>$ with intensity 1V/m

3.6 Green's Function for a Composite Plate with an Interphase Crack

To investigate the effects of coupled mechanical and electric fields in piezoceramic bimorphs with defects of a gap-phase type it is necessary to use Green's function for a compound (piecewise) piezoceramic plate with an interphase crack.

Consider unlimited medium consisting of two heterogeneous piezoceramic halfplanes continuously connected along $x_3 = 0, |x_1| \geq a$. On the area prescribed by $x_3 = 0, |x_1| < a$ there is a gap which may be considered as an interphase crack. Let a concentrated force $\overrightarrow{\mathbf{P}} = (P_1, P_3)$ or concentrated electric charge ρ (Fig. 3.22) be adjusted at the point (x_{10}, x_{30}) of the upper halfspace.

Ascribing the upper index "r" ($r = 1, 2$) to the quantities referring to the r-th halfplane we write down the mechanical and electric boundary conditions

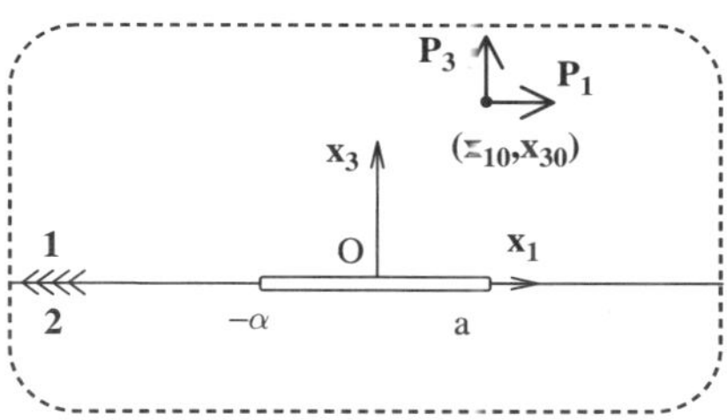

Fig. 3.22. The bimorph scheme with an interphase crack

on x_1 axis in the form of

$$\begin{aligned} &\sigma_{i3}^{(1)} = \sigma_{i3}^{(2)}, \quad du_i^{(1)} = du_i^{(2)}, \quad |x_1| \geq a \quad (i = 1, 3), \\ &E_1^{(1)} = E_1^{(2)}, \quad D_3^{(1)} = D_3^{(2)}, \quad -\infty < x_1 < \infty, \\ &\sigma_{i3}^{(1)+} = 0, \quad \sigma_{i3}^{(2)-} = 0, \quad |x_1| < a, \end{aligned} \tag{3.57}$$

where σ_{i3}, u_1, E_1, D_3 are the corresponding components of the stress tensor, shear vector a and electric stress and induction vectors; $\sigma^{\pm}$ are the limiting values of function σ on the upper (sign "plus") and lower (sign "minus") crack edges.

Let us assume that the vector of preliminarily polarization of the piezoceramics in the upper and lower halfplane is directed along x_3 -axis. Then, according to (3.10) the components of vector $\overrightarrow{\mathbf{U}} = \{U_k\} = \{\sigma_{33}, -\sigma_{13}, u_1', u_3', E_1, -D_3\}$ read

$$\begin{aligned} &U_k = 2\mathrm{Re}\sum_{\nu=1}^{3} c_{k\nu}\varPhi_\nu\left(z_\nu\right) \quad (k = 1, 2, \ldots, 6), \\ &c_{1\nu} = \gamma_\nu, \quad c_{2\nu} = \gamma_\nu\mu_\nu, \quad c_{3\nu} = p_\nu, \\ &c_{4\nu} = q_\nu, \quad c_{5\nu} = \lambda_\nu, \quad c_{6\nu} = r_\nu. \end{aligned} \tag{3.58}$$

Hence, we come to solve a boundary problem for the determination of six functions $\varPhi_\nu^{(r)}\left(z_\nu^{(r)}\right)$ each of which is analytical on its complex plane $z_\nu^{(r)} = x_1 + \mu_\nu^{(r)} x_3$ ($\nu = 1, 2, 3$; $r = 1, 2$) according to the conditions of conjugation (3.22) on area $|x_1| \geq a$ and the conditions $U_1^{\pm} = 0$, $U_2^{\pm} = 0$, $U_5^{(1)} = U_5^{(2)}$, $U_6^{(1)} = U_6^{(2)}$ on the cut edges $[-a, a]$.

The above problem can be solved by applying analytical expansion of the corresponding functions derived by subsequent reduction of it to Riemann problem [50]. In this case we use the fundamental solution for a composite plate without a gap (see Fig. 3.2b).

According to the above, the solution is given by

$$W_\nu^{(r)}\left(z_\nu^{(r)}\right) = \varPhi_\nu^{(r)}\left(z_\nu^{(r)}\right) + \Psi_\nu^{(r)}\left(z_\nu^{(r)}\right) \quad (\nu = 1, 2, 3; r = 1, 2), \tag{3.59}$$

where the functions $\Phi_\nu^{(r)}\left(z_\nu^{(r)}\right)$ are determined in (3.24)–(3.26), and function $\Psi_\nu^{(r)}\left(z_\nu^{(r)}\right)$ that depends on the disturbance introduced by an interphase crack must be determined.

At this point we use the concept of the analytical continuation $\Psi_\nu^{(r)}\left(z_\nu^{(r)}\right)$ in the following way:

Functions $\Psi_\nu^{(1)}(z)$ $\left(\Psi_\nu^{(2)}(z)\right)$ can be analytically continued through areas $|x_1| \geq a$ in the lower (upper) halfplane with the help of function $\overline{\Psi_\nu^{(1)}}(z)$

$\left(\overline{\Psi_\nu^{(2)}(z)}\right)$ which satisfy the relation $\overline{\Psi_\nu^{(1)+}(x_1)} = \overline{\Psi_\nu^{(1)-}}(x_1) \left(\overline{\Psi_\nu^{(2)-}(x_1)} = \overline{\Psi_\nu^{(2)+}}(x_1)\right)$. In this case we consider determination $\overline{\Psi}(z) = \overline{\Psi(\overline{z})}$, and by $\Psi^\pm(x_1)$ we imply the corresponding limiting values of function $\Psi(z)$. Based on the above, from (3.58) we get:

$$U_k^+(x_1) = \sum_{\nu=1}^{3} \left\{ c_{k\nu}^{(1)} \Psi_k^{(1)+}(x_1) + \overline{c_{k\nu}^{(1)} \Psi_k^{(1)-}}(x_1) \right\} + 2\mathrm{Re} \sum_{\nu=1}^{3} c_{k\nu}^{(1)} \Phi_\nu^{(1)}(x_1),$$

$$U_k^-(x_1) = \sum_{\nu=1}^{3} \left\{ c_{k\nu}^{(2)} \Psi_k^{(2)-}(x_1) + \overline{c_{k\nu}^{(2)} \Psi_k^{(2)+}}(x_1) \right\} + 2\mathrm{Re} \sum_{\nu=1}^{3} c_{k\nu}^{(2)} \Phi_\nu^{(2)}(x_1). \tag{3.60}$$

Let us introduce the function

$$F_k(z) = \begin{cases} f_k(z) = \sum\limits_{\nu=1}^{3} \left\{ c_{k\nu}^{(1)} \Psi_\nu^{(1)}(z) - \overline{c_{k\nu}^{(2)} \Psi_\nu^{(2)}}(z) \right\}, & x_3 > 0, \\ f_k(z) = \sum\limits_{\nu=1}^{3} \left\{ c_{k\nu}^{(2)} \Psi_\nu^{(2)}(z) - \overline{c_{k\nu}^{(1)} \Psi_\nu^{(1)}}(z) \right\}, & x_3 < 0, \end{cases} \tag{3.61}$$

$(k = 1, 2, \ldots, 6)$.

Due to the equality (3.58) and the infinite continuation of function U_k $(k = 1, 2, 5, 6)$ through the media interface we obtain

$$f_k^+(x_1) - f_k^-(x_1) = U_k^+(x_1) - U_k^-(x_1) = 0. \tag{3.62}$$

Hence, the function $F_k(z)$ $(k = 1, 2, 5, 6)$ with elements $f_k(z)$ at $\mathrm{Im}\, z > 0$ and $f_k(z)$ at $\mathrm{Im}\, z < 0$ are analytical on the full z - planes and they may be assumed to be equal to zero.

From relation (3.61) we get the following limiting equalities

$$\sum_{\nu=1}^{3} \left\{ c_{k\nu}^{(1)} \Psi_\nu^{(1)+}(x_1) - \overline{c_{k\nu}^{(2)} \Psi_\nu^{(2)+}}(x_1) \right\} = f_k^+(x_1), \tag{3.63}$$

$$\sum_{\nu=1}^{3} \left\{ \overline{c_{k\nu}^{(1)} \Psi_\nu^{(1)-}}(x_1) - c_{k\nu}^{(2)} \Psi_\nu^{(2)-}(x_1) \right\} = -f_k^-(x_1) \quad (k = 1, 2, \ldots, 6),$$

which due to (3.60), (3.62) and boundary conditions $U_k^+ = U_k^- = 0$ $(k = 1, 2)$ result to the following matrix Riemann problem on the interval $(-a, a)$

$$Bf^+(x_1) - \overline{B}f^-(x_1) = R(x_1), \tag{3.64}$$

$$B = \|b_{ij}\| = \left\| \begin{matrix} b_{11}, b_{12}, \ldots, b_{16} \\ b_{21}, b_{22}, \ldots, b_{26} \end{matrix} \right\| = AC_*^{-1}, \quad R(x_1) = \left\| \begin{matrix} N_1 \\ N_2 \end{matrix} \right\|,$$

$$A = \left\| \begin{matrix} c_{11}^{(1)}, c_{12}^{(1)}, c_{13}^{(1)}, \overline{c_{11}^{(2)}}, \overline{c_{12}^{(2)}}, \overline{c_{13}^{(2)}} \\ c_{21}^{(1)}, c_{22}^{(1)}, c_{23}^{(1)}, \overline{c_{21}^{(2)}}, \overline{c_{22}^{(2)}}, \overline{c_{23}^{(2)}} \end{matrix} \right\|, \quad N_k = -4\mathrm{Re} \sum_{\nu=1}^{3} c_{k\nu}^{(2)} \Phi_\nu^{(2)}(x_1),$$

$$C_* = \left\| \begin{matrix} c_{11}^{(1)}, c_{12}^{(1)}, c_{13}^{(1)}, -\overline{c_{11}^{(2)}}, -\overline{c_{12}^{(2)}}, -\overline{c_{13}^{(2)}} \\ c_{21}^{(1)}, c_{22}^{(1)}, c_{23}^{(1)}, -\overline{c_{21}^{(2)}}, -\overline{c_{22}^{(2)}}, -\overline{c_{23}^{(2)}} \\ \cdots\cdots\cdots\cdots\cdots\cdots \\ \cdots\cdots\cdots\cdots\cdots\cdots \\ c_{61}^{(1)}, c_{62}^{(1)}, c_{63}^{(1)}, -\overline{c_{61}^{(2)}}, -\overline{c_{62}^{(2)}}, -\overline{c_{63}^{(2)}} \end{matrix} \right\|, \quad f^{\pm}(x_1) = \left\| \begin{matrix} 0 \\ 0 \\ f_3^{\pm} \\ f_4^{\pm} \\ 0 \\ 0 \end{matrix} \right\|.$$

Now let us now choose the vectors (ρ_{k1}, ρ_{k2}) and the number λ_k $(k = 1, 2)$ so that they are the eigenvectors and the corresponding characteristic numbers of the following homogeneous system

$$\rho_{k1}\left(\bar{b}_{13} + \lambda_k b_{13}\right) + (-1)^{k-1} \rho_{k2}\left(\bar{b}_{23} + \lambda_k b_{23}\right) = 0 \quad (k = 1, 2), \tag{3.65}$$

$$\rho_{k1}\left(\bar{b}_{14} + \lambda_k b_{14}\right) + (-1)^{k-1} \rho_{k2}\left(\bar{b}_{24} + \lambda_k b_{24}\right) = 0 \quad (k = 1, 2). \tag{3.66}$$

Then, the relation (3.64) results to two scalar Riemann problems:

$$\rho_k^+(x_1) + \lambda_k \rho_k^-(x_1) = N_k^*(x_1), \quad |x_1| < a \quad (k = 1, 2), \tag{3.67}$$

where:

$$\rho_k(x_1) = \sum_{j=3}^{4} d_{kj} f_j(x_1), \quad d_{kj} = \frac{2i\mathrm{Im}\left(b_{1j}\bar{b}_{2j}\right)}{\bar{b}_{2j} + \lambda_k b_{2j}} \quad (j = 3, 4),$$

$$N_k^* = \rho_{k1} N_1(x_1) + (-1)^{k-1} \rho_{k2} N_2(x_1).$$

The characteristic numbers λ_1, λ_2 appearing in (3.67) can be found from the conditions of the non-trivial solvability of the system (3.65), (3.66). These conditions are brought to the following quadrature equations

$$\lambda_{1,2}^2 + 2r_1\lambda_{1,2} + r_2 = 0, \tag{3.68}$$

$$r_1 = \frac{\mathrm{Re}\left(\bar{b}_{13}b_{24} - \bar{b}_{23}b_{14}\right)}{b_{13}b_{24} - b_{14}b_{23}}, \quad r_2 = \frac{\bar{b}_{13}\bar{b}_{24} - \bar{b}_{23}\bar{b}_{14}}{b_{13}b_{24} - b_{14}b_{23}}.$$

It is obvious that the roots of (3.68) may be represented as

$$\lambda_1 = \frac{1}{R}e^{-i\vartheta}, \quad \lambda_2 = \mathrm{R}e^{-i\vartheta}, \quad \vartheta = -\frac{1}{2}\arg r_2, \tag{3.69}$$

$$\mathrm{Im}R = 0, \quad 0 \le \vartheta < 2\pi.$$

Solving the problem (3.67) we find

$$\rho_k(z) = \left\{D_k(z) + M_k\right\} X_k(z) \quad (k = 1, 2), \tag{3.70}$$

$$D_k(z) = \frac{1}{2\pi i}\int_{-a}^{a} \frac{N_k^*(x)\,dx}{(x - z)\,X_k(x)},$$

$$X_k(z) = (z + a)^{-\gamma_k}(z - a)^{\gamma_k - 1},$$

$$\gamma_1 = \frac{1}{2} - \frac{\vartheta}{2\pi} + \frac{i}{2\pi}\ell nR, \quad \gamma_2 = \overline{\gamma_1},$$

where $X_k(x) = X_k^+(x)$ are the values of the canonical function $X_k(z)$ on the upper cut edge, and M_k are arbitrary complex constants. To fix these constants it is necessary to render the shear vector continuous through $|x_1| \geq a$ (the conditions of conjugation consist of derivatives from displacements). This condition is fulfilled if M_1 and M_2 are connected by $\overline{M_2} = \lambda_1 M_1$.

Further, let us apply the conditions of the displacements homogeneity on a composite plane

$$\int_c du_i = -\int_{-a}^{a} d[u_i] = 0 \quad (i = 1, 2), \tag{3.71}$$

where c is a closed contour covering section $[-a, a]$ of x_1 -axis, $[u]$ is a jump of function u while crossing the cut. By introducing the expression $u_i'(x_1)$, using (3.58) and relations (3.59), (3.70) and after applying some transformations we find

$$M_i = \frac{2}{1+\lambda_i} \sum_{m=1}^{3} \left(A_m^{(1)} \Omega_{i1}^{(m)} + \overline{A_n^{(1)} \Omega_{i2}^{(m)}} \right) \quad (i = 1, 2), \tag{3.72}$$

$$\Omega_{11}^{(m)} = \Omega_{22}^{(m)} = -\sum_{\nu=1}^{3} \sum_{j=1}^{2} \rho_{1j} c_{j\nu}^{(2)} \alpha_{\nu+3,m}^{(1)} \quad (m = 1, 2, 3),$$

$$\Omega_{21}^{(m)} = \Omega_{12}^{(m)} = \sum_{\nu=1}^{3} \sum_{j=1}^{2} (-1)^j \rho_{2j} c_{j\nu}^{(2)} \alpha_{\nu+3,m}^{(1)}, \quad \rho_{k1} = 1,$$

$$\rho_{k2} = (-1)^k \left(\bar{b}_{13} + \lambda_k b_{13} \right) \left(\bar{b}_{23} + \lambda_k b_{23} \right)^{-1} \quad (k = 1, 2).$$

Now let us derive the sought-for functions by $f_k(z)$ from (3.61). We have

$$\Psi_\nu^{(1)}(z) = \sum_{j=3}^{4} l_{\nu j} f_j(z), \quad \mathrm{Im} z > 0 \quad (\nu = 1, 2, 3), \tag{3.73}$$

$$\Psi_\nu^{(2)}(z) = -\sum_{j=3}^{4} \bar{l}_{\nu+3,j} f_j(z), \quad \mathrm{Im} z < 0,$$

where l_{mj} $(m = 1, 2, \ldots, 6; j = 3, 4)$ are the elements of matrix C_*^{-1}.

Substituting the functions $f_3(z)$, $f_4(z)$, found from (3.70), fulfilling the necessary quadratures and using (3.72) we finally get

$$\Psi_\nu^{(r)}\left(z_\nu^{(r)}\right) = \sum_{m=1}^{3} \sum_{n=1}^{2} \left\{ \left[\frac{1 - X_n\left(z_\nu^{(r)}\right) X_n^{-1}\left(z_{m0}^{(1)}\right)}{z_\nu^{(r)} - z_{m0}^{(1)}} + X_n\left(z_\nu^{(r)}\right) \right] \right.$$

$$\Omega_{n1}^{(m)} \gamma_{n\nu}^{(r)} A_m^{(1)} +$$

$$+\left[\frac{1-X_n\left(z_\nu^{(r)}\right)X_n^{-1}\left(\overline{z_{m0}^{(1)}}\right)}{z_\nu^{(r)}-\overline{z_{m0}^{(1)}}}+X_n\left(z_\nu^{(r)}\right)\right]\overline{\Omega_{n2}^{(m)}}\gamma_{n\nu}^{(r)}\overline{A_m^{(1)}}, \tag{3.74}$$

$$\gamma_{n\nu}^{(r)}=\frac{2L_{n\nu}^{(r)}}{(1+\lambda_n)\,d_0},\qquad d_0=d_{13}d_{24}-d_{14}d_{23},$$

$$L_{n\nu}^{(1)}=\sum_{j=3}^{4}(-1)^{n-j}\,m_{nj}l_{\nu j},\quad L_{n\nu}^{(2)}=-\sum_{j=3}^{4}(-1)^{n-j}\,m_{nj}\bar{l}_{\nu+3,j},$$

$$m_{n3}=\frac{d_{14}d_{24}}{d_{n4}},\qquad m_{n4}=\frac{d_{13}d_{23}}{d_{n3}}\qquad(\nu=1,2,3;r=1,2)\,.$$

Thus, the Green's function, for a composite piezoceramic plane with an interphase crack, is determined in the explicit form by relations (3.59), (3.24) and (3.74). From the analytical representation of Green's function it follows that the mechanical stresses and also the components of the electric field on the tips of an interphase crack posses power singularities strengthened by oscillations in the small vicinity of the points $x_1=\pm a$. This phenomenon takes place also in the case of composite isotropic media [196].

To determine the stress intensity factors on the tips of interphase cracks we keep in expressions (3.74) only those terms that contain canonical functions $X_n(z)$. Then in the vicinity of points $x_1=\pm a$ we have:

at $x_1>a$

$$\left\{V_k^{(r)}\right\}=4\mathrm{Re}\sum_{m=1}^{3}\sum_{n=1}^{2}A_m^{(1)}\left\{\frac{(x_1-a)^{\gamma_n-1}}{(2a)^{\gamma_n}}\Omega_{1n}^{(m)}D_{kn}\left[\left(\frac{z_{m0}^{(1)}+a}{z_{m0}^{(1)}-a}\right)^{\gamma_n}+1\right]\right\}$$
$$+O(1)\,, \tag{3.75}$$

at $x_1<-a$

$$\left\{V_k^{(r)}\right\}=-4\mathrm{Re}\sum_{m=1}^{3}\sum_{n=1}^{2}A_m^{(1)}\left\{\frac{(x_1+a)^{-\gamma_n}}{(2a)^{1-\gamma_n}}\Omega_{1n}^{(m)}D_{kn}\left[\left(\frac{z_{m0}^{(1)}-a}{z_{m0}^{(1)}+a}\right)^{1-\gamma_n}+1\right]\right\}$$
$$+O(1)\,,$$

where:

$$V_1^{(r)}=\sigma_{33}^{(r)},\quad V_2^{(r)}=-\sigma_{13}^{(r)}\quad(r=1,2)\,,$$

$$D_{k1}^{(1)}=\sum_{\nu=1}^{3}\sum_{j=3}^{4}(-1)^{j-1}c_{k\nu}^{(1)}l_{\nu j}m_{1j},\quad D_{k2}=\overline{D}_{k1},\quad D_{k1}=D_{k1}^{(1)}d_0^{-1}.$$

As one would expect, the obtained expressions for $V_k^{(r)}$ do not depend on r.

The stress intensity factors are determined according to the above out asymptotic in the following way.

On the tip $x_1 = a$:

$$K_j^+ = \ell im\left\{\left(\frac{x_1}{a}-1\right)^{\frac{1}{2}+\frac{\vartheta}{2\pi}}\sigma_j\left(x_1\right)\right\} = \frac{\Lambda}{ah}\left\langle K_j^+\right\rangle, \quad x_1\to a \quad \left(x_1>a\right),$$

$$\sigma_1\left(x_1\right)=\sigma_{33}, \quad \sigma_2\left(x_1\right)=\sigma_{13}, \tag{3.76}$$

$$B_m^{(1)} = hA_m^{(1)}\left\{P_1^2+P_3^2+\left(\frac{\rho}{d_{33}^{(1)}}\right)^2\right\}^{-\frac{1}{2}},$$

$$\left\langle K_j^+\right\rangle = 4\mathrm{Re}\sum_{m=1}^{3}B_m^{(1)}\sum_{n=1}^{2}\frac{\Omega_{1n}^{(m)}D_{jn}}{2^{\gamma_n}}\left\{\left(\frac{z_{m0}^{(1)}+a}{z_{m0}^{(1)}-a}\right)^{\gamma_n}+1\right\} \quad (j=1,2).$$

On the tip $x_1 = -a$:

$$K_j^- = \ell im\left\{\left|\frac{x_1}{a}+1\right|^{\frac{1}{2}-\frac{\vartheta}{2\pi}}\sigma_j\left(x_1\right)\right\} = \frac{\Lambda}{ah}\left\langle K_j^-\right\rangle, \quad x_1\to -a \quad \left(x_1<-a\right),$$

$$\left\langle K_j^-\right\rangle = -4\mathrm{Re}\sum_{m=1}^{3}B_m^{(1)}\sum_{n=1}^{2}\frac{\Omega_{1n}^{(m)}D_{jn}}{2^{1-\gamma_n}}\left\{\left(\frac{z_{m0}^{(1)}-a}{z_{m0}^{(1)}+a}\right)^{1-\gamma_n}+1\right\}. \tag{3.77}$$

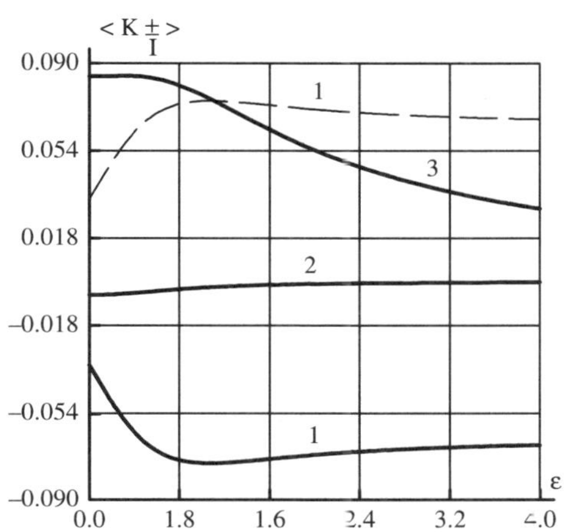

Fig. 3.23. Changes of $< K_I^{\pm} >$ as a function of parameter $\varepsilon = H/a$ for various types of loading (the material of the lower halfplane is PZT-5, the upper is $BaTiO_3$

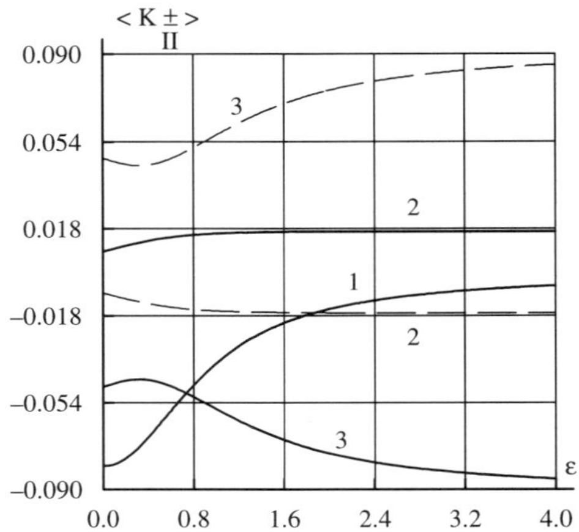

Fig. 3.24. Changes $< K_{II}^{\pm} >$ as a function of parameter $\varepsilon = H/a$ for various types of loading (the material of the upper halfplane is $BaTiO_3$, the lower is PZT-5)

In (3.76), (3.77) $\Lambda = P$ at $P = \sqrt{P_1^2 + P_3^2} > 0$, $\rho = 0$ and $\Lambda = \rho/d_{33}^{(1)}$ at $\rho \neq 0$, $P = 0$.

On calculating the given limits, as in [196], was assumed that $\mathrm{Im}\gamma_1 \cdot \ell n\,|x_1 \pm a| \approx 0$.

The calculations confirm the validity of this approximate equality. In the "microscopic" vicinity of the tip, their physical incorrectness of the solutions

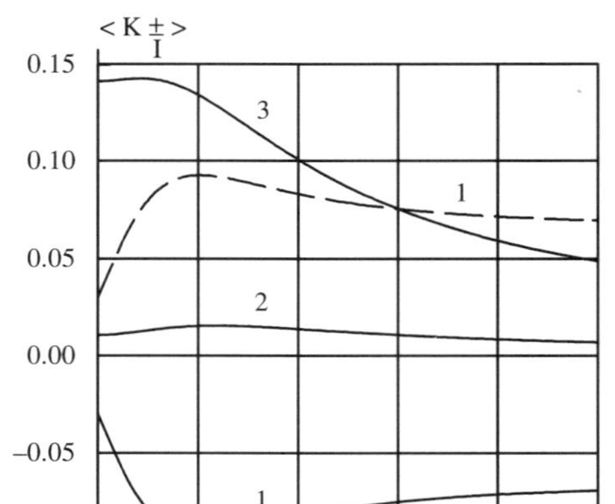

Fig. 3.25. Changes of $< K_I^{\pm} >$ as a function of parameter $\varepsilon = H/a$ for various types of loading (the material of the upper halfplane is $BaTiO_3$, the lower is PZT-5)

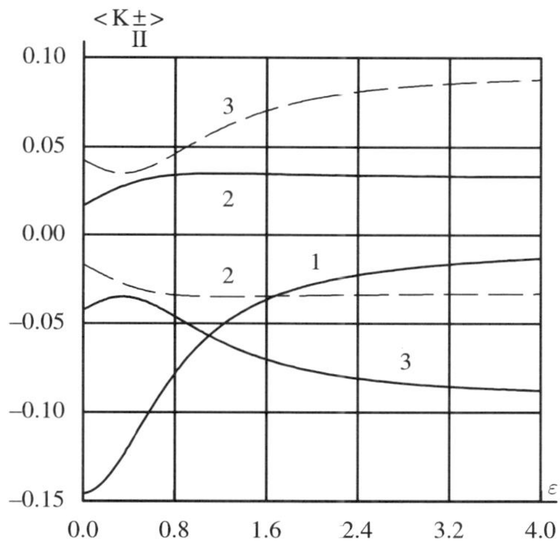

Fig. 3.26. Changes $< K_{II}^{\pm} >$ as a function of parameter $\varepsilon = H/a$ for various types of loading (the material of the upper halfplane is $BaTiO_3$, the lower is PZT-5)

connected with the fact that the stresses change the sign an infinite number of times.

As an example lets consider a pair of PZT-5 (the material of the upper halfplane), $BaTiO_3$ (the material of the lower halfplane) with interface crack $x_3 = 0, \ -a < x_1 < a$, subjected to a concentrated force (P_1, P_3) or to a charge ρ at point $x_{10} = 0, \ x_{30} = H > 0$. Changes of the relative stress intensity factors $\langle K_I^{\pm} \rangle$ and $\langle K_{II}^{\pm} \rangle$ in the function of parameter $\varepsilon = H/a$ are shown in Figs. 3.23 and 3.24, where curves 1, 2 and 3 correspond to $P_1 \neq 0, \ P_3 = \rho = 0; \ \rho \neq 0, \ P_1 = P_3 = 0$ and $P_3 \neq 0, P_1 = \rho = 0$, respectively. The solid curves correspond to $\langle K_I^- \rangle$, and dashed lines to $\langle K_I^+ \rangle$.

If we exchange the places of the pair components, the results change. Figs. 3.25 and 3.26 show the $\langle K_I^{\pm} \rangle$ and $\langle K_{II}^{\pm} \rangle$ for this case. In the calculations it was assumed that $P/h = 1, \ \rho/h = 1$.

3.7 A Case of an Inner Crack Reaching the Interphase

In this section we derive a system of integral equations when in the upper halfplane of the piezoceramic bimorph a crack-like defect L exists. Te defect L "touches" the crack that exists on the interphase (Fig. 3.27). Let us assume that there are no forces on the inner and interphase cracks and that stresses $< \sigma_{11}^{(1)} >, \ < \sigma_{13} >, \ < \sigma_{33} >$ are prescribed at infinity.

To obtain the integral representations of the complex potentials in this case let us use the built in Par. 3.6 Green's function for a compound piezoceramic

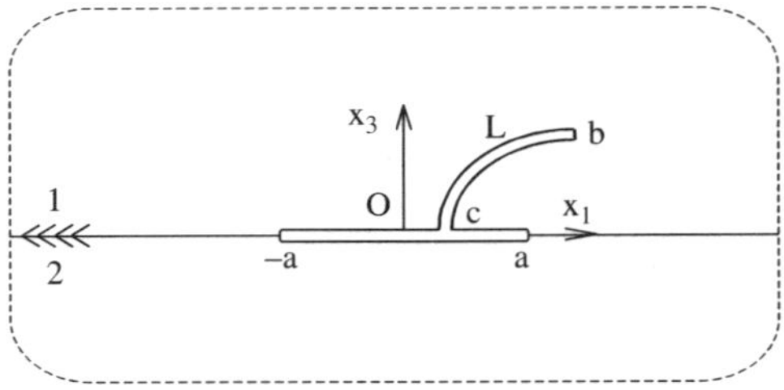

Fig. 3.27. The bimorph scheme with a crack reaching the interphase

plane with an interphase crack. Allowing for formulas (3.24), (3.28) and (3.74) at $r = 1$ we write

$$\Phi_{\nu}^{(1)}\left(z_{\nu}^{(1)}\right) = B_{\nu}^{(1)} + \frac{1}{2\pi i}\int\limits_{L}\frac{\omega_{\nu}(\zeta)\, d\zeta_{\nu}^{(1)}}{\zeta_{\nu}^{(1)} - z_{\nu}^{(1)}} + \sum_{m=1}^{3}\frac{\overline{\beta_{\nu m}^{(1)}}}{2\pi i}\int\limits_{L}\frac{\overline{\omega_m(\zeta)}\, d\overline{\zeta_m^{(1)}}}{\overline{\zeta_m^{(1)}} - z_{\nu}^{(1)}}, \qquad (3.78)$$

$$\Psi_{\nu}^{(1)}\left(z_{\nu}^{(1)}\right) = \frac{1}{2\pi i}\sum_{m=1}^{3}\sum_{n=1}^{2}\left\{\int\limits_{L}\omega_m(\zeta)\, G_n\left(z_{\nu}^{(1)}, \zeta_m^{(1)}\right)\Omega_{n1}^{(m)}\gamma_{n\nu}^{(1)}\, d\zeta_m^{(1)} - \right.$$

$$\left. - \int\limits_{L}\overline{\omega_m(\zeta)}\, G_n\left(z_{\nu}^{(1)}, \overline{\zeta_m^{(1)}}\right)\overline{\Omega_{n2}^{(m)}}\gamma_{n\nu}^{(1)}\, d\overline{\zeta_m^{(1)}}\right\},$$

$$G_n(x, y) = \frac{1 - X_n(x)\, X_n^{-1}(y)}{y - x} - X_n(x) \quad (\nu = 1, 2, 3).$$

The canonical functions $X_n(x)$ appearing in (3.78) are determined in (3.70); $\omega_{\nu}(\zeta)$ are the functions that have to be calculated the constants $B_{\nu}^{(1)}$ reflect the prescribed conditions at infinity.

Thus, representations (3.78) provide the automatic fulfillment of the conjugation conditions on the media interphase.

The mechanical and electric boundary conditions on L may be represented in the form of (3.23), however instead of functions $\Phi_{\nu}^{(1)}\left(\zeta_{\nu}^{(1)}\right)$ $(\nu = 1, 2, 3)$ it is recommended to take the limiting values of functions $W_{\nu}^{(1)}\left(z_{\nu}^{(1)}\right) = \Phi_{\nu}^{(1)}\left(z_{\nu}^{(1)}\right) + \Psi_{\nu}^{(1)}\left(z_{\nu}^{(1)}\right)$, determined allowing for representations (3.78).

By substituting of the limiting values of function $W_{\nu}^{(1)}\left(z_{\nu}^{(1)}\right)$ in conditions (3.23) we get a mixed system of real singular and algebraic equations L

$$\operatorname{Re}\sum_{m=1}^{3}\int\limits_{L}\omega_m(\zeta)\, g_{mp}(\zeta, \zeta_0)\, ds = f_p(\zeta_0) \quad (p = 1, 2), \qquad (3.79)$$

$$\operatorname{Re}\sum_{m=1}^{3} b_{nm}(\zeta)\, \omega_m(\zeta) = 0 \quad (n = 1, 2, \ldots, 4),$$

$$g_{mp}(\zeta,\zeta_0)=\frac{1}{\pi i}\left\{\frac{b^0_{pm}}{\zeta^{(1)}_m-\zeta^{(1)}_{m0}}+\sum_{\nu=1}^{3}\sum_{k=1}^{2}b^0_{p\nu}\Omega^{(m)}_{k1}\gamma^{(1)}_{k\nu}G_k\left(\zeta^{(1)}_{\nu0},\zeta^{(1)}_m\right)-\right.$$

$$\left.-\sum_{\nu=1}^{3}\frac{\overline{b^0_{p\nu}}\beta^{(1)}_{\nu m}}{\zeta^{(1)}_m-\overline{\zeta^{(1)}_{\nu0}}}-\sum_{\nu=1}^{3}\sum_{k=1}^{2}\overline{b^0_{p\nu}}\Omega^{(m)}_{k2}\overline{\gamma^{(1)}_{k\nu}G_k\left(\zeta^{(1)}_{\nu0},\overline{\zeta^{(1)}_m}\right)}\right\}a^{(1)}_m(\psi),$$

$$f_1(\zeta_0)=-<\sigma^{(1)}_{11}>\cos\psi_0-<\sigma_{13}>\sin\psi_0,$$

$$f_2(\zeta_0)=<\sigma_{13}>\cos\psi_0+<\sigma_{33}>\sin\psi_0$$

$$b^0_{n\nu}=b_{n\nu}(\zeta_0),\quad \zeta^{(1)}_m=\mathrm{Re}\zeta+\mu^{(1)}_m\mathrm{Im}\zeta;\quad \zeta,\zeta_0\in L.$$

Here the kernels $g_{mp}(\zeta,\zeta_0)$ are singular (of Cauchy type), ψ is the angle between the normal to the left edge L and axis Ox_1, the function $b_{nm}(\zeta)$ is determined in (3.23).

The stress at the tip of crack b has a singularity of quadrature root type and at tip c they are restricted. Therefore the numerical solution of integral and algebraic (3.79) should be found in the function class not bounded at tip a and bounded at tip c [121].

4

Diffraction of a Shear Wave on Tunnel Cracks in Media of Various Configurations (Antiplane Deformation)

In this chapter we investigate the diffraction of a shear wave on tunnel defects of crack type in piezoceramic media. Stationary and non-stationary dynamic problems of electro-elasticity are reduced, with the help of the corresponding Green's functions, to singular integral equations of the first kind. By asymptotic analysis of the integral representations of the solutions, the equations for the factors of stress shear intensity at the crack tips are obtained.

4.1 An Anisotropic Space

The increasing need for the development of modern and sophisticated technologies in the areas of material science and engineering stimulates the design and manufacturing of new composite anisotropic materials with respect to elastic and other physical properties. The anisotropy influences the strength of the constructive elements, as well as the parameters of failure of such elements if crack-like defects exist.

In order to investigate the influence of a) the character of anisotropy, b) the configuration of the crack and its orientation (to the main axes of anisotropy on) the stress intensity factor we consider the stationary dynamic problem in the framework of elasticity for a piecewise-homogeneous anisotropic medium.

Referring to Cartesian coordinates $Ox_1x_2x_3$ let an elastic anisotropic media be weakened by tunnel along axis x_3 curvilinear cracks-cuts $L_j\ (j = 1, 2, \ldots, k)$.

Let us assume that a plane monochromatic shear wave radiates from infinity to the cracks and on their surfaces a harmonically changing with time, independent on coordinate x_3 stress vector $X_{1n}^{\pm} = X_{2n}^{\pm} = 0, X_{3n}^{\pm} = \mathrm{Re}\left(X_3^{\pm} e^{-i\omega t}\right)$ is possibly given

Assume that at each point of the medium, there exists a plane of elastic symmetry perpendicular to x_3 -axis; L_j are the simple disconnected ares ($\cap L_j = \emptyset$) with curvatures satisfying the Hölder condition, and $X_3 = X_3^{+} = -X_3^{-}$ is the function of H -class on $L = \bigcup\limits_{j=1}^{k} L_j$.

Under the given conditions in an anisotropic medium, a stationary wave process corresponding to the state of antiplane deformation takes place. The full system of the equations of the stated boundary problem includes the following basic relations [134, 135]

$$\partial_1\sigma_{13} + \partial_2\sigma_{23} = \rho\frac{\partial^2 u_3}{\partial t^2}, \quad \partial_m = \frac{\partial}{\partial x_m} \quad (m = 1, 2), \tag{4.1}$$

$$\sigma_{23} = c_{44}\partial_2 u_3 + c_{45}\partial_1 u_3, \quad \sigma_{13} = c_{45}\partial_2 u_3 + c_{55}\partial_1 u_3, \tag{4.2}$$

$$(\sigma_{13}\cos\psi + \sigma_{23}\sin\psi)^{\pm} = X_{3n}^{\pm}. \tag{4.3}$$

Here (4.1) is the *motion equation*, (4.2) are the *equations of medium state*, (4.3) is the *condition on the surface*; c_{ij} are the moduli of the material elasticity; ρ is the density of the medium; ψ is the angle between the normal to the left edge L_j (when it moves from the beginning a_j to the end b_j) and Ox_1 -axis (Fig. 4.1).

The wave displacement field in a medium with cracks consists from the field of incoming monochromatic wave $\left(u_3^{(0)}\right)$ and the field of the dispersed defects (u_3^*). Therefore, we write

$$u_3 = u_3^{(0)} + u_3^*, \quad u_3^* = \mathrm{Re}\left(U_3^* e^{-i\omega t}\right), \tag{4.4}$$

where the displacement amplitude $U_3^{(0)}$ in the wave is determined by

$$\begin{aligned} u_3^{(0)} &= \mathrm{Re}\left(U_3^{(0)} e^{-i\omega t}\right), \quad U_3^{(0)} = \tau e^{i(\alpha_1 x_1 + \alpha_2 x_2)}, \\ \alpha_1 &= -\gamma\cos\beta, \quad \alpha_2 = -\gamma\left(\mathrm{Re}\mu\cos\beta + \mathrm{Im}\ \mu\ \sin\beta\right), \\ \gamma &= \frac{\omega}{c}, \quad c = \sqrt{\frac{\Delta}{\rho c_{44}}}, \quad \Delta = c_{44}c_{55} - c_{45}^2 > 0, \\ \mu &= \frac{-c_{45} + i\sqrt{\Delta}}{c_{44}}. \end{aligned} \tag{4.5}$$

Here β is the angle between the normal to the front wave and Ox_1 -axis; γ is the wave number. As it follows from (4.5) the characteristic number μ is imaginary.

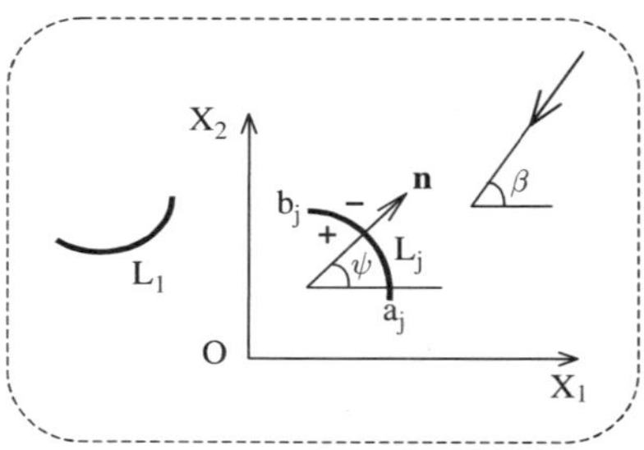

Fig. 4.1. A space containing curvilinear cracks

From (4.1), (4.2) the differential equation for the amplitude of shear displacement reads

$$c_{55}\partial_1^2 U_3 + 2c_{45}\partial_1\partial_2 U_3 + c_{44}\partial_2^2 U_3 + \rho\omega^2 U_3 = 0. \tag{4.6}$$

The boundary condition (4.3) for an arbitrary point $(\xi_{10}, \xi_{20}) \in L_j$ on the right and left edges of the cut has the following form.

$$\begin{aligned}
&a(\psi_0)\left(\frac{\partial U_3}{\partial \zeta_{10}}\right)^{\pm} - \overline{a(\psi_0)}\left(\frac{\partial U_3}{\partial \overline{\zeta}_{10}}\right)^{\pm} = \pm\frac{i}{\sqrt{\Delta}}X_3^{\pm}(\zeta_0),\\
&a(\psi_0) = \mu\cos\psi_0 - \sin\psi_0, \quad \psi_0 = \psi(\zeta_0),\\
&\zeta_{10} = \xi_{10} + \mu\xi_{20}, \quad \overline{\zeta}_{10} = \xi_{10} + \overline{\mu}\xi_{20}, \quad \xi_0 = \xi_{10} + i\xi_{20},\\
&\frac{\partial}{\partial z_1} = \frac{\overline{\mu}\partial_1 - \partial_2}{\overline{\mu} - \mu}, \quad \frac{\partial}{\partial \overline{z_1}} = \frac{\mu\partial_1 - \partial_2}{\mu - \overline{\mu}}, \quad z_1 = x_1 + \mu x_2.
\end{aligned} \tag{4.7}$$

The differential (4.6) and the boundary conditions (4.7) represent a boundary value problem for the determination of the displacement amplitude U_3^*. In order to bring the problem to an integral equation, the correct integral forms of the solutions must be constructed in such way that insures the existence of the displacement jump u_3^*, and also the continuity of the stress vector on crossing cut $L_j\,(j = 1, 2, \ldots, k)$.

Besides, these representations satisfy the *conditions of radiation* [69] at infinity.

To avoid the procedure of normalization of spreading on the contours of integral cuts we must construct the integral representations of the displacements derivatives.

Let us consider together with the main state of the system, the subsidiary state, characterized by the presence, at a certain point of area $(x_{10},\ x_{20})$ of a concentrated functional $\delta(x_1 - x_{10}, x_2 - x_{20})$. Calculating the sums of production of the differential (4.6) for the i-th state on the corresponding derivative from the displacement amplitude j-th state $i \neq j, \quad i, j = 1,\ 2$, we obtain the expression of the divergence type. Carrying out the integration in D occupied by the body, by subsequent application of Green's formulas [168] and after some transformations we finally find [118]

$$\begin{aligned}
\frac{\partial U_3^*}{\partial z_{10}} &= \int_L \left[c_{44}a(\psi)p(\zeta)\frac{\partial E}{\partial \zeta_1} + \frac{i}{2}\gamma^2\sqrt{\Delta}a(\psi)q(\zeta)E\right]ds,\\
\frac{\partial U_3^*}{\partial \overline{z}_{10}} &= \int_L \left[c_{44}a(\psi)p_1(\zeta)\frac{\partial E}{\partial \overline{\zeta}_1} - \frac{i}{2}\gamma^2\sqrt{\Delta}a(\psi)q(\zeta)E\right]ds,
\end{aligned} \tag{4.8}$$

where,

$$p(\zeta) = \bar{\mu}\left[\partial_1 U_3^*\right] - \left[\partial_2 U_3^*\right], \quad p_1(\zeta) = \mu\left[\partial_1 U_3^*\right] - \left[\partial_2 U_3^*\right],$$

$$q(\zeta) = \left[U_3^*\right], \quad [\Phi] = \Phi^+ - \Phi^-,$$

$$E = E(\xi_1 - x_{10}; \xi_2 - x_{20}) = -\frac{i}{4\sqrt{\Delta}} H_0^{(1)}(\gamma r),$$

$$r = |\zeta_1 - z_{10}|, \quad z_{10} = x_{10} + \mu x_{20}, \quad \bar{z}_{10} = x_{10} + \bar{\mu} x_{20},$$

$$\zeta_1 = \xi_1 + \mu \xi_2, \quad (\xi_1, \xi_2) \in L, \quad (x_{10}, x_{20}) \in D.$$

In (4.8) the functions $p(\zeta)$, $p_1(\zeta)$ and $q(\zeta)$ are the "*densities*" to be determined; $H_\nu^{(1)}(x)$ is the ν-*th* order Hankel's function [22, 109, 166] of the first kind, $E(\xi_1 - x_{10}; \xi_2 - x_{20})$ are the fundamental solutions [43] of the differential (4.6) while ds is the element of the arc-length of contour L

The integral representations (4.8) are obtained by using the methods of the theory of potentials [98]. In the same way we obtain the representation for the displacement amplitude, given by

$$U_3^*(x_{10}, x_{20}) = c_{44} \int\limits_L \left\{ E\left(\mu a(\psi) \left[\frac{\partial U_3^*}{\partial \zeta_1}\right] + \overline{\bar{\mu} a(\psi)} \left[\frac{\partial U_3^*}{\partial \overline{\zeta_1}}\right] \right) + \right. \tag{4.9}$$

$$\left. + \left[U_3^*\right] \left(\bar{\mu} a(\psi) \frac{\partial E}{\partial \zeta_1} + \overline{\mu a(\psi)} \frac{\partial E}{\partial \overline{\zeta_1}} \right) \right\} ds.$$

Here we must note that (4.9) actually guarantees the existence of the displacement jump on L_j $(j = 1, 2, \ldots, k)$ and also the fulfillment of the radiation conditions.

Equations (4.8) must satisfy the requirements of continuity of the stress vector when crossing L_j. Let us write them in the form of a sum of singular and regular parts [175] as follows:

$$\begin{aligned} &\frac{\partial}{\partial z^n} H_0^{(1)}(\gamma r) = \left(-\frac{\gamma}{2}\right)^n e^{-in\alpha} H_n^{(1)}(\gamma r), \\ &\frac{\partial^n}{\partial \bar{z}^n} H_0^{(1)}(\gamma r) = \left(-\frac{\gamma}{2}\right)^n e^{in\alpha} H_n^{(1)}(\gamma r), \\ &\frac{\partial^2}{\partial z \partial \bar{z}} H_0^{(1)}(\gamma r) = -\frac{\gamma^2}{4} H_0^{(1)}(\gamma r), \quad \gamma = \text{const} > 0, \\ &r = |z - z_0|, \quad \alpha = \arg(z - z_0), \\ &z = x_1 + i x_2, \quad z_0 = x_{10} + i x_{20}. \end{aligned} \tag{4.10}$$

As a result we get

$$\frac{\partial U_3^*}{\partial z_{10}} = \frac{c_{44}}{4\pi\sqrt{\Delta}} \int\limits_L \frac{p(\zeta)\, d\zeta_1}{\zeta_1 - z_{10}} - \tag{4.11}$$

$$- \frac{\gamma}{8i} \int\limits_L \left[\frac{c_{44}}{\sqrt{\Delta}} a(\psi)\, e^{-i\alpha} p(\zeta)\, H_1(\gamma r) - i\gamma q(\zeta)\, \overline{a(\psi)} H_0^{(1)}(\gamma r) \right] ds,$$

$$\frac{\partial U_3^*}{\partial \overline{z_{10}}} = \frac{c_{44}}{4\pi\sqrt{\Delta}} \int\limits_L \frac{p_1(\zeta)\, d\zeta_1}{\bar{\zeta}_1 - \bar{z}_{10}} -$$

$$- \frac{\gamma}{8i} \int\limits_L \left[\frac{c_{44}}{\sqrt{\Delta}} \overline{a(\psi)} e^{i\alpha} p_1(\zeta)\, H_1(\gamma r) + i\gamma q(\zeta)\, a(\psi)\, H_0^{(1)}(\gamma r) \right] ds,$$

$$H_1(\gamma r) = \frac{2i}{\pi\gamma r} + H_1^{(1)}(\gamma r), \quad \alpha = \arg(\zeta_1 - z_{10}).$$

By applying the formulas of Sohotsky-Plemmely for the generalized integrals of Cauchy type, the limiting values of the derivatives (4.11) at $z_{10} \to \zeta_{10} \in L'$ (L' is the image L on the affine z_1-plane) read

$$\left[\frac{1}{2\pi i} \int\limits_{L'} \frac{\omega(\zeta)\, d\zeta_1}{\zeta_1 - z_{10}} \right]^{\pm} = \pm \frac{\omega(\zeta_0)}{2} + \frac{1}{2\pi i} \int\limits_{L'} \frac{\omega(\zeta)\, d\zeta_1}{\zeta_1 - \zeta_{10}}, \tag{4.12}$$

$$\left[\frac{1}{2\pi i} \int\limits_{L'} \frac{\omega(\zeta)\, d\bar{\zeta}_1}{\bar{\zeta}_1 - \bar{z}_{10}} \right]^{\pm} = \mp \frac{\omega(\zeta_0)}{2} + \frac{1}{2\pi i} \int\limits_{L'} \frac{\omega(\zeta)\, d\bar{\zeta}_1}{\bar{\zeta}_1 - \bar{\zeta}_{10}}.$$

By calculating the difference of the harmonic expressions (4.7) which correspond to the left and the right edges L under the condition $X_3 = X_3^+ = -X_3^-$ we get

$$p(\zeta) = -\frac{i\mathrm{Im}\,\mu}{a(\psi)} \left[\frac{dU_3^*}{ds} \right], \quad p_1(\zeta) = \frac{i\mathrm{Im}\,\mu}{\overline{a(\psi)}} \left[\frac{dU_3^*}{ds} \right]. \tag{4.13}$$

Hence, using expressions (4.13) in representations (4.8) a single "*density*" appears, with respect to the displacement jump amplitude $[U_3^*]$ on L. Thus, satisfying the boundary condition (4.7) on any of the edges L (for example, on the left one) we derive a singular integro-differential equation of the first kind with respect to function $[U_3^*]$

$$\int\limits_L \left[\frac{dU_3^*}{ds} \right] g(\zeta, \zeta_0)\, ds + \int\limits_L [U_3^*] G(\zeta, \zeta_0)\, ds = N(\zeta_0),$$

$$g(\zeta, \zeta_0) = \frac{1}{\pi i} \mathrm{Re} \frac{a(\psi_0)}{\zeta_1 - \zeta_{10}} + \frac{\gamma}{2} H_1(\gamma r_0)\, \mathrm{Re}\left[a(\psi_0)\, e^{-i\alpha_0} \right], \tag{4.14}$$

$$G(\zeta,\zeta_0)=\frac{\gamma^2}{2}H_0^{(1)}(\gamma r_0)\,\mathrm{Re}\left[a(\psi_0)\,\overline{a(\psi)}\right],$$

$$N(\zeta_0)=\frac{2i}{\sqrt{\Delta}}X_3(\zeta_0)-2\gamma\tau\mathrm{Im}\left[a(\psi_0)\,e^{-i\beta}\right]e^{i(\alpha_1\xi_{10}+\alpha_2\xi_{20})},$$

$$[U_3^*]=\int\limits_0^s\left[\frac{dU_3^*}{ds}\right]ds,\quad r_0=|\zeta_1-\zeta_{10}|,\quad \alpha_0=\arg(\zeta_1-\zeta_{10}).$$

where the point $(\xi_{10},\ \xi_{20})\in L_j\ (j=1,\ 2,\dots,k)$; the kernel $G(\zeta,\zeta_0)$ has a logarithmic singularity, the kernel $g(\zeta,\zeta_0)$ consists of a singular component (Kernels of Cauchy type) and of a component due to the assumptions about L not more than a weak singularity.

For the unique solvability of (4.14) in a function class with derivatives not limited in the vicinity of the ends of cracks L_j [81, 121] it is necessary to add the following additional conditions to it:

$$\int\limits_{L_j}\left[\frac{dU_3^*}{ds}\right]ds=0\quad(j=1,2,\dots,k).\tag{4.15}$$

Conditions (4.15) regulate the equality to zero of the displacement jumps on tips $L=\cup L_j$.

In order to construct the asymptotic of stresses we will proceed from the formulas determining the behavior of integrals of Cauchy type in the vicinity of ends L in case when the density has a power singularity [50]

$$\frac{1}{2\pi i}\int\limits_L\frac{\lambda(\zeta)\,d\zeta}{\zeta-z}=\begin{cases}\dfrac{e^{\sigma\pi i}\lambda^*(a)(z-a)^{-\sigma}}{2i\sin\pi\sigma}+\Phi_1(z), & z\in O(a),\\[2ex] -\dfrac{e^{-\sigma\pi i}\lambda^*(b)(z-b)^{-\sigma}}{2i\sin\pi\sigma}+\Phi_2(z), & z\in O(b),\end{cases}\tag{4.16}$$

$$\lambda(\zeta)=\frac{\lambda^*(\zeta)}{(z-c)^\sigma},\quad \sigma=\kappa_1+i\kappa_2,\quad 0\le\kappa_1\le1.$$

For the function $\Phi_m(z)$ the following relations hold:

$$\lim_{z\to a}\Phi_1(z)(z-a)^\sigma=0,\quad \lim_{z\to b}\Phi_2(z)(z-b)^\sigma=0.$$

From asymptotic analysis of the integral (4.14) in the vicinity of the cut tip on integrating line L it is not difficult to deduce that $\sigma=1/2$. Therefore, by introducing the parameterization of the crack contour as $\zeta=\zeta(\delta)$, it is possible to assume that

$$\left[\frac{dU_3^*}{ds}\right]=\frac{p^*(\zeta)}{\sqrt{(\zeta-a)(\zeta-b)}}=\frac{\Omega_0(\delta)}{s'(\delta)\sqrt{1-\delta^2}},\tag{4.17}$$

$$s'(\delta)=\frac{ds}{d\delta}>0,\quad \Omega_0\in H[-1,1].$$

On the basis of formulas (4.16), (4.17) the asymptotics of the generalized integrals of Cauchy type are given by

$$\frac{1}{2\pi i}\int_{L'}\frac{\left[\frac{dU_3^*}{ds}\right]d\zeta_1}{\zeta_1 - z_{10}} = \begin{cases} \frac{\Omega_0(-1)\,R(-1)}{\sqrt{z_{10}-a_1}} + \Phi_1(z_{10}), & z_{10}\in O(a_1),\\ \frac{\Omega_0(1)\,R(1)}{\sqrt{b_1 - z_{10}}} + \Phi_2(z_{10}), & z_{10}\in O(b_1), \end{cases} \tag{4.18}$$

$$\frac{1}{2\pi i}\int_{L'}\frac{\left[\frac{dU_3^*}{ds}\right]d\overline{\zeta_1}}{\overline{\zeta_1} - \overline{z}_{10}} = \begin{cases} -\frac{\Omega_0(-1)\overline{R(-1)}}{\sqrt{\overline{z}_{10}-\overline{a_1}}} + \overline{\Phi_1^*(z_{10})}, & z_{10}\in O(a_1),\\ -\frac{\Omega_0(1)\overline{R(1)}}{\sqrt{\overline{b_1} - \overline{z}_{10}}} + \overline{\Phi_2^*(z_{10})}, & z_{10}\in O(b_1), \end{cases}$$

$$R(\pm 1) = \frac{1}{2}\sqrt{\frac{a(\psi(\pm 1))}{2s'(\pm 1)}},\quad \zeta_1 = \zeta_1(\delta),\quad a_1 = \zeta_1(-1),\quad b_1 = \zeta_1(1).$$

Allowing for (4.11), (4.13) and (4.18) and applying the equation of state (4.2) we can express the main asymptotic shear stresses in the vicinity of the cut tips as follows

$$\sigma_{13} = \begin{cases} \frac{c_{44}\,\mathrm{Im}\mu}{2\sqrt{2s'(-1)}}\,\mathrm{Im}\,\frac{\mu\,\mathrm{Re}\left[e^{-i\omega t}\Omega_0(-1)\right]}{\sqrt{a[\psi(-1)]\,(z_{10}-a_1)}}, & z \in O(a),\\ -\frac{c_{44}\,\mathrm{Im}\mu}{2\sqrt{2s'(1)}}\,\mathrm{Re}\,\frac{\mu\,\mathrm{Re}\left[e^{-i\omega t}\Omega_0(1)\right]}{\sqrt{a[\psi(1)]\,(z_{10}-b_1)}}, & z \in O(b), \end{cases} \tag{4.19}$$

$$\sigma_{23} = \begin{cases} -\frac{c_{44}\,\mathrm{Im}\mu}{2\sqrt{2s'(-1)}}\,\mathrm{Im}\,\frac{\mathrm{Re}\left[e^{-i\omega t}\Omega_0(-1)\right]}{\sqrt{a[\psi(-1)]\,(z_{10}-a_1)}}, & z \in O(a),\\ \frac{c_{44}\,\mathrm{Im}\mu}{2\sqrt{2s'(1)}}\,\mathrm{Re}\,\frac{\mathrm{Re}\left[e^{-i\omega t}\Omega_0(1)\right]}{\sqrt{a[\psi(1)]\,(z_{10}-b_1)}}, & z \in O(b). \end{cases}$$

The shear stress on the continuation over the crack tips have the form of

$$\sigma_n = \sigma_{13}\cos\psi(\pm 1) + \sigma_{23}\sin\psi(\pm 1) = \pm\frac{\sqrt{\Delta}\mathrm{Re}\left[e^{-i\omega t}\Omega_0(\pm 1)\right]}{2\sqrt{2r^* s'(\pm 1)}}. \tag{4.20}$$

Here $r^* = |z - c|$, where c is the tip; the lower sign refers to the beginning of the cut $c = a$, the upper one to the end $c = b$.

The stress intensity factor K_{III} is determined by [138, 159]

$$K_{III} = \lim_{r^*\to 0}\sqrt{2\pi r^*}\sigma_n. \tag{4.21}$$

From the expressions (4.20), (4.21) it follows that the tip K_{III} is the functional determined by the solution of the integro-differential (4.14). This condition permits the solution of the inverse (optimization) problem of the control of breakage parameters in the bodies with cracks, using the solution of the corresponding direct problems [172].

As an example let us consider an anisotropic space containing a singular crack L_1, the contour of which has the form of a parabola ($\xi_1 = \delta,\ \xi_2 = p_*\delta^2$, $break - 1 \leq \delta \leq 1$) or a sinusoid ($\xi_1 = (1+\delta)/2,\ \xi_2 = 0.2 \sin \nu\pi\xi_1$) for the case where a shear wave comes from infinity, along the negative direction of axis x_2 ($\beta = \pi/2$). In Table 4.1 are given the approximate values referring to stress intensity factors $\langle K_{III}^{\pm} \rangle = \sqrt{\Delta}\,|\Omega_0(\pm 1)| / \left(2S_{23}^0 \sqrt{\ell s'(\pm 1)}\right)$ for different relations of the elastic modules of the fictitious materials as functions of normalized wave number $\gamma\ell$.

Assuming that the quantity $\langle K_{III}^{\pm} \rangle$ is known the stress intensity factor may be determined by

$$K_{III}^{\pm} = \pm\sqrt{\pi\ell}\ S_{23}^0 \langle K_{III}^{\pm} \rangle \cos(\omega t - \arg \Omega_0(\pm 1)),$$

where $S_{23}^0 = |\tau|\,\gamma c_{44}$ is the modulus of the stress amplitude σ_{23} in the shear wave.

Figure 4.2 shows changes of $\langle K_{III}^{\pm} \rangle$ as a function of $\gamma\ell$ in the case of the diffraction of a shear wave ($\beta = \pi/2$) on a sinusoidal crack (the value of parameter ν corresponds to the curve number).

Figure 4.3 demonstrates illustrates graphs of $\langle K_{III}^{\pm} \rangle = \sqrt{\Delta}\,|\Omega_0(\pm 1)| / \left(2\,|X_3|\,\sqrt{\ell s'(\pm 1)}\right)$ in the case where (a) there is no wave and, (b) harmonically changing time-dependent shear forces act on the edges of the sinusoidal crack ($X_3 = \text{const}, \tau = 0$). The parameter values of the anisotropic material were taken to be equal $c_{45}/c_{44} = 0, c_{55}/c_{44} = 3$. The solid and dashed lines correspond to the tips of cracks a_1 and b_1 respectively.

The dependence of the factor K_{III} on the orientation angle of crack φ has characteristic extremal properties. Fig. 4.4 illustrates the behaviour

Table 4.1. The approximate values for the stress intensity factors referring to different relations of the elastic modules of fictional materials as a function the normalized wave number

$\gamma\ell$	$c_{44} = 1, c_{55} = 4, c_{45} = 0$		$c_{44} = 1,\ c_{55} = 6,\ c_{45} = 0.5$			
	$p^* = -0.5$	$p^* = 0.5$	$p^* = 0.5$		$p^* = -0.5$	
			$\langle K_{III}^- \rangle$	$\langle K_{III}^+ \rangle$	$\langle K_{III}^- \rangle$	$\langle K_{III}^+ \rangle$
0	0.968	0.968	1.153	0.876	0.876	1.153
0.1	0.978	0.978	1.164	0.885	0.886	1.165
0.2	1.003	1.000	1.187	0.910	0.912	1.192
0.3	1.042	1.034	1.217	0.948	0.955	1.235
0.4	1.097	1.077	1.251	1.000	1.018	1.297
0.5	1.171	1.128	1.287	1.069	1.107	1.385
0.6	1.267	1.187	1.318	1.158	1.228	1.505
0.7	1.386	1.248	1.334	1.264	1.383	1.662
0.8	1.522	1.299	1.311	1.370	1.555	1.842
0.9	1.652	1.316	1.218	1.429	1.690	1.988
1.0	1.732	1.268	1.052	1.382	1.699	2.007

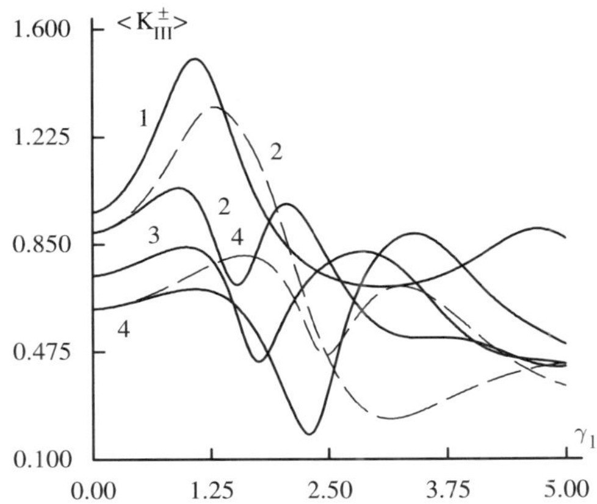

Fig. 4.2. Changes of quantity $< K^{\pm}_{III} >$ as a function of normalized wave number γl for various forms of sinusoidal crack under the influence of the shear wave

of the quantity $\langle K^{\pm}_{III} \rangle$ in the case of a linear crack $\xi_1 = \delta \cos\varphi$, $\xi_2 = \delta \sin\varphi \,(-1 \leq \delta \leq 1)$ when $\tau = 0, X_3 = \text{const}$. The curve indexed by the number m is constructed for $\gamma\ell = 1 + 0.3\,(m-1)$, $c_{45}/c_{44} = 0$, $c_{55}/c_{44} = 5$.

From the analysis, it follows that the intensity factor depends strongly on the frequency of the harmonic excitation, the anisotropy of materials, the crack curvature and also its orientation on the main axes of the anisotropy.

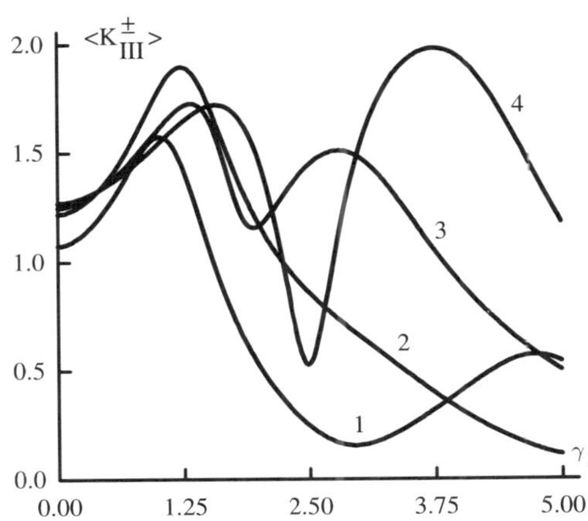

Fig. 4.3. Change of quantity $< K^{\pm}_{III} >$ as a function of normalized wave number γl for various forms of sinusoidal crack under the influence of shear forces

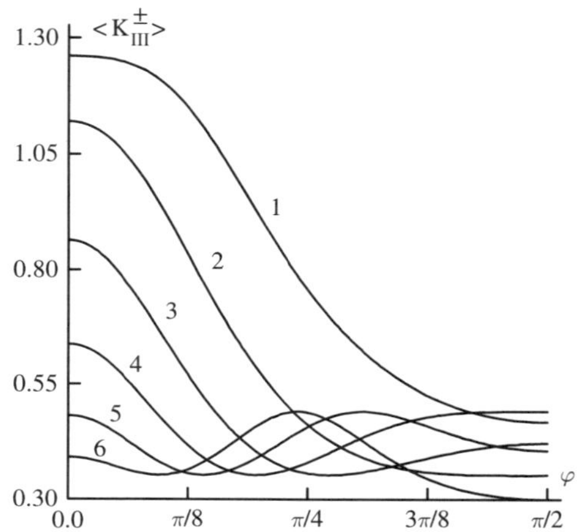

Fig. 4.4. Changes of quantity $< \mathrm{K}^{\pm}_{\mathrm{III}} >$ as a function of the orientation angle of crack ϕ for various values of parameter γl

4.2 A Piezoceramic Space

Let us now consider an unrestricted piezoceramic medium (a piezoelectric hexagonal class of $6mm$ [25]) in the Cartesian $Ox_1x_2x_3$ coordinates weakened by tunnel curvilinear cracks along the axis of symmetry x_3, symbolized as L_j $(j = 1, 2, \ldots, k)$. Assume that on the surface of cracks, a constant, along x_3- axis and harmonically changing with time, stress vector of the form $X^{\pm}_{1n} = X^{\pm}_{2n} = 0$, $X^{\pm}_{3n} = \mathrm{Re}\left(X^{\pm}_3 e^{-i\omega t}\right)$ is given and from infinity a plane monochromatic shear wave $u^{(0)}_3 = \mathrm{Re}\left(U^{(0)}_3 e^{-i\omega t}\right)$ possibly radiates.

The same assumptions made in Sect. 4.1 about cuts L_j and functions $X_{3n} = X^{+}_{3n} = -X^{-}_{3n}$ hold. It is necessary to determine the wave electric field under the conditions of anti-plane deformation of a piezoelectric medium with cracks.

The full system of equations includes the equation of motion (4.1), the material equations of the medium

$$\sigma_{m3} = c^{E}_{44}\partial_m u_3 - e_{15}E_m, \quad D_m = e_{15}\partial_m u_3 + \ni^{\varepsilon}_{11} E_m \quad (m = 1, 2) \tag{4.22}$$

and Maxwell-Lorentz equations [104, 129, 167] which in the given case read

$$\partial_1 E_2 - \partial_2 E_1 + \mu\frac{\partial H_3}{\partial t} = 0, \quad \partial_1 D_1 + \partial_2 D_2 = 0, \tag{4.23}$$

$$\partial_2 H_3 = \frac{\partial D_1}{\partial t}, \quad \partial_1 H_3 = -\frac{\partial D_2}{\partial t}.$$

Where E_1, E_2, H_3 and D_1, D_2 are the corresponding components of the strength vector of electric and magnetic fields, and also of electric induction; c^{E}_{44}, e_{15} and $\ni^{\varepsilon}_{11}$ are shear moduli measured at constant values of the electric field, while the piezoelectric constant and dielectric permeability, are measured at fixed deformations, respectively; μ is the magnetic permeability of the medium. It is assumed that there are no external charges and that the specific conductance of the medium is equal to zero.

The electromagnetic edge conditions on the cut boundaries read [115, 152]

$$E_s^+ = E_s^-, \quad D_n^+ = D_n^-, \quad H_s^+ = H_s^-, \quad B_n^+ = B_n^-, \quad \overrightarrow{B} = \mu \overrightarrow{H}, \tag{4.24}$$

where E_s and H_s are the tangential components of the vectors of the electric and magnetic strength, and D_n and B_n are the normal components of the vectors of electric and magnetic inductions. Conditions (4.24) express the circumstances under which the corresponding components of an electromagnetic field do not undergo jumps when crossing the cuts L_j.

Further, all the calculations are performed in the *electromagnetic system of units*.

Let us introduce the function Φ defined by the formulas [139]

$$\begin{aligned} E_1 &= -\frac{e_{15}}{\text{э}_{11}^{\varepsilon}} \partial_1 u_3 + \partial_2 \Phi, \quad E_2 = -\frac{e_{15}}{\text{э}_{11}^{\varepsilon}} \partial_2 u_3 - \partial_1 \Phi, \\ H_3 &= \text{э}_{11}^{\varepsilon} \frac{\partial \Phi}{\partial t}. \end{aligned} \tag{4.25}$$

Substituting from (4.22) the expressions for the stresses σ_{13}, σ_{23} into the equation of motion (4.1) and considering (4.25) we get the *wave equation*

$$\begin{aligned} &\nabla^2 u_3 - \frac{1}{c_*^2} \frac{\partial^2 u_3}{\partial t^2} = 0, \\ &c_* = \sqrt{\frac{c_{44}^E}{\rho} (1 + k_{15}^2)}, \quad k_{15}^2 = \frac{e_{15}^2}{\text{э}_{11}^{\varepsilon} \, c_{44}^E}. \end{aligned} \tag{4.26}$$

Here c_* is the shear wave velocity in a piezoelectric medium, k_{15} is the factor of electromagnetic coupling [115].

In the same way from the first (4.23) we obtain

$$\nabla^2 \Phi - \frac{1}{c_\alpha^2} \frac{\partial^2 \Phi}{\partial t^2} = 0, \quad c_\alpha = \frac{1}{\sqrt{\mu \, \text{э}_{11}^{\varepsilon}}}. \tag{4.27}$$

It should be noted that the quantity c_α which is the velocity of the electromagnetic wave propagation in a piezoelectric medium is much higher than the velocity of the acoustic waves. Therefore in the case of not very large cuts (the length of a cut is considerably smaller than the length of electromagnetic waves) the second compound in (4.27) may be neglected. The determination of the function Φ using the quasi-static approximation described in [140] is achieved with the help of the equation

$$\nabla^2 \Phi = 0. \tag{4.28}$$

Due to (4.22) and (4.25) we have

$$\begin{aligned} \sigma_{13} &= c_{44}^E\left(1+k_{15}^2\right)\partial_1 u_3 - e_{15}\partial_2\Phi, \quad D_1 = \ni_{11}^{\varepsilon}\,\partial_2\Phi, \\ \sigma_{23} &= c_{44}^E\left(1+k_{15}^2\right)\partial_2 u_3 + e_{15}\partial_1\Phi, \quad D_2 = -\ni_{11}^{\varepsilon}\,\partial_1\Phi, \\ u_3 &= u_3^{(0)} + u_3^*, \quad D_3 = 0. \end{aligned} \tag{4.29}$$

Here the quantity u_3^* characterizes disturbances of the displacement field stipulated by the presence of the cuts. Assuming that the following relations hold

$$\begin{aligned} &u_3^* = \operatorname{Re}\left(U_3^* e^{-i\omega t}\right), \quad U_3 = U_3^{(0)} + U_3^*, \\ &\Phi = \operatorname{Re}\left(F e^{-i\omega t}\right), \quad U_3^{(0)} = \tau e^{i(\alpha_1 x_1 + \alpha_2 x_2)}, \\ &\alpha_1 = -\gamma\cos\beta, \quad \alpha_2 = -\gamma\sin\beta, \quad \gamma = \frac{\omega}{c_*}, \end{aligned} \tag{4.30}$$

(β is the angle between the normal to the front wave and axis Ox_1) we can write (4.26) and (4.28) using the amplitudes, as follows

$$\nabla^2 U_3 + \gamma^2 U_3 = 0, \quad \nabla^2 F = 0. \tag{4.31}$$

The connection between functions F, U_3 and the potential amplitude of electric field ϕ_* are ascertained below.

In the same way, (4.3), (4.24) may be represented by the edge conditions on the boundaries L_j in the following form

$$\begin{aligned} &c_{44}^E\left(1+k_{15}^2\right)\left\{e^{i\psi}\left(\frac{\partial U_3}{\partial\zeta}\right)^{\pm} + e^{-i\psi}\left(\frac{\partial U_3}{\partial\overline{\zeta}}\right)^{\pm}\right\} - \\ &- ie_{15}\left\{e^{i\psi}\left(\frac{\partial F}{\partial\zeta}\right)^{\pm} - e^{-i\psi}\left(\frac{\partial F}{\partial\overline{\zeta}}\right)^{\pm}\right\} = X_3^{\pm}, \\ &i\frac{e_{15}}{\ni_{11}^{\varepsilon}}\left\{e^{i\psi}\left[\frac{\partial U_3}{\partial\zeta}\right] + e^{-i\psi}\left[\frac{\partial U_3}{\partial\overline{\zeta}}\right]\right\} + e^{i\psi}\left[\frac{\partial F}{\partial\zeta}\right] + e^{-i\psi}\left[\frac{\partial F}{\partial\overline{\zeta}}\right] = 0, \\ &e^{i\psi}\left[\frac{\partial F}{\partial\zeta}\right] - e^{-i\psi}\left[\frac{\partial F}{\partial\overline{\zeta}}\right] = 0, \\ &\zeta = \xi_1 + i\xi_2, \quad \overline{\zeta} = \xi_1 - i\xi_2, \quad [\eta] = \eta^+ - \eta^-, \quad \zeta\in L_j. \end{aligned} \tag{4.32}$$

Here the upper sign refers to left boundary L_j ($j = 1, 2, \ldots, k$), if its motion occurs from its beginning a_j to the end b_j; ψ is the angle between the normal to the left boundary and Ox_1- axis

Thus, the stated boundary problem is reduced to the definition of functions U_3 and F from (4.31) and boundary conditions (4.32). The integral equations for the problem under study are obtained by using general representations of the functions U_3^* and F. It is easy to obtain such representations for the derivatives from displacement amplitude U_3^* if in formulas (4.8) we assume that $c_{45} = 0, c_{55} = c_{44}^E, c_{44} = c_{44}^E \; (\mu = i)$. We have

$$\frac{\partial U_3^*}{\partial z} = c_{44}^E \int\limits_L \left\{ ip(\zeta)\, e^{i\psi} \frac{\partial E}{\partial \zeta} + \frac{\gamma^2}{2} e^{-i\psi} E q(\zeta) \right\} ds, \tag{4.33}$$

$$\frac{\partial U_3^*}{\partial \overline{z}} = c_{44}^E \int\limits_L \left\{ -ip_1(\zeta)\, e^{-i\psi} \frac{\partial E}{\partial \overline{\zeta}} + \frac{\gamma^2}{2} e^{i\psi} E q(\zeta) \right\} ds,$$

$$p(\zeta) = -2i \left[\frac{\partial U_3^*}{\partial \zeta} \right], \quad p_1(\zeta) = 2i \left[\frac{\partial U_3^*}{\partial \overline{\zeta}} \right], \quad q(\zeta) = [U_3^*],$$

$$E = E(\xi_1 - x_1, \xi_2 - x_2) = -\frac{i}{4c_{44}^E} H_0^{(1)}(\gamma r), \quad r = |\zeta - z|,$$

$$z = x_1 + ix_2, \quad \zeta = \xi_1 + i\xi_2 \in L_j \quad (j = 1, 2, \ldots, k),$$

where the derivatives of the sought-for function are determined in an internal point of the area $(x_1, \; x_2)$.

Formulas (4.33) match the integral representation for the displacement amplitude

$$U_3^*(x_1, x_2) = c_{44}^E \int\limits_L \left\{ q(\zeta) \left(e^{i\psi} \frac{\partial E}{\partial \zeta} + e^{-i\psi} \frac{\partial E}{\partial \overline{\zeta}} \right) - \right.$$
$$\left. - \frac{i}{2} E \left(p(\zeta)\, e^{i\psi} - p_1(\zeta)\, e^{-i\psi} \right) \right\} ds. \tag{4.34}$$

The equality (4.34) follows from (4.9) in the same way.

The composite harmonic function F may be represented in the following way

$$F(x_1, x_2) = \frac{1}{2\pi i} \int\limits_L \left\{ f(\zeta)\, \ell n(\zeta - z)\, d\zeta - \overline{f_1(\zeta)} \ell n \left(\overline{\zeta} - \overline{z} \right) d\overline{\zeta} \right\}. \tag{4.35}$$

The physical meaning of the functions f and f_1 will be clear below.

Differentiating (4.35) and applying the Sohotsky-Plemmely formulas of (4.12) at $z \to \zeta_0 \in L$, we find

$$\left\{ \frac{\partial F}{\partial z} \right\}^{\pm} = \mp \frac{f(\zeta_0)}{2} - \frac{1}{2\pi i} \int\limits_L \frac{f(\zeta)\, d\zeta}{\zeta - \zeta_0}, \tag{4.36}$$

$$\left\{ \frac{\partial F}{\partial \overline{z}} \right\}^{\pm} = \mp \frac{\overline{f_1(\zeta_0)}}{2} + \frac{1}{2\pi i} \int\limits_L \frac{\overline{f_1(\zeta)}\, d\overline{\zeta}}{\overline{\zeta} - \overline{\zeta_0}}.$$

Substituting the limiting values of the functions (4.33) at $z \to \zeta_0 \in L$ in the boundary conditions (4.32) allowing for formulas (4.36), we obtain

$$p(\zeta) = -e^{-i\psi}\left[\frac{dU_3^*}{ds}\right], \quad p_1(\zeta) = -e^{i\psi}\left[\frac{dU_3^*}{ds}\right], \tag{4.37}$$
$$f(\zeta) = -\frac{e_{15}}{2\,\text{э}_{11}^{\varepsilon}}p(\zeta), \quad \overline{f_1(\zeta)} = -\frac{e_{15}}{2\,\text{э}_{11}^{\varepsilon}}p_1(\zeta).$$

Actually, formulas (4.37) provide the continuation of the corresponding electromagnetic quantities through cuts and the vector of shear stress. In reality the continuation of quantity E_s, D_n, σ_n follows from the above-given constructions. To prove the continuation of the component of magnetic stress H_3 let us use the formulas (4.25) and (4.35). We have

$$H_3 = -i\omega\,\text{э}_{11}^{\varepsilon}\,\mathrm{Re}\left\{F(x_1,x_2)\,e^{-i\omega t}\right\}.$$

The jump $F(x_1, x_2)$ on L due to (4.35) reads

$$[F(\zeta_0)] = \int_a^{\zeta_0} f(\zeta)d\zeta + \int_a^{\zeta_0} \overline{f_1(\zeta)}d\bar{\zeta}.$$

Taking into account relation (4.37) we have $[F(\zeta_0)] = 0$, which proves the continuation of H_3.

Thus, as it is seen from (4.37), all "*densities*" appearing in the integral representations (4.33)–(4.35) are expressed by a jump of the derivative $[dU_3^*/ds]$. In that way the above representations acquire physical meaning.

The mechanical condition in (4.32) allowing for (4.37) induces the following singular integro-differentiated equation

$$\int_L \left[\frac{\partial U_3^*}{ds}\right] g(\zeta,\zeta_0)\,ds + \int_L [U_3^*] G(\zeta,\zeta_0)\,ds = N(\zeta_0),$$

$$g(\zeta,\zeta_0) = -\frac{1}{\pi i}\mathrm{Im}\frac{e^{i\psi_0}}{\zeta-\zeta_0} + \frac{\gamma}{2}\left(1+k_{15}^2\right)H_1(\gamma r_0)\sin(\alpha_0-\psi_0),$$

$$G(\zeta,\zeta_0) = \frac{\gamma^2}{2}\left(1+k_{15}^2\right)H_0^{(1)}(\gamma r_0)\cos(\psi-\psi_0), \tag{4.38}$$

$$N(\zeta_0) = \frac{2i}{c_{44}^E}X_3(\zeta_0) - 2\gamma\tau\left(1+k_{15}^2\right)e^{i(\alpha_1\xi_{10}+\alpha_2\xi_{20})}\cos(\psi_0-\beta),$$

$$H_1(\gamma r) = \frac{2i}{\pi\gamma r} + H_1^{(1)}(\gamma r),\ r_0 = |\zeta-\zeta_0|,\ \alpha_0 = \arg(\zeta-\zeta_0),$$

$$\psi_0 = \psi(\zeta_0), \quad [U_3^*] = \int_0^s \left[\frac{dU_3^*}{ds}\right]ds, \quad \zeta_0 = \xi_{10} + i\xi_{20} \in L_j\ (j=1,2,\ldots,k).$$

Here the kernel $g(\zeta,\zeta_0)$ is singular (Cauchy type), and the kernel $G(\zeta,\zeta_0)$ has a logarithmic singularity at the point $\zeta = \zeta_0$.

For the unique solution of (4.38) in the class function with derivatives not limited in the vicinity of crack tips L_j, it is necessary to add the conditions of the following type:

$$\int\limits_{L_j} \left[\frac{dU_3^*}{ds}\right] ds = 0 \quad (j = 1, 2, \ldots, k) . \tag{4.39}$$

Reasoning in the same way as in Sect. 4.1 we find that the asymptotic of the shear stresses on the continuation over the crack tips and stress intensity factor is given by

$$\sigma_{13} + i\sigma_{23} = \pm c_{44}^E e^{i\psi(\pm 1)} \frac{\operatorname{Re}\left[\Omega_0(\pm 1)\, e^{-i\omega t}\right]}{2\sqrt{2r^* s'(\pm 1)}}, \tag{4.40}$$

$$K_{III}^{\pm} = \lim_{r^* \to 0} \sqrt{2\pi r^*}\sigma_n = \pm \frac{c_{44}^E}{2} \sqrt{\frac{\pi}{s'(\pm 1)}} \operatorname{Re}\left[\Omega_0(\pm 1)\, e^{-i\omega t}\right],$$

$$\left[\frac{dU_3^*}{ds}\right] = \frac{\Omega_0(\delta)}{s'(\delta)\sqrt{1-\delta^2}}, \quad s'(\delta) = \frac{ds}{d\delta} > 0, \quad \Omega_0(\delta) \in H[-1,1] .$$

The asymptotic of a normal compound of electric induction vector on the continuation over the crack tip reads

$$D_n = D_1 \cos\psi(\pm 1) + D_2 \sin\psi(\pm 1) = \pm e_{15} \frac{\operatorname{Re}\left[\Omega_0(\pm 1)\, e^{-i\omega t}\right]}{2\sqrt{2r^* s'(\pm 1)}} . \tag{4.41}$$

The other electromagnetic quantities are limited. In reality, from the equations of state (4.22) we have

$$\sigma_n = c_{44}^E \frac{\partial u_3}{\partial n} - e_{15} E_n, \quad D_n = e_{15} \frac{\partial u_3}{\partial n} + \ni_{11}^{\varepsilon} E_n, \tag{4.42}$$

Where D_n is the normal component of electric induction on arc L' which is vey close to L. As $[\sigma_n] = [D_n] = 0$ and the determinant of system (4.42) differs from zero, we find $[\partial u_3/\partial n] = [E_n] = 0$. Thus, the electric field $\overrightarrow{E}$ continues through the cut and that implies that it is continuous everywhere.

After the solution of integral (4.38) functions F, U_3^* and $\partial_i U_3^*$ $(i = 1, 2)$ are restored by formulas (4.33)–(4.35). Then we can calculate the tangential stress, the strength and induction of the electric field in any point of the area by using relations (4.25), (4.29). The amplitude of the electric potential ϕ_* may be calculated as follows

Condition of Cauchy-Riemann [103]:

$$\partial_1 \left[\phi_* - \frac{e_{15}}{\ni_{11}^{\varepsilon}} U_3\right] = -\partial_2 F, \quad \partial_2 \left[\phi_* - \frac{e_{15}}{\ni_{11}^{\varepsilon}} U_3\right] = \partial_1 F.$$

Consequently, based on (4.35), (4.37) we get:

$$\phi_* (x_1, x_2) = \frac{e_{15}}{\ni_{11}^{\varepsilon}} \left\{ U_3 (x_1, x_2) + \frac{1}{2\pi} \int_L \left[\frac{\partial U_3^*}{ds} \right] \arg (\zeta - z) \, ds \right\}. \tag{4.43}$$

Due to condition (4.39) the integral appearing here is the unique function of the complex variable z, and the displacement amplitude $U_3 (x_1, x_2)$ and is calculated by using (4.34).

As an example consider the space of piezoceramics PZT–4 (See: Appendix A, Table A.1), containing a crack of parabolic or sinusoidal form. Table 4.2 gives the values of relative stress intensity factor $< K_{III}^+ >= c_{44}^E |\Omega_0(1)|/(2\Lambda\sqrt{\ell s'(1)})$ for a parabolic crack ($\xi_1 = \delta,\ \xi_2 = p_* \delta^2$) under the action of shear wave ($\beta = \pi/2$) in *piezoceramic* and in corresponding *piezopassive* media ($e_{15} = 0$). It is obvious that the influence of the interaction of the coupled fields on $K_{III}^+ = \sqrt{\pi\ell}\Lambda \left\langle K_{III}^+ \right\rangle \cos (\omega t - \arg \Omega_0 (1))$ $(\Lambda = S_{23}^0 = \gamma \left|\tau\right| c_{44}^E \left(1 + k_{15}^2\right)$ is the modulus of stress amplitude σ_{23} in a shear wave) in the space is not of great importance.

Figures 4.5 and 4.6 illustrate the behaviors of quantity $< K_{III}^+ >$ as a function of the normalized wave number $\gamma^* \ell = \gamma\ell\sqrt{1 + k_{15}^2}$ under the influence of harmonic forces ($\Lambda = |X_3| = const.$) and shear wave ($\beta = \pi/2$), in a space with a crack of sinusoidal form respectively. The number of the curve conforms ν in parametric conditions L:$\xi_1 = (1 + \delta)/2,\ \xi_2 = 0,\ 2 \sin \nu\pi\xi_1\ (\delta \in [-1, 1])$. The solid and dashed lines correspond to α and b, respectively.

Table 4.2.

$\gamma^* \ell$	$p^* = 0.5$		$p^* = -0.5$	
	PZT-4	$e_{15} = 0$	PZT-4	$e_{15} = 0$
0	0.857	0.857	0.857	0.857
0.1	0.864	0.864	0.865	0.864
0.2	0.884	0.881	0.884	0.882
0.3	0.912	0.906	0.913	0.907
0.4	0.950	0.937	0.953	0.941
0.5	0.996	0.973	1.002	0.981
0.6	1.050	1.014	1.061	1.028
0.7	1.112	1.056	1.130	1.081
0.8	1.177	1.097	1.207	1.136
0.9	1.240	1.131	1.285	1.189
1.0	1.289	1.150	1.354	1.232

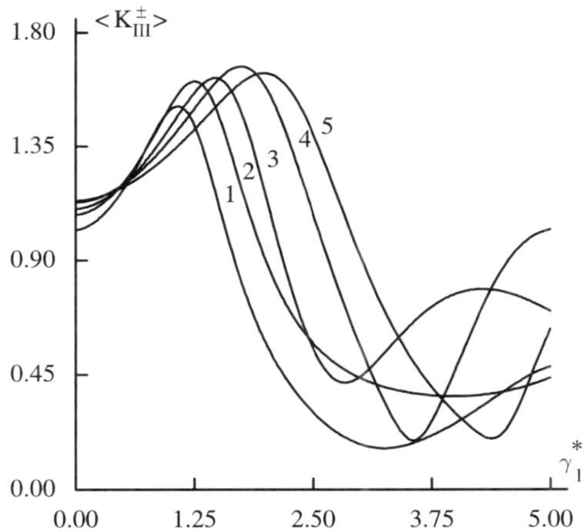

Fig. 4.5. Behaviour of $< K_{III}^{\pm} >$ as a function of normalized wave number $\gamma^* l$ for various forms of a sinusoidal crack under the influence of shear forces

Figure 4.7 shows the contour lines of the modulus of displacement amplitude $U_3(x_1, x_2)$ in the vicinity of a linear crack with the length of 2ℓ loaded with shear forces ($X_3 = \text{const.}$) ($\gamma^*\ell = 2$). Fig. 4.8 illustrates analogous results for the moduli of amplitude of electric potential $\phi_*(x_1, x_2)$, obtained by using formulas (4.43) ($\gamma^*\ell = 0.5$). The lighter zones correspond to maximal values.

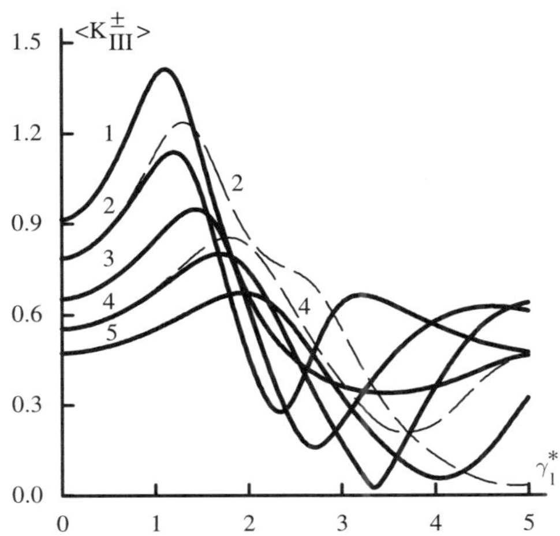

Fig. 4.6. Behaviour of $< K_{III}^{\pm} >$ as a function of normalized wave number $\gamma^* l$ for various sinusoidal cracks under the influence of a shear wave

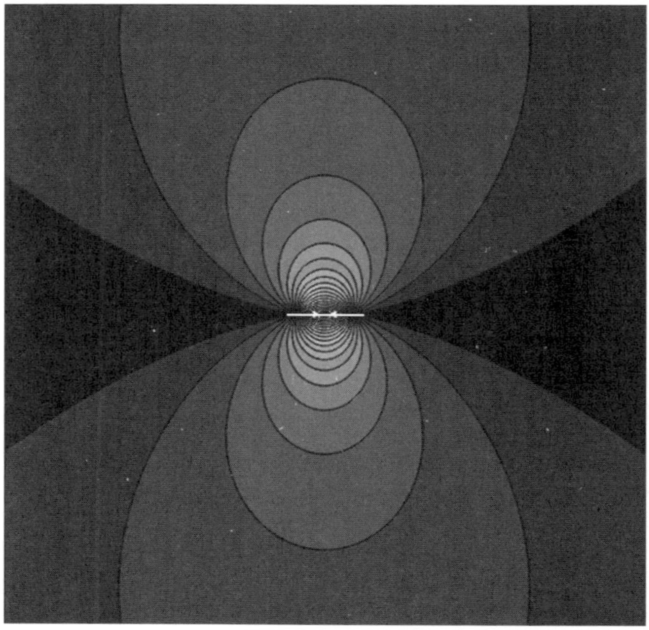

Fig. 4.7. The contour lines of the modulus of displacement amplitude $|U_3|$ in a space with a linear crack

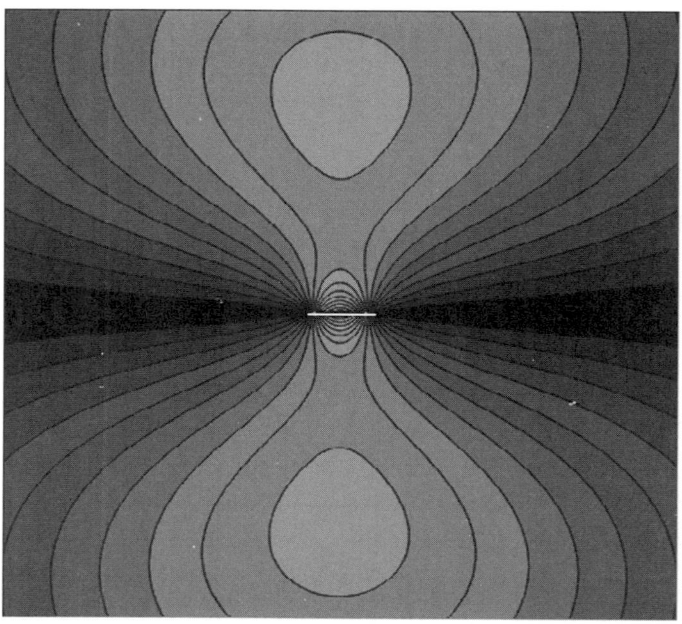

Fig. 4.8. The contour lines of the modulus of the electric potential amplitude in a space with a linear crack

4.3 A Piezoceramic Halfspace. Free Boundary and Rigid Fixture Problems

We have already considered the interaction of shear waves with curvilinear cracks-cuts in infinite media. In order to take into account the influence of the boundary let us study the analogous problem on a piezoceramic halfspace. To construct the integral representations of the solutions we use Green's function in a halfspace.

Consider the piezoceramic halfspace $x_2 \geq 0$, weakened by tunnel cuts, along x_3 -axis, that are curvilinear and noted as L_j $(j = 1, 2, \ldots, k)$. On the boundaries L_j let us assign independent on coordinates x_3, continuous by Hölder shear loadings $X_{3n} = X_{3n}^{+} = -X_{3n}^{-} == \mathrm{Re}\left(X_3 e^{-i\omega t}\right)$. The presence of radiation from infinity in the form of a plane monochromic shear wave (see Fig. 4.9) is not excluded.

Let us suppose that on the boundary of the half-space the following two types of edge conditions are possible:

a) a halfspace is rigidly fixed and covered along the boundary by a grounded electrode

$$u_3 = 0, \quad E_1 = 0 \quad (x_2 = 0)\,; \tag{4.44}$$

b) a halfspace free from forces and bounded with vacuum

$$\sigma_{23} = 0, \quad D_2 = 0 \quad (x_2 = 0)\,. \tag{4.45}$$

Under the given conditions in a halfspace, stationary waves appeared described by quantities $(m = 1, 2)$

$$\begin{aligned} \sigma_{m3} &= \mathrm{Re}\{S_{m3}(x_1, x_2)e^{-iwt}\}, \quad & E_m &= \mathrm{Re}\{E_{m^*}(x_1, x_2)e^{-iwt}\}, \\ D_m &= \mathrm{Re}\{D_{m^*}(x_1, x_2)e^{-iwt}\}, \quad & H_3 &= \mathrm{Re}\{H_{3*}(x_1, x_2)e^{-iwt}\}. \end{aligned}$$

The full system of equations of the stated boundary problem has the form of (4.1), (4.22), (4.23) and (4.32). Here we must also augment the above system with the mechanical and electric conditions on the boundary of a halfspace.

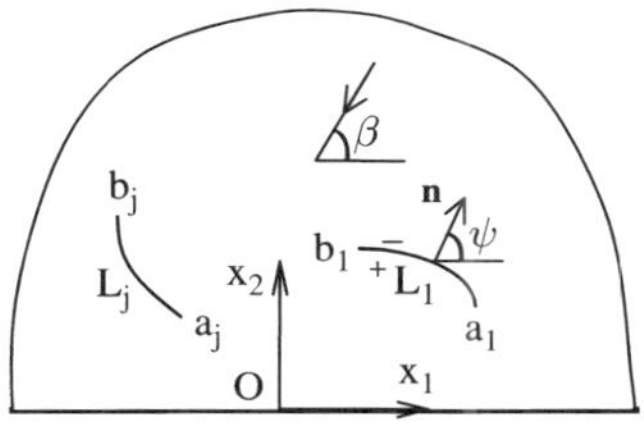

Fig. 4.9. The scheme of a halfspace with a crack

Proceeding in the same way as we did in Sect. 4.2 we come to the following differential equations corresponding to the displacement amplitude U_3 and the function F introduced into (4.25)

$$\nabla^2 U_3 + \gamma^2 U_3 = 0, \quad \nabla^2 F = 0. \tag{4.46}$$

Thus, the problem is reduced to the definition of functions U_3 and F from (4.46), of boundary conditions (4.32) on the cuts and also on mechanical and electric conditions on the boundary of a halfspace. Having constructed the subsidiary function of Green's (4.46) for a half-space it is better to fulfill the latter. Applying the method of images we find

$$\begin{aligned} &G\left(z_0, z\right) = H_0^{(1)}\left(\gamma r\right) - AH_0^{(1)}\left(\gamma r_1\right), \\ &H\left(z_0, z\right) = \ell n\left(z_0 - z\right) + A\ell n\left(z_0 - \overline{z}\right), \\ &z = x_1 + ix_2, \quad z_0 = x_{10} + ix_{20}, \quad r = \left|z - z_0\right|, \quad r_1 = \left|z - \overline{z_0}\right|. \end{aligned} \tag{4.47}$$

Here the function $G\left(z_0, z\right)$ is in conformity with Helmhöltz equation and $H\left(z_0, z\right)$ with Laplace equation. The value $A = -1$ corresponds to a free from force halfspace bounded with vacuum; $A = 1$ corresponds to a fixed and covered with grounded electrodes halfspace.

The wave displacement field in a halfspace with defects is combined from a direct, a reflected -from the boundary of a halfspace- wave and of a field of dispersed defects. According to this, the sought-for function is constructed in the following form [181]

$$\begin{aligned} &U_3\left(x_1, x_2\right) = -\frac{i}{4}\int\limits_L \left[U_3\right]\frac{\partial G\left(\zeta, z\right)}{\partial n_\zeta} ds + U_3^{(0)} - AU_3^{(1)}, \\ &F\left(x_1, x_2\right) = \frac{1}{2\pi i}\int\limits_L \left\{f\left(\zeta\right)H\left(\zeta, z\right)d\zeta - \overline{f_1\left(\zeta\right)H\left(\zeta, z\right)d\zeta}\right\}, \\ &U_3^{(1)} = \tau e^{i\left(\alpha_1 x_1 - \alpha_2 x_2\right)}, \quad \zeta = \xi_1 + i\xi_2 \in L, \end{aligned} \tag{4.48}$$

where $U_3^{(0)}, U_3^{(1)}$ denote the displacement amplitudes in falling and reflected from the boundary of a halfspace waves, respectively $\left[U_3\right]$, f and f_1 are sought-for "*densities*"; function G and H are defined in (4.47); differential operator $\partial/\partial n_\zeta$ defines the derivative by the normal to contour L, ds is the element of the arc length L.

The integral representations (4.48) conform to (4.46), mechanical and electric boundary conditions of the mentioned type on the boundary of halfspace $x_2 = 0$ and also the conditions of radiation, respectively. At $A = 0$ we obtain the representations of the solutions for an unlimited piezoceramic condition with cracks.

The procedure of reducing the considered boundary problem to the integral is as follows. Using formulas (4.25), (4.29) and (4.48) we obtain integral representations of electric and mechanical quantities. Substituting the corresponding limiting values of these quantities at $z \to \zeta_0 \in L$ in boundary conditions (4.32) we come to the dependence (4.37) between functions f, f_1 and $[dU_3/ds]$ and the singular integro-differential equation:

$$\int\limits_L p'(\zeta) g_1(\zeta,\zeta_0)\, ds + \int\limits_L p(\zeta) g_2(\zeta,\zeta_0)\, ds = N(\zeta_0), \tag{4.49}$$

$$g_1(\zeta,\zeta_0) = \mathrm{Im}\frac{e^{i\psi_0}}{\zeta-\zeta_0} + Ak_{15}^2 \mathrm{Im}\frac{e^{-i\psi_0}}{\zeta-\overline{\zeta}_0[6pt]},$$

$$g_2(\zeta,\zeta_0) = \frac{\pi i\gamma^2}{4}\left(1+k_{15}^2\right)\Big[AH_0^{(1)}(\gamma r_{10})\cos(\psi+\psi_0)$$

$$-H_0^{(1)}(\gamma r_0)\cos(\psi-\psi_0)+$$

$$+H_2(\gamma r_0)\cos(\psi+\psi_0-2\alpha_0) - AH_2^{(1)}(\gamma r_{10})\cos(\psi-\psi_0+2\alpha_{10})\Big],$$

$$N(\zeta_0) = \frac{\pi}{c_{44}^E}X_3(\zeta_0) + \pi i\tau\gamma\left(1+k_{15}^2\right)\Big[e^{i(\alpha_1\xi_{10}+\alpha_2\xi_{20})}\cos(\psi_0-\beta) -$$

$$-A\, e^{i(\alpha_1\xi_{10}-\alpha_2\xi_{20})}\cos(\psi_0+\beta)\Big],$$

$$H_2(\gamma r) = \frac{4i}{\pi\gamma^2 r^2} + H_2^{(1)}(\gamma r), \quad r_0 = |\zeta-\zeta_0|, \quad r_{10} = \left|\overline{\zeta}-\zeta_0\right|,$$

$$\alpha_0 = \arg(\zeta_0-\zeta), \quad \alpha_{10} = \arg\left(\zeta_0-\overline{\zeta}\right),$$

$$\psi = \psi(\zeta), \quad \psi_0 = \psi(\zeta_0), \quad p(\zeta) = \frac{[U_3]}{2}, \quad p'(\zeta) = \frac{dp(\zeta)}{ds},$$

$$\zeta_0 = \xi_{10} + i\xi_{20}; \quad \zeta,\zeta_0 \in L = \cup L_j.$$

Here, the case $A = -1$ corresponds to a free from forces halfspace bounded with vacuum; $A = 1$ corresponds to a halfspace covered with grounded electrodes. At $A = 0$ we obtain the integral equation corresponding to unlimited space with cracks.

It is necessary to add the additional condition of the following type to the integral (4.49)

$$\int\limits_{L_j} p'(\zeta) ds = 0 \quad (j = 1, 2, \ldots, k). \tag{4.50}$$

The expression for the stress intensity factor K_{III} may be obtained by asymptotic analysis of stresses in the vicinity of the crack tip. Using representations (4.48) we find

$$K_{III}^{\pm} = \pm c_{44}^{E} \sqrt{\frac{\pi}{s'(\pm 1)}} \operatorname{Re} \left\{ \Omega_0 (\pm 1) e^{-i\omega t} \right\}, \tag{4.51}$$

$$p'(\zeta) = \frac{\Omega_0(\delta)}{s'(\delta)\sqrt{1-\delta^2}}, \quad \Omega_0(\delta) \in H[-1,1].$$

The numerical solution of singular integro-differential (4.43) was carried out by the method of quadrature (see: Appendix 12.8, Par. B.1). Fig. 4.10 shows the contours of absolute values of the amplitude of electric potential $\phi_*(x_1, x_2)$ the area covering a linear crack for different types of mechanical and electric edge conditions on the boundary of a halfspace with $(A = 1, \gamma^* \ell = 0.5)$, Fig. 4.11 with $(A = -1, \gamma^* \ell = 1)$ and Fig. 4.12 with $(A = -1, \gamma^* \ell = 0.5)$. As loading we consider shear forces acting on the crack boundary. Calculations were carried out by using the following formula:

$$\phi_*(x_1, x_2) = \frac{e_{15}}{\ni_{11}^{\varepsilon}} \left\{ U_3(x_1, x_2) + \frac{1}{\pi} \int_L p'(\zeta) \left[\arg(\zeta - z) \right. \right.$$
$$\left. \left. + A \arg\left(\overline{\zeta} - z\right) \right] \right\}, \zeta \in L. \tag{4.52}$$

The lines of the modulus of displacement amplitude $U_3(x_1, x_2)$ for the same type of loading are given in Fig. 4.13 with $(A = -1, \gamma^* \ell = 1)$.

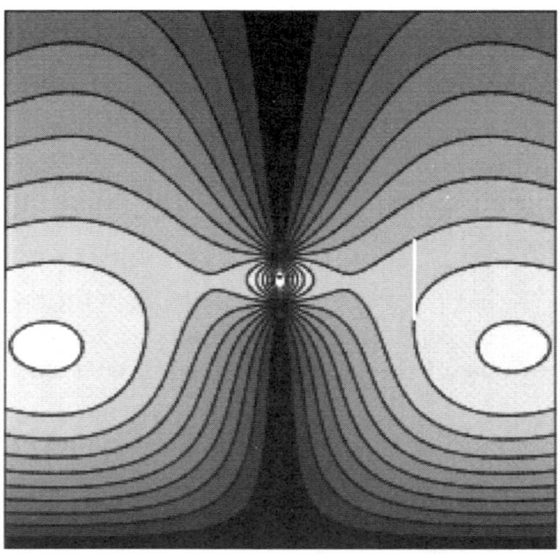

Fig. 4.10. The contour lines of the modulus of the displacement amplitude in the vicinity of a crack under various conditions on the boundary of a halfspace under the influence of shear forces

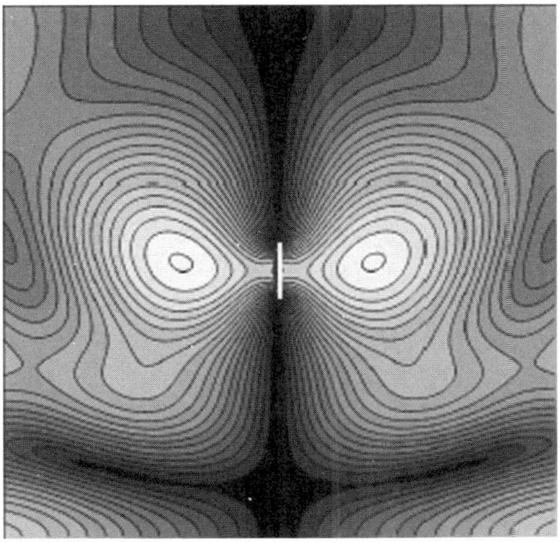

Fig. 4.11. The contour lines of the modulus of the displacement amplitude in the vicinity of a crack under various conditions on the boundary of a halfspace under the influence of shear forces

To investigate the influence of the cut curve and the frequencies of the harmonic excitation on the behavior of the stress intensity factor we considered the diffraction of shear wave ($\beta = \pi/2$) in a halfspace from piezoceramics PZT-4 with a parabolic crack parameterization of type $\xi_1 + i\xi_2 = \delta\ e^{i\varphi}\,(p_1 + ip_2\delta) + ih,\ \ \delta \in [-1, 1]$ (φ is the angle characterizing the crack orientation in a halfspace).

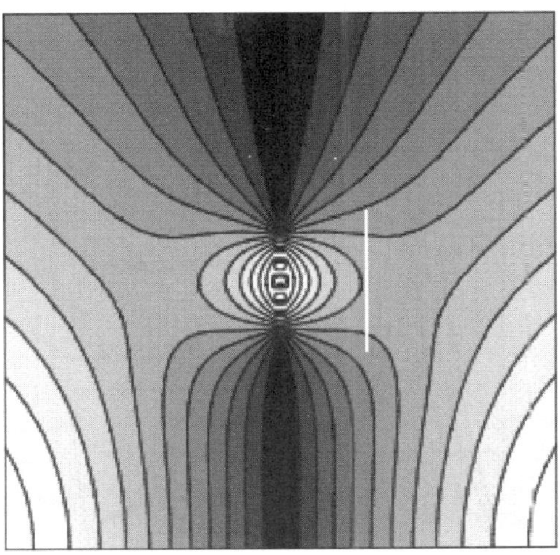

Fig. 4.12. The contour lines of the modulus of the displacement amplitude in the vicinity of a crack under various conditions on the boundary of a halfspace under the influence of shear forces

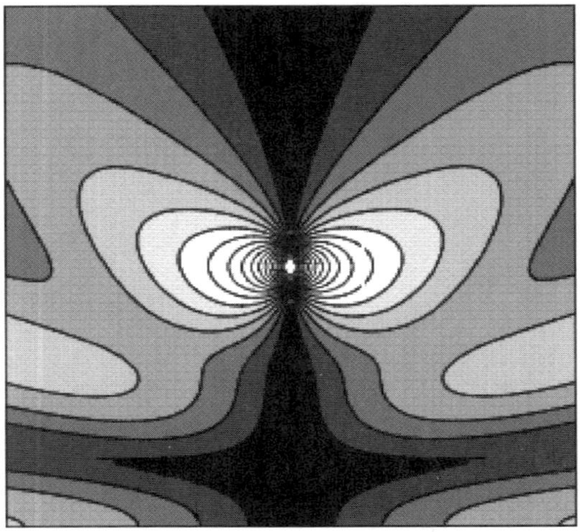

Fig. 4.13. The contour lines of the modulus of the electric potential amplitude in a free halfspace which is bounded with vacuum and is under the influence of shear forces

Figures 4.14 and 4.15 illustrate the behavior of the relative quantities $\langle K^{+}_{III}\rangle = c^{E}_{44}\left|\Omega_0\left(1\right)\right| / \left(S^{0}_{23}\sqrt{\ell s'\left(1\right)}\right)$ and $\langle K^{+}_{III}\rangle = c^{E}_{44}\left|\Omega_0\left(1\right)\right| / \left(\left|X_3\right|\sqrt{\ell s'\left(1\right)}\right)$ as a function of $\gamma^{*}\ell$ in the case of a free and conjugated with vacuum ($A = -1$), respectively. Curves 1, 2 and 3 are constructed for the values $h/p_1 = 2$, $\varphi = 0$, $p_2/p = 0, -0.6, -1.2$, respectively.

Analogous results for fixed halfspace ($A = 1$) are shown in Fig. 4.16 ($\tau = \text{const}$) and Fig. 4.17 ($X_3 = \text{const}$) and curves 1–4 relate to values $p_2/p = 0, -0.3, -0.6, -0.9$ and $h/p_1 = 3$, $\varphi = 0$. As it follows from Fig. 4.16,

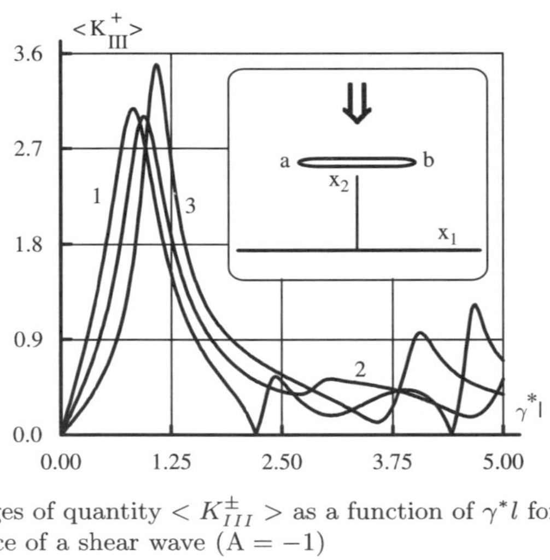

Fig. 4.14. Changes of quantity $< K^{\pm}_{III} >$ as a function of $\gamma^{*}l$ for a parabolic crack under the influence of a shear wave (A $= -1$)

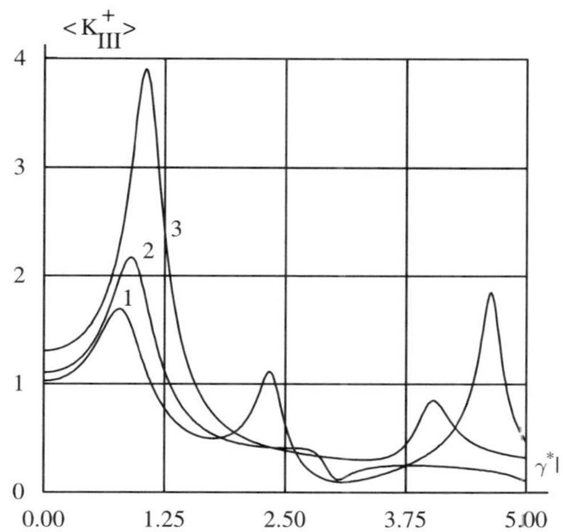

Fig. 4.15. Changes of quantity $< K^{\pm}_{III} >$ as a function of $\gamma^* l$ for a parabolic crack under the influence of shear forces $(A = -1)$

the quantity $\langle K_{III} \rangle$ may exceed its statistic analogue more than by 4–5 times.

Figures 4.18–4.21 illustrate the changes of the quantity $\langle K^{\pm}_{III} \rangle$ with respect to the orientation angle φ of linear cracks $(p_2 = 0)$ in the fixed and free halfspaces for both types of loading. In Fig. 4.18, Fig. 4.19 the lines with number m $(X_3 \neq 0)$ and Fig. 4.19 $(\tau \neq 0)$ correspond to the values of parameters $A = 1$, $h/p_1 = 3/2$, $\gamma^*\ell = 5 + (m-1)/4$ and $\gamma^*\ell = 1 + (m-1)/4$, respectively; in Fig. 4.20 $(X_3 \neq 0)$, Fig. 4.21 $(\tau \neq 0)$ shows the results for

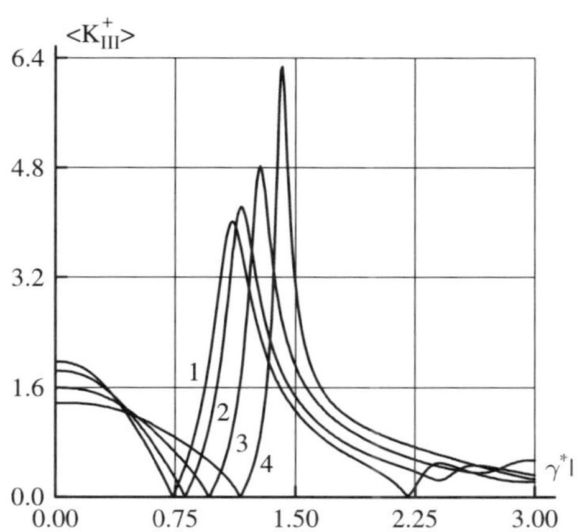

Fig. 4.16. Changes of quantity $< K^{\pm}_{III} >$ as a function of $\gamma^* l$ for a parabolic crack under the influence of a shear wave (A = 1)

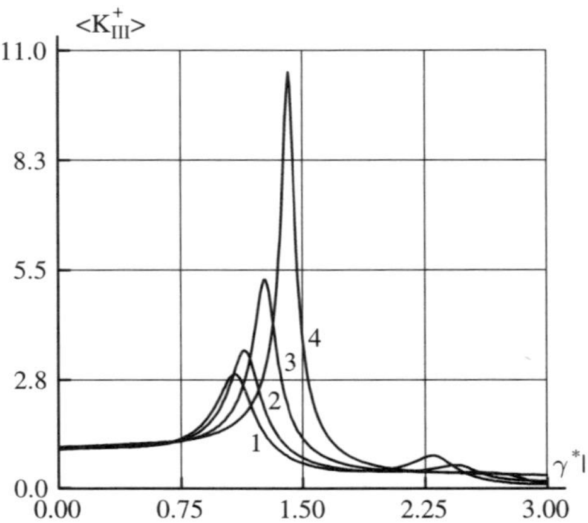

Fig. 4.17. Changes of quantity $< K^{\pm}_{III} >$ as a function of $\gamma^* l$ for a parabolic crack under the influence of shear forces $(A = -1)$

$A = -1$, $h/p_1 = 4$, $\gamma^*\ell = 0.8 + (m-1)/5$. In the case of a linear crack $(\varphi = 0)$ the propagation angle of wave $\beta = 0$ in the free halfspace of factor K_{III} is equal to zero.

Assuming that the quantity $\langle K^{\pm}_{III} \rangle$ is known, the intensity factor may be defined by $K^{\pm}_{III} = \pm\sqrt{\pi\ell}\Lambda \langle K^{\pm}_{III} \rangle \cos(\omega t - \arg\Omega_0(\pm 1))$, where $\Lambda = |X_3|$ in case $\tau = 0$, $X_3 \neq 0$ and $\Lambda = S^0_{23}$, if $\tau \neq 0$, $X_3 = 0$.

From the results it follows that in a halfspace with defect of the prescribed configuration, a series of values of a wave number exist where the factor K_{III}

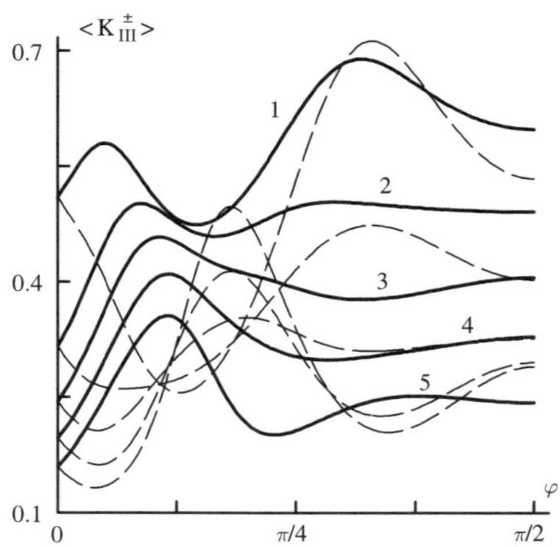

Fig. 4.18. Changes of quantity $< K^{\pm}_{III} >$ as a function of the orientation angle φ with a linear crack $(X_3 \neq 0,\ A = 1)$

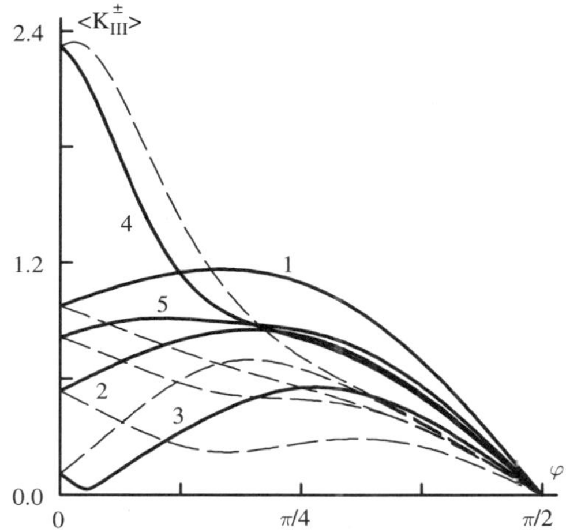

Fig. 4.19. Changes of quantity $< K^{\pm}_{III} >$ as a function of the orientation angle φ with a linear crack ($\tau \neq 0,\ A = 1$)

is close to zero. Similarly, we may note a series of values of a wave number, at which the curve of quantity K_{III} has local maximums.

The constructed algorithm corresponds to the situation where the cracks do not cross the boundary of the halfspace.

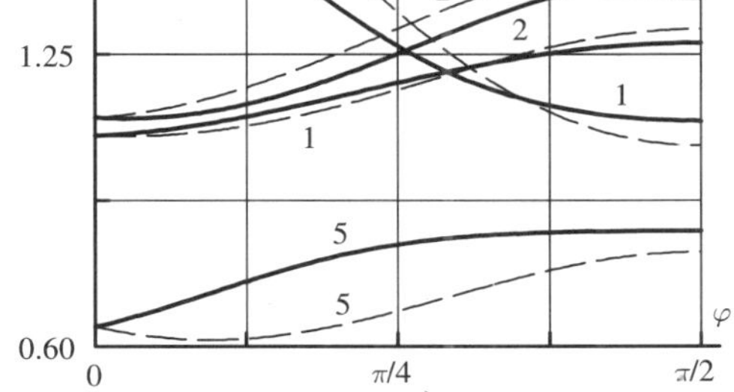

Fig. 4.20. Changes of quantity $< K^{\pm}_{III} >$ as a function of the orientation angle φ with a linear crack ($X_3 \neq 0,\ A = -1$)

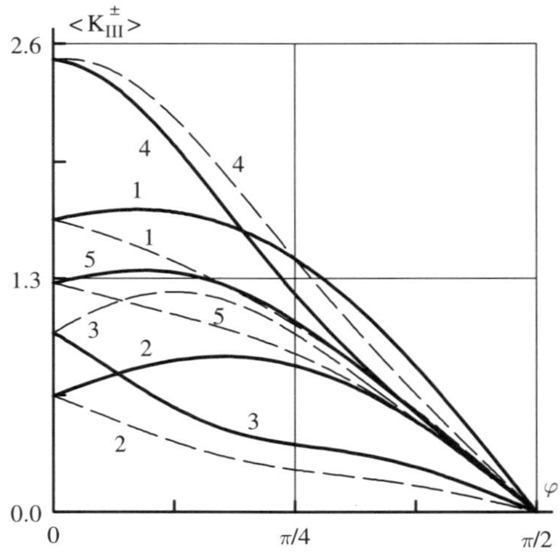

Fig. 4.21. Changes of quantity $< K_{III}^{\pm} >$ as a function of the orientation angle φ with a linear crack ($\tau \neq 0,\ A = -1$)

4.4 A Halfspace with a Crack Reaching the Boundary

Let the beginning of cut $\zeta = a$ lie on the boundary of halfspace $x_2 = 0$. In that case it is obvious that the representations (4.48) hold. Therefore, differentiating them with subsequent integration by parts of the main singular terms we obtain

$$
\begin{aligned}
\frac{\partial U_3}{\partial z} &= \frac{1}{2\pi i}\int\limits_L \frac{dp(\zeta)}{\zeta - z} + \frac{A}{2\pi i}\int\limits_L \frac{dp(\zeta)}{\overline{\zeta} - z} + \\
&+ \int\limits_L p(\zeta)\, K_1(\zeta, z) ds + \frac{p(a)}{2\pi i}\left(\frac{1}{a - z} + \frac{A}{\overline{a} - z}\right) + \frac{\partial}{\partial z}\left(U_3^{(0)} - AU_3^{(1)}\right), \\
\frac{\partial U_3}{\partial \overline{z}} &= -\frac{1}{2\pi i}\int\limits_L \frac{dp(\zeta)}{\overline{\zeta} - \overline{z}} - \frac{A}{2\pi i}\int\limits_L \frac{dp(\zeta)}{\zeta - \overline{z}} - \\
&- \int\limits_L p(\zeta)\, K_2(\zeta, z) ds - \frac{p(a)}{2\pi i}\left(\frac{1}{\overline{a} - \overline{z}} + \frac{A}{a - \overline{z}}\right) + \frac{\partial}{\partial \overline{z}}\left(U_3^{(0)} - AU_3^{(1)}\right), \\
\frac{\partial F}{\partial z} &= -\frac{1}{2\pi i}\int\limits_L \frac{f(\zeta)\, d\zeta}{\zeta - z} + \frac{A}{2\pi i}\int\limits_L \frac{\overline{f_1(\zeta)} d\overline{\zeta}}{\overline{\zeta} - z}, \\
\frac{\partial F}{\partial \overline{z}} &= \frac{1}{2\pi i}\int\limits_L \frac{\overline{f_1(\zeta)} d\overline{\zeta}}{\overline{\zeta} - \overline{z}} - \frac{A}{2\pi i}\int\limits_L \frac{f(\zeta)\, d\zeta}{\zeta - \overline{z}},
\end{aligned}
\tag{4.53}
$$

$$K_1(\zeta,z)=\frac{i\gamma^2}{8}e^{i\psi}\left[AH_0^{(1)}(\gamma r_1^*)+H_2(\gamma r)e^{-2i\alpha}\right]-$$
$$-\frac{i\gamma^2}{8}e^{-i\psi}\left[H_0^{(1)}(\gamma r)+AH_2(\gamma r_1^*)e^{-2i\alpha_1^*}\right],$$
$$K_2(\zeta,z)=\frac{i\gamma^2}{8}e^{i\psi}\left[AH_2(\gamma r_1^*)e^{2i\alpha_1^*}+H_0^{(1)}(\gamma r)\right]-$$
$$-\frac{i\gamma^2}{8}e^{-i\psi}\left[AH_0^{(1)}(\gamma r_1^*)+H_2(\gamma r)e^{2i\alpha}\right],$$
$$\alpha=\arg(z-\zeta),\quad \alpha_1^*=\arg\left(z-\overline{\zeta}\right),\quad r=|z-\zeta|,\quad r_1^*=\left|z-\overline{\zeta}\right|.$$

Substituting the limited values of the derivatives (4.53) in the boundary condition (4.32) on L, after some transformations we get a singular integro-differential equation for the function $p(\zeta)$ of the type (4.49), where instead of $H_2^{(1)}(\gamma r_{10})$ we have $H_2(\gamma r_{10})$ and

$$g(\zeta,\zeta_0)=\mathrm{Im}\frac{e^{i\psi_0}}{\zeta-\zeta_0}+A\mathrm{Im}\frac{e^{i\psi_0}}{\overline{\zeta}-\zeta_0} \tag{4.54}$$

For the derivation of the above equation it was taken into account that for a fixed halfspace is $p(a)=0$, and under this assumption the non-integral terms are vanished. In the case of a free halfspace these terms also vanish. In (4.54) the first component has mobile singularity of Cauchy type, and in the second-immobile singularity at point $\zeta=\zeta_0=a$. The theory of singular integral equations with mobile singularity is developed in [66].

The function $p'(\zeta)$ in the vicinity of tip b exhibits power singularity (the order of singularity $\sigma=1/2$). In the vicinity of tip a let us present this function in the form ($\varphi\in H$ on arc $[a,b)$)

$$p'(\zeta)=\frac{dp}{ds}=\frac{\varphi(\zeta)}{(\zeta-a)^{\sigma}},\quad \mathrm{Im}\sigma=0,\quad 0\le\sigma\le1. \tag{4.55}$$

Applying the formulas for asymptotic values of the integral of Cauchy type on the line of integrating [50] we find

$$\int_L p'(\zeta)\mathrm{Im}\frac{e^{i\psi_0}}{\zeta-\zeta_0}d\zeta=\varphi_1(\zeta_0)-\frac{\pi\varphi(a)}{(\zeta_0-a)^{\sigma}}\mathrm{ctg}\,\pi\sigma, \tag{4.56}$$
$$\int_L p'(\zeta)\mathrm{Im}\frac{e^{i\psi_0}}{\overline{\zeta}-\zeta_0}d\zeta=\varphi_2(\zeta_0)+\frac{\pi\varphi(a)\cos\left[2(\sigma-1)\psi(a)\right]}{(\zeta_0-a)^{\sigma}\sin\pi\sigma}.$$

where $\psi(a)$ is the value of $\psi(\zeta)$ at point $\zeta=a$; the functions $\varphi_i(\zeta)$ may exhibit singularity at point $\zeta=a$, but more weak than $(\zeta-a)^{-\sigma}$. Substituting (4.56) into (4.49) considering (4.54), multiplying the left and right part of the obtained equation with $(\zeta_0-a)^{\sigma}$ and satisfying the limiting transition at $\zeta_0\to a$, we obtain

$$A\cos\left[2(\sigma-1)\psi(a)\right]-\cos\pi\sigma=0. \tag{4.57}$$

The analysis of solution of this equation for case $A = 1$ gives

$$\sigma = \begin{cases} \psi(a)\left[\psi(a) + \pi/2\right]^{-1}, & 0 \le \psi(a) < \pi/2, \\ \psi(a)\left[\psi(a) - \pi/2\right]^{-1}, & -\pi/2 < \psi(a) \le 0. \end{cases} \tag{4.58}$$

At $A = -1$ there are no solutions of transcendental (4.57) satisfying condition $0 < \sigma < 1$.

Hence, in the case of an isotropic halfspace with a crack reaching the boundary [175], the stress in the free from forces piezoceramic halfspace are limited in the vicinity of reaching the tip boundary; in the case of a fixed halfspace a power singularity of order σ exists, determined by formula (4.58).

4.5 Harmonic Excitation of a Halfspace by External Sources

In real systems (*electroacoustic transducer*) the sources of oscillations in many cases may be considered linear. If a linear source is located along the symmetry axis of a piezoelectric medium with hexagonal symmetry, in its interior connected electric and acoustic oscillations corresponding to the antiplane deformation will be excited. The initial system of equations for the displacement amplitude and electric potential has the following form [52]

$$\begin{aligned} & c_{44}^{E}\nabla^2 U_3 + e_{15}\nabla^2\phi_* + \rho\omega^2 U_3 = -P_0\delta(x_1 - x_{10}, x_2 - x_{20}), \\ & e_{15}^{\nabla 2} U_3 - \ni_{11}^{\varepsilon}\nabla^2\phi_* = Q_0\delta(x_1 - x_{10}, x_2 - x_{20}). \end{aligned} \tag{4.59}$$

Here P_0 and Q_0 are the linear densities of the concentrated forces and charges acting at the point $z_0 = x_{10} + ix_{20}$ of the medium; $\delta(x, y) = \delta(x)\delta(y)$ is the Dirac delta function. The solution of (4.59) is found simply by

$$\begin{aligned} U_3(x_1, x_2) &= \left(\frac{k_{15}^2 Q_0}{4ie_{15}(1 + k_{15}^2)} - \frac{P_0}{4ic_{44}^{E}(1 + k_{15}^2)} \right) H_0^{(1)}(\gamma r), \\ \phi_*(x_1, x_2) &= -\frac{Q_0}{2\pi \ni_{11}^{\varepsilon}}\ell n r + \frac{i}{4(1 + k_{15}^2)\ni_{11}^{\varepsilon}}\left[\frac{e_{15}P_0}{c_{44}^{E}} - k_{15}^2 Q_0\right] H_0^{(1)}(\gamma r), \\ r &= |z - z_0|, \quad z = x_1 + ix_2. \end{aligned} \tag{4.60}$$

In the case where the source is concentrated on the line $x_1 = x_{10}$, $x_2 = 0$, $-\infty < x_3 < \infty$ on the boundary of a free halfspace, the solution may be found with the help of the method of images. If we assume that the external medium is vacuum, we get equations of type (4.60), where r is substituted by $r^* = |z - x_{10}|$ and also the amplitudes P_0 and Q_0 are doubled.

To investigate the influence of external sources on stress intensity factor K_{III} in a halfspace with cracks we will calculate the corresponding right part of integral (4.49). Considering combination $\sigma_n = \sigma_{13}\cos\psi + \sigma_{23}\sin\psi$ in

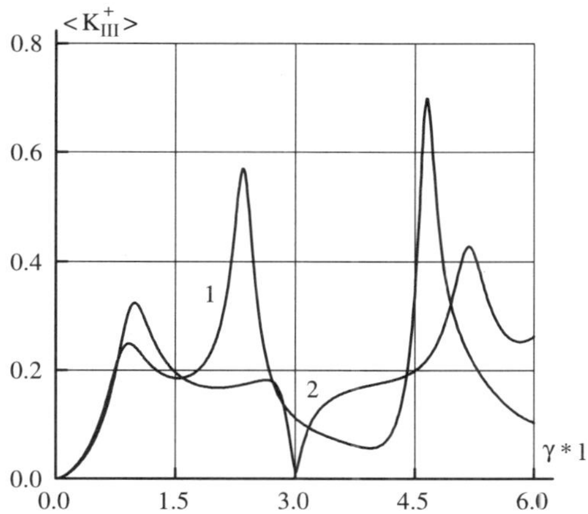

Fig. 4.22. Changes of quantity $< K_{III}^{\pm} >$ as a function of $\gamma^{*}l$ in a space with a parabolic crack under the influence of concentrated electric charges

the edge condition (4.3) allowing for constitutive relation (4.22) and formula (4.60) we find [180]

$$N(\zeta_0) = \left[\frac{\pi i \gamma P_0}{2c_{44}^{E}} H_1^{(1)}(\gamma r_0^{*}) + \frac{\pi k_{15}^{2} Q_0}{2e_{15}} \left(i\gamma H_1^{(1)}(\gamma r_0^{*}) - \frac{2}{\pi r_0^{*}} \right) \right] \cos(\psi_0 - \alpha_0^{*}),$$

$$r_0^{*} = |\zeta_0 - x_{10}|, \quad \alpha_0^{*} = \arg(\zeta_0 - x_{10}), \quad \psi_0 = \psi(\zeta_0). \tag{4.61}$$

Thus, the algorithm for solving the problem is reduced to the definition of the function $p(\zeta)$ from integral (4.49) where the right part is given by

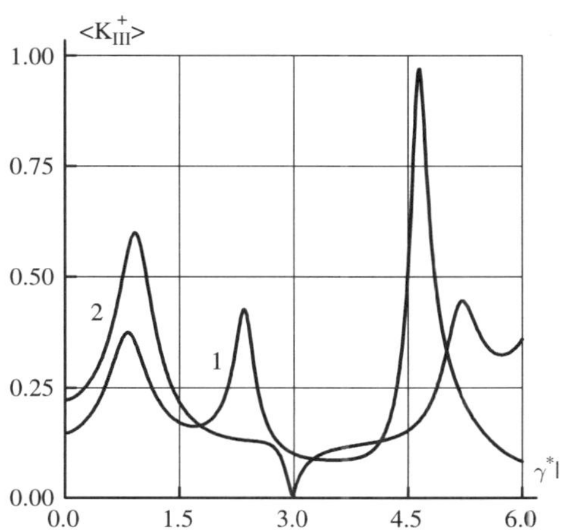

Fig. 4.23. Changes of quantity $< K_{III}^{\pm} >$ as a function of $\gamma^{*}l$ in a space with a parabolic crack under the influence of concentrated shear forces

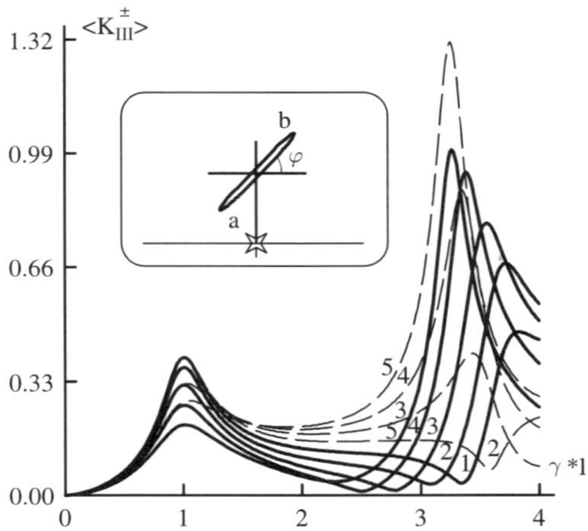

Fig. 4.24. Behavior of $< K_{III}^{\pm} >$ as a function of $\gamma^* l$ for various orientation angles of crack in a halfspace, ($Q_0 \neq 0$)

(4.61). The stress intensity factor K_{III} is found by the formula (4.51). If it is necessary to calculate the values of field quantities of internal area we should use integral representations (4.48) and formulas (4.25), (4.29).

As an example let us investigate the situation where the piezoceramic halfspace from ceramics PZT-4 is weakened by a parabolic crack ($\xi_1 = p_1\delta,\ \xi_2 = p_2\delta^2 + h, -1 \leq \delta \leq 1$). Figs. 4.22 and 4.23 illustrate changes of the quantity $\langle K_{III}^{+} \rangle = e_{15}\sqrt{\ell}\,|\Omega_0(1)| \big/ \left(|Q_0|\sqrt{s'(1)}\right)$ and $\langle K_{III}^{+} \rangle = c_{44}^{E}\sqrt{\ell}\,|\Omega_0(1)|$

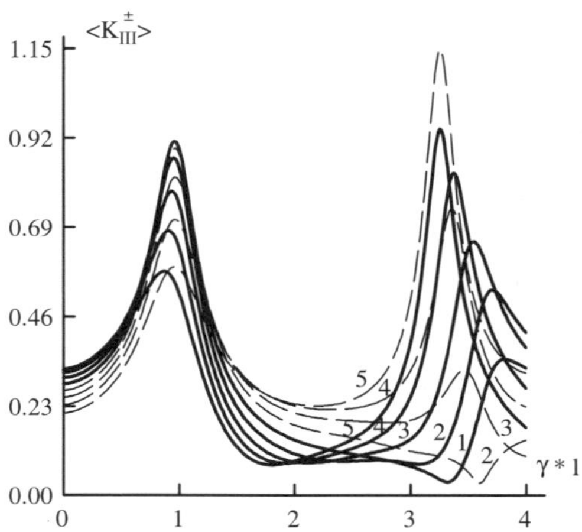

Fig. 4.25. Behaviour of $< K_{III}^{\pm} >$ as a function of $\gamma^* l$ for various orientation angles of a crack in a halfspace ($P_0 \neq 0$)

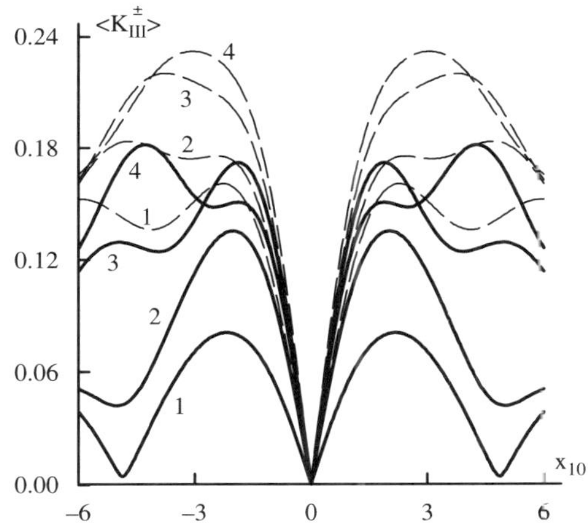

Fig. 4.26. Changes of quantity $< K^{\pm}_{III} >$ according to the coordinates of application of electric charges on the boundary of a halfspace for various values of $\gamma^* l$

$/\left(|P_0|\sqrt{s'(1)}\right)$, respectively, with respect to the normalized wave number $\gamma^* \ell = \gamma \ell \sqrt{1 + k_{15}^2}$ in the case where on the boundary of halfspace ($x_{10} = 0$) a linear source of electric charge ($Q_0 = \text{const}$, $P_0 = 0$) type and shear force ($P_0 = \text{const}$, $Q_0 = 0$) occur. Curves 1, 2 are constructed for the parameter values $h/p_1 = 2$, $p_*/p = 0, -0.6$.

Figures 4.24 and Fig. 4.25 For a parabolic crack orientated at angle φ to axis x_1, the behavior of $\langle K_{III} \rangle$ is shown in Figs. 4.24 ($Q_0 \neq 0$) and 4.25 ($P_0 \neq 0$). The line with number m corresponds to the parameters $h/p_1 =$

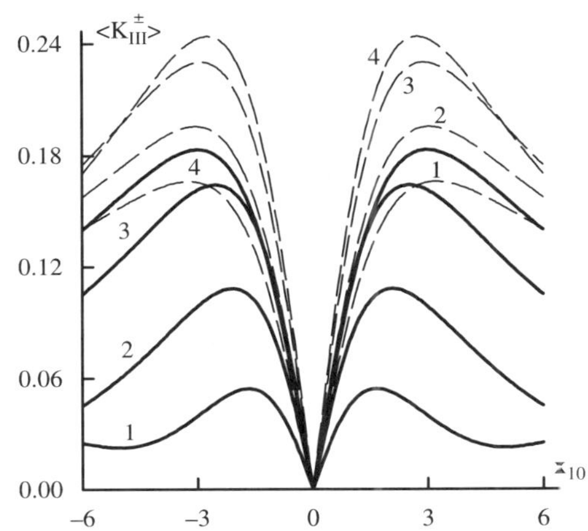

Fig. 4.27. Changes of quantity $< K^{\pm}_{III} >$ according to the coordinates of application of shear forces on the boundary of a halfspace for various values of $\gamma^* l$

1.5, $p_2/p_1 = -0.5$, $\varphi = 0.05\pi\,(m-1)$. The graphs characterizing the changes of $\langle K^{\pm}_{III}\rangle$ as a function of the coordinate of application of the current source of a charge type and shear force are represented in Figs. 4.26 and 4.27, respectively. Curves 1–4 correspond to parameters $p_*/p = 0$, $h/p = 3$ and $\gamma^*\ell = 2, 2.3, 2.6, 2.9$.

In Figs. 4.24–4.27 The solid and dashed lines are constructed for crack tips a and b, respectively.

The stress intensity factor K_{III} for both types of loading is determined by

$$K^{\pm}_{III} = \pm\sqrt{\frac{\pi}{\ell}}\frac{c^E_{44}}{e_{15}}\,|Q_0|\,\langle K^{\pm}_{III}\rangle \cos\left[\omega t - \arg\left(\Omega_0\left(\pm 1\right)\right)\right],$$

$$K^{\pm}_{III} = \pm\sqrt{\frac{\pi}{\ell}}\,|P_0|\,\langle K^{\pm}_{III}\rangle \cos\left[\omega t - \arg\left(\Omega_0\left(\pm 1\right)\right)\right].$$

The results shown in Figs. 4.22 and 4.23 confirm the conclusion that the crack curve in dynamic priblems may influence the behavior of the stress intensity factor.

4.6 Arbitrary with Time Excitation of a Halfspace

The analysis of electroacoustic fields in piezoelectric bodies with cracks under the action of dynamic loading changing with time arbitrarily are of great interest as it helps to realize deeply the processes preceding the brittle fracture. As in the case of harmonic effects, the problems of shock loading of bodies with cracks due to complexity of the appearing problems are solved analytically only in case of some idealized statements. These problems also include problems for infinite areas. The stress intensity factor in the case of a crack with a finite length loaded by a shock impulse of longitudinal shear is determined in [235].

Below we investigate the behavior of the stress intensity factor K_{III} under the dynamic influence on a piezoceramic halfspace containing a stationary crack of an infinite length [183].

Let on the line $x_1 = x_{10}$, $x_2 = 0, -\infty < x_3 < \infty$ of a boundary free from a bounded with vacuum halfspace to act shear forces $P\left(x_1, t\right) = P_0\left(t\right)\delta\left(x_1 - x_{10}\right)$, or electric charges $Q\left(x_1, t\right) = Q_0\left(t\right)\delta\left(x_1 - x_{10}\right)$ arbitrarily changing with time. In this case the solution i is the superposition of "elementary" solutions along the whole spectrum of frequencies. Applying Fourier's integral transform [165]

$$\{\widehat{U_3}\left(x_1, x_2, \omega\right), \hat{\phi}\left(x_1, x_2, \omega\right)\} = \frac{1}{\sqrt{2\pi}}\int\limits_0^\infty \{u_3\left(x_1, x_2, t\right), \phi\left(x_1, x_2, t\right)\}e^{i\omega t}dt,$$

$$\{u_3\left(x_1, x_2, t\right), \phi\left(x_1, x_2, t\right)\} = \sqrt{\frac{2}{\pi}}\mathrm{Re}\int\limits_0^\infty \left\{\widehat{U_3}\left(x_1, x_2, \omega\right), \hat{\phi}\left(x_1, x_2, \omega\right)\right\}e^{-i\omega t}d\omega, \tag{4.62}$$

we first consider the situation where on the boundary of a homogeneous piezoelectric halfspace (without defects) at point $x_1 = x_{10}$ acts a linear source. By considering the same assumptions of Sect. 4.5 we find that the corresponding displacement transformants and electric potentials are given by

$$\widehat{U_3^0}(x_1, x_2, \omega) = \left(\frac{k_{15}^2 \widehat{Q}_0(\omega)}{2ie_{15}(1+k_{15}^2)} - \frac{\widehat{P}_0(\omega)}{2ic_{44}^E(1+k_{15}^2)} \right) H_0^{(1)}(\gamma r^*), \tag{4.63}$$

$$\widehat{\phi^0}(x_1, x_2, \omega) = -\frac{\widehat{Q}_0(\omega)}{\pi_{11}^\varepsilon} \ell n r^* +$$

$$+ \frac{i}{2_{11}^\varepsilon(1+k_{15}^2)} \left[\frac{e_{15}\widehat{P}_0(\omega)}{c_{44}^E} - k_{15}^2 \widehat{Q}_0(\omega) \right] H_0^{(1)}(\gamma r^*),$$

$$r^* = |z - x_{10}|, \quad z = x_1 + ix_2.$$

In the case of a halfspace with cracks where a source on its boundary acts we obtain a integro-differential equation of type (4.49), the right part of which has the following form

$$N(\zeta_0, \omega) = \left[\frac{\pi i \gamma \widehat{P_0}(\omega)}{2c_{44}^E} H_1^{(1)}(\gamma r_0^*) + \frac{\pi k_{15}^2 \widehat{Q}_0(\omega)}{2e_{15}} \left(i\gamma H_1^{(1)}(\gamma r_0^*) - \frac{2}{\pi r_0^*} \right) \right]$$

$$\cos(\psi_0 - \alpha_0^*), \tag{4.64}$$

$$r_0^* = |\zeta_0 - x_{10}|, \quad \alpha_0^* = \arg(\zeta_0 - x_{10}).$$

In order to calculate the stress intensity factor K_{III} we derive the main asymptotic of shear stress transformant on the continuation across the crack tip. iIt must be taken into account that for the function $\hat{U}_3^*(x_1, x_2, \omega)$ corresponding to the defect dispersed shear field, the representation (4.48) (when non-integral compounds are lacking) is valid at $A = -1$. Due to (4.22) we have (leaving the compounds causing asymptotic)

$$\widehat{\sigma}_n = c_{44}^E \frac{\partial \widehat{U}_3^*}{\partial n} + \ldots = c_{44}^E \left[e^{i\psi_c} \frac{\partial \widehat{U}_3^*}{\partial z} + e^{-i\psi_c} \frac{\partial \widehat{U}_3^*}{\partial \bar{z}} \right] + \ldots, \tag{4.65}$$

where c is the cut tip, $\psi_c = \psi(c)$.

Allowing for expressions (4.53) let us write out the main part of function (4.65)

$$\widehat{\sigma}_n^0 = \frac{c_{44}^E}{\pi} \int_L p'(\zeta) \operatorname{Im} \frac{e^{i\psi_c}}{\zeta - z} ds. \tag{4.66}$$

Applying the asymptotic formulas for the integrals and contour parameterization $L\left(\zeta=\zeta\left(\delta\right)\right)$ appearing in (4.66) we find

$$\widehat{\sigma}_n^0 = \pm c_{44}^E \frac{\Omega_0\left(\pm 1,\omega\right)}{\sqrt{2rs'\left(\pm 1\right)}}, \quad p'\left(\zeta\right) = \frac{\Omega_0\left(\delta,\omega\right)}{s'\left(\delta\right)\sqrt{1-\delta^2}}, \tag{4.67}$$

$$s'\left(\pm 1\right) = \left.\frac{ds}{d\delta}\right|_{\delta=\pm 1} \qquad \left(-1 \le \delta \le 1\right).$$

where $r=|z-c|$, the lower sign corresponds to the tip $c=a$, and the upper one to $c=b$.

Let us introduce the "*standard*" solution Ω_1 and Ω_2 of the integral (4.49) with the right part (4.64) given by the formula

$$\Omega_0\left(\delta,\omega\right) = \frac{\hat{P}_0\left(\omega\right)}{c_{44}^E}\Omega_1\left(\delta,\omega\right) + \frac{\hat{Q}_0\left(\omega\right)}{e_{15}}\Omega_2\left(\delta,\omega\right). \tag{4.68}$$

The functions $\Omega_j\left(\delta,\omega\right), (j=1,2)$ corresponds to the following right parts of the integral (4.49)

$$N_1\left(\zeta_0,\omega\right) = \frac{\pi i\gamma}{2}H_1^{(1)}\left(\gamma r_0^*\right)\cos\left(\psi_0-\alpha_0^*\right), \tag{4.69}$$

$$N_2\left(\zeta_0,\omega\right) = \frac{\pi k_{15}^2}{2}\left(i\gamma H_1^{(1)}\left(\gamma r_0^*\right) - \frac{2}{\pi r_0^*}\right)\cos\left(\psi_0-\alpha_0^*\right).$$

According to (4.67), (4.68) we have

$$\widehat{K}_{III}^{\pm}\left(\omega\right) = \lim_{r\to 0}\sqrt{2\pi r}\,\widehat{\sigma}_n^0 = \tag{4.70}$$

$$= \pm\sqrt{\frac{\pi}{s'\left(\pm 1\right)}}\left[\widehat{P}_0\left(\omega\right)\Omega_1\left(\pm 1,\omega\right) + \frac{c_{44}^E}{e_{15}}\hat{Q}_0\left(\omega\right)\Omega_2\left(\pm 1,\omega\right)\right].$$

Using the initial notation from (4.62) we find stress intensity factor

$$K_{III}^{\pm}\left(t\right) = \pm\sqrt{\frac{2}{s'\left(\pm 1\right)}}\operatorname{Re}\int_0^\infty \left[\hat{P}_0\left(\omega\right)\Omega_1\left(\pm 1,\omega\right) + \right.$$

$$\left. + \frac{c_{44}^E}{e_{15}}\hat{Q}_0\left(\omega\right)\Omega_2\left(\pm 1,\omega\right)\right]e^{-i\omega t}d\omega. \tag{4.71}$$

If in the integral (4.49) with the right part (4.64) we substitute ω by $-\omega$, the solution of this new equation $\Omega_0^*\left(\omega\right) = \Omega_0\left(-\omega\right)$ becomes equal to $\overline{\Omega_0\left(\omega\right)}$. This follows from the equation [57]

$$H_\nu^{(1)}\left(e^{i\pi}z\right) = -e^{-i\pi\nu}H_\nu^{(2)}\left(z\right), \quad \overline{H_\nu^{(2)}}\left(z\right) = H_\nu^{(2)}\left(\overline{z}\right),$$

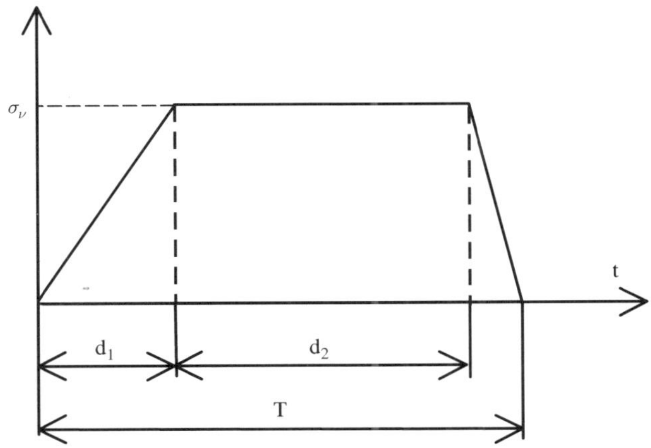

Fig. 4.28. The scheme of a trapezoidal impulse

and also from the relation in (4.49) connecting $H_2^{(1)}(\gamma r)$ with $H_2(\gamma r)$. Based on the above the quantity K_{III} may be represented in the form

$$K_{III}^{\pm}(t) = \pm\sqrt{\frac{1}{2s'(\pm 1)}} \int\limits_{-\infty}^{\infty} \left[\hat{P}_0(\omega)\,\Omega_1(\pm 1,\omega) + \right. \tag{4.72}$$

$$\left. + \frac{c_{44}^E}{e_{15}}\hat{Q}_0(\omega)\,\Omega_2(\pm 1,\omega)\right] e^{-i\omega t} d\omega.$$

Example. Let an electroelastic field excited by an impulse loading of trapezoidal form in a halfspace (space) weakened by a single linear crack with parametric equations $\xi_1 = \delta, \xi_2 = h\,(-1 \le \delta \le 1)$ (see Fig. 4.28). The corresponding spectral function of the given impulse has the form

$$R_\nu(\omega) = \frac{\sigma_\nu}{\omega^2\sqrt{2\pi}}\left[d_1^{-1}\left(e^{i\mu_1}-1\right) - \mu_2\left(e^{i\omega T} - e^{i\mu_3}\right)\right], \tag{4.73}$$

$$R_1(\omega) = \widehat{P}_0(\omega), \quad R_2(\omega) = \widehat{Q}_0(\omega), \quad \mu_1 = \omega d_1,$$

$$\mu_2 = (T - d_1 - d_2)^{-1}, \quad \mu_3 = \omega(d_1 + d_2) \quad (\nu = 1, 2).$$

In case wheh $d_2 = 0$ we deduce the spectral functions of loading given in the form of a triangle impulse.

At this point it is helpful to introduce the dimensionless time parameters (c_* are the velocity of a shear wave in a piezoelectric medium)

$$t^* = c_* t h^{-1}, \quad d_\nu^* = c_* d_\nu h^{-1}, \quad T^* = c_* T h^{-1} \quad (\nu = 1, 2).$$

Substituting (4.73) into (4.71) we obtain the formula for the stress intensity factor $K_{III}(t)$. Appearing in it "*standard*" solutions were calculated from the system of (4.49), (4.50) using (4.69) by the scheme of the method of quadrature (see: Appendix B 12.8, Par. B.1). A half infinite interval of integrating in (4.71) was substituted by finite section $[0, \omega^*]$; the quantity ω^* was determined numerically by minimizing the error.

Figure 4.29 illustrates the changes of quantities $\lambda = \sqrt{\pi \ell}\, K_{III}^{+}/\Lambda$ as a function of t^* for the case where the shear force (curve 1) or the electric charge (curve 2) act in a piezoceramic space with a linear crack. It was assumed that the material under study was PZT-4 ceramics parallel to the axis of the crack abscissa, is free from force and its center is at a distance of $h/\ell = 3$ (2ℓ is the crack length) from the line of the source action ($x_{10} = 0$). The values of the impulse parameters were set to $T^* = 10, d_1^* = 1, d_2^* = 8, \sigma_1^* = 1N/\left(s \cdot m\right), \sigma_2^* = 1cal\left(s \cdot m\right) \sigma_\nu^* = c_0 \sigma_\nu h^{-1} (v = 1,\ 2)$. $\Lambda = \sigma_1$ if $P_0(t) \neq 0, Q_0(t) = 0$ and $\Lambda = \sigma_2\, c_{44}^E/e_{15}$ if $P_0\left(t\right) \neq 0,\ Q_0\left(t\right) = 0$ and $\Lambda = \sigma_2 c_{44}^E/e_{15}$ if $P_0\left(t\right) = 0, Q_0\left(t\right) \neq 0$.

As it follows from Fig. 4.29 the waves of stress don't have the time to reach the crack tips under the action of shear forces and for time $t < c_*^{-1}\sqrt{h^2 + \ell^2}$, therefore $K_{III} = 0$. In the time interval $c_*^{-1}\sqrt{h^2 + \ell^2} < t < 2.22 c_*^{-1} h$ the quantity λ increases rapidly up to their global maximum $\lambda_{\max} \approx 0.28$, and then, some time later tends to the "static" value of $\lambda_0 \approx 0.16$, corresponding to the time interval $6c_*^{-1} h < t < c_*^{-1}\left(9h + \sqrt{h^2 + \ell^2}\right)$. After the impulse $\left(t > 10 c_*^{-1} h\right)$ goes away the quantity λ decreases to the value $\lambda_{\min} \approx -0.12$ and in the consequent time the stress intensity factor tends to zero due to the energy dispersion at infinity. Under the action of an electric charge we get a quite different picture. Due to the quasistatic character of the process appearing at the initial moment of time $t = 0$ the electric field of the charge influences the crack instantaneously causing stresses in its vicinity. It is interesting to note that in time period $6c_*^{-1} h < t < 9c_*^{-1} h$, corresponding to "static" behavior of electroacoustic fields the stress intensity factor is close to

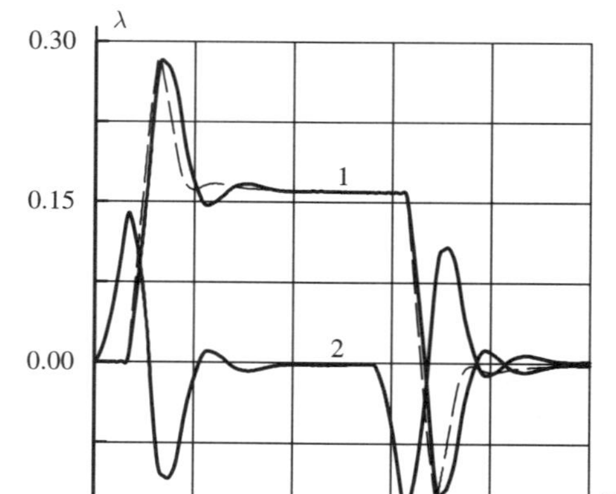

Fig. 4.29. Changes of quantity $\lambda = \sqrt{\pi l K_{III}^{+}}/\Lambda$ as a function of t^* in a space with a linear crack

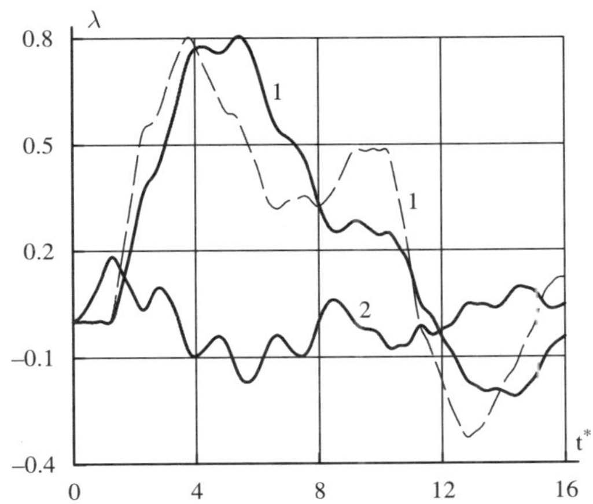

Fig. 4.30. Behaviour of quantity λ as a function of t^* in a halfspace with a parabolic crack

zero (Fig. 4.29). This situation is conditioned by the fact that in statics due to condition of antiplane deformation the electric excitation of the piezomedium does not causes stresses in it.

Figure 4.30 depicts the behavior of quantity λ in a halfspace with a parabolic crack $\left(\xi_1 = p_1\delta,\ \xi_2 = p_2\delta^2 + h, -1 \leq \delta \leq 1\right)$ in the case where the line of the source action is in the parabola focus $(h/p_1 = 1,\ p_2/p_1 = -0.25)$. The rather chaotic state appearing in the above figure is due to the appearence of the reflected, from the boundary, halfspace waves.

To estimate the influence of the piezoelectric effect on the stress intensity factor in Figs. 4.29 and 4.30, dashed lines corresponding to piezopassive media $(e_{15} = 0)$, in case when $P_0(t) \neq 0,\ Q_0(t) = 0$ are given.

At this point we should mention that the developed approach to the analysis of non-stationary dynamic electroelastic fields in the medium with cracks are effective and may be generalized for several cracks of various configurations.

4.7 A Layer

Let the piezoceramic layer $0 \leq x_1 \leq a, -\infty < x_2 < \infty, -\infty < x_3 < \infty$ defined in the Cartesian system of coordinates with a tunnel along x_3-axis; the cracks $L_j\ (j = 1, 2, \ldots, k)$ are excited by radiating at infinity a monochromic shear wave (see Fig. 4.31). On the boundaries of the cracks there may also act shear stress $X_{3n} = X_{3n}^+ = -X_{3n}^- = \mathrm{Re}\left(X_3 e^{-i\omega t}\right)$ harmonically changing with time and constant along axis x_3. Let us suppose that the main layer is not subjected to any force and is bounded with vacuum (the direction cf polarization of the ceramics is parallel to axis x_3). It is required to determine the electroacoustic wave fields and stress intensity factor K_{III}.

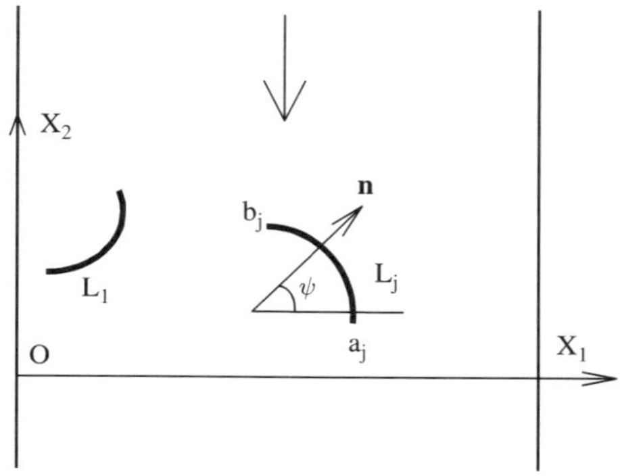

Fig. 4.31. The scheme of a layer with curvilinear cracks

As it was mentioned above, the given problem includes the integration of (4.46) at the corresponding boundary conditions on the cuts of (4.32) type, and also at mechanical and electric conditions on the basis of a layer. The latter may be formally represented in the form of

$$\sigma_{13} = 0, \quad D_1 = 0 \quad (x_1 = 0, a)\,. \tag{4.74}$$

Let us now construct the Green's function for a piezoelectric layer. The Boundary problems (4.46) and (4.74) with respect to the amplitudes using (4.29) are given by

$$\nabla^2 U_3 + \gamma^2 U_3 = 0, \quad \partial_1 U_3 = 0 \quad (x_1 = 0, a)\,, \tag{4.75}$$

$$\nabla^2 F = 0, \quad \partial_2 F = 0 \quad (x_1 = 0, a)\,. \tag{4.76}$$

Green's functions corresponding to problems (4.75), (4.76) read [147]

$$G(\zeta, z) = \sum_{\nu=0}^{\infty} b_\nu (x_2 - \xi_2) \cos \alpha_\nu \xi_1 \cos \alpha_\nu x_1, \tag{4.77}$$

$$E(\zeta, z) = \sum_{\nu=1}^{\infty} d_\nu (x_2 - \xi_2) \sin \alpha_\nu \xi_1 \sin \alpha_\nu x_1,$$

$$\nabla^2 G + \gamma^2 G = \delta(x_1 - \xi_1, x_2 - \xi_2)\,, \quad \alpha_\nu = \frac{\pi\nu}{a},$$

$$\nabla^2 E = \delta(x_1 - \xi_1, x_2 - \xi_2) = \delta(x_1 - \xi_1)\,\delta(x_2 - \xi_2)\,,$$

$$z = x_1 + i x_2, \quad \zeta = \xi_1 + i\xi_2.$$

where $\delta(x)$ is $2a$ - periodical Dirac δ - function.

Using the following expansion

$$\delta\left(x_1-\xi_1\right)=\frac{1}{a}+\frac{2}{a}\sum_{\nu=1}^{\infty}\cos\alpha_\nu\xi_1\cos\alpha_\nu x_1, \tag{4.78}$$

$$\delta\left(x_1-\xi_1\right)=\frac{2}{a}\sum_{\nu=1}^{\infty}\sin\alpha_\nu\xi_1\sin\alpha_\nu x_1,$$

If we separate the variables in (4.75), (4.76) and then apply the procedure of the fundamental solution of a usual differential equation we find

$$\begin{aligned} b_\nu &= -\frac{1}{a\lambda_\nu}e^{-\lambda_\nu|x_2-\xi_2|}, \quad b_0=\frac{1}{2ia\gamma}e^{i\gamma|x_2-\xi_2|},\\ d_\nu &= -\frac{1}{a\alpha_\nu}e^{-\alpha_\nu|x_2-\xi_2|},\\ \lambda_\nu &= \begin{cases}\sqrt{\alpha_\nu^2-\gamma^2}, & \gamma<\alpha_\nu\\ -i\sqrt{\gamma^2-\alpha_\nu^2}, & \gamma>\alpha_\nu\end{cases} \qquad (\nu=1,2,\ldots). \end{aligned} \tag{4.79}$$

A series of function $E(\zeta,\ z)$ in (4.77) allowing for (4.79) is easily summed using equality

$$\sum_{m=1}^{\infty}\frac{e^{-m|x|}}{m}\cos\ my=\frac{|x|}{2}-\frac{1}{2}\ell n\left[2\left(\operatorname{ch}x-\cos y\right)\right] \tag{4.80}$$

and has the form

$$\begin{aligned} E\left(\zeta,z\right) &= \frac{1}{2\pi}\ell n\left|\sin\frac{\pi\left(\zeta-z\right)}{2a}\sec\frac{\pi\left(\zeta+\overline{z}\right)}{2a}\right|,\\ z &= x_1+ix_2,\quad \overline{z}=x_1-ix_2. \end{aligned} \tag{4.81}$$

To single out the main part of function $G(\zeta,z)$ we write down Green's function G_0 of the senior operator in Helmhöltz (4.77). Summing up the corresponding series by using (4.80) we obtain

$$\begin{aligned} G_0 &= -\frac{1}{a}\sum_{m=1}^{\infty}a_m\left(x_1,\xi_1\right)e^{-\alpha_m|x_2-\xi_2|}=-\frac{|x_2-\xi_2|}{2a}+\\ &+\frac{1}{2\pi}\ell n\left|4\sin\frac{\pi\left(\zeta-z\right)}{2a}\sin\frac{\pi\left(\zeta+\overline{z}\right)}{2a}\right|,\\ a_m &= \frac{\cos\alpha_m\xi_1\cos\alpha_m x_1}{\alpha_m}. \end{aligned} \tag{4.82}$$

Allowing for (4.79), (4.82) let us represent function G in (4.77) in its final form

$$G(\zeta, z) = G_0 + G_1, \tag{4.83}$$

$$G_1 = \frac{1}{2ia\gamma} e^{i\gamma|x_2-\xi_2|} - \frac{1}{a}\sum_{m=1}^{\infty} c_m(x_2-\xi_2)\cos\alpha_m\xi_1\cos\alpha_m x_1,$$

$$c_m(x_2-\xi_2) = \frac{1}{\lambda_m} e^{-\lambda_m|x_2-\xi_2|} - \frac{1}{\alpha_m} e^{-\alpha_m|x_2-\xi_2|} \quad (m = 1, 2, \ldots).$$

Thus, the functions $E(\zeta, z)$ and $G(\zeta, z)$ determined by formulas (4.81) – (4.83) are the Green's functions of the boundary problems (4.75), (4.76) for a strip. The conditions of radiation in problem (4.75) and damping in problem (4.76) are satisfied. After we examine the main singularities in (4.77) the general compound of the series in (4.83) damps at point $z = \zeta$ as m^{-3}.

Summarized displacement field U_3 will be found similarly to [147]

$$U_3(x_1, x_2) = \int_L [U_3] \frac{\partial G(\zeta, z)}{\partial n_\zeta} ds + U_3^{(0)}, \quad \zeta \in L. \tag{4.84}$$

Here the unknown quantity $[U_3]$ has the meaning of a jump of displacement amplitude U_3 on L, the integral term reflects the cut dispersed field and compound $U_3^{(0)} = \tau \exp(-i\gamma x_2)$ corresponds to a falling shear wave.

The function F is represented in the following way

$$F(x_1, x_2) = \int_L f(\zeta) E(\zeta, z)\, ds, \tag{4.85}$$

Where ds is the element of arc contour length L.

In order to give the physical meaning of "*density*" $f(\zeta)$ in (4.85) let us first calculate the derivatives $\partial U_3/\partial z$, $\partial U_3/\partial \overline{z}$. As a result, after some transformations to improve the convergence of slowly converging series we obtain

$$\frac{\partial U_3}{\partial z} = \frac{\pi}{8ia^2}\int_L p(\zeta)\,\mathrm{cosec}^2\frac{\pi(\zeta - z)}{2a} d\zeta +$$
$$+ \int_L p(\zeta)\left(R_1 e^{i\psi} + R_3 e^{-i\psi}\right) ds - \frac{\tau\gamma}{2} e^{-i\gamma x_2},$$

$$\frac{\partial U_3}{\partial \overline{z}} = -\frac{\pi}{8ia^2}\int_L p(\zeta)\,\mathrm{cosec}^2\frac{\pi\left(\overline{\zeta} - \overline{z}\right)}{2a} d\overline{\zeta} +$$
$$+ \int_L p(\zeta)\left(R_2 e^{i\psi} + R_4 e^{-i\psi}\right) ds + \frac{\tau\gamma}{2} e^{-i\gamma x_2},$$

$$
\begin{aligned}
R_1(\zeta, z) &= \frac{i\gamma}{4a} e^{i\gamma|x_2-\xi_2|} - \frac{1}{2a}\left[A_1 - B_2 - i(A_2 + B_1)\right], \\
R_2(\zeta, z) &= -\frac{i\gamma}{4a} e^{i\gamma|x_2-\xi_2|} - \frac{\pi}{8a^2}\operatorname{cosec}^2 \frac{\pi(\zeta+\overline{z})}{2a} - \\
&- \frac{1}{2a}\left[A_1 + B_2 + i(A_2 - B_1)\right], \\
R_3(\zeta, z) &= -\frac{i\gamma}{4a} e^{i\gamma|x_2-\xi_2|} - \frac{\pi}{8a^2}\operatorname{cosec}^2 \frac{\pi(\overline{\zeta}+z)}{2a} - \\
&- \frac{1}{2a}\left[A_1 + B_2 - i(A_2 - B_1)\right], \qquad (4.86)\\
R_4(\zeta, z) &= \frac{i\gamma}{4a} e^{i\gamma|x_2-\xi_2|} - \frac{1}{2a}\left[A_1 - B_2 + i(A_2 + B_1)\right], \\
A_1 &= \frac{a\gamma^2}{4\pi}\ell n\left|\sin\frac{\pi(\zeta+\overline{z})}{2a}\sec\frac{\pi(\zeta-z)}{2a}\right| + \\
&+ \sum_{\nu=1}^{\infty}\left\{\frac{\alpha_\nu^2}{\lambda_\nu}e^{-\lambda_\nu|x_2-\xi_2|} - \left(\alpha_\nu + \frac{\gamma^2}{2\alpha_\nu}\right)e^{-\alpha_\nu|x_2-\xi_2|}\right\}\sin\alpha_\nu\xi_1\sin\alpha_\nu x_1, \\
A_2 &= \sum_{\nu=1}^{\infty}\alpha_\nu\beta_\nu^-\operatorname{sign}(x_2-\xi_2)\sin\alpha_\nu\xi_1\cos\alpha_\nu x_1, \\
B_1 &= -\sum_{\nu=1}^{\infty}\alpha_\nu\beta_\nu^-\operatorname{sign}(x_2-\xi_2)\cos\alpha_\nu\xi_1\sin\alpha_\nu x_1, \\
B_2 &= \frac{\gamma^2}{4}|x_2-\xi_2| - \frac{a\gamma^2}{4\pi}\ell n\left|4\sin\frac{\pi(\zeta-z)}{2a}\sin\frac{\pi(\zeta+\overline{z})}{2a}\right| + \\
&+ \sum_{\nu=1}^{\infty}\left\{\left(\alpha_\nu - \frac{\gamma^2}{2\alpha_\nu}\right)e^{-\alpha_\nu|x_2-\xi_2|} - \lambda_\nu e^{-\lambda_\nu|x_2-\xi_2|}\right\}\cos\alpha_\nu\xi_1\cos\alpha_\nu x_1, \\
\beta_\nu^- &= e^{-\lambda_\nu|x_2-\xi_2|} - e^{-\alpha_\nu|x_2-\xi_2|}, \quad p(\zeta) = \frac{[U_3]}{2}.
\end{aligned}
$$

Calculating the limiting values of derived functions (4.85) at $z \to \zeta_0 \in L$ we find

$$
\left[\frac{\partial F}{\partial \zeta}\right] = -\frac{e^{-i\psi}}{2}f(\zeta), \quad \left[\frac{\partial F}{\partial \overline{\zeta}}\right] = -\frac{e^{i\psi}}{2}f(\zeta). \tag{4.87}
$$

Hence the last condition in (4.32) is satisfied automatically while the one above it allowing for (4.86), (4.87) results to equality expressing the connection of function $f(\zeta)$ from (4.85) with a derivative from the displacement jump

$$
f(\zeta) = \frac{2e_{15}}{\ni_{11}^{\varepsilon}}p'(\zeta), \quad p'(\zeta) = \frac{dp(\zeta)}{ds}. \tag{4.88}
$$

Now substituting the limiting functions (4.86) and the derivatives $\partial F/\partial z$, $\partial F/\partial \bar{z}$ at $z \to \zeta_0$ into mechanical boundary condition (4.32) on one of the edges L and using relations (4.88) we get a singular integrodifferential equation referring to the amplitude of the displacement jump on L:

$$\int_L p'(\zeta)\, g_1(\zeta,\zeta_0)\, ds + \int_L p(\zeta)\, g_2(\zeta,\zeta_0)\, ds = N(\zeta_0), \tag{4.89}$$

$$g_1(\zeta,\zeta_0) = \frac{1}{a}\mathrm{Im}\left\{ e^{i\psi_0}\left(\mathrm{ctg}\frac{\pi(\zeta-\zeta_0)}{2a} - k_{15}^2\mathrm{ctg}\frac{\pi(\overline{\zeta}+\zeta_0)}{2a}\right)\right\},$$

$$g_2(\zeta,\zeta_0) = 2\left(1+k_{15}^2\right)\left\{e^{i\psi_0}\left(e^{i\psi}R_1^0 + e^{-i\psi}R_3^0\right) + e^{-i\psi_0}\left(e^{i\psi}R_2^0 + e^{-i\psi}R_4^0\right)\right\},$$

$$N(\zeta_0) = \frac{2}{c_{44}^E}X_3 + 2i\gamma\tau\left(1+k_{15}^2\right)e^{-i\gamma\xi_{20}}\sin\psi_0,$$

$$\psi_0 = \psi(\zeta_0), \quad \zeta_0 = \xi_{10} + i\xi_{20} \in L_j \quad (j = 1,2,\ldots,k).$$

where the Kernel $g_1(\zeta,\zeta_0)$ is singular (of Hilbert type); the Kernel $g_2(\zeta,\zeta_0)$ on the assumptions concerning L may have just a slight singularity; functions $R_m^0 = R_m(\zeta,\zeta_0)$ are determined from (4.86).

The equation (4.89) should be considered together with the additional condition of (4.50) type on L_j $(j = 1,2,\ldots,k)$.

The asymptotic analysis of stresses on the continuation across cut tip L brings to formulas (4.51).

As an example let us consider a layer (ceramics PZT-4) with a straight crack oriented at angle φ to axis x_1. The center of the crack of length 2ℓ is located in the middle of the layer. Figs. 4.32, 4.33 illustrate the behavior relevant to stress intensity factor $\left\langle K_{III}^{\pm}\right\rangle = c_{44}^E\left|\Omega_0(\pm 1)\right| / \left(|\Lambda|\sqrt{\ell s'(\pm 1)}\right)$ in the cases of loading $X_3 = \mathrm{const}$, $\tau = 0$ and $X_3 = 0$, $\tau = \mathrm{const}$, respectively. The solid lines correspond to a, the dashed to b. The curve with number m

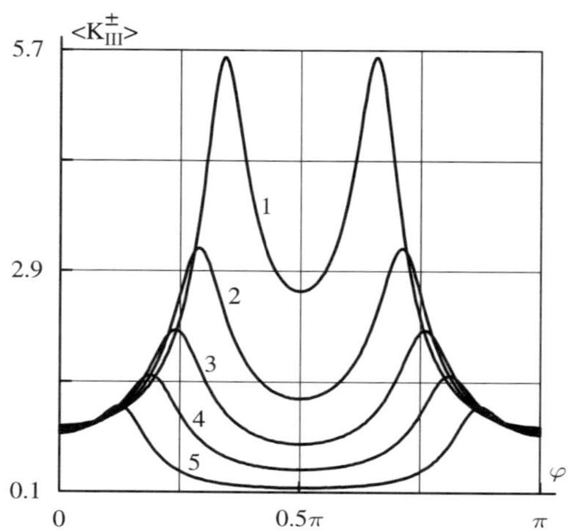

Fig. 4.32. Behaviour of $< K_{III}^{\pm} >$ as a function of orientation angle ϕ of the crack, in layer, for various values of $\gamma^*\alpha$ $(X_3 = 0, \tau = const)$

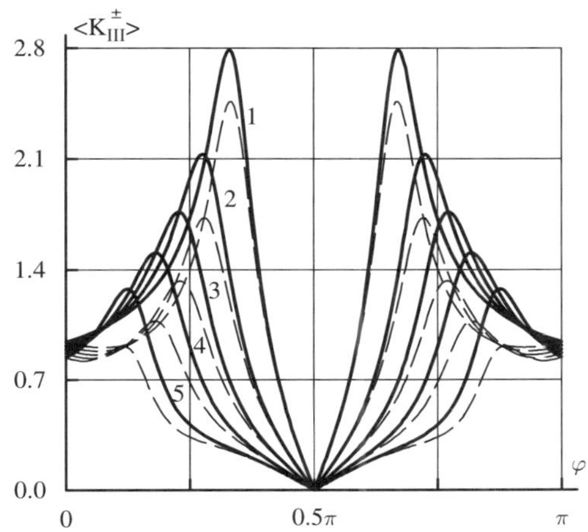

Fig. 4.33. Behaviour of $< K_{III}^{\pm} >$ as a function of orientation angle ϕ of the crack, in layer, for various values of $\gamma^* \alpha$ $(X_3 \neq 0, \tau = 0)$

is constructed for the value of normalized wave number $\gamma^* a = \gamma a \sqrt{1 + k_{15}^2} = 3.5 + 0.2\,(m - 1)\,\big(m = \overline{1,5}\big)$ and $\ell/a = 0.3$.

The changes of quantity $\langle K_{III}^{+} \rangle$ in a layer with parabolic crack $(\xi_1 + i\xi_2 = \delta\,(p_1 + ip_2\delta) + p_3)$ as a function of parameter $\gamma^* a$ for the variants of loading $\tau = 0,\ X_3 \neq 0$ and $X_3 = 0,\ \tau \neq 0$ is shown in Figs. 4.34 and 4.35, respectively. Curves $1, 2, \ldots, 4$ correspond to values $p_1/a = 1/5,\ p_3/a = 1/2,\ p_2/a =$

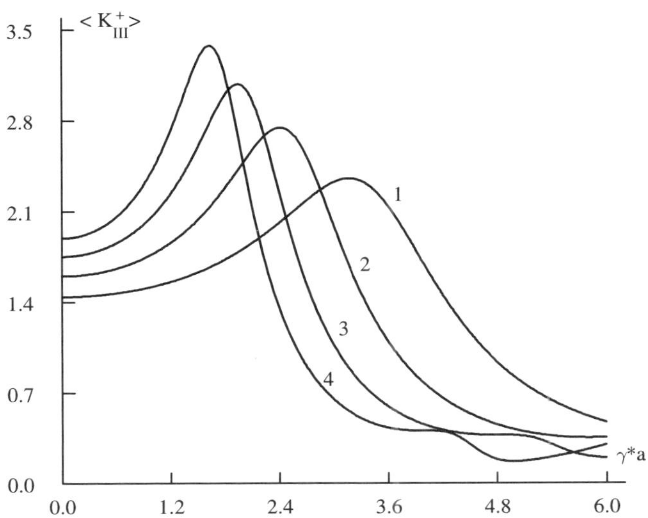

Fig. 4.34. Behaviour of $< K_{III}^{\pm} >$ as a function of normalized wave number $\gamma^* \alpha$ in a layer with a parabolic crack $(X_3 \neq 0, \tau = 0)$

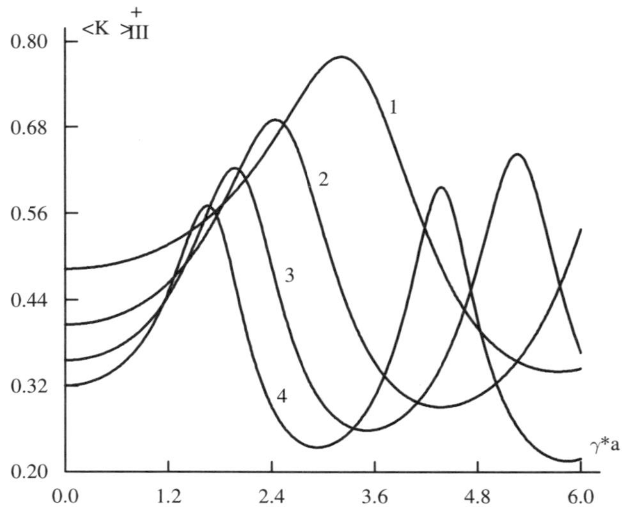

Fig. 4.35. Behaviour of $< K_{III}^{\pm} >$ as a function of normalized wave number $\gamma^{*}\alpha$ in a layer with a parabolic crack ($\tau \neq 0, X_3 = 0$)

$0.5, 0.7, 0.9$ and 1.1. Let us mention that if the crack is parallel to x_1 -axis with its center located in the middle of the layer, quantity $\langle K_{III}^{+} \rangle$ (in the considered interval of frequencies) is changing slightly.
The stress intensity factor K_{III} may be determined by the formula

$$K_{III}^{\pm} = \pm \Lambda \sqrt{\pi \ell} \left\langle K_{III}^{\pm} \right\rangle (\cos \omega t - \arg \Omega_0 (\pm 1)),$$
$$p'(\zeta) = \frac{\Omega_0(\delta)}{s'(\delta)\sqrt{1-\delta^2}}, \qquad \delta \in [-1, 1],$$

Where $\Lambda = |X_3|$ and $\Lambda = \gamma |\tau| c_{44}^{E} \left(1 + k_{15}^{2}\right)$ in the corresponding cases of the harmonic loading.

It must be noted that when the wave number γ exceeds value $\alpha_m = m\pi/a$ $(m = 1, 2, \ldots)$ instability of the solution is observed, due to the existence a new running wave carrying the power along the waveguide from the defect to infinity.

4.8 A Halflayer. Various Variants of Boundary Conditions

With the help of the method of images the Green's function we found above for a piezoceramic layer may be generalized for the case of a strip $(0 \leq x_1 \leq a,\ 0 \leq x_2 \leq \infty,\ -\infty \leq x_3 \leq \infty)$. Let us assume that the side bases of the strip are free from forces and is bound with vacuum. On the boundary $x_2 = 0$ the following types of mechanical and electric conditions may occur.

a) Lack of forces, fixed with vacuum

$$\sigma_{23} = 0, \quad D_2 = 0; \tag{4.90}$$

b) Rigid constraint, the boundary covered with electrodes and grounded

$$u_3 = 0, \quad E_1 = 0. \tag{4.91}$$

It is not difficult to conclude that Green's functions in that case are determined by formulas (4.77) where factors b_ν, d_ν have the form

$$b_\nu = -\frac{1}{a\lambda_\nu}\left[e^{-\lambda_\nu|x_2-\xi_2|} - Ae^{-\lambda_\nu(x_2+\xi_2)}\right],$$

$$b_0 = \frac{1}{2ia\gamma}\left[e^{i\gamma|x_2-\xi_2|} - Ae^{i\gamma(x_2+\xi_2)}\right],$$

$$d_\nu = -\frac{1}{a\alpha_\nu}\left[e^{-\alpha_\nu|x_2-\xi_2|} + Ae^{-\alpha_\nu(x_2+\xi_2)}\right] \quad (\nu = 1, 2, \ldots). \tag{4.92}$$

Summarizing the corresponding series in (4.77) allowing for (4.92) we obtain

$$G^*(\zeta, z) = G(\zeta, z) + A\left\{\frac{i}{2a\gamma}\, e^{i\gamma(x_2+\xi_2)} + \right.$$

$$+ \frac{x_2+\xi_2}{2a} - \frac{1}{2\pi}\ell n\left|4\sin\frac{\pi(\zeta+z)}{2a}\sin\frac{\pi(\zeta-\overline{z})}{2a}\right| + \tag{4.93}$$

$$\left. + \frac{1}{a}\sum_{m=1}^{\infty} c_m^*(x_2+\xi_2)\cos\alpha_m\xi_1\cos\alpha_m x_1\right\},$$

$$E^*(\zeta, z) = E(\zeta, z) +$$

$$+ \frac{A}{2\pi}\ell n\left|\sin\frac{\pi(\zeta-\overline{z})}{2a}\sec\frac{\pi(\zeta+z)}{2a}\right|,$$

$$c_m^*(x_2+\xi_2) = \frac{1}{\lambda_m}e^{-\lambda_m(x_2+\xi_2)} - \frac{1}{\alpha_m}e^{-\alpha_m(x_2+\xi_2)} \quad (m = 1, 2, \ldots),$$

$$z = x_1 + ix_2, \quad \zeta = \xi_1 + i\xi_2.$$

Here the case $A = -1$ corresponds to a free boundary with vacuum halflayer; the case $A = 1$ – corresponds to a fixed and covered with grounded electrodes along boundary $x_2 = 0$ halflayer; the functions G and E are determined in (4.81)–(4.83).

The integral representation of solutions of initial (4.75) and (4.76) in the given case has the form similar to (4.84), (4.85)

$$U_3(x_1,x_2)=\int_L [U_3]\frac{\partial G^*(\zeta,z)}{\partial n_\zeta}ds+U_3^{(0)}-AU_3^{(1)}, \tag{4.94}$$

$$F(x_1,x_2)=\int_L f(\zeta)E^*(\zeta,z)\,ds,\quad z=x_1+ix_2,\quad \zeta\in L,$$

where $U_3^{(1)}=\tau\exp(i\gamma x_2)$ corresponds to the reflected from the boundary halfspace $x_2=0$ shear wave.

Carrying out the same procedure allowing for expressions (4.93), (4.94) as in Sect. 4.7 we get the follwoing integro-differentiated equation

$$\int_L p'(\zeta)\,g_1^*(\zeta,\zeta_0)ds+\int_L p(\zeta)\,g_2^*(\zeta,\zeta_0)ds=N(\zeta_0), \tag{4.95}$$

$$g_1^*(\zeta,\zeta_0)=g_1(\zeta,\zeta_0)-\frac{Ak_{15}^2}{a}\mathrm{Im}\left[e^{i\psi_0}\left(\mathrm{ctg}\frac{\pi(\zeta+\zeta_0)}{2a}+\mathrm{ctg}\frac{\pi(\overline{\zeta}-\zeta_0)}{2a}\right)\right],$$

$$g_2^*(\zeta,\zeta_0)=2\left(1+k_{15}^2\right)\left\{e^{i\psi_0}\left(R_1^*e^{i\psi}+R_3^*e^{-i\psi}\right)+e^{-i\psi_0}\left(R_2^*e^{i\psi}+R_4^*e^{-i\psi}\right)\right\},$$

$$N(\zeta_0)=\frac{2}{c_{44}^E}X_3+2i\gamma\tau\left(1+k_{15}^2\right)\sin\psi_0\left(e^{-i\gamma\xi_{20}}+Ae^{i\gamma\xi_{20}}\right), \tag{4.96}$$

$$R_1^*=R_1^0+\frac{i\gamma A}{4a}e^{i\gamma(x_2+\xi_2)}+\frac{A\pi}{8a^2}\cos\mathrm{ec}^2\frac{\pi(\zeta+z)}{2a}-\frac{1}{2a}\left[A_1^*-B_2^*-i\left(A_2^*+B_1^*\right)\right],$$

$$R_2^*=R_2^0-\frac{i\gamma A}{4a}e^{i\gamma(x_2+\xi_2)}-\frac{A\pi}{8a^2}\cos\mathrm{ec}^2\frac{\pi(\zeta-\overline{z})}{2a}-\frac{1}{2a}\left[A_1^*+B_2^*+i\left(A_2^*-B_1^*\right)\right],$$

$$R_3^*=R_3^0-\frac{i\gamma A}{4a}e^{i\gamma(x_2+\xi_2)}-\frac{A\pi}{8a^2}\cos\mathrm{ec}^2\frac{\pi(\overline{\zeta}-z)}{2a}-\frac{1}{2a}\left[A_1^*+B_2^*-i\left(A_2^*-B_1^*\right)\right],$$

$$R_4^*=R_4^0+\frac{i\gamma A}{4a}e^{i\gamma(x_2+\xi_2)}+\frac{A\pi}{8a^2}\cos\mathrm{ec}^2\frac{\pi(\overline{\zeta}+\overline{z})}{2a}-\frac{1}{2a}\left[A_1^*-B_2^*+i\left(A_2^*+B_1^*\right)\right],$$

$$A_1^*=A_1-A\sum_{\nu=1}^{\infty}\left\{\frac{\alpha_\nu^2}{\lambda_\nu}e^{-\lambda_\nu(x_2+\xi_2)}-\alpha_\nu e^{-\alpha_\nu(x_2+\xi_2)}\right\}\sin\alpha_\nu\xi_1\sin\alpha_\nu x_1,$$

$$A_2^*=A_2-A\sum_{\nu=1}^{\infty}\alpha_\nu\beta_\nu^+\sin\alpha_\nu\xi_1\cos\alpha_\nu x_1,$$

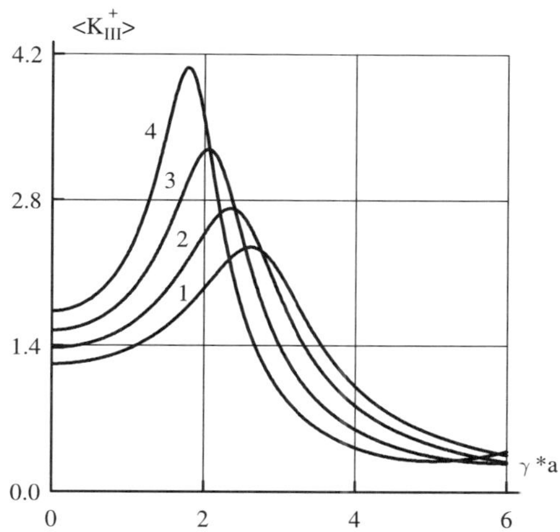

Fig. 4.36. Changes of $< K_{III}^{\pm} >$ as a function parameter $\gamma^* \alpha$ in a halflayer with a parabolic crack ($X_3 = 0,\ A = -1$)

$$B_1^* = B_1 - A \sum_{\nu=1}^{\infty} \alpha_\nu \beta_\nu^+ \cos\alpha_\nu \xi_1 \sin \alpha_\nu x_1,$$

$$B_2^* = B_2 + A \sum_{\nu=1}^{\infty} \left\{ \alpha_\nu e^{-\alpha_\nu (x_2+\xi_2)} - \lambda_\nu e^{-\lambda_\nu (x_2+\xi_2)} \right\} \cos\alpha_\nu \xi_1 \cos \alpha_\nu x_1,$$

$$\beta_\nu^+ = e^{-\lambda_\nu (x_2+\xi_2)} - e^{-\alpha_\nu (x_2+\xi_2)}, \quad p(\zeta) = \frac{[U_3]}{2} \quad (\nu = 1, 2, \ldots).$$

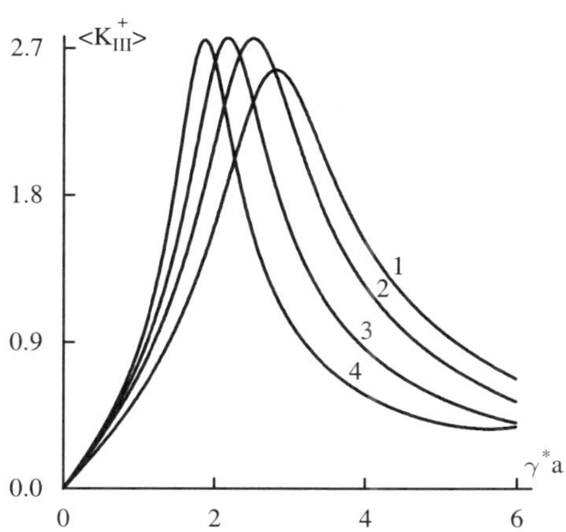

Fig. 4.37. Changes of $< K_{III}^{\pm} >$ as a function of parameter $\gamma^* \alpha$ in a halflayer with a parabolic crack ($\tau \neq 0,\ A = -1$)

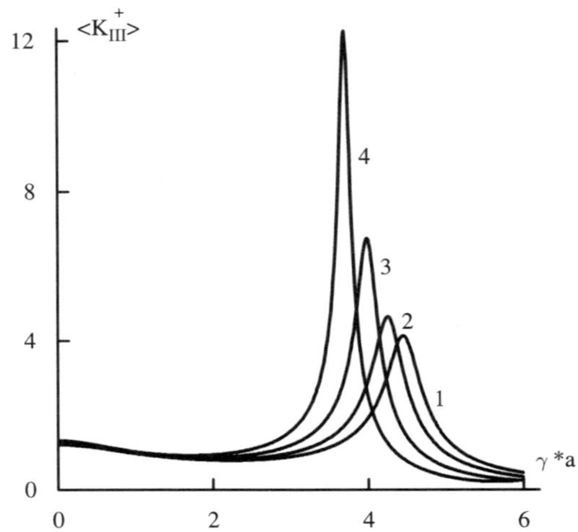

Fig. 4.38. Changes of $< K_{III}^{\pm} >$ as a function of parameter $\gamma^{*}\alpha$ in a halflayer with a parabolic crack $(X_3 \neq 0, A = 1)$

The functions $g_1(\zeta, \zeta_0)$, A_i, B_i and $R_i^0 = R_i(\zeta, \zeta_0)$ appearing in the above expressions are determined in (4.89), (4.86). Equation (4.95) should be solved together the additional conditions of (4.50).

Figures 4.36–4.43 show results form the numerical solution of integro-differential (4.95) for a halflayer with parabolic crack $\left(\xi_1 + i\xi_2 = \delta e^{i\varphi}(p_1 + ip_2\delta) + p_3 + ih, \delta \in [-1, 1]\right)$.

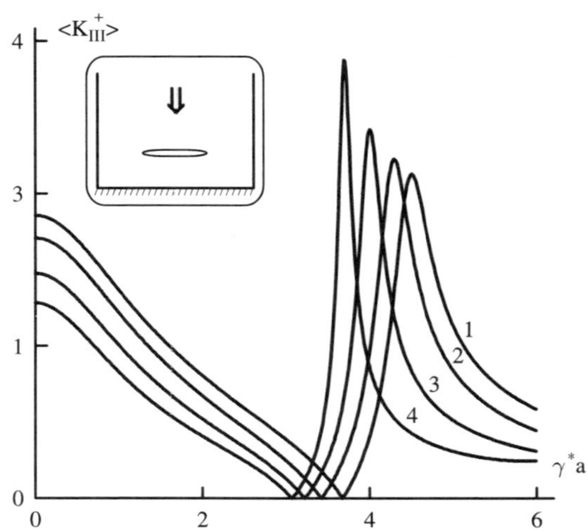

Fig. 4.39. Changes of $< K_{III}^{\pm} >$ as a function of parameter $\gamma^{*}\alpha$ in a halflayer with a parabolic crack $(\tau \neq 0, A = 1)$

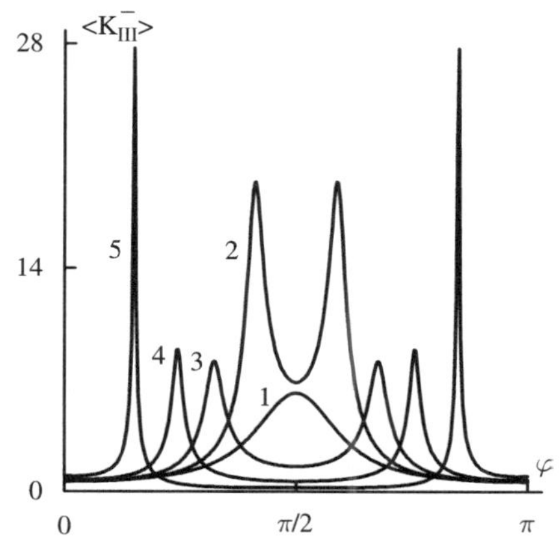

Fig. 4.40. Changes of $< K_{III}^{\pm} >$ as a function of ϕ for various values of $\gamma^* \alpha$ $(X_3 \neq 0, A = -1)$

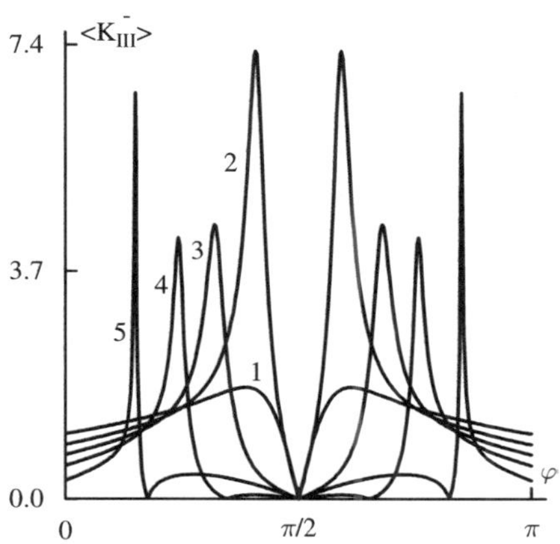

Fig. 4.41. Changes of $< K_{III}^{\pm} >$ as a function of ϕ for various values of $\gamma^* \alpha$ $(\tau \neq 0, A = -1)$

The graphs in Figs. 4.36 ($X_3 \neq 0$) and Fig. 4.37 ($\tau \neq 0$) are constructed for $A = -1$, $\varphi = 0$, $p_3/a = 1/2$, $p_1/a = 0.3$, $h/a = 0.3$, $p_2/a = 0.1\,(m-1)$ (m is the curve number); Figs. 4.38, 4.39 show the same things but for $A = 1$, $h/a = 0.6$.

The curves in Figs. 4.40 ($X_3 \neq 0$) and Fig. 4.41 ($\tau \neq 0$) correspond to $A = -1$, $p_3/a = 0.5$, $p_1/a = 0.3$, $p_2/a = 0$, $h/a = 1$, $\gamma^* a = 3 + 0.3\,(m-1)$; Fig. 4.42 ($X_3 \neq 0$) and Fig. 4.43 ($\tau \neq 0$) correspond to

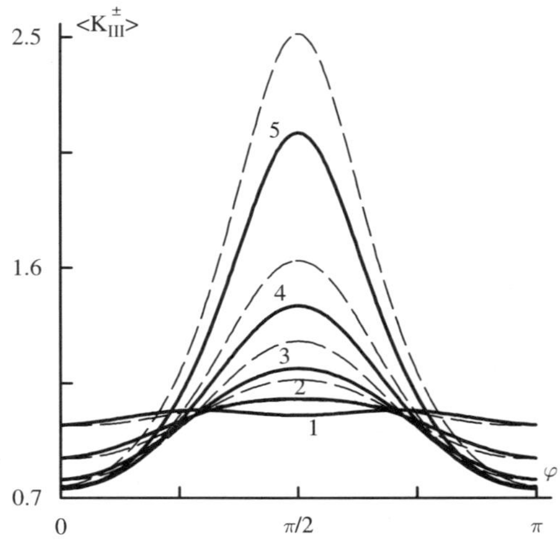

Fig. 4.42. Changes of $< K^{\pm}_{III} >$ as a function of ϕ for various values of $\gamma^{*}\alpha$ $(X_3 \neq 0, A = 1)$

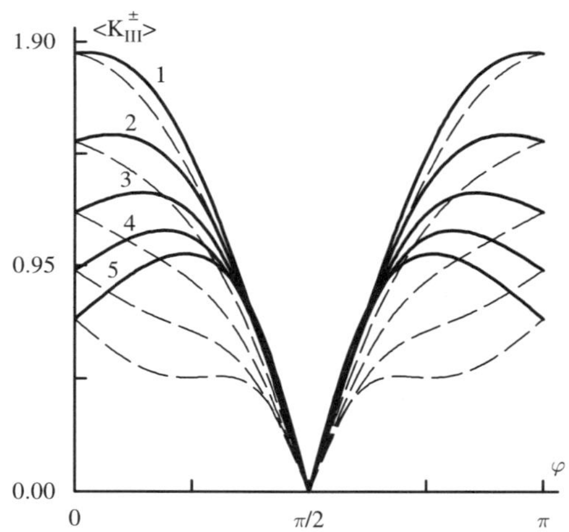

Fig. 4.43. Changes of $< K^{\pm}_{III} >$ as a function of ϕ for various values of $\gamma^{*}\alpha$ $(\tau \neq 0, A = 1)$

the same values of parameters but for $A = 1$, $h/a = 0.5$ and $\gamma^{*}a = 1 + 0.5\,(m-1)\,,\ (m = 1, 2, \ldots, 5)$.

5

Scattering of a Shear Wave by Cylindrical Inhomogeneities in Piezoceramic Media of Various Configurations (Antiplane Deformation)

In this chapter we study antiplane stationary and non-stationary dynamic problems of electroelasticity on the diffraction of a shear wave in tunnel cavities and inhomogeneous cylindrical inclusions in piezoceramic media. We derive algorithms in the form of singular integral and integrodifferential equations for the investigation of the concentration of stresses in the vicinity of inhomogeneities.

5.1 A Space and a Half-Space with Tunnel Openings

Let us start by considering a half-infinite piezoceramic medium ($x_2 \geq 0$) in the Cartesian system of coordinates, $Ox_1x_2x_3$, containing tunnel openings $\Gamma_m\,(m = 1, 2, \ldots, n)$ with generators parallel to optic x_3- axis. A constant shear loading $Z_n = \operatorname{Re}\left(Ze^{-i\omega t}\right)$ acts along x_3- axis on the surfaces of the cavities and let a plane monochromic shear wave radiating under angle β to x_1 - axis from infinity:

$$u_3^{(0)} = \operatorname{Re}\left(U_3^{(0)} e^{-i\omega t}\right),\ U_3^{(0)} = \tau e^{-i(\alpha_1 x_1 + \alpha_2 x_2)},$$
$$\alpha_1 = \gamma\cos\beta,\ \alpha_2 = \gamma\sin\beta,\ \gamma = \frac{\omega}{c_*}, \tag{5.1}$$
$$c = \left[\frac{c_{44}^E\left(1 + k_{15}^2\right)}{\rho}\right]^{1/2},\ k_{15}^2 = \frac{e_{15}^2}{\ni_{11}^{\varepsilon}\, c_{44}^E}.$$

The boundary of the halfspace $x_2 = 0$ is taken to be fixed or free of loading. In the case of a fixed halfspace we consider that the boundary is electroded with a zero potential ($E_1 = 0$); for a free halfspace bounded with vacuum it is assumed that $D_2 = 0$ at the boundary.

Let us assume that the electric potential is equal to zero on the surfaces of the cavities, indicating that the plating of the cavities is grounded electrodes.

We consider that the curvature of contour the $\Gamma_m\left(\cap\Gamma_m=\emptyset\right)$ and the amplitudes of shear forces $Z=\left\{Z_m\left(\zeta\right),\zeta\in\Gamma_m\right\}$ satisfy Hölder's condition.

Within this framework of analysis, the body (area Ω) is subject to the coupled wave fields of stresses $\sigma_{13}\left(x_1,x_2\right)$, $\sigma_{23}\left(x_1,x_2\right)$ as well as electric and magnetic stresses $E_1\left(x_1,x_2\right)$, $E_2\left(x_1,x_2\right)$ and $H_3\left(x_1,x_2\right)$, respectively.

Following the analysis presented in Sect. 4.2, the full system of equations of the assumed antiplane problem of electroelasticity in the quasistatic approximation is reduced to the Helmhöltz and Laplace (4.46) with respect to the functions $U_3\left(x_1,x_2\right)$ and $F\left(x_1,x_2\right)$. The only difference is the recording of mechanical and electric boundary conditions on the surface of openings $\Gamma_m\left(m=1,2,\ldots,n\right)$. In that case, due to (4.25), (4.29) these conditions are represented in the form

$$
\begin{aligned}
&c_{44}^{E}\left(1+k_{15}^{2}\right)\left[e^{i\psi}\frac{\partial U_3}{\partial\zeta}+e^{-i\psi}\frac{\partial U_3}{\partial\bar{\zeta}}\right]-ie_{15}\left[e^{i\psi}\frac{\partial F}{\partial\zeta}-e^{-i\psi}\frac{\partial F}{\partial\bar{\zeta}}\right]=Z, \quad (5.2)\\
&\frac{ie_{15}}{\ni_{11}^{\varepsilon}}\left[e^{i\psi}\frac{\partial U_3}{\partial\zeta}-e^{-i\psi}\frac{\partial U_3}{\partial\bar{\zeta}}\right]+e^{i\psi}\frac{\partial F}{\partial\zeta}+e^{-i\psi}\frac{\partial F}{\partial\bar{\zeta}}=0.
\end{aligned}
$$

Here derivatives $\partial U_3/\partial\zeta$, $\partial U_3/\partial\bar{\zeta}$, $\partial F/\partial\zeta$, $\partial F/\partial\bar{\zeta}$ are evaluated at the limit of $z=x_1+ix_2\to\zeta\in\Gamma$ with the transition taking place from the medium to the hole, while ψ is the angle between the contour of the cavity and Ox_1 -axis (Fig. 5.1).

Thus, the initial boundary problem is reduced to the definition of functions U_3 and F from differential (4.46), boundary conditions (5.2) and also the conditions on the boundary halfspace. For the derivation of the corresponding integral equations it is necessary to have the general representations of the formulas U_3 and F.The wave field in a halfspace with cavities is the result of the sum incoming and reflected (from the boundary) waves as well

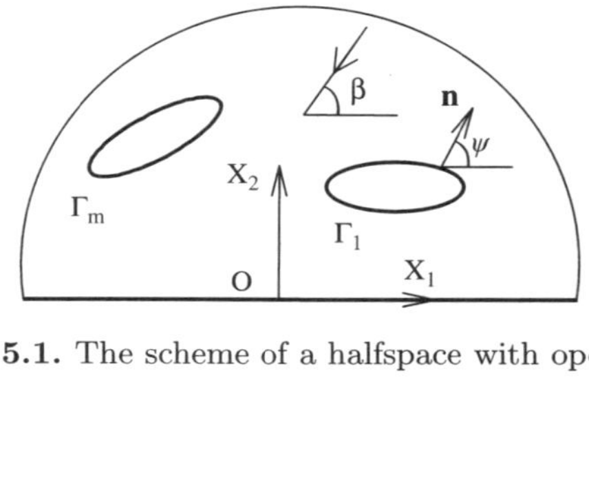

Fig. 5.1. The scheme of a halfspace with openings

as the ones scattered by the cavities, so let us represent these functions in the form [183]

$$U_3(x_1,x_2) = U_3^{(0)} - AU_3^{(1)} + \int\limits_\Gamma p(\zeta)\left[H_0^{(1)}(\gamma r) - AH_0^{(1)}(\gamma r_1)\right]ds,$$

$$F(x_1,x_2) = \frac{1}{\pi}\int\limits_\Gamma f(\zeta)(\ell n r + A\ell n r_1)ds, \tag{5.3}$$

$$r = |\zeta - z|,\ r_1 = \left|\bar{\zeta} - z\right|,\ z = x_1 + ix_2 \in \Omega,\ \zeta \in \Gamma = \cup\Gamma_m.$$

Here $p(\zeta)$ and $f(\zeta)$ are the unknown "*densities*"; $H_\nu^{(1)}(x)$ is the Hankel's function of the first kind and of ν-th order; the quantity $U_3^{(1)} = \tau\exp(-i(\alpha_1 x_1 - \alpha_2 x_2))$ is the amplitude of the displacement of a wave reflected from the boundary of a homogeneous halfspace. The value of $A = -1$ corresponds to the free, bounded with vacuum halfspace. The value of $A = 1$ corresponds to the fixed and covered by grounded electrodes halfspace.

The integral representations (5.3) satisfy (4.46), the mechanical and electric conditions on the boundary halfspace $x_2 = 0$ and also the conditions of radiation.

Substituting the limiting values of derivatives $\partial U_3/\partial z$, $\partial U_3/\partial \bar{z}$, $\partial F/\partial z$, $\partial F/\partial \bar{z}$, at $z \to \zeta_0 \in \Gamma_m\ (m = 1,2,\ldots,n)$ in the boundary conditions (5.2) we get a system of two singular integral equations of the second kind with respect to the functions $p(\zeta)$ and $f(\zeta)$

$$p(\zeta_0) + \int\limits_\Gamma p(\zeta)g_1(\zeta,\zeta_0)\,ds + \int\limits_\Gamma f(\zeta)\,g_2(\zeta,\zeta_0)ds = N_1(\zeta_0),$$

$$-f(\zeta_0) + \int\limits_\Gamma f(\zeta)g_3(\zeta,\zeta_0)\,ds + \int\limits_\Gamma p(\zeta)\,g_4(\zeta,\zeta_0)ds = N_2(\zeta_0),$$

$$g_1(\zeta,\zeta_0) = -\frac{1}{\pi}\mathrm{Re}\frac{e^{i\psi_0}}{\zeta_0 - \zeta} - \frac{i\gamma}{2}\left[H_1(\gamma r_0)\cos(\psi_0 - \alpha_0) - \right.$$
$$\left. -AH_1^{(1)}(\gamma r_{10})\cos(\psi_0 - \alpha_{10})\right],$$

$$g_2(\zeta,\zeta_0) = \frac{ie_{15}}{2c_{44}^E(1 + k_{15}^2)}\mathrm{Im}G(\zeta,\zeta_0),\ g_3(\zeta,\zeta_0) = \mathrm{Re}G(\zeta,\zeta_0),$$

$$G(\zeta,\zeta_0) = \frac{1}{\pi}\left[e^{i\psi_0}\left(\frac{1}{\zeta - \zeta_0} + \frac{A}{\bar{\zeta} - \zeta_0}\right)\right],$$

$$g_4(\zeta,\zeta_0) = \frac{c_{44}^E k_{15}^2}{e_{15}}\left\{\frac{2}{\pi i}\mathrm{Im}\frac{e^{i\psi_0}}{\zeta - \zeta_0} + \gamma\left[H_1(\gamma r_0)\sin(\psi_0 - \alpha_0) - \right.\right.$$
$$\left.\left. -AH_1^{(1)}(\gamma r_{10})\sin(\psi_0 - \alpha_{10})\right]\right\}, \tag{5.4}$$

$$N_1(\zeta_0) = \frac{Z(\zeta_0)}{2ic_{44}^E(1+k_{15}^2)} + \frac{\gamma\tau}{2}\left[e^{-i(\alpha_1\xi_{10}+\alpha_2\xi_{20})}\cos(\psi_0-\beta) - \right.$$

$$\left. -Ae^{-i(\alpha_1\xi_{10}-\alpha_2\xi_{20})}\cos(\psi_0+\beta)\right],$$

$$N_2(\zeta_0) = \frac{ie_{15}\gamma\tau}{\ni_{11}^{\varepsilon}}\left[e^{-i(\alpha_1\xi_{10}+\alpha_2\xi_{20})}\sin(\psi_0-\beta) - \right.$$

$$\left. -Ae^{-i(\alpha_1\xi_{10}-\alpha_2\xi_{20})}\sin(\psi_0+\beta)\right],$$

$$H_1(x) = \frac{2i}{\pi x} + H_1^{(1)}(x), \; r_0 = |\zeta-\zeta_0|, \; r_{10} = |\bar{\zeta}-\zeta_0|,$$

$$\alpha_0 = \arg(\zeta-\zeta_0), \; \alpha_{10} = \arg(\bar{\zeta}-\zeta_0), \quad \psi_0 = \psi(\zeta_0),$$
$$\zeta_0 = \xi_{10} + i\xi_{20}; \; \zeta, \zeta_0 \in \Gamma.$$

Here, the functions $g_2(\zeta,\zeta_0)$, $g_4(\zeta,\zeta_0)$ represent singular kernels of Cauchy type; the value $A=1$ corresponds to a fixed halfspace, and that of $A=-1$ corresponds to a free one. At $A=0$ we have an unrestricted space with cavities.

Here, we can show that the summed index in (5.4) is equal to zero and, hence, it is uniquely solvable under the assumption that the corresponding homogeneous system of integral equations has only trivial solution [41].

It should be also mentioned that at $e_{15}=0$ the system of integral (5.4) is divided into two independent equations with respect to the functions $p(\zeta)$ and $f(\zeta)$. In this case the integral equation with the "*density*" $p(\zeta)$ corresponds to a piezopassive (isotropic) halfspace with cavities.

Remark. It is possible, considering the harmonic oscillations of a multi-coupled piezoelectric cylinder with tunnel cavities $\Gamma_m\,(m=1,2,\ldots,n)$, excited by harmonic shear forces, to apply the obtained system of integral (5.4). For that it is necessary to set $A=0$, $\Gamma=\cup\Gamma_j\,(j=0,1,2,\ldots,n)$, where Γ_0 denotes the outer contour of a cylinder; the terms corresponding to shear waves should be omitted. The direction of tracing of all contour Γ_m while calculating (5.4) should be fulfilled in such a way that the area of the body is left on the left.

To determine the concentration of stresses in a halfspace with cavities we have to calculate the shear stress $\sigma_\vartheta, = \sigma_{23}\cos\psi - \sigma_{13}\sin\psi$ on the surface Γ:

$$\sigma_\vartheta = \mathrm{Re}\left(Te^{-i\omega t}\right), \; T = T(\xi_{10},\xi_{20}) = ic_{44}^E\left(1+k_{15}^2\right)\left(e^{i\psi_0}\frac{\partial U_3}{\partial\zeta_0} - \right. \tag{5.5}$$

$$\left. -e^{-i\psi_0}\frac{\partial U_3}{\partial\bar{\zeta}_0}\right) + e_{15}\left(e^{i\psi_0}\frac{\partial F}{\partial\zeta_0} + e^{-i\psi_0}\frac{\partial F}{\partial\bar{\zeta}_0}\right), \; \zeta_0 \in \Gamma_m \;\; (m=1,2,...,n),$$

Where the partial derivatives are calculated at $z \to \zeta_0 \in L$ of area Ω; ϑ is the angle between the positive tangential to the contour of an opening at the point ζ_0 and Ox_1 -axis.

It is more convenient to represent formula (5.5) in another, excluding all its terms containing function F. Substituting the second boundary condition into (5.2) and the limiting values of the functions $\partial U_3/\partial z$, $\partial U_3/\partial \overline{z}\,(z \to \zeta_0 \in \Gamma)$ calculated using (5.3) we find

$$T(\zeta_0) = \frac{e_{15}}{k_{15}^2}\left[N_2(\zeta_0) - \int_\Gamma p(\zeta)\, g_4(\zeta, \zeta_0)\, ds\right]. \tag{5.6}$$

The functions $g_4(\zeta, \zeta_0)$ and $N_2(\zeta_0)$ are defined in (5.4).

In Table 5.1 we give the approximated values of $\max\left|T/S_{23}^0\right|$ (S_{23}^0 is the modulus of amplitude of stress σ_{23} in a shear wave) on the contour of elliptic cavity ($\xi_1 = R_1 \cos\varphi,\ \xi_2 = R_2 \sin\varphi$) in a space made by of PZT-4 ceramics, for different values of parameter $\gamma^* R = \gamma R\sqrt{1 + k_{15}^2}$ ($\beta = \pi/2$, $R = 0.5\,(R_1 + R_2)$).

Figures 5.2–5.4 illustrate the behaviour of the relative tangential stress $\delta = \left|T/S_{23}^0\right|$ at the point $\varphi = 0$ of the contour of an elliptic opening according to normalized wave number $\gamma^* R$, respectively, for the values of parameter $A = 0, -1$ and 1. The curve with number k corresponds to $h/R_1 = 2,\ \beta = \pi/2$ and $R_2/R_1 = 0.5 + 0.25\,(k - 1)$. The dashed line in Fig. 5.2 corresponds to the piezopassive space with the singular cavity. The results obtained by the method of series in [65] are represented by circles.

Figure 5.5 and Fig. 5.7 depict changes of the quantity δ on the contour of the opening under the influence of shear ($\beta = \pi/2$) for $A = 0,\ -1$ and 1, respectively. In Fig. 5.5 the curve k is calculated for $\gamma^* R = 0.4,\ R_2/R_1 = 0.25k$, while in Figs. 5.6 and 5.7 for $\gamma^* R = 0.4,\ h/R_1 = 2$ and $\gamma^* R = 0.5,\ h/R_1 = 5,\ R_2/R_1 = 0.25k$, respectively.

Here, we should note that the above results correspond to the diffraction of a plane monochromatic wave in the cavity. In practice, what interests to

Table 5.1. Approximate values of max $\mid T/S_{23}^0 \mid$ on the contour of the elliptic cavity in a space made by PZT-4 ceramic for differenct values of the parameter $\gamma^* R$

$\gamma^* R$	$R_1/R_2 = 2$		$R_1/R_2 = 1$	
	PZT-4	$e_{15} = 0$	PZT-4	$e_{15} = 0$
0	3	3	2	2
0.1	3.072	3.049	2.065	2.033
0.2	3.222	3.139	2.184	2.079
0.3	3.406	3.229	2.290	2.095
0.4	3.570	3.282	2.314	2.060
0.5	3.636	3.264	2.221	1.975
0.6	3.536	3.157	2.059	1.668
0.7	3.265	2.974	1.907	1.757
0.8	2.902	2.745	1.808	1.660
0.9	2.536	2.508	1.768	1.576
1.0	2.221	2.288	1.773	1.498

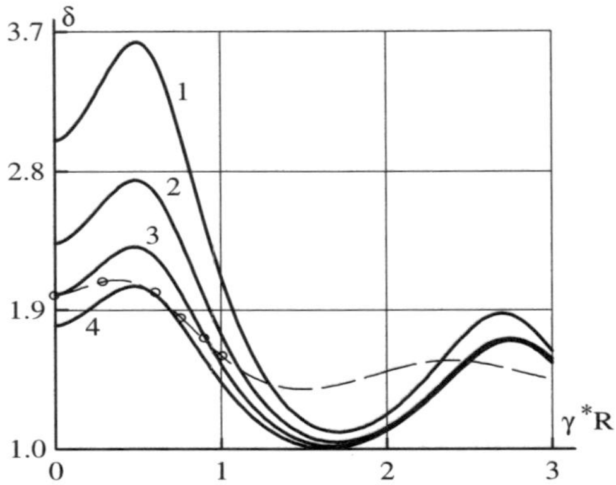

Fig. 5.2. Changes of the quantity $\delta = |T/S_{23}^0|$ at point $\varphi = \pi$ of the contour of an elliptic opening with respect to the parameter $\gamma^* R$ in a space during the radiation of a shear wave

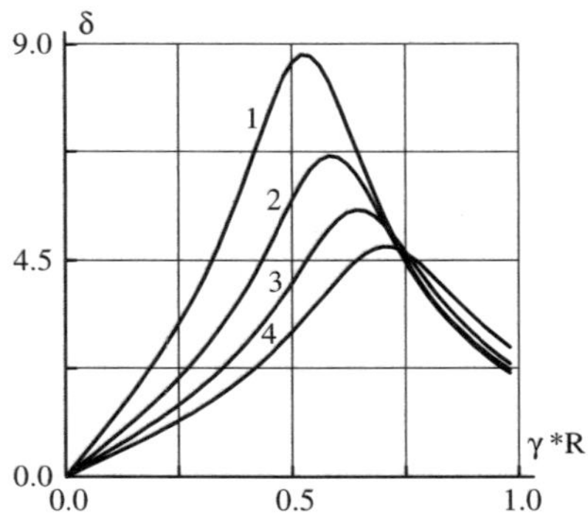

Fig. 5.3. Changes of the quantity $\delta = |T/S_{23}^0|$ at the point $\varphi = \pi$ of the contour of an elliptic opening with respect to the parameter $\gamma^* R$ in a halfspace (A = −1)

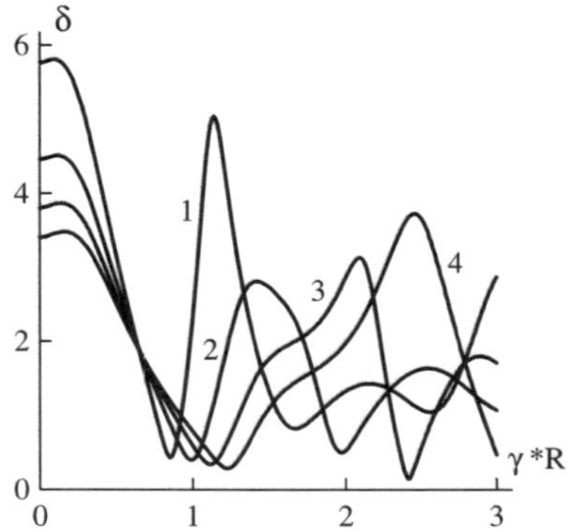

Fig. 5.4. Changes of the quantity $\delta = |T/S_{23}^0|$ at the point $\varphi = \pi$ of the contour of an elliptic opening with respect to parameter $\gamma^* R$ in a halfspace (A = 1)

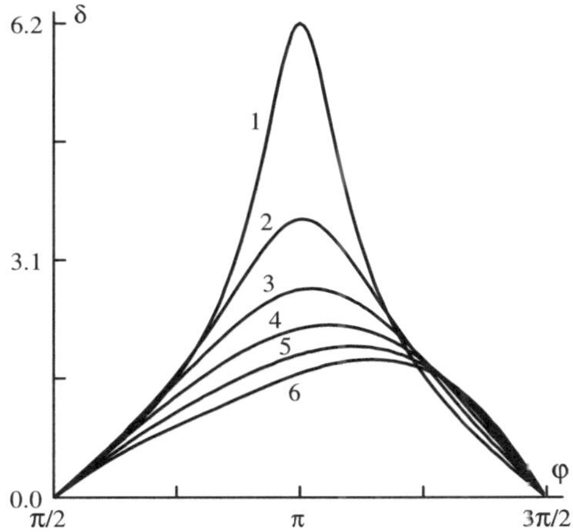

Fig. 5.5. Changes of the quantity δ on the contour of an opening under the influence of the shear wave in a space

study is the interaction of cylindrical waves with defects. As reflectors of a *cylindrical* shear wave let us study linear sources, that is either electric charges or shear forces, concentrated on line $x_1 = x_{10},\ x_2 = 0,\ -\infty < x_3 < \infty$ of the boundary of a free halfspace (space). The operator in the system of integral

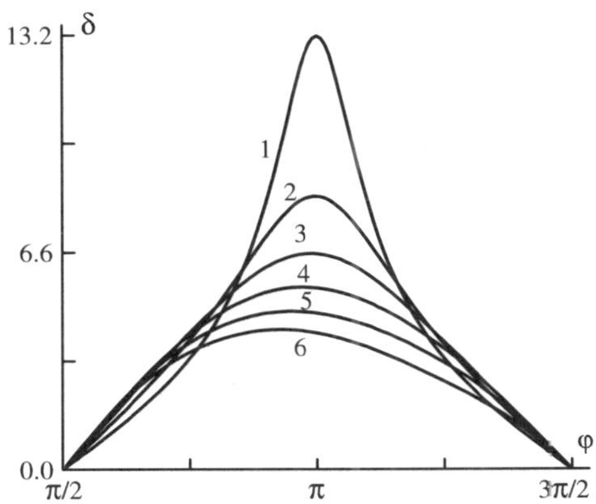

Fig. 5.6. Changes of the quantity δ on the contour of an opening under the influence of a shear wave in a halfspace (A = −1)

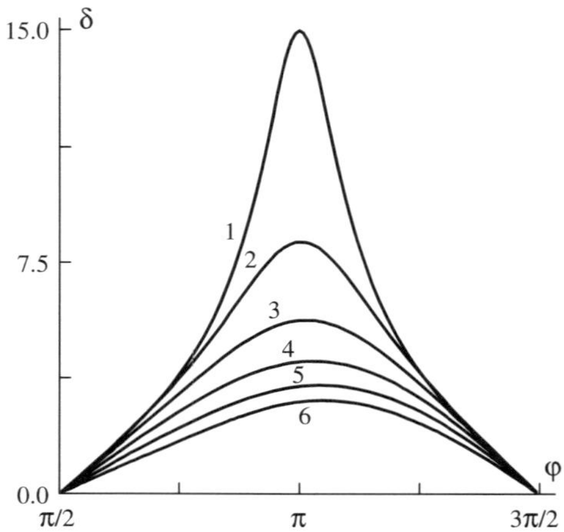

Fig. 5.7. Changes of quantity δ on the contour of an opening under the influence of a shear wave in a halfspace (A = 1)

(5.4) remains valid and the right-hand sides are determined, using formulas (4.60) and boundary conditions (5.2), as follows

$$N_1(\zeta_0) = \frac{1-A}{8c_{44}^E(1+k_{15}^2)}\cos(\psi_0-\alpha_0^*)\left\{\gamma P_0 H_1^{(1)}(\gamma r_0^*) - \right.$$

$$\left. -\frac{e_{15}Q_0}{\ni_{11}^{\varepsilon}}\left[\gamma H_1^{(1)}(\gamma r_0^*) + \frac{2i}{\pi r_0^*}\right]\right\},$$

$$N_2(\zeta_0) = \frac{1-A}{4\ni_{11}^{\varepsilon}}\sin(\psi_0-\alpha_0^*)\times\left\{P_0\frac{ie_{15}\gamma}{c_{44}^E(1+k_{15}^2)}H_1^{(1)}(\gamma r_0^*) - \right.$$

$$\left. -Q_0\left[\frac{i\gamma k_{15}^2}{1+k_{15}^2}H_1^{(1)}(\gamma r_0^*) - \frac{2}{\pi r_0^*}\right]\right\}, \tag{5.7}$$

$$r_0^* = |\zeta_0 - x_{10}|\,,\ \alpha_0^* = \arg(\zeta_0 - x_{10})\,.$$

where the parameter A may take the values 0 and -1.

Figures 5.8 and 5.9 demonstrate the results obtained by the numerical solution of integral (5.4) with right part (5.7) for the case of space ($A = 0$) with an elliptic cavity under the influence of charges and shear forces. Therein it was assumed that the center of the opening is at $h/R_1 = 2$ distance from the line of the source action. Curves 1 and 2 refer to relative tangential stress $\delta = |T/\Lambda|$ at contour point $\varphi = 0$ and values $R_2/R_1 = 0.5$ and 1, respectively; $\Lambda = P_0$, if $P_0 \neq 0$, $Q_0 = 0$ and $\Lambda = Q_0 c_{44}^E/c_{15}$ if

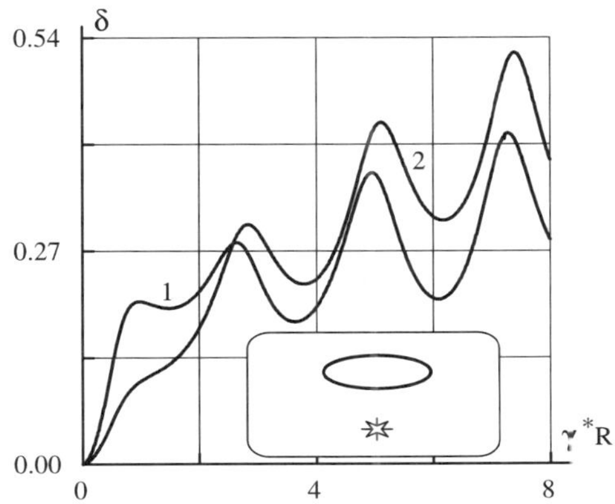

Fig. 5.8. Changes of the quantity δ at the point $\varphi = \pi$ on the contour of an elliptic cavity, as a function of $\gamma^* \mathrm{R}$ in a space under the influence of electric charges

$P_0 = 0$, $Q_0 \neq 0$. The dashed lines in Fig. 5.9 correspond to the piezopassive condition ($e_{15} = 0$).

The graphs of the quantity δ for a halfspace for the same values of parameters and in the same correspondence are presented in Figs. 5.10 and 5.11.

Comparing Figs. 5.9 and 5.11, it follows that the influence of the effect of interaction of the coupled fields with the concentration of stresses is more substantial in case of a halfspace than in a full space.

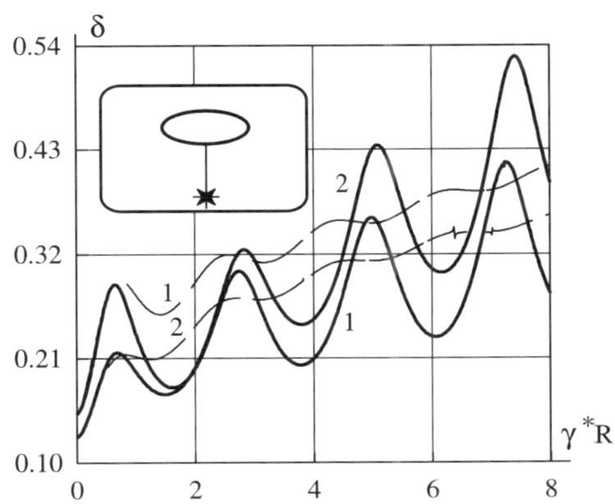

Fig. 5.9. Changes of the quantity δ at the point $\varphi = \pi$ on the contour of an elliptic cavity, as a function of $\gamma^* R$ in a space under the influence of shear forces

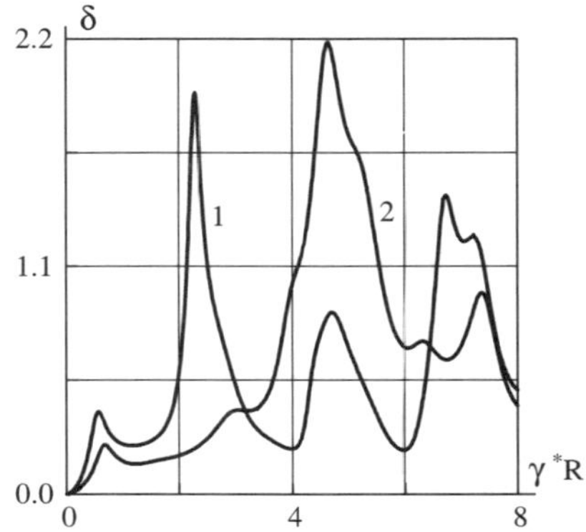

Fig. 5.10. Changes of the quantity δ at the point $\varphi = \pi$ on the contour of an elliptic cavity, as a function of $\gamma^* R$, in a halfspace under the influence of electric charges (A = -1)

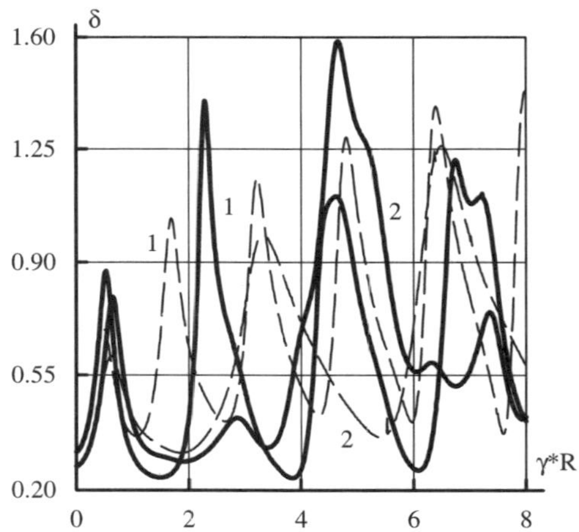

Fig. 5.11. Changes of the quantity δ at point $\varphi = \pi$ on the contour of an elliptic cavity, as a function of $\gamma^* R$, in a halfspace under the influence of shear forces (A=–1)

5.2 Impulse Excitation of a Half-Space with Openings

Let us consider the half-infinite piezoceramic medium $-\infty < x_1 < \infty$, $0 \leq x_2 < \infty$, $-\infty < x_3 < \infty$ in the Cartesian sytem of coordinates $Ox_1x_2x_3$, which is weakened by tunnel openings $\Gamma_m \, (m = 1, 2, \ldots, n)$ with generators parallel to x_3- axis. Let on line $x_1 = x_{10}$, $x_2 = 0$, $-\infty < x_3 < \infty$ of the free boundary of a halfspace shear forces $P\left(x_1, t\right) = P_0\left(t\right) \delta\left(x_1 - x_{10}\right)$, or electric charges $Q(x_1, t) = Q_0(t)\,\delta(x_1 - x_{10})$, arbitrary changing with time, act where $\delta(x)$ is Dirac delta function; t is the time.

Assume that the halfspace is bounded by vacuum; the surfaces of cavities Γ_m are free of mechanical loading and covered by grounded electrodes (the x_3 - axis coincides with the direction of the electric fields' force lines of the ceramics' preliminarily polarization.

The initial relations in quasistatic approximation are reduced to a system of two differential (4.26) and (4.28) for the displacement u_3 and the function Φ. Applying the integral Fourier transform with respect to time we obtain the equations of Helmhöltz and Laplace type (4.46) for the transformant functions u_3 and Φ. In Fourier transformants U_3, F the boundary conditions on the contour Γ_m have the form of (5.2) which allows us to use the system of integral (5.4) with the right part (5.7) written for function transformant $P_0(t)$, $Q_0(t)$.

We will determine tangential stress σ_ϑ on the surface of cavity Γ_m. Using (5.7) and because of the linear character of the problem and the validity of the superposition principle [196], proceeding from (5.5) we find [19]

$$\sigma_\vartheta(\zeta_0) = \sqrt{\frac{2}{\pi}}\mathrm{Re}\int\limits_0^\infty T(\zeta_0, \omega)e^{-i\omega t}d\omega,$$

$$T(\zeta_0, \omega) = T^* - \frac{e_{15}}{k_{15}^2}\int\limits_{\Gamma_m} p(\zeta)\, g_4(\zeta, \zeta_0)ds, \quad \zeta_0 \in \Gamma_m, \tag{5.8}$$

$$T^* = \frac{i\gamma(1-A)}{4(1+k_{15}^2)\ni_{11}^\varepsilon}\left[\ni_{11}^\varepsilon\hat{P}_0(\omega) - e_{15}\hat{Q}_0(\omega)\right]\sin(\psi_0 - \alpha_0^*)\,H_1^{(1)}(\gamma r_0^*).$$

The functions $g_4(\zeta, \zeta_0)$, r_0^* and α_0^* in (5.8) are defined in (5.4), (5.7); $\overline{P}_0(\omega)$, $\overline{Q}_0(\omega)$ are the Fourier transforms of the functions $P_0(t)$ and $Q_0(t)$.

As an example, let us consider a space from PZT-4 weakened by tunnel cavity of circular cross-section exposed to an impulse loading of trapezoidal form (Fig. 5.28). The spectral function of the impulse has form of (4.73).

Figures 5.12 and 5.13 illustrate the change $\lambda = \sigma_\vartheta R/\Lambda$ of the opening on the contour for various values of the dimensionless time $t^* = c_* t h^{-1}$ under the action of the shear forces and electric charges, respectively. Curves $1, 2, \ldots, 6$ are constructed for the values of dimensionless time $t^* = 1.2, 1.5, 2.3, 10.3, 10.8$ and 11.5 at $T^* = 10$, $d_1^* = 1$, $d_2^* = 8$, $\sigma_1^* = c_*\sigma_1 h^{-1} N/(sm)$, $h^{-1} = 1N/(s\cdot M)$, $\sigma_2^* = 1cal/s\cdot m$, $h/R = 4$ (R is the opening radius, h is the distance from the center of the opening to the line of the source action). Here, $\Lambda_1 = \sigma_1$ if $p(t) \neq 0$, $q(t) = 0$ and $\Lambda = \sigma_2 c_{44}^E/e_{15}$, if $p_0(t) = 0$, $q_0(t) \neq 0$. For comparison purposes the behaviour λ in the case of piezopassive medium is also shown (dashed lines).

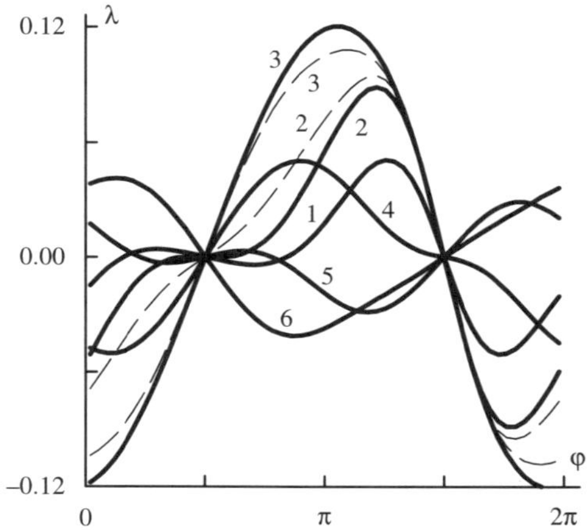

Fig. 5.12. Changes of the quantity λ on the contour of an opening for various values t^* under the impulse influence of shear forces, in a space

As it follows from the presented results the redistribution of stress σ_ϑ on the contour of the cavity occurs under a dynamic loading. The influence of the connected mechanical and electric fields on the surface of the cavity in a piezoceramic medium can be explained by the existence of Guliaev-Bluestein wave [140].

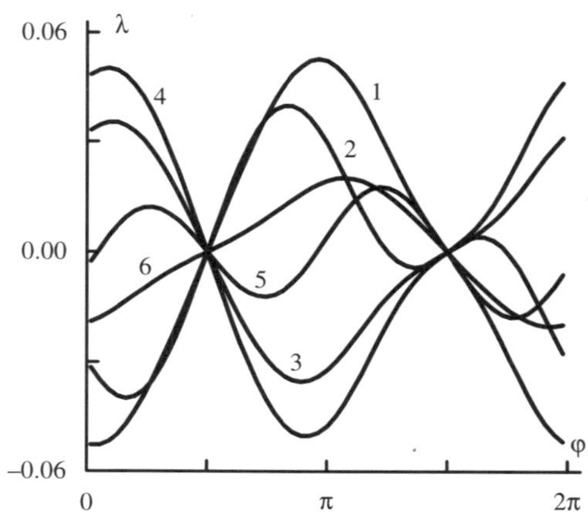

Fig. 5.13. Changes of the quantity λ on the contour of an opening for various values t^* under the impulse influence of electric charges, in a space

5.3 Stress Concentration in a Layer with Openings

Consider a piezoceramic layer $(0 \leq x_1 \leq a, -\infty \leq x_2 \leq \infty, -\infty \leq x_3 \leq \infty)$ in the Cartesian system of coordinates x_1, x_2, x_3 containing tunnel cavities $\Gamma_m \, (m = 1, 2, \ldots, n)$ along the x_3 -axis. On the surface of cavities which are considered to change harmonically with time, a shear loading $Z_n = \mathrm{Re}\left(Ze^{-i\omega t}\right)$, independent of $x_3 =$ coordinate, acts and a reflection of a plane shear wave may occur at infinity:

$$u_3^0 = \mathrm{Re}\left(U_3^0 e^{-i\omega t}\right), \; U_3^0 = \tau e^{-i\gamma x_2}, \; \gamma = \frac{\omega}{c_*}, \tag{5.9}$$

where the wave velocity of shear c_* is defined in (5.1).

Assume that the bases of the layer are free from forces and bounded by vacuum, while the surfaces of cavities Γ_m are electroded and grounded. The usual form of these conditions is displayed in (4.74) and (5.2). For the curvatures of contours Γ_m and the amplitudes Z we impose the same, as in Sect. 5.1, assumptions.

The stated boundary problem in a quasistatic approximation is described by (4.46). To reduce it to integral equations let us present the sought-for functions U_3 and F in form [145]

$$\begin{gathered} U_3\left(x_1, x_2\right) = U_3^{(0)} + \int\limits_{\Gamma} p\left(\zeta\right) G\left(\zeta, z\right) ds, \\ F\left(x_1, x_2\right) = \int\limits_{\Gamma} f\left(\zeta\right) E\left(\zeta, z\right) ds, \\ z = x_1 + i x_2 \in \Omega, \; \zeta = \xi_1 + i\xi_2 \in \Gamma = \cup \Gamma_m. \end{gathered} \tag{5.10}$$

where, the function $p\left(\zeta\right), \; f\left(\zeta\right)$ are to be determined, ds is the arc- element of the contour Γ and Ω is the area occupied by the body. For a piezoceramic layer Green's functions G and F are defined in (4.81)–(4.83). Thus, the integral representations of solutions (5.10) automatically satisfy the mechanical and electric boundary conditions on the bases of the layer and also the conditions of radiation and damping at infinity.

Substituting the corresponding derivatives at $z \to \zeta_0 \in \Gamma_m \; (m = 1, 2, \ldots, n)$ in the boundary conditions (5.2) we get the following system of singular integral equations for the determination of $p\left(\zeta\right)$ and $f\left(\zeta\right)$

$$\begin{gathered} p\left(\zeta_0\right) + \int\limits_{\Gamma} p\left(\zeta\right) g_1\left(\zeta, \zeta_0\right) ds + \int\limits_{\Gamma} f\left(\zeta\right) g_2\left(\zeta, \zeta_0\right) ds = N_1\left(\zeta_0\right), \\ f\left(\zeta_0\right) + \int\limits_{\Gamma} f\left(\zeta\right) g_3\left(\zeta, \zeta_0\right) ds + \int\limits_{\Gamma} p\left(\zeta\right) g_4\left(\zeta, \zeta_0\right) ds = N_2\left(\zeta_0\right), \end{gathered}$$

$$
\begin{aligned}
&g_1(\zeta,\zeta_0)=\frac{1}{2a}\mathrm{Re}\left[e^{i\psi_0}ctg\frac{\pi(\zeta_0-\zeta)}{2a}\right]+P_1e^{i\psi_0}+P_2e^{-i\psi_0},\\
&g_2(\zeta,\zeta_0)=\frac{e_{15}}{2ac_{44}^E(1+k_{15}^2)}\mathrm{Im}g,\ g_3(\zeta,\zeta_0)=\frac{\mathrm{Re}g}{2a},\\
&g_4(\zeta,\zeta_0)=\frac{ik_{15}^2c_{44}^E}{e_{15}}\left\{\frac{i}{2a}\mathrm{Im}\left[e^{i\psi_0}ctg\frac{\pi(\zeta_0-\zeta)}{2a}\right]+P_1e^{i\psi_0}-P_2e^{-i\psi_0}\right\},\\
&g=e^{i\psi_0}\left\{ctg\frac{\pi(\zeta_0-\zeta)}{2a}-ctg\frac{\pi\left(\zeta_0+\bar{\zeta}\right)}{2a}\right\},\\
&A_0=\sum_{k=1}^{\infty}\beta_{1k}\alpha_k\cos\alpha_k\xi_1\sin\alpha_k\xi_{10},\\
&B_0=\sum_{k=1}^{\infty}\beta_{0k}sign(\xi_{20}-\xi_2)\cos\alpha_k\xi_1\cos\alpha_k\xi_{10},\\
&\beta_{mk}=\frac{1}{\alpha_k^m}e^{-\alpha_k|\xi_2-\xi_{20}|}-\frac{1}{\lambda_k^m}e^{-\lambda_k|\xi_2-\xi_{20}|},\\
&P_1=P+S-\frac{1}{a}(A_0-iB_0),\ P_2=\bar{P}-S-\frac{1}{a}(A_0+iB_0), \qquad (5.11)\\
&P=\frac{1}{4a}ctg\frac{\pi\left(\bar{\zeta}+\zeta_0\right)}{2a},\ S=\frac{1}{2ia}sign(\xi_2-\xi_{20})\left(1-e^{i\gamma|\xi_2-\xi_{20}|}\right),\\
&N_1(\zeta_0)=\frac{2Z(\zeta_0)}{c_{44}^E(1+k_{15}^2)}+2i\gamma\tau e^{-i\gamma\xi_{20}}\sin\psi_0,\\
&N_2(\zeta_0)=\frac{2i\gamma\tau e_{15}}{\ni_{11}^{\varepsilon}}e^{-i\gamma\xi_{20}}\cos\psi_0,\\
&\psi_0=\psi(\zeta_0),\ \zeta_0=\xi_{10}+i\xi_{20}\in\Gamma_m\ (m=1,2,...,n).
\end{aligned}
$$

Here $g_2(\zeta,\zeta_0)$, $g_4(\zeta,\zeta_0)$ are singular kernels (of Hilbert type); the functions $g_1(\zeta,\zeta_0)$, $g_3(\zeta,\zeta_0)$ may posses not more than one weak singularity. If we set $e_{15}=0$ into the system (5.11) the equation with "*density*" $p(\zeta)$ corresponds to a piezopassive piecewise-homogeneous layer giving us the possibility to investigate the influence of the piezoelectric effect on the concentration of stresses.

The calculation of tangential stress $\sigma_\vartheta=\mathrm{Re}\left(Te^{-i\omega t}\right)$ with the use of formula (5.5) reads

$$
T(\zeta_0)=\frac{e_{15}}{k_{15}^2}\left[\int_\Gamma p(\zeta)\,g_4(\zeta,\zeta_0)\,ds-N_2(\zeta_0)\right], \qquad (5.12)
$$

where the functions $g_4(\zeta,\zeta_0)$ and $N_2(\zeta_0)$ are determined in (5.11).

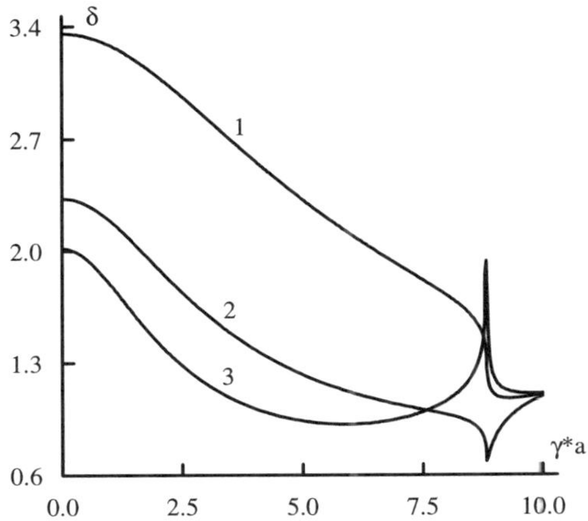

Fig. 5.14. Changes of $\delta = |T/S_{23}^0|$ at the point of contour $\varphi = \pi$ of a symmetrically positioned elliptic opening in a layer as a function of $\gamma^*\alpha$ under the influence of a shear wave

Figure 5.14 illustrates the changes of $\delta = \left|T/S_{23}^0\right|$ at the point of contour $\varphi = \pi$ of the elliptic opening with a center, located in the middle of a layer from ceramics PZT-4, with respect to γ^*a for $R_1/a = 0.2$, $R_2/a = 0.1, 0.2$ and 0.3 (curves 1, 2 and 3, respectively) under the action of the shear wave. Calculations were carried out by applying (5.12). Analogous results for the case of an opening shifted from the axis of the layer symmetry at distance of $0.1a$ to the left are given in Fig. 5.15.

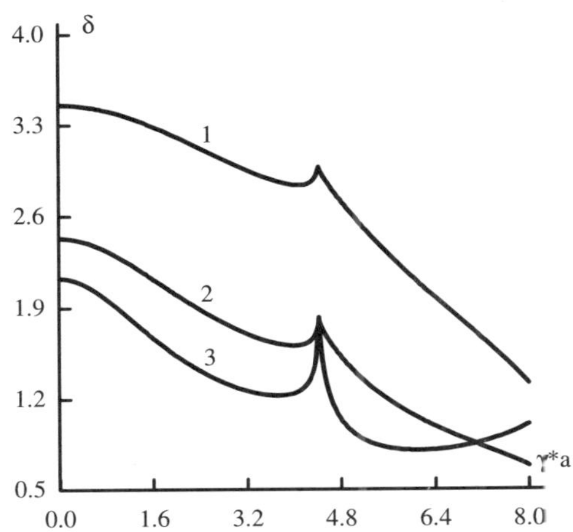

Fig. 5.15. Changes of δ at the point of contour $\varphi = \pi$ of a α symmetric elliptic opening as a function of $\gamma^*\alpha$ under the influence of a shear wave

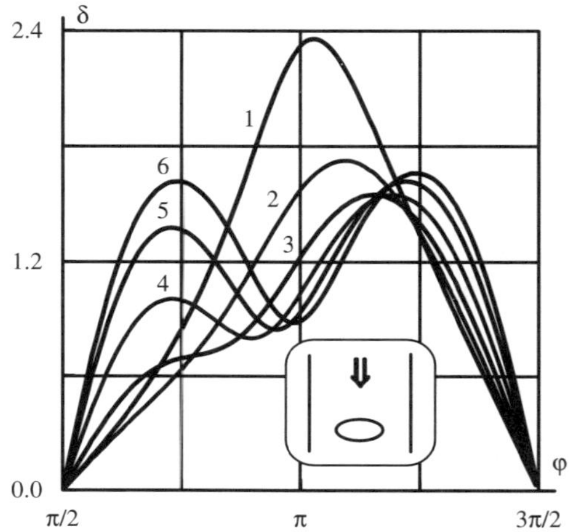

Fig. 5.16. Changes of δ on the contour of a symmetrically positioned opening in a layer, as a function of $\gamma^*\alpha$, for various values of parameter R_2/α under the influence of a shear wave

Changes of δ on the contour of the symmetrically located openings are represented in Fig. 5.16. The curve with number k corresponds to parameters $R_2/a = 0,1 + 0.05\,(k-1),\ \gamma^* a = 5$.

5.4 A Half layer with Openings

Applying the obtained in Sect. 4.8 Green's function for the piezoceramic half-layer ($0 \leq x_1 \leq a,\ 0 < x_2 < \infty,\ -\infty < x_3 < \infty$) we may in the same way construct the algorithm for the investigation of the diffraction of a shear wave on tunnel openings in a half-layer. Assume that the sidewise cases of the half-layer are free of forces and are bounded by vacuum and on the boundary $x_2 = 0$ either edge conditions (4.45) (lack of forces, contact with vacuum), or (4.44) (rigid restraint, the boundary covered by electrodes and grounded) hold.

In this case the integral representations of solutions (5.10) are somewhat generalized reading [148]

$$U_3(x_1, x_2) = U_3^0 - AU_3^{(1)} + \int\limits_\Gamma p(\zeta) G^*(\zeta, z) ds,$$

$$F(x_1, x_2) = \int\limits_\Gamma f(\zeta) E^*(\zeta, z) ds, \tag{5.13}$$

$$z = x_1 + ix_2 \in \Omega,\ \zeta = \xi_1 + i\xi_2\ \in \Gamma = \cup \Gamma_m$$

Here the term $u_3^{(1)} = \tau \exp(ijx_2)$ corresponds to shear waves reflected from the boundary of a half-layer $x_2 = 0$; the expressions for functions G^* and E^* are represented in (4.93). The case of $A = -1$ corresponds to the free, bounded with vacuum, half-layer while the case $A = 1$ to the fixed and covered with grounded electrodes along boundary $x_2 = 0$ half-layer.

Using (5.13) and boundary conditions (5.2) we get the system of singular integral equations of (5.11) type, where we should use the following relations

$$g = e^{i\psi_0}\left\{ctg\frac{\pi(\zeta_0-\zeta)}{2a} - ctg\frac{\pi(\zeta_0+\overline{\zeta})}{2a} + A\left[ctg\frac{\pi(\zeta_0-\overline{\zeta})}{2a} - ctg\frac{\pi(\zeta_0+\zeta)}{2a}\right]\right\}$$

$$P = \frac{1}{4a}\left\{ctg\frac{\pi(\zeta_0+\overline{\zeta})}{2a} - A\left[ctg\frac{\pi(\zeta_0+\zeta)}{2a} - ctg\frac{\pi(\overline{\zeta}-\zeta_0)}{2a}\right]\right\} \tag{5.14}$$

$$S = \frac{1}{2ia}\left\{sign(\xi_2-\xi_{20})\left[1-e^{i\gamma|\xi_2-\xi_{20}|}\right] + A\left[1-e^{i\gamma(\xi_2-\xi_{20})}\right]\right\},$$

$$A_0 = \sum_{k=1}^{\infty}(\beta_{1k}^- - A\beta_{1k}^+)a_k\cos a_k\xi_1\cos a_k\xi_{10}$$

$$B_0 = \sum_{k=1}^{\infty}\left[\beta_{0k}^- sign(\xi_{20}-\xi_2) - A\beta_{0k}^+\right]\cos a_k\xi_1\cos a_k\xi_{10}$$

$$\beta_{mk}^{\pm} = \frac{1}{a_k^m}e^{-a_k|\xi_2\pm\xi_{20}|} - \frac{1}{\lambda_k^m}e^{-\lambda_k|\xi_2\pm\xi_{20}|}$$

$$N_1(\zeta_0) = \frac{2Z(\zeta_0)}{c_{44}^E(1+k_{15}^2)} + 2i\gamma\tau\sin\psi_0(e^{-i\gamma\xi_{20}} + Ae^{i\gamma\xi_{20}}),$$

$$N_2(\zeta_0) = \frac{2i\gamma\tau e_{15}}{\underset{11}{\varepsilon}}\cos\psi_0(e^{-i\gamma\xi_{20}} + Ae^{i\gamma\xi_{20}})$$

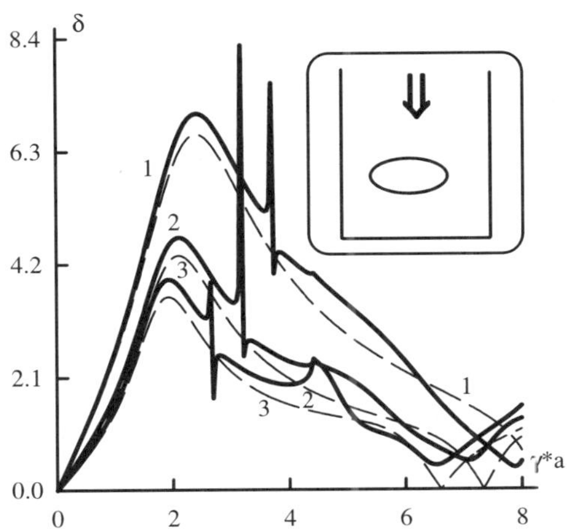

Fig. 5.17. Changes of δ at point $\varphi = \pi$ of the contour of an elliptic cavity as a function of $\gamma^*\alpha$ under the influence of a shear wave ($A = -1$)

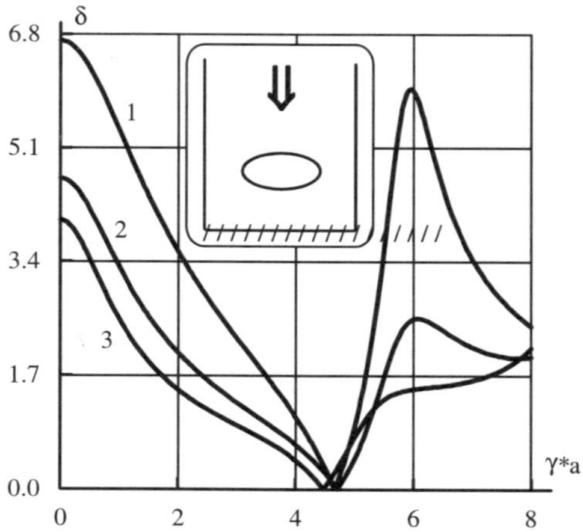

Fig. 5.18. Changes of δ at point $\varphi = \pi$ of the contour of an elliptic cavity as a function of $\gamma^*\alpha$ under the influence of a shear wave ($A = 1$)

Equation (5.12) are still valid for determination of tangential stress σ_θ, but now the functions $g_4(\zeta, \zeta_0)$ and $N_2(\zeta, \zeta_0)$ should be calculated with the help of equalities (5.11) and (5.14).

The calculations were carried out for a case of a half-layer with an elliptic opening. Figs. 5.17 and 5.18 show the behavior of δ at the point $\varphi = \pi$ of contour as a function of $\gamma^*\alpha$ at radiation of the shear wave for the values of parameters $A = -1$ and 1, respectively. The solid lines

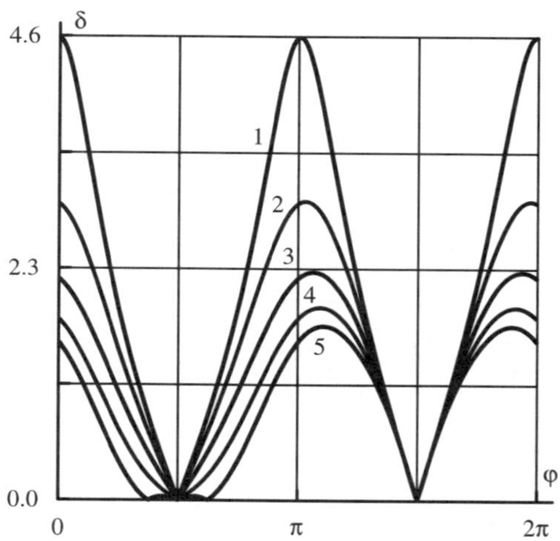

Fig. 5.19. Behaviour of quantity δ on the contour of an elliptic opening in halflayer ($A = -1$)

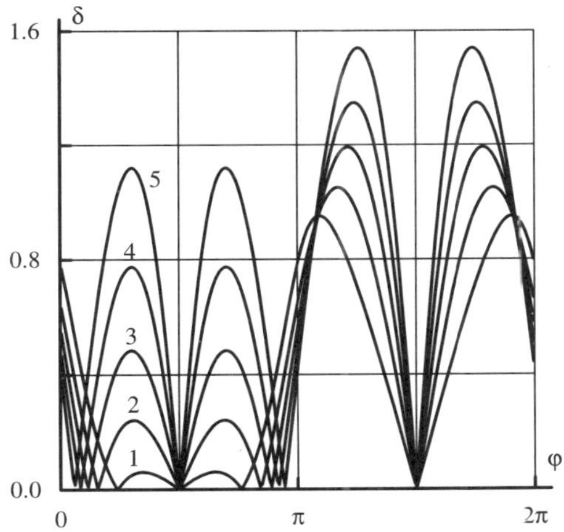

Fig. 5.20. Behaviour of quantity δ on the contour of an elliptic opening in halflayer ($A = 1$)

correspond to the opening shifted from the axis of symmetry at distance *0.1α* to the left; the dashed lines to the symmetrically positioned opening. Curves 1, 2 and 3 correspond to $R_1/a = 0.2$, $h/a = 0.5$, $R_2/a = 0.1$, 0.2 and 0.3. Changes of δ on the contour of a symmetrically positioned opening at $A = -1$ and 1 are shown in Figs. 5.19 and 5.20, respectively. The curve with number k is constructed for $\gamma^*\alpha = 3$, $h/\alpha = 0.7$, $R_2/a = 0.1 + 0.06(k-1)$.

5.5 A Space and a Halfspace with Cylindrical Inclusions. Integrodifferential Equations of a Boundary Problem

Consider a half-infinite $x_2 \geq 0$ medium in the Cartesian system of coordinates $Ox_1x_2x_3$ containing heterogeneous cylindrical inclusions (areas Ω_j) continuously connected with the matrix along contour C_j $(j = 1, 2, \ldots, k)$ with the generators parallel to x_3 -axis (see Fig. 5.21). Let us assume that the material

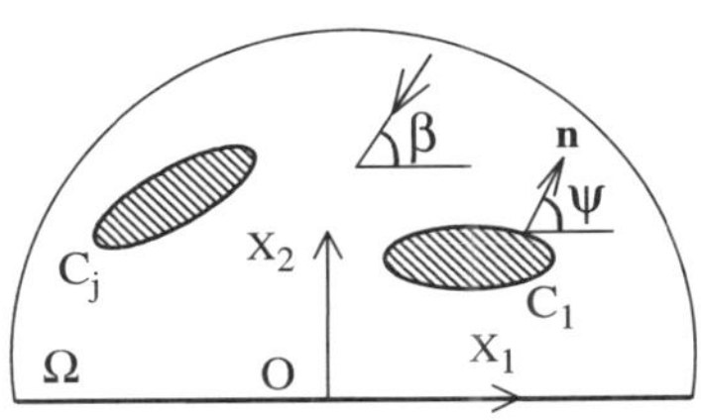

Fig. 5.21. The scheme of a halfspace with cylindrical inclusions

of the matrix and the inclusions is a piezoelectric of a hexagonal symmetry with the symmetry axis coinciding with x_3-axis. A plane monochromatic shear wave is reflected from infinity under angle β to axis Ox_1, and the component of shear vector $u_3^{(0.0)}$ and electric potential ϕ_0 are

$$u_3^{(0,0)} = \mathrm{Re}\left(U_3^{(0,0)} e^{-i\omega t}\right), \; U_3^{(0,0)} = \tau e^{-i(\alpha_1 x_1 + \alpha_2 x_2)}, \tag{5.15}$$
$$\phi_0 = \mathrm{Re}\left(\Phi_0 e^{-i\omega t}\right), \quad \Phi_0 = \frac{\tau e_{15}}{\ni_{11}^{\varepsilon}} e^{-i(\alpha_1 x_1 + \alpha_2 x_2)},$$

where α_i is determined in (5.1), when $\tau = const.$

It is assumed that the boundary of the halfspace is either fixed and covered with grounded electrodes ($u_3 = 0, \; E_1 = 0$) or free from forces and conjugated with vacuum ($\tau_{23} = 0, \; D_2 = 0$). On the surfaces of conjugation of the matrix (area Ω) with inclusions it is necessary to require the fulfillment of conditions of continuity of the displacement and stress vectors at crossing across C_j. The electric conditions should be stated as conditions of conjugation on the boundary of two dielectrics ($D_n^+ = D_n^-, \; E_s^+ = E_s^-$).

Assume that the curves of contour C_j satisfy Hölder's condition and besides ($C_i \cap C_j = \emptyset \, (i, j = 1, 2, \ldots, k)$).

In a quasistatic approximation the initial relations are reduced to the system differential (4.46). The amplitudes of stresses and electric quantities are expressed by functions U_3 and F calculated in (4.25), (4.29)

The index j is assigned to all the quantities referring to the j-th inclusion. Using (4.29) the conditions of coupling read

$$c_{44}^E \left(1 + k_{15}^2\right) \Lambda_0^+ - ie_{15}\Gamma_0^- = c_{44}^{E(j)} \left[1 + \left(k_{15}^{(j)}\right)^2\right] \Lambda_j^+ - ie_{15}^{(j)} \Gamma_j^-,$$
$$U_3 - U_3^{(j)} = 0, \; -\frac{ie_{15}}{\ni_{11}^{\varepsilon}} \Lambda_0^- - \Gamma_0^+ = -\frac{ie_{15}^{(j)}}{\ni_{11}^{\varepsilon(j)}} \Lambda_j^- - \Gamma_j^+, \tag{5.16}$$
$$\ni_{11}^{\varepsilon} \Gamma_0^- = \ni_{11}^{\varepsilon(j)} \Gamma_j^-,$$

where the differential operators $\Lambda_m^{\pm}$ and $\Gamma_m^{\pm}$ are determined by

$$\Lambda_m^{\pm} = \left[e^{i\psi_0} \frac{\partial U_3^{(m)}}{\partial \zeta_0} \pm e^{-i\psi_0} \frac{\partial U_3^{(m)}}{\partial \bar{\zeta}_0}\right], \; \zeta_0 = \xi_{10} + i\xi_{20},$$
$$\Gamma_m^{\pm} = \left[e^{i\psi_0} \frac{\partial F^{(m)}}{\partial \zeta_0} \pm e^{-i\psi_0} \frac{\partial F^{(m)}}{\partial \bar{\zeta}_0}\right], \; \bar{\zeta}_0 = \xi_{10} - i\xi_{20},$$
$$U_3^{(0)} = U_3, \; F^{(0)} = F \; (m = 0, 1, ..., k).$$

In (5.16) by quantities $U_3^{(j)}, \partial U_3^{(j)}/\partial\zeta_0, \partial U_3^{(j)}/\partial\bar{\zeta}_0$, $\partial F^{(j)}/\partial\zeta_0$ and $\partial F^{(j)}/\partial\bar{\zeta}_0$ are considered their limiting values at $z = x_1 + ix_2 \to \zeta_0 \in C_j$ where the transition takes place from the body area to the contours Ω_j, Ψ_0 is the angle between the positive normal to the inclusion contour and axis Ox_1 at point ζ_0.

The total wave field of the displacement in a halfspace with inclusions is summed by the falling and reflected from the boundary of a halfspace wave and also the field scattered by heterogeneities. According to it let us write down the integral representations of the wanted functions as [20]

$$U_3(z) = U_3^{(0,0)} - AU_3^{(0,1)} + \int_C p(\zeta)\left\{\frac{\partial G_0(\zeta,z)}{\partial\zeta}d\zeta - \frac{\partial G_0(\zeta,z)}{\partial\bar{\zeta}}d\bar{\zeta}\right\}, \quad z \in \Omega,$$

$$U_3^{(j)}(z) = U_3^{(j,0)} - AU_3^{(j,1)} + \int_C p_j(\zeta)\left\{\frac{\partial G_j(\zeta,z)}{\partial\zeta}d\zeta - \frac{\partial G_j(\zeta,z)}{\partial\bar{\zeta}}d\bar{\zeta}\right\}, \quad z \in \Omega_j, \tag{5.17}$$

$$F(z) = \frac{1}{\pi \ni_{11}^{\varepsilon}}\int_C f(\zeta)E(\zeta,z)\,ds, \quad z \in \Omega,$$

$$F^{(j)}(z) = \frac{1}{\pi \ni_{11}^{(j)}}\int_C f(\zeta)E(\zeta,z)\,ds, \quad z \in \Omega_j,$$

$$G_m(\zeta,z) = H_0^{(1)}(\gamma_m r) - AH_0^{(1)}(\gamma_m r_1), \quad E(\zeta,z) = \ell n r - A\ell n r_1,$$

$$U_3^{(j,k)} = \tau \exp\left[-i\left(\alpha_1^{(j)}x_1 + (-1)^k\alpha_2^{(j)}x_2\right)\right],$$

$$\alpha_1^{(j)} = \gamma_j\cos\beta, \quad \alpha_2^{(j)} = \gamma_j\sin\beta, \quad r = |\zeta - z|, \quad r_1 = |\bar{\zeta} - z|,$$

$$\gamma_0 = \gamma, \quad \zeta \in C = \cup C_j \quad (j = 1, 2, ..., k).$$

Here p, p_j and f are "*densities*" to be determined; ds is the arc-length element of contour C; $H_\nu^{(1)}(x)$ is the Hankel function of the first kind and of ν-th order. The parameter A identifies the type of the boundary conditions; at $A = -1$ we have a free bounded by vacuum halfspace; at $A = 1$ we have a fixed covered with grounded electrodes halfspace. The value $A = 0$ corresponds to the piezoelectric space with inclusions.

The integral representations (5.17) satisfy (4.46) in the corresponding areas Ω and Ω_j, mechanical and electric conditions of the mentioned types on the boundary of halfspace $x_2 = 0$ and also the conditions of reflection. Writing functions F and $F^{(j)}$ in the form of (5.17) gives the possibility to automatically satisfy the last condition in (5.16) (continuation of D_n at crossing C_j).

Substituting the limiting values of functions (5.17) and their derivatives into the conjugation conditions (5.16) we get a system of integrodifferential equations of the second kind:

$$
\begin{aligned}
&\int_C \left\{G_1 p'(\zeta) + G_2^{(j)} p_j'(\zeta) + G_3 p(\zeta) + G_4^{(j)} p_j(\zeta) + G_5^{(j)} f(\zeta)\right\} ds = N(\zeta_0),\\
&p(\zeta_0) + p_j(\zeta_0) + \int_C \left\{G_6 p(\zeta) + G_7^{(j)} p_j(\zeta)\right\} ds = M(\zeta_0),\\
&2e_{15} p'(\zeta_0) + 2e_{15}^{(j)} \lambda_j p_j'(\zeta_0) + (1+\lambda_j) f(\zeta_0) + \int_C \{G_8 p'(\zeta) + \\
&+ G_9^{(j)} p_j'(\zeta) + G_{10} p(\zeta) + G_{11}^{(j)} p_j(\zeta) + G_{12}^{(j)} f(\zeta)\Big\} ds = L(\zeta_0),
\end{aligned}
\tag{5.18}
$$

where

$$
\begin{aligned}
G_1 =& -\frac{2}{\pi} c_{44}^E \left(1 + k_{15}^2\right) \operatorname{Im} g,\ G_2^{(j)} = \frac{2}{\pi} c_{44}^{E(j)} \left(1 + \left(k_{15}^{(j)}\right)^2\right) \operatorname{Im} g,\\
G_3 =& c_{44}^E \left(1 + k_{15}^2\right) \left(e^{-i\psi_0} K_{20} - e^{i\psi_0} K_{10}\right),\ g = \frac{e^{i\psi_0}}{\zeta - \zeta_0},\\
G_4^{(j)} =& c_{44}^{E(j)} \left(1 + k_{15}^2\right) \left(e^{i\psi_0} K_{1j} - e^{-i\psi_0} K_{2j}\right),\ \lambda_j = \frac{\ni_{11}^\varepsilon}{\ni_{11}^{\varepsilon(j)}},\\
G_5^{(j)} =& \left(\frac{e_{15}^{(j)}}{\ni_{11}^{\varepsilon(j)}} - \frac{e_{15}}{\ni_{11}^\varepsilon}\right) \pi^{-1} \operatorname{Im} g_0,\ g_0 = g + A \frac{e^{i\psi_0}}{\bar{\zeta} - \zeta_0},\\
G_6 =& 0.5 i\gamma \left[H_1^{(1)}(\gamma r_0) \cos(\psi - \alpha_0) - A H_1^{(1)}(\gamma r_{10}) \cos(\psi + \alpha_{10})\right],\\
G_7^{(j)} =& 0.5 i\gamma_j \left[A H_1^{(1)}(\gamma_j r_{10}) \cos(\psi + \alpha_{10}) - H_1^{(1)}(\gamma_j r_0) \cos(\psi - \alpha_0)\right],\\
G_8 =& -\frac{2}{\pi} e_{15} \operatorname{Re} g,\ G_9^{(j)} = \frac{2}{\pi} e_{15}^{(j)} \lambda_j \operatorname{Re} g,\\
G_{10} =& -ie_{15} \left(e^{i\psi_0} K_{10} + e^{-i\psi_0} K_{20}\right),\\
G_{11}^{(j)} =& ie_{15}^{(j)} \lambda_j \left(e^{i\psi_0} K_{1j} + e^{-i\psi_0} K_{2j}\right),\ G_{12}^{(j)} = \frac{1}{\pi} (\lambda_j - 1) \operatorname{Re} g_0,\\
K_{1m} =& \frac{i\gamma_m^2}{4} \left\{e^{i\psi} \left[A H_0^{(1)}(\gamma_m r_{10}) + H_2(\gamma_m r_0) e^{-2i\alpha_0}\right] - \right.\\
&\left. -e^{-i\psi} \left[H_0^{(1)}(\gamma_m r_0) + A H_2^{(1)}(\gamma_m r_{10}) e^{-2i\alpha_{10}}\right]\right\},\\
K_{2m} =& \frac{i\gamma_m^2}{4} \left\{e^{i\psi} \left[A H_2^{(1)}(\gamma_m r_{10}) e^{2i\alpha_{10}} + H_0^{(1)}(\gamma_m r_0)\right] - \right.\\
&\left. -e^{-i\psi} \left[H_2(\gamma_m r_0) e^{2i\alpha_0} + A H_0^{(1)}(\gamma_m r_{10})\right]\right\},
\end{aligned}
$$

$$N(\zeta_0) = c_{44}^{E(j)} \left(1 + \left(k_{15}^{(j)}\right)^2\right) d_j(\zeta_0) - c_{44}^{E}\left(1 + k_{15}^2\right) d_0(\zeta_0),$$

$$M(\zeta_0) = 0.5\left(W_0^{(j)} - AW_1^{(j)} - W_0 + AW_1\right),$$

$$L(\zeta_0) = e_{15}^{(j)} \lambda_j e_j(\zeta_0) - e_{15} e_0(\zeta_0),$$

$$d_m(\zeta_0) = -i\gamma_m \left[B_m^{+}\cos(\psi_0 - \beta) - AB_m^{-}\cos(\psi_0 + \beta)\right],$$

$$e_m(\zeta_0) = i\gamma_m \left[B_m^{+}\sin(\psi_0 - \beta) - AB_m^{-}\sin(\psi_0 + \beta)\right],$$

$$H_2(x) = \frac{4i}{\pi x^2} + H_2^{(1)}(x), \ p_j'(\zeta) = \frac{dp_j}{ds}, \ p_j'(\zeta_0) = \frac{dp_j}{ds_0},$$

$$B_m^{\pm} = \tau \exp\left[-i\left(\alpha_1^{(m)}\xi_{10} \pm \alpha_2^{(m)}\xi_{20}\right)\right], \ \psi = \psi(\zeta), \ \psi_0 = \psi(\zeta_0),$$

$$r_0 = |\zeta_0 - \zeta|, \ r_{10} = \left|\zeta_0 - \bar{\zeta}\right|, \alpha_0 = \arg(\zeta_0 - \zeta), \ \alpha_{10} = \arg\left(\zeta_0 - \bar{\zeta}\right),$$

$$s_0 = s(\zeta_0), \ \zeta = \xi_1 + i\xi_2 \in C, \ \zeta_0 \in C_j \ \ (j = 1, 2, ..., k).$$

In the system (5.18) the kernels G_1, $G_2^{(j)}$, $G_5^{(j)}$ are singular, while the other kernels, due to the assumptions referring to contour C_j, may have not more than weak singularity.

It should be mentioned that at $e_{15} = e_{15}^{(j)} = 0$ the system (5.18) is divided into the independent equations with respect to functions $p_m(\zeta)$ and $f(\zeta)$. In this case the first two equations (with "*densities*" $p_m(\zeta)$) correspond to a piezopassive composition.

To investigate the stress concentration in a halfspace with inclusions let us calculate the stresses $\sigma_n = \sigma_{13}\cos\psi_0 + \sigma_{23}\sin\psi_0$ and $\sigma_s = \sigma_{23}\cos\psi_0 - \sigma_{13}\sin\psi_0$ at the contour point $\zeta_0 \in C$. Using equalities (5.17), (4.29) the expressions for the amplitude of tangential stresses read

$$T_n(\zeta_0) = \int_C \left[p'(\zeta) G_1(\zeta, \zeta_0) + p(\zeta) G_3(\zeta, \zeta_0)\right] ds +$$
$$+ c_{44}^{E}\left(1 + k_{15}^2\right) d_0(\zeta_0) - \frac{e_{15}}{\pi \ni_{11}^{\varepsilon}} \int_C f(\zeta) \mathrm{Im} g_0 ds,$$

$$T_s^{-}(\zeta_0) = c_{44}^{E}\left(1 + k_{15}^2\right)\left\{2p'(\zeta_0) - \frac{2}{\pi}\int_C p'(\zeta)\,\mathrm{Re} g ds - \right.$$
$$\left. -i\int_C p(\zeta)\left(e^{i\psi_0}K_{10} + e^{-i\psi_0}K_{20}\right) ds + e_0(\zeta_0)\right\} +$$
$$+ \frac{e_{15}}{\ni_{11}^{\varepsilon}}\left\{f(\zeta_0) - \frac{1}{\pi}\int_C f(\zeta)\,\mathrm{Re} g_0 ds\right\},$$

$$T_s^+(\zeta_0) = -c_{44}^{E(j)}\left(1+\left(k_{15}^{(j)}\right)^2\right)\left\{2p_j'(\zeta_0)+\frac{2}{\pi}\int\limits_C p_j'(\zeta)\,\mathrm{Re}g\,ds+\right.$$

$$\left.+i\int\limits_C p_j(\zeta)\left(e^{i\psi_0}K_{1j}+e^{-i\psi_0}K_{2j}\right)ds - e_j(\zeta_0)\right\}-$$

$$-\frac{e_{15}^{(j)}}{\ni_{11}^{\varepsilon(j)}}\left\{f(\zeta_0)+\frac{1}{\pi}\int\limits_C f(\zeta)\,\mathrm{Re}g_0\,ds\right\}. \tag{5.19}$$

The functions G_m, K_{im}, d_m, e_m, g, g_0 are determined from (5.18), the "plus" and "minus" signs show the limiting values of stresses T_s from areas of j-th inclusions and matrix, respectively.

As an example let us consider a halfspace with inclusions of elliptic cross-section with parametric equations

$$\xi_1 = R_1\cos\varphi,\ \xi_2 = R_2\sin\varphi + h\ (0 \le \varphi \le 2\pi). \tag{5.20}$$

The material of the inclusions and of the matrix corresponded to piezoceramics PZT-4 and $BaTiO_3$.

Using (5.20) the integrodifferential (5.18) can be reduced to a system of linear algebraic equations using the scheme of the method of quadrature (see: Appendix B, B.3).

Figure 5.22 shows the graphs of $\lambda = |T_n|\,R/\left(c_{44}^E|\tau|\right)$ at point $\varphi = \pi$ of the contour of an elliptic inclusion as function $\gamma^*R = R\sqrt{\gamma\gamma_1\eta}$ for values $\beta = \pi/2$, $R_2/R_1 = 2$ and 1 (curves 1 and 2, respectively).

Figures 5.23 and 5.24 illustrate the behaviour of $\mu^- = |T_s^-|\,R/\left(c_{44}^E|\tau|\right)$ and $\lambda = |T_n|\,R/\left(c_{44}^E|\tau|\right)$ on the inclusion contour in space ($A = 0$). Lines 1,

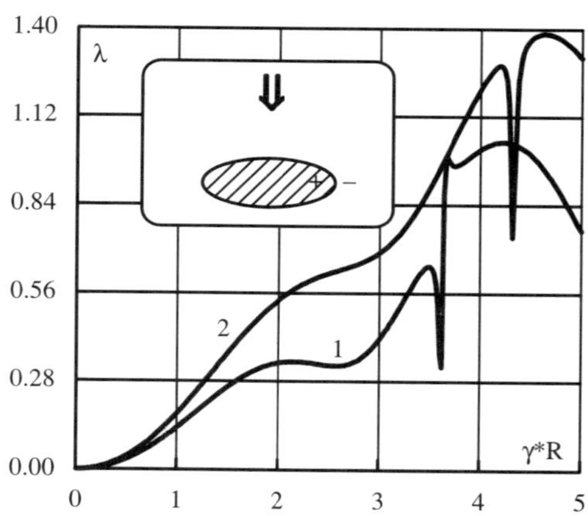

Fig. 5.22. Changes of quantity λ at the point $\varphi = \pi$ of the contour of an elliptic inclusion as a function of γ^*R during the radiation of a shear wave in a space

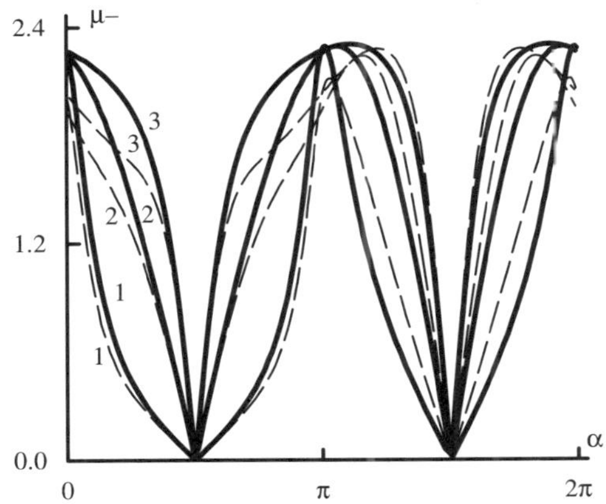

Fig. 5.23. Behaviour of μ^- on the contour of an inclusion in a halfspace ($\gamma^* R = 2$)

2 and 3 are constructed for $\beta = \pi/2$, $\gamma^* R = 2$, $R_2/R_1 = 0.5, 1$ and 1.5, respectively. Here, $\eta = \sqrt{(1 + k_{15}^2)\left(1 + \left(k_{15}^{(1)}\right)^2\right)}$, $R = (R_1 + R_2)/2$. The dashed lines correspond to the piezopassive composition.

Figures 5.25 and 5.26 demonstrate the change of λ on the inclusion contour in a halfspace as a function of the vectorial angle α for (A $= -1$) and (A $= 1$) respectively. Curves 1, 2 and 3 are constructed for the values of parameters $\gamma^* R = 2$, $\beta = \pi/2$, $R_2/R_1 = 0.5, 1$ and 1.5, respectively. The distance from the boundary of the halfspace to the center of the inclusion is assumed equal to $h = 2.5R_1$ (Fig. 5.25) and $h = 2R_1$ (Fig. 5.26).

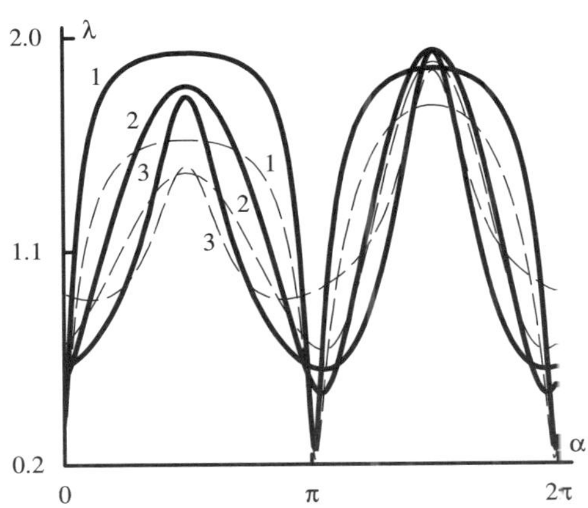

Fig. 5.24. The behaviour of λ on the contour of an inclusion in a halfspace ($\gamma^* R = 2$)

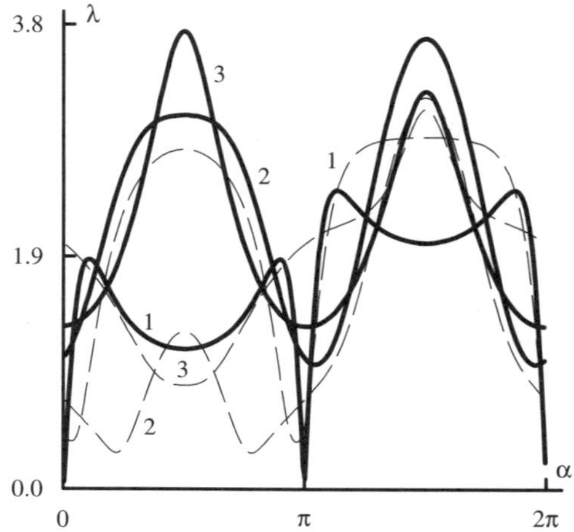

Fig. 5.25. Changes of λ on the contour of an inclusion in a halfspace (A = −1)

From the results it follows that the influence of coupling of the mechanical and electric fields on the stress state is considerably strong in the vicinity of extremal values of tangential stresses on the interphase of media. This circumstance should be taken into account for the calculation of the strength of the glue layer on the boundary of "*fibre-matrix*", and also for the estimation of the appearance interphase cracks. Not taking into account a piezoeffect may bring to (in some range of changes of vector angle α) a considerable decrease of calculated stresses σ_n and σ_s in comparison with real ones. The characteristic

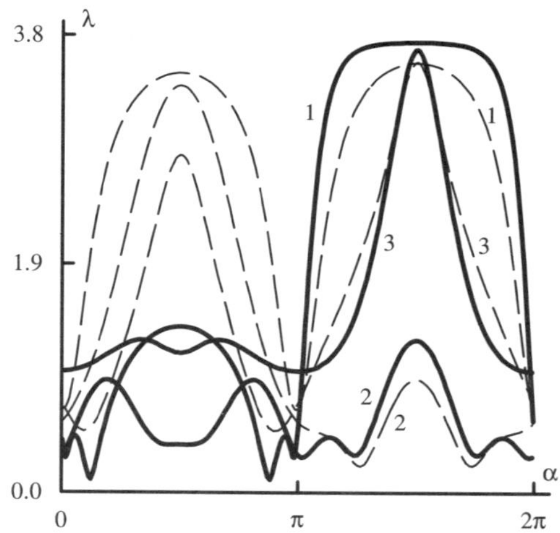

Fig. 5.26. Changes of λ on the contour of an inclusion in a halfspace (A = 1)

curves of the lines in Fig. 5.22 correspond to the release of the energy from the inclusion to the matrix.

5.6 Interaction of Openings and Cracks in a Space

In the Cartesian coordinate system O, x_1, x_2, x_3, consider a piezoceramic space weakened by a tunnel in the direction of x_3 -axis, L_j $(j = 1, 2, \ldots, k)$ cracks and openings C_m $(m = 1, 2, \ldots, n)$. A monochrome plane shear wave radiates from infinity, at an angle β with the x_1-axis, with a displacement component $u_3^{(0)} = \mathrm{Re}\left(U_3^{(0)} e^{-i\omega t}\right)$ (Fig. 5.27). We assume that the surfaces of the cavities are electroded and grounded, while stresses of longitudinal shear $Z_n = \mathrm{Re}\left(Z e^{-i\omega t}\right)$ may be applied (independent of the x_3 -coordinate). On the edges of the cracks may also act self-balanced shear forces $X_{3n} = \mathrm{Re}\left(X_3 e^{-i\omega t}\right)$ harmonic with time.

Assume that the curvature of contours C_m and L_j, and also the amplitudes of forces $Z(x_1,\ x_2)$ and $X_3(x_1,\ x_2)$ are the functions of Hölder's class and besides $\cap L_j = \emptyset,\ C_m \cap L_j = \emptyset$.

The total wave field of the displacement satisfying Helmhöltz (4.46) may be represented in the form

$$u_3 = \mathrm{Re}\left(U_3 e^{-i\omega t}\right), \quad U_3 = U_3^{(0)} + U_3^{*} \tag{5.21}$$

Here U_3^{*} is the displacement conditioned by the presence of tunnel heterogeneities, while the quantity $U_3^{(0)}$ is defined in (4.30).

The electric boundary conditions on the surface of the openings and cracks read

$$\phi = 0 \ \ (E_s = -\partial\phi/\partial s = 0) \ \text{ on } C_m \, (m = 1, 2, \ldots, n)\,, \tag{5.22}$$

$$E_s^{+} = E_s^{-}\,,\ D_n^{+} = D_n^{-} \ \text{ on } L_j \, (j = 1, 2, \ldots, k)\,. \tag{5.23}$$

Here ϕ is the potential of an electric field, E_s and D_n are tangential components of the vector of electric strength and the normal component of the vector of electric induction, s is the arc-length abscissa.

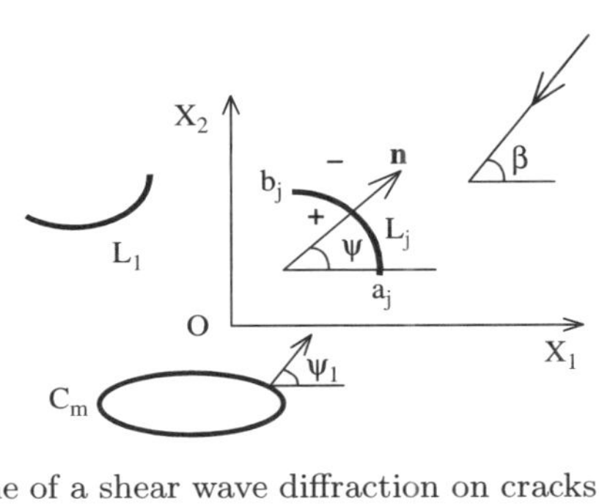

Fig. 5.27. The scheme of a shear wave diffraction on cracks and opening in a space

Based on expressions (4.25), (4.29) for the components of an electroelastic field corresponding to the state of an antiplane deformation let us write down the sought-for functions U_3 and E as a superposition of the integral representations of type (5.3) and (4.48)

$$U_3(x_1, x_2) = U_3^{(0)} + \int_C p(\zeta^*) H_0^{(1)}(\gamma r_1^*)\, ds - \frac{1}{4} \int_L [U_3] \frac{\partial H_0^{(1)}(\gamma r^*)}{\partial n_\zeta} ds,$$

$$F(x_1, x_2) = \frac{e_{15}}{\pi \ni_{11}^{\varepsilon}} \left\{ \frac{1}{2} \int_L \left[\frac{dU_3}{ds} \right] \ell n r^* ds + \int_C f(\zeta^*) \ell n r_1^* ds \right\}, \tag{5.24}$$

$$r^* = |\zeta - z|, \ r_1^* = |\zeta^* - z|, \ z = x_1 + ix_2, \ \zeta \in L = \cup L_j, \ \zeta^* \in C = \cup C_m.$$

Here $H_n^{(1)}(x)$ is the Hankel's function of first kind and of n-th order, the square brackets denote a jump of the corresponding quantity on L, ds is the element of a contour arc over which the integration is carried out.

The integral representations of solutions (5.24) are correct in the sense that they satisfy differential (4.46), provide the existence of displacement jump U_3^* on the cracks, the continuity of the strength vector through L_j, and also the automatic satisfaction of electric conditions (5.23). Appearing into (5.24) densities $[U_3]$, $p(\zeta)$ and $f(\zeta^*)$ should be determined using the mechanical boundary conditions on contours C and L, and also electrical conditions (5.22). The latter taking into account (4.29), may be represented as follows

$$c_{44}^E (1 + k_{15}^2) \left\{ e^{i\psi_1} \left(\frac{\partial U_3}{\partial \zeta^*} \right)^+ + e^{-i\psi_1} \left(\frac{\partial U_3}{\partial \bar{\zeta}^*} \right)^+ \right\}$$
$$- ie_{15} \left\{ e^{i\psi_1} \left(\frac{\partial F}{\partial \zeta^*} \right)^+ + e^{-i\psi_1} \left(\frac{\partial F}{\partial \bar{\zeta}^*} \right)^+ \right\} = Z,$$
$$\frac{ie_{15}}{\varepsilon_{11}} \left\{ e^{i\psi_1} \left(\frac{\partial U_3}{\partial \zeta^*} \right)^+ - e^{-i\psi_1} \left(\frac{\partial U_3}{\partial \bar{\zeta}^*} \right)^+ \right\} + e^{i\psi_1} \left(\frac{\partial F}{\partial \zeta^*} \right)^+ + e^{-i\psi_1} \left(\frac{\partial F}{\partial \bar{\zeta}^*} \right)^+ = 0, \tag{5.25}$$
$$c_{44}^E (1 + k_{15}^2) \left\{ e^{i\psi} \left(\frac{\partial U_3}{\partial \zeta} \right)^+ + e^{-i\psi} \left(\frac{\partial U_3}{\partial \bar{\zeta}} \right)^+ \right\}$$
$$- ie_{15} \left\{ e^{i\psi} \left(\frac{\partial F}{\partial \zeta} \right)^+ - e^{-i\psi} \left(\frac{\partial F}{\partial \bar{\zeta}} \right)^+ \right\} = X_3.$$

Here the corresponding variables are calculated at their limiting values at $z \to \zeta^* \in C$ and $z \to \zeta \in L$ from the body area; the functions $\psi = \psi(\zeta)$ and $\psi_1 = \psi_1(\zeta^*)$ denote the angles between axes x_1 and the normals to contour L and C, respectively. The second condition in (5.25) expresses the equality to zero of the tangential component of the vector of electric strength on the outer electroded boundary of the cylinder. Due to the fact that representations (5.24) provide the continuity of the stress vector on the cracks it suffices to satisfy the mechanical boundary condition on one of the edges (for example, on the left one).

Substituting the limiting values of function (5.24) in to boundary condition (5.25) we get a system of three integrodifferential equations of the second kind

$$\begin{aligned}
&\int_L \left\{ \left[\frac{dU_3}{ds} \right] g_1(\zeta,\zeta_0) + [U_3]\, g_2(\zeta,\zeta_0) \right\} ds + \\
&+ \int_C \{ p(\zeta^*)\, g_3(\zeta^*,\zeta_0) + f(\zeta^*)\, g_4(\zeta^*,\zeta_0) \} ds = N_1(\zeta_0), \\
&- p(\zeta_0^*) + \int_L \left\{ \left[\frac{dU_3}{ds} \right] g_5(\zeta,\zeta_0^*) + [U_3]\, g_6(\zeta,\zeta_0^*) \right\} ds + \\
&+ \int_C \{ p(\zeta^*)\, g_7(\zeta^*,\zeta_0^*) + f(\zeta^*)\, g_8(\zeta^*,\zeta_0^*) \} ds = N_2(\zeta_0), \\
&f(\zeta_0^*) + \int_L \left\{ \left[\frac{dU_3}{ds} \right] g_9(\zeta,\zeta_0^*) + [U_3]\, g_{10}(\zeta,\zeta_0^*) \right\} ds + \\
&+ \int_C \{ p(\zeta^*)\, g_{11}(\zeta^*,\zeta_0^*) + f(\zeta^*)\, g_{12}(\zeta^*,\zeta_0^*) \} ds = N_3(\zeta_0),
\end{aligned} \tag{5.26}$$

where the kernels g_m $(m = 1, 2, \ldots, 12)$ and the right parts are determined by the expressions

$$\begin{aligned}
g_1(\zeta,\zeta_0) &= \frac{1}{2}\mathrm{Im}\frac{e^{i\psi_0}}{\zeta-\zeta_0}, \quad g_8(\zeta^*,\zeta_0^*) = \frac{\kappa}{2\pi i}\mathrm{Im}\frac{e^{i\psi_{10}}}{\zeta^*-\zeta_0^*}, \\
g_2(\zeta,\zeta_0) &= \frac{i\pi}{8}\gamma^2\left(1+k_{15}^2\right)\left[H_2(\gamma r_0)\cos(\psi+\psi_0-2\alpha_0)\right. \\
&\quad \left. -H_0^{(1)}(\gamma r_0)\cos(\psi-\psi_0)\right], \\
g_3(\zeta^*,\zeta_0) &= \pi\left(1+k_{15}^2\right)\gamma H_1^{(1)}(\gamma r_1)\cos(\psi_0-\alpha_1), \\
g_4(\zeta^*,\zeta_0) &= -k_{15}^2\mathrm{Im}\frac{e^{i\psi_0}}{\zeta^*-\zeta_0}, \quad g_5(\zeta,\zeta_0^*) = \frac{\kappa}{4\pi i}\mathrm{Im}\frac{e^{i\psi_{10}}}{\zeta^*-\zeta_0^*}, \\
g_6(\zeta,\zeta_0^*) &= -\frac{\gamma^2}{16}\left[H_2^{(1)}(\gamma r_2)\cos(\psi+\psi_{10}-2\alpha_2) - H_0^{(1)}(\gamma r_2)\cos(\psi-\psi_{10})\right], \\
g_7(\zeta^*,\zeta_0^*) &= -\frac{\gamma}{2i}H_1(\gamma r_3)\cos(\psi_{10}-\alpha_3) + \frac{1}{\pi}\mathrm{Re}\frac{e^{i\psi_{10}}}{\zeta^*-\zeta_0^*}, \\
g_9(\zeta,\zeta_0^*) &= -\frac{1}{2\pi}\mathrm{Re}\frac{e^{i\psi_{10}}}{\zeta-\zeta_0^*}, \quad g_{12}(\zeta^*,\zeta_0^*) = -\frac{1}{\pi}\mathrm{Re}\frac{e^{i\psi_{10}}}{\zeta^*-\zeta_0^*},
\end{aligned}$$

$$g_{10}(\zeta,\zeta_0^*) = -\frac{i\gamma^2}{8}\left[H_2^{(1)}(\gamma r_2)\sin(\psi+\psi_{10}-2\alpha_2) + H_0^{(1)}(\gamma r_2)\sin(\psi-\psi_{10})\right],$$

$$g_{11}(\zeta^*,\zeta_0^*) = -\gamma H_1(\gamma r_3)\sin(\psi_{10}-\alpha_3) + \frac{2i}{\pi}\mathrm{Im}\frac{e^{i\psi_{10}}}{\zeta^*-\zeta_0^*},$$

$$H_1(x) = \frac{2i}{\pi x} + H_1^{(1)}(x),\ H_2(x) = \frac{4i}{\pi x^2} + H_2^{(1)}(x),$$

$$N_1(\zeta_0) = \frac{\pi}{c_{44}^E}X_3(\zeta_0) + i\pi\gamma\tau\left(1+k_{15}^2\right)e^{i(\alpha_1\xi_{10}+\alpha_2\xi_{20})}\cos(\psi_0-\beta),$$

$$N_2(\zeta_0^*) = -\frac{Z}{2ic_{44}^E(1+k_{15}^2)} - \frac{\gamma\tau}{2}e^{i(\alpha_1\xi_{10}^*+\alpha_2\xi_{20}^*)}\cos(\psi_{10}-\beta),\ \kappa = \frac{k_{15}^2}{1+k_{15}^2},$$

$$N_3(\zeta_0^*) = -i\gamma\tau e^{i(\alpha_1\xi_{10}^*+\alpha_2\xi_{20}^*)}\sin(\psi_{10}-\beta),$$

$$r = |\zeta-\zeta_0|,\ r_1 = |\zeta^*-\zeta_0|,\ r_2 = |\zeta-\zeta_0^*|,\ r_3 = |\zeta^*-\zeta_0^*|,$$

$$\alpha = \arg(\zeta-\zeta_0),\ \alpha_1 = \arg(\zeta^*-\zeta_0),\ \alpha_2 = \arg(\zeta-\zeta_0^*),\ \alpha_3 = \arg(\zeta^*-\zeta_0^*),$$

$$\psi = \psi(\zeta),\ \psi_0 = \psi(\zeta_0),\ \psi_{10} = \psi_1(\zeta_0^*);$$

$$\zeta = \xi_1 + i\xi_2,\ \zeta_0 = \xi_{10} + i\xi_{20};\ \zeta^* = \xi_1^* + i\xi_2^*,\ \zeta_0^* = \xi_{10}^* + i\xi_{20}^*;$$

$$\zeta,\zeta_0 \in L = \cup L_j,\ \zeta^*,\zeta_0^* \in C = \cup C_m.$$

To complete the algorithm it is necessary to augment the system (5.26) with the additional conditions

$$\int\limits_{L_j}\left[\frac{dU_3}{ds}\right]ds = 0\ \ (j = 1,2,\ldots,k), \tag{5.27}$$

The above conditions express the absence of displacement jumps U_3 on the tips of the cracks.

Here we should note that when $k_{15} = 0$ the two first integrodifferential conditions (5.26) correspond to a piezopassive (isotropic) space with defects.

To determine the stress intensity factor let us introduce the parameterization of contour L

$$\zeta = \zeta(\delta),\ \left[\frac{dU_3}{ds}\right] = \frac{\Omega(\delta)}{s'(\delta)\sqrt{1-\delta^2}},\ s'(\delta) = \frac{ds}{d\delta} > 0,\ \delta \in [-1,1]. \tag{5.28}$$

The asymptotic analysis of quantity $\partial_i U_3$, $\partial_i F\ (i = 1,\ 2)$ given by (5.24) permits us to find the main singular part of shear stress on their continuation beyond the tips of the cracks by applying formulas (4.29), (5.28)

$$\tau_n^0 = \pm c_{44}^E\frac{\mathrm{Re}\left[e^{-i\omega t}\Omega(\pm 1)\right]}{2\sqrt{2r_* s'(\pm 1)}},\ s'(\pm 1) = \left.\frac{ds}{d\delta}\right|_{\delta=\pm 1}. \tag{5.29}$$

Here the lower sign corresponds to tip a, the upper one to tip b, r_* is the distance from the point to the tip. From (5.28) we obtain the following expression for the intensity stress

$$K_{III}^{\pm} = \lim_{r\to 0}\sqrt{2\pi r_*}\tau_n^0 = \pm c_{44}^E\frac{\mathrm{Re}\left[e^{-i\omega t}\Omega(\pm 1)\right]}{2\sqrt{s'(\pm 1)}}. \tag{5.30}$$

To determine the concentration of stresses in a medium with cracks let us evaluate tangential stress $\tau_{\vartheta} = \tau_{23}\cos\psi - \tau_{13}\sin\psi$ on C. We have

$$T(\zeta_0^*) = ic_{44}^E\left(1+k_{15}^2\right)\left(e^{i\psi_{10}}\frac{\partial U_3}{\partial \zeta_0^*} - e^{-i\psi_{10}}\frac{\partial U_3}{\partial \overline{\zeta_0^*}}\right) + e_{15}\left(e^{i\psi_{10}}\frac{\partial F}{\overline{\partial \zeta_0^*}} + e^{-i\psi_{10}}\frac{\partial F}{\overline{\partial \zeta_0^*}}\right), \tag{5.31}$$

$$\tau_{\vartheta} = \mathrm{Re}\left(Te^{-i\omega t}\right), \ \psi_{10} = \psi_1(\zeta_0^*), \ \zeta_0^* \in C.$$

In (5.31) the partial derivatives of the corresponding quantities are calculated at their limiting values at $z \to \zeta_0^*$ of the medium area. By applying formula (5.24) we find

$$T(\zeta_0^*) = c_{44}^E\left(1+k_{15}^2\right)\left\{\int_L [U_3]\, g_{10}(\zeta,\zeta_0^*)ds - \int_C p(\zeta^*)g_{11}(\zeta^*,\zeta_0^*)\,ds + S(\zeta_0^*)\right\}$$
$$- \frac{c_{44}^E k_{15}^2}{\pi}\left\{\frac{1}{2}\int_L\left[\frac{dU_3}{ds}\right]\mathrm{Re}\frac{e^{i\psi_{10}}}{\zeta-\zeta_0^*}ds - \pi f(\zeta_0^*) + \int_C f(\zeta^*)\,\mathrm{Re}\frac{e^{i\psi_{10}}}{\zeta^*-\zeta_0^*}ds\right\} \tag{5.32}$$

For quantity $S(\zeta_0^*)$ we have the following cases: 1) $S(\zeta_0^*) = 0$ when shear forces act on the opening or cracks and 2) $S(\zeta_0^*) = i\gamma\tau e^{i(\alpha_1\xi_{10}^*+\alpha_2\xi_{20}^*)}\sin(\psi_{10}-\beta)$ when we have wave radiation.

As an example consider a medium from ceramics PZT-4 weakened by an opening of an elliptic cross-section and rectangular crack. The parametric equations of contours C and L have, respectively, the following form

$$\mathrm{Re}\zeta^* = R_1\cos\varphi, \ \mathrm{Im}\zeta^* = R_2\sin\varphi \ \ (0 \le \varphi < 2\pi),$$
$$\mathrm{Re}\zeta = p_1\delta + p_0, \ \mathrm{Im}\zeta = h \ \ (-1 \le \beta \le 1). \tag{5.33}$$

The solution of the system of integrodifferential (5.26) together with (5.27) with the help of (5.28), (5.33) was obtained numerically using the method of quadratures (see Appendix 12.8, Par. B1, Par. B2). The stress intensity factor and the stress of a cavity contour were calculated by the formulas (5.30), (5.32).

Figures 5.28 and 5.29 show the changes of relative intensity factor $\langle K_{III}^+\rangle = c_{44}^E\,|\Omega(1)|\,/\left(2S_{23}^0\sqrt{\ell s'(1)}\right)$ as a function of the normalized wave number $\gamma^* R = \gamma R\sqrt{1+k_{15}^2}$ in the process of radiation of a shear wave ($Z = X_3 = 0,\ R = 0.5\,(R_1+R_2)$). In Fig. 5.28 the curves 1 and 2 correspond to the parameter values $\beta = \pi/2,\ R_1/R_2 = 1,\ p_1/R_1 = 1,\ p_0/R_1 = 0,\ h/R_1 = 0.6$ and 0.4, respectively; in Fig. 5.29 for the same values and $h/R_1 = -0.6$ and -0.4. The dashed line 1 satisfies the piezopassive piecewise-homogeneous medium

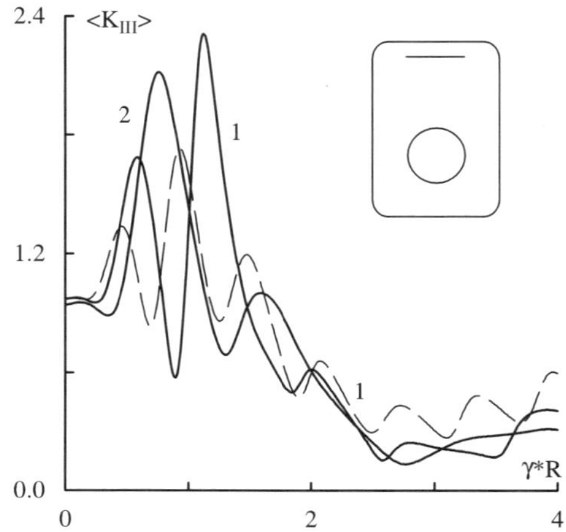

Fig. 5.28. Behaviour of $< K_{III}^{+} >$ as a function of parameter $\gamma^* R$ for various positions of an opening and a crack in a space

($k_{15} = 0$). Assuming that we have already at hand the value of the quantity $\langle K_{III}^{\pm} \rangle$ we may define the corresponding stress intensity factor by formula

$$K_{III}^{\pm} = \pm\sqrt{\pi \ell} S_{23}^{0} \langle K_{III}^{\pm} \rangle \cos\left\{\omega t - \arg \Omega\left(\pm 1\right)\right\}. \tag{5.34}$$

Here $S_{23}^{0} = \tau \gamma c_{44}^{E} \left(1 + k_{15}^{2}\right)$ is the modulus of the stress amplitude τ_{23} in the shear wave, 2ℓ is the length of the crack.

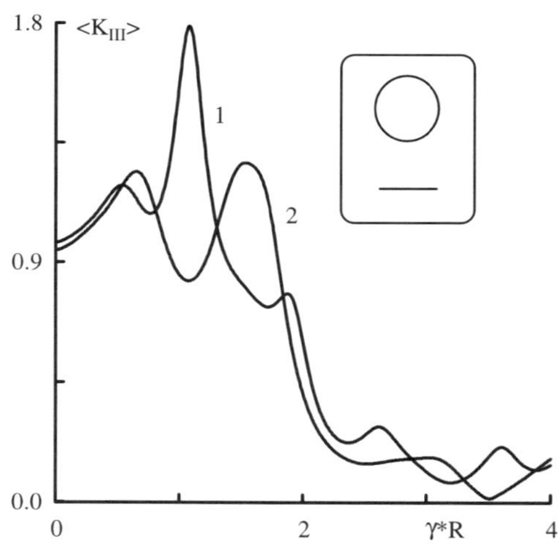

Fig. 5.29. Behaviour of $< K_{III}^{+} >$ as a function of parameter $\gamma^* R$ for various positions of an opening and a crack in a space

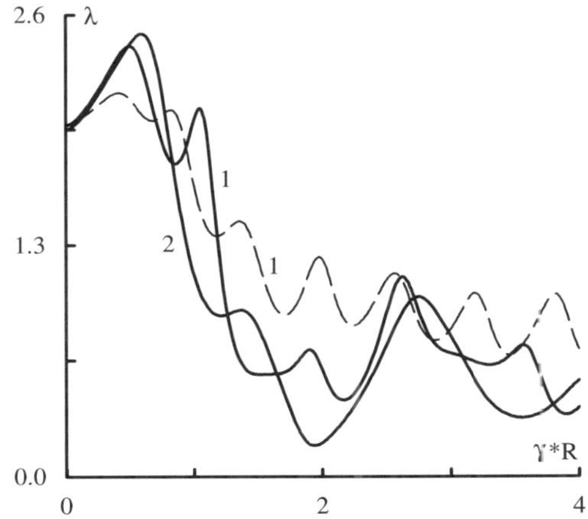

Fig. 5.30. Behaviour of relative shear stresses $\lambda = |T/S_{23}^0|$ at the point of contour $\varphi = \pi$ as function of γ^*R for various positions of an opening and a crack in a space

Figures 5.30 and 5.31 demonstrate the behavior of the relative shear stress $\lambda = \left|T/S_{23}^0\right|$ at the point of contour $\varphi = \pi$ as the function $\gamma^* R$ for the same parameters, using the same relations as the ones in Figs. 5.28 and 5.29.

From the results we conclude that the parameters of rigidity and fracture depend significantly on the frequency of the harmonic loading and the mutual location of the defects. For example, as it follows from Fig. 5.28, the quantity

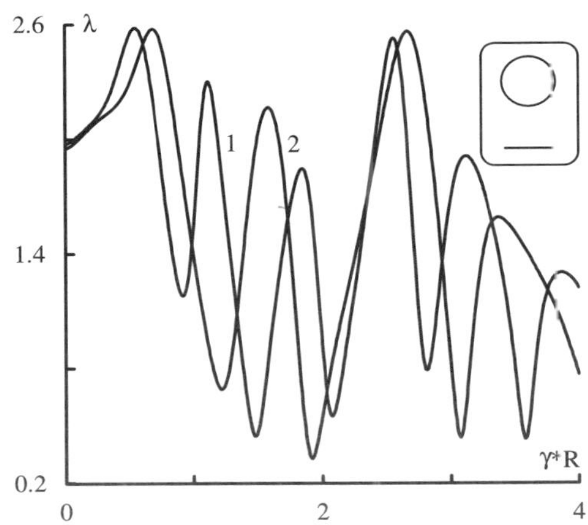

Fig. 5.31. Behaviour of relative shear stress $\lambda = |T/S_{23}^0|$ at the point of contour $\varphi = \pi$ as function of $\gamma^* R$ for various positions of an opening and a crack in a space

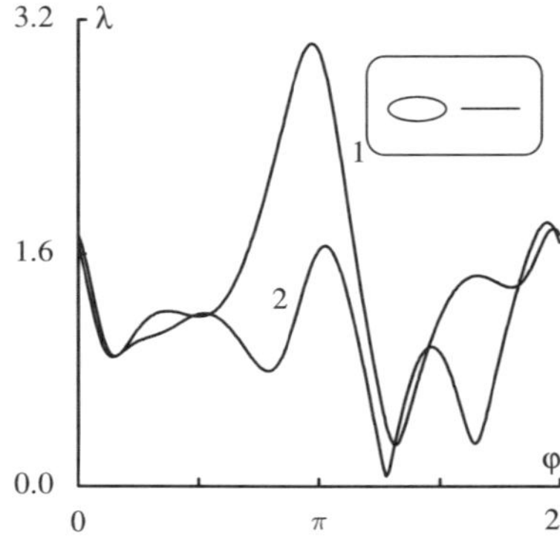

Fig. 5.32. Changes of λ on the contour of an elliptic cavity with an inclusion crack under the influence of a shear wave ($\beta = \pi/4$)

$\langle K_{III}^{+} \rangle$ may overcome its static analogue approximately 2.3 times (Curve 1). If the wave falls in such a way that the crack is in the "*shadow*" of the opening (Fig. 5.29), the maximum value $\langle K_{III}^{+} \rangle$ in the considered frequency range is smaller than in the situation, corresponding to Fig. 5.28.

Figure 5.32 illustrates the changes of λ on the contour of an elliptic cavity in the vicinity of an inclusion under the influence of shear waves ($\beta = \pi/4$). Curves 1 and 2 satisfy values $R_1/R_2 = 2$, $p_0/R_1 = 2.5$, $p_1/R_1 = 1$, $h/R_1 = 0$ and $\gamma^* R = 1$ and 2, respectively.

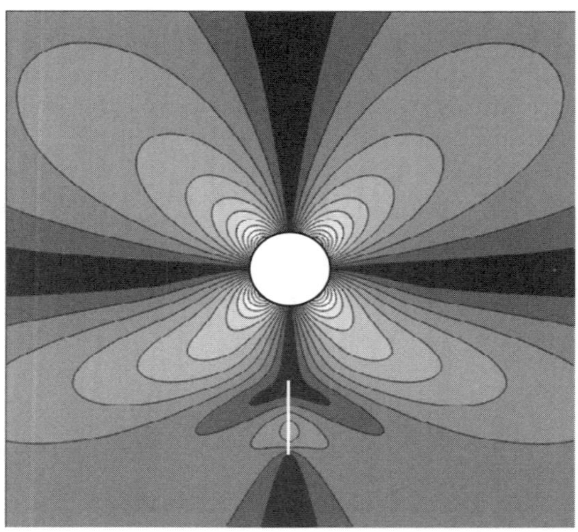

Fig. 5.33. The contour lines of the modulus of the displacement amplitude in the area with an opening and a crack under the influence of shear forces $Z = \sin(2\varphi)$ on the opening

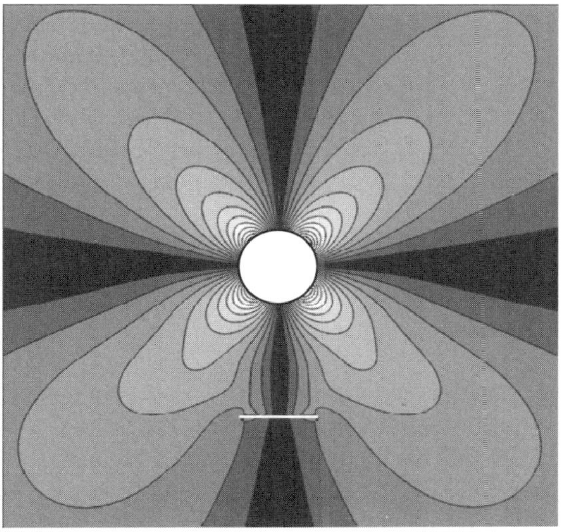

Fig. 5.34. The contour lines of the modulus of the displacement amplitude in the area with an opening and a crack under the influence of shear forces Z = $\sin(2\varphi)$ on the opening

The influence of the effect of connection of mechanical and electric fields on the behaviour of the investigated quantities in some frequency ranges appears to be significant (see Figs. 5.28 and 5.30).

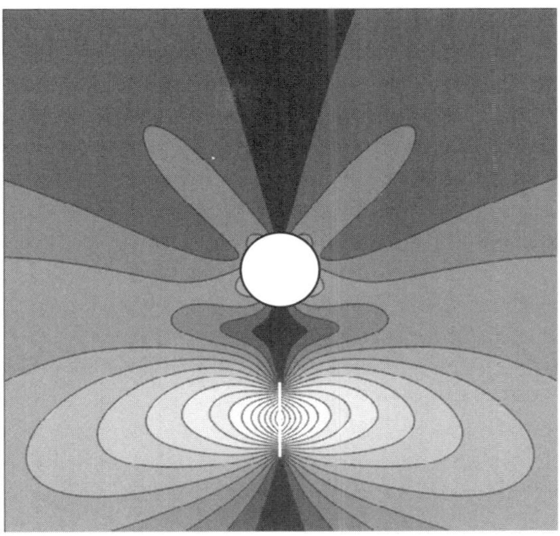

Fig. 5.35. The contour lines of the modulus of the displacement amplitude in the area with an opening and a crack, under the influence of shear forces X_3 = const, on the opening

Figures 5.33 and 5.34 illustrate the contour lines of the displacement amplitude modulus in the area covering an opening and a crack of different orientation when on the surface of the opening there act shear forces $Z = \sin(2\varphi)$ $(R_1/R_2 = 1,\ p_1/R_1 = 1,\ p_0/R_1 = 0,\ h/R_1 = -4,\ \gamma^* R = 2)$. The analogous result for the same parameter values are given in Fig. 5.35 when the shear forces of constant intensity ($X_3 = const$) act on the crack edges.

5.7 Fundamental Solution for a Composite Anisotropic Space

In this section we consider a composite anisotropic space referring to Cartesian coordinates x_1, x_2, x_3 and effected by, the concentrated on line $x_1 = \xi_1,\ x_2 = \xi_2 > 0, -\infty < x_3 < \infty$ harmonically changing with time shear forces $q = \mathrm{Re}\left(Q\delta\left(x_1 - \xi_1, x_2 - \xi_2\right)e^{-i\omega t}\right)$ of constant along axis x_3 intensity (t is the time, ω is the circular frequency, $\delta(x, y)$ is Dirac-delta function). It is assumed that the materials of the composite space with respect to the elastic properties are orthotropic.

In the given conditions in a composite space a steady wave process occurs, corresponding to the state of antiplane deformation. The system of equations of the problem includes the following relations [134]

$$\partial_1\sigma_{13}^{(r)} + \partial_2\sigma_{23}^{(r)} = \rho_r\frac{\partial^2 u_3^{(r)}}{\partial t^2} - \delta_r^1 q,\ \partial_j = \partial/\partial x_j, \tag{5.35}$$

$$\sigma_{23}^{(r)} = c_{44}^{(r)}\partial_2 u_3^{(r)},\ \sigma_{13}^{(r)} = c_{55}^{(r)}\partial_1 u_3^{(r)}\ \ (r = 1, 2) \tag{5.36}$$

Here (5.35) is the equation of movement, (5.36) are the equations of the medium state, c_{ij} are the moduli of the material elasticity; ρ is the material density, δ_r^j is the Kronecker delta. Index "r" $(r = 1, 2)$ for all the quantities referring to r-th half-space; $x_2 > 0$ if $r = 1$ and $x_2 \leq 0$ if $r = 2$.

On the boundary of interphase $x_2 = 0$ it is necessary to require the performance of conditions of an ideal mechanical

$$u_3^{(1)} = u_3^{(2)},\ \ \sigma_{23}^{(1)} = \sigma_{23}^{(2)} \tag{5.37}$$

From (5.35), (5.36) we obtain Helmhöltz equations for the displacement amplitude in a composite space

$$\begin{aligned} &\nabla^2 U_3^{(1)} + \gamma_1^2 U_3^{(1)} = -\frac{Q}{\sqrt{\Delta_1}}\delta\left(x_1 - \xi_1, \eta_1 - \chi\right)\ (x_2 > 0) \\ &\nabla_2^2 U_3^{(2)} + \gamma_2^2 U_3^{(2)} = 0\ \ (x_2 \leq 0) \end{aligned} \tag{5.38}$$

$$\nabla_r^2 = \frac{\partial^2}{\partial x_1^2} + \frac{\partial^2}{\partial \eta_r^2}$$

$$u_3^{(r)} = \mathrm{Re}\left(U_3^{(r)} e^{-i\omega t}\right), \ \gamma_r = \frac{\omega}{c_r}, \ c_r = \sqrt{\frac{\Delta_r}{\rho_r c_{44}^{(r)}}}, \ \Delta_r = c_{44}^{(r)} c_{55}^{(r)}$$

$$\eta_r = x_2 \ \mathrm{Im}\ \mu_r, \ \zeta = \xi_2 \ \mathrm{Im}\ \mu_1 > 0, \ \mu_r = i\frac{\sqrt{\Delta_r}}{c_{44}^{(r)}} \ (r = 1, 2)$$

Thus, the problem is reduced to the definition of the function from differential (5.38), conjugation (5.37) and also from the conditions at infinity.

To solve the problem let us apply integral Fourier transform to (5.38)

$$F(p) = \frac{1}{\sqrt{2\pi}} \int_{-\infty}^{\infty} f(x_1) e^{-ipx_1} dx_1, \ f(x_1) = \frac{1}{\sqrt{2\pi}} \int_{-\infty}^{\infty} F(p) e^{ipx_1} dp \tag{5.39}$$

As a result we come to ordinary differential equations with respect to the spectral displacement functions

$$\begin{aligned} &\frac{d^2 \hat{U}_3^{(1)}}{d\eta_1^2} + \left(\gamma_1^2 - p^2\right) \hat{U}_3^{(1)} = -\frac{Q}{\sqrt{2\pi}\sqrt{\Delta_1}} e^{-ip\xi_1} \delta(\eta_1 - \chi) \ (x_2 > 0) \\ &\frac{d^2 \hat{U}_3^{(2)}}{d\eta_2^2} + \left(\gamma_2^2 - p^2\right) \hat{U}_3^{(2)} = 0 \ (x_2 \le 0) \end{aligned} \tag{5.40}$$

The general solutions of (5.40) providing the performance of radiation conditions [43] at infinity may be represented as follows

$$\begin{aligned} \hat{U}_3^{(1)} &= A e^{-\lambda_1 \eta_1} + \frac{Q e^{-ip\xi_1}}{2\sqrt{2\pi}\sqrt{\Delta_1}\lambda_1} e^{-\lambda_1 |\eta_1 - \zeta|} \ (x_2 > 0) \\ \hat{U}_3^{(2)} &= B e^{\lambda_2 \eta_2} \ (x_2 \le 0) \\ \lambda_r &= \begin{cases} -i\sqrt{\gamma_r^2 - p^2}, \ \gamma_r > |p| \\ \sqrt{p^2 - \gamma_r^2}, \ \gamma_r < |p| \end{cases} \quad (r = 1, 2) \end{aligned} \tag{5.41}$$

Constants A and B are determined from conditions (5.37) on the boundary of conjugation of media $x_2 = 0$, which allowing for (5.36) in Fourier transformants have the following form

$$\begin{aligned} &\hat{U}_3^{(1)} = \hat{U}_3^{(2)} \\ &c_{44}^{(1)} \ \mathrm{Im}\ \mu_1 \frac{d\hat{U}_3^{(1)}}{d\eta_1} = c_{44}^{(2)} \ \mathrm{Im}\ \mu_2 \frac{d\hat{U}_3^{(2)}}{d\eta_2} \end{aligned} \tag{5.42}$$

Proceeding from (5.41), (5.42) we obtain the expressions for the spectral functions of the displacement amplitude

$$\begin{aligned}
\hat{U}_3^{(1)} &= \frac{Qe^{-ip\xi_1}}{2\sqrt{2\pi}\sqrt{\Delta_1\lambda_1}}e^{-\lambda_1|\eta_1-x_1|} + \frac{Q\alpha(p)}{2\sqrt{2\pi}\sqrt{\Delta_1\lambda_1}}e^{-ip\xi_1}e^{-\lambda_1(\eta_1+\chi)} \quad (x_2>0)\\
\hat{U}_3^{(2)} &= \frac{Q\left[1+\alpha(p)\right]}{2\sqrt{2\pi}\sqrt{\Delta_1\lambda_1}}e^{-ip\xi_1}e^{-\lambda_1\chi}e^{\lambda_2\eta_2} \quad (x_2\le 0) \qquad (5.43)\\
\alpha(p) &= \frac{\lambda_1\sqrt{\Delta_1}-\lambda_2\sqrt{\Delta_2}}{\lambda_1\sqrt{\Delta_1}+\lambda_2\sqrt{\Delta_2}}
\end{aligned}$$

Using the initial notation, according to (5.39), we find

$$\begin{aligned}
U_3^{(1)}(x_1,x_2;\xi_1,\xi_2) &= \frac{Q}{2\pi\sqrt{\Delta_1}}\int\limits_0^{\infty}\frac{\cos p(x_1-\xi_1)}{\lambda_1}e^{-\lambda_1|x_2-\xi_2|\mathrm{Im}\mu_1}dp+ \qquad (5.44)\\
&+\frac{Q}{4\pi\sqrt{\Delta_1}}\int\limits_{-\infty}^{\infty}\frac{\alpha(p)}{\lambda_1}e^{ip(x_1-\xi_1)}e^{-\lambda_1(x_2+\xi_2)\mathrm{Im}\mu_1}dp \quad (x_2>0)\\
U_3^{(2)}(x_1,x_2;\xi_1,\xi_2) &= \frac{Q}{4\pi\sqrt{\Delta_1}}\int\limits_{-\infty}^{\infty}\frac{1+\alpha(p)}{\lambda_1}e^{ip(x_1-\xi_1)}e^{-\lambda_1\xi_2\mathrm{Im}\mu_1}e^{\lambda_2 x_2\mathrm{Im}\mu_2}dp\\
&(x_2\le 0)
\end{aligned}$$

Integral in (5.44) prescribed on a semi-infinite interval according to the radiation conditions will be understood in the general meaning [43]

$$\begin{aligned}
I &= \frac{Q}{2\pi\sqrt{\Delta_1}}\int\limits_0^{\infty}\frac{\cos p(x_1-\xi_1)}{\lambda_1}e^{-\lambda_1|x_2-\xi_2|\mathrm{Im}\mu_1}dp=\\
&= \frac{Q}{2\pi\sqrt{\Delta_1}}\lim_{\varepsilon\to+0}\int\limits_0^{\infty}\frac{e^{-\sqrt{p^2+\gamma_*^2}|x_2-\xi_2|\mathrm{Im}\mu_1}}{\sqrt{p^2+\gamma_*^2}}\cos p(x_1-\xi_1)\,dp\\
\gamma_* &= -i\sqrt{\gamma_1^2+i\varepsilon}, \qquad \varepsilon>0,\ \mathrm{Re}\gamma_*>0
\end{aligned}$$

Using the value of the integral :

$$\int\limits_0^{\infty}\frac{e^{-c\sqrt{x^2+z^2}}}{\sqrt{x^2+z^2}}\cos bx\,dx = K_0\left(z\sqrt{b^2+c^2}\right) \quad (b,\mathrm{Re}c,\mathrm{Re}z>0)$$

and the connection between functions of MacDonald and Hankel [109]

$$K_0(-ix) = \frac{\pi i}{2} H_0^{(1)}(x)$$

we find

$$I = \frac{iQ}{4\sqrt{\Delta_1}} H_0^{(1)}(\gamma_1 \rho_*) \tag{5.45}$$

$$\rho_* = |z_1 - z_*|, \; z = x_1 + \mu_1 x_2, \; z_* = \xi_1 + \mu_1 \xi_2$$

Thus, the expression for displacement amplitude when $x_2 > 0$ may be represented in the following form

$$U_3^{(1)}(x_1, x_2; \xi_1, \xi_2) = I + \frac{Q}{4\pi\sqrt{\Delta_1}} \int_{-\infty}^{\infty} \frac{\alpha(p)}{\lambda_1} e^{ip(x_1-\xi_1)} e^{-\lambda_1(x_2+\xi_2)\mathrm{Im}\mu_1} dp \tag{5.46}$$

where quantity I is the fundamental solution for a homogeneous anisotropic space. Using the equations of state (5.36) and formulas (5.44)–(5.46) we will determine the expressions for the stress amplitudes in a composite anisotropic space. Taking into account the formulas of differentiation (4.10) and equality

$$\partial_1 = \frac{\partial}{\partial z_1} + \frac{\partial}{\partial \overline{z_1}}, \; \partial_2 = \mu_1 \frac{\partial}{\partial z_1} + \bar{\mu}_1 \frac{\partial}{\partial \bar{z}_1}$$

we will have

$$\sigma_{kj}^{(r)} = \mathrm{Re}\left(S_{kj}^{(r)} e^{-i\omega t}\right) \quad (r = 1, 2) \tag{5.47}$$

$$S_{23}^{(1)}(x_1, x_2; \xi_1, \xi_2) = -\frac{iQ}{4}\gamma_1 H_1^{(1)}(\gamma_1\rho_*)\sin\beta$$
$$-\frac{Q}{4\pi}\int_{-\infty}^{\infty} \alpha(p) e^{ip(x_1-\xi_1)} e^{-\lambda(x_2+\xi_2)\mathrm{Im}\mu_1} dp$$

$$S_{23}^{(2)}(x_1, x_2; \xi_1, \xi_2) = -\frac{Q}{4\pi}\sqrt{\frac{\Delta_2}{\Delta_1}} \int_{-\infty}^{\infty} \frac{\lambda_2}{\lambda_1}[1+\alpha(p)] e^{ip(x_1-\xi_1)} e^{-\lambda_1\xi_2\mathrm{Im}\mu_1} e^{\lambda_2 x_2 \mathrm{Im}\mu_2} dp$$

$$S_{13}^{(1)}(x_1, x_2; \xi_1, \xi_2) = -\frac{iQc_{55}^{(1)}}{4\sqrt{\Delta_1}}\gamma_1 H_1^{(1)}(\gamma_1\rho_*)\cos\beta$$
$$+\frac{iQc_{55}^{(1)}}{4\sqrt{\Delta_1}}\int_{-\infty}^{\infty} \frac{p\alpha(p)}{\lambda_1} e^{ip(x_1-\xi_1)} e^{-\lambda_1(x_2+\xi_2)\mathrm{Im}\mu_1} dp$$

$$S_{13}^{(2)}(x_1, x_2; \xi_1, \xi_2) = \frac{iQc_{55}^{(2)}}{4\pi\sqrt{\Delta_1}} \int_{-\infty}^{\infty} \frac{p}{\lambda_1}[1+\alpha(p)] e^{ip(x_1-\xi_1)} e^{-\lambda_1\xi_2\mathrm{Im}\mu_1} e^{\lambda_2 x_2 \mathrm{Im}\mu_2} dp$$

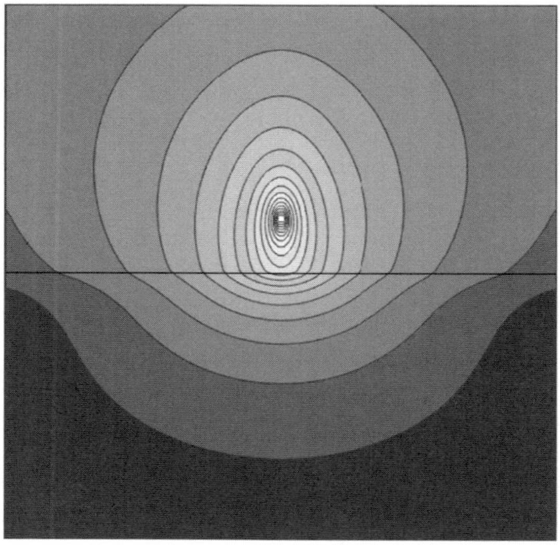

Fig. 5.36. The contour lines of the modulus of the displacement amplitude in composite space ($\gamma_2 = 4, c_{44}^{(1)} = 4, c_{44}^{(2)} = 1, c_{55}^{(1)} = 1, c_{55}^{(2)} = 2$)

It should be mentioned here that in the process of determining $S_{kj}^{(r)}$ in (5.47), we used the procedure of differentiation over the parameter with the sign of an improper integral, which is not admitted in this case.

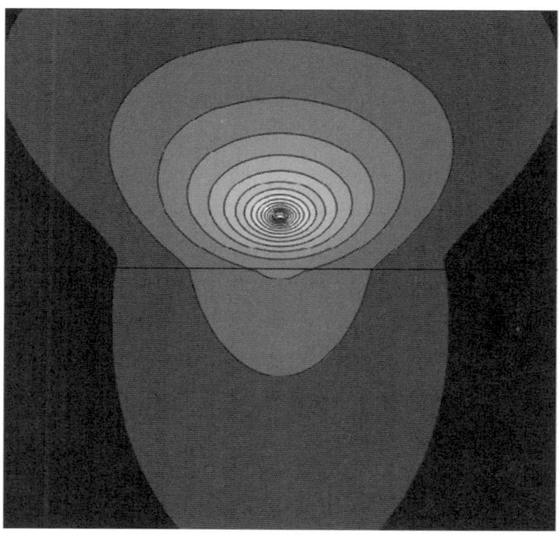

Fig. 5.37. The contour lines of the modulus of the displacement amplitude in composite space ($\gamma_2 = 4, c_{44} = 1, c_{44}^{(2)} = 4, c_{55}^{(1)} = 2, c_{55}^{(2)} = 1$)

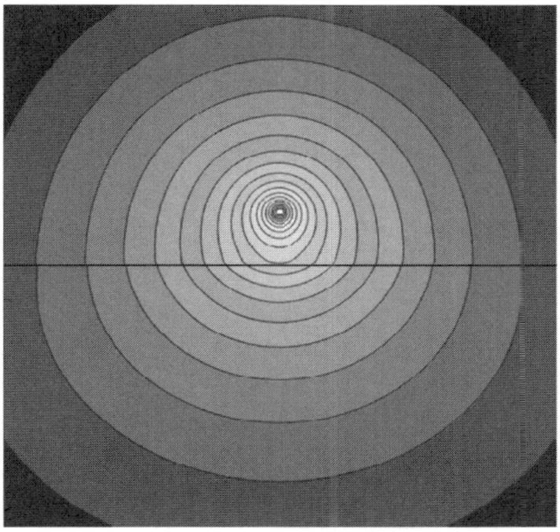

Fig. 5.38. The contour lines of the modulus of displacement amplitude in composite isotropic space ($\gamma_2 = 1, c_{44}^{(1)} = 3, c_{44}^{(2)} = 1, c_{55}^{(1)} = 3, c_{55}^{(2)} = 1$)

Let us investigate the distribution of elastic displacements and stresses in a composite anisotropic space under the influence of concentrated forces of shear depending on the character of the anisotropic material and frequency of harmonic loading. The contour line of the absolute values of displacements in the area covering point $(\xi_1, \ \xi_2)$ for different relations of the moduli of elasticity of the material are represented in Figs. 5.36–5.39. Figures 5.40 and 5.41 illustrate the behaviour of the contour lines of the moduli of the amplitude of stresses

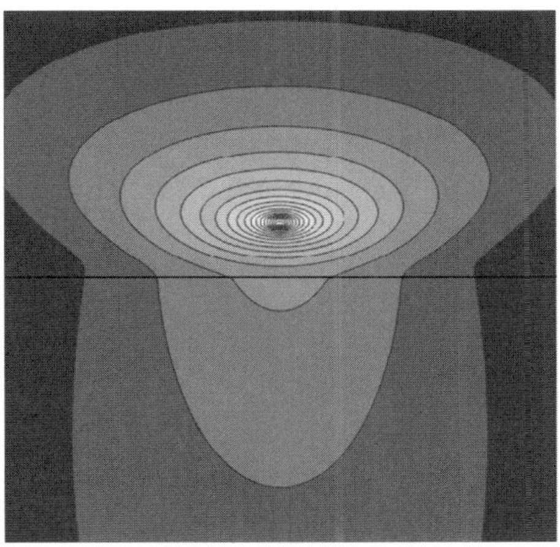

Fig. 5.39. The contour lines of the modulus of displacement amplitude in composite isotropic space ($\gamma_2 = 3, c_{44} = 1, c_{44}^{(2)} = 5, c_{55}^{(1)} = 5, c_{55}^{(2)} = 1$)

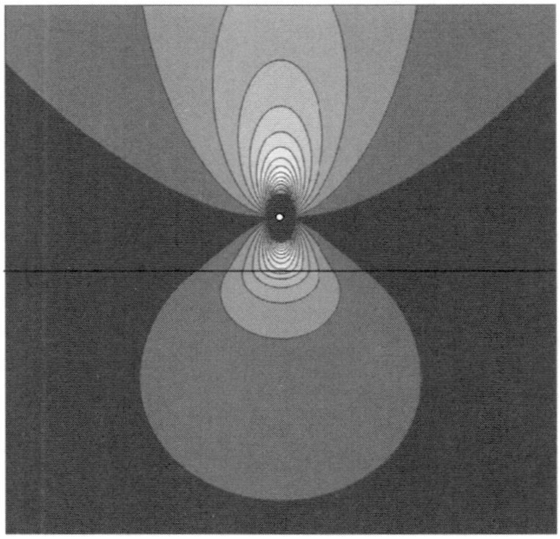

Fig. 5.40. The contour lines of the modulus of stress amplitude σ_{23} in a composite space ($\gamma_2 = 2, c_{44} = 3, c_{44}^{(2)} = 1, c_{55}^{(1)} = 1, c_{55}^{(2)} = 2$)

σ_{23} and σ_{13}, respectively. In calculations we assumed that the beginning of the system of coordinates is in the center of the considered square area and $\xi = 0,\ \xi_2/a = 0.1,\ p_1 = p_2$ (a is the length of a side of a square). More light regions correspond to maximum values of the investigated value.

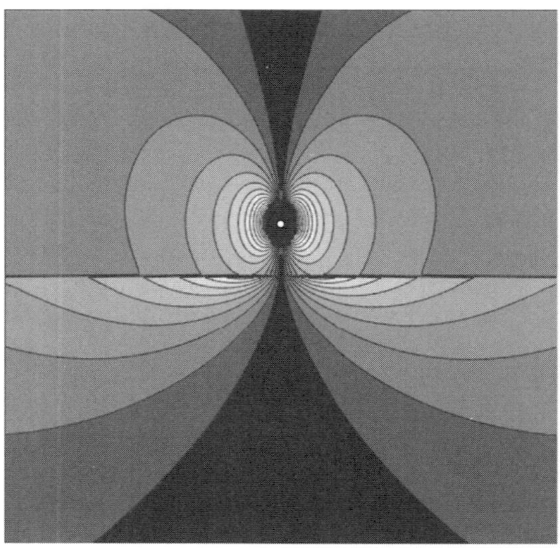

Fig. 5.41. The contour lines of the modulus of stress amplitude σ_{13} in a composite space ($\gamma_2 = 2, c_{44} = 3, c_{44}^{(2)} = 1, c_{55}^{(1)} = 1, c_{55}^{(2)} = 2$)

5.8 An Anisotropic Bimorph with Tunnel Openings

Let us use the constructed above fundamental solution for investigation of the stationary wave process in an orthotropic bimorph with heterogeneities of tunnel opening types.

In the Cartesian system of coordinates $x_1,\ x_2,\ x_3$ let a composite anisotropic space (bimorph) be weakened in the upper half-space $x_2 > 0$ by tunnel along axis x_3 openings $\Gamma_j\ (j = 1, 2, \ldots, n)$. Along line $x_1 = \xi_1,\ x_2 = \xi_2 > 0,\ -\infty < x_3 < \infty$ a concentrated, horizontal, changing with time, not depending on coordinate x_3 shear force $q = \operatorname{Re}\left(Q\delta\left(x_1 - \xi_1, x_2 - \xi_2\right) e^{-i\omega t}\right)$ acts, and on the surface opening shear stress $X_{3n} = \operatorname{Re}\left(X_3 e^{-i\omega t}\right)$ possibly appears. It is assumed that the elastic properties of the materials of the composite space are orthotropic, the curves of contour Γ_j and function X_3 belong to the class of functions infinite by Hölder $(\cap \Gamma_j = \emptyset)$.

Under the given conditions the composite space is a state of antiplane deformation described by the system of (5.35), (5.36). It is necessary to add coupling condition (5.37) to them, and also the boundary conditions on the surface of the openings at point $\zeta = \varepsilon_1 + i\varepsilon_2 \in \Gamma_j\ (j = 1, 2, \ldots, n)$

$$\sigma_{13}^{(1)} \cos\psi + \sigma_{23}^{(1)} \sin\psi = X_{3n}. \tag{5.48}$$

Here ψ is the angle between the normal to contour Γ_j at point ζ and axis Ox_1 (Fig. 5.42).

Assuming $u_3^{(r)} = \operatorname{Re}\left(U_3^{(r)} e^{-i\omega t}\right)$ we will write boundary condition (5.48) with respect to the amplitude of displacement as follows

$$a(\psi) \frac{\partial U_3^{(1)}}{\partial \zeta_1} - \overline{a(\psi)} \frac{\partial U_3^{(1)}}{\partial \bar{\zeta}_1} = \frac{iX_3}{\sqrt{\Delta_1}} \tag{5.49}$$
$$a(\psi) = \mu_1 \cos\psi - \sin\psi,\ \zeta_1 = \varepsilon_1 + \mu_1 \varepsilon_2,\ \xi_2 > 0,\ (\varepsilon_1, \varepsilon_2) \in \Gamma.$$

Here, by partial derivatives from the amplitude of displacement we imply their limiting values when the inner point of region z tends to boundary point $\zeta \in \Gamma$.

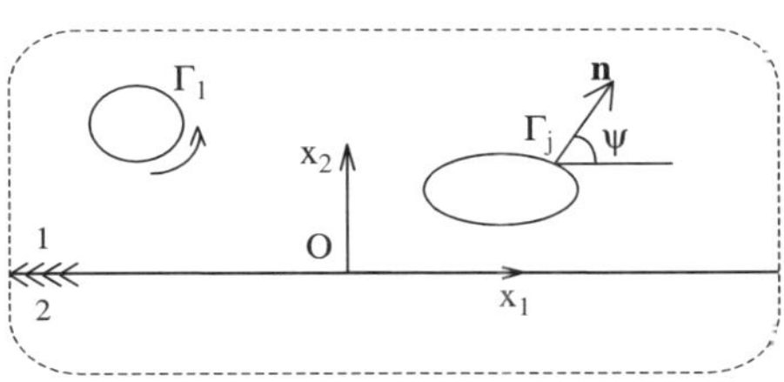

Fig. 5.42. A composite space with tunnel openings

Thus, it is required to determine the amplitudes of displacements in the bimorph from differential equations of Helmohöltz (5.38), coupling conditions (5.37), boundary conditions (5.49), and also the conditions of radiations at infinity.

The wave field of displacement in the bimorph with heterogenities is composed of a field caused by the action of a harmonic source and a field diffused by tunnel openings. Using the fundamental solution determined by formulas (5.44)–(5.46) we can represent the amplitude of displacement in region $x_2 > 0$ in the following way

$$U_3^{(1)}(z) = \int_\Gamma f(\zeta) K(\zeta, z)\, ds + U_3^*(z), \tag{5.50}$$

$$K(\zeta, z) = \frac{i}{4\sqrt{\Delta_1}} H_0^{(1)}(\gamma_1 \rho_*) + \frac{i}{4\pi\sqrt{\Delta_1}} \int_{-\infty}^{\infty} \frac{\alpha(p)}{\lambda_1} e^{ip(x_1-\varepsilon_1)} e^{-\lambda_1(x_2+\varepsilon_2)\mathrm{Im}\mu_1} dp$$

$$\rho_* = |z_1 - \zeta_1|,\ \zeta_1 = \varepsilon_1 + \mu_1\varepsilon_2,\ \zeta = \varepsilon_1 + i\varepsilon_2 \in \Gamma,\ z = x_1 + ix_2$$

Here, displacement $U_3^*(z)$ caused by the action of a source is determined by formulas (5.45), (5.46); ds is the element of the length of the arc of contour Γ.

Taking into account (5.44) we obtain the analogous representation in region $x_2 \le 0$

$$U_3^{(2)}(z) = \int_\Gamma f(\zeta) L(\zeta, z)\, ds + U_3^{**}(z), \tag{5.51}$$

$$L(\zeta, z) = \frac{1}{4\pi\sqrt{\Delta_1}} \int_{-\infty}^{\infty} \frac{1+\alpha(p)}{\lambda_1} e^{ip(x_1-\varepsilon_1)} e^{-\lambda_1\varepsilon_2\mathrm{Im}\mu_1} e^{\lambda_2 x_2 \mathrm{Im}\mu_2} dp,$$

where function $U_3^{**}(z)$ is determined by the second formula in (5.44).

Substituting the limiting values of derivative functions $U_3^1(z)$ at $z \to \zeta_0 \in \Gamma$ in boundary conditions (5.49) we come to the integral equation of the second kind

$$-\frac{f(\zeta_0)}{2} + \int_\Gamma f(\zeta) G(\zeta, \zeta_0)\, ds = N(\zeta_0), \tag{5.52}$$

$$\begin{aligned} G(\zeta, \zeta_0) = {} & \frac{\gamma_1}{4i} H_1^{(1)}(\gamma_1 P_*) \,\mathrm{Im}\left[w^{-i\beta_0} a(\psi_0)\right] \\ & + \frac{1}{8\pi}\left[a(\psi_0) I_1(\zeta, \zeta_0) - \overline{a(\psi_0)} I_2(\zeta, \zeta_0)\right] \end{aligned}$$

$$I_k(\zeta, \zeta_0) = \int_{-\infty}^{\infty} \left[\frac{p}{\lambda_1} - (-1)^k\right] \alpha(p)\, e^{ip(\varepsilon_{10}-\varepsilon_1)} e^{-\lambda_1(\varepsilon_{20}+\varepsilon_2)\mathrm{Im}\mu_1} dp,$$

$$N(\zeta_0) = X_3(\zeta_0) - QG^*(z_*, \zeta_0),$$

$$G^{*}\left(z_{*},\zeta_{0}\right)=\frac{\gamma_{1}}{4i}H_{1}^{(1)}\left(\gamma_{1}P_{*}^{0}\right)\operatorname{Im}\left[e^{-i\beta_{*}}a\left(\psi_{0}\right)\right]$$
$$+\frac{1}{8\pi}\left[a\left(\psi_{0}\right)M_{1}\left(z_{*},\zeta_{0}\right)-\overline{a\left(\psi_{0}\right)}M_{2}\left(z_{*},\zeta_{0}\right)\right]$$

$$M_{k}\left(z_{*},\zeta_{0}\right)=\int\limits_{-\infty}^{\infty}\left[\frac{p}{\lambda_{1}}-(-1)^{k}\right]\alpha\left(p\right)e^{ip(\varepsilon_{10}-\xi_{1})}e^{-\lambda_{1}(\varepsilon_{20}+\xi_{2})\operatorname{Im}\mu_{1}}dp,$$

$$P_{*}=\left(\zeta_{10}-\zeta_{1}\right),\ P_{*}^{0}=\left|\zeta_{10}-z_{*}\right|,$$
$$\beta_{0}=\arg\left|\zeta_{10}-z_{*}\right|,\ \beta_{*}=\arg\left(\zeta_{10}-z_{*}\right),$$
$$\psi_{0}=\psi\left(\zeta_{0}\right),\ \zeta_{0}=\varepsilon_{10}+i\varepsilon_{20},\ \zeta_{10}=\varepsilon_{10}+\mu_{1}\varepsilon_{20},$$
$$\left(\varepsilon_{10},\varepsilon_{20}\right)\in\Gamma=\cup\Gamma_{j}\ \ (j=1,2,\ldots,n).$$

Here $z_{*}=\xi_{1}+\mu_{1}\xi_{2}$ ($\xi_{2}>0$) is the point of application of the concentrated forces of shear.

To investigate the concentration of stresses at the opening, the influence of the interphase of media on it, the parameters of the anisotropy of the materials it is necessary to calculate stresses $\sigma_{s}=\sigma_{23}\cos\psi-\sigma_{13}\sin\psi$ on the boundary. We have

$$\sigma_{s}=\operatorname{Re}\left(T_{s}e^{-i\omega t}\right), \tag{5.53}$$
$$T_{s}\left(\zeta_{0}\right)=i\sqrt{\Delta_{1}}\left\{e\left(\psi_{0}\right)\frac{\partial U_{3}^{(1)}}{\partial\zeta_{10}}-\overline{e\left(\psi_{0}\right)}\frac{\partial\bar{U}_{3}^{(1)}}{\partial\bar{\zeta}_{10}}\right\},$$
$$e\left(\psi_{0}\right)=\cos\psi_{0}+\mu_{1}\sin\psi_{0}.$$

Here, by the partial derivatives we imply their limiting values at $z\to\zeta_{0}\in\Gamma$.

Using integral representation(5.50) we finally obtain

$$T_{s}\left(\zeta_{0}\right)=\frac{1}{2}q\left(\zeta_{0}\right)\operatorname{Re}\left[\frac{e\left(\psi_{0}\right)}{a\left(\psi_{0}\right)}\right]+\int\limits_{\Gamma}q\left(\zeta\right)H\left(\zeta,\zeta_{0}\right)ds+H_{*}\left(z_{*},\zeta_{0}\right) \tag{5.54}$$
$$H\left(\zeta,\zeta_{0}\right)=\frac{i\gamma_{1}}{4}H_{1}^{(1)}\left(\gamma_{1}P_{*}\right)\operatorname{Im}\left[e\left(\psi_{0}\right)e^{-i\beta_{0}}\right]-\frac{1}{8\pi}\left[e\left(\psi_{0}\right)I_{1}-\overline{e\left(\psi_{0}\right)}I_{2}\right]$$
$$H_{*}\left(z_{*},\zeta_{0}\right)=\frac{i\gamma_{1}}{4}H_{1}^{(1)}\left(\gamma_{1}P_{*}^{0}\right)\operatorname{Im}\left[e\left(\psi_{0}\right)e^{-i\beta_{*}}\right]-\frac{1}{8\pi}\left[e\left(\psi_{0}\right)M_{1}-\overline{e\left(\psi_{0}\right)}M_{2}\right]$$

Quantities $I_{k},\ M_{k}$ are determined in (5.52).

Let us investigate the concentration of stresses in bimorph containing in the upper half space $x_{2}>0$ a single tunnel opening of an elliptic cross-section with parametric equations of type

$$\varepsilon_{1}=R_{1}\cos\varphi,\ \varepsilon_{2}=R_{2}\sin\varphi+h\ \ (0\leq\varphi<2\pi) \tag{5.55}$$

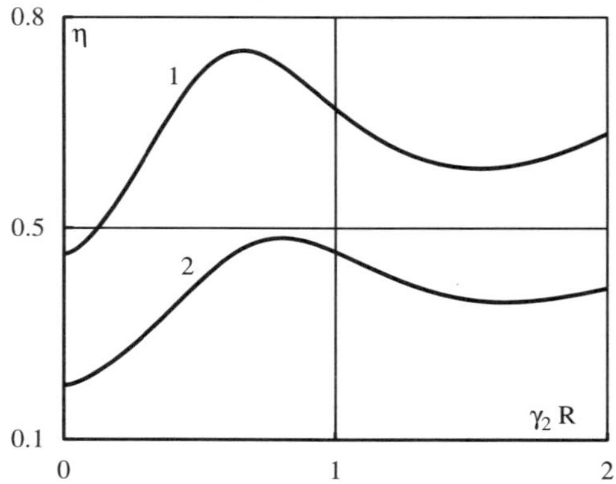

Fig. 5.43. Changes of relative shear stress $\eta = |T_S R/Q|$ at point $\varphi = \pi$ of the contour of an elliptic cavity as a function of normalized wave number $\gamma_2 R$

For this case integral (5.52) was realized by the numerical method of quadratures. The calculations were carried out with the help of formula (5.54) allowing for (5.55). In calculations we assumed that $\xi_1 = 0$, $\xi_2/R_1 = 4$, $h/R_1 = 2$, $\rho_1 = \rho_2$.

In Figs. 5.43 and 5.44 there are given the changes of quantities $\eta = |T_s R/Q|$ and $\sigma = |T_s/Z|$ in the function of normalized wave number $\gamma_2 R$ at point of contour $\varphi = \pi$ under the influence of the concentrated shear forces

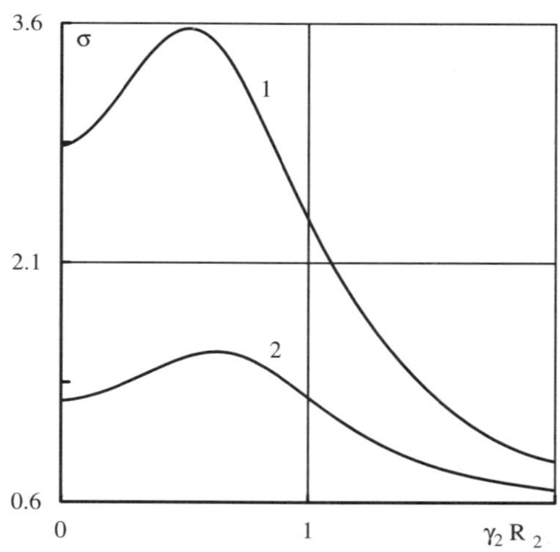

Fig. 5.44. Changes of relative shear stress $\sigma = |T_S/Z|$ at point $\varphi = \pi$ of the contour of an elliptic cavity as a function of normalized wave number $\gamma_2 R$

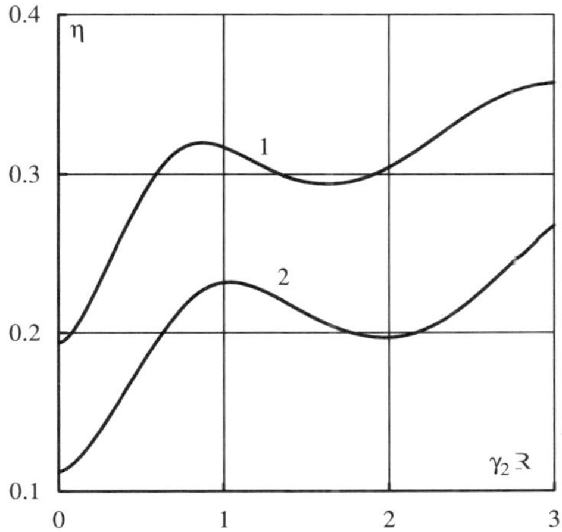

Fig. 5.45. Changes of relative shear stress $\eta = |T_S R/Q|$ at point $\varphi = \pi$ of the contour of a circular cavity as a function of $\gamma_2 R$

and stress $X_3 = Z \sin \varphi$, respectively. Curves 1 and 2 correspond to values $R_1/R_2 = 2$, $c_{44}^{(1)} = 2$, $c_{44}^{(2)} = 1$, $c_{55}^{(1)} = 1$, $c_{55}^{(2)} = 1$ and $c_{44}^{(1)} = 2$, $c_{44}^{(2)} = 1$, $c_{55}^{(1)} = 2$, $c_{55}^{(2)} = 1$. The similar results for the same parameters and in the same correspondence at $R_1/R_2 = 1$ are shown in Figs. 5.45 and 5.46.

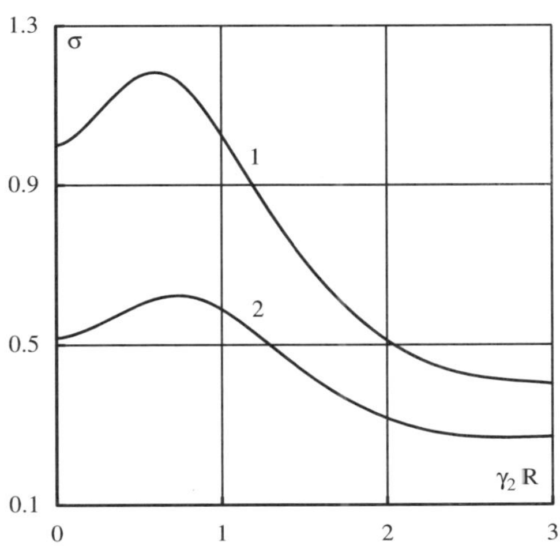

Fig. 5.46. Changes of relative shear stress $\sigma = |T_S/Z|$ at point $\varphi = \pi$ of the contour of a circular cavity as a function of $\gamma_2 R$

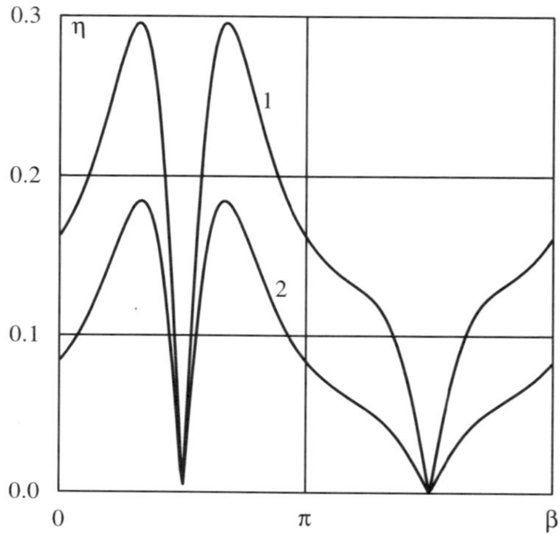

Fig. 5.47. Behaviour of quantity $\eta = |T_S R/Q|$ on the contour of a circular opening for various values of $\gamma_2 R$

Figure 5.47 illustrates the changes of quantity η on the contour of circular cavity at $c_{44}^{(1)} = 1$, $c_{44}^{(2)} = 2$, $c_{55}^{(1)} = 2$, $c_{55}^{(2)} = 1$. Curve 1 is constructed for the static case of loading ($\gamma_2 R = 0$), curve 2 is for $\gamma_2 R = 1.5\,(R = 0.5\,(R_1 + R_2))$.

The represented results illustrate the influence of the dynamic effect and configuration of the opening on the concentration of stresses for various relations of moduli of elasticity of the material components of the bimorph.

6

Mixed Dynamic Problems of Electroelasticity for Piezoelectric Bodies with Surface Electrodes

In this chapter, we investigate some aspects of the application of boundary integral equations and integral transformations methods, used for the solution of mixed dynamic problems of electroelasticity. We describe the antiplane boundary problems of electric and mechanical loading of piezoceramic bodies with electrodes, and also the plane problems of excitation of Lamb and Rayleigh waves by the system of surface electrodes..

6.1 An Unbounded Medium with a Tunnel Opening. Direct and Inverse Piezoelectric Effect

Let us consider in the Cartesian system of coordinates $Ox_1x_2x_3$ an unbounded piezoceramic medium with a tunnel along the axis of symmetry x_3 opening, the cross-section of which is restrained by the smooth contour C. On the free from mechanical stresses surface of the opening $2n$ infinitely long (in the direction of axis x_3) thin electrodes with prescribed difference of the electric potential are located, and the unelectrodized regions of the surface of the cavity are bounded with vacuum (air). Let the boundaries of the k-th electrode be determined by quantities the β_{2k-1} and β_{2k} $\left(k=\overline{1,2n}\right)$. $\phi_k^*=\mathrm{Re}\left(\Phi_k^* e^{-i\omega t}\right)$ (Fig. 6.1a). It is assumed that the electrodes are weightless and have negligible rigidity.

The state of antiplane deformation is formulated under the conditions of an inverse piezoelectric effect in the medium with a tunnel opening [140]. Under the quasistatic approximation, the system of equations of the antiplane boundary system of electroelasticity is reduced to a system of differential equations with respect to the displacement $u_3=\mathrm{Re}\left(U_3 e^{-i\omega t}\right)$ and the electric potential $\phi=\mathrm{Re}\left(\Phi e^{-i\omega t}\right)$.

$$c_{44}^E\nabla^2 u_3+e_{15}\nabla^2\phi=\rho\frac{\partial^2 u_3}{\partial t^2},$$
$$e_{15}\nabla^2 u_3-\ni_{11}^{\varepsilon}\nabla^2\phi=0. \qquad (6.1)$$

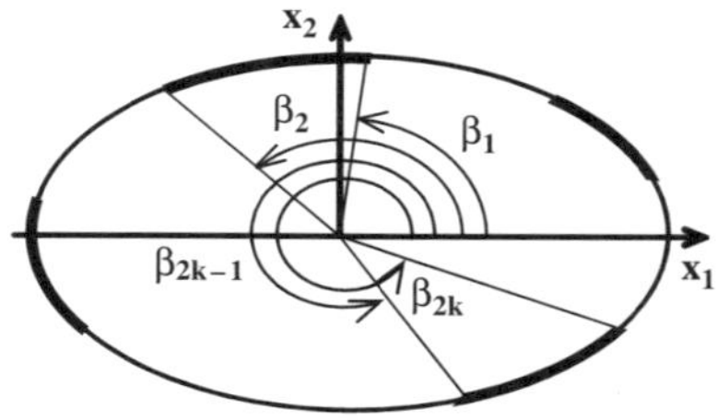

Fig. 6.1a. Tunnel opening with a system of surface electrodes

here $c_{44}^E, \ni_{11}^\varepsilon, e_{15}$ and ρ are the shear modulus measured at a constant electric field, the dielectric permeability measured at constant deformation, the piezoelectric constant and the mass density of the material, respectively; t is the time.

From the system (6.1) we obtain the following relations

$$\nabla^2 u_3 - c^{-2}\frac{\partial^2 u_3}{\partial t^2} = 0,\ \nabla^2 F = 0,$$

$$\phi = \frac{e_{15}}{\ni_{11}^\varepsilon} u_3 + F,\ c = \sqrt{\frac{c_{44}^E\left(1+k_{15}^2\right)}{\rho}},\ k_{15} = \frac{e_{15}}{\sqrt{c_{44}^E \ni_{11}^\varepsilon}}, \tag{6.2}$$

where c is the velocity of the shear wave in the piezomedium, k_{15} is the electromechanic coupling coefficient.

The components of the electric field are expressed as functions of u_3 and F

$$\sigma_{13} - i\sigma_{23} = 2\frac{\partial}{\partial z}\left[c_{44}^E\left(1+k_{15}^2\right)u_3 + e_{15}F\right],$$

$$D_1 - iD_2 = -2\ni_{11}^\varepsilon \frac{\partial F}{\partial z},$$

$$E_1 - iE_2 = -2\frac{\partial}{\partial z}\left(F + \frac{e_{15}}{\ni_{11}^\varepsilon}u_3\right),\ z = x_1 + ix_2. \tag{6.3}$$

Here σ_{ij} is the stress of a longitudinal shear, D_j and E_j are the components of the induction vectors and the electric field strength, respectively.

Taking into account (6.2), (6.3) the mechanical and electric boundary problems on the surface of the opening may be written down in the form

$$\frac{\partial}{\partial n}\left\{c_{44}^E\left(1+k_{15}^2\right)u_3 + e_{15}F\right\} = 0 \text{ on } C,$$

$$\phi(\zeta,t) = F + \frac{e_{15}}{\ni_{11}^\varepsilon}u_3 = \phi^*(\zeta,t),\quad \zeta\in C_\phi, \tag{6.4}$$

$$\text{“}D_n = -\ni_{11}^\epsilon\frac{\partial F}{\partial n} = 0 \text{ on } C\backslash C_\phi,$$

where C_ϕ is the part of contour C, corresponding to the electrodized surface of the cavity, $\phi^*(\zeta,t)$ are the given function, t is the time.

Equation (6.2), recorded with respect to amplitude functions U_3 and F, we obtain the form

$$\nabla^2 U_3 + \gamma^2 U_3 = 0, \ \nabla^2 F^* = 0,$$
$$\Phi = \frac{e_{15}}{\ni_{11}^{\varepsilon}} U_3 + F^*, \ \gamma = \frac{\omega}{c}, \tag{6.5}$$

where γ is the wave number.

The task here is to define displacement amplitude U_3 and harmonic function F^* from differential (6.5) and boundary conditions (6.4).

The solution of the stated mixed boundary problem of electroelasticity is obtained using the method of boundary integral equations. Under this framework we seek for a solution of the following form [190]

$$U_3(x_1, x_2) = \frac{1}{c_{44}^E} \int_C p(\zeta) H_0^{(1)}(\gamma r)\, ds,$$
$$F^*(x_1, x_2) = \int_C f(\zeta) \frac{\partial}{\partial n_\zeta} \ell n r ds, \ r = |\zeta - z|, \ \zeta \in C. \tag{6.6}$$

where $H_\nu^{(1)}(x)$ is the Hankel's function of the second kind and of ν-th order; ds is the element of the arc-length of contour C; the differential operator $\partial/\partial n$ defines the derivative in the direction of the normal to contour C. The representation of $U_3(x_1, x_3)$ in (6.6) satisfies the radiation conditions of the wave field of displacement at infinity.

Substituting the limiting values of functions (6.6) at $z \to \zeta_0 \in C$ into conditions (6.4) and integrating by parts the divergence integrals; we derive a system of singular integrodifferential equations of the second kind

$$2ip(\zeta_0) + \int_C \{p(\zeta) g_1(\zeta, \zeta_0) + f'(\zeta) g_2(\zeta, \zeta_0)\} ds = N_1(\zeta_0), \ \zeta_0 \in C,$$

$$-\pi f(\zeta_0) + \int_C \{p(\zeta) g_3(\zeta, \zeta_0) + f(\zeta) g_4(\zeta, \zeta_0)\} ds = N_2(\zeta_0), \ \zeta_0 \in C_\phi,$$

$$\int_C f'(\zeta) g_5(\zeta, \zeta_0)\, ds = 0, \ \zeta_0 \in C \backslash C_\phi,$$

$$g_1(\zeta, \zeta_0) = \frac{2}{\pi i} \mathrm{Re} \frac{e^{i\psi_0}}{\zeta - \zeta_0} + \gamma H_1(\gamma r_0) \cos(\psi_0 - \alpha_0),$$

$$g_2(\zeta, \zeta_0) = \frac{e_{15}}{1 + k_{15}^2} g_5(\zeta, \zeta_0), \ g_5(\zeta, \zeta_0) = \mathrm{Im} \frac{e^{i\psi_0}}{\zeta - \zeta_0},$$

$$g_3(\zeta, \zeta_0) = \frac{k_{15}^2}{e_{15}} H_0^{(1)}(\gamma r_0), \ g_4(\zeta, \zeta_0) = \mathrm{Re} \frac{e^{i\psi}}{\zeta - \zeta_0},$$

$$N_1(\zeta_0) = 0, \ N_2(\zeta_0) = \Phi^*(\zeta_0),$$

$$f'(\zeta) = \frac{df}{ds}, \ H_1(x) = \frac{2i}{\pi x} + H_1^{(1)}(x),$$
$$r_0 = |\zeta - \zeta_0|, \ \alpha_0 = \arg(\zeta - \zeta_0), \ \psi_{10} = \psi(\zeta_0). \tag{6.7}$$

Here ψ is the angle between the normal to contour C and Ox_1-axis; $\Phi^*(\zeta_0)$ is the piecewise-constant function prescribing the peak values of potentials on the electrodized regions.

We find functions $p(\zeta)$ and $f(\zeta)$ from system (6.7) based on (6.2), (6.3) using representations (6.6) so afterwards we can calculate all components of the wave electroelastic field in the medium.

Parameterizing contour C with the help of $\zeta = \zeta(\beta), \zeta_0 = \zeta(\beta_0)$ $(0 \leq \beta, \beta_0 \leq 2\pi)$, we can find an expression giving the amplitude of distribution density of electric charges $q_k(\beta)$ on k-th electrode. Taking into account the fact that the opening is bounded by vacuum, we can write

$$q_k(\beta) = D_n^{(k)}(\beta), \ \beta_{2k-1} < \beta < \beta_{2k} \ \left(k = \overline{1, 2n}\right). \tag{6.8}$$

Here $D_n^{(k)}(\beta)$ represents the amplitude of the normal component of the electric induction vector on the surface of the opening covered with k-th electrodes. Using the integral representation (6.6) for the function $F^*(x_1, x_2)$ and taking into account (6.3), (6.8) we find

$$q_k(\beta_0) = -\ni_{11}^{\varepsilon} \int_C f'(\zeta) \mathrm{Im} \frac{e^{i\psi_0}}{\zeta - \zeta_0} ds, \ \zeta_0 \in C_{\phi_k}, \tag{6.9}$$

where C_{ϕ_k} is the part of contour C on which the k-th electrode is placed.

Integrating expression (6.9) with respect to the variable β_0 over the limits from β_{2k-1} to β_{2k}, we obtain the peak value of the total charge Q_k of the k-th electrode and equal to the unity of its length. We can determine the current passing through the given electrode that is equal to the conduction current in the generator circuit, using formula

$$I_k(t) = Re\left\{ i\omega e^{-i\omega t} \int_{\beta_{2k-1}}^{\beta_{2k}} q_k(\beta_0) s'(\beta_0) \, d\beta_0 \right\}, \ s'(\beta_0) = \frac{ds}{d\beta_0}. \tag{6.10}$$

Let us now apply the above described approach to the situation where a piezoceramic medium with a tunnel opening is used as a generator of electric energy. The mechanical excitation in that case comes from two plane monochromatic shear waves propagating in the positive and negative directions of x_2-axes. Considering the values of the displacement amplitude u_3 and electric potential ϕ, to be known we get

$$U_3^{(1)} = \tau_1 e^{-i\gamma x_2}, \ U_3^{(2)} = \tau_2 e^{i\gamma x_2},$$
$$\Phi^{(j)} = \frac{e_{15}}{\ni_{11}^{\varepsilon}} U_3^{(j)} \ (j = 1, 2). \tag{6.11}$$

We also assume that the cross-section of the cavity is symmetrical to the coordinate axes and that on its surface there are two infinitely long electrodes placed symmetrically to x_2- axis (Fig. 6.1b). To obtain the difference of electric potentials $2V(t)$ under the medium deformation, appearance of electric charges of opposite signs on the electrodized surfaces is necessary. This requires the compliance of the displacement amplitude in monochromatic waves. Therefore in (6.11) we write $\tau_1 = -\tau_2 = \tau$.

The generated energy is consumed on the outer electric circuit connecting the electrodes, and may be modeled as by losses on the element with conduction Y (Fig. 6.1b). In this case both the value of the differences of potentials on electrodes $2V(t)$, and the current $I(t)$ in the circuit are unknown [115]. The electric boundary condition of the considered problem is obtained by applying the Ohm's law in the outer circuit [167]

$$I(t) = 2YV(t). \tag{6.12}$$

In this case the construction of the solution of the boundary problem is derived by prescribing the unknown difference of electric potentials $2V(t)$ on the electrodes, i.e. the use of the boundary conditions (6.4) under the influence of harmonic waves. Thus, from equalities (6.9), (6.10) and (6.12) we can determine the unknown potential amplitude $V(t)$ on the electrode

$$\begin{gathered} V^*(\omega) = \frac{i\tau\omega \ni_{11}^{\varepsilon} B_1}{2Y - i\omega \ni_{11}^{\varepsilon} B_2}, \\ B_m = \int\limits_{\beta_1}^{\beta_2} A_m(\beta_0)\, s'(\beta_0)\, d\beta_0 \quad (m = 1, 2), \\ A_m(\beta_0) = -\int\limits_C f'_m(\zeta)\, \mathrm{Im}\frac{e^{i\psi_0}}{\zeta - \zeta_0} ds. \end{gathered} \tag{6.13}$$

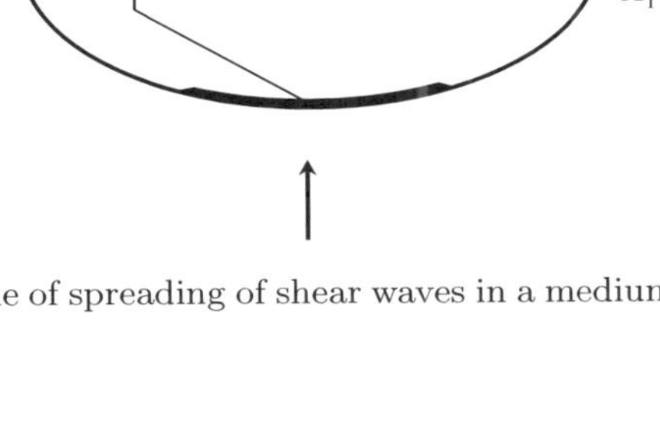

Fig. 6.1b. The scheme of spreading of shear waves in a medium with an electrodized opening

Here functions $f_m(\zeta)$ $(m = 1, 2)$ represent "*standard*" solutions of system (6.7) in which the right parts are:

$$\begin{aligned} N_1^{(1)}(\zeta_0) &= 2i\gamma c_{44}^E \sin\psi_0 \cos\gamma\xi_{20}, \\ N_2^{(1)}(\zeta_0) &= \frac{2e_{15}}{\ni_{11}^{\varepsilon}} i \sin\gamma\xi_{20}, \\ N_1^{(2)}(\zeta_0) &= 0, \\ N_2^{(2)}(\zeta_0) &= \begin{cases} 1, & \beta_1 < \beta_0 < \beta_2, \\ -1, & \beta_3 < \beta_0 < \beta_4, \end{cases} \qquad \zeta_0 = \xi_{10} + i\xi_{20} \in C, \end{aligned} \tag{6.14}$$

where quantities β_k $\left(k = \overline{1,4}\right)$ define the positions of the electrodes.

From (6.14) we get two limiting cases for the disconnected $(Y = 0)$ and short-connected $(Y \to \infty)$ circuits. In the first case the total charge on the electrodes does not change under the deformation of the medium, and in the second it is obvious that $V(t) = 0$.

As an example consider the piezoceramic medium (of $PZT-4$), weakened by the cavity of elliptic cross-section (R_1 and R_2 are the half-axes of the ellipse) with two and four active surface electrodes. For these cases, the system (6.7) was solved numerically by the method of quadratures (see: Appendix B, Par. B.3).

Figure 6.2 illustrates the change of quantity $Q^* = |Q/(\ni_{11}^{\varepsilon}\,\Phi^*)|$ (Q is the amplitude of the total charge on the electrode) as a function of the normalized wave number γR due to the potential differences $2\Phi^*$ on two electrodes. Curves 1, 2 and 3 are constructed for $R_1/R_2 = 1$, $\beta_1 = \pi/6$, $\beta_2 = 5\pi/6$, $\beta_3 = 7\pi/6$, $\beta_4 = 11\pi/6$; $\beta_k = (2k-1)\pi/4$ $\left(k = \overline{1,4}\right)$ and $\beta_1 = 5\pi/14$, $\beta_2 = 9\pi/14$, $\beta_3 = 19\pi/14$, $\beta_4 = 23\pi/14$ $(R = 0.5(R_1 + R_2))$.

The graphs characterizing the change of $\delta = c_{44}^E \left|U_3/\Phi^*\right|$ as a function of the polar angle β on the contour C is given in Fig. 6.3. Curves 1–4 are

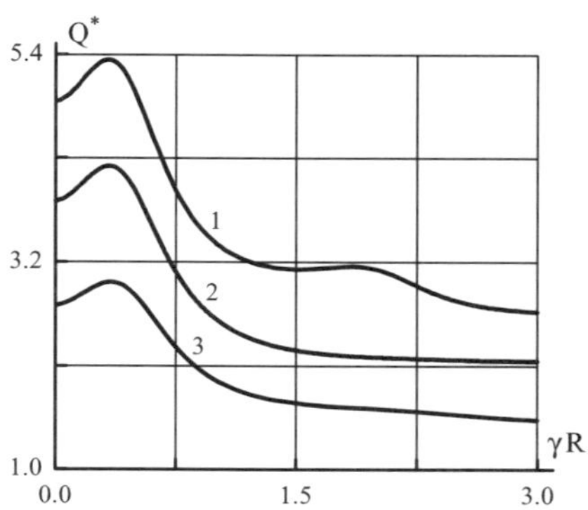

Fig. 6.2. Changes of summarized electric charge on an electrode as a function of the normalized wave number

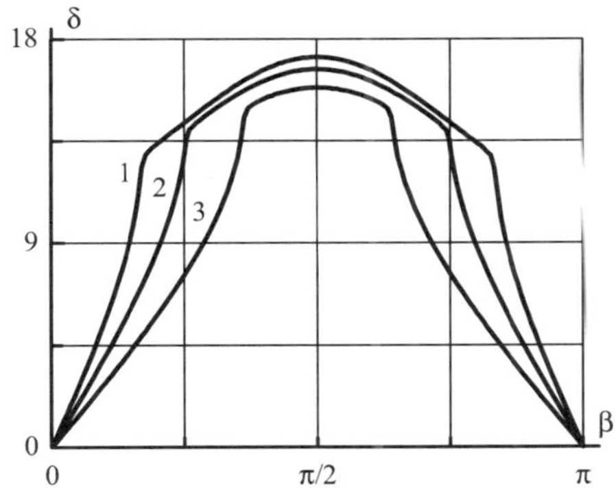

Fig. 6.3. Changes of quantity $\delta = c_{44}^E/u_3$ on the contour of an opening for various areas of the electrodized coating

constructed for $R_1/R_2 = 2$, $\gamma R = 0.5$, $\beta_1 = 5\pi/14$, $\beta_2 = 9\pi/14$, $\beta_3 = 19\pi/14$, $\beta_4 = 23\pi/14$; $\beta_k = (2k-1)\,\pi/4$ $\left(k = \overline{1,4}\right)$ and $\beta_1 = \pi/6$, $\beta_2 = 5\pi/6$, $\beta_3 = 7\pi/6$, $\beta_4 = 11\pi/6$.

Figure 6.4. illustrates the behavior of δ on the contour of the circular cavity for $\gamma R = 0, 1, 2$ and 3 at $\beta_1 = 5\pi/14$, $\beta_2 = 9\pi/14$, $\beta_3 = 19\pi/14$, $\beta_4 = 2\pi/14$ (shown by curves 1-4, respectively).

Figure 6.5 shows the change of the amplitude modulus with respect to the electric potential $\langle V^* \rangle = |\ni_{11}^{\varepsilon}\, V^*/\tau e_{15}|$ on an electrode as a function of γR under the influence of the harmonic waves of (6.11) type for a circular opening. The calculations were carried out using (6.13) for the "*no-load* " mode (disconnected electrodes). Curves 1–3 correspond to the following variants of the distribution of the electrodes: $\beta_1 = 5\pi/14$, $\beta_2 = 9\pi/14$, $\beta_3 = 19\pi/14$, $\beta_4 = 23\pi/14$; $\beta_k = (2k-1)\pi/4 \left(k = \overline{1,4}\right)$ and $\beta_1 = \pi/6$, $\beta_2 = 5\pi/6$, $\beta_3 = 7\pi/6$, $\beta_4 = 11\pi/6$.

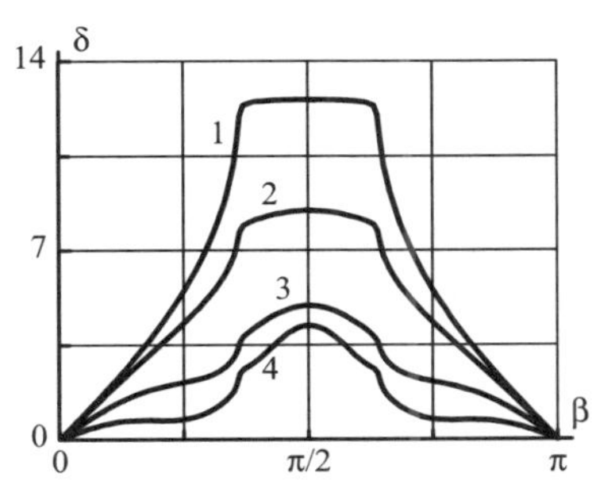

Fig. 6.4. Changes of $\delta = c_{44}^E|u_3/\Phi^*|$ on the contour of an opening for various values of parameter γR

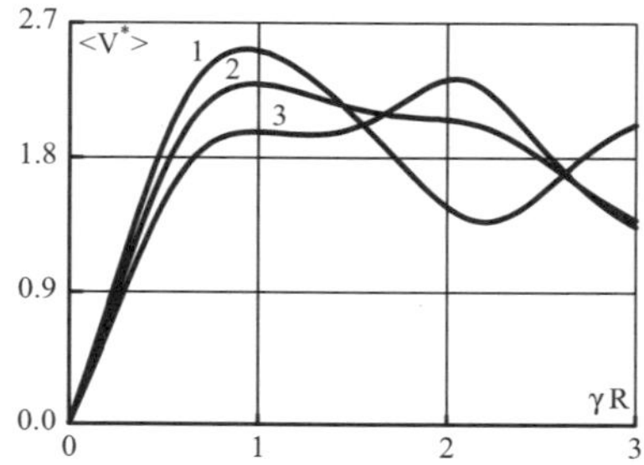

Fig. 6.5. Changes of the electric potential amplitude on an electrode as a function of the normalized wave number for various sizes of electrodized areas

Figures 6.6 and 6.7 show the contour line of the modulus of the displacement amplitude in the nearest and farthest zones of the opening, for the cases of two and four active electrodes, respectively; we used the following values: $\gamma R = 1$, $\beta_1 = 5\pi/14$, $\beta_2 = 9\pi/14$, $\beta_3 = 19\pi/14$, $\beta_4 = 23\pi/14$ (Fig. 6.6) and $\beta_k = (2k-1)\pi/8$ $\left(k = \overline{1,8}\right)$ (Fig. 6.7). In each of the four electrodes the electric potential was assumed to be equal $\Phi_j^* = (-1)^{j+1}B\left(j = \overline{1,4}\right)$.

From the results it follows that the inertial effect influences considerably the behavior of the electric quantities both in the direct and the inverse piezoeffect. For example, from Fig. 6.2 we conclude that the quantity Q^* characterizing the total electric charge on an electrode, under the dynamic loading may exceed its static analogue by 9%. The distribution of the displacements

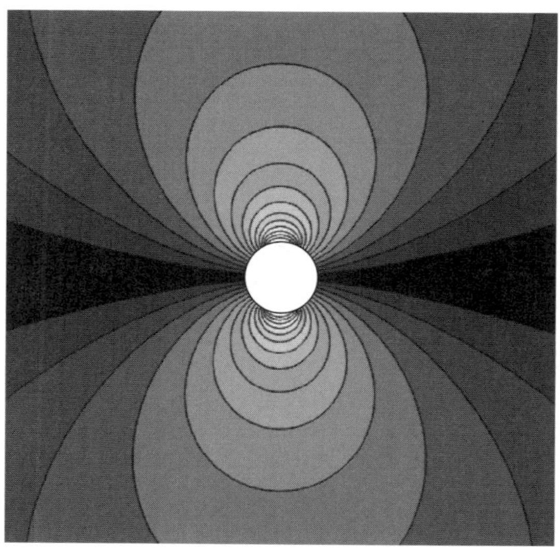

Fig. 6.6. Distribution of the displacement amplitude in a medium with circular openings with four active electrodes at *$\gamma R = 1$*

Fig. 6.7. Distribution of the displacement amplitude in a medium with circular openings with four active electrodes at $\gamma R = 1$

in the medium region is considerably changed, with respect to the number of the electrodes and their location on the surface of the opening (see Figs. 6.6 and 6.7). Also the zones of the maximum displacements are opposite to the electrodized regions of the surface.

6.2 Interaction of Two Openings in an Unbounded Medium

Let us mow consider a piezoceramic medium weakened by two tunnel along the axis of the symmetry of material x_3 openings C_1 and C_2 in the Cartesian system of coordinates $Ox_1x_2x_3$ (Fig. 6.8). On the free from mechanical stresses surfaces of openings C_1 and C_2, $2n_1$ and $2n_2$ infinitely long (in the direction of axis x_3) electrodes are placed, with prescribed differences of the electric potentials. The unelectrodized regions are bounded with vacuum (air). The boundaries of the k-th electrode located on contour C_m are determined by the quantities $\beta_{2k-1}^{(m)}$ and $\beta_{2k}^{(m)}$ ($k = \overline{1, 2n_1}$, if $m = 1$; $k = \overline{1, 2n_2}$, if $m = 2$), and the electric potential on it, is prescribed by the quantity $\phi_k^{(m)} = \mathrm{Re}\left(\Phi_k^{(m)} e^{-i\omega t}\right)$. The configuration of the opening and the location of coupled electrodes cannot be quite arbitrary; it should be such that the reflected waves from the boundary can introduce similar charges (over the absolute quantity) on these electrodes.

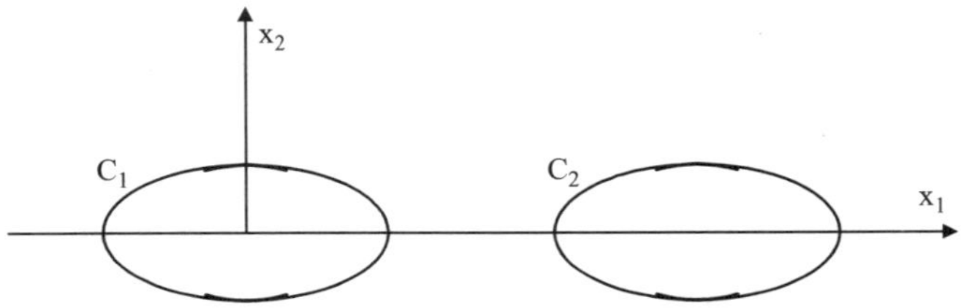

Fig. 6.8. The scheme of a space with two openings

The mechanical and electric boundary conditions on the openings have the form of (6.4), where we assume $C = C_1 \cup C_2$.

Let us represent the amplitudes of solutions of (6.5) in the form

$$\begin{aligned} U_3(x_1, x_2) &= \int\limits_{C_1} p_1(\zeta) H_0^{(1)}(\gamma r)\, ds + \int\limits_{C_2} p_2(\zeta^*) H_0^{(1)}(\gamma r^*)\, ds; \\ F^*(x_1, x_2) &= \int\limits_{C_1} f_1(\zeta) \frac{\partial}{\partial n} \ell n r ds + \int\limits_{C_2} f_2(\zeta^*) \frac{\partial}{\partial n} \ell n r^* ds; \\ r &= |\zeta - z|,\ r^* = |\zeta^* - z|,\ \zeta \in C_1,\ \zeta^* \in C_2, \end{aligned} \tag{6.15}$$

where $H_\nu^{(1)}(x)$ is the Hankel's function of the first kind and of ν-th order.

The integral representations (6.15) satisfy the conditions of radiation at infinity. Substituting the limiting values of functions (6.15) at $z \to \zeta_0 \in C_1$ and $z \to \zeta_0^* \in C_2$ into the boundary conditions (6.4) we arrive at a system of singular integrodifferential equations

$$\begin{aligned} &2ip_1(\zeta_0) + \int\limits_{C_1} \{p_1(\zeta) g_1(\zeta, \zeta_0) + f_1'(\zeta) g_2(\zeta, \zeta_0)\} ds + \\ &+ \int\limits_{C_2} \{p_2(\zeta^*) g_1(\zeta^*, \zeta_0) + f_2'(\zeta^*) g_2(\zeta^*, \zeta_0)\} ds = 0,\ \zeta_0 \in C_1; \\ &2ip_2(\zeta_0^*) + \int\limits_{C_1} \{p_1(\zeta) g_1(\zeta, \zeta_0^*) + f_1'(\zeta) g_2(\zeta, \zeta_0^*)\} ds + \\ &+ \int\limits_{C_2} \{p_2(\zeta^*) g_1(\zeta^*, \zeta_0^*) + f_2'(\zeta^*) g_2(\zeta^*, \zeta_0^*)\} ds = 0,\ \zeta_0^* \in C_2; \\ &-\pi f_1(\zeta_0) + \int\limits_{C_1} \{p_1(\zeta) g_3(\zeta, \zeta_0) + f_1(\zeta) g_4(\zeta, \zeta_0)\} ds + \\ &+ \int\limits_{C_2} \{p_2(\zeta^*) g_3(\zeta^*, \zeta_0) + f_2(\zeta^*) g_4(\zeta^*, \zeta_0)\} ds = \Phi_1^*(\zeta_0),\ \zeta_0 \in C_{1\phi}; \end{aligned}$$

$$
\begin{aligned}
&- \pi f_2(\zeta_0^*) + \int\limits_{C_1} \{p_1(\zeta) g_3(\zeta, \zeta_0^*) + f_1(\zeta) g_4(\zeta, \zeta_0^*)\} ds + \\
&+ \int\limits_{C_2} \{p_2(\zeta^*) g_3(\zeta^*, \zeta_0^*) + f_2(\zeta^*) g_4(\zeta^*, \zeta_0^*)\} ds = \Phi_2^*(\zeta_0^*), \ \zeta_0^* \in C_{2\phi}; \\
&\int\limits_{C_1} f_1'(\zeta) g_5(\zeta, \zeta_0) ds + \int\limits_{C_2} f_2'(\zeta^*) g_5(\zeta^*, \zeta_0) ds = 0, \ \zeta_0 \in C_1 \backslash C_{1\phi}; \\
&\int\limits_{C_1} f_1'(\zeta) g_5(\zeta, \zeta_0^*) ds + \int\limits_{C_2} f_2'(\zeta^*) g_5(\zeta^*, \zeta_0^*) ds = 0, \ \zeta_0^* \in C_2 \backslash C_{2\phi}; \\
&g_1(\xi, \eta) = \frac{2}{\pi i} Re \frac{e^{i\psi_0}}{\xi - \eta} + \gamma H_1(\gamma r_0) \cos(\psi_0 - \alpha_0); \\
&g_2(\xi, \eta) = \frac{e_{15}}{c_{44}^E (1 + k_{15}^2)} g_5(\xi, \eta), \ g_3(\xi, \eta) = \frac{e_{15}}{\ni_{11}^\varepsilon} H_0^{(1)}(\gamma r_0); \\
&g_5(\xi, \eta) = Im \frac{e^{i\psi_0}}{\xi - \eta}, \ H_1(x) = \frac{2i}{\pi x} + H_1^{(1)}(x); \\
&g_4(\xi, \eta) = Re \frac{e^{i\psi}}{\xi - \eta}, \ r_0 = |\xi - \eta|, \ f_m'(\zeta) = \frac{df_m}{ds}; \\
&\alpha_0 = \arg(\xi - \eta), \ \psi_0 = \psi(\eta), \ \psi = \psi(\xi).
\end{aligned} \tag{6.16}
$$

Here ψ is the angle between the normal to contour C and x_1-axis, $\Phi_m^*(\eta)$ is the piecewise-constant function, giving the values of the electric potential on the electrodes. We calculate functions $p_m(\zeta)$ and $f_m(\zeta)$ as defined in (6.16), from formulas (6.3) using representations (6.15) so afterwards we may determine all field quantities in the medium.

Let us now find the expression for the density amplitude of the distribution of electric charges $q_k^{(m)}(\beta^{(m)})$ on the k-the electrode located on the contour C_m $(m = 1, 2)$. Parameterizing the contour C_m with the help of $\zeta = \zeta(\beta^{(1)})$, $\zeta^* = \zeta^*(\beta^{(2)})$ $(0 \le \beta^{(m)} \le 2\pi)$ and taking into account the fact that, the openings are bounded by vacuum, we can write

$$
q_k^{(m)}\left(\beta^{(m)}\right) = D_n^{(m,k)}\left(\beta^{(m)}\right), \ \beta_{2k-1}^{(m)} < \beta^{(m)} < \beta_{2k}^{(m)} \ (m = 1, 2). \tag{6.17}
$$

Here $D_n^{(m,k)}(\beta^{(m)})$ represents the amplitude of the normal component of the electric induction vector on the corresponding electrodized region of the contour C_m.

Using (6.15) f and taking into account (6.17) we find

$$
q_k^{(m)}\left(\beta_0^{(m)}\right) = - \ni_{11}^\varepsilon \left\{ \int\limits_{C_1} f_1'(\zeta) \operatorname{Im} \frac{e^{i\psi_0}}{\zeta - \eta} ds + \int\limits_{C_2} f_2'(\zeta^*) \operatorname{Im} \frac{e^{i\psi_0}}{\zeta^* - \eta} ds \right\}, \tag{6.18}
$$

where $\eta\left(\beta_0^{(m)}\right) \in C_{m\phi_k}$, $C_{m\phi_k}$ are the parts of contour C_m, where the k-th electrode is located.

Integrating (6.18) from $\beta_{2k-1}^{(m)}$ to $\beta_{2k}^{(m)}$, we obtain the peak value of the total charge $Q_k^{(m)}$ of the k-th electrode on contour C_m with respect to the unity of its length. The current flow through the given electrodes may be determined by

$$I_k^{(m)}(t) = Re\left\{ i\omega e^{-i\omega t} \int\limits_{\beta_{2k-1}^{(m)}}^{\beta_{2k}^{(m)}} q_k^{(m)}\left(\beta_0^{(m)}\right) s'\left(\beta_0^{(m)}\right) d\beta_0^{(m)} \right\},$$

$$s'\left(\beta_0^{(m)}\right) = \frac{ds}{d\beta_0^{(m)}}. \tag{6.19}$$

As an example let us consider a medium from ceramics $PZT-4$ with two circular openings determined by

$$\zeta = Re^{i\beta^{(1)}} - a,\ \zeta^* = Re^{i\beta^{(2)}} + a,\ \beta^{(m)} \in [0, 2\pi]. \tag{6.20}$$

Two infinite electrodes are located on the surface of each cavity C_m, $\left(\beta_1^{(m)} = 5\pi/14,\ \beta_2^{(m)} = 9\pi/14,\ \beta_3^{(m)} = 19\pi/14,\ \beta_4^{(m)} = 23\pi/14\right)$.

Figure 6.9 shows the change of quantity $Q^* = \left|Q_1^{(1)}/\left(\ni_{11}^{\varepsilon}\,\Phi^*\right)\right|$, as a function of the normalized wave number γR. Curves 1 and 2 are constructed for the cases of loading $\Phi_1^{(1)} = \Phi_1^{(2)} = \Phi^*$, $\Phi_2^{(1)} = \Phi_2^{(2)} = -\Phi^*$ and $\Phi_1^{(1)} = -\Phi_1^{(2)} = \Phi^*$, $\Phi_2^{(1)} = -\Phi_2^{(2)} = -\Phi^*$, respectively. The solid lines correspond to $a/R = 2$, the dashed ones to $a/R = 3$.

Figures 6.10a, b and 6.11a, b illustrate the contour lines of the modulus of the displacement amplitude for various variants of loading at $\gamma R = 0.5$ and 2 respectively (more light zones correspond to the maximum value of quantity $|U_3|$).

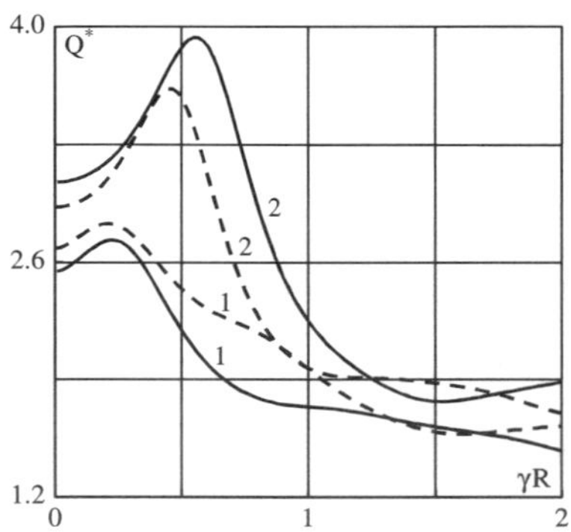

Fig. 6.9. Changes of relative total electric charge $Q^* = |Q_1^{(1)}/(\varepsilon_{11}^{s}\Phi^*)|$ on an electrode as a function of the normalized wave number

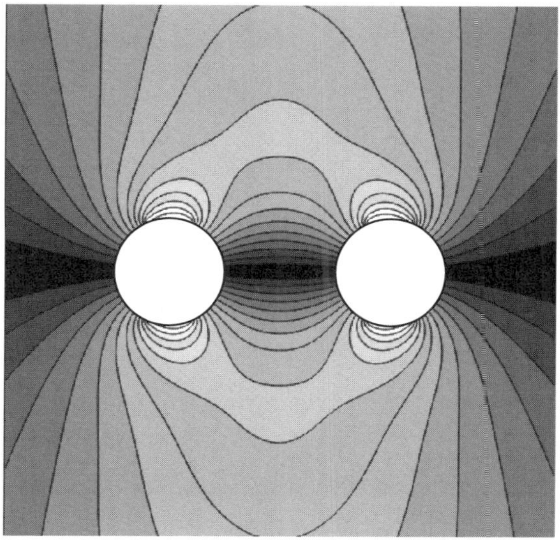

Fig. 6.10a. The contour lines of the modulus of the displacement amplitude in a medium with two openings ($\gamma R = 0,5, \Phi_1^{(1)} = \Phi_1^{(2)} = \Phi^*,\ \Phi_2^{(1)} = \Phi_2^{(2)} = -\Phi^*$)

The analysis of the results shows that the behavior of the components of the electric field on the boundary and in the area of a piecewise-homogeneous medium depends considerably on the frequencies of harmonic excitation, the disposition of the openings and the values of the potentials prescribed on the

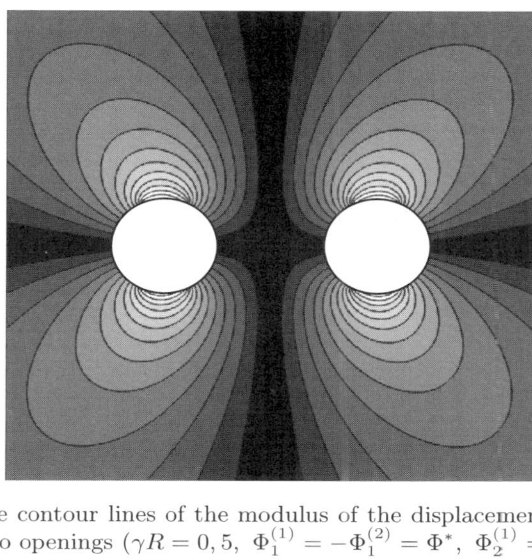

Fig. 6.10b. The contour lines of the modulus of the displacement amplitude in a medium with two openings ($\gamma R = 0,5,\ \Phi_1^{(1)} = -\Phi_1^{(2)} = \Phi^*,\ \Phi_2^{(1)} = \Phi_2^{(2)} = -\Phi^*$)

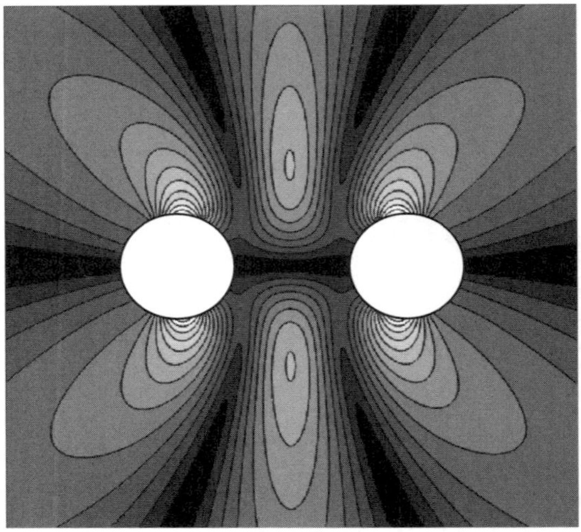

Fig. 6.11a. The contour lines of the modulus of the displacement amplitude in a medium with two openings ($\gamma R = 2,\ \Phi_1^{(1)} = \Phi_1^{(2)} = \Phi^*,\ \Phi_2^{(1)} = \Phi_2^{(2)} = -\Phi^*$)

system of electrodes. As it follows from Fig. 6.9 due to the dynamic effect, the quantity Q^* may exceed its static analogue by 27%.

In the case of antiplane deformation the stresses of longitudinal shear on a free from mechanical loading surface do not have singularities on the edges of the electrodes. At the same time considering the singular integral (6.16) and expressions (6.17), (6.18) we conclude that the components of the

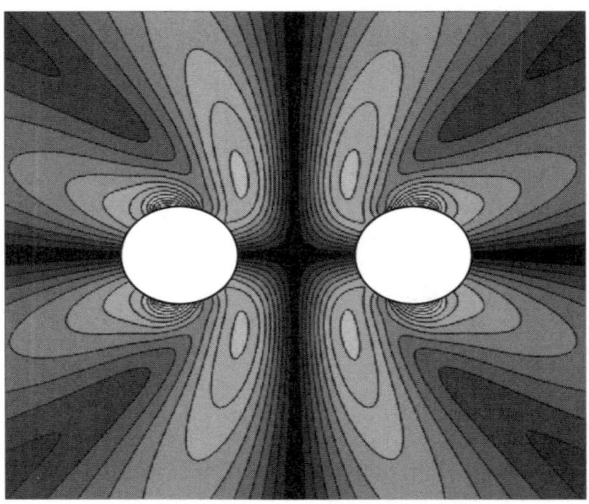

Fig. 6.11b. The contour lines of the modulus of the displacement amplitude in a medium with two openings ($\gamma R = 2,\ \Phi_1^{(1)} = -\Phi_1^{(2)} = \Phi^*,\ \Phi_2^{(1)} = \Phi_2^{(2)} = -\Phi^*$)

electric induction vector have singularities on the edges of the electrodes of the root type.

The presented approach used for a coupled electroelastic field in a medium with openings may be also implemented for the estimation of the characteristics of the perforated piezoelectric transducers of various geometry under electric loading with the help of multi-electroded systems, and also for the estimations of piezoelectric stress generators, where the direct transformation of mechanical energy into electric en energy occurs under the influence of mechanical load. The numerical scheme given in Appendix 12.8, Par. B.3 for the solution of system (6.16) permits us to investigate various variants of the electric excitation of the fields without any principle complication of the calculation algorithm.

6.3 Excitation of a Medium with an Opening by an Electric Impulse

The above described methodology for the harmonic loading of a medium with tunnel openings may be generalized to deal with the case of a non-stationary (impulse) change with time of the differences of the electric potential. In this case the solution of the dynamic problem is the superposition of the "*elementary*" solutions over the whole spectrum of frequencies. Using the Fourier integral transform

$$\mu(x,t) = \frac{1}{\sqrt{2\pi}} \int_{-\infty}^{\infty} M(x,\omega) e^{-i\omega t} d\omega,$$

$$M(x,\omega) = \frac{1}{\sqrt{2\pi}} \int_{-\infty}^{\infty} \mu(x,t) e^{i\omega t} dt, \ M(x,-\omega) = \overline{M(x,\omega)} \tag{6.21}$$

the initial mixed problem is reduced to equations of (6.4) type with respect to displacement transformant u_3 and functions F under the corresponding boundary conditions. The following procedure for the solution of the problem is analogous to Par. 6.1. In conclusion we deduce the solution according to the formula of inversion of Fourier integral transform (6.21).

Parametrizing contour C with the help of $\zeta = \zeta(\beta), \zeta_0 = \zeta(\beta_0)\,(0 \leq \beta, \beta_0 \leq 2\pi)$, we find the expression for the spectral function of the distribution density of electric charges $q_k(\beta,\omega)$ on the k-th electrode. Taking into account the fact that the opening is bounded by vacuum, we may write

$$q_k(\beta,\omega) = D_n^{(k)}(\beta,\omega), \ \beta_{2k-1} < \beta < \beta_{2k}. \tag{6.22}$$

Here $D_n^{(k)}(\beta,\omega)$ represents the transformant of the normal component of the electric induction vector on the surface of the opening covered with the k-th electrode. From the integral representation (6.6) for the function $F^*(x_1, x_2, \omega)$ and using $D_n = - \ni_{11}^{\varepsilon} \partial F/\partial n$ we find

$$q_k(\beta_0, \omega) = - \ni_{11}^{\varepsilon} \int\limits_C f'(\zeta) \mathrm{Im} \frac{e^{i\psi_0}}{\zeta - \zeta_0} ds, \;\; \zeta_0(\beta_0) \in C_{\phi_k}, \tag{6.23}$$

where C_{ϕ_k} is the part of contour C where the k-th electrode is positioned.

Integrating expression (6.23) with respect to β_0 from β_{2k-1} to β_{2k}, we obtain the spectral function for the total charge of the k-th electrode corresponding to the unity of its length. Finally, using the initial notation, we find

$$Q_k(t) = \sqrt{\frac{2}{\pi}} r Re \int\limits_0^{\infty} Y_k(\omega) e^{-i\omega t} d\omega,$$

$$Y_k(\omega) = \int\limits_{\beta_{2k-1}}^{\beta_{2k}} q_k(\beta_0, \omega) s'(\beta_0) d\beta_0. \tag{6.24}$$

As an example, in Fig. 5.28 we give the calculations of the time relation of the full charge Q on the electrode for the medium from $PZT-4$ with a circular opening under the influence of electric impulse $2\phi^*(t)$ of the trapezoidal form. Fig. 6.12 demonstrate the change of $Q^* = Q/(\sigma \ni_{11}^{\varepsilon})$ $(\sigma = \max \phi^*(t))$, for each of the two active electrodes as a function of dimensionless time $t^* = c_* t R^{-1}$ (R is the radius of the opening, c_* is the velocity of the shear wave in the piezomedium). The values of the parameters of the impulse were set to $T^* = c_* T R^{-1} = 10$, $d_1^* = c_* d_1 R^{-1} = 1$, $d_2^* = c_* d_2 R^{-1} = 8$, $\sigma^* = c_* \sigma R^{-1} = 1V/s$. Curves 1 and 2 are constructed for $\beta_1 = \pi/6$, $\beta_2 = 5\pi/6$, $\beta_3 = 7\pi/6$, $\beta_4 = 11\pi/6$ and $\beta_1 = 5\pi/14$, $\beta_2 = 9\pi/14$, $\beta_3 = 19\pi/14$, $\beta_4 = 23\pi/14$, respectively. The quantity $Q(t)$ is calculated with the help of (6.23), (6.24), and the function $f'(\zeta)$ is calculated from system (6.7) for each value of ω

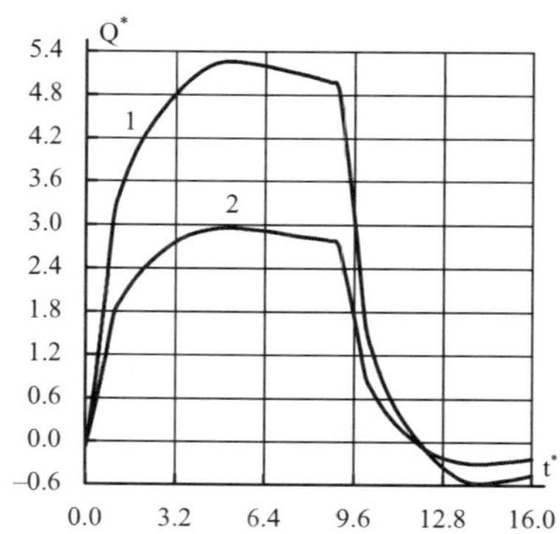

Fig. 6.12. Changes of total electric charge Q^* on an electrode during the impulse change of the difference of the electric potential

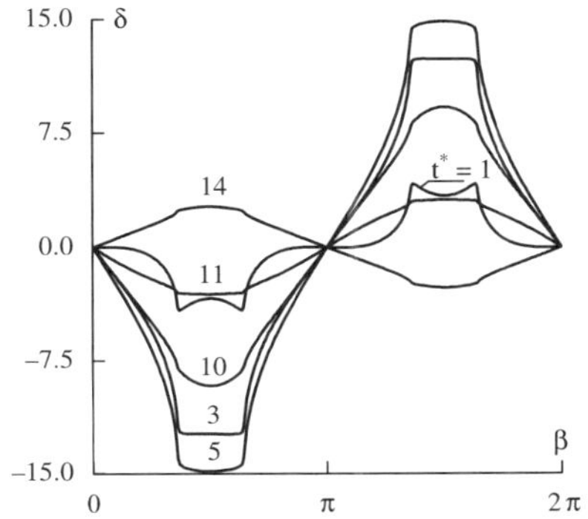

Fig. 6.13. Changes of quantity δ as a function of polar angle β for various values t^*

Figure 6.13 illustrates the evolutionary development of the shear displacement on the boundary of the circular cavity under the influence of an electric impulse of trapezoidal form. The graphs characterizing the change of quantity $\delta = c_{44}^E u_3/\sigma$ as a function of polar angle β, are represented for various values of dimensionless time t^*.

Here we should mention that the curves in Fig. 6.12 have characteristic breaks, corresponding to the same breaks of the graph of a trapezoidal impulse. This situation is conditioned by the quasistatic character of the behavior of an electric field which immediately reacts to the changes of the electric field potential $\phi^*(t)$.

6.4 Excitation of Shear Waves in an Infinite Cylinder with an Arbitrary System of Electrodes

Consider an antiplane mixed boundary problem of electroelasticity for a piezoceramic cylinder with a system of active surface electrodes, exciting its oscillation. The same problem for a circular cylinder with a system of two symmetrically positioned electrodes by the method of series is investigated in [140].

In the Cartesian system of coordinates $Ox_1x_2x_3$ let us have an infinite, along x_3-axis, a piezoceramic cylinder with axial polarization, the cross-section of which is bounded by smooth the contour C (Fig. 6.14). On the free from mechanical forces surface of the cylinder, $2n$ infinitely long along axis x_3 electrodes are positioned. The differences of the electric potential on them are given while the unelectrodized areas of the cylinder are bounded by vacuum. The boundary of the k-th electrode is determined by β_{2k-1} and β_{2k} $\left(k = \overline{1, 2n}\right)$, and the electric potential on it is prescribed by $\phi_k^*(t) = \mathrm{Re}\left(\Phi_k^* e^{-i\omega t}\right)$.

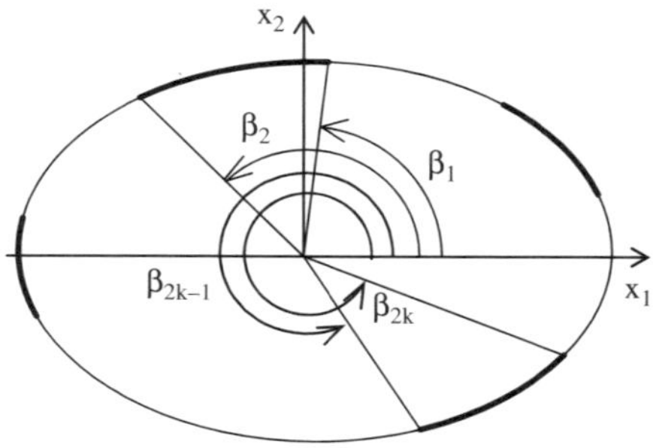

Fig. 6.14. The scheme of the cross-section of a partially electrodized cylinder

The above problem unlike the one studied in Sect. 6.1 represents an inside *boundary problem* for the cylinder. Therefore the differential (6.5) and boundary conditions (6.4) are valid here too. As a result we obtain the system of integrodifferential (6.7), where the signs before the integrated summand should be changed to the opposite ones. Here we must note that in this case the solution of system (6.7) exists for any frequency ω not coinciding with its eigenfrequency.

As an example we will investigate the behavior of the components of an electric field in a circular cylinder (from ceramics $PZT-4$) with radius R excited by two symmetrically positioned electrodes with a difference of potential amplitudes equal to $2\Phi^*$.

The graphs giving the change $\delta = c_{44}^E \left|U_3/\Phi^*\right|$ as a function of the polar angle β on the contour C are given in Fig. 6.15. Curves 1–4 are constructed for $\gamma R = 1, 2, 2.5$ and 3.5, respectively

At certain values of frequency ω the cylinder enters the state of resonance. Fig. 6.16 illustrates this phenomenon; it is shown the change of the amplitude

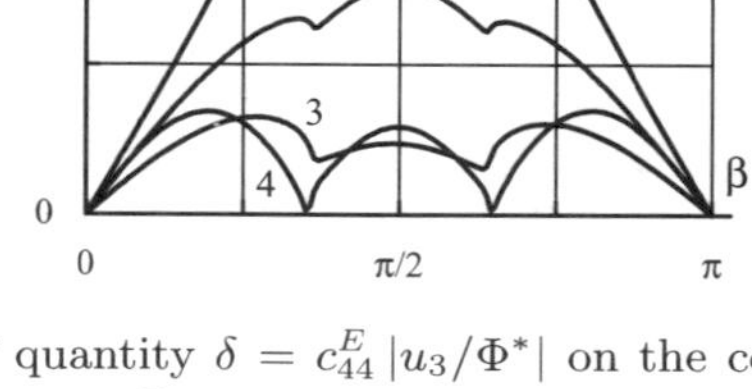

Fig. 6.15. Changes of quantity $\delta = c_{44}^E \left|u_3/\Phi^*\right|$ on the contour of an opening for various values of parameter γR

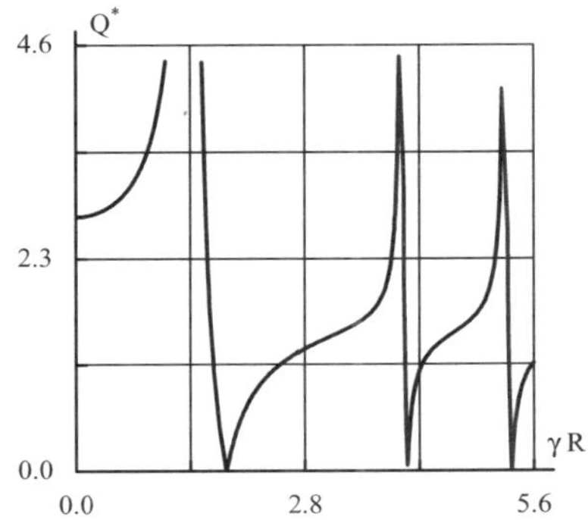

Fig. 6.16. Changes of the amplitude of relative total charge $Q^* = |Q/(\ni_{11}^{\varepsilon}\ \Phi^*)|$ on an electrode as a function of γR

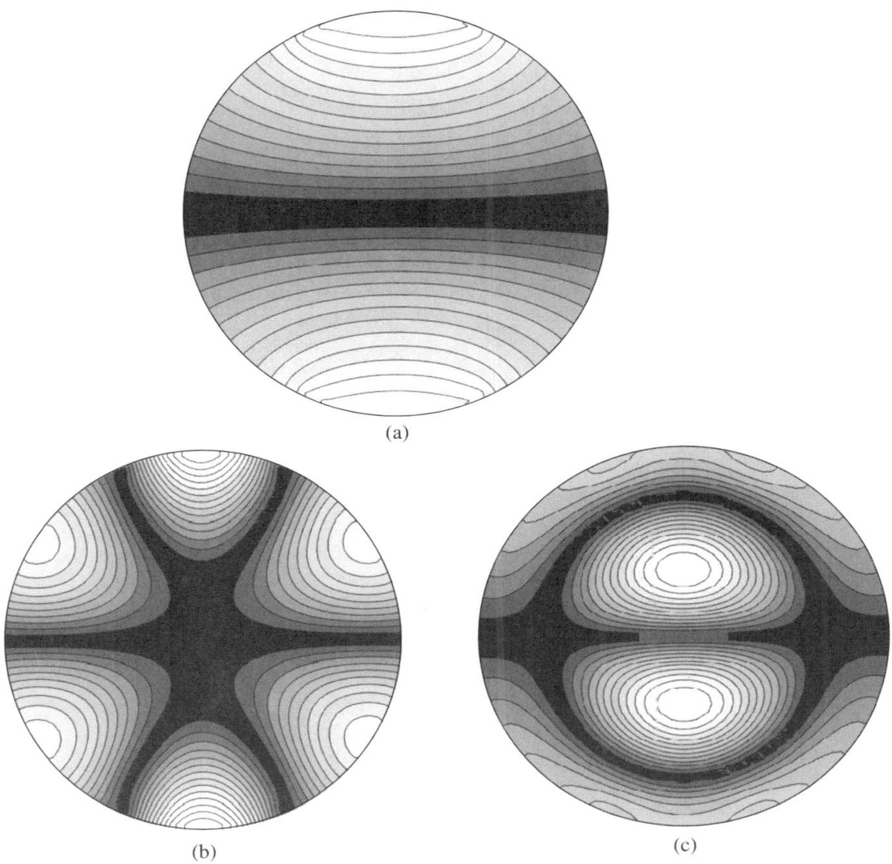

Fig. 6.17a,b and c. The contour lines of quantity $|u_3|$ at the area of a cylinder in the vicinity of the first three natural frequencies of oscillation

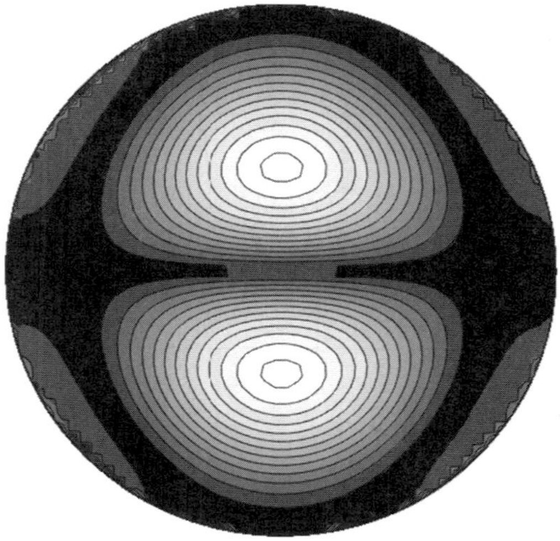

Fig. 6.18. Equipotentials in the area of a cylinder in the vicinity of the third eigenfrequency

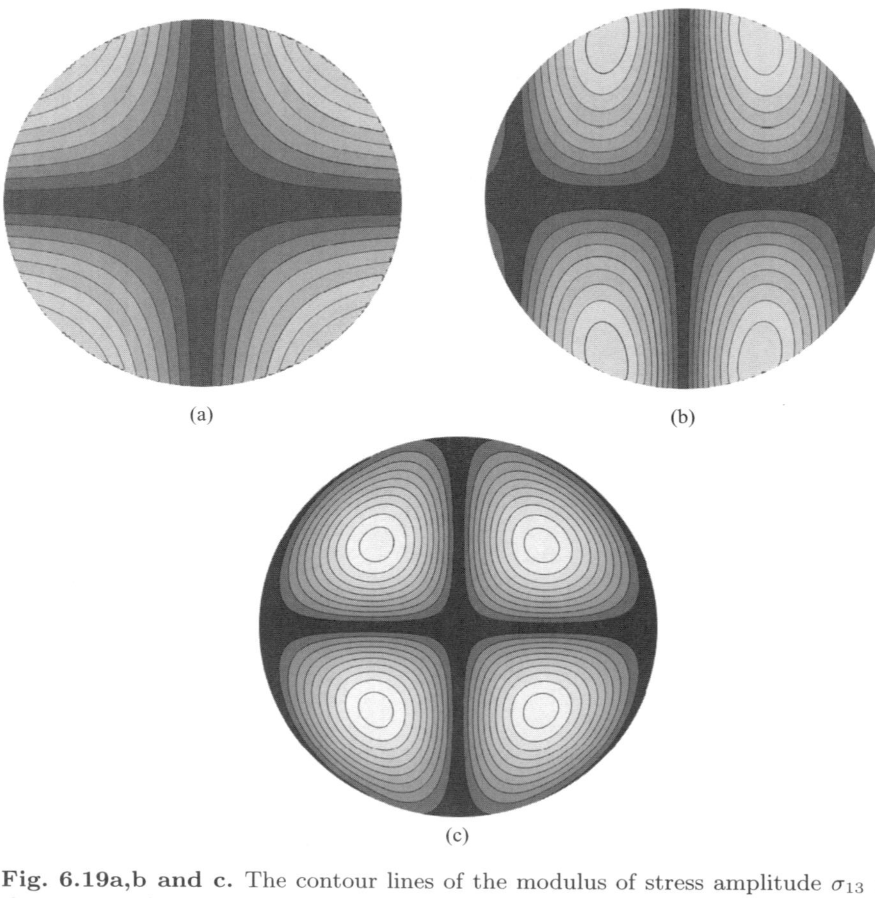

Fig. 6.19a,b and c. The contour lines of the modulus of stress amplitude σ_{13} in the vicinity of the first three eigenfrequencies

of the total charge $Q^* = |Q/(\ni_{11}^{\varepsilon}\,\Phi^*)|$ on an electrode as a function of the normalized wave number γR in the vicinity of the first three natural frequencies of oscillations. The calculations were carried out with the help of (6.9).

Figure 6.17a,b and c show the contour lines of quantity $|U_3|$ in the cylinder area and in the vicinity of three natural frequencies of oscillation $\left(\gamma^{(1)}R \approx 1.42\,,\ \gamma^{(2)}R \approx 3.98, \gamma^{(3)}R \approx 5.25\right)$, respectively. Figure 6.18 illustrates the exponentials in the vicinity of the third natural frequency. In all the calculations we have set $\beta_1 = 5\pi/14,\ \beta = 9\pi/14,\ \beta_3 = 19\pi/14,\ \beta_4 = 23\pi/14$.

Figure 6.19a,b and c and 6.20a,b and c show the contour line of the modulus of stress amplitudes σ_{13} and σ_{23} in the vicinity of the first three eigenfrequencies of oscillations. Figure 6.21a and b illustrates contour lines of the modulus of amplitude of electric induction component D_1, and Fig. 6.21a and

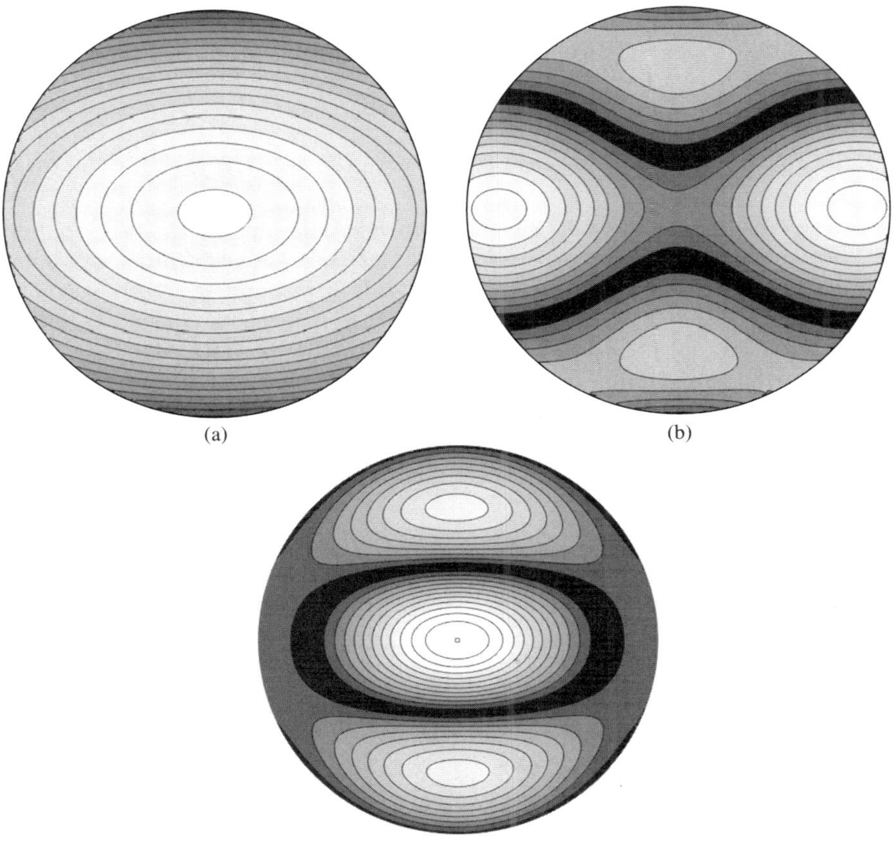

Fig. 6.20a,b and c. The contour lines of the modulus of stress amplitude σ_{23} in the vicinity of the first three eigenfrequencies of oscillations

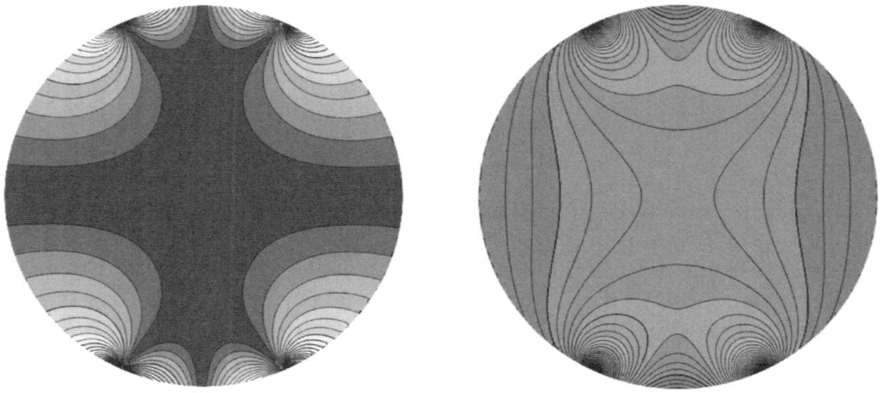

Fig. 6.21a and b. The contour lines of the modulus of the electric induction component D1 (Fig. 6.21a), D2 (Fig. 6.21b) amplitude in the vicinity of the third eigenfrequency

b show the behavior of components D_2 in the vicinity of the third natural frequency.

The dependence of total charge $Q^* = |Q/(2\pi\, \ni_{11}^{\varepsilon}\, \Phi^*)|$ on an electrode on parameter $\lambda = \cos\alpha$ characterizing the area of the electrodized coating is represented in Fig. 6.22. Curves 1, 2 and 3 correspond to $\beta_1 = -\alpha, \beta_2 = \alpha, \beta_3 = \pi - \alpha, \beta_4 = \pi + \alpha, \gamma R = 0, 1.2$ and 1.3, respectively. The analogous result in statics, obtained in Fig. 6.22, is represented by dots [140].

Here, what interests is the consideration of the oscillations of the cylinder of a complicated form. Figure 6.23a and 6.23b show the distribution $|U_3|$ in the case of an excitation by the electrodes, positioned on the rectangular areas of the boundary, for the normalized wave numbers $\gamma R = 2, 5$ and 5 respectively;

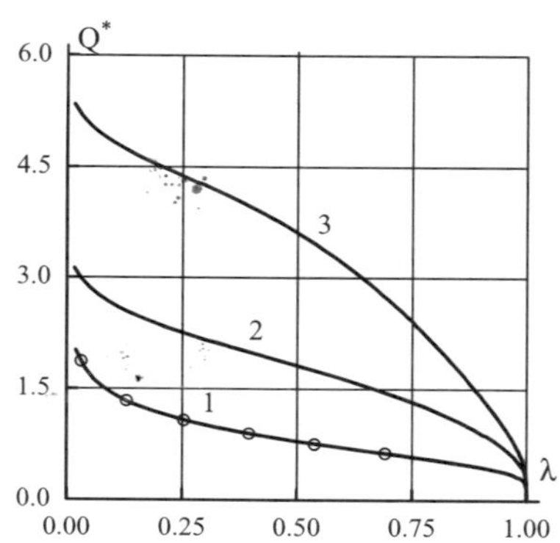

Fig. 6.22. Dependence of the total charge $Q^* = |Q/(2\pi\, \ni_{11}^{\varepsilon}\, \Phi^*)|$ on an electrode, on parameter $\lambda = \cos a$, characterizing the area of an electrodized coating

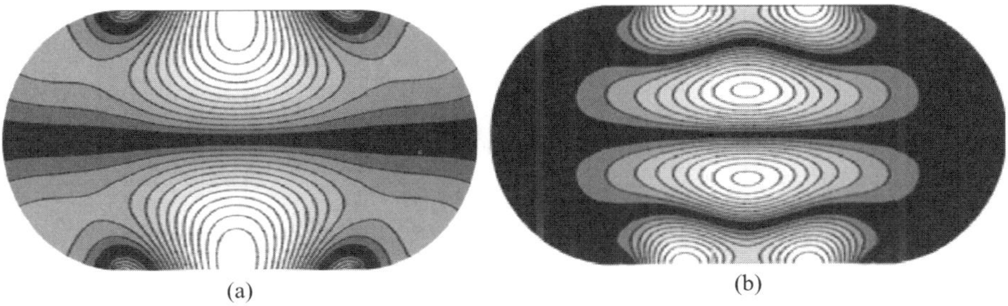

Fig. 6.23 a and b. Distribution $|u_3|$ in the area of a cylinder with cross-section in the form of a "stadium" for various values of a normalized wave number

the cylinder area has cross-section in the form of a "*stadium*" with parametric equations

$$\zeta(\beta) = \begin{cases} R\left(1 + e^{2i\beta}\right), & -\pi/4 \le \beta \le \pi/4 \\ R\left(i + ctg\beta\right), & \pi/4 < \beta \le 3\pi/4 \\ -R\left(1 + e^{2i\beta}\right), & 3\pi/4 < \beta \le 5\pi/4 \\ -R\left(i + ctg\beta\right), & 5\pi/4 < \beta \le 7\pi/4 \end{cases}$$

Comparing Figs. 6.23a and 6.23b it is obvious that the displacement of the zones of the local extremums of quantity U_3 is related to the frequency ω.

6.5 A Hollow Cylinder

Let us now generalize the procedure presented in Sect. 6.4 for the calculation of coupled fields in a continuous cylinder for the case of a hollow cylinder, the cross-section of which is restrained by two arbitrary smooth contours C_1 and C_2 (see Fig. 6.24). On the free from mechanical forces outer and inner surfaces of the cylinder $2n_1$ and $2n_2$ electrodes are positioned with prescribed differences of the electric potential; the unelectrodized areas of the cylinder are bounded with air. The boundaries of the k-th electrode positioned on the contour C_m $(m = 1, 2)$ are determined by $\beta_{2k-1}^{(m)}$ and $\beta_{2k}^{(m)}$ $(k = \overline{1, 2n_1}$, if

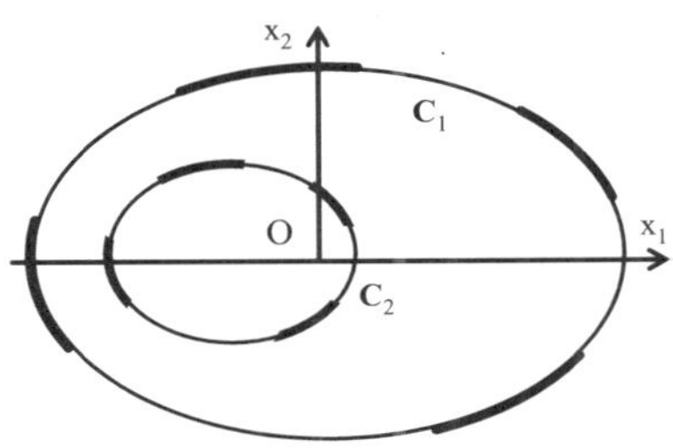

Fig. 6.24. The cross-section of a hollow cylinder with electrodes

$m = 1$; $k = \overline{1, 2n_2}$, if $m = 2$), and the electric potential on it is prescribed by $\phi_k^{(m)} = \mathrm{Re}\left(\Phi_k^{(m)} e^{-i\omega t}\right)$.

The boundary conditions for mechanical and electric quantities on the surface of the hollow cylinder have the form of (6.23), where in this case C_ϕ defines the part of united contour $C = C_1 \cup C_2$, corresponding to the electrodized surface.

The integral representations of the calculations are found in (6.15). Applying the standard procedure, and using (6.15) and boundary conditions (6.24), we come to the system of integrodifferential (6.16), where the signs before the integrated members in the equations given on outer contour C_1, should be switched to the opposite ones.

The expression for the determination of the displacement current on the k-th electrode has the form of (6.19). With the help of (6.19) we may determine the anti-resonance frequencies, at $I_k^{(m)}(t) = 0$.

As an example let us consider a hollow cylinder (of $PZT-4$), the cross-section of which is bounded by two circular contours $\left(\zeta = R_1 e^{i\beta},\ \zeta^* = R_2 e^{i\beta} + a, \beta \in [0, 2\pi]\right)$. The excitation of the cylinder is carried out by four electrodes, positioned in pairs on its outer and inner surfaces.

In Fig. 6.25 we present the amplitude-frequency characteristics of $Q^* = \left|Q_1^{(1)} / \left(\ni_{11}^{\varepsilon}\, \Phi_1^{(1)}\right)\right|$, characterizing the amplitude of total charge $Q_1^{(1)}$ on electrode ($m = k = 1$). Curves 1 and 2 are constructed using $a = 0$, $R_2/R_1 = 0.5$ and 0.8, respectively. The potentials on the electrodes were set to $\Phi_1^{(1)} = \Phi_2^{(2)} = 1V$, $\Phi_2^{(1)} = \Phi_1^{(2)} = -1V$. The position of the electrodes was fixed by

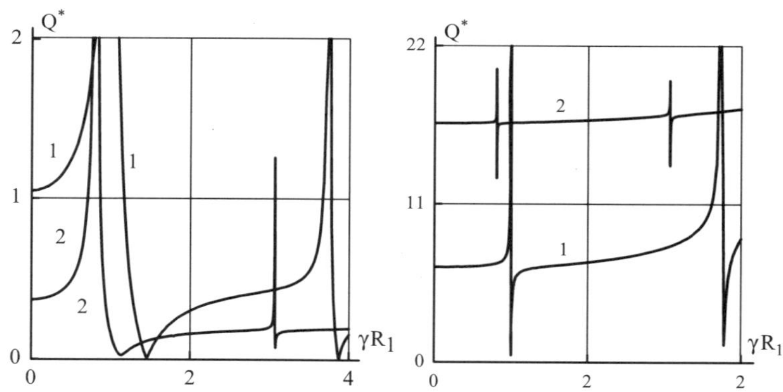

Figs. 6.25 and 6.26. Changes of the amplitude of relative total charge on an electrode as a function of a normalized wave number, for a hollow cylinder of circular cross-section

$$(6.25: \Phi_1^{(1)} = \Phi_2^{(2)} = \Phi^*, \Phi_2^{(1)} = \Phi_1^{(2)} = -\Phi^*)$$
$$(6.26: \Phi_1^{(1)} = \Phi_1^{(2)} = \Phi_2^{(1)} = \Phi_2^{(2)} = -\Phi*)$$

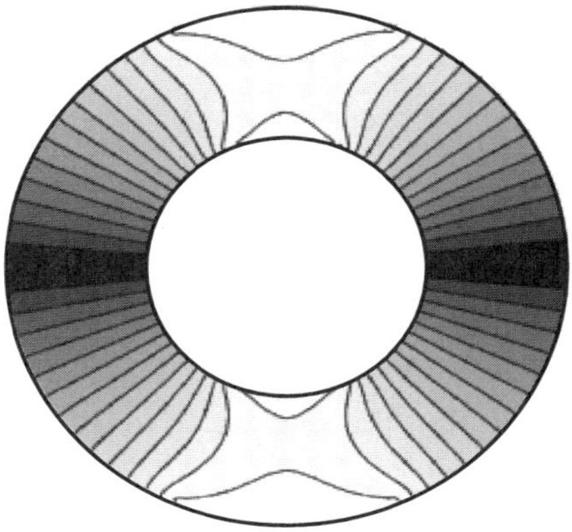
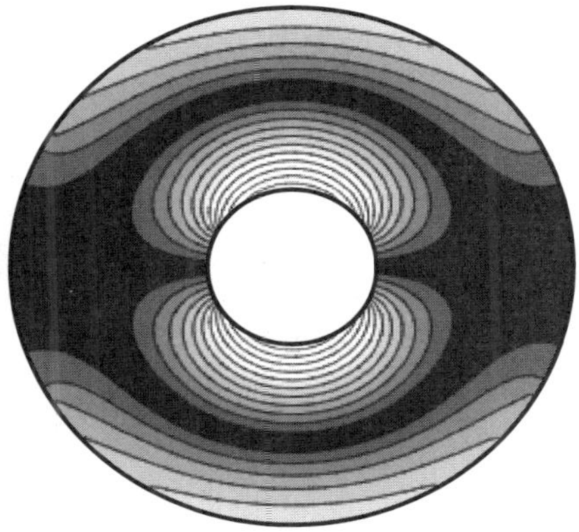

Figs. 6.27 and 6.28. The contour lines of the modulus of the displacement

$\beta_1^{(m)} = 5\pi/14$, $\beta_2^{(m)} = 9\pi/14$, $\beta_3^{(m)} = 19\pi/14$, $\beta_4^{(m)} = 23\pi/14$ $(m = 1, 2)$. The analogous results for the case of $\Phi_1^{(1)} = \Phi_1^{(2)} = 1V$, $\Phi_2^{(1)} = \Phi_2^{(2)} = -1V$ are given in Fig. 6.26.

Here what interests is the investigation of the influence of electric loading on the distribution of the mechanical quantities in the area of a hollow cylinder. Figures 6.27, 6.28 illustrate the contour lines of the modulus of the displacement amplitude with respect to $\beta_1^{(m)} = 5\pi/14$, $\beta_2^{(m)} = 9\pi/14$, $\beta_3^{(m)} = 19\pi/14$, $\beta_4^{(m)} = 23\pi/14$ and $\beta_1^{(m)} = \pi/6$, $\beta_2^{(m)} = 5\pi/6$, $\beta_3^{(m)} = 7\pi/6$, $\beta_4^{(m)} = 11\pi/6$ $(m = 1, 2)$. The values of electric potentials were set to $\Phi_1^{(1)} = \Phi_1^{(2)} = 1V$, $\Phi_2^{(1)} = \Phi_2^{(2)} = -1V$, and $\gamma R_1 = 1$.

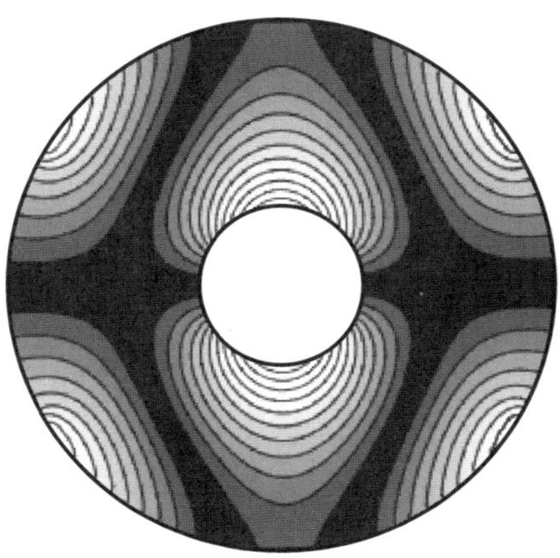
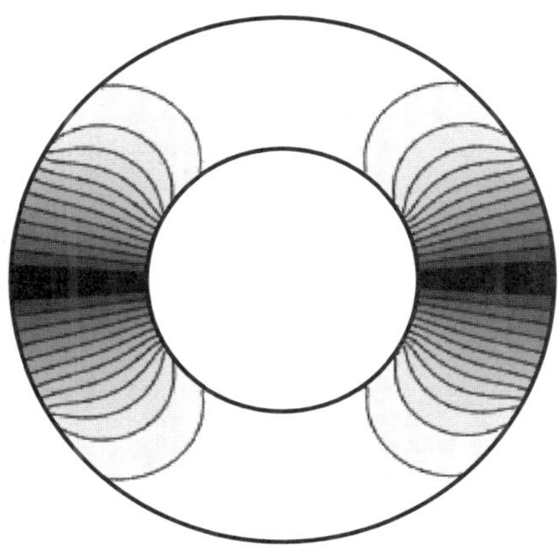

Figs. 6.29 and 6.30. The contour lines of quantity $|U_3|$

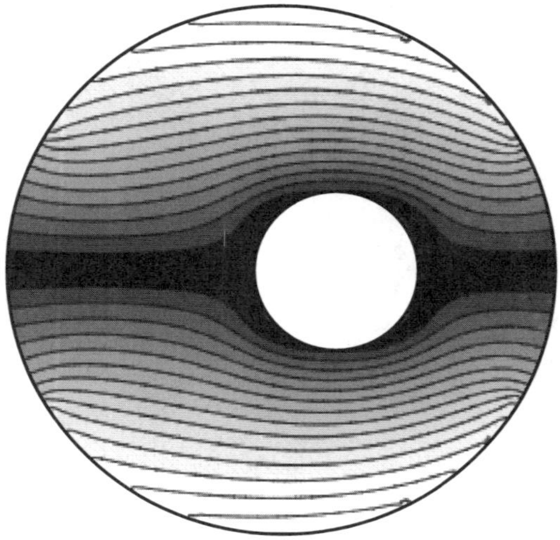

Fig. 6.31. The contour lines of quantity $|U_3|$ in the area of an eccentric cylinder

The contour lines of quantity $|U_3|$ for cases $\Phi_1^{(1)} = -1V$, $\Phi_1^{(2)} = 5V$, $\Phi_2^{(1)} = 1V$, $\Phi_2^{(2)} = -5V$ and $\Phi_1^{(1)} = -5V$, $\Phi_1^{(2)} = 1V$, $\Phi_2^{(1)} = 5V$, $\Phi_2^{(2)} = -1V$ are represented in Fig. 6.29 and 6.30 for $\beta_1^{(m)} = \pi/6$, $\beta_2^{(m)} = 5\pi/6$, $\beta_3^{(m)} = 7\pi/6$, $\beta_4^{(m)} = 11\pi/6$, $\gamma R_1 = 3$ $(m = 1, 2)$.

Figure 6.31 illustrates contour lines $|U_3|$ in the area of an eccentric cylinder at $\gamma R_1 = 1$ for the values of potential $\Phi_1^{(1)} = \Phi_2^{(2)} = 1V$, $\Phi_2^{(1)} = \Phi_1^{(2)} = -1V$ (the position of the electrodes is the same as in Fig. 6.27 and 6.28).

Figure 6.32a and b show the contour lines of the modulus of displacement amplitude $|U_3|$ in the case where the outer contour has and elliptic form, for $\beta_1^{(m)} = 5\pi/14$, $\beta_2^{(m)} = 9\pi/14$, $\beta_3^{(m)} = 19\pi/14$, $\beta_4^{(m)} = 23\pi/14$ and $\beta_1^{(m)} = \pi/6$, $\beta_2^{(m)} = 5\pi/6$, $\beta_3^{(m)} = 7\pi/6$, $\beta_4^{(m)} = 11\pi/6$ $(m = 1, 2)$ respetively. Here we set $\gamma R = 1$, $R = 0.5\,(R_1 + R_2)$, where R_1 and R_2 are the horizontal and vertical half-axes of an ellipse. The values of the potentials were chosen as $\Phi_1^{(1)} = \Phi_1^{(2)} = 1V$, $\Phi_2^{(1)} = \Phi_2^{(2)} = -1V$.

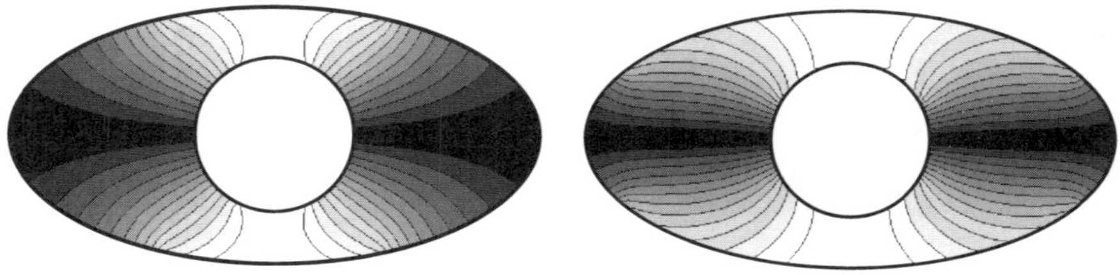

Fig. 6.32a and b. The contour lines of $|U_3|$ in an elliptic cylinder with circular opening

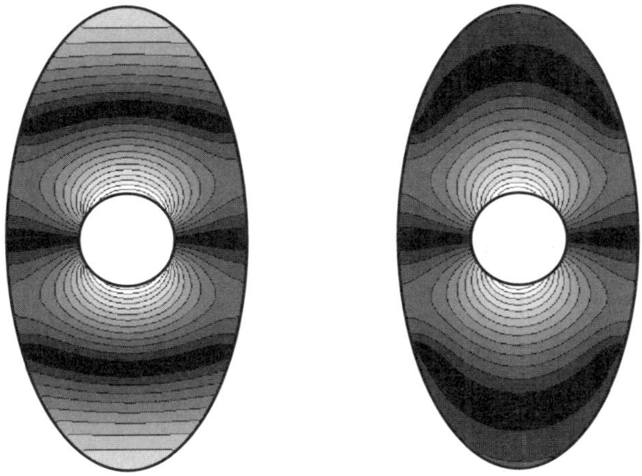

Fig. 6.33a and b. The contour lines of $|U_3|$ in an elliptic cylinder with a circular opening

Figure 6.33a and b show analogous results for a cylinder with a cross-section in the form of an extended along x_2 -axis ellipse with a cut-out circle (for the same position of electrodes, as in Fig. 6.32a and b). The calculations were carried out for parameters $\gamma R = 2$ at $\Phi_1^{(1)} = -1V$, $\Phi_1^{(2)} = 10V$, $\Phi_2^{(1)} = 1V$, $\Phi_2^{(2)} = -10V$ (Fig. 6.33a and b) and $\Phi_1^{(1)} = -10V$, $\Phi_1^{(2)} = 1V$, $\Phi_2^{(1)} = 10V$, $\Phi_2^{(2)} = -1V$ (Fig. 6.33a and b).

From results it follows that the distribution of the displacement in the cylinder is influenced strongly by the frequency of excitation, sizes and position of the electrodes, and also the values of the given on them potentials.

6.6 A Halfspace with Tunnel Openings

In order to investigate the influence of the boundary area on the behavior of the components of the electric field let us consider the stationary dynamic problem of electroelasticity for a piezoceramic halfspace weakened by a tunnel cavity with a system of active surface electrodes.

Let in the Cartesian system of coordinates $Ox_1x_2x_3$ a piezoceramic halfspace weakened by a tunnel along the symmetry x_3 axis of the material opening, the cross-section of which is bounded by arbitrary smooth contour C (Fig. 6.34a). The boundaries of the k-th electrode are determined by β_{2k-1} and β_{2k} $\left(k = \overline{1, 2n}\right)$, and the electric potential on it is given by $\phi_k^* = Re\left(\Phi_k^* e^{-i\omega t}\right)$. It is assumed that the cross-section of the cavity has a symmetry with respect to x_2, and the electrodes are weightless and their rigidity is negligible. The position of the electrodes cannot be quite arbitrary; the required conditions of matching will be indicated below.

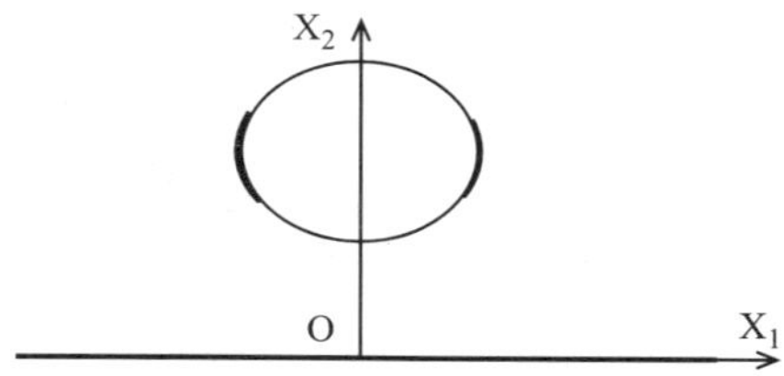

Fig. 6.34a. A halfspace with a partially electrodized tunnel cavity

Consider two types of the boundary conditions on the boundary of half-space $x_2 = 0$:

a) a halfspace rigidly fixed and covered by grounded electrodes along the boundary

$$u_3 = 0, \ \phi = 0; \tag{6.25}$$

b) a halfspace free from forces and bounded with vacuum

$$\sigma_{23} = 0, \ D_2 = 0. \tag{6.26}$$

Hence, the boundary problem of electroelasticity is reduced to the determination of functions U_3 and F^* from the differential equations of Helmhöltz and Laplace (6.5) and the boundary conditions (6.4), (6.25) or (6.26)

To solve the above problem it is recommended to have integral representations of the solutions, satisfying conditions (6.25) or (6.26) and conditions of radiation at infinity. Using the method of images let us represent the solution of the form

$$\begin{aligned} U_3(x_1, x_2) &= \frac{1}{c_{44}^E} \int\limits_C p(\zeta) \left[H_0^{(1)}(\gamma r) - A H_0^{(1)}(\gamma r_1) \right] ds, \\ F^*(x_1, x_2) &= \int\limits_C f(\zeta) \frac{\partial}{\partial n_\zeta} (\ell n r - A \ell n r_1)\, ds, \\ r &= |\zeta - z|, \ r_1 = \left|\bar{\zeta} - z\right|, \ \zeta \in C. \end{aligned} \tag{6.27}$$

Here $H_\nu^{(1)}(x)$ is the Hankel's function of the first kind and of ν-th order, ds is the element of an arc-length of contour C. The value $A = -1$ corresponds to the free- from forces halfspace-bounded by vacuum; $A = 1$ corresponds to the fixed and covered by grounded electrodes of a half-space. For $A = 0$ we have an unbounded space with a tunnel cavity.

Substituting the limiting values of functions (6.27) at $z \to \zeta_0 \in C$ into the boundary conditions (6.4) and integrating by parts we come to the system ofsingular integrodifferential equations of the second type (6.7), where the kernel g_m $(m = 1, 2, \ldots, 5)$ and the right parts are determined by the expressions

$$
\begin{aligned}
g_1(\zeta, \zeta_0) &= \frac{2}{\pi i} Re \frac{e^{i\psi_0}}{\zeta - \zeta_0} + \gamma \left[H_1(\gamma r_0) \cos(\psi_0 - \alpha_0) \right. \\
&\quad \left. - A H_1^{(1)}(\gamma r_{10}) \cos(\psi_0 - \alpha_{10}) \right], \\
g_2(\zeta, \zeta_0) &= \frac{e_{15}}{1 + k_{15}^2} g_5(\zeta, \zeta_0), \ g_3(\zeta, \zeta_0) = \frac{k_{15}^2}{e_{15}} \left[H_0^{(1)}(\gamma r_0) - A H_0^{(1)}(\gamma r_{10}) \right], \\
g_4(\zeta, \zeta_0) &= \mathrm{Re} \left\{ \frac{e^{i\psi}}{\zeta - \zeta_0} - \frac{A e^{i\psi}}{\zeta - \bar{\zeta}_0} \right\}, \ f'(\zeta) = \frac{df}{ds}, \qquad (6.28) \\
g_5(\zeta, \zeta_0) &= \mathrm{Im} \left\{ \frac{e^{i\psi_0}}{\zeta - \zeta_0} + \frac{A e^{i\psi_0}}{\bar{\zeta} - \zeta_0} \right\}, \\
& N_1(\zeta_0) = 0, \ N_2(\zeta_0) = \Phi^*(\zeta_0), \\
& r_0 = |\zeta - \zeta_0|, \ \alpha_0 = \arg(\zeta - \zeta_0), \ r_{10} = |\bar{\zeta} - \zeta_0|, \ \alpha_{10} = \arg(\bar{\zeta} - \zeta_0), \\
& \psi = \psi(\zeta), \ \psi_0 = \psi(\zeta_0), \ \zeta, \zeta_0 \in C.
\end{aligned}
$$

Here ψ is the angle between the normal to contour C and x_1 -axis $\Phi^*(\zeta_0)$ is the piecewise-constant function, determining the electric potential values on the system of electrodes. The kernels $g_2(\zeta, \zeta_0)$, $g_5(\zeta, \zeta_0)$ are singular; the other kernels, due to the assumption that contour C is smooth, can not posses more than a weak singularity.

In this case the amplitude of the density distribution of electric charges $q_k(\beta)$ on the k-th electrode reads

$$
q_k(\beta_0) = - \ni_{11}^{\varepsilon} \int_C f'(\zeta) \mathrm{Im} \left\{ \frac{e^{i\psi_0}}{\zeta - \zeta_0} + \frac{A e^{i\psi_0}}{\bar{\zeta} - \zeta_0} \right\} ds, \ \zeta_0 \in C_{\phi_k}, \qquad (6.29)
$$

where C_{ϕ_k} is the part of contour C, where the k-th electrode is positioned.

By analogy, using (6.3) we find the expression for the peak values of the electric field component in the area of a piecewise-homogeneous halfspace. We have

$$
\begin{aligned}
\sigma_{13}^* &= \left(1 + k_{15}^2\right) \gamma \int_C p(\zeta) \left\{ H_1^{(1)}(\gamma r) \cos\alpha - A H_1^{(1)}(\gamma r_1) \cos\alpha_1 \right\} ds + \\
&\quad + e_{15} \int_C f'(\zeta) \mathrm{Im} \left[\frac{1}{\zeta - z} + \frac{A}{\bar{\zeta} - z} \right] ds, \\
\sigma_{23}^* &= \left(1 + k_{15}^2\right) \gamma \int_C p(\zeta) \left\{ H_1^{(1)}(\gamma r) \sin\alpha - A H_1^{(1)}(\gamma r_1) \sin\alpha_1 \right\} ds +
\end{aligned}
$$

$$+ e_{15} \int_C f'(\zeta) \mathrm{Re} \left[\frac{1}{\zeta - z} + \frac{A}{\bar{\zeta} - z} \right] ds,$$

$$E_1^* = -\frac{k_{15}^2 \gamma}{e_{15}} \int_C p(\zeta) \left\{ H_1^{(1)}(\gamma r) \cos\alpha - A H_1^{(1)}(\gamma r_1) \cos\alpha_1 \right\} ds -$$

$$- \int_C f'(\zeta) \mathrm{Im} \left[\frac{1}{\zeta - z} + \frac{A}{\bar{\zeta} - z} \right] ds,$$

$$E_2^* = -\frac{k_{15}^2 \gamma}{e_{15}} \int_C p(\zeta) \left\{ H_1^{(1)}(\gamma r) \sin\alpha - A H_1^{(1)}(\gamma r_1) \sin\alpha_1 \right\} ds -$$

$$- \int_C f'(\zeta) \mathrm{Re} \left[\frac{1}{\zeta - z} + \frac{A}{\bar{\zeta} - z} \right] ds,$$

$$D_1^* = - \ni_{11}^{\varepsilon} \int_C f'(\zeta) \mathrm{Im} \left[\frac{1}{\zeta - z} + \frac{A}{\bar{\zeta} - z} \right] ds,$$

$$D_2^* = - \ni_{11}^{\varepsilon} \int_C f'(\zeta) \mathrm{Re} \left[\frac{1}{\zeta - z} + \frac{A}{\bar{\zeta} - z} \right] ds,$$

$$\alpha = \arg(\zeta - z), \ \alpha_1 = \arg(\bar{\zeta} - z), \ \zeta \in C. \tag{6.30}$$

Now consider the situation where the fixed and grounded boundary $x_2 = 0$ of a piezoceramic halfspace with tunnel openings is used as a generator of electric energy. In this case the mechanical excitation comes from two plane monochromatic shear waves, propagating in the positive and negative directions of x_1 -axis having the following values of displacement amplitude u_3 and electric potential ϕ:

$$U_3^{(1)} = \tau_1 \left(e^{-i\gamma x_1} - A e^{-i\gamma x_1} \right), \ U_3^{(2)} = \tau_2 \left(e^{i\gamma x_1} - A e^{-i\gamma x_1} \right),$$
$$\Phi^{(j)} = \frac{e_{15}}{\ni_{11}^{\varepsilon}} U_3^{(j)} \ (j = 1, 2). \tag{6.31}$$

Here the value $A = 1$ corresponds to the fixed halfspace with zero potential on the boundary, while the value $A = 0$ corresponds to the full space.

For assume that the cross-section of the cavity has vertical and horizontal axes of symmetry and that on its surface there are two symmetrically positioned infinitely long electrodes (Fig. 6.34b). To impose the differences of electric potentials $2V(t)$ under the deformation of the medium in (6.31) it is necessary to assume $\tau_1 = -\tau_2 = \tau$.

The generated energy is consumed in the outer electric circuit connecting the electrodes and as a model may can be modeled by losses on an element with conductivity Y (Fig. 6.34b). Following the procedurepresented

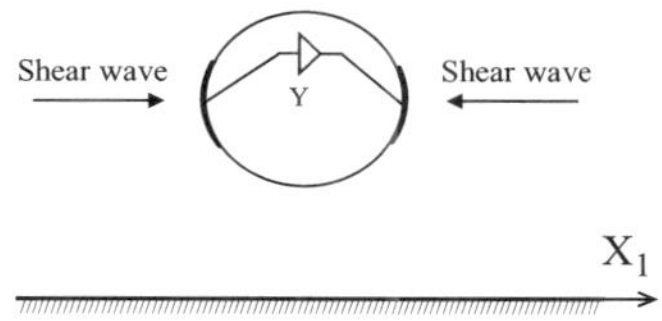

Fig. 6.34b. The scheme of generation of the differences of electric potentials on electrodes under the influence of monochromatic shear waves in a fixed and grounded along the boundaries halfspace

in Sect. 6.1 we determine the unknown amplitude of potential $V(t)$ on an electrode as

$$V^*(\omega) = \frac{i\tau\omega \ni_{11}^{\varepsilon} B_1}{2Y - i\omega \ni_{11}^{\varepsilon} B_2},$$
$$B_m = \int_{\beta_1}^{\beta_2} A_m(\beta_0)\, s'(\beta_0)\, d\beta_0 \quad (m = 1, 2), \tag{6.32}$$
$$A_m(\beta_0) = -\int_C f_m'(\zeta)\, \mathrm{Im}\left\{\frac{e^{i\psi_0}}{\zeta - \zeta_0} + \frac{Ae^{i\psi_0}}{\bar{\zeta} - \zeta_0}\right\} ds.$$

Here the functions $f_m(\zeta)$ $(m = 1, 2)$ represent "*standard*" solutions of the system (6.7) with kernels (6.28) where the right parts of (6.7) are:

$$N_1^{(1)}(\zeta_0) = 2i\gamma c_{44}^{E}(1 + A)\cos\psi_0 \cos\gamma\xi_{10},$$
$$N_2^{(1)}(\zeta_0) = \frac{2e_{15}}{\ni_{11}^{\varepsilon}} i(1 + A)\sin\gamma\xi_{10},$$
$$N_1^{(2)}(\zeta_0) = 0, \tag{6.33}$$
$$N_2^{(2)}(\zeta_0) = \begin{cases} 1, \beta_1 < \beta_0 < \beta_2, \\ -1, \beta_3 < \beta_0 < \beta_4, \end{cases} \quad \zeta_0 = \xi_{10} + i\,\xi_{20} \in C,$$

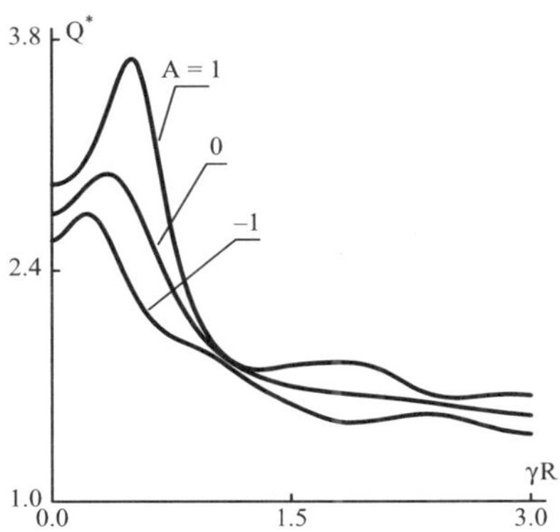

Fig. 6.35. Changes of the amplitude of relative total charge on an electrode as a function of a normalized wave number for various types of boundary conditions on the boundary of a halfspace

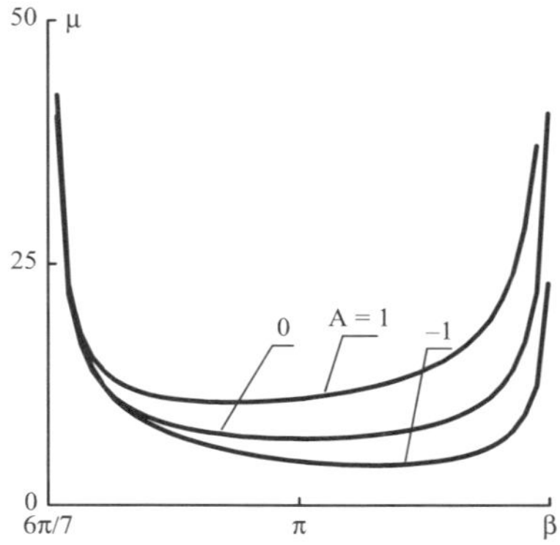

Fig. 6.36. The distribution intensity of electric charges on an electrode for various types of boundary conditions on the boundary of a halfspace

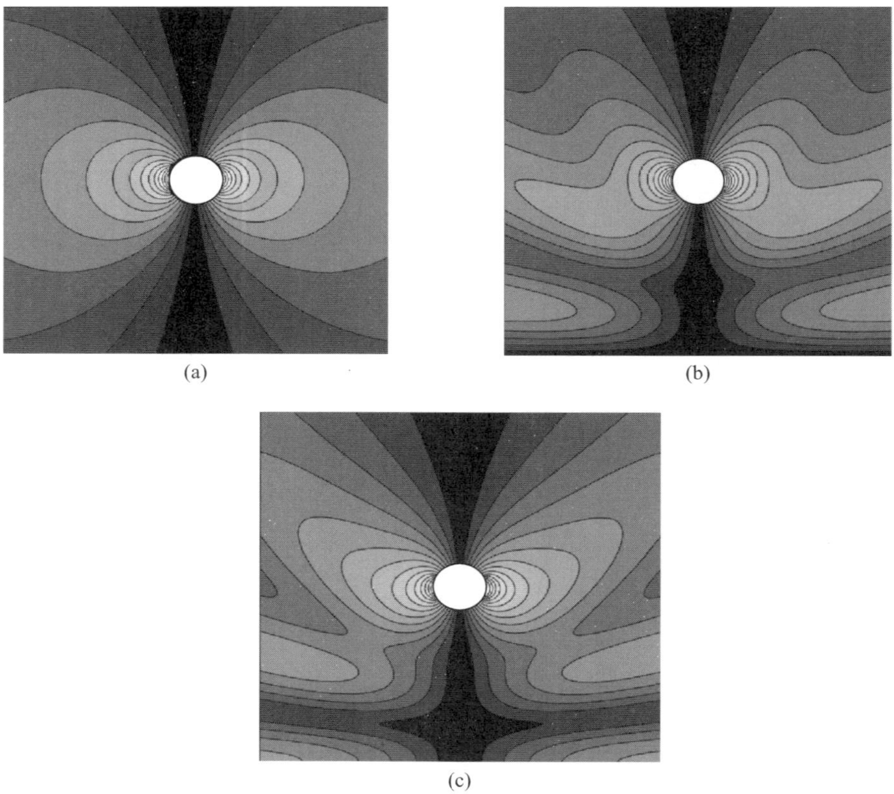

Fig. 6.37a,b and c. The contour lines of the modulus of the shear displacement amplitude in the vicinity of a circular opening with two electrodes in: a) a space ($A = 0$), b) a free halfspace ($A = -1$) and c) a fixed halfspace ($A = -1$)

where $\beta_k \left(k = \overline{1,4}\right)$ denotes the position of the electrodes.

As a first example consider a halfspace (of $PZT-4$ ceramics) with circular opening $\zeta = \mathrm{R}e^{i\beta} + ih \, (\beta \in [0, 2\pi])$, excited by two electrodes with the difference of the amplitude of electric potentials $2\Phi^*$ positioned symmetrically with respect to x_2-axis ($\beta_1 = -\pi/7,\ \beta_2 = \pi/7,\ \beta_3 = 6\pi/7,\ \beta_4 = 8\pi/7$).

Figure 6.35, shows the change of quantity $Q^* = |Q_1/(\ni_{11}^{\varepsilon}\ \Phi^*)|$ characterizing the amplitude of total electric charge Q_1 on the electrode as a function of normalized wave number γR for various variants of the edge conditions on the boundary of halfspace ($h/R = 2.5$). It is seen that in the case of a fixed halfspace ($A = 1$) the quantity Q^* may exceed its static analogue by 26%. The influence of the inertial effect in the space is negligible.

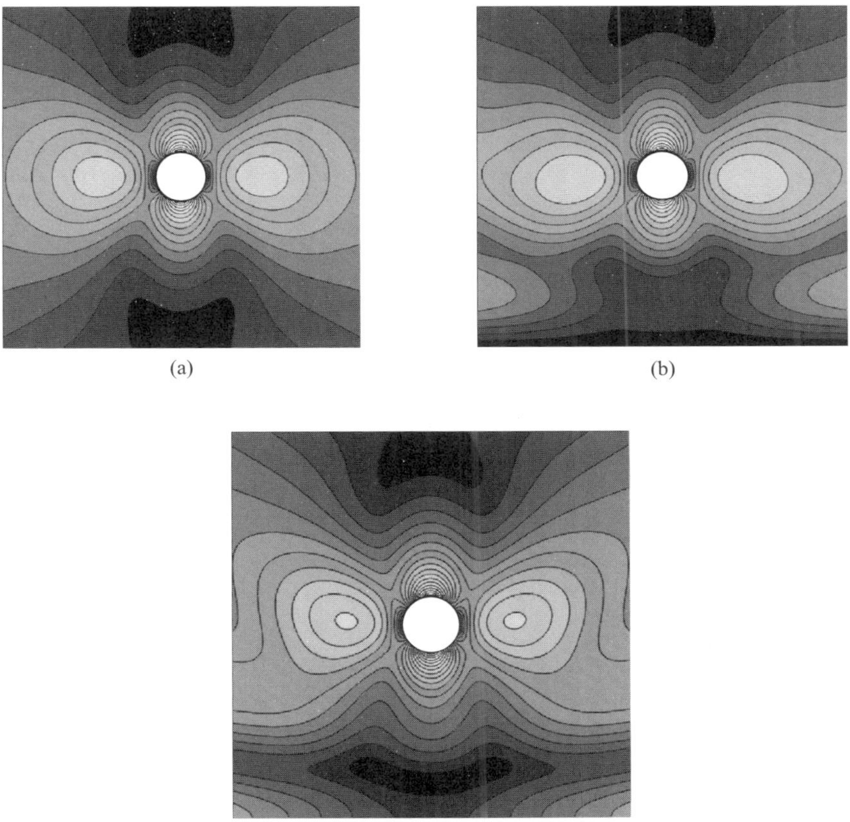

Fig. 6.38a,b and c. The contour lines of $|\sigma_{13}^*|$ in the vicinity of a circular opening with two electrodes in: a) a space (A $= 0$) b) a free halfspace ($A = -1$) and c) a fixed halfspace ($A = 1$)

Figure 6.36 depicts the behavior of the quantity $\mu = |q_2(\beta) / (\ni^{\varepsilon}_{11}\, \Phi^*)|$ on the electrode at $h/R = 1.5, \gamma R = 1$ for various values of the identificator of boundary conditions A. As it follows from the last singular equation in (6.7) and expression (6.29), the intensity of distribution of the charges (the normal component of the electric induction vector) has singularities of root type on the edges of the electrodes, which is also confirmed by the curves in Fig. 6.36.

Figure 6.37a,b and c illustrates the contour lines of the modulus of displacement amplitude $|U_3|$ in the area covering the opening, for various conditions on boundary $x_2 = 0$ at $\gamma R = 1,\ h/R = 7.5$. The lighter zones correspond to the maximum values of quantity $|U_3|$. Figure 6.37a,b and c show analogous results for the values of parameter $A = 0,\ 1$ and -1, respectively.

The distribution of the modulus of stress amplitudes $|\sigma^*_{13}|$ and $|\sigma^*_{23}|$ in the close and far regions at $\gamma R = 1,\ h/R = 7.5$ for values $A = 0, 1$ and -1 are

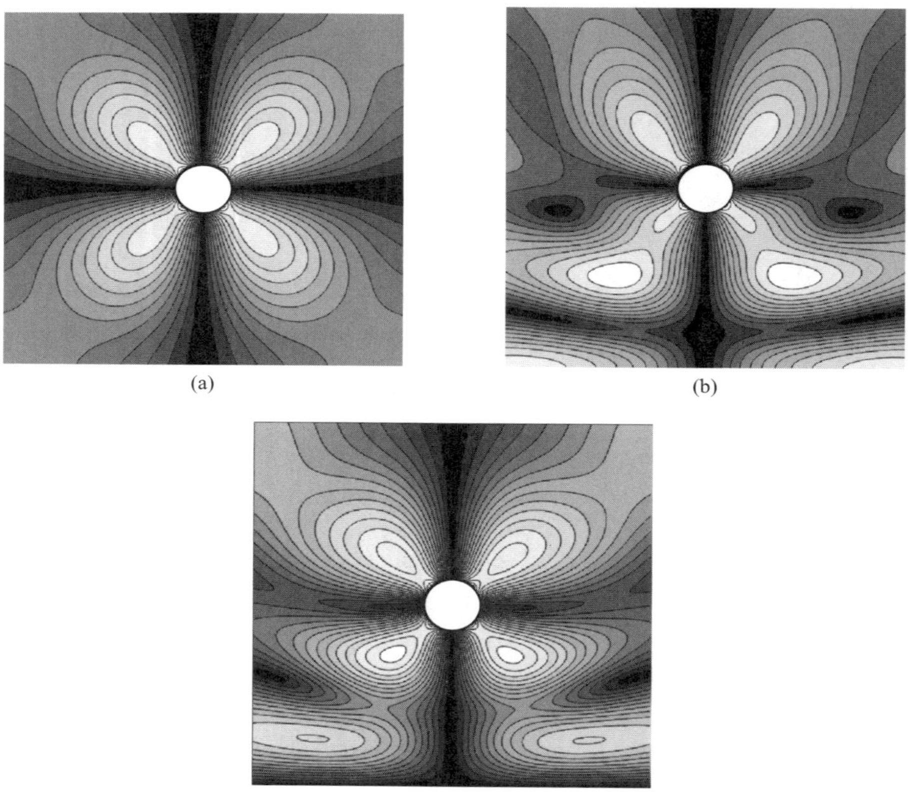

(a) (b) (c)

Fig. 6.39a,b and c. The contour lines of $|\sigma^*_{23}|$ in the vicinity of a circular opening with two electrodes in: a) a space ($A = 0$), b) a free halfspace ($A = -1$) and c) a fixed halfspace ($A = 1$)

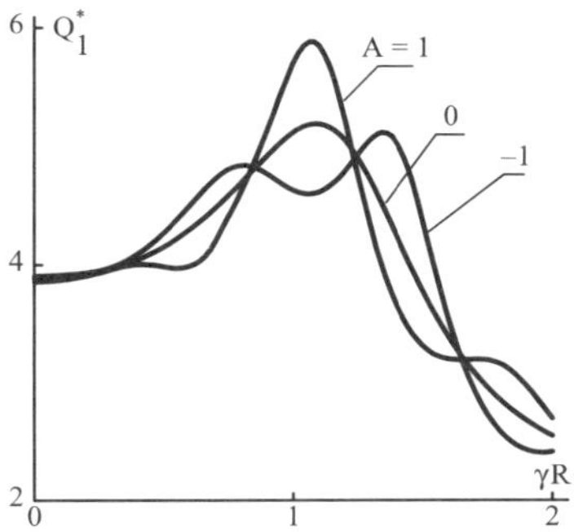

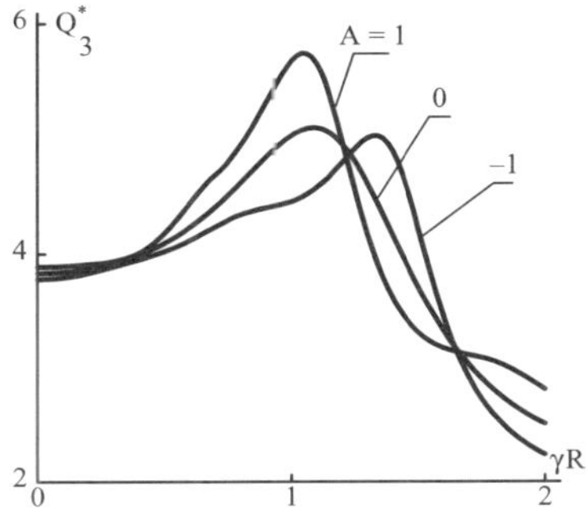

Fig. 6.40a and b. Changes of the amplitude of relative total charge $Q_1^* = |Q_1/(\ni_{11}^{\varepsilon}\ V)|$ and $Q_3^* = |Q_3/(\ni_{11}^{\varepsilon}\ V)|$ as a function of the normalized wave number for various types of boundary conditions on the boundary of the halspace $\Phi_1^* = V, \Phi_2^* = -V, \Phi_3 = 100V, \Phi_4^* = -100V$

(a)

(b)

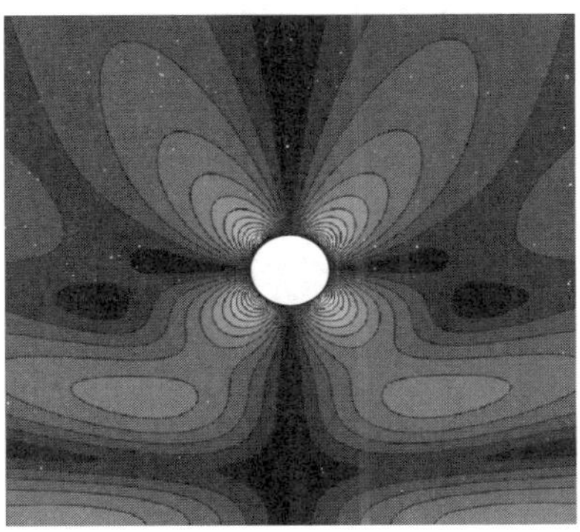

(c)

Fig. 6.41a,b and c. The contour lines of the modulus of the shear displacement amplitude in the vicinity of a circular opening with four electrodes in: a) a space ($A = 0$), b) a free halfspace ($A = -1$) and c) a fixed halfspace ($A = 1$)

represented in Figs. 6.38a,b and c and 6.39 a, b, c, respectively. Here it should be mentioned, that under static conditions ($\omega = 0$) the electric loading of the medium under an antiplane deformation do not cause mechanical stresses in it.

Now, consider the case of excitation of coupled fields by four electrodes, the position of which is fixed by quantities $\beta_k = (2k-1)\pi/8 \ (k = 1, 8)$. Figures 6.40a and b and 6.40a and b show respectively the behavior of quantities $Q_1^* = |Q_1/(\ni_{11}^{\varepsilon} V)|$, and $Q_3^* = |Q_3/(\ni_{11}^{\varepsilon} V)|$ far and near to the boundary of the halfspace electrodes, with respect to γR for various variants of the edge conditions on the boundary of halfspace ($h/R = 2.5$). On the electrodes the potentials were prescribed as follows $\Phi_1^* = V$, $\Phi_2^* = -V$, $\Phi_3^* = V$, $\Phi_4^* = -V$.

The results for the distribution of the contour lines of quantities $|U_3|$, $|\sigma_{13}^*|$ and $|\sigma_{23}^*|$ in the vicinity of a circular opening are given in Figs. 6.41a, b, c; 6.42a, b, c and 6.43a, b, c at $\gamma R = 1$, $h/R = 7.5$, $\Phi_1^* = V$, $\Phi_2^* = -V$, $\Phi_3^* = -V$, $\Phi_4^* = V$. Figure 6.44 illustrates the contour lines of quantities $|\sigma_{13}^*|$ and $|\sigma_{23}^*|$ for a free halfspace in the case of $\Phi_1^* = V$, $\Phi_2^* = -V$, $\Phi_3^* = -V$, $\Phi_4^* = V$ at $\gamma R = 1$.

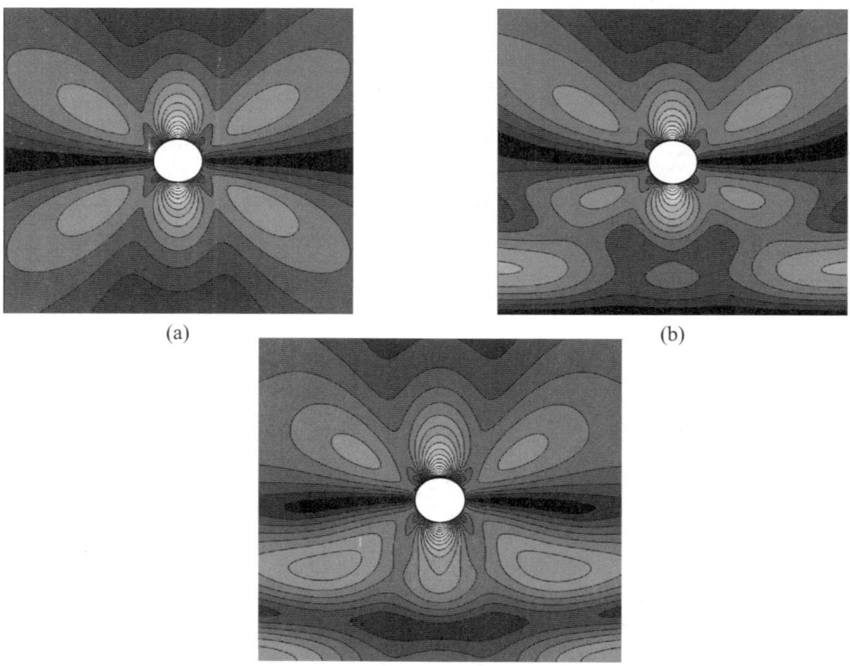

(a) (b) (c)

Fig. 6.42a,b and c. The contour lines of $|\sigma_{13}^*|$ in the vicinity of a circular hole with four electrodes in a) a space (A = 0) b) a free halfspace (A = −1) and c) a fixed halfspace (A = 1)

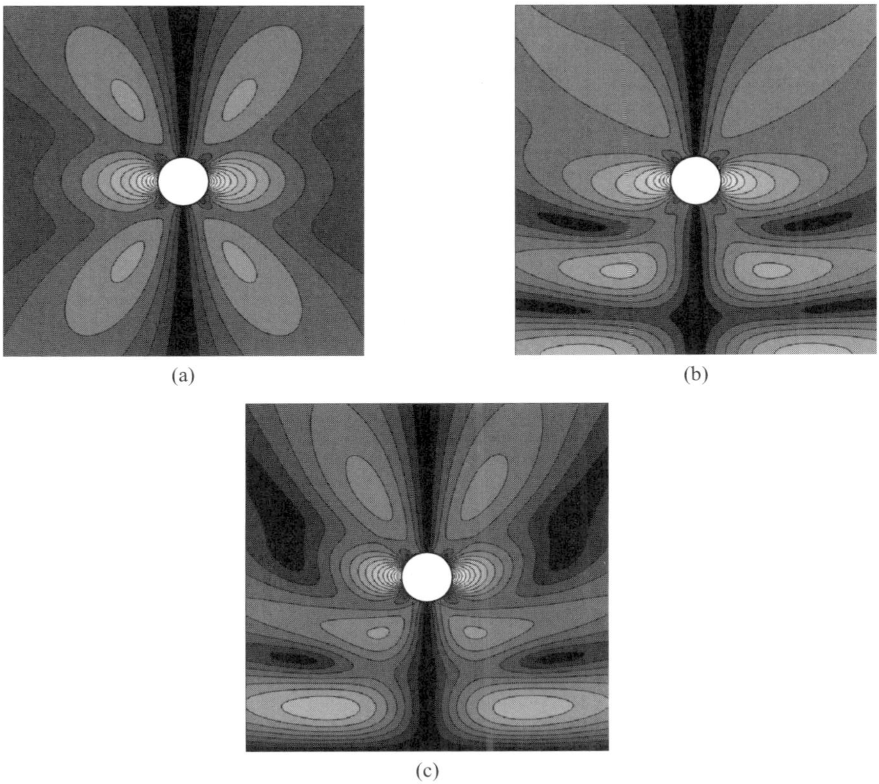

Fig. 6.43a,b and c. The contour lines of $|\sigma_{23}^*|$ in the vicinity of a circular hole with four electrodes in a) a space (A = 0) b) a free halfspace (A = −1) and c) a fixed halfspace (A = 1)

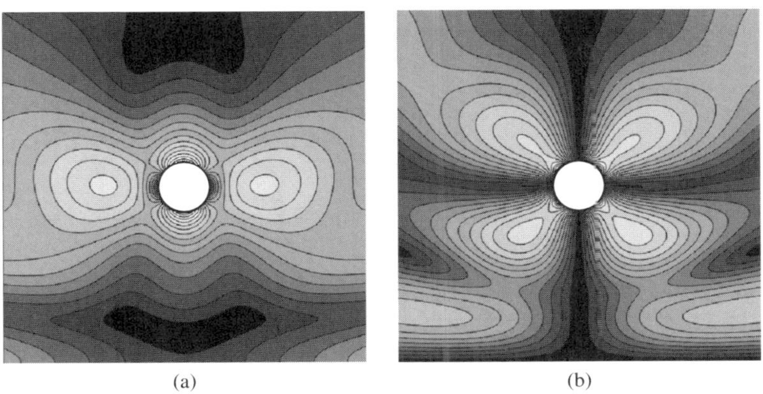

Fig. 6.44 a and b. Contour lines of $|\sigma_{13}^*|$ and $|\sigma_{23}^*|$ respectively in the vicinity of a circular hole containing four electodes for a free halfspace for ($\Phi_1^* = V$, $\Phi_2^* = -V$, $\Phi_3^* = -V$, $\Phi_4^* = V$)

The changes of the modulus of amplitude on the electrode with respect to the electrical potential $\langle V^* \rangle = |\ni_{11}^{\varepsilon} V^*/\tau e_{15}|$ as a function of γR under the influence of the harmonic waves of (6.31) type are given for the values of $A = 0$ and 1, respectively, in Fig. 6.45a and 6.45b ($h/R = 2.5$). The calculations were carried out with the help of (6.32) for the "*no-load*" mode (in that case the electrodes are disconnected). Curves 1–3 correspond to the following variants of the distribution of the electrodes: $\beta_1 = -\pi/7$, $\beta_2 = \pi/7$, $\beta_3 = 6\pi/7$, $\beta_4 = 8\pi/7$; $\beta_1 = -\pi/4$, $\beta_2 = \pi/4$, $\beta_3 = 3\pi/4$, $\beta_4 = 5\pi/4$ and $\beta_1 = -\pi/3$, $\beta_2 = \pi/3$, $\beta_3 = 2\pi/3$, $\beta_4 = 5\pi/3$.

The analysis of the results show that the most efficient electroacoustic transformation of the energy is observed at the smallest area of the electrodized coating, and in a halfspace it is considerably higher than in the full space. For example, comparing curves 1 in Fig. 6.45 and 6.45b we observe that the maximum value of quantity $\langle V^* \rangle$ in a halfspace exceeds almost two times the similar value in case of a full space.

From the results it follows that under the condition of reverse piezoelectric effect the distribution of mechanical quantities in a halfspace changes considerably according to the boundary conditions on and the prescribed electric potentials on the system of the electrodes.

Here it should be mentioned that, as the reflected (by the boundary of the halfspace) coupled wave field from the boundary of a half-space introduces additional charges on the paired electrodes (connected to a separate generator), the latter should be positioned symmetrically with respect to axis x_2 (a case when the centers of the electrodes lie on this axis, are obviously excluded). In that case the system of integrodifferential equations becomes unsolvable.

The constructed algorithm may be generalized in the case of n-tunnel openings C_m $(m = \overline{1,n})$ with cross-sections of canonic form, if their centers of symmetry are positioned on x_2 - axis, and the electrodes are also positioned

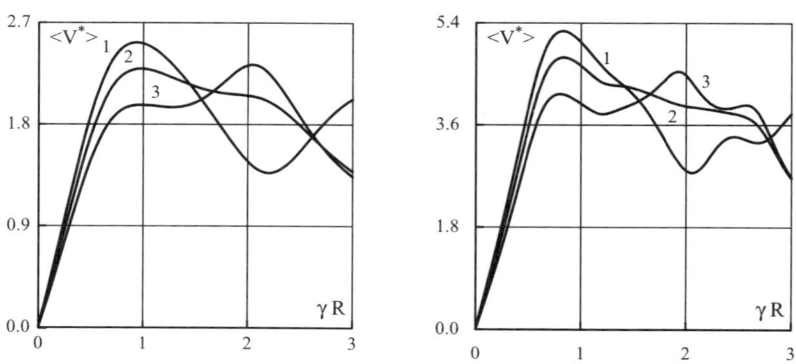

Fig. 6.45a and b. Changes of the modulus of relative electric potential on an electrode according to the normalized wave number in case of: a) a space ($A = 0$) b) a space ($A = 1$)

symmetrically with respect to this axis. For this purpose, in (6.7), we assume $p(\zeta) = \{p_m(\zeta), \zeta \in C_m\}$, $f(\zeta) = \{f_m(\zeta), \zeta \in C_m\}$, $C = \bigcup_{m=1}^{n} C_m$.

6.7 A Layer

Let a piezoceramic layer ($0 \leq x_1 \leq a, -\infty < x_2 < \infty, -\infty < x_3 < \infty$) in the Cartesian system of coordinates $Ox_1x_2x_3$ weakened by a tunnel along x_3-axis opening, the cross-section of which is bounded by smooth contour C (Fig. 6.46a). Assume that the basis of the layer are free from forces and bounded with vacuum (the direction of polarization of the ceramics is parallel to x_3 -axis). On the free from mechanical stresses surfaces of the opening-$2n$ thin electrodes are positioned, with prescribed differences of the electric potential, and the unelectrodized areas of the opening are bounded by vacuum (air). The boundaries of the k-th electrode are determined by β_{2k-1} and β_{2k} $\left(k = \overline{1, 2n}\right)$, and the electric potential on it is prescribed by quantity $\phi_k^* = \mathrm{Re}\left(\Phi_k^* e^{-i\omega t}\right)$. The position of the electrodes and the configuration of the cavity cannot be arbitrary and the imposed on them requirements will be given below.

The mechanical and electric boundary conditions at the bases of the layer are formally represented in the form

$$\sigma_{13} = 0, \ D_1 = 0 \ (x_1 = 0, a). \tag{6.34}$$

The boundary conditions (6.4) remain valid on the surface of the cavity.

In order to solve the stated problem it is recommended to have integral representations of the solutions satisfying conditions (6 34) as well the conditions of radiation at infinity. Having that in mind, let us write the Green's functions for a homogeneous piezoceramic layer.

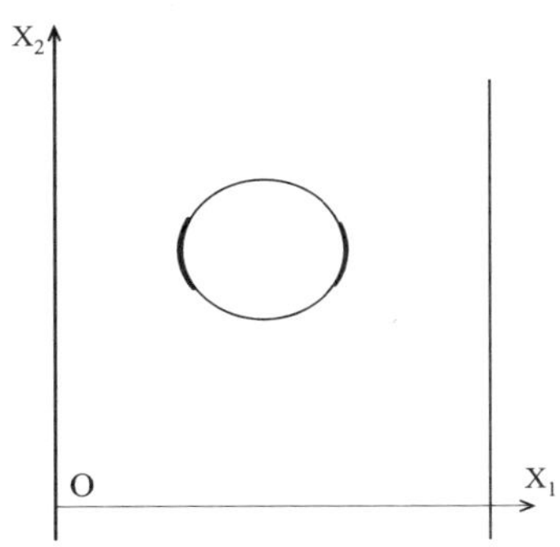

Fig. 6.46a. A layer with a partially electrodized tunnel cavity

The boundary problems (6.5), (6.34) may be written as

$$\nabla^2 U_3 + \gamma^2 U_3 = 0;\ \partial_1 U_3 = 0\ \ (x_1 = 0, a)\,, \tag{6.35}$$

$$\nabla^2 F = 0,\ \partial_1 F^* = 0,\ (x_1 = 0, a)\,. \tag{6.36}$$

The Green's function corresponding to problem (6.35) is represented by formulas (4.82), (4.83). The expression for the Green's function of problem (6.36) is changed and written in the form of

$$\begin{aligned} &E(\zeta, z) = -\frac{|x_2 - \xi_2|}{2a} + \frac{1}{2\pi}\ell n \left| 4\sin\frac{\pi(\zeta - z)}{2a}\sin\frac{\pi(\zeta + \bar{z})}{2a} \right|, \\ &\nabla^2 E = \delta(x_1 - \xi_{11}, x_2 - \xi_2) = \delta(x_1 - \xi_1)\,\delta(x_2 - \xi_2)\,, \\ &z = x_1 - i x_2,\ \zeta = \xi_1 + i\xi_2, \end{aligned} \tag{6.37}$$

where $\delta(x) - 2a$ is the periodical Dirac' δ-function.

Using the Green's function we write the integral representation of the solution in the form

$$\begin{aligned} U_3(x_1, x_2) &= \int_C p(\zeta) G(\zeta, z)\, ds, \\ F^*(x_1, x_2) &= \int_C f(\zeta) \frac{\partial E(\zeta, z)}{dn_\zeta} ds,\ \zeta \in C. \end{aligned} \tag{6.38}$$

Here ds is the element of the arc length of contour C. Expression (6.38) satisfy both the boundary conditions (6.34) on the bases of the layer, and the conditions of radiation at infinity.

Taking into account (6.37) the representation for function $F^*(x_1, x_2)$ is transformed into the form

$$\begin{aligned} &F^*(x_1, x_2) = \int_C f(\zeta) K(\zeta, z) ds, \\ &K(\zeta, z) = \frac{\sin\psi}{2a} sign(x_2 - \xi_2) + \frac{1}{4a}\mathrm{Re}\left\{ e^{i\psi}\left[ctg\frac{\pi(\zeta - z)}{2a} + ctg\frac{\pi(\zeta + \bar{z})}{2a} \right] \right\}. \end{aligned} \tag{6.39}$$

Here $\psi = \psi(\zeta)$ is the angle between the normal to contour C and Ox_1-axis, at $\zeta = C$.

Using Sohotsky-Plemmely formulas and the decomposing into partial fractions

$$ctg\pi x = \frac{1}{\pi x} + \frac{2x}{\pi}\sum_{m=1}^{\infty} \frac{1}{x^2 - m^2}$$

the expression for the limiting values of the appearing in (6.39) integrals with the kernels of Hilbert type at $z \to \zeta_0 \in C$ read

$$\left\{\int_C f(\zeta)\, ctg\frac{\pi(z-\zeta)}{2a}d\zeta\right\}^{\pm} = \pm 2iaf(\zeta_0) + \int_C f(\zeta)\, ctg\frac{\pi(\zeta_0-\zeta)}{2a}d\zeta, \quad (6.40)$$

$$\left\{\int_C f(\zeta)\, ctg\frac{\pi(\bar{z}-\bar{\zeta})}{2a}d\bar{\zeta}\right\}^{\pm} = \pm 2iaf(\zeta_0) + \int_C f(\zeta)\, ctg\frac{\pi(\bar{\zeta}_0-\bar{\zeta})}{2a}d\bar{\zeta}.$$

Differentiating function $F^*(x_1, x_2)$ in (6.38) we get

$$\frac{\partial F^*}{\partial z} = \frac{\pi}{16a^2}\int_C f(\zeta)\left\{e^{i\psi}\sec^2\frac{\pi(\zeta-z)}{2a} - e^{-i\psi}\sec^2\frac{\pi(\bar{\zeta}+z)}{2a}\right\}ds, \quad (6.41)$$

$$\frac{\partial F^*}{\partial \bar{z}} = \frac{\pi}{16a^2}\int_C f(\zeta)\left\{e^{-i\psi}\sec^2\frac{\pi(\bar{\zeta}-\bar{z})}{2a} - e^{i\psi}\sec^2\frac{\pi(\zeta+\bar{z})}{2a}\right\}ds.$$

At $z \to \zeta_0 \in C$ the integrals in (6.41) become diverging. To regulate them it is necessary to integrate by parts allowing for the conditions of periodicity function $f(z)$.

Substituting the limiting values of functions (6.38), (6.39) and their derivatives at $z \to \zeta_0 \in C$ under the boundary conditions (6.4) and taking into account (6.40) we get a system of singular integrodifferential equations of the second kind

$$p(\zeta_0) + \int_C p(\zeta)\, g_1(\zeta,\zeta_0)\, ds + \int_C f'(\zeta) g_2(\zeta,\zeta_0)\, ds = N_1(\zeta_0),$$

$$-\frac{1}{2}f(\zeta_0) + \int_C \{p(\zeta)\, g_3(\zeta,\zeta_0) + f(\zeta)\, g_4(\zeta,\zeta_0)\}\, ds = N_2(\zeta_0), \ \zeta \in C_\phi,$$

$$\int_C f'(\zeta) g_5(\zeta,\zeta_0)\, ds = 0, \ \zeta_0 \in C \backslash C_\phi,$$

$$g_1(\zeta,\zeta_0) = \frac{1}{2a}\mathrm{Re}\left\{e^{i\psi_0}\left[ctg\frac{\pi(\zeta_0-\zeta)}{2a} + ctg\frac{\pi(\zeta_0+\bar{\zeta})}{2a}\right]\right\} + P_1 e^{i\psi_0} + P_2 e^{-i\psi_0},$$

$$g_2(\zeta,\zeta_0) = \frac{2e_{15}}{c_{44}^E(1+k_{15}^2)} g_5(\zeta,\zeta_0),$$

$$g_3(\zeta,\zeta_0) = \frac{e_{15}}{\text{э}_{11}^{\varepsilon}}\left\{-\frac{|\xi_{20}-\xi_2|}{2a} + \frac{1}{2\pi}\ell n\left|4\sin\frac{\pi(\zeta-\zeta_0)}{2a}\sin\frac{\pi(\zeta+\bar{\zeta}_0)}{2a}\right| + \right.$$

$$\left. + \frac{1}{2ia\gamma}e^{i\gamma|\xi_{20}-\xi_2|} - \frac{1}{a}\sum_{m=1}^{\infty} c_m(\xi_{20}-\xi_2)\cos\alpha_m\xi_1\cos\alpha_m\xi_{10}\right\},$$

$$
\begin{aligned}
g_4(\zeta,\zeta_0) &= \frac{1}{4a}\mathrm{Re}\left\{e^{i\psi}\left[ctg\frac{\pi(\zeta-\zeta_0)}{2a}+ctg\frac{\pi(\zeta+\bar{\zeta}_0)}{2a}\right]\right\} \\
&\quad+\frac{\sin\psi}{2a}sign(\xi_{20}-\xi_2),\\
g_5(\zeta,\zeta_0) &= \frac{1}{4a}\mathrm{Im}\left\{e^{i\psi_0}\left[ctg\frac{\pi(\zeta-\zeta_0)}{2a}+ctg\frac{\pi(\bar{\zeta}+\zeta_0)}{2a}\right]\right\},\\
P_1 &= S-\frac{1}{a}(A_0-iB_0),\ P_2=-S-\frac{1}{a}(A_0-iB_0),\\
S &= \frac{1}{2ia}sign(\xi_2-\xi_{20})\left(1-e^{i\gamma|\xi_2-\xi_{20}|}\right),\\
A_0 &= \sum_{k=1}^{\infty}\beta_{1k}\alpha_k\cos\alpha_k\xi_1\sin\alpha_k\xi_{10},\\
B_0 &= \sum_{k=1}^{\infty}\beta_{0k}sign(\xi_{20}-\xi_2)\cos\alpha_k\xi_1\cos\alpha_k\xi_{10},\\
\beta_{mk} &= \frac{1}{\alpha_k^m}e^{-\alpha_k|\xi_2-\xi_{20}|}-\frac{1}{\lambda_k^m}e^{-\lambda_k|\xi_2-\xi_{20}|},\\
c_m(\xi_{20}-\xi_2) &= \frac{1}{\lambda_m}e^{-\lambda_m|\xi_{20}-\xi|}-\frac{1}{\alpha_m}e^{-\alpha_m|\xi_{20}-\xi_2|},\\
N_1(\zeta_0) &= 0,\ N_2(\zeta_0)=\Phi^*(\zeta_0),\\
\psi &= \psi(\zeta),\ \psi_0=\psi(\zeta_0),\ \zeta,\zeta_0\in C.
\end{aligned}
\tag{6.42}
$$

Here $\Phi^*(\zeta_0)$ is a piecewise-constant function, determining the value of electric potential on the system of electrodes. The kernels $g_2(\zeta,\zeta_0)$, $g_5(\zeta,\zeta_0)$ are singular (of Hilbert type).

Here it should be mentioned, that the appearing in the process of oscillation, shear waves reflected from boundary $x_1=0$ and $x_1=a$ cause additional charges on the active electrodes. Therefore the configuration of the section of cavity, its position, and also the position of paired electrodes (powered from a separate generator) should have certain symmetry with respect to the bases, i.e. the applied on the indicated electrodes charges over the absolute quantity should be the same. If this requirement is violated the system (6.42) becomes unsolvable.

The amplitude of the density distribution of electric charges $q_k(\beta)$ on the k-th electrode reads

$$
q_k(\beta_0)=-\frac{\mathfrak{s}_{11}^{\varepsilon}}{4a}\int_C f'(\zeta)\mathrm{Im}\left\{e^{i\psi_0}\left[ctg\frac{\pi(\zeta-\zeta_0)}{2a}+ctg\frac{\pi(\bar{\zeta}+\zeta_0)}{2a}\right]\right\}ds
$$
$$
\zeta_0\in C_{\phi_k}, \tag{6.43}
$$

where C_{ϕ_k} is the part of contour C, where the k-th electrode is positioned.

Now consider the case, where a piezoceramic layer with a tunnel opening is used as a generator of electric energy. In this case, the mechanic excitation comes from two plane monochromatic shear waves, propagating in the positive and negative directions of x_2 -axis having the following values of displacement amplitude u_3 and electric potential ϕ, respectively:

$$U_3^{(1)} = \tau_1 e^{-i\gamma x_2},\ U_3^{(2)} = \tau_2 e^{i\gamma x_2},$$
$$\Phi^{(j)} = \frac{e_{15}}{\ni_{11}^{\varepsilon}} U_3^{(j)} \quad (j = 1, 2)\,. \tag{6.44}$$

We assume that the cross-section of the cavity has a vertical axis of symmetry and on its surface two symmetrically positioned infinitely long electrodes are positioned (Fig. 6.46b). In order to sustain the difference of electric potential $2V\,(t)$ under the deformation of the medium electric charges of opposite signs should appear on the electrodized coatings, which require the matching of the displacement amplitudes in monochromatic waves. Due to this in (6.44) we have to set $\tau_1 = -\tau_2 = \tau$.

As a result the unknown potential amplitude $V\,(t)$ on the electrode read

$$V^*\,(\omega) = \frac{i\tau\omega \ni_{11}^{\varepsilon} B_1}{2Y - i\omega \ni_{11}^{\varepsilon} B_2},$$
$$B_m = \int_{\beta_1}^{\beta_2} A_m\,(\beta_0)\, s'\,(\beta_0)\, d\beta_0 \quad (m = 1, 2)\,, \tag{6.45}$$
$$A_m\,(\beta_0) = -\frac{1}{4a} \int_C f_m'\,(\zeta)\, \mathrm{Im} \left\{ e^{i\psi_0} \left[ctg \frac{\pi\,(\zeta - \zeta_0)}{2a} + ctg \frac{\pi\,(\overline{\zeta} - \zeta_0)}{2a} \right] \right\} ds.$$

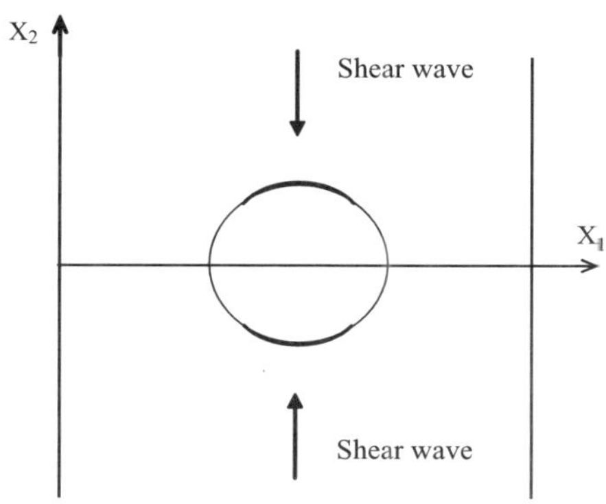

Fig. 6.46b. The generation of the difference of electric potentials on electrodes under the influence of monochromatic shear waves in a piece-wise homogeneous layer

where the functions $f_m(\zeta)\,(m=1,2)$ are "*standard*" solutions of system (6.42) where the right parts are given by:

$$
\begin{aligned}
&N_1^{(1)}(\zeta_0) = 4i\gamma\cos\gamma\xi_{20}\sin\psi_0,\\
&N_2^{(1)}(\zeta_0) = \frac{2ie_{15}}{\ni_{11}^{\varepsilon}}\sin\gamma\xi_{20},\\
&N_1^{(2)}(\zeta_0) = 0,\\
&N_2^{(2)}(\zeta_0) = \begin{cases} 1, \beta_1 < \beta_0 < \beta_2, \\ -1, \beta_3 < \beta_0 < \beta_4, \end{cases} \qquad \zeta_0 = \xi_{10} + i\xi_{20} \in C,
\end{aligned} \tag{6.46}
$$

where the quantities $\beta_k\,\left(k=\overline{1,4}\right)$ prescribe the position of the electrodes.

As a first example consider a layer (of $PZT-4$ ceramics) with a circular cavity of radius R excited by two electrodes, the centers of which are positioned on its vertical diameter ($\beta_1 = 5\pi/14,\ \beta_2 = 9\pi/14,\ \beta_3 = 19\pi/14,\ \beta_4 = 23\pi/14$).

Figure 6.47 depicts the the change of quantity $Q^* = \left|Q/\left(\ni_{11}^{\varepsilon}\,\Phi^*\right)\right|$, characterizing the amplitude of total electric charge Q on an electrode is shown, with respect to the normalized wave numbers γa ($2\Phi^*$ is the difference of the amplitude of electric potential on electrodes). Curve 1 corresponds to the opening, displaced from the axis of symmetry of the layer at a distance of $0.1a$; curve 2 corresponds to the symmetrically positioned opening ($R/a = 0.1$). In the first case it is seen that due to the inertial effect, the quantity Q^* may exceed its static analogue by 16%. Here we should note that on exceeding wave number with values $\alpha_m = m\pi/a\ (m = 1, 2, \ldots)$ the solution is destabilized to a new running wave, carrying the energy along the wave guide from the heterogeneity to infinity. This circumstance provides a characteristic "breakwise" form of the curves in the vicinity of points $\gamma = \pi$ and $\gamma = 2\pi\ (a = 1)$.

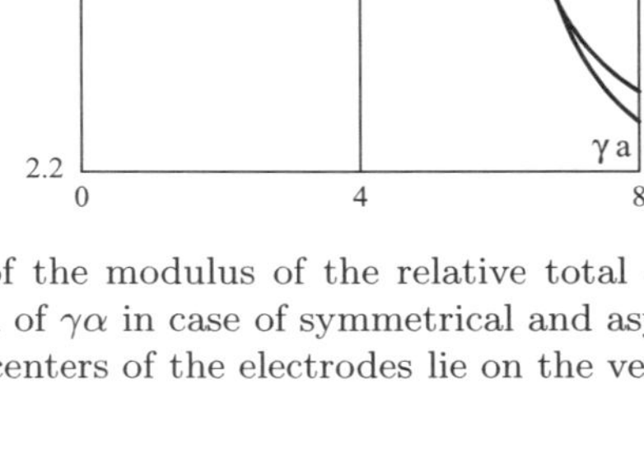

Fig. 6.47. Changes of the modulus of the relative total charge amplitude on an electrode as a function of $\gamma\alpha$ in case of symmetrical and asymmetrical position of a circular opening (the centers of the electrodes lie on the vertical diameter)

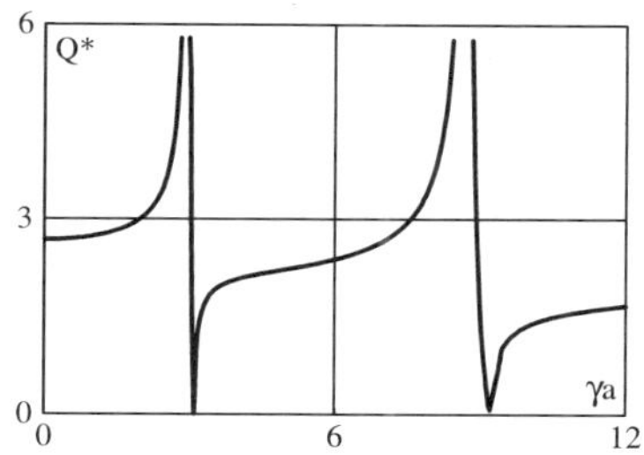

Fig. 6.48. Amplitude frequency characteristics of quantity Q^* at $\beta_1 = -\pi/7$, $\beta_2 = \pi/7$, $\beta_3 = 6\pi/7$ and $\beta_4 = 8\pi/7$ (the centers of the electrodes lie on the horizontal diameter of the opening)

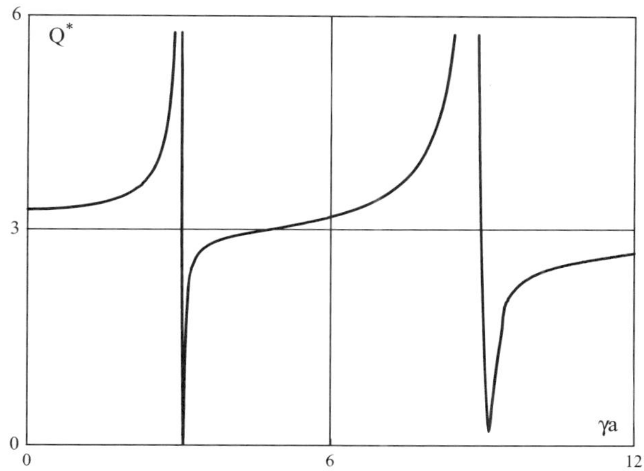

Fig. 6.49. Amplitude frequency characteristics of quantity Q^* at $\beta_1 = -3\pi/28$, $\beta_2 = 11\pi/28$, $\beta_3 = 17\pi/28$ and $\beta_4 = 25\pi/28$

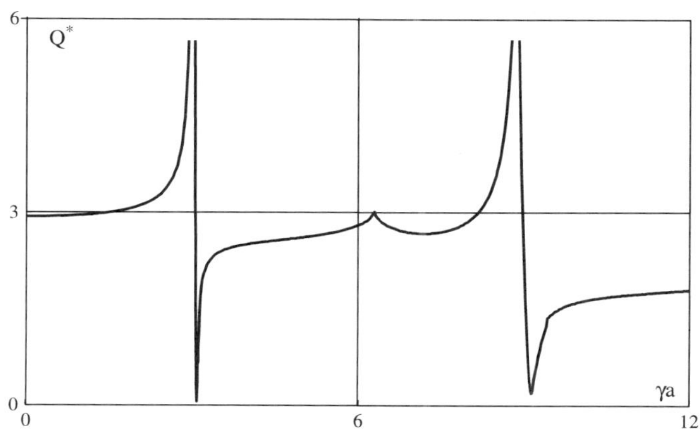

Fig. 6.50. Amplitude frequency characteristics of quantity Q^* at $\beta_1 = 3\pi/28$, $\beta_2 = 11\pi/28$, $\beta_3 = 31\pi/28$ and $\beta_4 = 39\pi/28$

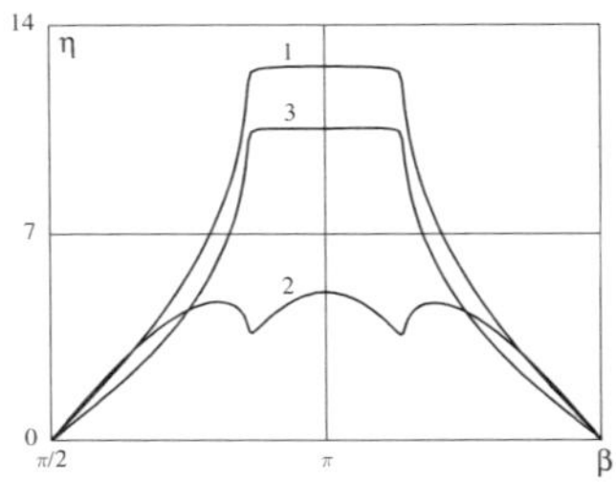

Fig. 6.51. Changes of quantity $\eta = c_{44}^E |u_3/\Phi^*|$ on the contour of symmetrically positioned circular cavity for various $\gamma\alpha$

Quite different picture is observed if the centers of two active electrodes lie on the horizontal diameter of symmetrically positioned openings ($\beta_1 = -\pi/7$, $\beta_2 = \pi/7$, $\beta_3 = 6\pi/7$, $\beta_4 = 8\pi/7$). In this case, as can be seen in Fig. 6.48, we observe the phenomenon of resonance. The values of the normalized wave numbers corresponding to the first and second natural frequencies of oscillations are approximately equal: $\gamma_{(1)}a \approx 2.95$ and $\gamma_{(2)}a \approx 8.69$. The anti-resonance frequency, at which the current in the generator circuit is zero, is equal $\gamma^*_{(1)}a \approx 3.1$. In the calculations we set $R/a = 0.1$.

The resonances also occur in situations where the centers of the electrodes are positioned on straight lines, inclined to the horizontal diameter

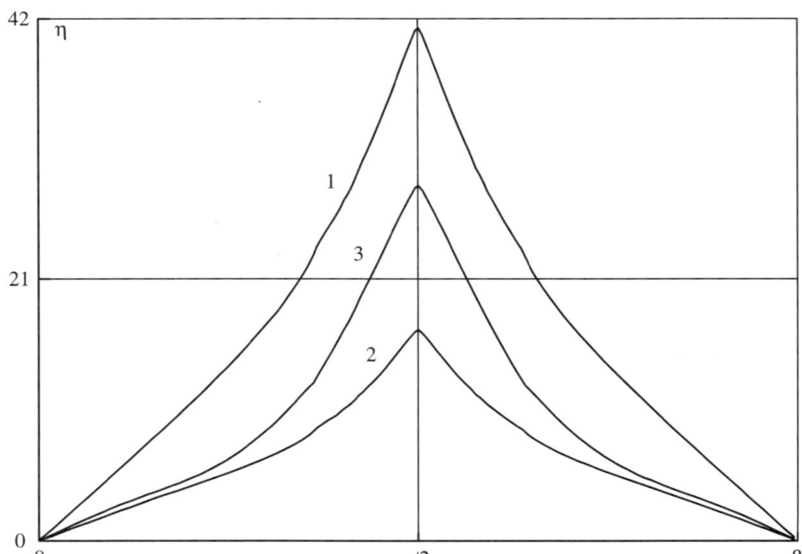

Fig. 6.52. Changes of quantity $\eta = c_{44}^E |u_3/\Phi^*|$ on the contour of symmetrically positioned circular cavity for various $\gamma\alpha$

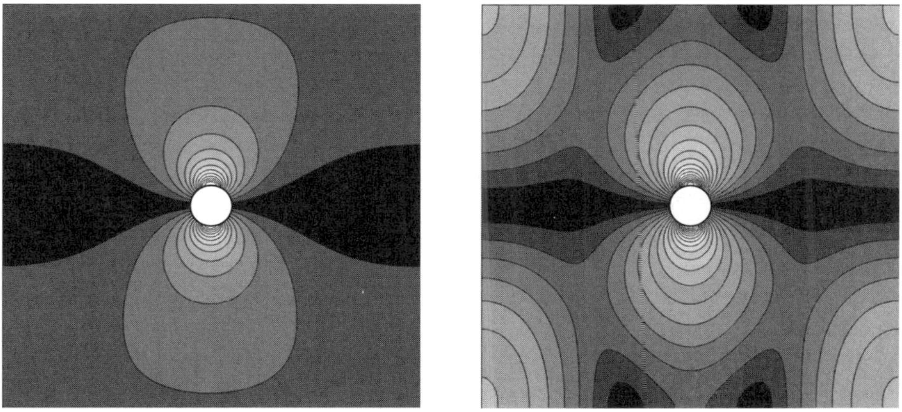

(a) (b)

Fig. 6.53 a and b. The contour lines of the modulus of the displacement amplitude in a layer with a circular opening

at an angle of 45°. The graphs of the quantities Q^* at $R/a = 0.1$, $\beta_1 = 3\pi/28$, $\beta_2 = 11\pi/28$, $\beta_3 = 17\pi/28$, $\beta_4 = 25\pi/28$, $\beta_1 = 3\pi/28$, $\beta_2 = 11\pi/28$, $\beta_3 = 31\pi/28$, $\beta_4 = 39\pi/28$ are represented in Figs. 6.49 and 6.50, respectively. In these cases the resonance frequencies are determined at $\gamma_{(1)}a \approx 3.01, \gamma_{(2)}a \approx 8.79$ and $\gamma_{(1)}a \approx 3.02$, $\gamma_{(2)}a \approx 8.85$, respectively.

Figures 6.51 and 6.52 illustrate the change of $\eta = c_{44}^E\,|U_3/\Phi^*|$ on the contour of symmetrically positioned circular cavity ($R/a = 0.1$). Curves 1–3 in Fig. 6.51 were obtained setting $\beta_1 = -\pi/7$, $\beta_2 = \pi/7$, $\beta_3 = 6\pi/7$, $\beta_4 = 8\pi/7$, $\gamma a = 0, 3.14$ and 6.28, respectively; in Fig. 6.52 they were obtained for the same values γa and $\beta_1 = 5\pi/14$, $\beta_2 = 9\pi/14$, $\beta_3 = 19\pi/14$, $\beta_4 = 23\pi/14$.

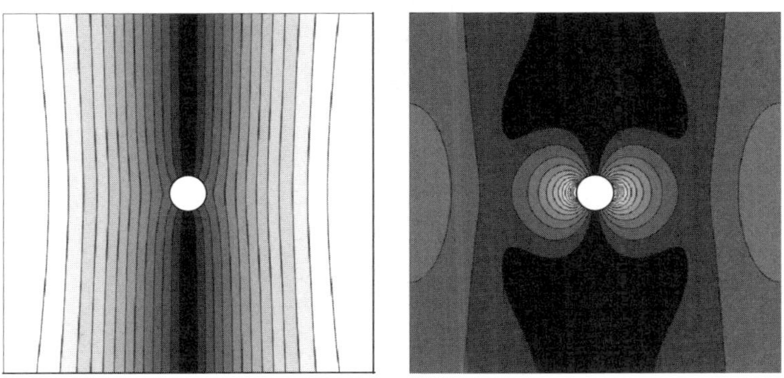

(a) (b)

Fig. 6.54 a and b. The contour lines of the modulus of the displacement amplitude in a layer with circular opening

The investigation of mechanical quantities in the area of a piecewise-homogeneous layer at electric excitation of an electroelastic field is of interest. Figure 6.53a and 6.53b show the contour lines of the modulus of the displacement amplitude in a layer with a circular cavity for $R/a = 0.05$, $\gamma a = 3.14$ and 6.28, respectively ($\beta_1 = 5\pi/14$, $\beta_2 = 9\pi/14$, $\beta_3 = 19\pi/14$, $\beta_4 = 23\pi/14$). The analogous results for the same values of parameters and $\beta_1 = -\pi/7$, $\beta_2 = \pi/7$, $\beta_3 = 6\pi/7$, $\beta_4 = 8\pi/7$ are given in Fig. 6.54a and b. Fig. 6.55a, b and c depict the distribution $|U_3|$ in the vicinity of resonance frequencies ($\gamma_{(1)}a = 2.95$, $\gamma_{(2)}a = 8.69$) and antiresonance frequencies $\gamma^*_{(1)}a = 3.1$ at $R/a = 0.1$, respectively for the same case of disposition of the electrodes.

In Figs. 6.56 and 6.57 the contour lines (U_3) are constructed for $\beta_1 = 3\pi/28$, $\beta_2 = 11\pi/28$, $\beta_3 = 17\pi/28$, $\beta_4 = 25\pi/28$ and $\beta_1 = 3\pi/28$,

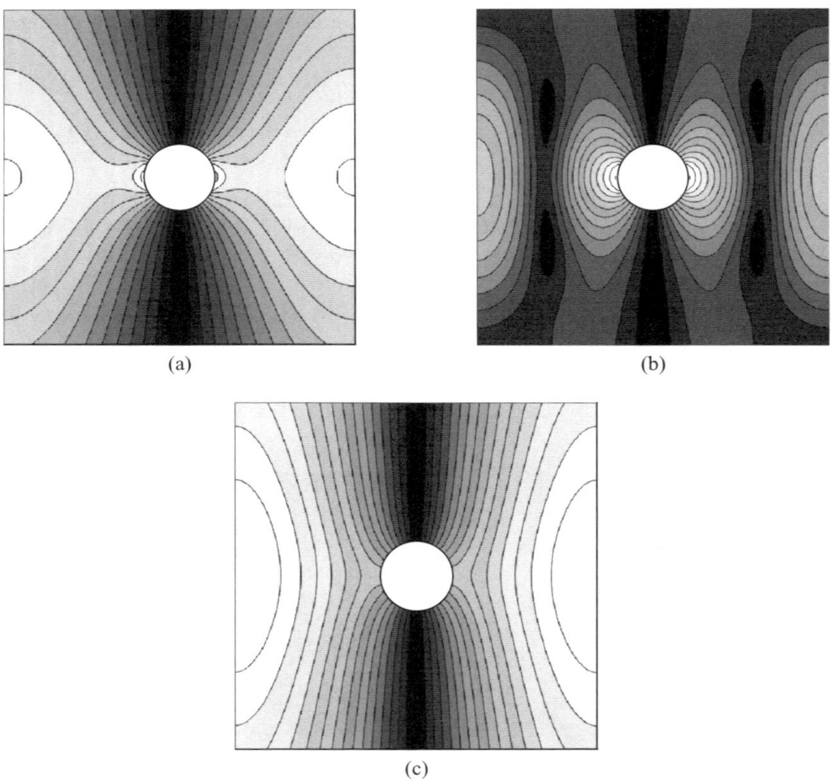

Fig. 6.55a, b and c. The contour lines of the modulus of the displacement amplitude in a layer with a circular opening in the vicinity of resonance frequency

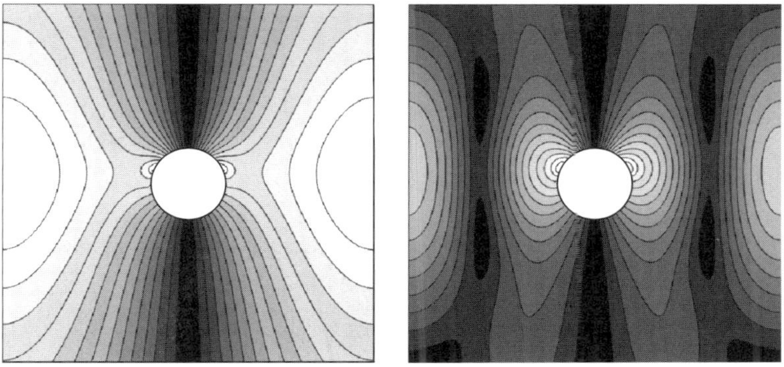

(a) (b)

Fig. 6.56. The contour lines of the modulus of the displacement amplitude in a layer with a circular opening in the vicinity of the first resonance frequency

$\beta_2 = 11\pi/28$, $\beta_3 = 31\pi/28$, $\beta_4 = 39\pi/28$ for the first two resonance frequencies respectively.

Now consider the case of excitation of oscillation by four electrodes, the displacement of which is fixed by $\beta_k = (2k-1)\,\pi/8$ $\left(k = \overline{1,8}\right)$.

In Fig. 6.58 the curves 1, 2 and 3, characterizing the change of $\eta = c_{44}^E\,|u_3/V|$ on the contour of the symmetrically positioned elliptic cavity, are constructed for $\gamma a = 0, 6.28$ and 10, respectively, at $R_1/a = 0.1$, $R_1/R_2 = 2$ and $\Phi_1^* = V$, $\Phi_2^* = -V$, $\Phi_3^* = -V$, $\Phi_4^* = V$ (R_1 and R_2 are horizontal and vertical half axes of an ellipse).

Figure 6.59a and 6.59b show the distribution $|U_3|$ in a layer with an opening of elliptic cross-section for values of potentials $\Phi_1^* = V$, $\Phi_2^* = -V$,

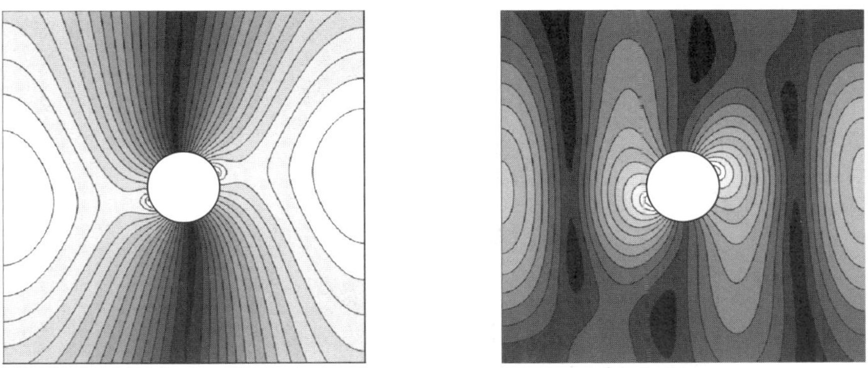

(a) (b)

Fig. 6.57a and b. The contour lines of the displacement amplitude in a layer with a circular opening in the vicinity of the first resonance frequency

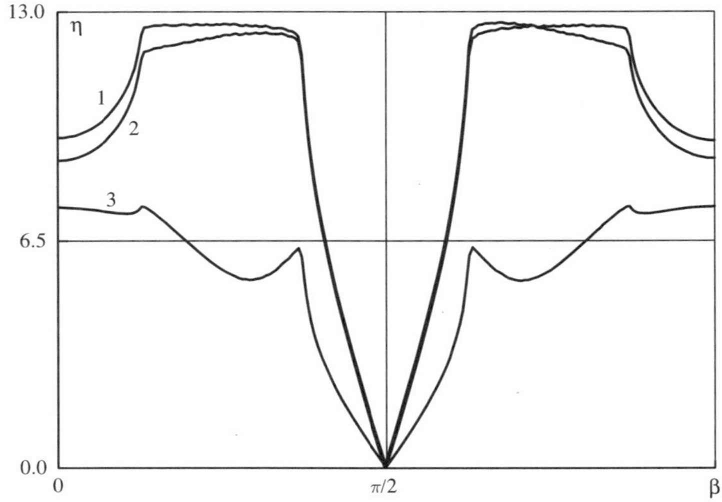

Fig. 6.58. Changes of quantity $\eta = c_{44}^E|u_3/\Phi^*|$ on the surface of an elliptic cavity for various values of $\gamma\alpha$ at excitation of the fields by four electrodes

$\Phi_3^* = V$, $\Phi_4^* = -V$ and $\Phi_1^* = V$, $\Phi_2^* = -V$, $\Phi_3^* = -V$, $\Phi_4^* = V$, respectively. We set $\gamma a = 9$, $R_1 = 0.1$, $R_1/R_2 = 2$. Figure 6.60 depicts the case of an asymmetrically positioned opening ($\gamma a = 13$, $R_2/a = 0.1$, $R_1/R_2 = 0.5$).

Figure 6.61a illustrates the changes of the modulus of the amplitude, concerning electric potential $\langle V^* \rangle = |\ni_{11}^{\varepsilon} V^*/\tau e_{15}|$ on the electrode as a function of γa in case of diffraction of monochromic waves of (6.44) type on the symmetrically positioned elliptic cavity ($R_1/R_2 = 2$). The calculations were

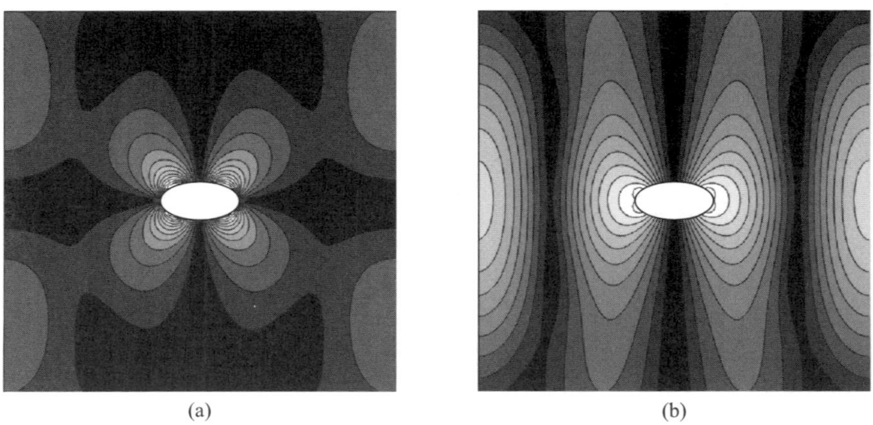

Fig. 6.59a and b. The contour lines $|u_3|$ of the modulus in a piece-wise homogeneous layer at excitation of the oscillations by four electrodes

Fig. 6.60. The contour lines $|u_3|$ of the modulus in a layer with asymmetrically positioned opening at excitation of the oscillations by four electrodes

carried out with the help of (6.45) for the "*no-load*" mode (the electrodes are disconnected). Curves 1–3 satisfy the following variants of the disposition of electrodes: $\beta_1 = 5\pi/14$, $\beta_2 = 9\pi/14$, $\beta_3 = 19\pi/14$ $\beta_4 = 23\pi/14$; $\beta_k = (2k-1)\,\pi/4\,\left(k=\overline{1,4}\right)$, and $\beta_1 = \pi/6$, $\beta_2 = 5\pi/6$, $\beta_3 = 7\pi/6$, $\beta_4 = 11\pi/6$. The analogous results for a case with an opening displaced from the axis

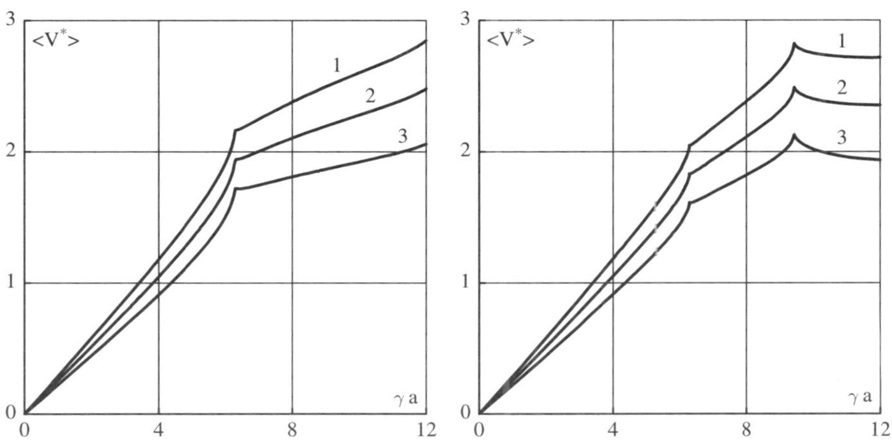

Fig. 6.61a and b. Changes of generalized on electrodes potentials $\langle V* \rangle = |\ni^{\varepsilon}_{11}\, V^*/\tau e_{15}|$ as a function of the normalized wave number for various areas of an electrodized coating at symmetrically and assymetrically positioned elliptic opening in a layer

of symmetry of the layer at distance $0.1a$, are given in Fig. 6.61a and b ($R_1/R_2 = 2$). The analysis of the results shows that the most efficient electroacoustic transform of energy in the considered interval of frequencies is observed at the smallest area of the electrodized coating.

6.8 Interaction of a Partially Electrodized Opening and Crack

Up to now the behaviour of the dynamic stress intensity factor (SIF) was investigated on the tip of a crack under the mechanical loading (direct piezoeffect). Below we will investigate the behaviour of the SIF at the harmonic with time electric excitation by the system of electrodes.

In the Cartesian system of coordinates $Ox_1x_2x_3$ consider a piezoceramic space containing tunnel along axis x_3 opening and curvilinear crack L. On the free from the mechanical forces opening surface there are placed $2n$ infinitely long in the direction of axis x_3 thin electrodes with given differences of the electric potential. The unelectrodized portions of the opening surface are bounded with vacuum (air) boundaries of the k-th electrode are determined by quantities β_{2k-1} and β_{2k} $\left(k = \overline{1, 2n}\right)$, and the electric potential on it is prescribed by quantity $\phi_k^* = \mathrm{Re}\left(\Phi_k^* e^{-i\omega t}\right)$. The relative position of the heterogeneities, their configurations, and also the position of the surface electrodes should satisfy certain requirements which will be indicated below.

Assuming that the edges of the crack are free from mechanical stresses, we will represent the mechanical and electric boundary conditions on contour L as usually

$$(\sigma_{13}\cos\psi + \sigma_{23}\sin\psi)^{\pm} = 0. \tag{6.47}$$

$$E_s^+ = E_s^-, \; D_n^+ = D_n^-. \tag{6.48}$$

Here E_s and D_n are a tangential component of the vector of electric strength and a normal component of the vector of electric induction, respectively; ψ is the angle between the normal to left edge L and axis Ox_1; the signs "plus" and "minus" refer to the left and right edges of the section at its movement from its beginning a to end b (Fig. 6.62).

Taking into account complex representations (6.3) the boundary conditions on the surfaces of the opening and crack we will represent in the form

$$\begin{aligned}
&\frac{\partial}{\partial n}\left\{c_{44}^E\left(1+k_{15}^2\right)U_3 + e_{15}F^*\right\} = 0 \text{ on } C,\\
&F^* + \frac{e_{15}}{\ni_{11}^{\varepsilon}}U_3 = \phi^*\left(\zeta^*\right), \quad \zeta^* \in C_\phi,\\
&D_n^* = -\ni_{11}^{\varepsilon}\frac{\partial F^*}{\partial n} = 0 \text{ on } C\backslash C_\phi,\\
&\frac{\partial}{\partial n}\left\{c_{44}^E\left(1+k_{15}^2\right)U_3 + e_{15}F^*\right\}^{\pm} = 0 \text{ on } L,
\end{aligned} \tag{6.49}$$

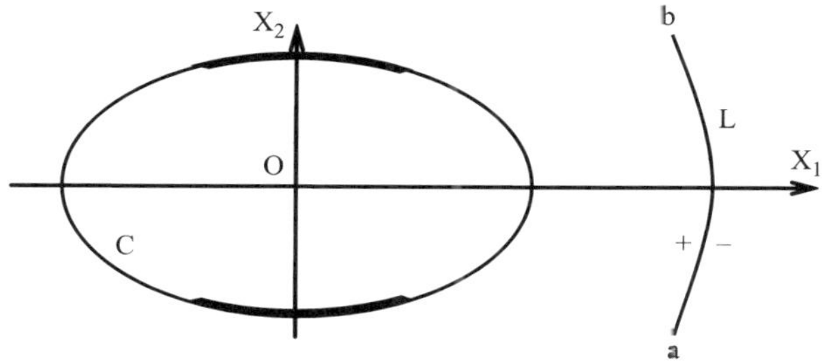

Fig. 6.62. A body with a partially electrodized hole and a curvilinear crack

where C_ϕ is the part of contour C, corresponding to the electrodized surface of the opening.

Now assume

$$U_3(z) = \int_C p(\zeta^*) H_0^{(1)}(\gamma r_1)\, ds - \frac{i}{4}\int_L [U_3] \frac{\partial H_0^{(1)}(\gamma r)}{\partial n_\zeta} ds,$$
$$F^*(z) = \int_C f(\zeta^*) \frac{\partial}{\partial n_{\zeta^*}} \ell n r_1 ds + \frac{e_{15}}{2\pi \ni_{11}^{\varepsilon}} \int_L \left[\frac{dU_3}{ds}\right] \arg(z-\zeta)\, ds, \qquad (6.50)$$
$$r = (z-\zeta), r_1 = (\zeta * -\zeta), \zeta \in L, \zeta^* \in C.$$

Integral representations (6.50) provide a displacement jump and continuation of the stress vector on L, and also the automatic satisfaction of electric conditions (6.48).

Substituting the limiting values of function (6.50) and their derivatives at $z \to \zeta_0 \in L$ and $z \to \zeta_0^* \in C$ under boundary conditions (6.49) we come to the system of singular integrodifferential equations of the second kind

$$2ip(\zeta_0^*) + \int_C p(\zeta^*) g_1(\zeta^*, \zeta_0^*)\, ds + \int_C f'(\zeta^*) g_2(\zeta^*, \zeta_0^*)\, ds +$$
$$+ \int_L \left[\frac{dU_3}{ds}\right] g_3(\zeta, \zeta_0^*)\, ds + \int_L [U_3] g_4(\zeta, \zeta_0^*)\, ds = 0,$$
$$-\pi f(\zeta_0^*) + \int_C p(\zeta^*) g_5(\zeta^*, \zeta_0^*)\, ds + \int_C f(\zeta^*) g_6(\zeta^*, \zeta_0^*)\, ds +$$
$$+ \int_C \left[\frac{dU_3}{ds}\right] g_7(\zeta, \zeta_0^*)\, ds + \int_C [U_3] g_8(\zeta, \zeta_0^*)\, ds = \Phi^*(\zeta_0^*), \ \zeta_0^* \in C_\phi,$$
$$\int_C f'(\zeta^*)\, g_9(\zeta^*, \zeta_0^*)\, ds + \int_L \left[\frac{dU_3}{ds}\right] g_{10}(\zeta, \zeta_0^*) ds = 0, \ \zeta_0^* \in C \backslash C_\phi,$$

$$\int_C p(\zeta^*)\, g_{11}(\zeta^*, \zeta_0)\, ds + \int_C f'(\zeta^*)\, g_{12}(\zeta^*, \zeta_0)\, ds + \int_L \left[\frac{dU_3}{ds}\right] g_{13}(\zeta, \zeta_0) ds +$$
$$+ \int_L [U_3] g_{14}(\zeta, \zeta_0)\, ds = 0, \tag{6.51}$$

where kernels g_m $(m = \overline{1,14})$ are determined by the following expressions

$$g_1(\zeta^*, \zeta_0^*) = \frac{2}{\pi i} Re \frac{e^{i\psi_{10}}}{\zeta^* - \zeta_0^*} + \gamma H_1(\gamma r_0)\cos(\psi_{10} - \alpha_{10}),$$
$$g_2(\zeta^*, \zeta_0^*) = \frac{e_{15}}{c_{44}^E(1 + k_{15}^2)} \mathrm{Im} \frac{e^{i\psi_{10}}}{\zeta^* - \zeta_0^*},\ g_3(\zeta, \zeta_0^*) = \frac{1}{2\pi(1 + k_{15}^2)} \mathrm{Im} \frac{e^{i\psi_{10}}}{\zeta - \zeta_0^*},$$
$$g_4(\zeta, \zeta_0^*) = \frac{i\gamma^2}{8}\left[H_2(\gamma r_{20})\cos(\psi + \psi_{10} - 2\alpha_{20}) - H_0^{(1)}(\gamma r_{20})\cos(\psi - \psi_{10})\right],$$
$$g_5(\zeta^*, \zeta_0^*) = \frac{e_{15}}{\text{э}_{11}^{\varepsilon}} H_0^{(1)}(\gamma r_{10}),\ g_6(\zeta^*, \zeta_0^*) = \mathrm{Re}\frac{e^{i\psi_1}}{\zeta^* - \zeta_0^*},$$
$$g_7(\zeta, \zeta_0^*) = \frac{e_{15}}{2\pi\ \text{э}_{11}^{\varepsilon}}\alpha_{20},\ g_8(\zeta, \zeta_0^*) = \frac{ie_{15}}{4\ \text{э}_{11}^{\varepsilon}}\gamma H_1^{(1)}(\gamma r_{20})\cos(\psi - \alpha_{20}),$$
$$g_9(\zeta, \zeta_0^*) = \mathrm{Im}\frac{e^{i\psi_{10}}}{\zeta^* - \zeta_0^*},\ g_{10}(\zeta, \zeta_0^*) = -\frac{e_{15}}{2\pi_{11}^{\varepsilon}}\mathrm{Im}\frac{e^{i\psi_{10}}}{\zeta - \zeta_0^*},$$
$$g_{11}(\zeta^*, \zeta_0) = \frac{2}{\pi i}\mathrm{Re}\frac{e^{i\psi_0}}{\zeta^* - \zeta_0} + \gamma H_1(\gamma r_{30})\cos(\psi_0 - \alpha_{30}),$$
$$g_{12}(\zeta^*, \zeta_0) = \frac{e_{15}}{c_{44}^E(1 + k_{15}^2)}\mathrm{Im}\frac{e^{i\psi_0}}{\zeta^* - \zeta_0},\ g_{13}(\zeta, \zeta_0) = \frac{1}{2\pi(1 + k_{15}^2)}\mathrm{Im}\frac{e^{i\psi_0}}{\zeta - \zeta_0},$$
$$g_{14}(\zeta, \zeta_0) = \frac{i\gamma^2}{8}\left[H_2(\gamma r_0)\cos(\psi + \psi_0 - 2\alpha_0) - H_0^{(1)}(\gamma r_0)\cos(\psi - \psi_0)\right],$$
$$H_1(x) = \frac{2i}{\pi x} + H_1^{(1)}(x),\ H_2(x) = \frac{4i}{\pi x^2} + H_2^{(1)}(x),$$
$$r_0 = (\zeta_0 - \zeta), \alpha_0 = \arg(\zeta_0 - \zeta), r_{10} = (\zeta * -\zeta_0^*), \alpha_{10} = \arg(\zeta^* - \zeta_0^*),$$
$$r_{20} = (\zeta_0^* - \zeta), \alpha_{20} = \arg(\zeta_0^* - \zeta), r_{30} = (\zeta * -\zeta_0), \alpha_{30} = \arg(\zeta^* - \zeta_0),$$
$$\psi = \psi(\zeta), \psi_1 = \psi(\zeta^*), \psi_0 = \psi(\zeta_0), \psi_{10} = \psi(\zeta_0^*), \zeta, \zeta_0 \in L, \zeta^*, \zeta_0^* \in C.$$

Here ψ and ψ_1 are the angles between the normals contours L and C and axis Ox_1, respectively; $\Phi^*(\zeta_0^*)$ are piecewise-constant functions, assigning the amplitude values of the electric potentials on the electrodes.

For the simple solvability of system (6.51) in the class of functions with derivatives unbounded in the vicinity of the tips of crack L, it is necessary to consider it together with the additional condition.

$$\int_L \left[\frac{dU_3}{ds}\right] ds = 0, \tag{6.52}$$

expressing the equality to zero of the jumps of displacement in the tips L. Besides, condition (6.52) provides the uniqueness of the integral representations of function $F^*(z)$ in (6.50).

It is necessary to mention here that as the appearing in the process of oscillation reflected from the crack electroelastic waves introduce additional charges on the paired electrodes, the position of the latters, and also the relative position of the opening and the crack and their configuration should provide the similarity of the additional charges according to the absolute value. Otherwise system (7.51) becomes unsolvable.

Having determined functions $[U_3]$, $p(\zeta^*)$ and $f(\zeta^*)$ from system (6.51), according to formulas (6.3) with the help of representations (6.50) we may determine all the components of the electroelastic field in a piecewise-homogeneous space.

Introducing the parameterization of contour C with the help of equality $\zeta^* = \zeta^*(\beta)$, $\zeta_0^* = \zeta^*(\beta_0)$ $(0 \le \beta, \beta_0 \le 2\pi)$ we will find the expression for the density amplitude of distribution of electric charges $q_k(\beta)$ on the k-th electrode. Taking into account the fact that the opening is bounded with vacuum we will write

$$q_k(\beta) = D_n^{(k)^*}(\beta), \ \beta_{2k-1} < \beta < \beta_{2k}. \tag{6.53}$$

Here $D_n^{(k)^*}(\beta)$ is the amplitude of the normal component of the vector of electric induction on the position of the opening surface covered with k-th electrode. Using integral representations (6.50) for function $F^*(z)$ we find

$$q_k(\beta_0) = -\ni_{11}^{\varepsilon} \int_C f'(\zeta^*) \mathrm{Im} \frac{e^{i\psi_{10}}}{\zeta^* - \zeta_0^*} ds, \ \zeta_0^* \in C_{\phi_k}, \tag{6.54}$$

where C_{ϕ_k} is the part of contour C, where the k-th electrode is positioned.

Integrating expression (6.54) over variable β_0 in the limits from β_{2k-1} to β_{2k}, we will obtain the amplitude value of total charge Q_k of k-th electrode, referring to the unity of its length. He current flowing through the given electrode and equal to the conductance current in the generator circuit, we may be determined over formula (6.10).

To determine stress intensity factor K_{III} we will obtain the main asymptotics of the shear stress on the continuation across the crack tip. We proceed from formula (4.16), determining the behaviour of the integrals of Cauchy type in the vicinity of the ends L in case when the density has the power singularity.

From the asymptotic analysis of the latter in (6.51) of the singular integrodifferential equation in the vicinity of tip ℓ it follows that $\sigma = 1/2$. Therefore, introducing the parameterization of crack contour $\zeta = \zeta(\delta)$ we may have

$$\left[\frac{dU_3}{ds}\right] = \frac{\Omega_0(\delta)}{s'(\delta)\sqrt{1-\delta^2}}, \ s'(\delta) = \frac{ds}{d\delta} > 0, \ -1 \le \delta \le 1, \tag{6.55}$$

where function $\Omega_0(\delta)$ is continuous according to Hölder.

Here we have (leaving only terms that contribute ε to asymptotics)

$$\sigma_n = \mathrm{Re}\left(S_n e^{-i\omega t}\right), \; S_n = c_{44}^E \frac{\partial U_3}{\partial n} + \ldots = c_{44}^E \left[e^{i\psi_c}\frac{\partial U_3}{\partial z} + e^{-i\psi_c}\frac{\partial U_3}{\partial \bar{z}}\right] + \ldots, \tag{6.56}$$

where c is the cut tip, $\psi_c = \psi(c)$.

On the basis of (6.50) we will write out the main part of function (6.56)

$$S_n^0 = \frac{c_{44}^E}{2\pi}\int_L \left[\frac{dU_3}{ds}\right]\mathrm{Im}\frac{e^{i\psi_c}}{\zeta - z}ds. \tag{6.57}$$

Using asymptotic formulas (4.16) from (6.57) we find

$$S_n^0 = \pm c_{44}^E \frac{\Omega_0(\pm 1)}{3\sqrt{2r^* s'(\pm 1)}}, \; s'(\pm 1) = \left.\frac{ds}{d\delta}\right|_{\delta=\pm 1}. \tag{6.58}$$

Here $r^* = (z - c)$, the lower sign refers to tip $c = a$, the upper one to $c = b$.

Proceeding from (6.58) we find stress intensity factor

$$K_{III}^{\pm} = \lim_{r^* \to 0} \sqrt{2\pi r^*}\sigma_n^0 = \pm\frac{c_{44}^E}{2}\sqrt{\frac{\pi}{s'(\pm 1)}}\mathrm{Re}\left[\Omega_0(\pm 1)e^{-i\omega t}\right]. \tag{6.59}$$

The asymptotics of the normal component of the vector of electric induction on the continuation across the cut tip are as follows

$$D_n = D_1\cos\psi(\pm 1) + D_2\sin\psi(\pm 1) = \pm e_{15}\frac{\mathrm{Re}\left[\Omega_0(\pm 1)e^{-i\omega t}\right]}{2\sqrt{2r^* s'(\pm 1)}}. \tag{6.60}$$

The other electric values in the vicinity of L are limited.

As an example consider a piezoceramic space (the material is $PZT-4$), containing an opening of elliptic cross-section and a crack, the contour of which is parabolic. Assume, that the excitation of the space is carried out by two electrodes with the difference of potential amplitude $2\Phi^*$. The parametric equation of contours L and C, respectively, have the form

$$\begin{aligned} &\zeta = \delta e^{i\vartheta}(p_1 + ip_2\delta) + h, \; \delta \in [-1, 1], \\ &\zeta^* = R_1\cos\beta + iR_2\sin\beta, \; \beta \in [0, 2\pi], \end{aligned} \tag{6.61}$$

where ϑ is the angle characterizing the orientation of the crack in the space.

The solution of the system of integrodifferential (6.51) together with (6.52) taking into account (6.61) was carried out over the scheme of the method of quadratures (Appendix 12.8, Par. B.1, Par. B.2).

The behaviour of quantity $Q^* = \left(Q/\left(\ni_{11}^{\varepsilon}\,\Phi^*\right)\right)$ ((Q is the amplitude of the total charge on an electrode)) in the function of normalized wave number γR are represented in Fig. 6.63. Curve 1 is constructed for the values of parameters

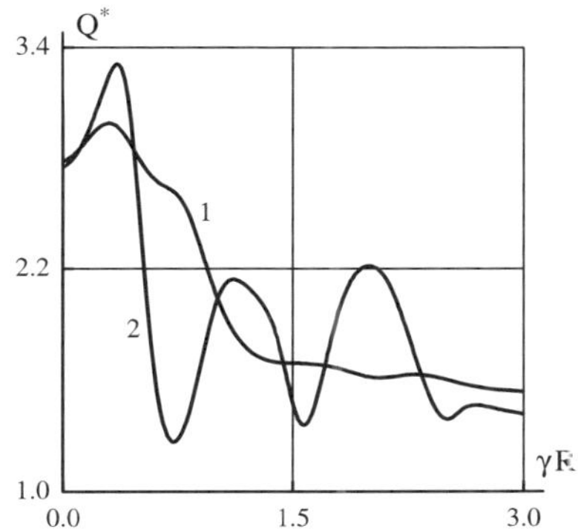

Fig. 6.63. The relative total charge as a function of the normalized wave number ($\beta_1 = 5\pi/14, \beta_2 = 9\pi/14,\ \beta_3 = 19\pi/14,\ \beta_4 = 23\pi/14$)

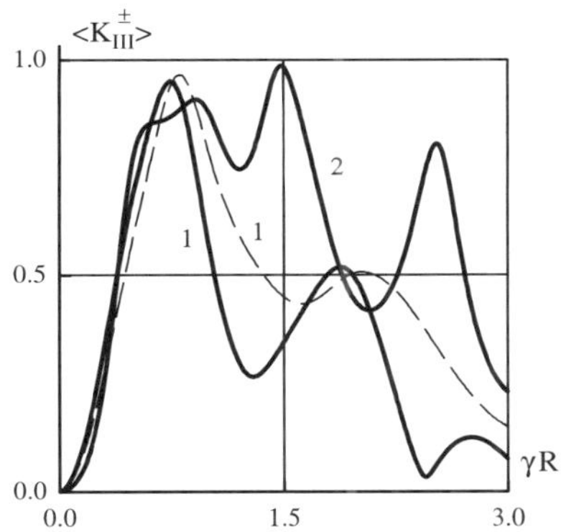

Fig. 6.64. The relative stress intensity factor as a function of the normalized wave number ($\beta_1 = 5\pi/14,\ \beta_2 = 9\pi/14,\ \beta_3 = 19\pi/14,\ \beta_4 = 23\pi/14$)

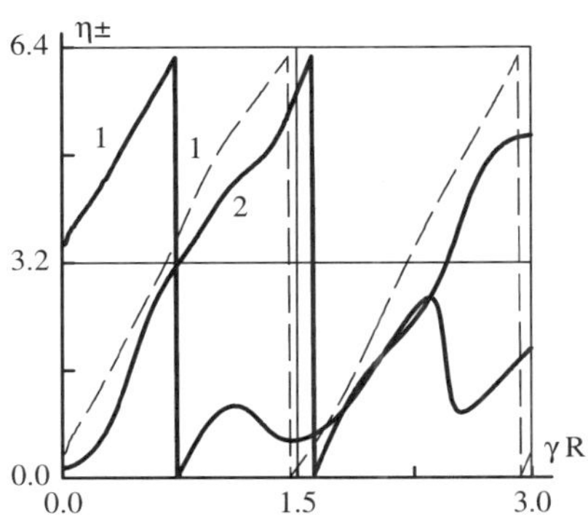

Fig. 6.65. Quantity $\eta^{\pm} = \arg(\Omega_0(\pm 1))$ as a function of the normalized wave number ($\beta_1 = 5\pi/14,\ \beta_2 = 9\pi/14,\ \beta_3 = 19\pi/14,\ \beta_4 = 23\pi/14$)

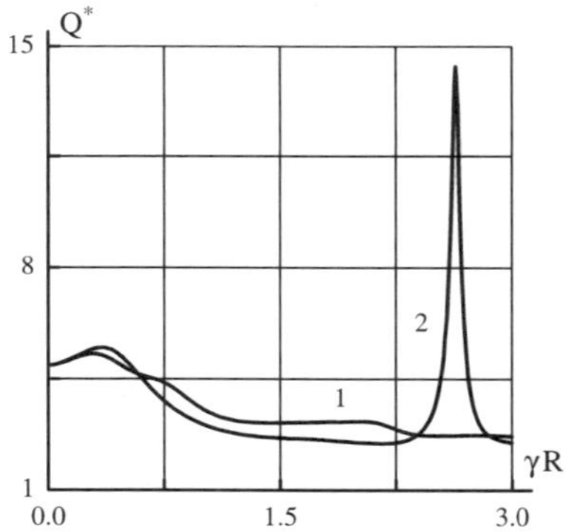

Fig. 6.66. The relative total charge as a function of the normalized wave number ($\beta_1 = \pi/6,\ \beta_2 = 5\pi/6,\ \beta_3 = 7\pi/6,\ \beta_4 = 11\pi/6$)

$R_1/R_2 = 1,\ p_1/R_1 = 1,\ p_2/R_1 = 0,\ h/R_1 = 3,\ \vartheta = 0,\ \beta_1 = 5\pi/14,\ \beta_2 = 9\pi/14,\ \beta_3 = 19\pi/14,\ \beta_4 = 23\pi/14$; curve 2 is constructed for those values besides $p_1/R_1 = 3,\ \vartheta = \pi/2$ ($R = 0.5\,(R_1 + R_2)$, 2ℓ is the cut length).

In Figs. 6.64 and 6.65 are given the relations of the relative stress intensity factor $< K^{\pm}_{III} >= c^E_{44}\sqrt{\pi\ell/s'\,(\pm 1)}\,(\Omega_0\,(\pm 1)\,(/\,(2e_{15}(\Phi^*))$ and $\eta^{\pm} = \arg(\Omega_0\,(\pm 1))$ from γR for the same values of the parameters and in the same correspondence as in Fig. 6.62. The solid lines correspond to tip a, the dashed ones to tip b.

Knowing quantity $(K^{\pm}_{III})$ the stress intensity factor may be determined by formula

$$K^{\pm}_{III} = \pm\frac{e_{15}(\Phi^*)}{\sqrt{\ell}}\left\langle K^{\pm}_{III}\right\rangle \cos\left(\omega t - \arg\Omega_0\,(\pm 1)\right). \tag{6.62}$$

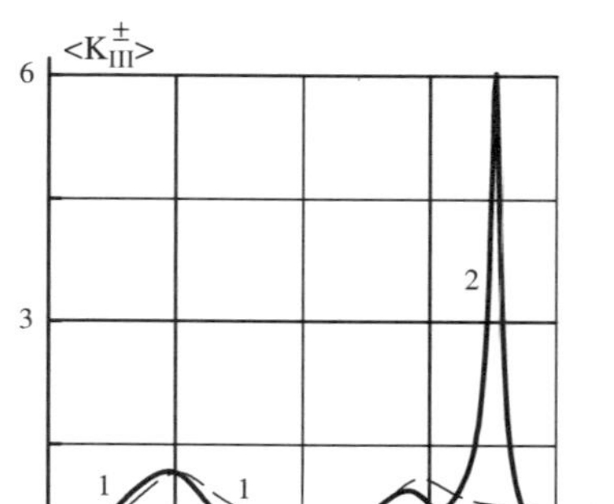

Fig. 6.67. The relative stress intensity factor as a function of the normalized wave number ($\beta_1 = \pi/6,\ \beta_2 = 5\pi/6,\ \beta_3 = 7\pi/6,\ \beta_4 = 11\pi/6$)

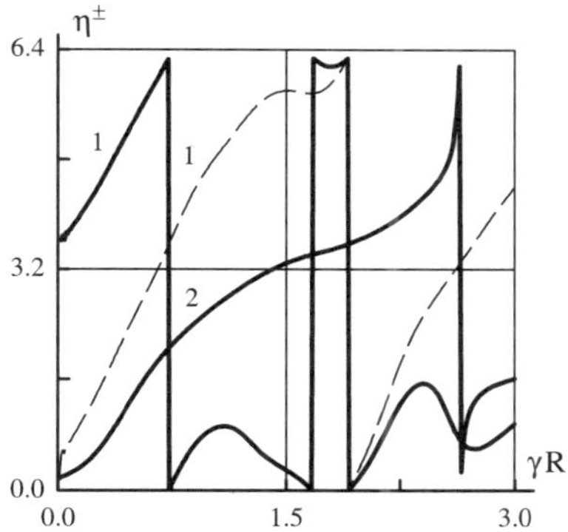

Fig. 6.68. Quantity $\eta^{\pm} = \arg(\Omega_0(\pm 1))$ as a function of the normalized wave number $(\beta_1 = \pi/6, \beta_2 = 5\pi/6, \beta_3 = 7\pi/6, \beta_4 = 11\pi/6)$

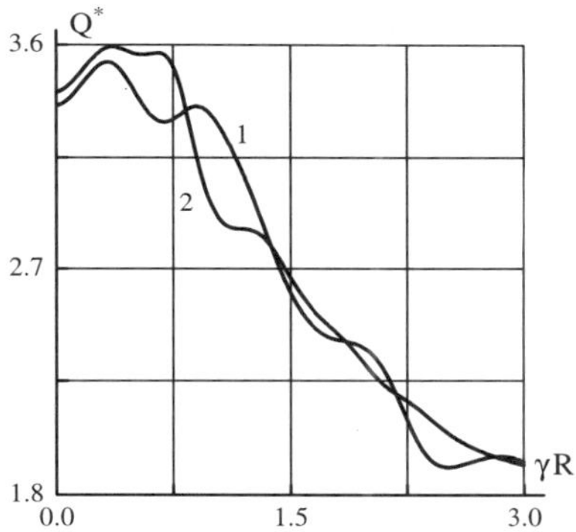

Fig. 6.69. The relative total electric charge on the electrodes as a function of the normalized wave number for various electrode positions

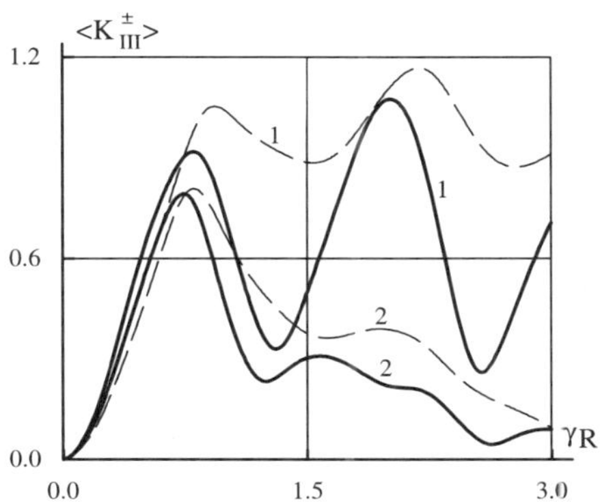

Fig. 6.70. The relative stress intensity factor as a function of the normalized wave number for various electrode positions

For the case of the larger area of the electrodized coating ($\beta_1 = \pi/6$, $\beta_2 = 5\pi/6$, $\beta_3 = 7\pi/6$, $\beta_4 = 11\pi/6$) the analogous results of the calculations in Figs. 6.66–6.68. Curves 1 and 2 correspond to parameters $R_1/R_2 = 1$, $p_1/R_1 = 1$, $p_2/R_1 = 0$, $h/R_1 = 3$, $\vartheta = 0$ and $\vartheta = \pi/2$.

Changes of Q^* and $< K_{III}^{\pm} >$ as a function of γR in case, when the centers of two active electrodes do not lie on axis x_2 are shown in Figs. 6.69 and 6.70. Curves 1 and 2 are constructed for parameters $R_2/R_1 = 1$, $p_1/R_1 = 1$, $p_2/R_1 = 0$, $h/R_1 = 3$, $\vartheta = 0$, $\beta_1 = 3\pi/28$, $\beta_2 = 11\pi/28$, $\beta_3 = 45\pi/28$, $\beta_4 = 53\pi/28$ and $\beta_1 = 17\pi/28$, $\beta_2 = 25\pi/28$, $\beta_3 = 31\pi/28$, $\beta_4 = 39\pi/28$, respectively.

Figure 6.71 illustrates the changes of value $\lambda = c_{44}^E \left([U_3] / \Phi^* \right)$, characterizing the jump of the displacement on the crack for various values of the curve parameter at $R_2/R_1 = 0.5$, $p_1/R_1 = 2$, $h/R_1 = 3$, $\gamma R = 1$, $\vartheta = \pi/2$ ($\beta_1 = 5\pi/14$, $\beta_2 = 9\pi/14$, $\beta_3 = 19\pi/14$, $\beta_4 = 23\pi/14$). Curves 1, 2 and 3 satisfy values $p_2/R_1 = 0$, 2 and 4, respectively.

The behaviour of the mechanical quantities in the area of a piecewise-homogeneous space is of interest. The line of the equal modulus of the displacement amplitude in the area covering the crack and the opening are represented in Fig. 6.72 ($R_2/R_1 = 1$, $p_1/R_1 = 0.5$, $h/R_1 = 2$, $\gamma R = 1$), Fig. 6.73 ($R_2/R_1 = 2$, $p_1/R_1 = 2$, $h/R_1 = 4$, $\gamma R = 1$) and Fig. 6.74 ($R_2/R_1 = 0.5$, $p_1/R_1 = 1$, $h/R_1 = 3$, $\gamma R = 0.6$). In the calculations we assumed $\beta_1 = 5\pi/14$, $\beta_2 = 9\pi/14$, $\beta_3 = 19\pi/14$, $\beta_4 = 23\pi/14$.

From the represented results of the calculations it follows that the behaviour of the electric and mechanical quantity substantially depends on the frequencies of the harmonic loadings, relative position and configuration of

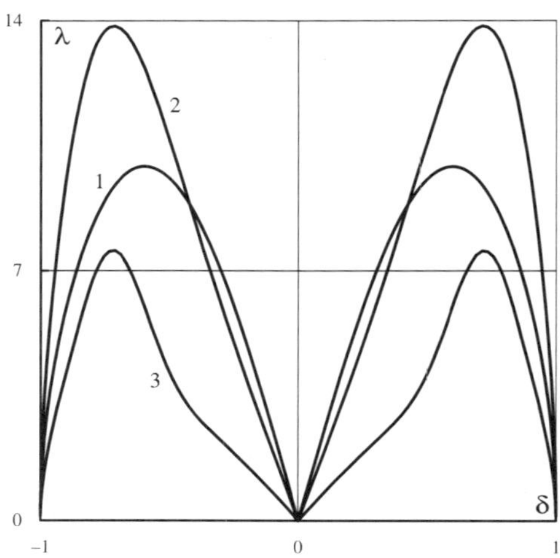

Fig. 6.71. The changes of quantity $\lambda = c_{44}^E |[U_3]/\Phi^*|$, that characterizes the displacement jump on the crack for various values of crack curvature

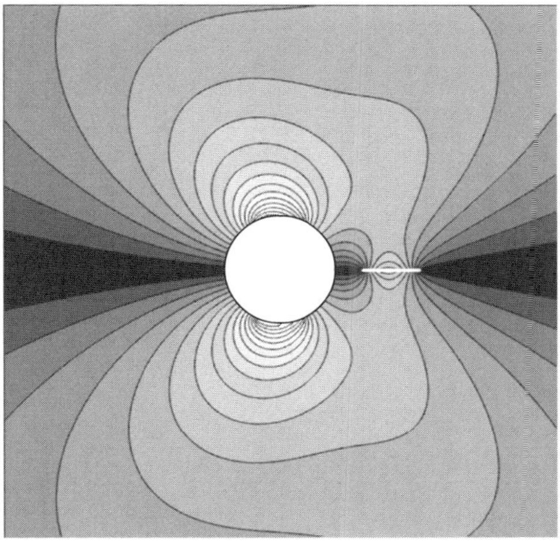

Fig. 6.72. The contour lines of the amplitude of displacements in a partially homogenous space for various values of frequency as well as relative positions of the defects

the heterogeneities, and also on the position and sizes of the surface electrodes. The availability of the crack may considerably increase the dynamic effect. For example, as it follows from Fig. 6.66, quantity Q^*, characterizing the total electric charge on the electrode, may exceed its static analogue 2.9 times as large (curve 2). In the absence of the crack this excess is only 9

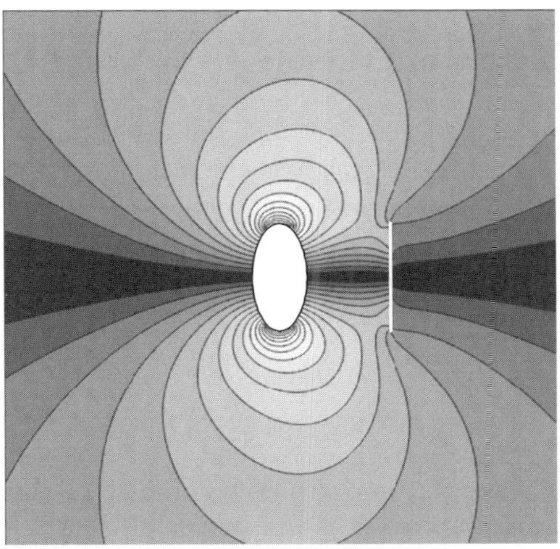

Fig. 6.73. The contour lines of the amplitude of displacements in a partially homogenous space for various values of frequency as well as relative positions of the defects

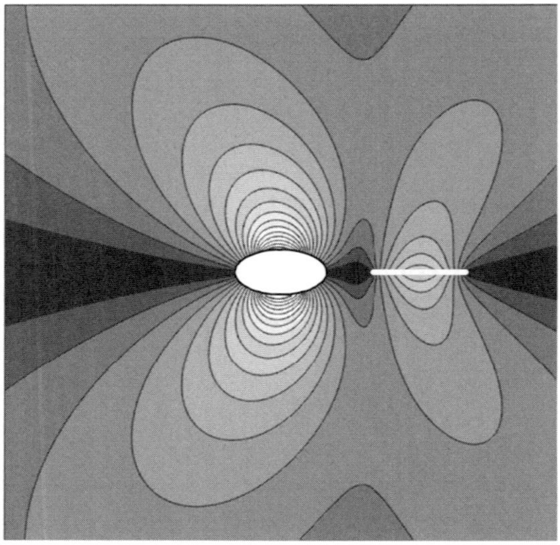

Fig. 6.74. The contour lines of the amplitude of displacements in a partially homogenous space for various values of frequency as well as relative positions of the defects

per cent. It should be mentioned that in statics ($\omega = 0$) the electric loading of the piezoceramic medium under the condition of antiplane deformation does not cause in it mechanical stresses, and therefore factor K_{III} in this case is equal to zero.

6.9 An Opening Strengthened by a Rigid Stringer

Consider the analogous problem for a medium containing tunnel opening C and thin rigid inclusion L (stringer). The excitation of the oscillation is carried out by the difference of electric potential applied on 2ℓ electrodes, positioned on the opening surface.

Assuming that the stringer is fixed, let us represent the mechanical and electric boundary conditions on contour L as follows

$$u_3^{\pm} = 0. \tag{6.63}$$

$$E_s^{+} = E_s^{-}, D_n^{+} = D_n^{-}. \tag{6.64}$$

Here E_s and D_n are the tangential component of the vector of electric strength and normal component of the vector of electric induction, respectively, the signs "plus" and "minus" refer to the left and right edges of inclusion L at its movement from its beginning a to the end b (Fig. 6.75).

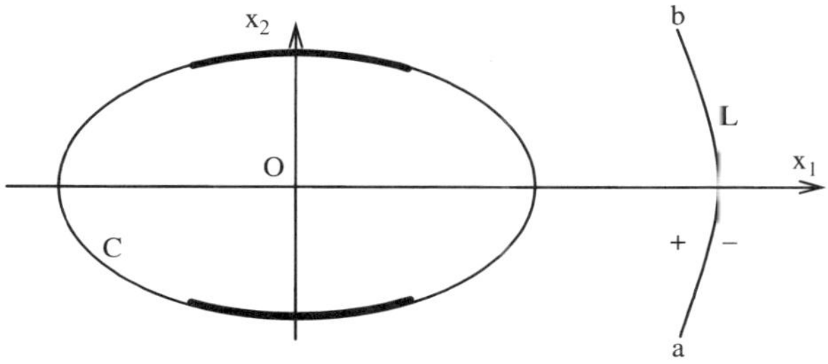

Fig. 6.75. A medium containing a partially electrodized hole and a curvilinear stringer

To obtain the efficient boundary condition (6.63) from the point of view of the numerical realization of the system of integral equations it is recommended to differentiate over arc abscissa s

$$\left(\frac{\partial u_3}{\partial s}\right)^{\pm} = 0 \tag{6.65}$$

The boundary condition on the surface of the partially electrodized opening have the form (6.4).

To bring the considered boundary problem of electroelasticity to the integral equations let us represent the sought-for functions u_3 and F in the form

$$U_3(x_1,x_2) = \frac{i}{4c_{44}^E\left(1+k_{15}^2\right)}\left\{\int_L q(\zeta)\,H_0^{(1)}(\gamma r)\,ds + \int_C p(\zeta^*)\,H_0^{(1)}(\gamma r_1)\,ds\right\},$$

$$F^*(x_1,x_2) = -\frac{1}{2\pi\,\ni_{11}^{\varepsilon}}\int_C f(\zeta^*)\frac{\partial}{\partial n}\ln r_1 ds, \tag{6.66}$$

$$r = |\zeta - z|\,,\ r_1 = |\zeta^* - z|\,,\ \zeta\in L,\ \zeta^*\in C.$$

Here $H_\nu^{(1)}(x)$ is the Hankel function of the first kind of order ν, ds is the element of the arc length contour over which the integration is carried out.

Integral representations (6.66) satisfy differential (6.5) and electric conditions (4.65) on insert L, and also support the fulfillment of condition $[U_3] = U_3^+ - U_3^- = 0$ in (6.64).

Substituting the limiting values of functions (6.67) and their derivatives at $z\to\zeta_0\in L$ and $z\to\zeta_0^*\in C$ into boundary condition (6.4), (6.65), we come to the system of integrodifferential equations of the second kind

$$\int_L q(\zeta)\,G_1(\zeta,\zeta_0)\,ds + \int_C p(\zeta^*)\,G_2(\zeta^*,\zeta_0)\,ds = 0,\ \zeta_0\in L,$$

$$+\int_C f'(\zeta^*)G_5(\zeta^*,\zeta_0^*)\,ds - 2p(\zeta_0^*) = 0,\ \zeta_0^*\in C,$$

$$\int_C f'(\zeta^*) G_6(\zeta^*, \zeta_0^*) ds = 0, \ \zeta_0^* \in C \backslash C_\phi,$$

$$\int_L q(\zeta) G_7(\zeta, \zeta_0^*) ds + \int_C p(\zeta^*) G_8(\zeta^*, \zeta_0^*) ds + \int_C f(\zeta^*) G_9(\zeta^*, \zeta_0^*) ds +$$

$$+ \frac{1}{2\,\text{э}_{11}^{\varepsilon}} f(\zeta_0^*) = \Phi(\zeta_0^*), \ \zeta_0^* \in C_\phi,$$

$$G_1(\zeta, \zeta_0) = \gamma H_1(\gamma r_0) \sin(\psi_0 - \alpha_0) - \frac{2i}{\pi} \mathrm{Im} \frac{e^{i\psi_0}}{\zeta - \zeta_0},$$

$$G_2(\zeta^*, \zeta_0) = \gamma H_1^{(1)}(\gamma r_{10}) \sin(\psi_0 - \alpha_{10}),$$

$$G_3(\zeta, \zeta_0^*) = i\gamma H_1^{(1)}(\gamma r_{20}) \cos(\psi_{10} - \alpha_{20}),$$

$$G_4(\zeta^*, \zeta_0^*) = i\gamma H_1(\gamma r_{30}) \cos(\psi_{10} - \alpha_{30}) + \frac{2}{\pi} \mathrm{Re} \frac{e^{i\psi_{10}}}{\zeta^* - \zeta_0^*},$$

$$G_5(\zeta^*, \zeta_0^*) = -\frac{2e_{15}}{\pi\,\text{э}_{11}^{\varepsilon}} \mathrm{Im} \frac{e^{i\psi_{10}}}{\zeta^* - \zeta_0^*},$$

$$G_6(\zeta^*, \zeta_0^*) = \mathrm{Im} \frac{e^{i\psi_{10}}}{\zeta^* - \zeta_0^*},$$

$$G_7(\zeta, \zeta_0^*) = \frac{ik_{15}^2}{4e_{15}(1 + k_{15}^2)} H_0^{(1)}(\gamma r_{20}),$$

$$G_8(\zeta^*, \zeta_0^*) = \frac{ik_{15}^2}{4e_{15}(1 + k_{15}^2)} H_0^{(1)}(\gamma r_{30}),$$

$$G_9(\zeta^*, \zeta_0^*) = -\frac{1}{2\pi\,\text{э}_{11}^{\varepsilon}} \mathrm{Re} \frac{e^{i\psi_1}}{\zeta^* - \zeta_0^*},$$

$$H_1(x) = \frac{2i}{\pi x} + H_1^{(1)}(x), \ r_0 = |\zeta - \zeta_0|, \ r_{10} = |\zeta^* - \zeta_0|,$$

$$r_{20} = |\zeta - \zeta_0^*|, \ r_{30} = |\zeta^* - \zeta_0^*|, \ \alpha_0 = \arg(\zeta - \zeta_0),$$

$$\alpha_{10} = \arg(\zeta^* - \zeta_0), \ \alpha_{20} = \arg(\zeta - \zeta_0^*), \ \alpha_{30} = \arg(\zeta^* - \zeta_0^*),$$

$$\psi_0 = \psi(\zeta_0), \ \psi_{10} = \psi_1(\zeta_0^*); \ \zeta, \zeta_0 \in L; \ \zeta^*, \zeta_0^* \in C. \tag{6.67}$$

Here ψ and ψ_1 are the angles between the normals to the contours L and C and the axis Ox_1, respectively; $\Phi^*(\zeta_0^*)$ is the piecewise-constant function assigning the value of the amplitude of electric potential on the electrodized portions of the opening surface.

It is necessary to mention that, as the appearing in the process of oscillation of reflected from the inclusion of electroelastic waves bring the additional charges on the paired electrodes, the position of the latter, and also the relative position of the opening and inclusion and their configuration should provide the similarity of the additional charges (according to absolute value). Otherwise system (6.67) becomes unsolvable.

Consider the behaviour of the electric field in the vicinity of the inclusion. From integral representation (6.66) for the displacement amplitude we obtain equality

$$q(\zeta) = c_{44}^{E}\left(1 + k_{15}^{2}\right)\left[\frac{\partial U_3}{\partial n}\right], \tag{6.68}$$

where the square brackets define a jump of the corresponding quantity on L. From equation of state (4.22) and relation (6.64) it follows

$$\begin{aligned} &\sigma_n = \mathrm{Re}\left(T_n e^{-i\omega t}\right), \\ &[T_n] = c_{44}^{E}\left[\frac{\partial U_3}{\partial n}\right] + e_{15}\left[\frac{\partial \Phi}{\partial n}\right], \quad \left[\frac{\partial \Phi}{\partial n}\right] = \frac{e_{15}}{\ni_{11}^{\varepsilon}}\left[\frac{\partial U_3}{\partial n}\right]. \end{aligned} \tag{6.69}$$

From expressions (6.68), (6.69) we obtain the equality

$$q(\zeta) = [T_n]. \tag{6.70}$$

Thus, on the basis of (6.70) function $q(\zeta)$ may be expressed as the intensity of contact forces of interaction of a rigid inclusion and piezomedium. Therefore, for the equilibrium of the inclusion there should be fulfilled the following equality

$$\int_L q(\zeta)\, ds = 0. \tag{6.71}$$

By asymptotic analysis of singular integral equations in the vicinity of the tops of the inclusion on the line of integration it is possible to show that on ends L function $q(\zeta)$ has singularities of root type. Thus, condition (6.71) must

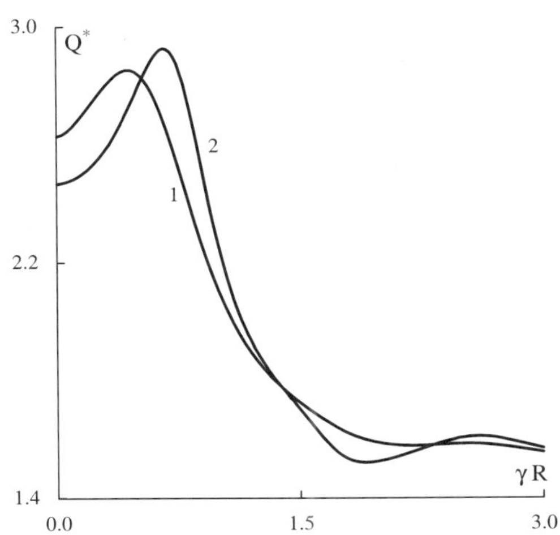

Fig. 6.76. The relative total electrical charge on the electrode as a function of the normalized wave number ($h/R_1 = 1,5$)

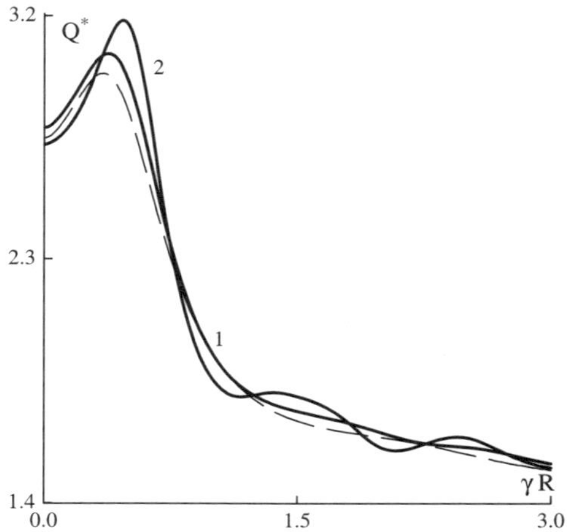

Fig. 6.77. The relative total electrical charge on the electrode as a function of the normalized wave number ($h/R_1 = 3$)

be considered as an additional during the solution of the system of singular integrodifferential (6.67) in the class of function, unbounded on ends L.

Due to (4.22), (6.64) and (6.65) we have

$$[D_s^*] = \ni_{11}^{\varepsilon} [E_s^*] = 0, \; [E_n^*] = -\frac{e_{15}}{\ni_{11}^{\varepsilon}} \left[\frac{\partial U_3}{\partial n}\right] = -\frac{k_{15}^2}{(1+k_{15}^2)\, e_{15}} q(\zeta), \qquad (6.72)$$

where the asterisk defines the amplitude of the corresponding value.

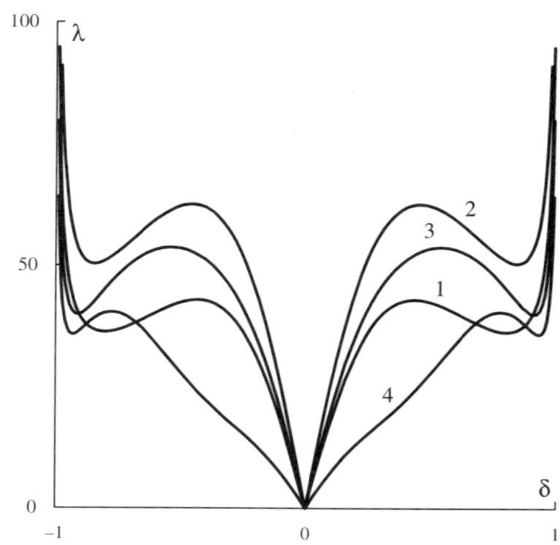

Fig. 6.78. Quantity λ (intensity of contact stresses) as a function of δ. Curves 1–4 represent different normalized wave numbers (γR) for the case of a straight inclusion

Proceeding from (6.72) we may conclude that the vector of electric induction $\overrightarrow{D}$ continue across L, and the vector of electric strength $\overrightarrow{E}$ undergoes breaks on the insert.

For example, let us find the density amplitude of position of electric charges $\rho_k(\beta)$ on the k-th electrode. Introducing the parameterization of the contour C with the help of the equalities $\zeta^* = \zeta^*(\beta)$, $\zeta_0^* = \zeta^*(\beta_0)$ $(0 \leq \beta, \beta_0 \leq 2\pi)$ and taking into account that the opening is bounded with vacuum we may write down

$$\rho_k(\beta) = D_n^{(k)}(\beta), \quad \alpha_{2k-1} < \beta < \alpha_{2k}. \tag{6.73}$$

Here $D_n^{(k)}(\beta)$ represents the amplitude of the normal component of the electric induction vector on the corresponding area of contour C covered with electrodes. Using integral representation (6.66) of function $F^*(x_1, x_2)$ we find

$$\rho_k(\beta_0) = -\ni_{11}^{\varepsilon} \int_C f'(\zeta^*) \mathrm{Im} \frac{e^{i\psi_{10}}}{\zeta^* - \zeta_0^*} ds, \ \zeta_0^* \in C_{\phi_k}. \tag{6.74}$$

Here C_{ϕ_k} is the part of the contour C where k-th electrode is placed.

To determine the concentrations of shear stresses near the opening we will calculate stress $\sigma_s = \sigma_{23}\cos\psi_1 - \sigma_{13}\sin\psi_1$ on its surface. Taking into account (4.22) and equations of electrostatics

$$\begin{aligned} div\overrightarrow{D} &= 0, \\ \overrightarrow{E} &= -\mathrm{grad}\phi. \end{aligned} \tag{6.75}$$

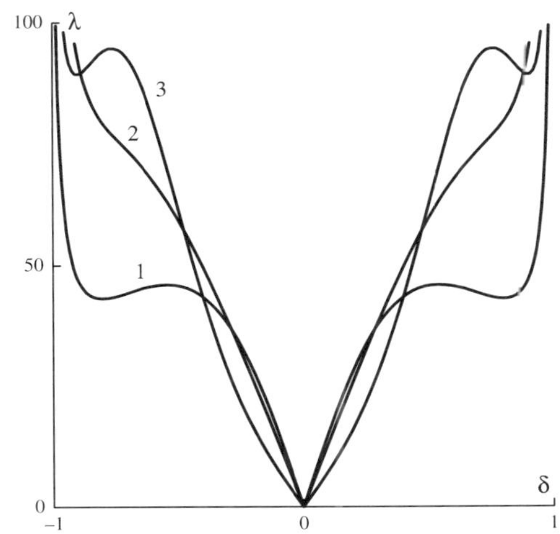

Fig. 6.79. Quantinty λ as a function of δ. Curves 1–4 represent different values of normalized wave numbers (γR) for the case of a parabolic inclusion

we will have

$$\sigma_s = \operatorname{Re}\left(T_s e^{-i\omega t}\right), \; T_s\left(\zeta_0^*\right) = c_{44}^E\left(1+k_{15}^2\right)\frac{\partial U_3}{\partial s} + e_{15}\frac{\partial F^*}{\partial s}. \tag{6.76}$$

Here, by partial derivatives from the corresponding quantities are understood their limiting values at $z \to \zeta_0^*$ from the area of the body. From (6.76) and representations (6.66) we find

$$\begin{aligned}
T_s\left(\zeta_0^*\right) &= \int_L q(\zeta) g_1\left(\zeta, \zeta_0^*\right) ds + \int_C p\left(\zeta^*\right) g_2\left(\zeta^*, \zeta_0^*\right) ds + \\
&+ \int_C f'\left(\zeta^*\right) g_3\left(\zeta^*, \zeta_0^*\right) ds + \frac{e_{15}}{2\,\text{э}_{11}^{\varepsilon}} f'\left(\zeta_0^*\right), \\
g_1\left(\zeta, \zeta_0^*\right) &= -\frac{i\gamma}{4} H_1^{(1)}\left(\gamma r_{20}\right) \sin\left(\psi_{10} - \alpha_{20}\right), \\
g_2\left(\zeta^*, \zeta_0^*\right) &= -\frac{i\gamma}{4} H_1^{(1)}\left(\gamma r_{30}\right) \sin\left(\psi_{10} - \alpha_{30}\right), \\
g_3\left(\zeta^*, \zeta_0^*\right) &= -\frac{e_{15}}{2\pi\,\text{э}_{11}^{\varepsilon}} \operatorname{Re}\frac{e^{i\psi_{10}}}{\zeta^* - \zeta_0^*}.
\end{aligned} \tag{6.77}$$

The appearing here quantities ψ_{10}, r_{20}, r_{30}, α_{20}, α_{30} are determined in (6.67).

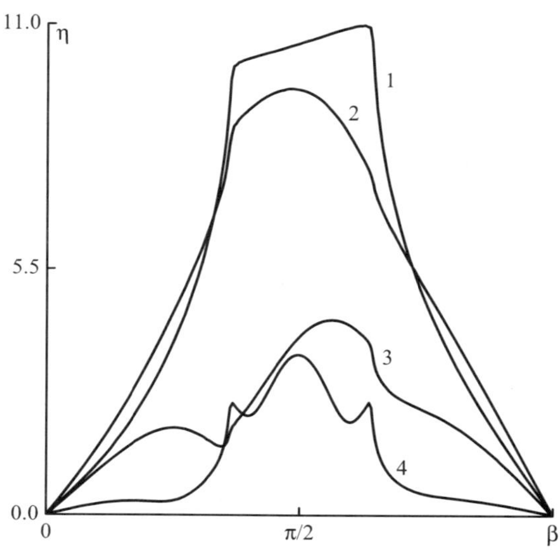

Fig. 6.80. Quantity η (module of displacement amplitude) on the contour of an elliptic hole with two electrodes as a function of β. Curves 1–4 are plotted for different values of normalized wave number (γR)

Formulas (6.77) allow investigating the concentration of stresses according to the frequency of the harmonic excitation, configuration of a cross-section opening and inserting, quantity and position of the active surface electrodes.

Consider a piezoceramic space (the material is $PZT - 4$), containing an opening of elliptic cross-section and an inclusion, the contour of which is parabolic. The parametric equations of contours L and C have the form

$$\begin{aligned} \zeta &= \delta e^{i\vartheta}\left(p_1 + ip_2\delta\right) + h, \ \ \delta \in [-1, 1], \\ \zeta^* &= R_1 \cos\beta + iR_2 \sin\beta, \ \ \beta \in [0, 2\pi], \end{aligned} \tag{6.78}$$

where ϑ is the angle characterizing the orientation of the stringer in a space.

Assume that the excitation of the medium is carried out by two surface electrodes with the difference of the amplitude of potentials $2\Phi^*$, the centers of which lie on axis ($\beta_1 = 5\pi/14$, $\beta_2 = 9\pi/14$, $\beta_3 = 19\pi/14$, $\beta_4 = 23\pi/14$).

Investigate the influence of the dynamic effect on the behaviour of the total electric charges on an electrode in case of an opening interacting with linear stringer ($p_2/R_1 = 0$). In Fig. 6.76 there is shown the behaviour of quantity $Q^* = \left(Q/\left(\text{э}^{\varepsilon}_{11}\,\Phi^*\right)\right)$ (Q is the total electric charge on an electrode) in the function of normalized wave number γR. Curves 1 and 2 are constructed for the values of parameters $R_2/R_1 = 1$, $h/R_1 = 1.5$, $\vartheta = \pi/2$, $p_1/R_1 = 1$ and 3; respectively, ($R = 0.5\left(R_1 + R_2\right)$). The analogous graphs for the same values of the parameters, beside $h/R_1 = 3$, are represented in Fig. 6.77. The dashed lines here refer to the case, when the inclusion is absent. Quantity Q was calculated with the help of formula (6.74).

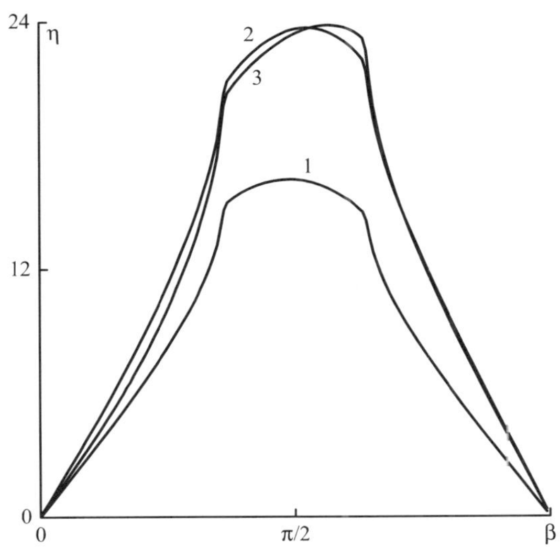

Fig. 6.81. Quantity η (module of displacement amplitude) on the contour of an elliptic hole with two electrodes as a function of β. Curves 1–4 are plotted for different values of curvature

Investigate the intensity of contact forces on an insert according to different parameters. In Fig. 6.78 there is represented the change of quantity $\lambda = (q(\delta)/\Phi^*)$ due to the prescribed on two surface electrodes the difference of potentials $2\Phi^*$. The calculation is carried out for elliptic opening $(R_1/R_2 = 2)$ with linear inclusion at $p_1/R_1 = 2,\ h/R_1 = 1.5,\ \vartheta = \pi/2$. Curves 1–4 are constructed for values $\gamma R = 0, 1, 2$ and 3, respectively.

Change λ on the contour of a parabolic inclusion is illustrated in Fig. 6.79 by curves 1,2,3 for values $p_2/R_1 = 0, 1.5, 3$, respectively at $\gamma R = 0.6,\ R_1/R_2 = 2,\ p_1/R_1 = 3,\ h/R_1 = 2.5,\ \vartheta = \pi/2$.

In Fig. 6.80 are represented the data of calculations characterizing the distribution of quantity $\eta = c_{44}^E|U_3/\Phi^*|$ on the opening contour with linear and parabolic inclusions, respectively. Curves 1–4 in Fig. 6.80 are constructed for the values of normalized wave number $\gamma R = 0, 1, 2$ and 3 at $R_1/R_2 = 2,\ p_2/R_1 = 2,\ h/R_1 = 1.5,\ \vartheta = \pi/2$. Lines 1–3 in Fig. 6.81 correspond to value $p_1/R_1 = 3,\ R_1/R_2 = 2,\ h/R_1 = 2.5,\ \gamma R = 0.6,\ \vartheta = \pi/2,\ p_2/R_1 = 0, 1.5$ and 3.

It should be mentioned here that at $\vartheta = 0$ a linear insert at the given symmetrically positioned two electrodes does not cause disturbance into the electroelastic state of the medium with an opening. The calculation shows that in this case $q(\zeta) \equiv 0$.

The investigation of the behaviour of the mechanical quantities at electric excitation of a piecewise-homogeneous medium is of interest. The lines of the equal modulus of the displacement amplitude in the area covering a circular tunnel opening and a linear stringer are represented in Figs. 6.82

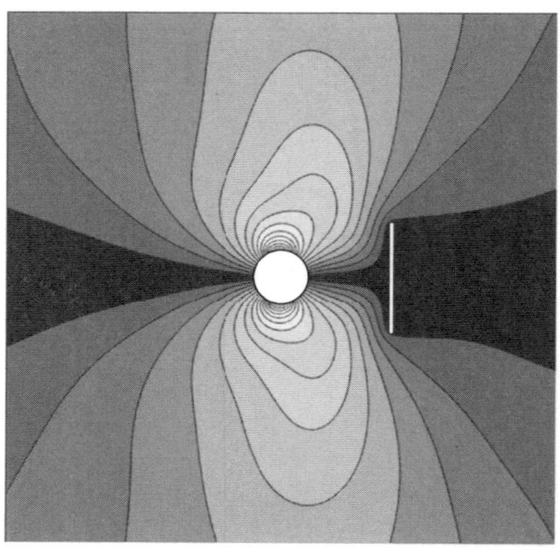

Fig. 6.82. The contour lines of the modulus of displacement amplitude in a medium containing a circular hole and a straight stringer ($\gamma R = 1,\ \beta_1 = 5\pi/14,\ \beta_2 = 9\pi/14,\ \beta_3 = 19\pi/14,\ \beta_4 = 23\pi/14$)

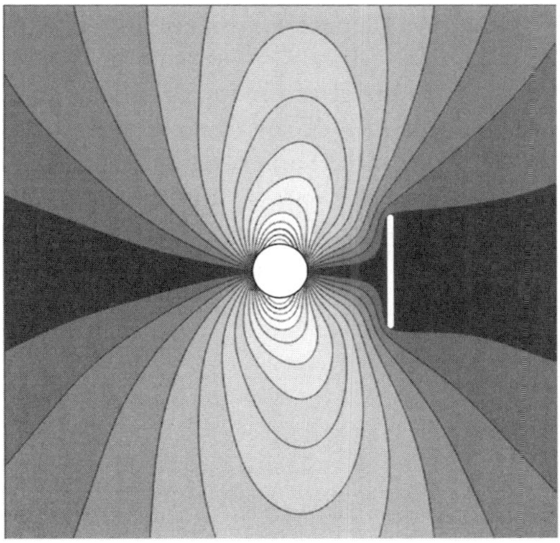

Fig. 6.83. The contour lines of the modulus of displacement amplitude in a medium containing a circular hole and a straight stringer ($\gamma R = 1$, $\beta_1 = \pi/6$, $\beta_2 = 5\pi/6$, $\beta_3 = 7\pi/6$, $\beta_4 = 11\pi/6$)

($\beta_1 = 5\pi/14$, $\beta_2 = 9\pi/14$, $\beta_3 = 19\pi/14$, $\beta_4 = 23\pi/14$) and 6.83 ($\beta_1 = \pi/6$, $\beta_2 = 5\pi/6$,$\beta_3 = 7\pi/6$, $\beta_4 = 11\pi/6$) at $p_1/R = 2$, $h/R = 4$, $\gamma R = 1$. The more light areas correspond to the maximum values of quantity $|U_3|$. Figure 6.84 illustrates contour lines $|U_3|$ in the vicinity of elliptic opening ($R_1/R_2 = 2$)

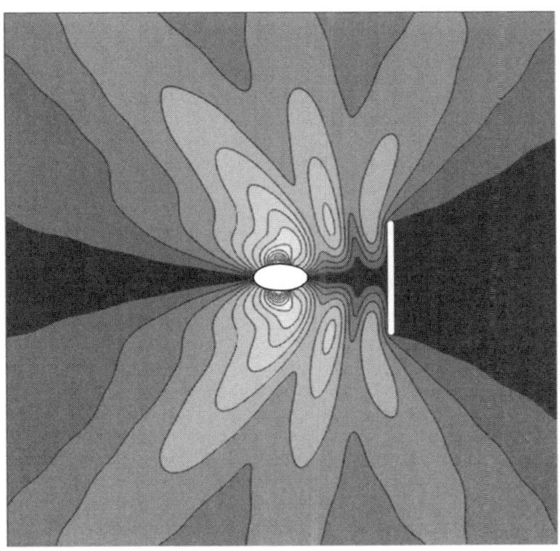

Fig. 6.84. The contour lines of the modulus of displacement amplitude in a medium containing a elliptic hole and a straight stringer ($\gamma R = 2$, $\beta_1 = 5\pi/14$, $\beta_2 = 9\pi/14$, $\beta_3 = 19\pi/14$, $\beta_4 = 23\pi/14$)

and a stringer for the values of parameters corresponding to Fig. 6.82 and $\gamma R = 2$.

From the represented results it is obvious that the availability of an inclusion may increase the influence of the dynamic effect on the behaviour of the total charge on the electrodes. For example, as it follows from Fig. 6.77 quantity Q^* may be greater than its static analogue by 17% (curve 2). In the absence of the inclusion the dynamic effect is only 9% per cent. As it follows from Figs. 6.82–6.84, the fixed rigid stringer prevents the propagation of the oscillations of the medium.

7

Harmonic Oscillations of Continuous Piezoceramic Cylinders with Inner Defects (Antiplane Deformation)

In the present chapter we deal with the problem of harmonic oscillations of continuous cylinders weakened by inner defects of crack and linear inclusion type under the conditions of direct and reverse piezoeffect. We derive analytical algorithms in the form of a system of singular integral and integro-differential equations, permitting the investigation of the amplitude-frequency characteristics of piecewise-homogeneous cylinders.

7.1 A Cylinder Weakened by Tunnel Cracks (Direct Piezoeffect)

In the Cartesian system of coordinates $Ox_1x_2x_3$ consider a continuous piezoceramic cylinder along x_3- axis, the cross-section of which is restrained by an outer contour C with a tunnel along the cracks elements, L_j $(j = 1, 2, \ldots, k)$. Assume that on the outer boundary of the cylinder, which is covered with thin grounded electrodes, a vector of stresses of longitudinal shear $Z_n = \mathrm{Re}\left(Ze^{-i\omega t}\right)$ (considered to be independent of the x_3 coordinate) is applied. On the edges of the cracks-cuts act also harmonic with time self-balanced shear forces $X_{3n}^{\pm} = \mathrm{Re}\left(X_3^{\pm}e^{-i\omega t}\right)$.

Providing that the direction of the electric field of preliminarily polarized ceramics is parallel to x_3 -axis, the curves of contours C and L_j, and also the amplitudes of shear forces $Z\left(x_1, x_2\right)$ and $X_3\left(x_1, x_2\right)$ satisfy the Hölder condition and besides $\cap L_j = \emptyset$, $C \cap L_j = \emptyset$.

In order to solve the antiplane electroelasticity problem we will use (4.25), (4.29), with the help of which the mechanical and electromagnetic quantities may be expressed by the displacement u_3 and the harmonic functions Φ. Under the quasistatic approximation, the problem is reduced to the determination of the displacement amplitude and the function Φ from the differential (4.46) augmented with the boundary conditions on the surface of the cylinder and the cracks.

The electric boundary conditions on the outer surface of the cylinder and cracks L_j are given by

$$\phi = 0\,(E_s = -\partial\phi/\partial s = 0) \text{ on } C, \tag{7.1}$$

$$E_s^+ = E_s^-, \quad D_n^+ = D_n^- \text{ on } L_j\,(j = 1, 2, \ldots, k)\,. \tag{7.2}$$

Here, ϕ is the potential field; E_s and D_n are the tangential component of the vector of electric strength and the normal component of the vector of electric induction, respectively; s is the arc abscissa. The "plus" and "minus" signs refer to the left and right edges of crack L_j during its movement from its beginning a_j to the end b_j (Fig. 7.1).

We seek the functions of U_3 and F in the following form [191]

$$U_3\,(x_1, x_2) = \int\limits_C p\,(\zeta^*) H_0^{(1)}\,(\gamma r_1)\, ds - \frac{1}{4}\int\limits_L [U_3] \left\{ \frac{\partial H_0^{(1)}\,(\gamma r)}{\partial\zeta} d\zeta \right.$$
$$\left. - \frac{\partial H_0^{(1)}\,(\gamma r)}{\partial\bar{\zeta}} d\bar{\zeta} \right\},$$

$$F\,(x_1, x_2) = \frac{e_{15}}{\pi\ni_{11}^{\varepsilon}} \left\{ \frac{1}{2}\int\limits_L \left[\frac{dU_3}{ds}\right] \ell n r ds + \int\limits_C f\,(\zeta^*)\,\ell n r_1 ds \right\}, \tag{7.3}$$

$$r = |\zeta - z|\,, \quad r_1 = |\zeta^* - z|\,, \quad z = x_1 + ix_2,$$
$$\zeta \in L = \cup L_j, \quad \zeta^* \in C,$$

where $H_\nu^{(1)}\,(x)$ is the Hankel function of the first kind and of order ν; the square brackets denote a jump of the corresponding quantity on L; ds is the element of the contour arc-length, over which the integration is carried out.

The integral representations (7.3) of the solutions are correct if they satisfy the differential (4.46) which provide existence of a displacement jump U_3 on the cracks, continuity of the stress vector during crossing L_j, and also the automatic fulfillment of conditions (7.2). The densities $[U_3]$, $p\,(\zeta)$ and $f\,(\zeta^*)$ appearing in (7.3) are determined from the mechanical boundary conditions

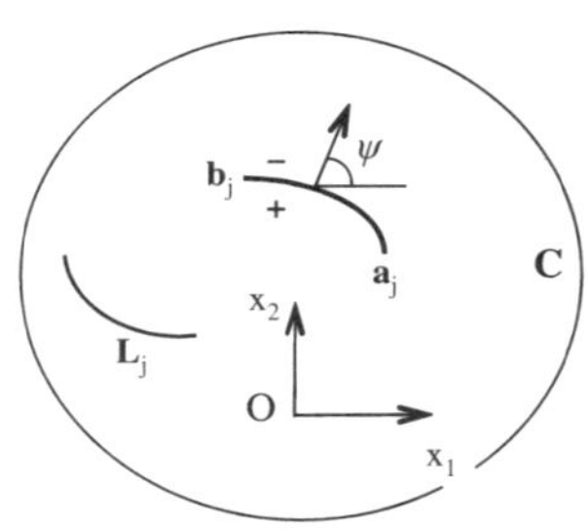

Fig. 7.1. The scheme of cross-section of a cylindrical with cracks

on the contour C and L_j, and also from the electric conditions (7.1). Taking into account (4.25), (4.29) the later reads

$$c_{44}^E\left(1+k_{15}^2\right)\left\{e^{i\psi}\left(\frac{\partial U_3}{\partial\zeta^*}\right)^+ + e^{-i\psi}\left(\frac{\partial U_3}{\partial\bar{\zeta}^*}\right)^+\right\} - ie_{15}\left\{e^{i\psi}\left(\frac{\partial F}{\partial\zeta^*}\right)^+\right.$$

$$\left.-e^{-i\psi}\left(\frac{\partial F}{\partial\bar{\zeta}^*}\right)^+\right\} = Z,\ \frac{ie_{15}}{\underset{11}{\varepsilon}}\left\{e^{i\psi}\left(\frac{\partial U_3}{\partial\zeta^*}\right)^+ - e^{-i\psi}\left(\frac{\partial U_3}{\partial\bar{\zeta}^*}\right)^+\right\} \tag{7.4}$$

$$+\left\{e^{i\psi}\left(\frac{\partial F}{\partial\zeta^*}\right)^+ + e^{-i\psi}\left(\frac{\partial F}{\partial\bar{\zeta}^*}\right)^+\right\} = 0,\ c_{44}^E\left(1+k_{15}^2\right)\left\{e^{i\psi}\left(\frac{\partial U_3}{\partial\zeta}\right)^+\right.$$

$$\left.+e^{-i\psi}\left(\frac{\partial U_3}{\partial\bar{\zeta}}\right)^+\right\} - ie_{15}\left\{e^{i\psi}\left(\frac{\partial F}{\partial\zeta}\right)^+ - e^{-i\psi}\left(\frac{\partial F}{\partial\bar{\zeta}}\right)^+\right\} = X_3.$$

Here the derivatives of the functions U_3 and F are calculated at their limiting values at $z\to\zeta^*\in C$ and $z\to\zeta\in L$ of the body; $\psi=\psi(\zeta)$ and $\psi_1=\psi_1(\zeta^*)$ denote the angles between x_1-axis and the normals to the contour L_j and C, respectively. The second condition in (7.4) is for the zero equality of the tangential component of the vector of the electric strength on the electrodized surface of the cylinder. Due to the fact that the forms (7.3) provide the continuity of the stress vector on the crack, the mechanical boundary condition may be satisfied only on one of the edges (e.g. on the left).

Substituting the limiting values of functions (7.3) into boundary conditions (7.4) we get a system of three integrodifferential equations of the second kind

$$\int\limits_L\left\{\left[\frac{dU_3}{ds}\right]g_1(\zeta,\zeta_0)+[U_3]\,g_2(\zeta,\zeta_0)\right\}ds+ \tag{7.5}$$

$$+\int\limits_C\{p(\zeta^*)\,g_3(\zeta^*,\zeta_0)+f(\zeta^*)\,g_4(\zeta^*,\zeta_0)\}ds=N_1(\zeta_0),$$

$$p(\zeta_0^*)+\int\limits_L\left\{\left[\frac{dU_3}{ds}\right]g_5(\zeta,\zeta_0^*)+[U_3]\,g_6(\zeta,\zeta_0^*)\right\}ds+$$

$$+\int\limits_C\{p(\zeta^*)\,g_7(\zeta^*,\zeta_0^*)+f(\zeta^*)\,g_8(\zeta^*,\zeta_0^*)\}ds=N_2(\zeta_0^*),$$

$$-f(\zeta_0^*)+\int\limits_L\left\{\left[\frac{dU_3}{ds}\right]g_9(\zeta,\zeta_0^*)+[U_3]\,g_{10}(\zeta,\zeta_0^*)\right\}ds+$$

$$+\int\limits_C\{p(\zeta^*)\,g_{11}(\zeta^*,\zeta_0^*)+f(\zeta^*)\,g_{12}(\zeta^*,\zeta_0^*)\}ds=0,$$

where the kernels g_m ($m = 1, 2, \ldots, 12$) and the right parts are determined by

$$g_1(\zeta, \zeta_0) = \frac{1}{2}\mathrm{Im}\frac{e^{i\psi_0}}{\zeta, \zeta_0}, \quad g_8(\zeta^*, \zeta_0^*) = \frac{\kappa}{2\pi i}\mathrm{Im}\frac{e^{i\psi_0}}{\zeta^* - \zeta_0^*}, \tag{7.6}$$

$$g_2(\zeta, \zeta_0) = \frac{i\pi}{8}\gamma^2\left(1 + k_{15}^2\right)\left[H_2(\gamma r_0)\cos(\psi + \psi_0 - 2\alpha_0)\right.$$
$$\left. - H_0^{(1)}(\gamma r_0)\cos(\psi - \psi_0)\right],$$

$$g_3(\zeta^*, \zeta_0) = \pi\left(1 + k_{15}^2\right)\gamma H_1^{(1)}(\gamma r_{10})\cos(\psi_0 - \alpha_1),$$

$$g_4(\zeta^*, \zeta_0) = -k_{15}^2\mathrm{Im}\frac{e^{i\psi_0}}{\zeta^* - \zeta_0}, \quad g_5(\zeta, \zeta_0^*) = \frac{\kappa}{4\pi i}\mathrm{Im}\frac{e^{i\psi_{10}}}{\zeta - \zeta_0^*},$$

$$g_6(\zeta, \zeta_0^*) = -\frac{\gamma^2}{16}\left[H_2^{(1)}(\gamma r_2)\cos(\psi + \psi_{10} - 2\alpha_2) - H_0^{(1)}(\gamma r_2)\cos(\psi - \psi_{10})\right],$$

$$g_7(\zeta^*, \zeta_0^*) = -\frac{\gamma}{2i}H_1(\gamma r_3)\cos(\psi_{10} - \alpha_3) + \frac{1}{\pi}\mathrm{Re}\frac{e^{i\psi_{10}}}{\zeta^* - \zeta_0^*},$$

$$g_9(\zeta, \zeta_0^*) = -\frac{1}{2\pi}\mathrm{Re}\frac{e^{i\psi_{10}}}{\zeta - \zeta_0^*}, \quad g_{12}(\zeta^*, \zeta_0^*) = -\frac{1}{\pi}\mathrm{Re}\frac{e^{i\psi_{10}}}{\zeta^*, \zeta_0^*},$$

$$g_{10}(\zeta, \zeta_0^*) = -\frac{i\gamma^2}{8}\left[H_2^{(1)}(\gamma r_2)\sin(\psi + \psi_{10} - 2\alpha_2) + H_0^{(1)}(\gamma r_2)\sin(\psi - \psi_{10})\right],$$

$$g_{11}(\zeta^*, \zeta_0^*) = -\gamma H_1(\gamma r_3)\sin(\psi_{10} - \alpha_3) + \frac{2i}{\pi}\mathrm{Im}\frac{e^{i\psi_{10}}}{\zeta^* - \zeta_0^*},$$

$$H_1(x) = \frac{2i}{\pi x} + H_1^{(1)}(x), \quad H_2(x) = \frac{4i}{\pi x^2} + H_2^{(1)}(x),$$

$$N_1(\zeta_0) = \frac{\pi}{c_{44}^E}X_3(\zeta_0), \quad N_2(\zeta_0) = -\frac{Z}{2ic_{44}^E\left(1 + k_{15}^2\right)},$$

$$\kappa = \frac{k_{15}^2}{1 + k_{15}^2}, \quad r_0 = |\zeta - \zeta_0|, \quad r_{10} = |\zeta^* - \zeta_0|,$$

$$r_2 = |\zeta - \zeta_0^*|, \quad r_3 = |\zeta^* - \zeta_0^*|, \quad \alpha_0 = \arg(\zeta - \zeta_0),$$

$$\alpha_1 = \arg(\zeta^* - \zeta_0), \quad \alpha_2 = \arg(\zeta - \zeta_0^*), \quad \alpha_3 = \arg(\zeta^* - \zeta_0^*),$$

$$\psi = \psi(\zeta), \quad \psi_0 = \psi(\zeta_0), \quad \psi_{10} = \psi_1(\zeta_0^*),$$

$$\zeta, \zeta_0 \in L = \cup L_j; \quad \zeta^*, \zeta_0^* \in C.$$

In order to solve the system (7.5) we have to augment it with the additional conditions

$$\int\limits_{L_j}\left[\frac{dU_3}{ds}\right]ds = 0 \quad (j = 1, 2, \ldots, k), \tag{7.7}$$

which express the absence of displacement jumps U_3 on the tips of the cracks.

The solution of the obtained system together with (7.7) is available for any frequency ω, not coinciding with its eigenfrequency. At $k_{15} = 0$ the first two integrodifferential equations in (7.5) correspond to the case of piezopassive cylinder with cracks.

To determine the stress intensity factor we introduce the parameterization of contour L using the expressions

$$\zeta = \zeta(\beta), \quad \left[\frac{dU_3}{ds}\right] = \frac{\Omega(\beta)}{s'(\beta)\sqrt{1-\beta^2}}, \tag{7.8}$$
$$s'(\beta) = \frac{ds}{d\beta} > 0, \quad \beta \in [-1,1].$$

Taking into account (7.3), (7.8) and performing asymptotic analysis of the quantities $\partial_i U_3$, $\partial_i F$ $(i = 1,2)$ we find the main singular part of the stress shear on the continuation of the tip of the crack to be

$$\sigma_n^0 = \pm c_{44}^E \frac{\mathrm{Re}\left[e^{-i\omega t}\Omega(\pm 1)\right]}{2\sqrt{2r^* s'(\pm 1)}}, \quad s'(\pm 1) = \left.\frac{ds}{d\beta}\right|_{\beta=\pm 1}. \tag{7.9}$$

Here the lower sign refers to tip a, the upper sign to tip b, r^* is the distance from the point to the tip. From (7.9) the stress intensity factor is given by

$$K_{III}^{\pm} = \lim_{r^*\to 0} \sqrt{2\pi r^*}\sigma_n^0 = \pm c_{44}^E \frac{\sqrt{\pi}\mathrm{Re}\left[e^{-i\omega t}\Omega(\pm 1)\right]}{2\sqrt{s'(\pm 1)}}. \tag{7.10}$$

To determine the stress concentration in the cylinder with cracks we should calculate as usual the stress $\sigma_s = \sigma_{23}\cos\psi - \sigma_{13}\sin\psi$ on C. We have

$$T_s(\zeta_0^*) = ic_{44}^E\left(1+k_{15}^2\right)\left(e^{i\psi_{10}}\frac{\partial U_3}{\partial \zeta_0^*} - e^{-i\psi_{10}}\overline{\frac{\partial U_3}{\partial \zeta_0^*}}\right) + \tag{7.11}$$
$$+ e_{15}\left(e^{i\psi_{10}}\frac{\partial F}{\partial \zeta_0^*} + e^{-i\psi_{10}}\overline{\frac{\partial F}{\partial \zeta_0^*}}\right),$$
$$\sigma_s = \mathrm{Re}\left(T_s e^{-i\omega t}\right), \quad \psi_{10} = \psi_1(\zeta_0^*), \quad \zeta_0^* \in C.$$

In (7.11) the derivatives are calculated at their limiting values at $z \to \zeta_0^*$ the transition is carried from the area of the body to the contour.

From (7.3) we find

$$T_s(\zeta_0^*) = c_{44}^E\left(1+k_{15}^2\right)\left\{\int_L [U_3]\, g_{10}(\zeta,\zeta_0^*)ds + \int_C p(\zeta^*)\, g_{11}(\zeta^*,\zeta_0^*)ds\right\} - \tag{7.12}$$
$$-\frac{c_{44}^E k_{15}^2}{\pi}\left\{\frac{1}{2}\int_L \left[\frac{dU_3}{ds}\right]\mathrm{Re}\frac{e^{i\psi_{10}}}{\zeta-\zeta_0^*}ds + \pi f(\zeta_0^*) + \int_C (\zeta^*)\mathrm{Re}\frac{e^{i\psi_{10}}}{\zeta^*-\zeta_0^*}ds\right\}.$$

where the functions $g_{10}(\zeta,\zeta_0^*)$, $g_{11}(\zeta^*,\zeta_0^*)$ defined in (7.6).

As an example consider a circular piezoceramic cylinder with radius R (by $PZT-4$ ceramics) with a horizontal linear crack of 2ℓ length and with the center lying on the axis of the cylinder.

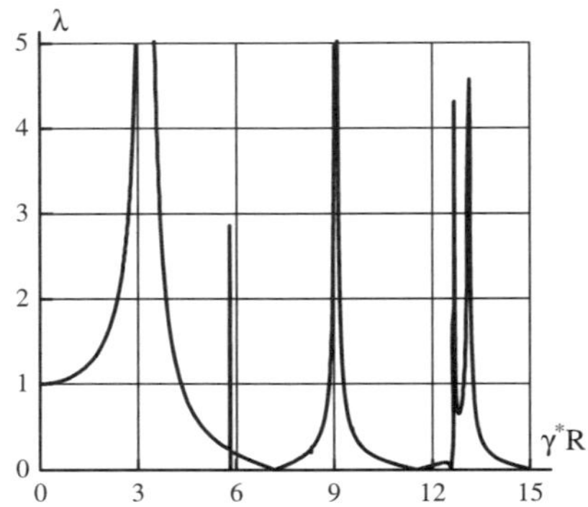

Fig. 7.2. Behaviour of quantity $\lambda = |T_s/Z_0|$ at point $\varphi = \pi$ as a function of normalized wave number of $\gamma^* R$ under the influence of forces $Z = Z_0 sin2\varphi$

The solution of (7.5), (7.7) is obtained by the method of quadrature (see: Appendix B, Par. B.1, Par. B.2). The maximum number of the division nodes on the intervals of integration $[0, 2\pi]$ and $[-1, 1]$ was assumed to be equal to $n_1 = 61$ and $n_2 = 40$, respectively; practically, a further increase of the parameters n_1, n_2 did not influence the accuracy of the results.

Figure 7.2 shows the behavior of the quantity $\lambda = |T_s/Z_0|$ at the point of contour $\varphi = \pi$ as a function of the normalized wave number $\gamma^* R = \gamma R\sqrt{1 + k_{15}^2}$. The calculation is carried out for the case, where the edges of the crack are free from loading, while shear stresses applied on the side surface of the cylinder. The shear stresses change according to $Z = Z_0 \sin 2\varphi$ (φ is the polar angle).

To evaluate the influence of the crack on the amplitude-frequency characteristics of the cylinder, we calculated the quantity $\lambda = \lambda\,(\gamma^* R)$ at $\varphi = \pi$ in the vicinity of the first three natural frequencies of oscillation of the piezoceramic cylinder (without a crack) (shown in Fig. 7.3). This is obtained with the help of the system of integral (5.4) where the signs before the integrated

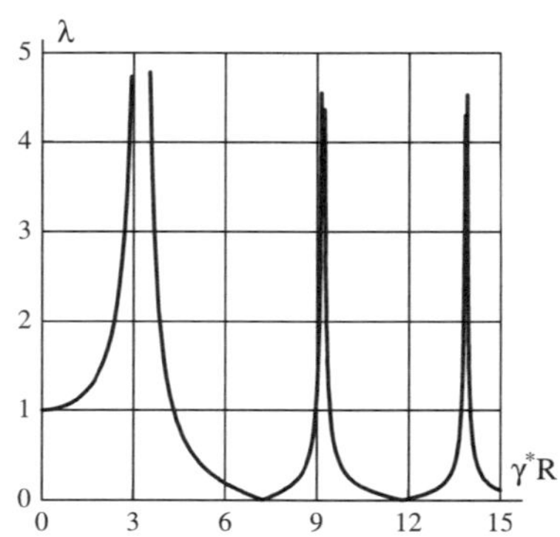

Fig. 7.3. The graph of quantity $\lambda = \lambda(\gamma^*\mathrm{R})$ at $\varphi = \pi$ in the vicinity of the first three natural frequencies of the oscillations of a homogenous piezoceramic cylinder under the influence of forces $Z = Z_0 sin2\varphi$

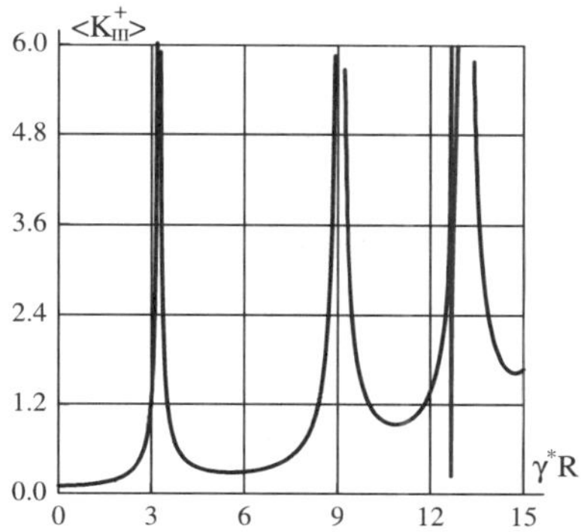

Fig. 7.4. Behaviour of the relative intensity factor $<K_{III}^{+}>$ according to γ^{*}R under the influence of forces $Z = Z_0 sin2\varphi$

summands should be changed to the opposite ones, and the terms corresponding to monochromatic waves should be omitted.

Comparing the graphs shown in Figs. 7.2 and 7.3 we conclude that the availability of the tunnel crack inside the cylinder may qualitatively change its amplitude-frequency characteristics.

Figure 7.4 depicts the behavior of relative intensity factor $\left\langle K_{III}^{+}\right\rangle = c_{44}^{E}\left|\Omega\left(1\right)\right| / \left(2\left|Z_0\right|\sqrt{\ell s'\left(1\right)}\right)$ with respect to $\gamma^{*}R$. Assuming that the quantity $\left\langle K_{III}^{+}\right\rangle$ is known, the stress intensity factor may be determined by $K_{III}^{\pm} = \pm\sqrt{\pi\ell}\left|Z_0\right| < \mathrm{K}_{\mathrm{III}}^{\pm} > \cos\left\{\omega t - \arg\Omega\left(\pm 1\right)\right\}$.

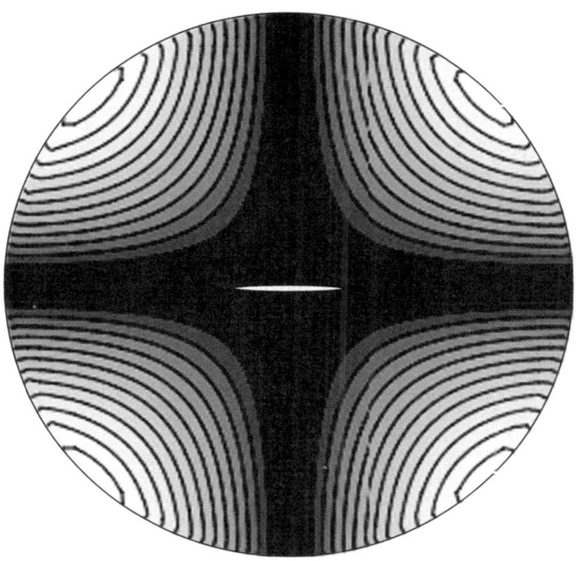

Fig. 7.5. The contour lines of the modulus of displacement amplitude U_3 in the vicinity of the first three natural excitation frequencies of a cylinder by shear forces $Z = Z_0 sin2\varphi$

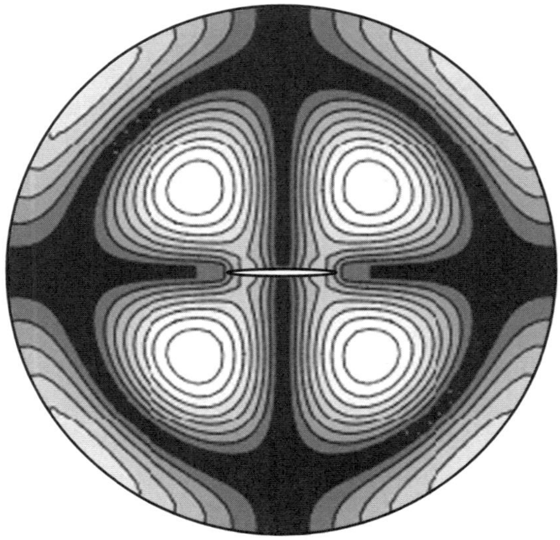

Fig. 7.6. The contour lines of the modulus of displacement amplitude U_3 in the vicinity of the first three natural excitation frequencies of a cylinder by shear forces $Z = Z_0 sin 2\varphi$

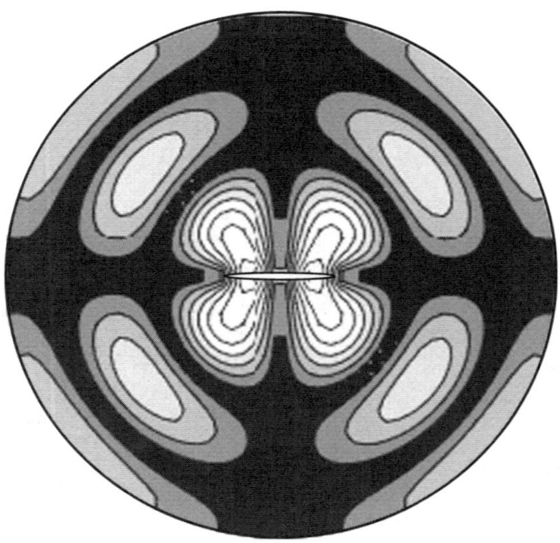

Fig. 7.7. The countour lines of the modulus of displacement amplitude U_3 in the vicinity of the first three natural excitation frequencies of a cylinder by shear forces $Z = Z_0 sin 2\varphi$

Figures 7.5–7.7 illustrate the contour lines of the modulus of displacement amplitude U_3 in the vicinity of the first three resonance frequencies of excitation of the cylinder by shear forces $Z = Z_0 \sin 2\varphi \, (X_3 = 0)$. The normalized wave number, corresponding to resonance frequencies, which are approximately equal to $\gamma^*_{(1)} R \approx 3.24$, $\gamma^*_{(2)} R \approx 9.06$ and $\gamma^*_{(3)} R \approx 13.125$. For our illustrations we used $\ell/R = 0.2$.

7.2 A Cylinder with a Thin Rigid Inclusion

Let us consider a piezoceramic cylinder continuous in the direction of x_3 -axis, containing a tunnel along the element rigid curvilinear insert L (stringer). Assuming that the vector of preliminary polarization of piezoceramics is directed along x_3- axis, there are two variants of the electric boundary conditions $Z_n = \mathrm{Re}\left(Ze^{-i\omega t}\right)$: (a) the cylinder surface is assumed to be electrodized and grounded (variant A); (b) the cylinder surface is not electrodized and coupled with vacuum (variant B).

In order to solve the above problem we use the complex representations of the components of electroelastic field (6.2), (6.3) in the case of antiplane deformation. In order to solve (6.1) we use (6.5) that refer to the amplitudes of the sought-for functions.

Assuming that the inclusion is fixed, the mechanical and electric boundary conditions on contour L read

$$U_3^{\pm} = 0, \tag{7.13}$$

$$E_s^{+} = E_s^{-}, \quad D_n^{+} = D_n^{-}. \tag{7.14}$$

Here E_s and D_n are the tangential component of the vector of electric strength and the normal component of the vector of electric induction, respectively; the plus and minus signs refer to the left and right edges of inclusion L at movement from its beginning a to the end b (Fig. 7.8).

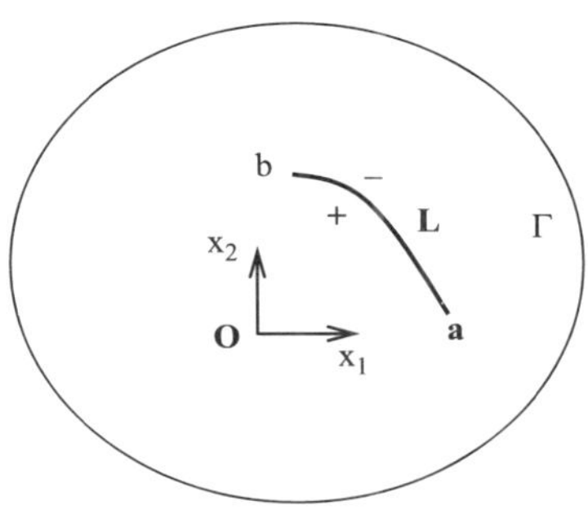

Fig. 7.8. The scheme of cross-section of a cylinder with inclusions

To obtain an efficient (with respect to numerical realization) system of integral equations, it is helpful to the differentiate boundary condition (7.13) over the arc coordinate s

$$\left(\frac{\partial U_3}{\partial s}\right)^{\pm} = 0. \tag{7.15}$$

The boundary conditions on the contour of cylinder Γ for the considered variants are given by

$$\frac{\partial}{\partial n}\left\{c_{44}^{E}\left(1+k_{15}^{2}\right)U_3 + e_{15}F^{*}\right\} = Z, \tag{7.16}$$

$$\phi_{*} = 0, \tag{7.17}$$

$$D_n = -\ni_{11}^{\varepsilon}\frac{\partial F^{*}}{\partial n} = 0. \tag{7.18}$$

The boundary equalities (7.17), (7.18) correspond to variants A and B, respectively. In what follows, instead of (7.17) we will use the condition

$$\frac{\partial}{\partial s}\left(F^{*} + \frac{e_{15}}{\ni_{11}^{\varepsilon}}U_3\right) = 0. \tag{7.19}$$

The question here is how, we can derive the functions U_3 and F^* from differential (6.5) and the boundary conditions (7.15), (7.16) and also (7.18) or (7.19).

The integral representations of functions U_3 and F^* is guided by the fundamental solution of the system of (6.1) in the case, where the dependence on time is harmonic. In this case, the system (6.1) may be represented in the form of (4.59) and its solution can be found by the formulas (4.60). We have

$$U_3\left(x_1,x_2\right) = \frac{i}{4c_{44}^{E}\left(1+k_{15}^{2}\right)}\left\{\int\limits_{L} q\left(\zeta\right)H_0^{1}\left(\gamma r\right)ds + \int\limits_{\Gamma} p\left(\zeta^{*}\right)H_0^{(1)}\left(\gamma r_1\right)ds\right\} + \tag{7.20}$$

$$+\frac{k_{15}^{2}}{4ie_{15}\left(1+k_{15}^{2}\right)}\int\limits_{\Gamma} f\left(\zeta^{*}\right)H_0^{(1)}\left(\gamma r_1\right)ds,$$

$$F^{*}\left(x_1,x_2\right) = -\frac{1}{2\pi\ni_{11}^{\varepsilon}}\int\limits_{\Gamma} f\left(\zeta^{*}\right)\ell n r_1 ds,$$

$$r = \left|\zeta - z\right|,\quad r_1 = \left|\zeta^{*} - z\right|,\quad \zeta\in L, \zeta^{*}\in\Gamma.$$

here $H_{\nu}^{(1)}(x)$ is the Hänkel function of the first kind and of order ν, ds is the element of the arc- length contour, over which the integration is carried out.

It is easy to realize that representations (7.20) automatically satisfy conditions (7.14) on L and provide the performance of equalities $[U_3] = U_3^{+} - U_3^{-} = 0$ in (7.13).

The unknown "*densities*" $q(\zeta)$, $p(\zeta^*)$ and $f(\zeta^*)$ are determined from the complex system of three integral equations, which are obtained as the result of substitution of the limiting values of derivatives of the functions in (7.20) at $z \to \zeta \in L$ and $z \to \zeta^* \in \Gamma$ in the corresponding boundary conditions.

The system is represented in the following form:

$$\int_L q(\zeta)G_1(\zeta,\zeta_0)\,ds + \int_\Gamma p(\zeta^*)G_2(\zeta^*,\zeta_0)\,ds + \int_\Gamma f(\zeta^*)G_3(\zeta^*,\zeta_0)\,ds = 0,$$

$$\frac{1}{2}p(\zeta_0^*) + \int_L q(\zeta)G_4(\zeta,\zeta_0^*)\,ds + \int_\Gamma p(\zeta^*)G_5(\zeta^*,\zeta_0^*)\,ds$$
$$+ \int_\Gamma f(\zeta^*)G_6(\zeta^*,\zeta_0^*)\,ds = Z(\zeta_0^*),\ \lambda f(\zeta_0^*) + \int_L q(\zeta)G_7(\zeta,\zeta_0^*)\,ds$$
$$+ \int_\Gamma p(\zeta^*)G_8(\zeta^*,\zeta_0^*)\,ds + \int_\Gamma f(\zeta^*)G_9(\zeta^*,\zeta_0^*)\,ds = 0,$$

$$G_1(\zeta,\zeta_0) = \frac{1}{c_{44}^E}\left\{\gamma H_1(\gamma r_0)\sin(\psi_0-\alpha_0) - \frac{2i}{\pi}\mathrm{Im}\frac{e^{i\psi_0}}{\zeta-\zeta_0}\right\}, \tag{7.21}$$
$$G_2(\zeta^*,\zeta_0) = \frac{\gamma}{c_{44}^E}H_1^{(1)}(\gamma r_{10})\sin(\psi_0-\alpha_1),$$
$$G_3(\zeta^*,\zeta_0) = -\frac{k_{15}^2\gamma}{e_{15}}H_1^{(1)}(\gamma r_{10})\sin(\psi_0-\alpha_1),$$
$$G_4(\zeta,\zeta_0^*) = \frac{i\gamma}{4}H_1^{(1)}(\gamma r_2)\cos(\psi_{10}-\alpha_2),$$
$$G_5(\zeta^*,\zeta_0^*) = \frac{1}{2\pi}\mathrm{Re}\frac{e^{i\psi_{10}}}{\zeta^*-\zeta_0^*} + \frac{i\gamma}{4}H_1(\gamma r_3)\cos(\psi_{10}-\alpha_3),$$
$$G_6(\zeta^*,\zeta_0^*) = \frac{c_{44}^E k_{15}^2}{4ie_{15}}\gamma H_1(\gamma r_3)\cos(\psi_{10}-\alpha_3),$$
$$H_1(x) = \frac{2i}{\pi x} + H_1^{(1)}(x), \quad \kappa = \frac{k_{15}^2}{1+k_{15}^2},$$
$$r_0 = |\zeta-\zeta_0|, \quad r_{10} = |\zeta^*-\zeta_0|, \quad r_2 = |\zeta-\zeta_0^*|,$$
$$r_3 = |\zeta^*-\zeta_0^*|, \quad \alpha_0 = \arg(\zeta-\zeta_0),$$
$$\alpha_1 = \arg(\zeta^*-\zeta_0), \quad \alpha_2 = \arg(\zeta-\zeta_0^*), \quad \alpha_3 = \arg(\zeta^*-\zeta_0^*),$$
$$\psi_0 = \psi(\zeta_0), \quad \psi_{10} = \psi_1(\zeta_0^*), \quad \zeta,\zeta_0 \in L, \quad \zeta^*,\zeta_0^* \in \Gamma.$$

When $\phi|_\Gamma = 0$ (variant A) we have

$$\lambda = 0, \quad G_7(\zeta,\zeta_0^*) = \frac{\kappa\gamma}{4ie_{15}}H_1^{(1)}(\gamma r_2)\sin(\psi_{10}-\alpha_2), \tag{7.22}$$
$$G_8(\zeta^*,\zeta_0^*) = \frac{\kappa}{4ie_{15}}\left\{\gamma H_1(\gamma r_3)\sin(\psi_{10}-\alpha_3) - \frac{2i}{\pi}\mathrm{Im}\frac{e^{i\psi_{10}}}{\zeta^*-\zeta_0^*}\right\},$$
$$G_9(\zeta^*,\zeta_0^*) = -\frac{1}{2\pi \ni_{11}^\varepsilon (1+k_{15}^2)}\mathrm{Im}\frac{e^{i\psi_0}}{\zeta^*-\zeta_0^*} - \frac{\kappa\gamma}{4i \ni_{11}^\varepsilon}H_1(\gamma r_3)\sin(\psi_{10}-\alpha_3).$$

To satisfy the boundary conditions $D_n|_\Gamma = 0$ in (7.21) (variant B) we have to use

$$\lambda = \frac{1}{2}, \quad G_7(\zeta, \zeta_0^*) = G_8(\zeta^*, \zeta_0^*) = 0, \qquad (7.23)$$
$$G_9(\zeta^*, \zeta_0^*) = \frac{1}{2\pi}\mathrm{Re}\frac{e^{i\psi_{10}}}{\zeta^* - \zeta_0^*}.$$

In (7.21)–(7.23) the quantities $\psi = \psi(\zeta)$ and $\psi_1 = \psi_1(\zeta^*)$ are the angles between x_1 -axis and the normal vectors to contours L and Γ, respectively.

Here, it should be mentioned, that the solution of the system of integral (7.21) can be obtained for any frequency ω not coinciding with its eigenfrequency. At $e_{15} = 0$ the system (7.21) corresponds to the piezopassive (isotropic) cylinder with inclusion.

To determine the shear stress $\sigma_s = \mathrm{Re}\left(T_s e^{-i\omega t}\right)$ on the boundary of the cylinder we use (6.76).

At $z \to \zeta_0^* \in \Gamma$, from (7.20), we can obtain the amplitude of shear stress T_s:

$$T_s(\zeta_0^*) = \frac{e_{15}}{\kappa}\left\{\int_\Gamma p(\zeta^*) G_8(\zeta^*, \zeta_0^*)\, ds + \int_L q(\zeta) G_7(\zeta, \zeta_0^*)\, ds\right\} - \qquad (7.24)$$
$$- \frac{e_{15}\gamma}{4i_{11}^{\varepsilon}} \int_\Gamma f(\zeta^*) \sin(\psi_{10} - \alpha_3) H_1(\gamma r_3)\, ds.$$

The functions $G_7(\zeta, \zeta_0^*)$, $G_8(\zeta, \zeta_0^*)$ appearing in (7.24) defined in (7.22).

Equation (7.24) permits us to investigate the concentration of stresses in the cylinder according to the excitation frequency, the position and configuration of the inclusion.

With quite similar assumptions to those that were stated in Sect. 6.9, we obtain

$$\int_L q(\zeta) ds = 0, \qquad (7.25)$$

This should be considered as an additional condition for the solution of the system of singular integral (7.21) in the class of functions, not restrained on tips, L.

As an example consider the circular cylinder with radius R (by $PZT-4$ ceramics), containing an inclusion of parabolic section orientated at angle ϑ to Ox_1 -axis. The parametric equations of contour L have the form

$$\mathrm{Re}\zeta = p\delta\cos\vartheta - p_*\delta^2\sin\vartheta,$$
$$\mathrm{Im}\zeta = p\delta\sin\vartheta + p_*\delta^2\cos\vartheta \quad (-1 \le \delta \le 1). \qquad (7.26)$$

Taking into account (7.26), the solution of system (7.21), (7.25) was carried out by the standard method. The inclusion is assumed to be orientated along

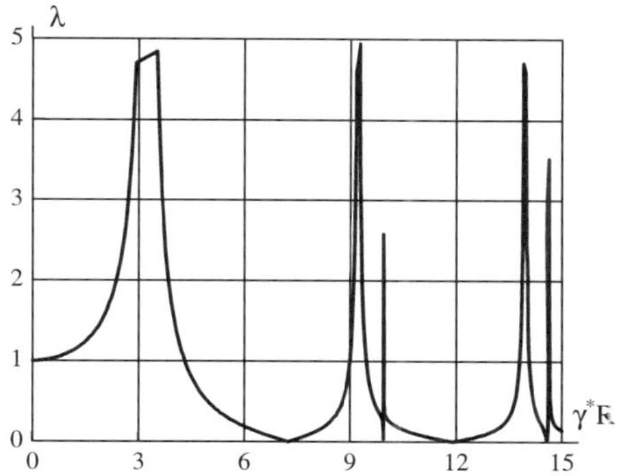

Fig. 7.9. Changes of $\lambda = |T_S/Z_0|$ at the point of contour $\beta = \pi$ in the vicinity of the first three eigenfrequencies of oscillations (variant A)

x_1 ($\vartheta = 0$)- axis and its center is on the axis of the cylinder, the boundary of which is electrodized and grounded. On the cylinder surface, the forces change according to the law $Z = Z_0 \sin 2\beta$ (β is the polar angle). The calculations show that in this case we have that $q(\zeta) \equiv 0$, and, thus, the existence of the inclusion does not change the electroelastic state of the cylinder.

The case where the linear inclusion is orientated at angle $\vartheta = \pi/4$ to Ox_1 axis, its influence under the above mentioned loading is already studied. Figures 7.9 and 7.10 illustrate changes of $\lambda = \lambda(\gamma^* R) = |T_s/Z_0|$ at the point of contour $\beta = \pi$ in the vicinity of the first three natural frequencies of the oscillations for variants A and B at $p/R = 0.2$. The influence of the electric boundary conditions on the contour of the cylinder is manifested by the fact that the resonance frequencies for case B are displaced to the right in comparison with variant A, and this displacement is reduced with the increase of the number of the resonance frequency. It is important to mention, that the

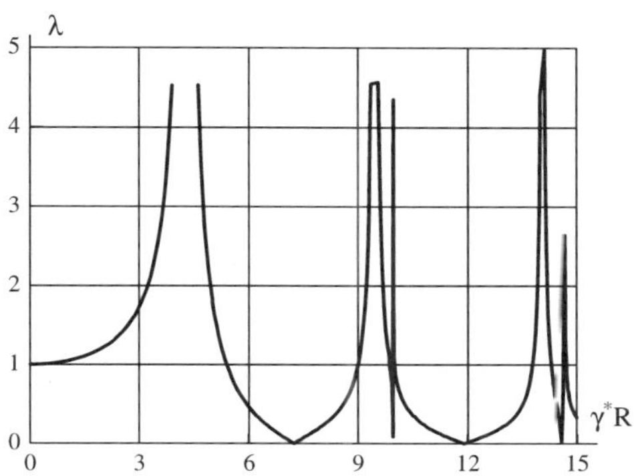

Fig. 7.10. Changes of $\lambda = |T_S/Z_0|$ at the point of contour $\beta = \pi$ in the vicinity of the first three eigenfrequencies of oscillations (variant B)

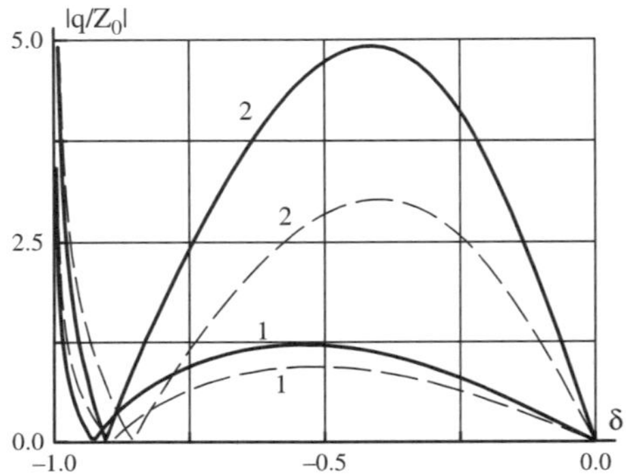

Fig. 7.11. Changes of the absolute intensity value of the contact force on a parabolic insert

resonance picks shown in Fig. 7.10 are wider than those in Fig. 7.9, revealing a more efficient electromechanical transformation of the energy at the contact of the cylinder surface with vacuum (air), than in the case of an electrodized boundary [115].

Figure 7.11 depicts the change of the absolute value of intensity of the contact force on a parabolic inclusion. Curves 1 and 2 are constructed for $\gamma^* R = 8$, $p/R = 0.4$, $\vartheta = 0$, $p_*/R = 0.1$ and 0.4. The solid and dashed lines are constructed for A and B variants, respectively. As it follows from the asymptotic analysis of system (7.21), the function $q(\zeta)$ has singularities of square root type in the vicinity of the ends of the inclusion ($\delta = \pm 1$).

Fig. 7.12. The contour lines of the modulus of the electric potential amplitude in the area of a cylinder with a straight insert (variant A)

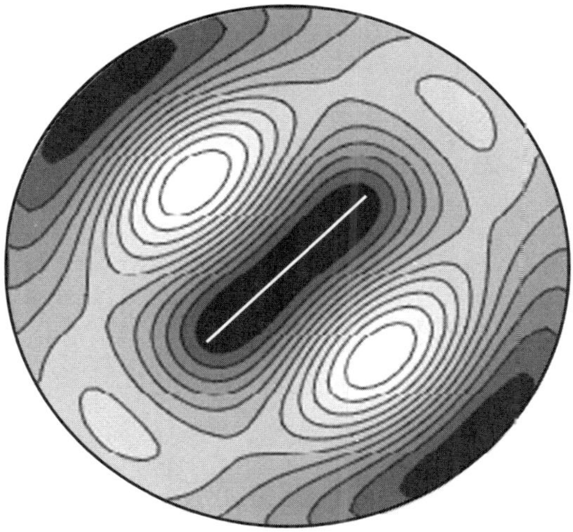

Fig. 7.13. The contour lines of the modulus of the displacement amplitude in the area of a cylinder with a straight insert (variant A)

Figures 7.12 and 7.13 show the contour line of the modulus of the amplitude of the electric potential and displacement, corresponding to variant A, in the area of the cylinder with a straight inclusion at $p/R = 0.4$, $\vartheta = \pi/4$, $\gamma^*R = 10$, respectively. Figures 7.14–7.16 show the results for the

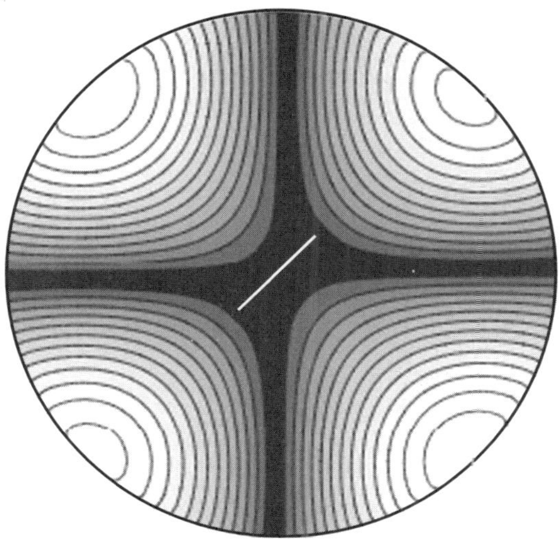

Fig. 7.14. The contour lines of the modulus of the displacement amplitude in the area of a cylinder with a straight insert in the vicinity of the first eigenfrequency of oscillations (variant B)

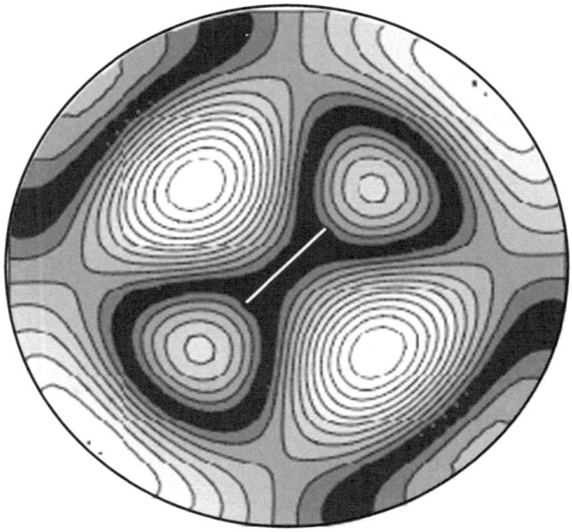

Fig. 7.15. The contour lines of the modulus of the displacement amplitude in the area of a cylinder with a straight insert in the vicinity of the first eigenfrequency of oscillations (variant B)

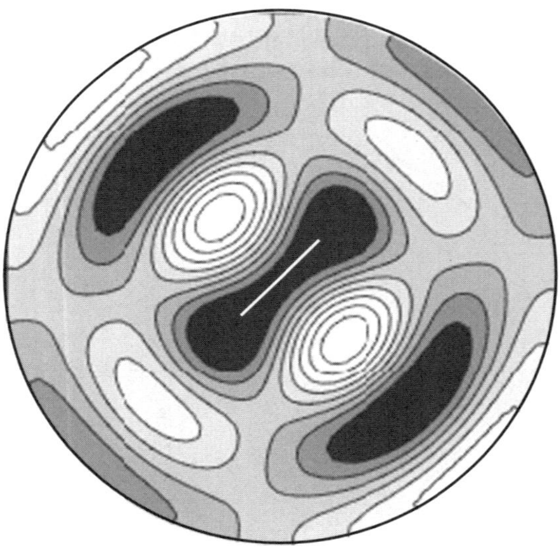

Fig. 7.16. The contour lines of the modulus of the displacement amplitude in the area of a cylinder with a straight insert in the vicinity of the first eigenfrequency of oscillations (variant B)

Fig. 7.17. Equipotentials in the vicinity of the main eigenfrequency of oscillations (variant A)

modulus of the amplitude of displacement for case B and for frequencies close to eigenfrequencies ($\gamma^* R \approx 4.3; 9.456$ and 14.06). In Fig. 7.17 the equipotentials for variant A in the vicinity of the main eigenfrequency of oscillations ($\gamma^* R \approx 3.263$) are given.

7.3 A Cylinder with a Crack Excited by a System of Electrodes

In the section we investigate the oscillations of a cylinder with a tunnel crack during electric excitation of an arbitrary system of electrodes (Fig. 7.18). Taking into account representations (6.2), (6.3) the boundary conditions on the

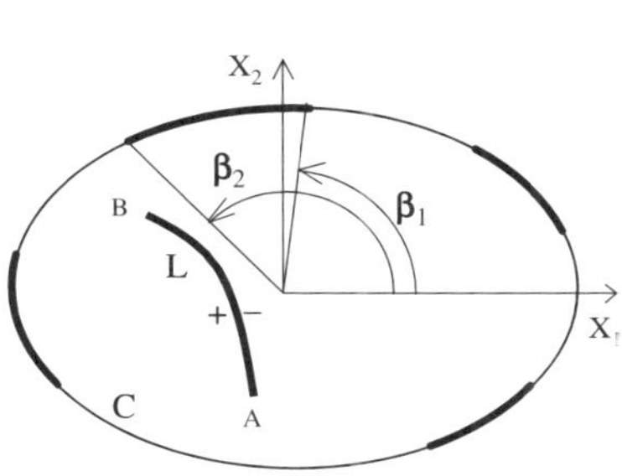

Fig. 7.18. The scheme of cross-section of an electrodized cylinder with a crack

surface of the opening and crack L may be represented in the form of (6.49). Working with the integral representations (6.50), providing the availability of a displacement jump, the continuity of the stress vector on L, and also the satisfaction of electric conditions $E_s^+ = E_s^-$, $D_n^+ = D_n^-$ on L, we get a system of integrodifferential (6.51), where before the integrated summands it is necessary to switch signs. This system should be augmented with the additional condition (6.52).

As an example consider a cylinder (made by $PZT-4$ ceramics) with an elliptic cross-section containing a linear crack orientated at an angle ϑ with respect to Ox_1 axis. The oscillation of the cylinder is generated by two electrodes with a difference of potential amplitude equal to $2\Phi^*$, the centers of which are positioned on x_2-axis. The parametric equations of contours L and C, respectively, read

$$\begin{aligned} \zeta &= \ell\delta\, e^{i\vartheta}, \quad \delta \in [-1,1] \\ \zeta^* &= R_1 \cos\beta + iR_2 \sin\beta, \quad \beta \in [0, 2\pi]\,. \end{aligned} \tag{7.27}$$

Figure 7.19 shows the dependence of the relative intensity factor $\langle K_{III}^+\rangle = c_{44}^E \sqrt{\pi\ell/s'(1)}\,|\Omega_0(1)|\,/\,(2e_{15}\,|\Phi^*|)$ as a function of the normalized wave number γR at $R_1/R_2 = 1$, $\ell/R = 0.2$, $\vartheta = 0$ ($R = 0.5\,(R_1+R_2)$, 2ℓ is the length of the cut). Figure 7.20 presents the behavior of the quantity $Q^* = |Q/(\ni_{11}^{\varepsilon}\,\Phi^*)|$ (Q is the amplitude of the total charge on an electrode) for the same parameter values. In this case the values of the first three natural frequencies of the oscillation are equal to $\gamma_{(1)}R \approx 1.35$, $\gamma_{(2)}R \approx 3.9$ and $\gamma_{(3)}R \approx 4.24$. Assuming that the quantity $\langle K_{III}^+\rangle$ is known, the stress intensity factor may be determined by (6.62).

Figure 7.21 illustrates the change of quantity $\lambda = c_{44}^E\,|[U_3]\,/\Phi^*|$, characterizing a displacement jump on the crack at various angles of its orientation in the circular cylinder at $\gamma R = 3$, $\ell/R = 0.4$. The curve with number m is

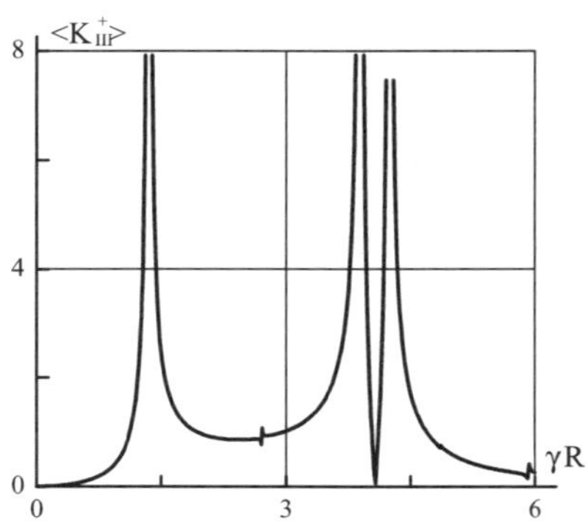

Fig. 7.19. Dependence of relative stress intensity factor $<\mathrm{K}_{\mathrm{III}}^{\pm}>$ on normalized wave number γR

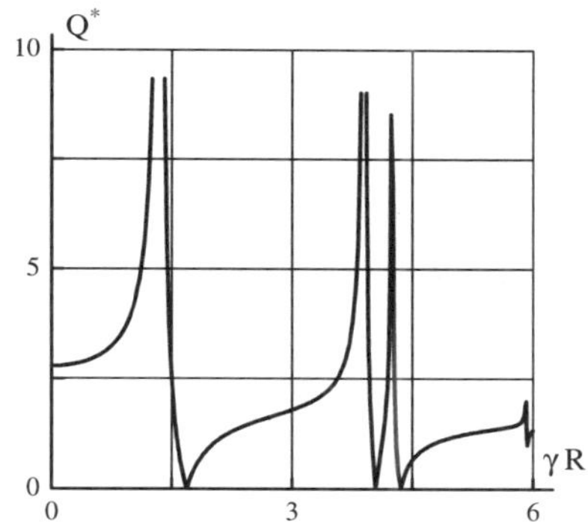

Fig. 7.20. Behaviour of relative total charge $Q^* = |Q/(\ni_{11}^{\varepsilon}\,\Phi^*)|$ on an electrode as a function of γR

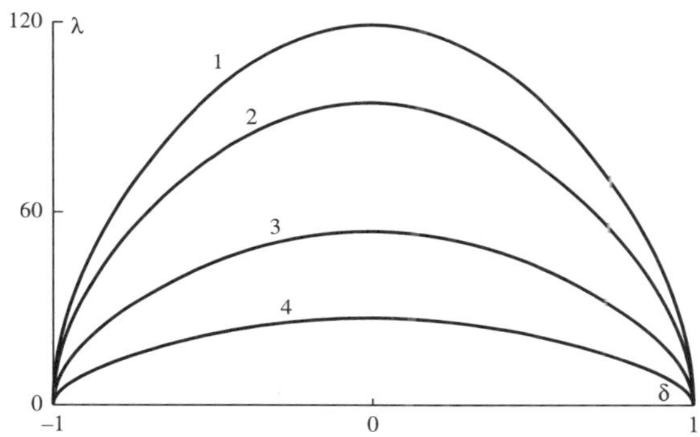

Fig. 7.21. Changes of quantity $\lambda = c_{44}^E\,|[U_3]\,/\Phi^*|$ on a crack for various angles of its orientation in a circular cylinder

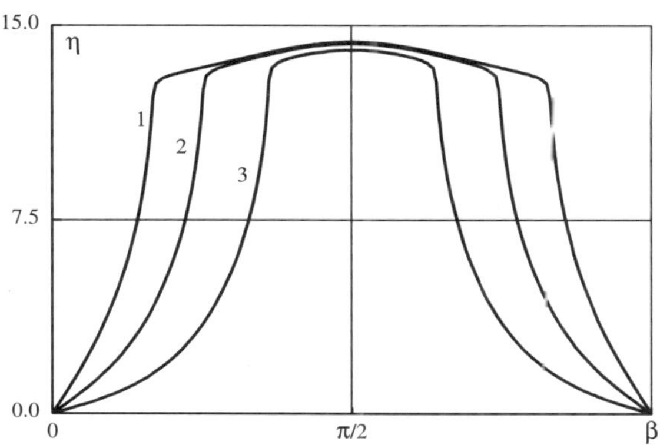

Fig. 7.22. Changes of quantity $\eta = c_{44}^E\,|[U_3]/\Phi^*|$ on the contour of a piece-wise homogenous cylinder for various areas of an electroded coating

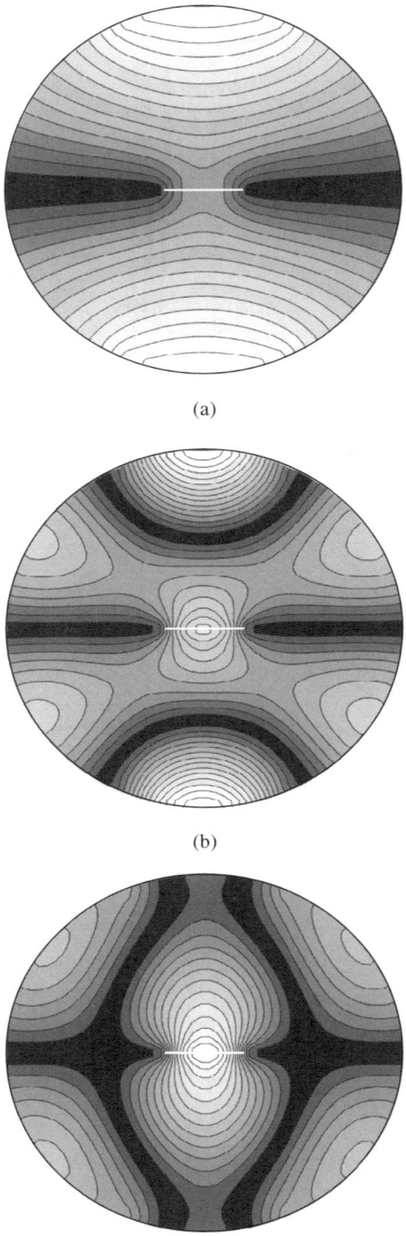

(a)

(b)

(c)

Fig. 7.23a,b and c. The contour lines of the modulus of the displacement in a cylinder with a crack in the vicinity of the first three eigenfrequencies of oscillation

constructed for $\vartheta = (m-1)\pi$. It should be mentioned, that at $\vartheta = \pi/2$ the availability of the crack does not change the electroelastic state of the cylinder.

We considered the parameters of the variants when the positions of the electrodes were prescribed as $\beta_1 = 5\pi/14, \beta_2 = 9\pi/14, \beta_3 = 19\pi/14, \ \beta_4 = 23\pi/14$.

Figure 7.22 depicts the distribution of $\eta = c_{44}^E \left|[U_3]/\Phi^*\right|$ on the contour of a piecewise-homogeneous cylinder for various areas of the electrodized coating at $R_1/R_2 = 2, \ell/R_1 = 0.4, \vartheta = 0, \gamma R = 0.5$. Curves 1, 2 and 3 are constructed for $\beta_1 = \pi/6, \beta_2 = 5\pi/6, \beta_3 = 7\pi/6, \beta_4 = 11\pi/6 \beta_k = (2k-1)\pi/4 \left(k = \overline{1,4}\right)$ and $\beta_1 = 5\pi/14, \beta_2 = 9\pi/14, \beta_3 = 19\pi/14, \ \beta_4 = 23\pi/14$.

What is of major interest here is the investigation of the behavior of the mechanical quantity in the area of a piecewise-homogeneous cylinder. Figure 7.23a,b and c illustrate lines of equal modulus of the displacement amplitude in the vicinity of the first three natural frequencies of oscillation $(\ell/R = 0.2)$ at $\beta_1 = 5\pi/14, \ \beta_2 = 9\pi/14, \ \beta_3 = 19\pi/14, \ \beta_4 = 23\pi/14$.

7.4 A Cylinder With an Inclusion Excited by a System of Electrodes

In the Cartesian system of coordinates $Ox_1x_2x_3$ consider a continuous piezoceramic cylinder oriented along the x_3-axis, containing a thin rigid inclusion L (see Fig. 7.24). The boundaries of the k-th electrode are determined by the quantities β_{2k-1} and β_{2k} $\left(k = \overline{1, 2\ell}\right)$, and the electric potential on it is prescribed by the quantity $\phi_k^* = \text{Re}\left(\Phi_k^* e^{-i\omega t}\right)$.

The boundary condition on the contours C and L are represented in the form of (6.63)–(6.65). Using the integral representations of the solution we arrive at a system of integrodifferential equations of the second kind of type (6.67), where the signs before the integrated summands should be switched. It is necessary to augment the system by the conditions (6.71).

We use formulas (6.76), (6.77) to determine the concentration of shear stresses in the cylinder with inclusion where the sign before the integrated summands should be switched.

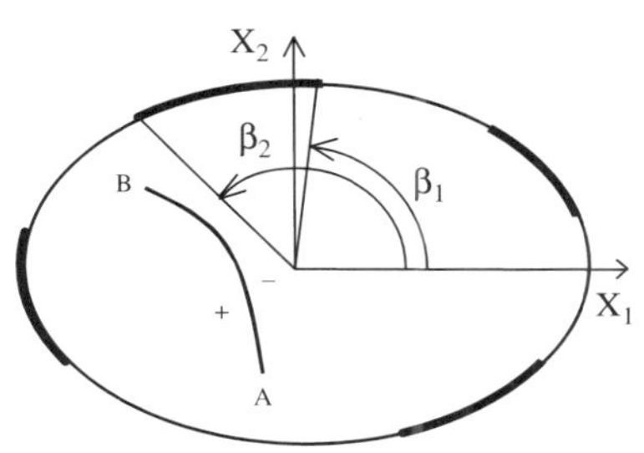

Fig. 7.24. The scheme of cross-section of an electrodized cylinder with an inclusion

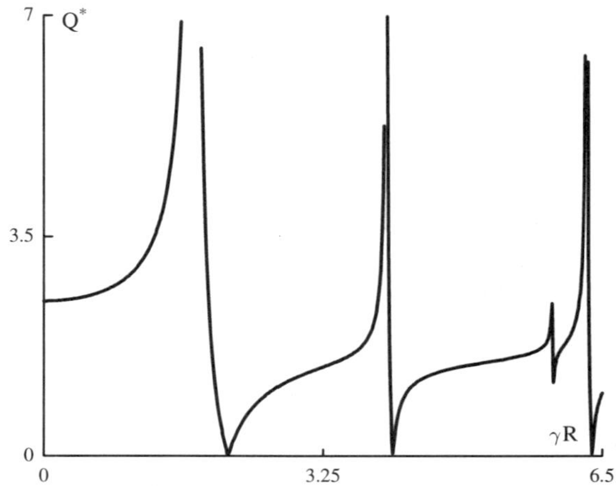

Fig. 7.25. Changes of the relative total electric charge on an electrode as a function of the normalized wave number in case of a cylinder of circular cross-section ($\beta_1 = 5\pi/14, \beta_2 = 9\pi/14, \beta_3 = 19\pi/14, \beta_4 = 25\pi/14$)

Consider a piezoceramic cylinder (the material is $PZT-4$) with an elliptic cross-section, containing a parabolic insert orientated at an angle ϑ with respect to Ox_1-axis. The parametric equations of contours L and C, read

$$\begin{aligned} \zeta &= \left(p_1\delta + ip_2\delta^2\right)e^{i\vartheta}, \quad \delta \in [-1,1] \\ \zeta^* &= R_1\cos\beta + iR_2\sin\beta, \quad \beta \in [0,2\pi]\,. \end{aligned} \tag{7.28}$$

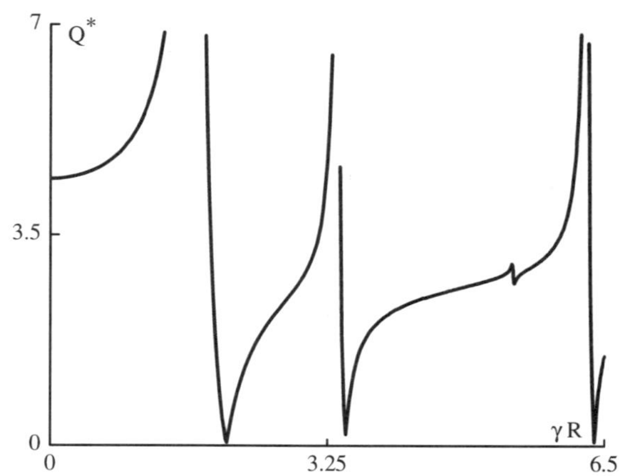

Fig. 7.26. Changes of the relative total electric charge on an electrode as a function of the normalized wave number in case of a cylinder of circular cross-section ($\beta_1 = \pi/6, \beta_2 = 5\pi/6, \beta_3 = 7\pi/6, \beta_4 = 11\pi/6$)

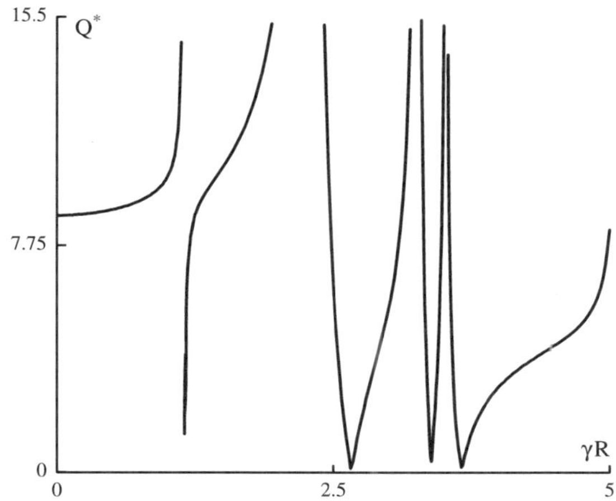

Fig. 7.27. Changes of the relative total electric charge on an electrode as a function of the normalized wave number in case of a cylinder of elliptic cross-section ($\beta_1 = \pi/6, \beta_2 = 5\pi/6, \beta_3 = 7\pi/6, \beta_4 = 11\pi/6$)

Our purpose is to investigate the amplitude-freqᴉency characteristics of a cylinder with linear inclusion ($p_2 = 0$), orientated along x_2 ($\vartheta = \pi/2$)-axis, at $R_1/R_2 = 1$ and $p_1/R = 0.4$ ($R = 0.5\,(R_1 + R_2)$). Figures 7.25 and 7.26 demonstrate the behavior of $Q^* = |Q/(\ni^{\varepsilon}_{11}\,\Phi^*)|$ (Q is the total charge on an electrode) as function of the normalized wave number γR due to the prescribed on two electrodes difference of potentials $2\Phi^*$. The

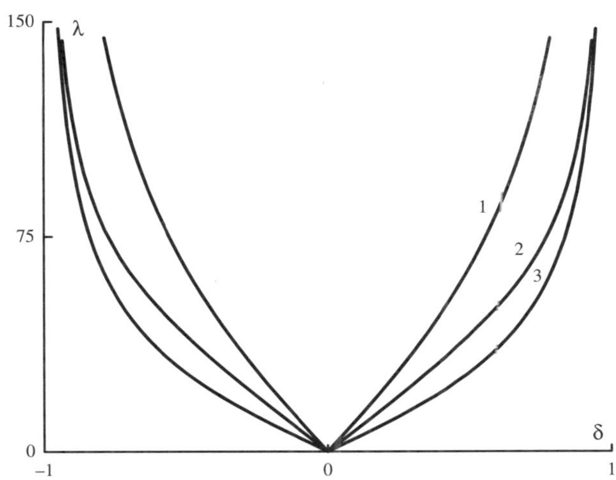

Fig. 7.28. Changes of the intensity of contact forces on a linear4 insert for the various values of normalized wave number γR

parameters determining the position of the electrodes were prescribed as follows: $\beta_1 = 5\pi/14$, $\beta_2 = 9\pi/14$, $\beta_3 = 19\pi/14$, $\beta_4 = 23\pi/14$ (Fig. 7.25) and $\beta_1 = \pi/6$, $\beta_2 = 5\pi/6$, $\beta_3 = 7\pi/6$, $\beta_4 = 11\pi/6$ (Fig. 7.26). For our illustrations we used the following approximate values of the first three natural frequencies of oscillation: $\gamma_{(1)}R = 1.74$, $\gamma_{(2)}R = 3.99$, $\gamma_{(3)}R = 6.32$ (Fig. 7.25); $\gamma_{(1)}R = 1.70$, $\gamma_{(2)}R = 3.36$, $\gamma_{(3)}R = 6.28$ (Fig. 7.26).

Analogous result for a cylinder with an elliptic section having an inclusion, orientated at an angle $\vartheta = \pi/4$ with respect to Ox_1 - axis, are given in Fig. 7.27 (setting $R_1/R_2 = 2$, $p_1/R_1 = 0.3$, $\beta_1 = \pi/6$, $\beta_2 = 5\pi/6$, $\beta_3 = 7\pi/6$, $\beta_4 = 11\pi/6$). In this case the values of the first three natural frequencies of oscillations are given by $\gamma_{(1)}R = 1.14$, $\gamma_{(2)}R = 2.30$, $\gamma_{(3)}R = 3.26$.

Let us investigate the intensity of contact forces on the insert according to the various parameters. In Fig. 7.28 shows the change of $\lambda = (q(\delta)/\Phi^*$ (due to the prescribed difference of potentials $2\Phi^*$ on the two surface electrodes. The calculation was carried out on a circular cylinder with a linear inclusion ($\vartheta = \pi/2$) at $p_1/R = 0.4$. Figure 7.25 depicts the results with respect to β_k $\left(k = \overline{1,4}\right)$. The curves 1, 2 and 3 are constructed for $\gamma R = 1, 3, 5$, respectively.

The changes of λ on the contour of an parabolic inclusion, orientated at an angle $\vartheta = \pi/2$ (Fig. 7.29) and $\vartheta = \pi/4$ (Fig. 7.30), are illustrated by curves 1, 2 and 3 for $p_2/R = 0$, 0.3, 0.5, respectively. The calculation was carried out setting $p_1/R = 0.4$, $\beta_1 = \pi/4$, $\beta_2 = 3\pi/4$, $\beta_3 = 5\pi/4$, $\beta_4 = 7\pi/4$. The graphs in Fig. 7.29 were constructed for $\gamma R = 2$; in Fig. 7.30 were constructed for $\gamma R = 4$.

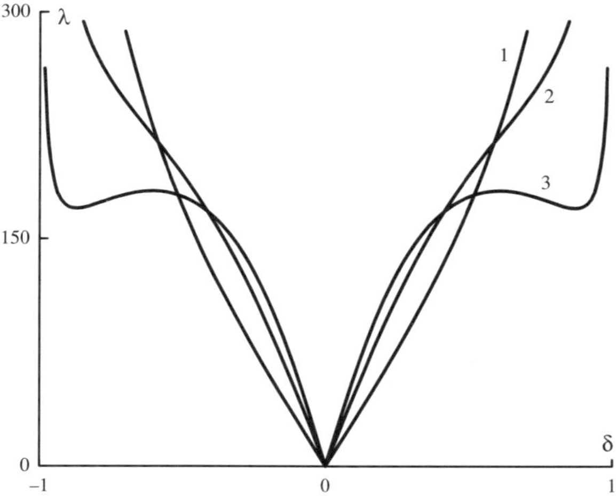

Fig. 7.29. Changes of intensity of the contact forces on a parabolic insert for the various parameters values of curvature ($\theta = \pi/2$)

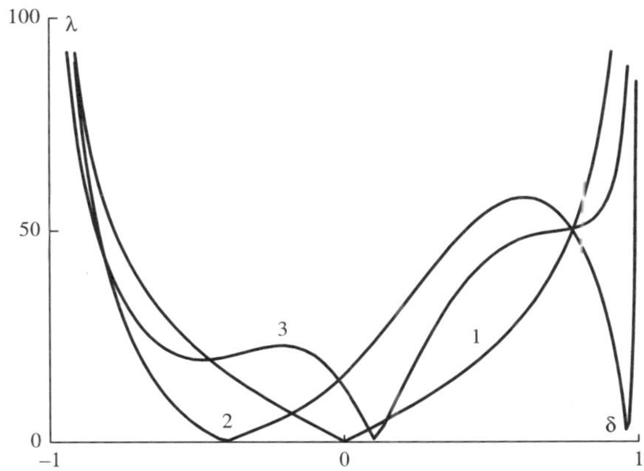

Fig. 7.30. Changes of intensity of the contact forces on a parabolic insert for various parameter values of curvature $\theta = \pi/4$

If a linear inclusion is orientated along x_1-axis, its center and the centers of the two electrodes are lying on x_2 -axis, it follows that $q(\zeta) \equiv 0$ (by the solutions of the integral equations). Therefore, in this case the availability of the inclusion does not change the state of electroelasticity of the cylinder.

Consider now the behavior of the displacement on the boundary of a piecewise-homogeneous cylinder according to the configuration of the inclusion, its orientation and the area of the electrodized coating.

Figures 7.31, 7.32 and 7.33 illustrate the distribution of $\eta = c_{44}^{E} \left| [U_3]/\Phi^* \right|$ on the contour of a circular cylinder with a linear inclusion, orientated along $x_2\,(\vartheta = \pi/2)$-axis (Fig. 7.31) and $x_1\,(\vartheta = 0)$ -axis (Figs. 7.32, 7.33) at $\gamma R = 1$. The values of the parameters, assigning the position of a pair

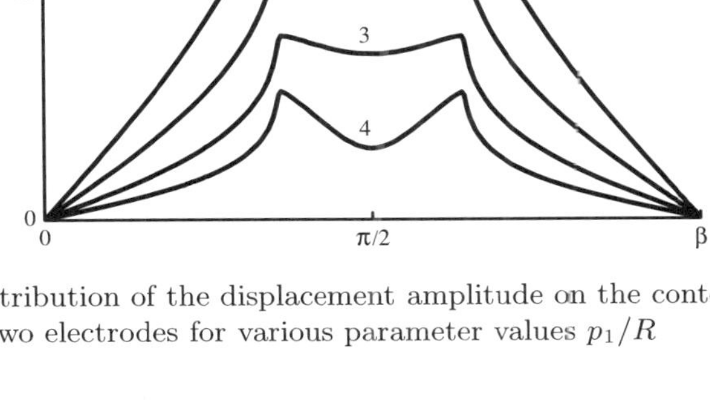

Fig. 7.31. Distribution of the displacement amplitude on the contour of a circular cylinder with two electrodes for various parameter values p_1/R

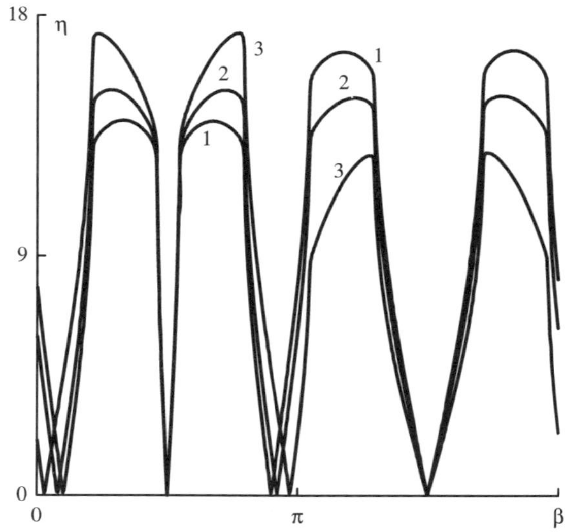

Fig. 7.32. Distribution of the displacement amplitude on the contour of a circular cylinder with four electrodes for various parameter values p_1/R ($\Phi_1^* = -\Phi_2^* = \Phi_3^* = -\Phi_4^* = \Phi$)

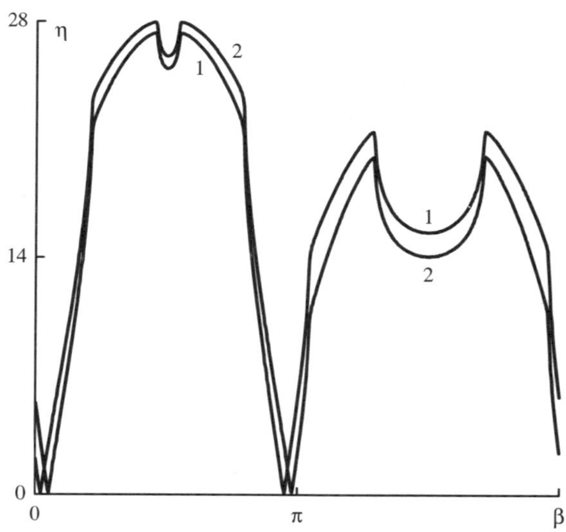

Fig. 7.33. Distribution of the displacement amplitude on the contour of a circular cylinder with four electrodes for various parameter values p_1/R ($\Phi_1^* = \Phi_2^* = -\Phi_3^* = -\Phi_4^* = \Phi*$)

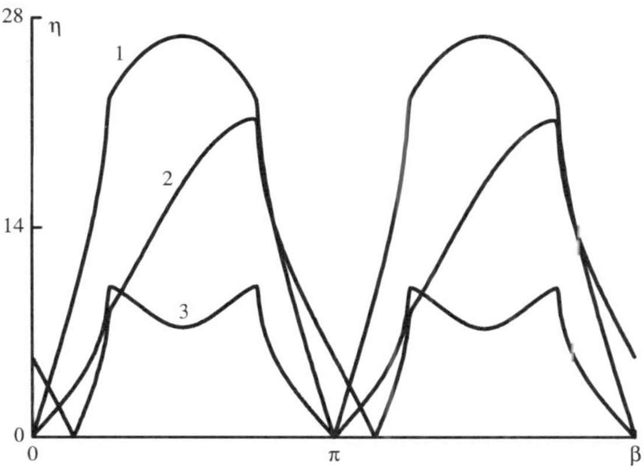

Fig. 7.34. Dependence of the displacement amplitude on the contour of a cylinder with two electrodes on the orientation angle of the insert

electrodes with the difference of potentials $2\Phi^*$, were: $\beta_1 = 5\pi/14, \beta_2 = 9\pi/14$, $\beta_3 = 19\pi/14$, $\beta_4 = 23\pi/14$ (Fig. 7.31). In Fig. 7.31 curves 1–4 are constructed for $p_1/R = 0.2, 0.4, 0.6, 0.8$, respectively. The graphs in Figs. 7.32 and 7.33 were obtained for a cylinder excited by four electrodes ($\beta_1 = 5\pi/24$, $\beta_2 = 11\pi/24$, $\beta_3 = 13\pi/24$, $\beta_4 = 19\pi/24$, $\beta_5 = 25\pi/24$, $\beta_6 = 31\pi/24$, $\beta_7 = 41\pi/24$, $\beta_8 = 47\pi/24$) with the prescribed electric potentials: $\Phi_1^* = -\Phi_2^* = \Phi_3^* = -\Phi_4^* = \Phi^*$ (Fig. 7.32) and $\Phi_1^* = \Phi_2^* = -\Phi_3^* = -\Phi_4^* = \Phi^*$ (Fig. 7.33). Curves 1, 2 and 3 in Fig. 7.32 correspond to $p_1/R = 0.2, 0.4, 0.8$; curves 1 and 2 in Fig. 7.33 to $p_1/R = 0.2$ and 0.8.

The dependence of the displacement on the cylinder contour on the orientation angle of the linear inclusion is illustrated in Fig. 7.34. Curves 1, 2 and 3

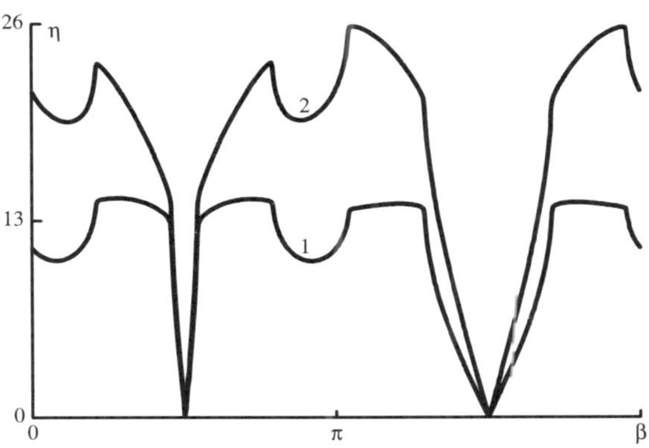

Fig. 7.35. Dependence of the displacement amplitude on the contour of a cylinder with four electrodes on the orientation angle of the insert ($\theta = 0$ and $\pi/4$)

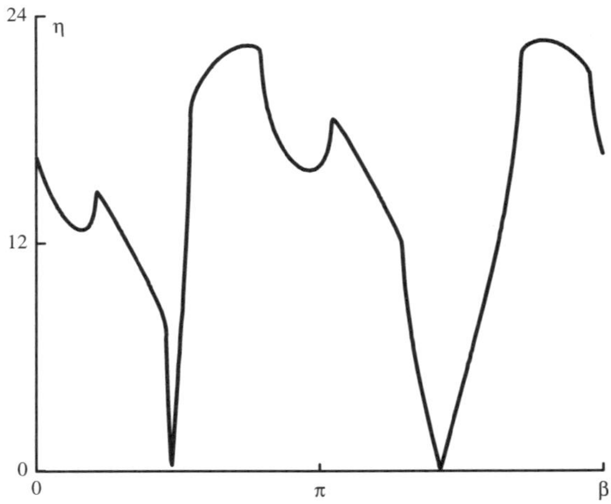

Fig. 7.36. Dependence of the displacement amplitude on the contour of a cylinder with four electrodes on the orientation angle of the insert $\theta = \pi/4$

represent the quantity η for $\vartheta = 0, \pi/4, \pi/2$ for and $p_1/R = 0.6$, respectively. The values of the parameters, determining the position of the two electrodes were set as follows: $\beta_1 = \pi/4,\ \beta_2 = 3\pi/4,\ \beta_3 = 5\pi/4,\ \beta_4 = 7\pi/4$.

Figures 7.35 and 7.36 depict the behavior of η on the contour of a circular cylinder, excited by four electrodes, for $\gamma R = 1,\ p_2/R = 0$ and $p_1/R = 0.4$. In Fig. 7.35 curves 1 and 2 correspond to $\vartheta = 0$ and $\vartheta = \pi/2$. The graphs in Fig. 7.36 are constructed at $\vartheta = \pi/4$. Figure 7.32 illustrate the results with

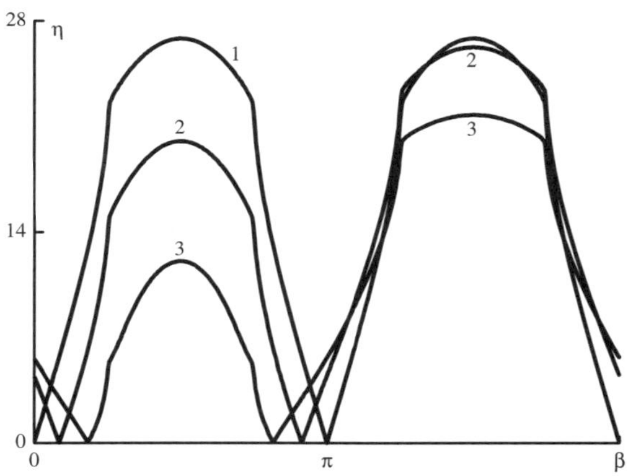

Fig. 7.37. Distribution of the displacement amplitude on the contour of a cylinder with two electrodes at various curvature of the parabolic inclusion

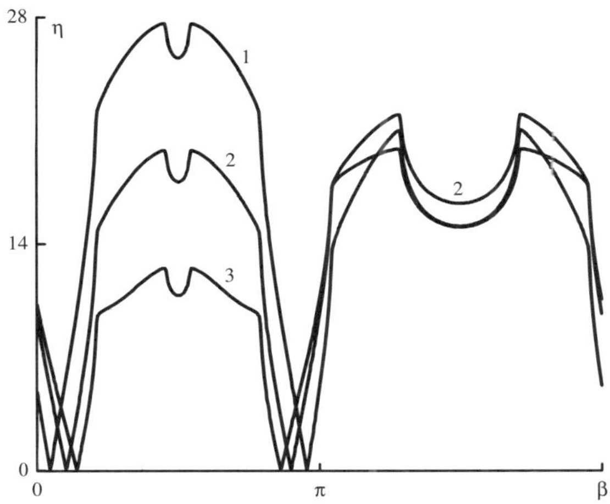

Fig. 7.38. Distribution of the displacement amplitude on the contour of a cylinder with four electrodes at various curvatures of the parabolic inclusion

respect to β_k $\left(k = \overline{1,8}\right)$. In our calculations we assumed that $\Phi_1^* = -\Phi_2^* = \Phi_3^* = -\Phi_4^* = \Phi^*$.

Figures 7.37 and 7.38 illustrate the graphs of quantity η on circular contour C for various curves of the parabolic inclusion. It was assumed that $\gamma R = 1$, $\vartheta = 0$ $p_1/R = 0.6$ (Fig. 7.37) and $p_1/R = 0.4$ (Fig. 7.38). Figure 7.37 corresponds to the case of the loading of a cylinder with two electrodes ($\beta_1 = \pi/4, \beta_2 = 3\pi/4, \beta_3 = 5\pi/4, \beta_4 = 7\pi/4$), and curves 1, 2 and 3 are constructed for $p_2/R = 0$, 0.3, 0.6, respectively. Figure 7.38 corresponds

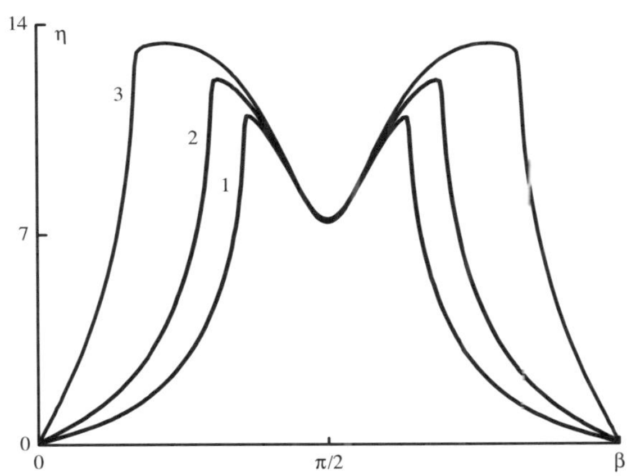

Fig. 7.39. Distribution of the displacement amplitude on the contour of a cylinder according to the area of the electrode coating

to the case of the loading of a cylinder with four electrodes (the parameters determining the position of the electrodes, correspond to Fig. 7.36). Curves 1, 2 and 3 were constructed for $p_2/R = 0,\ 0.3$ and 0.5, respectively.

Figure 7.39 presents the results characterizing the distribution of displacements on the contour of a cylinder of elliptic section with linear inclusion, orientated along x_2 -axis, with respect to the area of the electrodized coating.

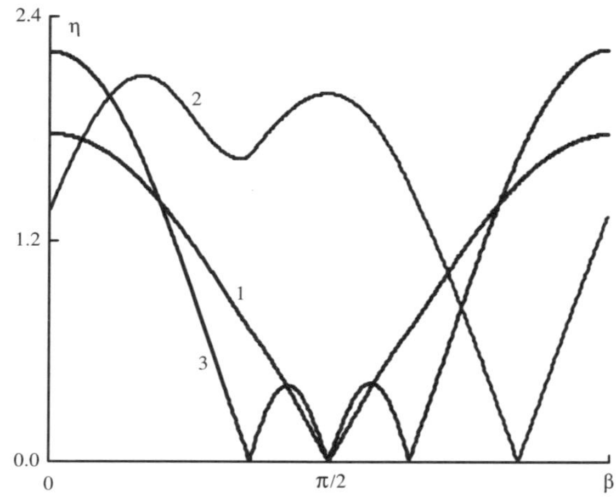

Fig. 7.40. Changes of the amplitude of the relative tangential stress on the contour of a cylinder with two electrodes according to the orientation angle of the linear inclusion

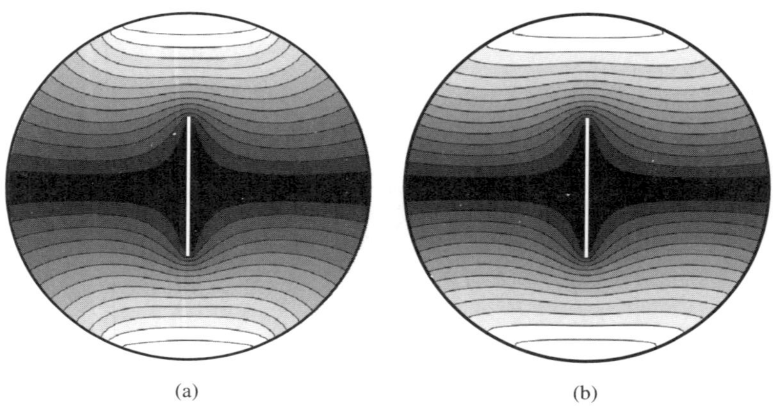

Fig. 7.41a and b. The lines of the equal modulus of the displacement amplitude in the area of a cylinder in the vicinity of the first (a) and second (b) eigenfrequencies of oscillations a) $\beta_1 = 5\pi/14,\ \beta_2 = 9\pi/14,\ \beta_3 = 19\pi/14,\ \beta_4 = 25\pi/14$ b) $\beta_1 = \pi/6,\ \beta_2 = 5\pi/6,\ \beta_3 = 7\pi/6,\ \beta_4 = 11\pi/6$

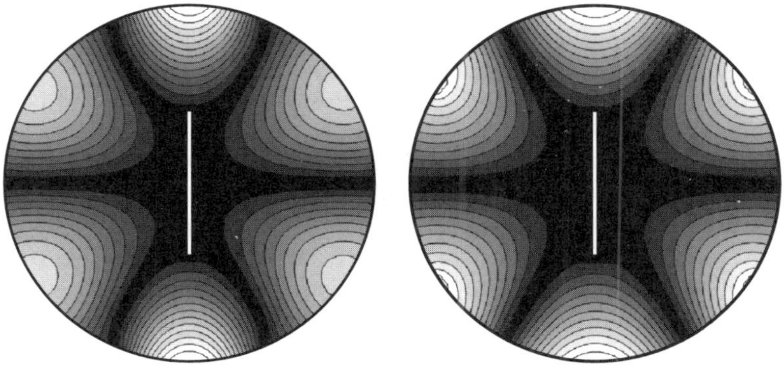

(a) (b)

Fig. 7.42a and b. The lines of the equal modulus of the displacement amplitude in the area of a cylinder in the vicinity of the first (a) and second (b) eigenfrequency of oscillations a) $\beta_1 = 5\pi/14,\ \beta_2 = 9\pi/14,\ \beta_3 = 19\pi/14,\ \beta_4 = 25\pi/14$. b) ($\beta_1 = \pi/6,\ \beta_2 = 5\pi/6,\ \beta_3 = 7\pi/6,\ \beta_4 = 11\pi/6$

It was assumed that $\gamma R = 1$, $R_1/R_2 = 2$ and $p_1/R_2 = 0.6$. Curves 1, 2 and 3 are constructed for: $\beta_1 = 5\pi/14,\ \beta_2 = 9\pi/14,\ \beta_3 = 19\pi/14,\ \beta_4 = 23\pi/14$; $\beta_1 = 3\pi/10,\ \beta_2 = 7\pi/10,\ \beta_3 = 13\pi/10,\ \beta_4 = 17\pi/10$ and $\beta_1 = \pi/6,\ \beta_2 = 5\pi/6,\ \beta_3 = 7\pi/6,\ \beta_4 = 11\pi/6$, respectively.

Figure 7.40 illustrates the distribution of relative shear stress $\tau = |T_s R/(e_{15}\Phi^*)|$ on the contour of a circular cylinder with two electrodes with respect to the orientation of the linear inclusionfor $p_1/R = 0.6$, $\gamma R = 2$ and $\beta_1 = 5\pi/14,\ \beta_2 = 9\pi/14,\ \beta_3 = 19\pi/14,\ \beta_4 = 23\pi/14$. Curves 1, 2 and 3 correspond to $\vartheta = 0,\ \pi/4$ and $\pi/2$, respectively.

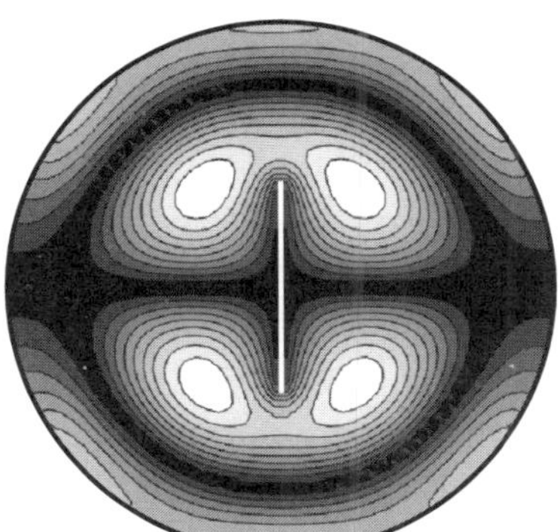

(c)

Fig. 7.43. The lines of the equal modulus of the displacement amplitude in the area of a cylinder in the vicinity of the third eigenfrequency of oscillations ($\beta_1 = \pi/6, \beta_2 = 5\pi/6, \beta_3 = 7\pi/6, \beta_4 = 11\pi/6$)

Figures 7.41–7.43 show the lines of the equal modulus of the displacement amplitude in the area of a piecewise-homogeneous cylinder in the vicinity of the first three natural frequencies of oscillations. Figures 7.41a and ba, 7.42a and ba illustrate results for $\beta_1 = 5\pi/14$, $\beta_2 = 9\pi/14$, $\beta_3 = 19\pi/14$, $\beta_4 = 23\pi/14$; Figs. 7.41a and bb, 7.42a and bb, 7.43 show results for $\beta_1 = \pi/6$, $\beta_2 = 5\pi/6$, $\beta_3 = 7\pi/6$, $\beta_4 = 11\pi/6$.

From the results it follows that according to the excitation frequency, the configuration of the inclusion, sizes and position of the electrodized portions, and also the prescribed differences of electric potentials on the electrodes we observe the redistribution of the displacements on the cylinder boundary. As it is seen in Figs. 7.41a and b–7.43 the rigid inclusion plays the role of a "fixing agent", preventing the propagation of the oscillations.

8

Electroacoustic Waves in Piezoceramic Media with Defects (Plane Deformation)

In the current chapter we study the interaction of waves of various types with defects of type of tunnel cracks and openings in a piezoelectric medium of the hexagonal class of symmetry with the help of the method of boundary integral equations. It is assumed that a crack in non-deformed state is a mathematical cut, and the edges of the crack do not contact during the harmonic with time effect.

8.1 Waves in a Homogeneous Medium

For the complete description of elastic and electromagnetic fields in a piezoelectric crystal it is necessary to use the equations of motion in the continuous medium

$$\partial_j \sigma_{ij} = \rho \frac{\partial^2 u_i}{\partial t^2} \tag{8.1}$$

and Maxwell-Lorentz equations without taking into account the external current and free charges

$$\begin{aligned} rot\overrightarrow{E} &= -\frac{\partial \overrightarrow{B}}{\partial t}, \quad div\overrightarrow{D} = 0 \\ rot\overrightarrow{H} &= \frac{\partial \overrightarrow{B}}{\partial t}, \quad div\overrightarrow{B} = 0 \end{aligned} \tag{8.2}$$

where the components σ_{ij}, D_i are connected with the deformations ε_{ij} and electric field stress E_i by the equations of medium state, $\overrightarrow{B} = \mu\overrightarrow{H}$.

As in (8.1) where stresses σ_{ij} are coupled with the electric field E_i, the system of (8.1), (8.2) describe coupled elastic electromagnetic waves in a piezoelectric medium. The coupling of elastic and electromagnetic fields is determined by the piezoelectric constants of the material. The solution of system (8.1), (8.2) are elastic waves which propagate with a velocity U and the electromagnetic waves propagating with a velocity $c_\alpha \cong 10^5 U$. For the elastic waves that propagate in an electric field we may neglect the magnetic field moving for velocities $U << c_\alpha$, and assume

$$rot\overrightarrow{E} \cong 0, \quad \overrightarrow{E} = -grad\varphi. \tag{8.3}$$

Here the first set of (8.1), (8.2) correspond to the equations of acoustic piezoelectric crystals in a quasi-static approximation

$$\begin{aligned} &\partial_j \sigma_{ij} = \rho \frac{\partial^2 u_i}{\partial t^2}, \\ &div\overrightarrow{D} = 0, \quad rot\overrightarrow{E} = 0, \quad \overrightarrow{E} = -grad\phi \end{aligned} \tag{8.4}$$

while the second set

$$rot\overrightarrow{H} = \frac{\partial \overrightarrow{D}}{\partial t}, \quad div\overrightarrow{H} = 0 \tag{8.5}$$

permits us to find the magnetic field of an electroacoustic wave that is determined from the solution of system (8.4).

Under this reprsesentation, a plane monocrhomatic wave may be represented in the form

$$u\left(\overrightarrow{x}, t\right) = u_0 \cos\omega \left(t - \frac{\overrightarrow{n} \cdot \overrightarrow{x}}{U}\right) = u_0 \cos\left(\omega t - \overrightarrow{k} \cdot \overrightarrow{x}\right).$$

u_0 is the wave amplitude; $\overrightarrow{k} = \omega\overrightarrow{n}/U = k\overrightarrow{n}$ is the wave vector; ω is the circular frequency; $k = \left|\overrightarrow{k}\right| = \omega/U = 2\pi/\lambda$ is the wave number; λ is the wave length, i.e. the distance, at which the disturbance is propagated for one period, $T = 2\pi/\omega$.

For a plane monochromatic (sinusoidal) wave the surfaces $\omega t - \overrightarrow{k}\overrightarrow{x} = const$ are planes perpendicular to wave vector $\overrightarrow{k}$. These planes move in the space with a velocity $U = \omega/k$, which is called the phase velocity of the wave. Accordingly, the vector $\overrightarrow{U} = U\overrightarrow{n}$ is called the *vector of phase velocity.* The plane monochromic wave is represented in the complex form,

$$U(\overrightarrow{x}, t) = U_0 e^{i\left(\overrightarrow{k} \cdot \overrightarrow{x} - \omega t\right)}, \qquad u(\overrightarrow{x}, t) = \mathrm{Re}U(\overrightarrow{x}, t), \qquad U_0 = U_0^{(1)} + iU_0^{(2)}.$$

In general case phase velocity U depends on quantity k and on the direction of propagation of plane wave $\overrightarrow{n}$. The medium, for which $U = U(k)$, is called isotropic dispersive; then, $f(\omega, \overrightarrow{k}) = 0$ is called dispersion equation and expresses the law of dispersion. If for all k and $\overrightarrow{n}$ the velocity U is constant, there is no dispersion, and in such a medium the signal having a certain form at $t = 0$ is propagated with velocity U without changing its shape (i.e. it is invariant).

It is known that in elastic, isotropic and infinite medium two types of main waves may appear: a) longitudinal waves, where the displacement of the particles of the medium is parallel to the wave propagation, b) transversal, shear waves, where the displacement of the particles is perpendicular to the wave vector. If the unrestrained medium is anisotropic, there will be propagation of three waves in any direction, and in the general case no one of them is pure longitudinal or pure transverse. In that case they are two *quasitransverse* waves and a *quasilongitudinal* wave (see Fig. 8.1). Polarization of these waves, i.e. the direction of the displacement the medium particles are always perpendicular.

Now consider the solution of (8.4), (8.5) for an unbcunded piezoelectric medium in the form of plane monochromatic waves, propagating in the direction of $\overrightarrow{n} = (n_1, n_2, n_3)$ (the vector $\overrightarrow{n}$ is perpendicular to the wave front). Substituting the expressions for the potential $\phi(\overrightarrow{x}, t)$ and the vector of the elastic displacement $u_j(\overrightarrow{x}, t)$

$$\phi(\overrightarrow{x}, t) = \Phi e^{i(\overrightarrow{n} \cdot \overrightarrow{x}/V - \omega t)}, \; u_j(\overrightarrow{x}, t) = U_j e^{i(\overrightarrow{n} \cdot \overrightarrow{x}/V - \omega t)} \quad (j = 1, 2, 3), \tag{8.6}$$

in (8.4), (8.5) and in the equations of the state of a piezoelectric medium we get a system of linear homogeneous equations with respect to the amplitude values of potential Φ and the elastic displacement U_j

$$\begin{aligned} &(c^E_{kips} n_l n_p - \rho U^2 \delta_{ks}) U_s + e_{skj} n_j n_s \Phi = 0, \\ &e_{skj} n_j n_s U_k - \ni^{\varepsilon}_{jk} n_j n_k \Phi = 0 \end{aligned} \tag{8.7}$$

Introducing the notation

$$\Gamma_{ks} = c^E_{kips} n_l n_p, \qquad \gamma_k = e_{skj} n_j n_s, \quad \ni = \ni^{\varepsilon}_{jk} n_j n_k,$$

the system of (8.7) reads

$$\left(\Gamma_{ks} - \rho U^2 \delta_{ks}\right) U_s + \gamma_k \Phi = 0, \qquad \gamma_k U_k - \ni \Phi = 0. \tag{8.8}$$

Eliminating the amplitude of the potential Φ from (8.8), we obtain

$$\left[\left(\Gamma_{ks} + \frac{\gamma_k \gamma_s}{\ni}\right) - \rho U^2 \delta_{ks}\right] U_s = 0 \qquad (k = 1, 2, 3). \tag{8.9}$$

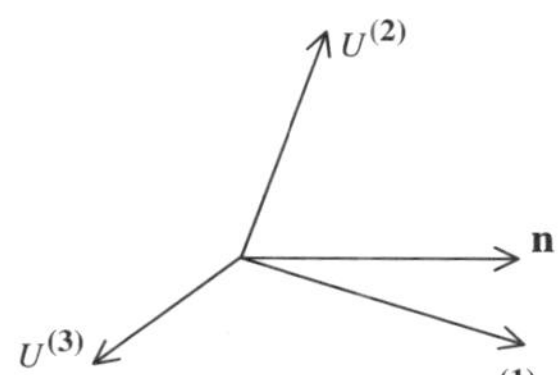

Fig. 8.1. The scheme illustrating the types of waves in an unbounded anisotropic medium

The existence condition of a non-trivial solution of system (8.9) is that its determinant has to be zero. Taking into the account the symmetry of tensor Γ_{ks} gets the form

$$\begin{vmatrix} \Gamma_{11}^{*} - \rho U^2 & \Gamma_{12}^{*} & \Gamma_{13}^{*} \\ \Gamma_{12}^{*} & \Gamma_{22}^{*} - \rho U^2 & \Gamma_{23}^{*} \\ \Gamma_{13}^{*} & \Gamma_{23}^{*} & \Gamma_{33}^{*} - \rho U^2 \end{vmatrix} = 0, \tag{8.10}$$

where $\Gamma_{ks}^{*} = \Gamma_{ks} + \gamma_k \gamma_s / \ni \;\; (k, s = 1, 2, 3)$.

Equality (8.10) allows us to determine the velocity of the plane waves, propagating along the direction of $\vec{n}$. This results to an equation of third order with respect to ρU^2. In the general case, for the given direction in the piezoelectric crystal three velocities of propagation ρU_j^2 $(j = 1, 2, 3)$ appear, which are determined as the positive roots of (8.10). Then, substituting each of U_j into (8.9), we find $U_s^{(j)}$ and $\Phi^{(j)} = \gamma_k U_k^{(j)} / \ni$, and due to the homogeneity of (8.9) we can find $U_s^{(j)}$, $\Phi^{(j)}$, multiplied by an arbitrary constant, which may be found with the help of the additional condition of normalization.

Thus, the problem of the determination of the velocity and the polarization of plane waves propagating along the arbitrary direction $\vec{n}$ of the piezocrystal is reduced to the determination of the characteristic values ρU_j^2 and the characteristic vectors $U_s^{(j)}$ of the tensor Γ_{ks}^{*}. Here, Iit should be mentioned that due to the symmetry of tensor Γ_{ks}^{*} its characteristic value are real quantities, and in addition because the quadrature form $U_k \Gamma_{ks} U_s$ is positive [12], the characteristic values of the tensor Γ_{ks}^{*} are also positive.

For any of three plane waves, propagating in a piezocrystal medium in the direction of $\vec{n}$, the displacement vector $\vec{\boldsymbol{U}}^{(j)}$ in the general case will not be parallel or perpendicular to the direction of propagation. However, the wave with vector $\vec{\boldsymbol{U}}^{(1)}$, forming the smallest angle in direction $\vec{n}$, is usually called quasilongitudinal and (Fig. 8.1) the two other waves with displacement vectors $\vec{U}^{(2)}$ and $\vec{U}^{(3)}$ are called quasi-transversal. Vectors $\vec{\boldsymbol{U}}^{(2)}$, $\vec{\boldsymbol{U}}^{(3)}$ $\vec{\boldsymbol{U}}^{(j)}$ $(j = 1, 2, 3)$ are perpendicular to each other.

In order to study the propagation of the waves along the directions connected with the elements of symmetry of a piezocrystal, the expressions for the components of tensor Γ_{ks}^{*} in the case of the piezoelectric medium of hexagonal class $6mm$ [68] are given by

$$\begin{aligned}
&\Gamma_{11} = c_{11}^E n_1^2 + c_{66}^E n_2^2 + c_{44}^E n_3^2, \qquad \Gamma_{12} = \left(c_{12}^E + c_{66}^E\right) n_1 n_2, \\
&\Gamma_{13} = \left(c_{13}^E + c_{44}^E\right) n_1 n_2, \qquad \Gamma_{22} = c_{66}^E n_1^2 + c_{11}^E n_2^2 + c_{44}^E n_3^2, \\
&\Gamma_{23} = \left(c_{13}^E + c_{44}^E\right) n_2 n_3, \qquad \Gamma_{33} = c_{44}^E \left(n_1^2 + n_2^2\right) + c_{33}^E n_3^2, \\
&\Gamma_{21} = \Gamma_{12}, \qquad \Gamma_{31} = \Gamma_{13}, \qquad \Gamma_{32} = \Gamma_{23}, \\
&\Gamma_{11}^{*} = \Gamma_{11} + \frac{\gamma_1^2}{\ni}, \qquad \Gamma_{12}^{*} = \Gamma_{12} + \frac{\gamma_1 \gamma_2}{\ni}, \qquad \Gamma_{13}^{*} = \Gamma_{13} + \frac{\gamma_1 \gamma_3}{\ni},
\end{aligned} \tag{8.11}$$

$$\Gamma_{22}^{*}=\Gamma_{22}+\frac{\gamma_2^2}{\ni},\qquad \Gamma_{23}^{*}=\Gamma_{23}+\frac{\gamma_2\gamma_3}{\ni},\qquad \Gamma_{33}^{*}=\Gamma_{33}+\frac{\gamma_3^2}{\ni},$$
$$\gamma_1=(e_{15}+e_{31})\,n_1n_2,\qquad \gamma_2=(e_{15}+e_{31})\,n_2n_3,$$
$$\gamma_3=e_{15}\left(n_1^2+n_2^2\right)+e_{33}n_3^2,$$

where $\ni=\ni_{11}^{\varepsilon}\left(n_1^2+n_2^2\right)+\ni_{33}^{\varepsilon}\,n_3^2$

Using (8.10) and expressions (8.11), we can determine the velocity of longitudinal and transversal waves, propagating along axes x_1,x_2,x_3 of the crystal of hexagonal class $6mm$ (axis x_3 coincides with the axis of symmetry of the sixth order). We examine the following cases:

1) The plane wave is propagating along x_1 -axis. Then

$$n_1=1,\qquad n_2=n_3=0,\qquad \Gamma_{11}^{*}=c_{11}^{E},\qquad \Gamma_{12}^{*}=\Gamma_{13}^{*}=\Gamma_{23}^{*}=0,$$
$$\Gamma_{22}^{*}=c_{66}^{E}=0.5\left(c_{11}^{E}-c_{12}^{E}\right),\qquad \Gamma_{33}^{*}=c_{44}^{E}+\frac{e_{15}^2}{\ni_{11}^{\varepsilon}}.$$

From (8.10) we find

$$\rho U_{11}^2=c_{11}^{E},\qquad \rho U_{12}^2=c_{66}^{E},\qquad \rho U_{13}^2=c_{44}^{E}+\frac{e_{15}^2}{\ni_{11}^{\varepsilon}},\tag{8.12}$$

where U_{11} is the velocity of the longitudinal wave propagating along x_1-axis; U_{12}, U_{13} are the velocities of the transversal waves. From now on we will make use of two indices, where the first index is for the direction of the wave propagation, and the second one for the direction of the amplitude vector (polarization).

2) The plane wave propagates along x_3 -axis. From (8.11) it follows

$$\Gamma_{11}^{*}=c_{44}^{E},\qquad \Gamma_{12}^{*}=\Gamma_{13}^{*}=\Gamma_{23}^{*}=0,$$
$$\Gamma_{22}^{*}=c_{44}^{E},\qquad \Gamma_{33}^{*}=c_{33}^{E}+\frac{e_{33}^2}{\ni_{33}^{\varepsilon}},$$

and from (8.10) we find the velocity U_{33} of the longitudinal velocity and velocities U_{31}, U_{32} of two transversal waves

$$\rho U_{33}^2=c_{33}^{E}+\frac{e_{33}^2}{\ni_{33}},\qquad \rho U_{32}^2=c_{44}^{E},\qquad \rho U_{31}^{2}=c_{44}^{E}.\tag{8.13}$$

3) The plane wave propagates along x_2-axis. Then

$$n_2=1,\qquad n_1=n_3=0,\qquad \Gamma_{11}^{*}=c_{66}^{E},\qquad \Gamma_{12}^{*}=\Gamma_{13}^{*}=\Gamma_{23}^{*}=0,$$
$$\Gamma_{22}^{*}=c_{11}^{E},\qquad \Gamma_{33}^{*}=c_{44}^{E}+\frac{e_{15}^2}{\ni_{11}^{\varepsilon}}$$

and the velocity U_{22} of the longitudinal waves, and also the velocities U_{21}, U_{23} of two transversal waves are determined by

$$\rho U_{22}^2 = c_{11}^E, \qquad \rho U_{21}^2 = c_{66}^E, \qquad \rho U_{23}^2 = c_{44}^E + \frac{e_{15}^2}{\ni_{11}^\varepsilon}. \tag{8.14}$$

4) The plane wave propagating in $x_1 = 0$ plane in a direction which forms an angle 45° with x_2-axis. In this case

$$n_1 = 0, \qquad n_2 = n_3 = \frac{1}{\sqrt{2}}, \qquad \Gamma_{11}^* = 0.5\left(c_{66}^E + c_{44}^E\right),$$

$$\Gamma_{12}^* = \Gamma_{13}^* = 0, \qquad \Gamma_{22}^* = 0.5\left(c_{13}^E + c_{44}^E\right) + \frac{0.5\left(e_{15} + e_{31}\right)^2}{\ni_{11}^\varepsilon + \ni_{33}^\varepsilon},$$

$$\Gamma_{23}^* = 0.5\left(c_{13}^E + c_{44}^E\right) + \frac{0.5\left(e_{15} + e_{31}\right)\left(e_{15} + e_{33}\right)}{\ni_{11}^\varepsilon + \ni_{33}^\varepsilon},$$

$$\Gamma_{33}^* = 0.5\left(c_{33}^E + c_{44}^E\right) + \frac{0.5\left(e_{15} + e_{33}\right)^2}{\ni_{11}^\varepsilon + \ni_{33}^\varepsilon}.$$

Solving (8.10) we obtain

$$\begin{aligned} \rho U_{(1)}^2 &= 0.5\left(c_{66}^E + c_{44}^E\right), \\ \rho U_{(2)}^2 &= 0.5\left(\Gamma_{33}^* + \Gamma_{22}^*\right) + \sqrt{0.25\left(\Gamma_{33}^* - \Gamma_{22}^*\right)^2 + \left(\Gamma_{23}^*\right)^2}, \\ \rho U_{(3)}^2 &= 0.5\left(\Gamma_{33}^* + \Gamma_{22}^*\right) - \sqrt{0.25\left(\Gamma_{33}^* - \Gamma_{22}^*\right)^2 + \left(\Gamma_{23}^*\right)^2}. \end{aligned} \tag{8.15}$$

8.2 General Representations of Coupled Fields in a Medium of the Hexagonal Class of Symmetry

Let a piezocrystal of hexagonal class of symmetry $6mm$, in the Cartesian system of coordinates x_1, x_2, x_3, be subjected to mechanical and electric loading harmonic in time, independent of the x_3-coordinate. If we suppose that the x_3-axis is the axis of symmetry of the sixth order, from the material equations [25] at $\varepsilon_{33} = 0$ given by

$$\begin{aligned} \sigma_{11} &= c_{11}^D \varepsilon_{11} + c_{12}^D \varepsilon_{22} - h_{31} D_3, \\ \sigma_{22} &= c_{12}^D \varepsilon_{11} + c_{11}^D \varepsilon_{22} - h_{31} D_3, \\ \sigma_{33} &= c_{13}^D \left(\varepsilon_{11} + \varepsilon_{22}\right) - h_{33} D_3, \\ \sigma_{12} &= \left(c_{11}^D + c_{12}^D\right) \varepsilon_{12}, \\ E_3 &= -h_{31}\left(\varepsilon_{11} + \varepsilon_{22}\right) + \beta_{33}^\varepsilon D_3, \\ \sigma_{13} &= 2c_{44}^D \varepsilon_{13} - h_{15} D_1, \qquad \sigma_{23} = 2c_{44}^D \varepsilon_{23} - h_{15} D_2, \\ E_1 &= -2h_{15}\varepsilon_{13} + \beta_{11}^\varepsilon D_1, \qquad E_2 = -2h_{15}\varepsilon_{23} + \beta_{11}^\varepsilon D_2, \end{aligned} \tag{8.16}$$

and also from (8.1) and (8.2) it follows, that the two-dimensional state of the piezocrystal can be divided into plane and antiplane deformations. In the first case we have the following components of the electromagnetic field

$$\begin{aligned}\sigma_{ij} &= \operatorname{Re}\{S_{ij}(x_1,x_2)\exp(-i\omega t)\},\\ E_3 &= \operatorname{Re}\{E_3^*(x_1,x_2)\exp(-i\omega t)\},\\ D_3 &= \operatorname{Re}\{D_3^*(x_1,x_2)\exp(-i\omega t)\},\\ H_i &= \operatorname{Re}\{H_i^*(x_1,x_2)\exp(-i\omega t)\} \qquad (i,j=1,2).\end{aligned} \tag{8.17}$$

In this case Maxwell's (8.2) read

$$\begin{aligned}\partial_1 E_3 &= \mu\frac{\partial H_2}{\partial t}, \qquad \partial_2 E_3 = -\mu\frac{\partial H_1}{\partial t},\\ \partial_1 H_2 - \partial_2 H_1 &= \frac{\partial D_3}{\partial t}, \qquad \partial_1 H_1 + \partial_2 H_2 = 0.\end{aligned} \tag{8.18}$$

Our aim is to construct general representations of the solutions of the system of (8.1), (8.16), (8.18). For this purpose we should introduce vector potential $\overrightarrow{A} = (0,0,A)$ by setting $\overrightarrow{H} = rot\overrightarrow{A}$. In that case, the last equation in (8.18) is fulfilled automatically, and taking into account (8.16) the other equations take the form of

$$\nabla^2 A = i\omega D_3^*, \qquad i\omega\mu A + h_{31}div\overrightarrow{U} - \beta_{33}^{\varepsilon} D_3^* = C. \tag{8.19}$$

Here $\overrightarrow{U}$ is the peak value of the displacement vector; C is a constant, which is assumed to be equal to zero according to the conditions of radiation.

Taking into account (8.16) and Cauchy relations the equations of motion (8.1) are given by

$$\begin{aligned}&\nabla^2\overrightarrow{U} + \sigma_1 graddiv\overrightarrow{U} + \gamma_2^2\overrightarrow{U} - \sigma_2 gradD_3^* = 0,\\ &\gamma_2^2 = \frac{2\rho\omega^2}{c_{11}^D - c_{12}^D}, \qquad \sigma_1 = \frac{c_{11}^D + c_{12}^D}{c_{11}^D - c_{12}^D}, \qquad \sigma_2 = \frac{2h_{31}}{c_{11}^D - c_{12}^D}.\end{aligned} \tag{8.20}$$

Eliminating D_3^*, the system of (8.19), (8.20) gives rise to two coupled equations with respect to the vector of mechanical displacement and vector potential $\overrightarrow{A} = (0,0,A)$

$$\begin{aligned}&\nabla^2\overrightarrow{U} + \sigma\cdot graddiv\overrightarrow{U} + \gamma_2^2\overrightarrow{U} + \chi\cdot gradA = 0\\ &\nabla^2 A + k^2 A = \alpha div\overrightarrow{U}.\end{aligned} \tag{8.21}$$

Here

$$\sigma = \frac{\gamma_2^2}{\gamma_1^2} - 1, \qquad \gamma_m = \frac{\omega}{c_m}, \qquad k = \frac{\omega}{c_\alpha} \qquad (m = 1, 2),$$

$$c_1 = \sqrt{\frac{c_{11}^D - \kappa_0^2}{\rho}}, \qquad c_2 = \sqrt{\frac{c_{11}^D - c_{12}^D}{2\rho}}, \qquad c_\alpha = \sqrt{\frac{\beta_{33}^\varepsilon}{\mu}},$$

$$\alpha = \frac{i\omega h_{31}}{\beta_{33}^\varepsilon}, \qquad \kappa_0^2 = \frac{h_{31}^2}{\beta_{33}^\varepsilon}, \qquad \chi = \frac{2k^2 h_{31}}{i\omega \left(c_{11}^D - c_{12}^D\right)}.$$

The quantity κ_0 characterizes the piezoelectric effect. The displacement vector reads [144]

$$\overrightarrow{U} = grad\Phi + rot\left(\overrightarrow{k}\,\Psi\right), \tag{8.22}$$

where Φ and Ψ are scalar functions to be determined; $\overrightarrow{k}$ is the unit vector. Substituting (8.22) into (8.21) we come to a system of three differential equations

$$\begin{aligned} &(1+\sigma)\,\nabla^2\Phi + \gamma_2^2\Phi + \chi A = 0, \qquad \nabla^2\Psi + \gamma_2^2\Psi = 0, \\ &\nabla^2 A + k^2 A = \alpha\nabla^2\Phi. \end{aligned} \tag{8.23}$$

Further, eliminating A from the first equation of (8.23) and substituting the outcome into the last equation we find

$$\begin{aligned} &\left(\nabla^2 + \beta_1^2\right)\cdot\left(\nabla^2 + \beta_2^2\right)\Phi = 0, \\ &\beta_1 = 0.5\left(\sqrt{(\gamma_1 + k)^2 + \delta_0^2} + \sqrt{(\gamma_1 - k)^2 + \delta_0^2}\right), \\ &\beta_2 = 0.5\left(\sqrt{(\gamma_1 + k)^2 + \delta_0^2} - \sqrt{(\gamma_1 - k)^2 + \delta_0^2}\right), \\ &\delta_0^2 = \frac{\kappa_0^2 k^2}{c_{11}^D - \kappa_0^2}. \end{aligned} \tag{8.24}$$

In the class of continuous functions the general solution of (8.24) may be written in the form

$$\Phi = \Phi_1 + \Phi_2, \tag{8.25}$$

The functions Φ_m $(m = 1, 2)$ represent the arbitrary solutions of Helmhöltz equations $\left(\nabla^2 + \beta_m^2\right)\Phi_m = 0$. According to this the integration of the last (8.23) allows us to determine function A

$$A = \alpha \sum_{m=1}^{2} \frac{\beta_m^2 \Phi_m}{\beta_m^2 - k^2}. \tag{8.26}$$

The other mechanical and electric quantities due to (8.16), (8.18) and (8.22) are expressed by three functions Ψ and Φ_m $(m = 1, 2)$ in the following way [146]

$$S_{11} + S_{22} = -\sum_{m=1}^{2} \beta_m^2 \left(c_{11}^D + c_{12}^D + \frac{2\kappa_0^2 \beta_m^2}{k^2 - \beta_m^2} \right) \Phi_m, \tag{8.27}$$

$$S_{22} - S_{11} + 2iS_{12} = -4\left(c_{11}^D - c_{12}^D\right) \frac{\partial^2}{\partial z^2} (\Phi + i\Psi),$$

$$S_{22} - S_{11} - 2iS_{12} = -4\left(c_{11}^D - c_{12}^D\right) \frac{\partial^2}{\partial \bar{z}^2} (\Phi - i\Psi),$$

$$E_3^* = h_{31} k^2 \sum_{m=1}^{2} \frac{\beta_m^2 \Phi_m}{k^2 - \beta_m^2}, \qquad D_3^* = \frac{h_{31}}{\beta_{33}^\varepsilon} \sum_{m=1}^{2} \frac{\beta_m^4 \Phi_m}{k^2 - \beta_m^2},$$

$$\vec{H} = rot \vec{A}, \qquad z = x_1 + ix_2, \qquad \bar{z} = x_1 - ix_2.$$

If from infinity, at angle β with axis Ox_1, travelling monochromatic waves are radiated, and functions Ψ^0 and Φ_ν^0 $(\nu = 1, 2)$ describe them. From (8.23) and (8.24) it is easy to deduce their forms

$$\Phi_1^0 = B_1 e^{-i\beta_1 \vec{x} \cdot \vec{n}}, \qquad \Phi_2^0 = B_2 h_{31} e^{-i\beta_2 \vec{x} \cdot \vec{n}}, \tag{8.28}$$

$$\Psi^0 = B_3 e^{-i\gamma_2 \vec{x} \cdot \vec{n}}, \qquad \vec{n} = (\cos\beta, \sin\beta), \qquad \vec{x} = (x_1, x_2).$$

Here Φ_ν^0 and Ψ^0 are the amplitude values of the corresponding potentials. The quantities β_ν are determined in (8.24), B_ν are complex constants. The multiplier h_{31} is introduced into the expression for Φ_2^0 to deal with the limiting case of an isotropic medium ($\kappa_0 = 0$). The electroacoustic magnetic wave field, corresponding to the three above-mentioned waves, are determined by functions (8.28) using (8.22), (8.26), (8.27).

From the analysis it follows that when the plane monochromatic waves are excited in the unbounded piezoelectric of class $6mm$, the shear wave does not cause a coupled electromagnetic wave. However, generally speaking "*pure*" *waves of expansion* and "*pure*" *electromagnetic waves* do not exist by themselves. Thus, when there is radiation of the transverse SV-wave and *quasilongitudinal wave* of "*type* β_1" and "*type* β_2", in each of them appear coupled mechanical and electromagnetic fields.

From (8.24) we obtain the asymptotic equalities

$$\beta_1 = \gamma_1 + O\left(k^2\right), \qquad \beta_2 = k + O\left(k^3\right). \tag{8.29}$$

Therefore, the wave of "type β_1" is propagated with a velocity close to the velocity of propagation of the wave of expansion - compression (P-waves), whole the wave of "type β_2" is propagated with velocity close to the velocity of electromagnetic wave c_α.

8.3 An Unbounded Medium with Tunnel Cracks. Integral Representations of Complex Potentials

Consider a transversal-isotropic piezoelectric medium, weakened by cracks L_j $(j = 1, 2, \ldots, n)$ with generators parallel to the axis of symmetry x_3. Plane monochromatic waves determined by equalities (8.28) are radiated from infinity to the cracks, and their edges L_j the constant, along axis x_3, loading $X_m(x_1, x_2, t) = \operatorname{Re}\left\{P_m(x_1, x_2)\, e^{-i\omega t}\right\}$ $(m = 1, 2)$ may act. Assume that P_m and the curves of contour L_j are ‘ functions of class H on $L = \cup L_j$. The problem is the determination of the components of electroelastic magnetic field (8.17) and the stress intensity factors K_I and K_{II}.

Considering the full boundary problem, i.e. the one arising without neglecting, as usual, the magnetic field. In this case we proceed using the general representations of mechanical and electromagnetic fields (8.27). The functions Ψ and Φ_m $(m = 1, 2)$, which represent arbitrary solutions of the corresponding equations of Helmhöltz-type, are determined from the mechanical boundary conditions on the edges of the cuts

$$(\sigma_{11} + \sigma_{22})^{\pm} - e^{2i\psi}(\sigma_{22} - \sigma_{11} + 2i\sigma_{12})^{\pm} = \pm 2e^{i\psi}\left(X_1^{\pm} - iX_2^{\pm}\right), \qquad (8.30)$$

and also from electromagnetic boundary conditions

$$E_3^{+} = E_3^{-}, \qquad H_1^{+} = H_1^{-}, \qquad H_2^{+} = H_2^{-}. \qquad (8.31)$$

Here the signs “plus” and “minus” correspond to the left and right edges L_j, respectively (at movement from beginning a_j to end b_j); ψ is the angle between the normal to the left edge and axis Ox_1. In what follows we will assume that $X_m^{+} = -X_m^{-}$.

To reduce the stated boundary problems to integral equations we will write the integral representations of sought-for functions Φ_ν and Ψ in the form [146]

$$\Phi_\nu(z) = \frac{(-1)^\nu\left(k^2 - \beta_\nu^2\right)}{\beta_\nu^2\left(\beta_2^2 - \beta_1^2\right)}\left\{\int_L q_1 H_0^{(1)}(\beta_\nu r)\, ds + \int_L \left(q_2 \frac{\partial H_0^{(1)}(\beta_\nu r)}{\partial \zeta} d\zeta + \right.\right.$$
$$\left.\left. + q_3 \frac{\partial H_0^{(1)}(\beta_\nu r)}{\partial \bar{\zeta}} d\bar{\zeta}\right)\right\} + \Phi_\nu^0(z), \qquad (8.32)$$

$$\Psi(z) = \frac{k^2}{i\beta_1^2\beta_2^2}\left\{\int_L q_4\, H_0^{(1)}(\gamma_2 r)\, ds + \int_L \left(q_2 \frac{\partial H_0^{(1)}(\gamma_2 r)}{\partial \zeta} d\zeta - \right.\right.$$
$$\left.\left. - q_3 \frac{\partial H_0^{(1)}(\gamma_2 r)}{\partial \bar{\zeta}} d\bar{\zeta}\right)\right\} + \Psi^0(z),$$

$$r = |\zeta - z|, \qquad \zeta = \xi_1 + i\xi_2, \qquad \bar{\zeta} = \xi_1 - i\xi_2,$$
$$\zeta \in L = \cup L_j, \qquad z = x_1 + ix_2 \qquad (\nu = 1, 2),$$

and sought-for *densities* q_m $\left(m = \overline{1,4}\right)$ are directly expressed by the jumps of displacements $[U_k]$ $(k = 1,2)$ as follows

$$q_m = q_m(\zeta) = \frac{\gamma_1^2}{8i}\left(e^{-i\psi}R_1 - (-1)^m e^{i\psi}R_2\right) \qquad (m = 1,4), \tag{8.33}$$

$$q_2 = q_2(\zeta) = \frac{i\gamma_1^2}{2\gamma_2^2}e^{-i\psi}R_1', \qquad q_3 = q_3(\zeta) = \frac{i\gamma_1^2}{2\gamma_2^2}e^{i\psi}R_2',$$

$$R_m = R_m(\zeta) = [U_1] - (-1)^m i[U_2], \qquad R_m' = \frac{dR_m}{ds} \qquad (m = 1,2).$$

In (8.32) the summands Φ_ν^0 and Ψ_ν^0 correspond to plane monochromatic waves, the integral terms express the dispersed by defects coupled wave fields; ds is the element of the length of contour L, $H_p^{(1)}(x)$ is the Hankel function of the first kind and of p - order.

It should be mentioned, that for the functions q_m $\left(m = \overline{1,4}\right)$ the following relations, derived by (8.33) are valid

$$q_1(\zeta) = \frac{\gamma_1^2}{4i}[U_n], \qquad q_4(\zeta) = \frac{\gamma_1^2}{4}[U_s], \tag{8.34}$$

$$q_2(\zeta) = \frac{i\gamma_1^2}{2\gamma_2^2}(V + iV^*), \qquad q_3(\zeta) = \frac{i\gamma_1^2}{2\gamma_2^2}(V - iV^*),$$

$$V = [U_n'] - \frac{[U_s]}{\rho_*}, \qquad V^* = [U_s'] + \frac{[U_n]}{\rho_*}.$$

Here U_n and U_s are the amplitudes of normal and transversal components of the displacements vector on L respectively; ρ_* is the radius of contour curve L at point ζ.

The construction of representations (8.32) is not at all a trivial operation and requires further explanation. Functions Ψ and Φ_ν should be the solutions of the corresponding Helmhöltz equations and satisfy the conditions of radiation. Besides, representations (8.32) should provide the availability of the jump of the displacement vector and the continuity of the stress vector, and also the fulfillment of electromagnetic conditions (8.31) on L. The indicated conditions due to (8.22), (8.27), (8.30) should have the form

$$\sum_{\nu=1}^{2}\alpha_\nu[\Phi_\nu] = -4\left(c_{11}^D - c_{12}^D\right)e^{2i\psi}\left[\frac{\partial^2}{\partial\zeta^2}(\Phi + i\Psi)\right], \tag{8.35}$$

$$\sum_{\nu=1}^{2}\alpha_\nu[\Phi_\nu] = -4\left(c_{11}^D - c_{12}^D\right)e^{-2i\psi}\left[\frac{\partial^2}{\partial\overline{\zeta}^2}(\Phi - i\Psi)\right],$$

$$\left[\frac{\partial}{\partial\zeta}\sum_{\nu=1}^{2}\frac{\beta_\nu^2\Phi_\nu}{k^2 - \beta_\nu^2}\right] = 0, \quad \left[\frac{\partial}{\partial\bar{\zeta}}\sum_{\nu=1}^{2}\frac{\beta_\nu^2\Phi_\nu}{k^2 - \beta_\nu^2}\right] = 0,$$

$$\sum_{\nu=1}^{2}\frac{\beta_\nu^2[\Phi_\nu]}{k^2 - \beta_\nu^2} = 0, \qquad \alpha_\nu = -\beta_\nu^2\left(c_{11}^D + c_{12}^D + \frac{2\kappa_0^2\beta_\nu^2}{k^2 - \beta_\nu^2}\right), \qquad \zeta \in L,$$

where symbol $[\bullet]$ denotes a jump of the corresponding quantity on L.

Representations (8.32) were constructed in such a way, that conditions (8.35) are automatically satisfied. In this case the remaining two complex functions $[U_n]$ and $[U_s]$ are required for the satisfaction of the other mechanical boundary conditions on the cracks.

It should be mentioned that under this approach, the solution of the dynamic problem of electroelasticity for bodies with cracks, the mechanical and electromagnetic quantities are expressed in terms of three functions, which are the solutions of the corresponding Helmhöltz equations. That is why the solution of such problems is a rather difficult task; functions Ψ and Φ_ν $(\nu = 1, 2)$ are solutions of second order differential equations and the boundary conditions on the cuts contain second order derivatives of these functions. As a result, if we use the method of integral equations, the obtained boundary integral equations contain non-integrated singularities. To overcome this difficulty when constructing integral representations (8.32) we used the approach presented in [187], due to which the combinations $\partial^2(\Phi + i\Psi)/\partial z^2$ and $\partial^2(\Phi - i\Psi)/\partial z^2$ appearing in the boundary condition (8.30) and the non-integrated singularities caused from *potential* and *vortex solutions* cancel each other.

8.4 Integrodifferential Equations of a Boundary Problem

Passing to the amplitude values in boundary conditions (8.30) we will write

$$\begin{aligned}
&(S_{11} + S_{22})^{\pm} - e^{2i\psi}\left(S_{22} - S_{11} + 2iS_{12}\right)^{\pm} = \pm 2e^{i\psi}\left(P_1 - iP_2\right), \\
&(S_{11} + S_{22})^{\pm} - e^{-2i\psi}\left(S_{22} - S_{11} + 2iS_{12}\right)^{\pm} = \pm 2e^{-i\psi}\left(P_1 + iP_2\right).
\end{aligned} \tag{8.36}$$

Due to the fact that integral representations (8.32) provide automatically the continuity of the stress vector on L, it is enough to satisfy each of the conditions (8.36) on one of the edges of the cuts, e.g. on the left one. Further, omitting the lower signs in (8.36) and using relation (8.27), we can express condition (8.36) directly in terms of solvable functions $(P_m^+ = -P_m^- = P_m; m = 1, 2)$

$$\begin{aligned}
&\sum_{\nu=1}^{2} \alpha_\nu \Phi_\nu^+ + 4\left(c_{11}^D - c_{12}^D\right) e^{2i\psi}\left\{\frac{\partial^2}{\partial z^2}\left(\Phi + i\Psi\right)\right\}^+ = 2e^{i\psi}\left(P_1 - iP_2\right), \\
&\sum_{\nu=1}^{2} \alpha_\nu \Phi_\nu^+ + 4\left(c_{11}^D - c_{12}^D\right) e^{-2i\psi}\left\{\frac{\partial^2}{\partial \bar{z}^2}\left(\Phi - i\Psi\right)\right\}^+ = 2e^{-i\psi}\left(P_1 - iP_2\right).
\end{aligned} \tag{8.37}$$

Substituting the limiting values in functions (8.37) we arrive at a system of two complex singular integrodifferential equations with respect to the jumps of the amplitude of displacements on L

$$
\begin{aligned}
&\sum_{m=1}^{2}\int_{L}\{f'_m(\zeta)\,G_{m\nu}(\zeta,\zeta_0)+f_m(\zeta)\,g_{m\nu}(\zeta,\zeta_0)\}ds=N_\nu(\zeta_0),\\
&f_1(\zeta)=[U_n],\quad f_2(\zeta)=[U_s],\qquad f'_m=\frac{df_m}{ds}\qquad(\nu=1,2),\\
&G_{1\nu}(\zeta,\zeta_0)=\frac{i\gamma_1^2}{\gamma_2^2}K_1\sin(\psi-\alpha_0)\pm\frac{c_{11}^D-c_{12}^D}{2}e^{\pm 2i\psi_0}\left(F_1\,e^{\pm i(\psi-3\alpha_0)}+\right.\\
&\qquad\left.+F_2e^{i\varepsilon_\nu(\psi+\alpha_0)}\right)+T_{1\nu}(\zeta,\zeta_0),\qquad\qquad(8.38)\\
&G_{2\nu}(\zeta,\zeta_0)=\frac{i\gamma_1^2}{\gamma_2^2}K_1\cos(\psi-\alpha_0)-\frac{c_{11}^D-c_{12}^D}{2i}e^{\pm 2i\psi_0}\left(F_1\,e^{\pm i(\psi-3\alpha_0)}-\right.\\
&\qquad\left.-F_2e^{i\varepsilon_\nu(\psi+\alpha_0)}\right)+T_{2\nu}(\zeta,\zeta_0),\\
&g_{1\nu}(\zeta,\zeta_0)=\frac{\gamma_1^2}{2i}K_0+\gamma_1^2\frac{c_{11}^D-c_{12}^D}{2i}\Phi_{22}e^{\pm 2i(\psi_0-\alpha_0)}+\frac{G_{2\nu}(\zeta,\zeta_0)}{\rho_*},\\
&g_{2\nu}(\zeta,\zeta_0)=\pm\gamma_2^2\frac{c_{11}^D-c_{12}^D}{2}H_2(\gamma_2r_0)\,e^{\pm 2i(\psi_0-\alpha_0)}-\frac{G_{1\nu}(\zeta,\zeta_0)}{\rho_*},\\
&T_{1\nu}(\zeta,\zeta_0)=T_\nu(\zeta,\zeta_0),\qquad T_{2\nu}(\zeta,\zeta_0)=\overline{T_\nu(\zeta,\zeta_0)},\\
&T_1(\zeta,\zeta_0)=\frac{\gamma_1^2}{\pi\gamma_2^2}\left(c_{11}^D+c_{12}^D-2\kappa_0^2\right)\left[2\mathrm{Im}\frac{e^{i\psi}}{\zeta-\zeta_0}-ie^{2i\psi_0}\frac{e^{-i\psi}+e^{i(\psi-2\alpha_0)}}{\zeta-\zeta_0}\right],\\
&T_2(\zeta,\zeta_0)=\frac{\gamma_1^2}{\pi\gamma_2^2}\left(c_{11}^D+c_{12}^D-2\kappa_0^2\right)\left[2\mathrm{Re}\frac{e^{i\psi}}{\zeta-\zeta_0}-e^{2i\psi_0}\frac{e^{-i\psi}-e^{i(\psi-2\alpha_0)}}{\zeta-\zeta_0}\right],\\
&K_0=c_2^*H_0^{(1)}(\beta_2r_0)-c_1^*H_0^{(1)}(\beta_1r_0),\qquad F_1=\frac{\gamma_1^2}{\gamma_2^2}\Phi_{33}+\gamma_2H_3(\gamma_2r_0),\\
&K_1=c_2^*\beta_2H_1(\beta_2r_0)-c_1^*\beta_1H_1(\beta_1r_0),\qquad F_2=\frac{\gamma_1^2}{\gamma_2^2}\Phi_{31}-\gamma_2H_1(\gamma_2r_0),\\
&\Phi_{mn}=\left(\beta_2^2-\beta_1^2\right)^{-1}\left[\beta_2^{m-2}\left(k^2-\beta_2^2\right)H_n(\beta_2r_0)-\beta_1^{m-2}\left(k^2-\beta_1^2\right)H_n(\beta_1r_0)\right],\\
&c_i^*=\left(\beta_2^2-\beta_1^2\right)^{-1}\left[\beta_i^2\left(c_{11}^D+c_{12}^D-2\kappa_0^2\right)-k^2\left(c_{11}^D+c_{12}^D\right)\right],\\
&H_1(x)=\frac{2i}{\pi x}+H_1^{(1)}(x),\qquad H_2(x)=\frac{4i}{\pi x^2}+H_2^{(1)}(x),\\
&H_3(x)=\frac{16i}{\pi x^3}+\frac{2i}{\pi x}+H_3^{(1)}(x),\qquad r_0=|\zeta-\zeta_0|,\qquad \alpha_0=\arg(\zeta-\zeta_0),\\
&N_\nu(\zeta_0)=4e^{\pm i\psi_0}(P_1+\varepsilon_\nu iP_2)+2\exp\left(-i\lambda\left(\xi_{10}\cos\beta+\xi_{20}\sin\beta\right)\right)\times
\end{aligned}
$$

$$\times \left\{ \sum_{j=1}^{2} \beta_j h_j B_j \left(c_{11}^D + c_{12}^D + \frac{2\kappa_0^2 \beta_j^2}{k^2 - \beta_j^2} \right) \right.$$

$$\left. + \lambda^2 \left(c_{11}^D - c_{12}^D \right) e^{\pm 2i(\psi_0 - \beta)} \sum_{j=1}^{3} h_j B_j \right\},$$

$$\psi_0 = \psi\left(\zeta_0\right), \qquad \zeta = \xi_1 + i\xi_2, \qquad \zeta_0 = \xi_{10} + i\xi_{20}; \qquad \zeta, \zeta_0 \in L,$$
$$\varepsilon_\nu = (-1)^\nu, \qquad h_1 = 1, \qquad h_2 = h_{31}, \qquad h_3 = \pm i.$$

Here, the upper signs correspond to index $\nu = 1$; the lower to $\nu = 2$; β is the angle between the normal to the front of the incident wave and Ox_1 -axis; the quantities α_ν $(\nu = 1, 2)$ are defined in (8.35). Kernels $G_{m\nu}$, $g_{m\nu}$ have singular character of Cauchy type at the expense of functions T_1 and T_2. Depending on the type of the monochromatic wave the parameter λ may take the values β_1, β_2 and γ_3.

It is interesting to note, that if the cracks are direct $(\rho_* = \infty)$, the kernels $g_{m\nu}$ are simplified and may posses not more than one logarithmic singularity. To complete the algorithm it is required to add the following additional conditions to the integral (8.38)

$$\int_{L_j} [U_n'] ds = 0, \qquad \int_{L_j} [U_s'] ds = 0 \qquad (j = 1, 2, \ldots, n), \tag{8.39}$$

expressing the equality to zero of the jumps of the displacements at tips L.

Thus, the constructed algorithm of solution of the boundary problem of a piezoceramic medium with cuts is reduced to the following. From system (8.38), (8.39) we determine the jumps of the amplitudes of normal and transversal displacements on the cuts. Further, by this functions with the help of representations (8.32) we restore potentials Φ_ν and Ψ. Then by formulas (8.22), (8.27) we calculate the components of the electroacoustic magnetic field in the medium.

It should be mentioned that the solution of the boundary problem should be constructed in such way, that we could obtain the limiting case of the isotropic medium at $h_{31} = 0$. Actually, if there is no radiation from infinity, and if a mechanical loading acts on the edges of the cuts, from systems (8.38), (8.39) we obtain the finite values of quantities $[U_n]$ and $[U_s]$ at $h_{31} = \kappa_0 = 0$. Substituting (8.32) into (8.27) and (8.26) we conclude that the components of the electromagnetic field are equal to zero, which is a requirement in the case of an isotropic medium. We come to the same conclusion, in those situations, where a *mechanical shear wave* or *a quasilongitudinal wave* of "*type* β_1" radiates from infinity.

We have a quite different picture when a wave of "*type* β_2" radiates from infinity. In that case the right sides of (8.38) are proportional to quantity κ_0 and hence, in the limits of $\kappa_0 \to 0$, $[U_n] = [U_s] = 0$. From (8.26), (8.27) and (8.32) we conclude, that in that case the mechanical stresses and displacements are equal to zero, and there is only an incident electromagnetic wave.

In order to determine the dynamic stress intensity factors K_I and K_{II} it is necessary to find the main asymptotics of stresses in the vicinity of the tips. It is recommended to consider the following combination

$$2\left(N-iT\right)=S_{11}+S_{12}-e^{2i\psi}\left(S_{22}-S_{11}+2iS_{12}\right), \tag{8.40}$$

$$2\left(N+iT\right)=S_{11}+S_{12}-e^{-2i\psi}\left(S_{22}-S_{11}-2iS_{12}\right),$$

where N and T are the normal and tangential stresses on the area which is the continuation of the crack beyond its tip; these combinations of the stress amplitude are expressed as potentials Φ_ν $(\nu=1,2)$ and Ψ by formulas (8.27). Leaving only the main terms (i.e. the integrals containing Cauchy kernels) in these expressions, we find as a result of transformations

$$S^0_{11}+S^0_{22}=-\frac{1}{\pi i}\left(c^D_{11}+c^D_{12}-2\kappa_0^2\right)\left\{\int\limits_L\frac{q_2d\zeta}{\zeta-z}+\int\limits_L\frac{q_3d\bar{\zeta}}{\bar{\zeta}-\bar{z}}\right\}, \tag{8.41}$$

$$S^0_{22}-S^0_{11}+2iS^0_{12}=\frac{1}{\pi i}\left(c^D_{12}-c^D_{11}\right)\left\{\int\limits_L\frac{dR_2}{\zeta-z}+\right.$$

$$\left.+\left(1-\frac{\gamma_2^2}{\gamma_1^2}\right)\int\limits_L\frac{q_2e^{-2i\alpha}d\zeta}{\zeta-z}+\left(1+\frac{\gamma_2^2}{\gamma_1^2}\right)\int\limits_L\frac{q_3e^{-2i\psi}d\zeta}{\zeta-z}\right\},$$

$$S^0_{22}-S^0_{11}-2iS^0_{12}=\frac{1}{\pi i}\left(c^D_{12}-c^D_{11}\right)\left\{-\int\limits_L\frac{dR_1}{\bar{\zeta}-\bar{z}}+\right.$$

$$\left.+\left(1-\frac{\gamma_2^2}{\gamma_1^2}\right)\int\limits_L\frac{q_3e^{2i\alpha}d\bar{\zeta}}{\bar{\zeta}-\bar{z}}+\left(1+\frac{\gamma_2^2}{\gamma_1^2}\right)\int\limits_L\frac{q_2e^{2i\psi}d\bar{\zeta}}{\bar{\zeta}-\bar{z}}\right\},\qquad \alpha=\arg\left(\zeta-z\right).$$

The "*densities*" R_m, q_m are defined in (8.33).

Introducing the parameterization of the crack contour $\zeta=\zeta(\delta)(-1\le\delta\le1)$ and taking into account, that the solution of integral (8.38), (8.39) are sought in the class of functions which are unbounded at tips L (with the order of singularity $\sigma=1/2$) we write

$$\left[U_n'\right]=\frac{\Omega_1\left(\delta\right)}{s'\left(\delta\right)\sqrt{1-\delta^2}},\qquad \left[U_s'\right]=\frac{\Omega_2\left(\delta\right)}{s'\left(\delta\right)\sqrt{1-\delta^2}}, \tag{8.42}$$

$$\Omega_m\left(\delta\right)\in H\left[-1,1\right]\qquad (m=1,2).$$

Due to the formulas describing the behavior of the integrals of Cauchy type in the vicinity of the ends of the integration line, and also the asymptotic relations [50]

$$\frac{1}{2\pi i}\int\limits_{L}\frac{\omega(\zeta)\,e^{-2i\alpha}d\zeta}{\zeta-z}=\pm\frac{e^{\pm i\gamma\pi}\omega^{*}(c)\,h(z)}{2i\sin\gamma\pi\,(z-c)^{\gamma}}+F_{0}(z), \tag{8.43}$$

$$\omega(\zeta)=\frac{\omega^{*}(\zeta)}{(\zeta-c)^{\gamma}},\quad 0<\mathrm{Re}\gamma<1$$

We conclude, that the integrals into (8.41) possess a singularity root at tip c. Using these formulas we can write out the main asymptotic combination of the amplitudes of stresses S_{ij}

$$\begin{aligned} S_{11}^{0}+S_{22}^{0} &= -\frac{\gamma_1^2}{\gamma_2^2}\left(c_{11}^{D}+c_{12}^{D}-2\kappa_0^2\right)\frac{\Omega_1(\pm1)}{\sqrt{2r^{*}s'(\pm1)}},\\ S_{22}^{0}-S_{11}^{0}+2iS_{12}^{0} &= -\frac{i\gamma_1^2}{\gamma_2^2}\left(c_{11}^{D}+c_{12}^{D}-2\kappa_0^2\right)\frac{e^{-2i\psi(\pm1)}\Omega_2(\pm1)}{\sqrt{2r^{*}s'(\pm1)}},\\ S_{22}^{0}-S_{11}^{0}-2iS_{12}^{0} &= \frac{i\gamma_1^2}{\gamma_2^2}\left(c_{11}^{D}+c_{12}^{D}-2\kappa_0^2\right)\frac{e^{2i\psi(\pm1)}\Omega_2(\pm1)}{\sqrt{2r^{*}s'(\pm1)}}. \end{aligned} \tag{8.44}$$

Here the lower sign corresponds to $c=a$, the upper one to $c=b$, and r^{*} is the distance from the point to the tip.

Substituting (8.44) in (8.40) we determine the stress intensity factors

$$K_{I}^{\pm}=\sqrt{2\pi r^{*}}\sigma_n=\sqrt{2\pi r^{*}}\left|N^{0}\right|\cos\left(\omega t-\arg\Omega_1(\pm1)\right), \tag{8.45}$$
$$K_{II}^{\pm}=\sqrt{2\pi r^{*}}\tau_n=\sqrt{2\pi r^{*}}\left|T^{0}\right|\cos\left(\omega t-\arg\Omega_2(\pm1)\right),\qquad (r^{*}\to0),$$

where

$$N^{0}=\frac{\gamma_1^2}{2\gamma_2^2}\left(c_{11}^{D}+c_{12}^{D}-2\kappa_0^2\right)\frac{\Omega_1(\pm1)}{\sqrt{2r^{*}s'(\pm1)}},$$
$$T^{0}=\frac{\gamma_1^2}{2\gamma_2^2}\left(c_{11}^{D}+c_{12}^{D}-2\kappa_0^2\right)\frac{\Omega_2(\pm1)}{\sqrt{2r^{*}s'(\pm1)}}.$$

Thus, the factor K_I is expressed as a normal component and the quantity K_{II} as a tangential component of the jump of the displacement vector on L.

8.5 Reducing to a Case of an Isotropic Medium

Returning to the main system of differential (8.21), in the second equation we can neglect the term k^2A when compared to the term ∇^2A, which corresponds to the quasistatic approximation. Taking into account (8.19) we obtain

$$D_3^{*}=\frac{h_{31}}{\beta_{33}^{\varepsilon}}div\vec{U}. \tag{8.46}$$

Substituting D_3^* in the equation of motion (8.20) we find

$$\nabla^2\vec{U} + \sigma graddiv\vec{U} + \gamma_2^2\vec{U} = 0, \tag{8.47}$$
$$\sigma = \frac{c_{11}^* + c_{12}^*}{c_{11}^* - c_{12}^*}, \qquad c_{11}^* = c_{11}^D - \kappa_0^2, \qquad c_{12}^* = c_{12}^D - \kappa_0^2.$$

Thus, we have a usual equation of motion (in amplitudes) of some fictitious isotropic medium with elastic parameters c_{ij}^*. According to relations (8.21) the wave numbers γ_i $(i = 1, 2)$, are also expressed in terms of these parameters. The mechanical meaning of the carried out simplifications is made clear when studying the general representations (8.27). Here in the expressions for mechanical quantities and for the term corresponding to $m = 1$, the quantity k^2 may be neglected with respect to the term β_1^2 $(\beta_1 \cong \gamma_1)$, and as a result in all the mechanical combinations are embedded only in c_{ij}^*.

The dynamic problem in the theory of elasticity of isotropic media with curvilinear cuts was considered in [175, 176]. The integral representations of the solutions used in [176] were constructed by the method of the theory of potential read (with negligible changes)

$$\partial_j U_m = -\int\limits_L \left\{ (-1)^j S_{\nu,3-j}^{(m)} \left[\frac{\partial U_\nu}{\partial s}\right] + \rho\omega^2 \left[\vec{U}\right] \cdot \vec{U}^{(m)} n_j \right\} ds, \tag{8.48}$$
$$\partial_j = \partial/\partial x_j, \qquad \vec{n} = (\cos\psi, \sin\psi) \qquad (j, \nu, m = 1, 2).$$

Here $[\bullet]$ denotes the jump of the corresponding quantity on L; $\vec{U} = (U_1, U_2)$ is the amplitude vector of the elastic displacement at point $\zeta \in L$; $\vec{U}^m$ and $S_{\nu j}^{(m)}$ are the displacement vector and stress tensor of the m-th state, respectively, corresponding to the action of the concentrated force at point $z_0 = (x_{10}, x_{20})$ in the direction x_m; ψ is the angle between the normal to light edge L (at motion from beginning a_j to end b_j) and Ox_1- axis; ρ is the material density; ω is the circular frequencies; ν is the index of summation. Taking into account that the stress vector is continuous across the cut the quantities $S_{kj}^{(m)}$, $\vec{U}^{(m)}$ are determined by the fundamental solution of the *biwave equation* [134]

$$E(z, z_0) = \frac{H_0^{(1)}(\gamma_2 r) - H_0^{(1)}(\gamma_1 r)}{2ic_{11}^*(c_{11}^* - c_{12}^*)(\gamma_2^2 - \gamma_1^2)},$$
$$z = x_1 + ix_2, \qquad r = |z - z_0|$$

using the formulas

$$U_\nu^{(m)} = A_{\nu m} E, \qquad S_{\nu j}^{(m)} = \frac{c_{11}^* - c_{12}^*}{2}\left(\partial_j U_\nu^{(m)} + \partial_\nu U_j^{(m)}\right) + c_{12}^* \delta_\nu^j U_i,$$
$$A_{mm} = c_{11}^* \nabla^2 - \frac{c_{11}^* + c_{12}^*}{2}\partial_m^2 + \rho\omega^2, \qquad A_{12} = A_{21} = -\frac{c_{11}^* + c_{12}^*}{2}\partial_1\partial_2.$$

In [176] the system of integrodifferential equations of the equations of the dynamic boundary problem for an isotropic medium with cuts is obtained using (8.48). This system can be written in the compact form of

$$
\begin{aligned}
&\sum_{m=1}^{2} \int_{L} \left\{ R'_m (\zeta) \, G_{m\nu} (\zeta, \zeta_0) + R_m (\zeta) \, g_m (\zeta, \zeta_0) \right\} ds = N_\nu (\zeta_0) \, , \\
&G_{mm} = \operatorname{Im} \frac{e^{i\psi_0}}{\zeta - \zeta_0} + \frac{\pi i}{2} F_{31} \sin (\psi_0 - \alpha_0) + (-1)^m \, \frac{\pi \gamma_2^2}{8} p_{1m} \Phi_{11} \times \\
&\quad \times \left(3 - \frac{\gamma_2^2}{\gamma_1^2} - 2\bar{p}_{1m}^2 \right) \quad (\nu, m = 1, 2) \, , \quad R'_m (\zeta) = \frac{dR_m}{ds} \, , \\
&G_{m\nu} = \frac{(-1)^m \, i p_{3m} \left(1 + p_{1m}^2 \right)}{2r} - \frac{\pi}{4} (-1)^m \, p_{3m} \left[F_{31} + F_{33} p_{1m}^2 \right. \\
&\quad \left. + (-1)^m \, \frac{\gamma_2^2}{2} \left(3 - \frac{\gamma_2^2}{\gamma_1^2} \right) \Phi_{11} \right] \qquad (\nu \neq m) \, , \\
&g_{mm} = \frac{\pi i \gamma_2^2 p_{1m}}{16 p_{2m}} \left[2\gamma_2^2 \Phi_{00} - \left(1 + \frac{\gamma_2^2}{\gamma_1^2} \right) \Phi_{20} + \left(1 - \frac{\gamma_2^2}{\gamma_1^2} \right) p_{2m}^2 \Phi \right. \\
&\quad {}_{22} \left. - \frac{2\bar{p}_{1m}^2 p_{2m}^2}{\gamma_2^2 - \gamma_1^2} \left(\left(\gamma_1^2 + \gamma_2^2 \right) \Phi_{20} - 2\gamma_1^2 \gamma_2^2 \Phi_{00} \right) \right] , \\
&g_{m\nu} = \frac{\pi i \gamma_2^2 p_{3m}}{16 p_{2m}} \left[2\gamma_2^2 \Phi_{00} - \left(1 + \frac{\gamma_2^2}{\gamma_1^2} \right) \Phi_{20} + \left(1 - \frac{\gamma_2^2}{\gamma_1^2} \right) p_{2m}^2 \Phi_{22} - 2p_{1m}^2 \Phi_{22} \right] \\
&\quad (\nu \neq m) \, , \\
&\Phi_{m\nu} = \frac{\gamma_2^m H_\nu^{(1)} (\gamma_2 r) - \gamma_1^m H_\nu^{(1)} (\gamma_1 r)}{\gamma_2^2 - \gamma_1^2} \, , \qquad F_{3\nu} = \Phi_{3\nu} + \frac{2i}{\pi r} \, , \\
&N_\nu = N_{\nu 0} + \sum_{k=1}^{2} \sum_{j=1}^{2} N_{\nu k}^{(j)} \, , \qquad N_{\nu 0} = \frac{2\pi \gamma_2^2}{\left(\gamma_2^2 - \gamma_1^2 \right) \left(c_{11}^* - c_{12}^* \right)} \left[P_1 - (-1)^\nu \, i P_2 \right] , \\
&N_{\nu k}^{(1)} = \frac{\pi i \gamma_2^2 \sigma_k}{\gamma_1} \exp \left[-i\gamma_1 \xi_{k0} + (-1)^\nu \, i\psi_0 \right] \times \\
&\quad \times \left\{ \exp \left[(-1)^{\nu+1} \, 2i\psi_0 \right] - \frac{(-1)^k \, \gamma_1^2}{\gamma_2^2 - \gamma_1^2} \right\} , \\
&N_{\nu k}^{(2)} = \frac{(-1)^\nu \, \pi \gamma_2^3 \tau_k}{\gamma_2^2 - \gamma_1^2} \exp \left(-i\gamma_2 \xi_{k0} \right) \exp \left[(-1)^\nu \, i\psi_0 \right] , \\
&p_{11} = \bar{p}_{12} = \exp \left[i \left(\psi_0 - \alpha_0 \right) \right] , \qquad p_{21} = \bar{p}_{22} = \exp \left[i \left(\psi - \alpha_0 \right) \right] , \\
&p_{32} = \bar{p}_{31} = \exp \left[i \left(\psi_0 + \alpha_0 \right) \right] , \qquad r = \left| \zeta - \zeta_0 \right| , \qquad \alpha_0 = \arg \left(\zeta - \zeta_0 \right) , \\
&\psi_0 = \psi \left(\zeta_0 \right) , \qquad \zeta = \xi_1 + i\xi_2 , \qquad \zeta_0 = \xi_{10} + i\xi_{20} , \qquad \zeta, \zeta_0 \in L = \cup L_j ,
\end{aligned}
\tag{8.49}
$$

where functions $R_m (\zeta)$ $(m = 1, 2)$ are defined in (8.33).

We add the following conditions to the system (8.49)

$$\int_{L_j} R'_m(\zeta)\, ds = 0 \qquad (m = 1, 2; j = 1, 2, \ldots, n). \tag{8.50}$$

It is of interest to mention, that based on integral representations (8.48), and taking into account relations (8.34) as well as equalities

$$R'_1 = \frac{dR_1}{ds} = e^{i\psi}(V + iV^*), \qquad R'_2 = \frac{dR_2}{ds} = e^{-i\psi}(V - iV^*)$$

the formulas for the stress intensity factors, coincide with formulas (8.45). Hence, the intensity factors K_I and K_{II} both in precise and in the approximate variant of the solutions of the problem are determined by the same expressions. However, in the first case, the functionals $\Omega_\nu(\pm 1)$ are calculated from the system of integral (8.38), (8.39), and in the second case from system (8.49), (8.50).

The relative stress intensity factors read

$$\langle K_m^{\pm} \rangle = \frac{\gamma_1^2 (c_{11}^* + c_{12}^*)}{4\gamma_2^2 \Lambda \sqrt{\ell s'(\pm 1)}} \left[e^{-i\psi(\pm 1)} \Omega_1(\pm 1) - (-1)^m e^{i\psi(\pm 1)} \Omega_2(\pm 1) \right],$$

$$R'_j(\zeta) = \frac{\Omega_j(\delta)}{s'(\delta)\sqrt{1 - \delta^2}}, \qquad \delta \in [-1, 1] \qquad (j, m = 1, 2).$$

as a function of the normalized wave number $\gamma_2 \ell$ (2ℓ is the length of the crack). In case of diffraction on a parabolic crack $(\xi_1 = \delta,\ \xi_2 = p_* \delta^2)$ in correspondence with quasilongitudinal and transversal waves propagating along the negative direction of axis x_2, is shown in Figs. 8.2 and 8.3. The lines with number m are constructed for $p_* = 0.2\,(m - 1)$. Assuming theknowledge of

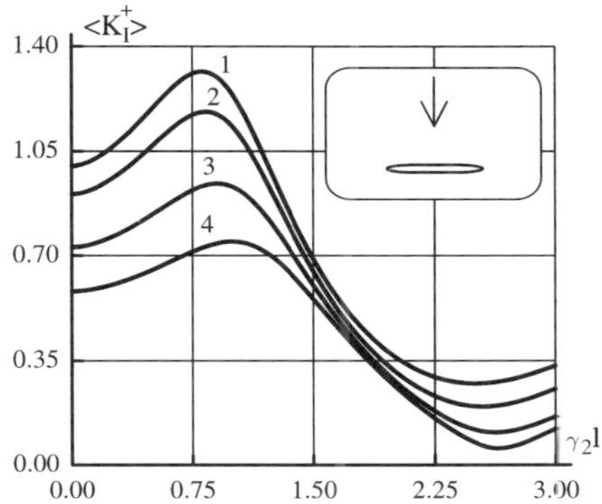

Fig. 8.2. Changes of quantity $< K_1^{\pm} >$ as a function of normalized wave number $\gamma_2 l$ at studying of a quasilongitudinal wave in the medium of a parabolic crack

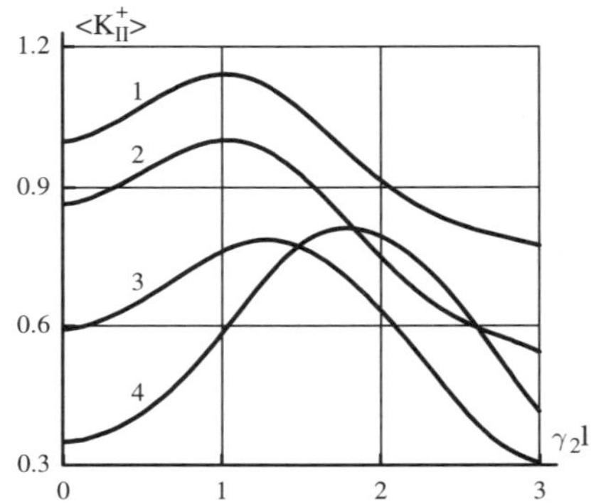

Fig. 8.3. Changes of quantity $\langle K_{II}^{+} \rangle$ as a function of γ_2l under the influence of a transverse wave in a medium with a parabolic crack

$\langle K_I \rangle$ and $\langle K_{II} \rangle$, the stress intensity factors may be determined by the formulas

$$K_m^{\pm} = \pm \Lambda \sqrt{\pi \ell} \left\langle K_m^{\pm} \right\rangle \cos\left(\omega t - \arg \Omega_m \left(\pm 1\right)\right) \qquad \left(m = 1, 2\right),$$

where the functions $\Omega_m(\delta)$ are calculated from the system of (8.49), (8.50), where quantity $\Lambda = |\tau| \gamma_1 \left(c_{11}^D - \kappa_0^2\right)$ is used in case of the radiation of a quasilongitudinal wave, and $\Lambda = |\tau| \gamma_2 \left(c_{11}^D - c_{12}^D\right)/2$ is used under the influence of the shear wave ($\tau = const$ is the amplitude of displacement in the wave).

$\left\langle K_I^{+} \right\rangle$ and $\left\langle K_{II}^{+} \right\rangle$ are shown in Figs. 8.4 and 8.5, respectively,, In the case where there are no radiations from infinity, and on the edges of the cracks there are prescribed normal ($P_1 = P \cos\psi_0,\ P_2 = P \sin\psi_0$) or tangential ($P_1 = P \sin\psi_0,\ P_2 = -P \cos\psi_0$) forces ($\Lambda = |P|$).

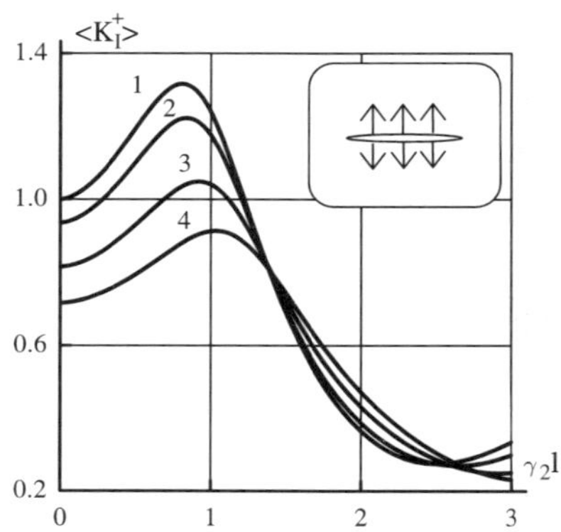

Fig. 8.4. Changes of $< K_I^{\pm} >$ as a function of $\gamma_2 l$ under the influence of normal forces on a parabolic crack

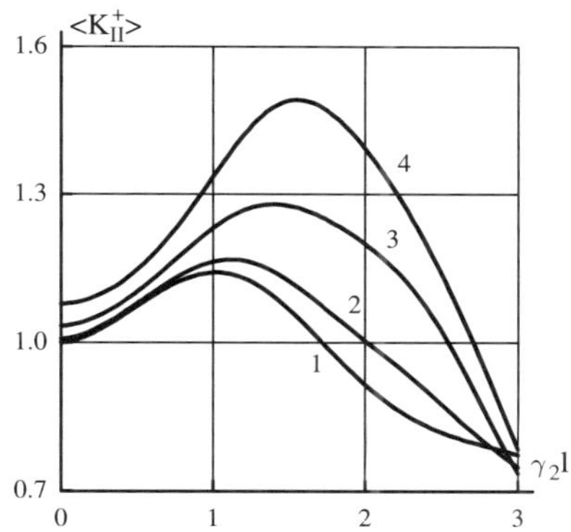

Fig. 8.5. Changes of $< K^{\pm}_{II} >$ as a function of $\gamma_2 l$ under the influence of tangential forces on a parabolic crack

From the results it follows that with increasing curve parameter p_* the peak values of quantity $\gamma_2\ell$ are displaced to the left, and the maximum of quantity $\langle K_I \rangle$ is decreased. As it is shown in Fig. 8.5, the increase of the cut curvature under the influence of transversal stress brings to an increase of maximum $\langle K_{II} \rangle$.

8.6 Effect of Mutual Hardening of Cracks

When two or more cracks interact in a body, there may be situations where hardening in a specified sense takes place, i.e. the absolute values of the corresponding stress intensity factors at the tips of one of a crack may decrease at the expense of an influence of other cracks.

We consider the numerical solution of (8.49), (8.50) in the case of a piezoelectric medium, containing two cracks. Here the algorithm is reduced to the solution of eight real integrodifferential equations (for a case of a singular crack they are four). In order to reduce the number of equations we will indicate below a scheme of numerical solution of the system of integral equations. The method successively decreases the dependence on the coupling in the considered area.

We set the cut L_1 as a main one and we introduce the integral (8.49) together with additional conditions (8.50) to the system of linear algebraic equations.

$$\sum_{j=1}^{2}\sum_{\nu=1}^{n} \alpha_{ij}\left(\delta_\nu, \delta^*_m\right)\Omega_{j\nu} = N_i\left(\delta^*_m\right) - \sum_{j=1}^{2}\sum_{\nu=1}^{n} \alpha_{ij}\left(\Delta_\nu, \delta^*_m\right)\Lambda_{j\nu}, \tag{8.51}$$

$$\sum_{j=1}^{2}\sum_{\nu=1}^{n} \alpha_{ij}\left(\delta_\nu, \Delta^*_m\right)\Omega_{j\nu} = N_i\left(\Delta^*_m\right) - \sum_{j=1}^{2}\sum_{\nu=1}^{n} \alpha_{ij}\left(\Delta_\nu, \Delta^*_m\right)\Lambda_{j\nu},$$

$$\sum_{\nu=1}^{n} \Omega_{i\nu} = 0, \qquad \sum_{\nu=1}^{n} \Lambda_{i\nu} = 0 \qquad (i = 1, 2; m = 1.2, \ldots, n-1),$$

$$\alpha_{ij}(x_\nu, y_m) = \frac{\pi}{n} G_{ij}(x_\nu, y_m) - \frac{2\pi}{n^2} \sum_{k=1}^{n} g_{ij}(x_k, y_m) S_{\nu k} S'(x_k) \sin\theta_k,$$

$$\frac{dR_j}{d\delta} = \frac{\Omega_j(\delta)}{\sqrt{1-\delta^2}}, \qquad \frac{dR_j}{d\Delta} = \frac{\Lambda_j(\Delta)}{\sqrt{1-\Delta^2}}, \quad \Omega_{j\nu} = \Omega_j(\delta_\nu), \quad \Lambda_{j\nu} = \Lambda_j(\Delta_\nu),$$

$$S_{\nu k} = \sum_{\ell=1}^{n-1} \frac{\cos\ell\theta_\nu \sin\ell\theta_k}{\ell}, \qquad \theta_\nu = \frac{2\nu - 1}{2n}\pi \quad (\nu = 1, 2, \ldots, n),$$

$$\Omega_j(\delta), \qquad \Lambda_j(\Delta) \in H[-1, 1] \qquad (j = 1, 2).$$

Here the contour L_1 is described by the parameter $\delta \in [-1, 1]$, contour L_2 by the parameter $\Delta \in [-1, 1]$; δ_ν, Δ_ν are the zeros of the Chebyshev polynomial of the first kind; $\delta_\nu^*, \Delta_\nu^*$ are the zeros of the Chebyshev polynomial of the first kind, n is the number of nodes of interpolation. Each fixed i corresponds to $2n$ of complex equations.

We will solve the first system in (8.51), with respect to the unknowns $\Omega_1(\delta_\nu)$, $\Omega_2(\delta_\nu)$, which correspond to the main cut. For this purpose we will introduce "*standard*" solutions $N_m^S, \gamma_{m\nu}^S$ and $\omega_{m\nu}^S$ $\left(\nu, m = \overline{1, n}\right)$ of the first system in order to fulfill the relations

$$\Omega_{km} = N_{m+i}^S + \sum_{\nu=1}^{N} \left(\gamma_{m+i,\nu}^s \Lambda_{1\nu} + \omega_{m+i,\nu}^S \Lambda_{2\nu}\right), \tag{8.52}$$

where $i = 0$ at $k = 1$ and $i = n$, if $k = 2$. If we apply (8.52) in (8.51), we come to a system of equations with the second group of unknowns $\Lambda_{j\nu}$. In fact such an algorithm is used for the consecutive solution of two uncoupled systems (of four real integral equations). As the numerical investigations show, the indicated procedure of "decreasing of the coupling of the area" may be efficiently used in those cases, where there are several cuts in the region.

As an example, consider a piezoceramic from $PZT-4$, which contains two cracks L_1 and L_2 of the parameterization of the form $\xi_1^{(1)} + i\xi_2^{(1)} = \delta \exp\left(i\varphi^{(1)}\right)\left(1 + ip_*^{(1)}\delta\right)$ and $\xi_1^{(2)} + i\xi_2^{(2)} = \Delta \exp\left(i\varphi^{(2)}\right)\left(1 + ip_*^{(2)}\Delta\right) + ih$ ($\varphi^{(j)}$ is the angle of orientation of crack L_j in the space).

For this case the number of interpolation during the solution procedure of system (8.51) was set to be equal $n = 20, 25, 33$, and a further increase of n does not practically improve the accuracy of the obtained results.

The relative stress intensity factors are given by

$$\left\langle K_{mj}^{\pm} \right\rangle = \frac{\gamma_1^2\left(c_{11}^* + c_{12}^*\right)}{4\gamma_2^2 \Lambda \sqrt{\ell_j s_j'(\pm 1)}} \left[e^{-i\psi^{(i)}(\pm 1)} \Pi_{1j}(\pm 1) - (-1)^m e^{i\psi^{(i)}(\pm 1)} \Pi_{2j}(\pm 1)\right],$$

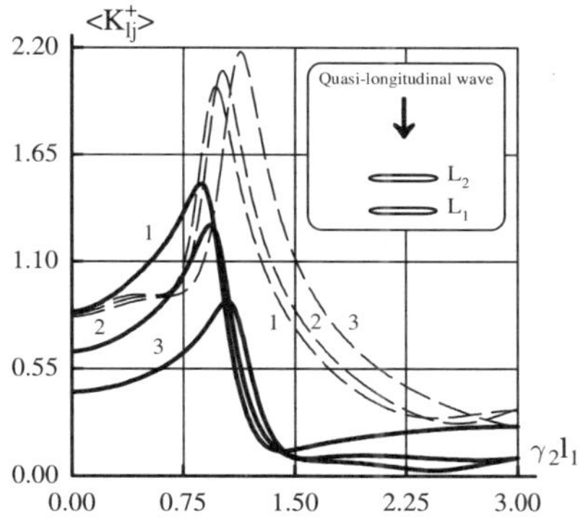

Fig. 8.6. Behaviour of quantity $< K_{II}^{+} >$ in a medium with two cracks according to parameter $\gamma_2 l_1$ at radiation of quasilongitudinal wave

where $\Pi_{m1} = \Omega_m$, $\Pi_{m2} = \Lambda_m$ $(m, j = 1, 2)$. Relative stress intensity factors are functions of the normalized wave number $\gamma_2\ell_1$ ($2\ell_j$ is the length of crack L_j) while a quasilongitudinal wave radiates along x_2-axis is shown in Figs. 8.6 and 8.7. The curves with number m are constructed for $p_*^{(2)}/\ell_2 = 0$, $p_*^{(1)}/\ell_2 = 0.3\,(m-1)$. It was assumed that $h/\ell_2 = 2$, $\varphi^{(j)} = 0$. The solid lines correspond to crack L_1, the dashed ones to L_2. It is obvious that $\langle K_{mj}^{+} \rangle = \langle K_{mj}^{-} \rangle$.

Figure 8.8 show $\left(K_{Ij}^{\pm} 1\right)$ as a function of the orientation angle of crack L_2 at $\gamma_2\ell_1 = 1$, $\varphi^{(1)} = 0$, $h/\ell_1 = 2$, $p_*^{(j)}/\ell_1 = 0$. Solid lines 1 and 2 are constructed for tips a_1 and a_2, the dashed ones for b_1 and b_2, respectively; curve 3 corresponds to the situations, where there is a single crack L_2 in the medium.

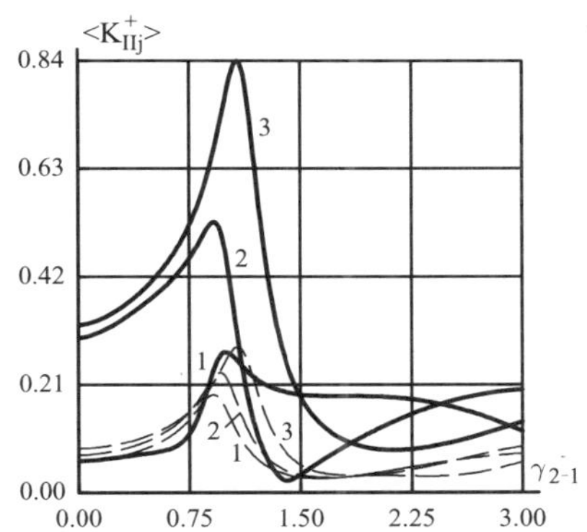

Fig. 8.7. Behaviour of quantity $< K_{III}^{+} >$ in a medium with two cracks according to parameter $\gamma_2 l_1$ at radiation of a quasilongitudinal wave

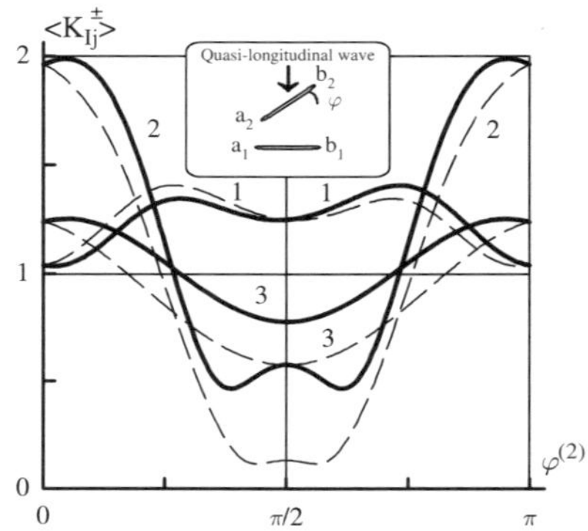

Fig. 8.8. Changes of quantity $< K_{ij}^{+} >$ as a function of the orientation angle of the crack L_2

From the results it follows, that during the interaction of two cracks the stress intensity factors depend considerably on the curvature of the cracks and on their relative position. In particular this is observed when monochromatic shear or expansion waves radiate on the cracks from infinity. It should be mentioned that for the certain values of quantities ω and the angles of orientation of the cracks it is observed an effect of their mutual hardening (Fig. 8.8).

8.7 An Inertial Effect in the Process of Impact Effect on a Crack

Here we investigate the behavior of stress intensity factors under conditions of impulse loading of a piezoceramic medium, weakened by tunnel cracks of rather arbitrary configuration. As loading we consider normal and tangential stresses acting on the edges of the crack.

Based on the analysis similar to that presented in Sect. 4.6 and taking into account formula (8.45) we come to the following expressions for the stress intensity factors

$$K_{I,II}^{\pm} = \pm \frac{\gamma_1^2 \left(c_{11}^D + c_{12}^D - 2\kappa_0^2\right)}{\gamma_2^2 \sqrt{2s'\left(\pm 1\right)}} \mathrm{Re} \int_0^\infty \Omega_{1,2}\left(\omega, \pm 1\right) e^{-i\omega t} d\omega. \tag{8.53}$$

Functions $\Omega_m\left(\omega, \pm 1\right)$ $(m = 1, 2)$ are determined from system of integrodifferential (8.38), (8.39), the right parts of which in the given case will be written down with respect to the spectral functions of loading.

Below we give some results for the influence of an inertial effect on stress intensity factors in a medium (ceramics $PZT{-}4$) weakened by one or two cracks. For definiteness, assume that the forces acting on the edges of the cracks are

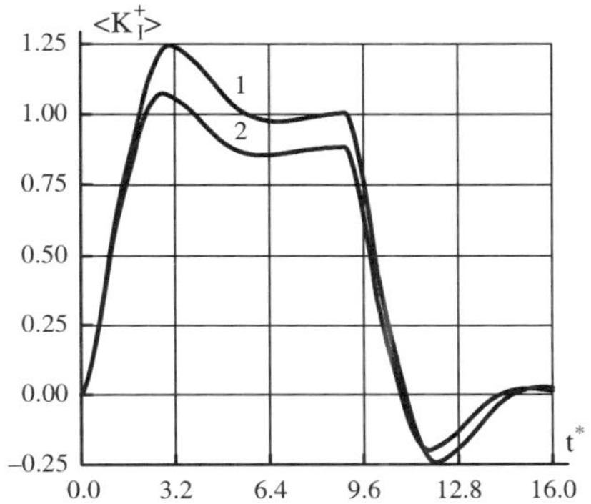

Fig. 8.9. Changes of $< K_I^+ >$ under the impulse loading of a crack by normal forces

of impulse trapezoidal form (Fig. 4.28). The spectral function of the given impulse is determined by formula (4.73), where the quantity $R_\nu(\omega)$ $(\nu = 1, 2)$ is now expressed by the spectral function of loading $P_\nu(\omega)$ appearing in integral (8.38) or (8.49).

Figure 8.9 show the behavior of the relative factor $\langle K_I^+ \rangle = K_I^+ / \left(\sigma_2 \sqrt{\pi \ell}\right)$ (2ℓ is the length of the crack), as a function of the dimensionless time $t^* = c_2 t \ell^{-1}$ (c_2 is the velocity of the shear wave in a piezoceramic) for the case where a parabolic crack $\left(\xi_1 = p\delta,\ \xi_2 = p_* \delta^2,\ \delta \in [-1, 1]\right)$ is evenly loaded by distributed normal conditions changing according to a trapezoidal impulse with parameters $T^* = c_2 T \ell^{-1} = 10$, $d_1^* = c_2 d_1 \ell^{-1} = 1$, $d_2^* = c_2 d_2 \ell^{-1} = 8$, $\sigma_2^* = c_2 \sigma_2 \ell^{-1} = 1N / \left(s \cdot m^2\right)$. Curves 1 and 2 are constructed for $p_*/p = 0$ and 0.3, respectively. The behavior of quantity $\langle K_{II}^+ \rangle = K_{II}^+ \big/ \left(\sigma_1 \sqrt{\pi \ell}\right)$ under the influence of analogous shear impulse is shown in Fig. 8.10.

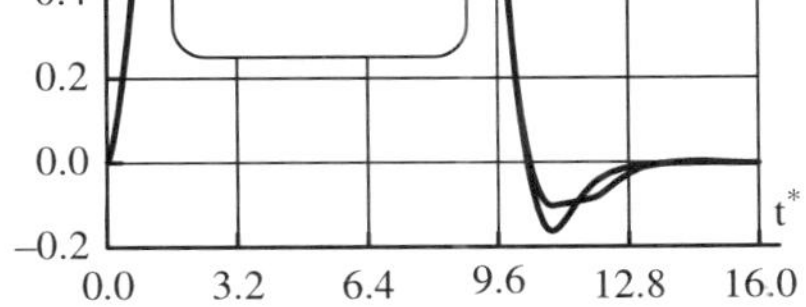

Fig. 8.10. Changes of $< K_{II}^+ >$ under the impulse loading of a crack by tangential forces

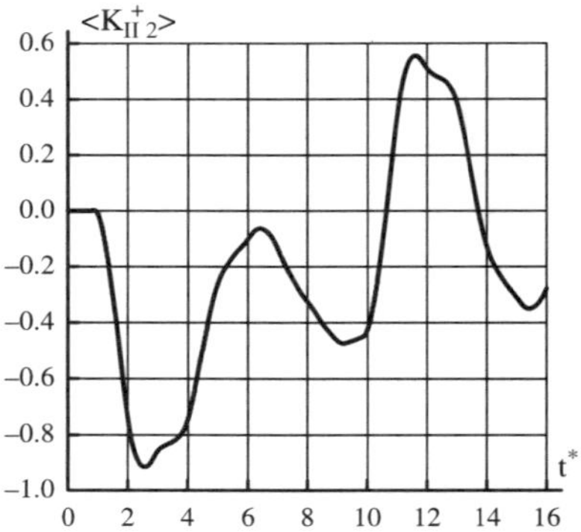

Fig. 8.11. Changes of $< K_{II2}^{+} >$ during the interaction of two cracks in a medium as a result of impulse changes of tangential forces on one of the cracks

Figure 8.11 shows the graph of quantity $\langle K_{II2}^{+} \rangle$ in the case of interaction of two collinear cracks L_1 and L_2 of different length with parametric equations $\xi_1^{(1)} = 3\delta,\ \xi_2^{(1)} = 0$ and $\xi_1^{(2)} = \delta,\ \xi_2^{(2)} = h$. It was assumed that the crack L_2 is free from loading, and a trapezoidal impulse of tangential stress acts on L_1 in the form of $T^* = c_2Th^{-1} = 10,\ d_1^* = c_2d_1h^{-1} = 1,\ d_2^* = c_2d_2h^{-1} = 8$, and $h/\ell_2 = 1,\ t^* = c_2th^{-1},\ \sigma_1^* = c_2\sigma_1h^{-1} = 1N/\left(s \cdot m^2\right)$.

As it follows from Fig. 8.9, due to the inertial effect of intensive dynamic stress the intensity factor $\langle K_I \rangle$ may exceed its static value $\langle K_1^0 \rangle = 1$ only by 25% (curve 1). The fact that, in the interval $11c_2^{-1}\ell < t < 14.6c_2^{-1}\ell$, quantity $< K_I >$ takes negative values expresses the tendency of the crack lips to approach one another, after some time since the load stops to exist (Fig. 8.10). From the calculations it also follows that during the interaction of two cracks the influence of the inertial effect on stress intensity factor is stronger than in a medium with one crack.

In concluding, the quantities $\langle K_I \rangle$ and $\langle K_{II} \rangle$ were calculated with the help of formulas (8.53) taking into account (4.73), where the semidimensionless interval was substituted by the final one, $[0, \omega^*]$; the functions Ω_m $(m = 1, 2)$ were found from system (8.38), (8.39) by the standard method, and in order to reach the satisfactory accuracy, the number of the nodes of interpolation on the crack contour did not exceed $n = 33$.

8.8 A Matrix of Fundamental Solutions of Two-Dimensional Equations of Electroelasticity

Consider a fundamental solution for the case of a set of oscillations in a piezoceramic medium. Consider the practically important and most complex from the mathematical point of view case, where a plane deformation of the medium is carried out on the plane parallel to the direction of an electric field of preliminarily polarization of the piezoceramics.

Let a piezoceramic medium be subjected to the influence of forces $X_k(x_1, x_3, t) = \delta(x_1, x_3)\,\mathrm{Re}\left(P_k e^{-i\omega t}\right)$ $(k = 1, 3)$, which are concentrated on line $x_1 = 0$, $-\infty < x_2 < \infty$, $x_3 = 0$ and harmonically change with time, or electric charges $q(x_1, x_3, t) = \delta(x_1, x_3)\,\mathrm{Re}\left(Qe^{-i\omega t}\right)$, where $\delta(x_1, x_3)$ is the Dirac δ-function. Assume, that the medium is preliminarily polarized in the direction of x_3 -axis of Cartesian system of coordinates $Ox_1x_2x_3$. In that case a plane deformation of a piezoelectric medium occurs in x_1Ox_3.

The problem under study is governed by the

- equations of motion of a medium

$$\partial_k \sigma_{ik} + X_i = \rho_* \partial^2 u_i / \partial t^2 \qquad (i, k = 1, 3); \tag{8.54}$$

- equations of electrostatics

$$div \vec{D} = q, \qquad \vec{E} = -grad\phi; \tag{8.55}$$

- material equations [140]

$$\begin{aligned} \sigma_{11} &= c_{11}\partial_1 u_1 + c_{13}\partial_3 u_3 - e_{31}E_3, \\ \sigma_{13} &= c_{44}(\partial_1 u_3 + \partial_3 u_1) - e_{15}E_1, \\ \sigma_{33} &= c_{13}\partial_1 u_1 + c_{33}\partial_3 u_3 - e_{33}E_3, \\ D_1 &= \ni_{11} E_1 + e_{15}(\partial_1 u_3 + \partial_3 u_1), \\ D_3 &= \ni_{33} E_3 + e_{31}\partial_1 u_1 + e_{33}\partial_3 u_3. \end{aligned} \tag{8.56}$$

Here σ_{ik}, u_i, E_i, D_i and ϕ are the mechanical stresses, displacements, strength, induction and potential of the electric field; $c_{ij} = c_{ij}^E$ is the elasticity modulus; $\ni_{jj} = \ni_{jj}^{\varepsilon}$ - is the dielectric permeability; e_{kj} is the piezoelectric modulus; ρ_* is the material density.

Assuming further,

$$u_j = \mathrm{Re}\left(U_j e^{-i\omega t}\right), \qquad \sigma_{jk} = \mathrm{Re}\left(T_{jk} e^{-i\omega t}\right), \qquad \phi = \mathrm{Re}\left(\Phi e^{-i\omega t}\right),$$

we can write the system (8.54)–(8.56) in matrix form, with respect to the amplitude values of displacements or electric potentials

$$\left\| \begin{matrix} L_{11} & L_{12} & L_{13} \\ L_{21} & L_{22} & L_{23} \\ L_{31} & L_{32} & L_{33} \end{matrix} \right\| \left\| \begin{matrix} U_1^{(1)} & U_1^{(2)} & U_1^{(3)} \\ U_3^{(1)} & U_3^{(2)} & U_3^{(3)} \\ \Phi^{(1)} & \Phi^{(2)} & \Phi^{(3)} \end{matrix} \right\| = \left\| \begin{matrix} -P_1\delta & 0 & 0 \\ 0 & -P_3\delta & 0 \\ 0 & 0 & Q\delta \end{matrix} \right\|. \tag{8.57}$$

Here the upper index $k = 1, 2, 3$ indicates the relation of quantity $U_j^{(k)}$ and $\Phi^{(k)}$ with respect to the concentrated loadings X_1, X_3 and q,and the differential operators L_{ij} are determined as follows

$$\begin{aligned} L_{11} &= c_{11}\partial_1^2 + c_{44}\partial_3^2 + \rho_*\omega^2, & L_{12} &= (c_{13} + c_{44})\,\partial_1\partial_3, \\ L_{13} &= (e_{31} + e_{15})\,\partial_1\partial_3, & L_{22} &= c_{44}\partial_1^2 + c_{33}\partial_3^2 + \rho_*\omega^2, \\ L_{23} &= e_{15}\partial_1^2 + e_{33}\partial_3^2, & L_{33} &= -\varepsilon_{11}\partial_1^2 - \varepsilon_{33}\partial_3^2. \end{aligned} \tag{8.58}$$

Applying a two-dimensional Fourier transform to the system (8.57) [165] over x_1 and x_3 coordinates

$$\begin{aligned} V_j^{(k)}(\xi,\omega) &= \frac{1}{2\pi}\int U_j^{(k)}(x,\omega)\,e^{i(\xi,x)}dx, & x &= (x_1, x_3), \\ U_j^{(k)}(x,\omega) &= \frac{1}{2\pi}\int V_j^{(k)}(\xi,\omega)\,e^{-i(\xi,x)}d\xi, & \xi &= (\xi_1, \xi_3). \end{aligned} \tag{8.59}$$

we get

$$\begin{gathered} V_j^{(k)} = \frac{\varepsilon_k R_k \Delta_j^{(k)}(\xi,\omega)}{2\pi\Delta(\xi,\omega)} \qquad \left(k, j = \overline{1,3}\right), \\ R_1 = P_1, \qquad R_2 = P_3, \qquad R_3 = Q, \qquad \varepsilon_1 = \varepsilon_2 = -1, \qquad \varepsilon_3 = 1. \end{gathered} \tag{8.60}$$

Here the functions $V_j^{(k)}$ $(j = 1, 2, 3)$ represent the transformations of the amplitude of the displacement U_1, U_3 and electric potential Φ, respectively.

Taking into account the transformation of the coordinates

$$\begin{aligned} \xi_1 &= \rho\cos\alpha, & x_1 &= r\cos\beta, \\ \xi_3 &= \rho\sin\alpha, & x_3 &= r\sin\beta, \end{aligned} \tag{8.61}$$

the quantities entering (8.60) may be represented in the form

$$\begin{aligned} &\Delta_j^{(k)}(\rho,\alpha,\omega) = b_j^{(k)}\rho^4 + c_j^{(k)}\gamma^2\rho^2 + \delta_j^3\delta_k^3\gamma^4 c_{44}^2, \\ &\Delta(\rho,\alpha,\omega) = c_{44}^3 A(\alpha)\,\rho^2\left(\rho^2 - \gamma^2\rho_1^2\right)\left(\rho^2 - \gamma^2\rho_2^2\right), \\ &A(\alpha) = c_{44}^{-3}\left(a_{11}a_{22}a_{33} + 2a_{12}a_{13}a_{23} - a_{11}a_{23}^2 - a_{22}a_{13}^2 - a_{33}a_{12}^2\right), \\ &A_1(\alpha) = \left(a_{11}a_{33} + a_{22}a_{33} - a_{23}^2 - a_{13}^2\right)/c_{44}^2, \qquad A_2(\alpha) = a_{33}/c_{44}, \\ &a_{11} = -c_{11}n_1^2 - c_{44}n_3^2, \qquad a_{13} = -(e_{31} + e_{15})\,n_1n_3, \\ &a_{12} = -(c_{13} + c_{44})\,n_1n_3, \qquad a_{22} = -c_{44}n_1^2 - c_{33}n_3^2, \qquad n_1 = \cos\alpha, \\ &a_{23} = -e_{15}n_1^2 - e_{33}n_3^2, \qquad a_{33} = \ni_{11} n_1^2 + \ni_{33} n_3^2, \qquad n_3 = \sin\alpha, \\ &b_1^{(1)} = a_{22}a_{33} - a_{23}^2, \qquad b_2^{(1)} = a_{13}a_{23} - a_{12}a_{33}, \\ &b_3^{(1)} = a_{12}a_{23} - a_{13}a_{22}, \qquad b_2^{(2)} = a_{11}a_{33} - a_{13}^2, \\ &b_3^{(2)} = a_{12}a_{13} - a_{11}a_{23}, \qquad b_3^{(3)} = a_{11}a_{22} - a_{12}^2, \qquad c_1^{(1)} = c_{44}a_{33}, \end{aligned} \tag{8.62}$$

$$c_3^{(1)} = -c_{44}a_{13}, \quad c_2^{(1)} = 0, \quad c_2^{(2)} = c_{44}a_{33}, \quad c_3^{(2)} = -c_{44}a_{23},$$
$$c_3^{(3)} = c_{44}\left(a_{11} + a_{22}\right), \quad c_k^{(j)} = c_j^{(k)}, \quad b_k^{(j)} = b_j^{(k)},$$
$$B_i\left(\alpha\right) = \frac{A_i\left(\alpha\right)}{A\left(\alpha\right)}, \quad \gamma = \frac{\omega}{c}, \quad c = \sqrt{\frac{c_{44}}{\rho_*}},$$
$$\rho_\nu^2\left(\alpha\right) = -\frac{B_1}{2} + (-1)^\nu\sqrt{\frac{B_1^2}{4} - B_2}.$$

Here δ_m^i is the Kronecker symbol, the quantity $\rho_\nu^2\left(a\right) > 0\,(\nu = 1, 2)$.

Taking into account (8.62) and decomposing expression (8.60) into partial fractions we obtain

$$V_j^{(k)} = \frac{\varepsilon_k R_k}{2\pi c_{44}^3 A\left(\alpha\right)}\left\{\sum_{\nu=1}^{2}\frac{\lambda_{\nu j}^{(k)}}{\rho^2 - \gamma^2\rho_\nu^2} + \delta_j^3\delta_k^3\frac{\lambda_0}{\rho^2}\right\} \qquad \left(j, k = \overline{1,3}\right), \tag{8.63}$$
$$\lambda_{\nu j}^{(k)} = \frac{\rho_\nu^2 b_j^{(k)} + c_j^{(k)} + \delta_j^3\delta_k^3 c_{44}^2/\rho_\nu^2}{(-1)^\nu\left(\rho_2^2 - \rho_1^2\right)}, \qquad \lambda_0 = \frac{c_{44}^2}{\rho_1^2\rho_2^2}.$$

Applying the inverse Fourier transform we find

$$W_j^{(k)}\left(r, \beta, \omega\right) = \frac{\varepsilon_k R_k}{2\pi^2 c_{44}^3}\int_0^\pi\int_0^\infty\left\{\sum_{\nu=1}^{2}\frac{\lambda_{\nu j}^{(k)}\rho\cos\left[r\rho\cos\left(\alpha - \beta\right)\right]}{A\left(\alpha\right)\left(\rho^2 - \gamma^2\rho_\nu^2\right)}\right.$$
$$\left.+\frac{\delta_j^3\delta_k^3\lambda_0\cos\left[r\rho\cos\left(\alpha - \beta\right)\right]}{A\left(\alpha\right)\rho}\right\}d\rho d\alpha,$$
$$W_1^{(k)} = U_1^{(k)}, \qquad W_2^{(k)} = U_3^{(k)}, \qquad W_3^{(k)} = \Phi^{(k)} \qquad \left(k = \overline{1,3}\right). \tag{8.64}$$

The improper integral in (8.64) according to the principle of limiting absorption [46] should be used in the following form [43, 57]

$$\int_0^\infty \frac{x\cos ax dx}{x^2 - b^2 - i0} = \lim_{\varepsilon\to+0}\int_0^\infty\frac{x\cos axdx}{x^2 + b_*^2} = \frac{1}{2}\left\{e^{-iab}E_1\left(-iab\right) - e^{iab}Ei\left(-iab\right)\right\},$$
$$b_* = -i\sqrt{b^2 + i\varepsilon}, \qquad \varepsilon > 0, \quad a > 0, \qquad \mathrm{Re}b_* > 0,$$

where $E_1\left(z\right)$, $E_i\left(z\right)$ are the integral exponential functions [22, 109]. Taking into account this observation, and the values of integral [57] ($C = 0.5772\ldots..$ is the Euler's constant)

$$\int_0^\infty\frac{\cos axdx}{x} = -\ell na - C \qquad (a > 0),$$

we obtain the formulas for the determination of the amplitude of displacements and electric potentials

$$W_j^{(k)}(r,\beta,\omega) = \frac{\varepsilon_k R_k}{2\pi^2 c_{44}^3} \int_0^{\pi} \left\{ \sum_{\nu=1}^{2} \lambda_{\nu j}^{(k)} \Psi(\theta_\nu) + \delta_j^3 \delta_k^3 \lambda_0 \Omega(\theta) \right\} \frac{d\alpha}{A(\alpha)}, \qquad (8.65)$$

$$\Psi(x) = \frac{\pi i}{2} e^{ix} - \cos x cix - \sin x six, \quad \Omega(x) = -\ell n x - C,$$

$$\theta_\nu = \gamma \rho_\nu r \left|\cos(\alpha-\beta)\right|, \quad \theta = r\left|\cos(\alpha-\beta)\right| \qquad \left(j,k=\overline{1,3}\right).$$

here *six, cix* are the corresponding integral sine and cosine [109].

Thus, the matrix of fundamental solutions is determined in a closed form over formula (8.65). The analogous result will be obtained in [39].

Here it should be mentioned that from the analysis of the homogeneous system of differential (8.57) it follows, that in the given case of a piezoceramic medium there may exist plane monochromatic waves of two types: *quasilongitudinal* and *quasitransverse ones.* The velocity of these waves depends on the direction, in which their propagation takes place. The amplitudes of displacements and electric potentials that refer to the waves with wave number γ are determined by expression

$$W_{j\nu}^0 = \tau_j \exp\left[-i\gamma_\nu(\varphi)(x_1\cos\vartheta + x_3\sin\varphi)\right]$$
$$(\nu = 1,2; j = 1,2,3), \qquad (8.66)$$
$$\gamma_\nu(\varphi) = \frac{\omega}{c_\nu(\varphi)}, \qquad c_\nu(\varphi) = \frac{\sqrt{c_{44}}}{\rho_\nu(\varphi)\sqrt{\rho_*}}.$$

Here φ is the angle between the normal to the front waves and x_1-axis; the functions $\rho_\nu(\varphi)$ are defined in (8.62). The amplitude of potential $W_{3\nu}^0$ in each of the electroacoustic waves is connected with the amplitude of displacements $W_{1\nu}^0$ and $W_{2\nu}^0$, by the relation

$$\tau_3 = \frac{0.5\tau_1(e_{15}+e_{31})\sin 2\phi + \tau_2\left(e_{15}\cos^2\varphi + e_{33}\sin^2\varphi\right)}{\ni_{11}\cos^2\varphi + \ni_{33}\sin^2\varphi}. \qquad (8.67)$$

Figures 8.12 and 8.13, illustrate the behavior of the electric fields in the transversal-isotropic with respect to elastic and electric properties of a piezoceramic medium on the line of peak values of displacements $U_3^{(3)}$ and $U_3^{(1)}$ under the influence of electric charges q and force X_1 in the centre of the considered area. Figure 8.14 shows the equipotentials for the case where an electric charge is the source of oscillations. The curves of the equal modulus of the amplitude of displacements $U_1^{(1)}$ under the influence of force X_1 are shown in Fig. 8.15. In the calculations of the physicalquantities with

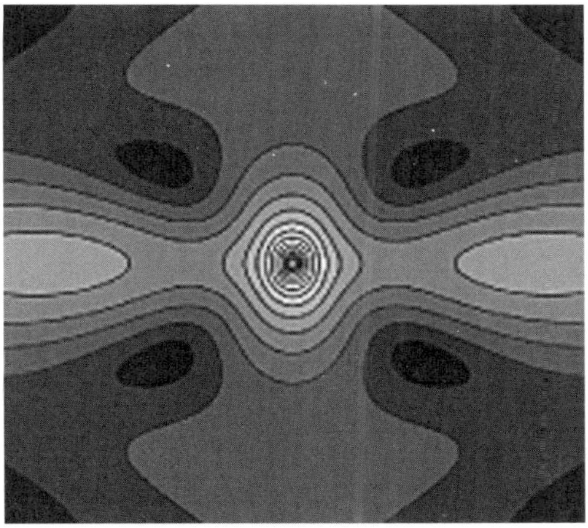

Fig. 8.12. The contour lines of the modulus of displacement amplitude $U_3^{(3)}$ under the influence of an electric charge

Fig. 8.13. The contour lines of the modulus of displacement amplitude $U_3^{(1)}$ under the influence of force X_1

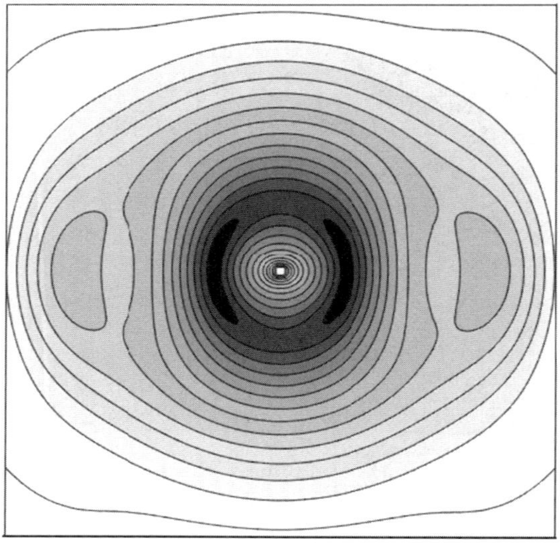

Fig. 8.14. The contour lines of the modulus of the electric potential amplitude under the influence of an electric charge

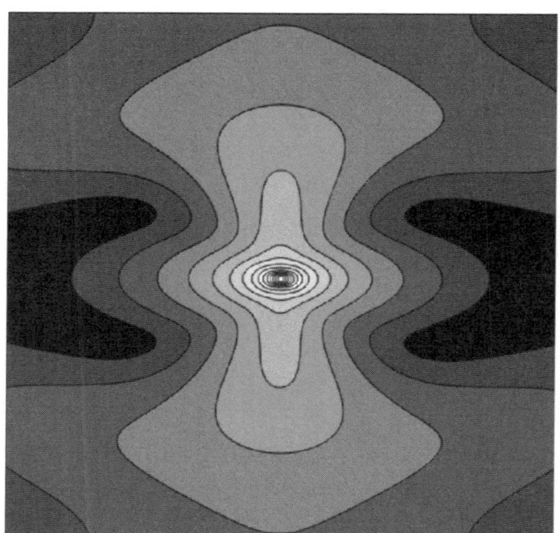

Fig. 8.15. The contour lines of the modulus of displacement amplitude $U_1^{(1)}$ under the influence of force X_1

respect to formula (8.65) it is recommended to use the quadrature formulas of trapezoids [137]

$$\frac{1}{\pi}\int_{0}^{\pi} f(\theta)\ell n\left|\cos\theta\right| d\theta \approx -\frac{1}{2N}\sum_{k=1}^{2N} f(\theta_k)S_k^N,$$

$$S_k^N = \ell n 2 + \sum_{m=1}^{N-1}\frac{(-1)^m\cos 2m\theta_k}{m}, \qquad \theta_k = \frac{(2k-1)\pi}{4N}.$$

8.9 An Unbounded Medium with Tunnel Openings

In the Cartesian system of coordinates x_1, x_2, x_3 let us assume a piezoceramic medium weakened by tunnel openings along-x_2 axis, the cross-section of which are bounded by contours Γ_m $(m = 1, 2, \ldots, n)$. Assume that the x_3-axis coincides with the direction of force lines of the electric field of the preliminarily polarization of the ceramics. On the surface of the cavities may act, stresses $X_m(x_1,\ x_3,\ t) = \mathrm{Re}\left\{P_m(x_1,\ x_3)\,e^{-i\omega t}\right\}$ $(m = 1, 3)$ harmonically changing with time and may radiate plane monochromatic waves of the corresponding types may occur from infinity.

Assumed that the curvatures of contours Γ_m and of functions P_m satisfy Hölder condition on $\Gamma = \cup\Gamma_m$ and, besides, $\cap\Gamma_m = \emptyset$.

In the stated problem, the medium with cavities is in state of plane deformation in x_1Ox_3. The full system of equations has the form of (8.54)–(8.56). It is necessary to add mechanical and electric boundary conditions on the surface of the cavities. Considering that the surface of the cavities are bounded with vacuum (air) we set

$$D_n = D_1\cos\psi + D_3\sin\psi = 0 \quad \text{on } \Gamma_m. \tag{8.68}$$

here ψ denotes the angle between the normal to contour Γ_m and axis Ox_1.

Using the matrix of fundamental solutions (8.65) we write down the integral representations of the amplitude of displacements and electric potentials, corresponding to the dispersed by cavities electroelastic field, in the form [212, 215]

$$W_j(z) = \sum_{k=1}^{3}\int_{\Gamma} p_k(\zeta)\, g_{kj}(\zeta, z)\, ds,$$

$$g_{kj}(\zeta, z) = \frac{\varepsilon_k}{2\pi^2 c_{44}^3}\int_{0}^{\pi}\left\{\sum_{\nu=1}^{2}\lambda_{\nu j}^{(k)}\Psi(\theta_\nu^*) + \delta_j^3\delta_k^3\lambda_0\Omega(\theta^*)\right\}\frac{d\alpha}{A(\alpha)}, \tag{8.69}$$

$$\theta_\nu^* = \gamma\rho_\nu r^*\left|\cos(\alpha-\beta^*)\right|, \qquad \theta^* = r^*\left|\cos(\alpha-\beta^*)\right|,$$
$$r^* = |\zeta - z|, \qquad \beta^* = \arg(\zeta - z), \qquad z = x_1 + ix_3, \qquad \zeta\in\Gamma.$$

Here $p_k(\zeta)$ are the unknown "*densities*" to be determined; ds is the element of the length of contour arc Γ. Integral representations (8.69) have the necessary completeness with respect to the considered boundary problem.

By differentiating functions (8.69) we find

$$\partial_\ell W_j(z) = \sum_{k=1}^{3} \int_\Gamma p_k(\zeta) G_{kj}^{(\ell)}(\zeta, z)\, ds \qquad (\ell = 1, 3), \tag{8.70}$$

$$G_{kj}^{(\ell)}(\zeta, z) = \frac{\varepsilon_k}{2\pi^2 c_{44}^3} \int_0^\pi \left\{ \sum_{\nu=1}^{2} \lambda_{\nu j}^{(k)} \Psi^{(\ell)}(\theta_\nu^*) + \delta_j^3 \delta_k^3 \lambda_0 \Omega^{(\ell)}(\theta^*) \right\} \frac{d\alpha}{A(\alpha)},$$

$$\Psi^{(\ell)}(\theta_\nu^*) = -\gamma \rho_\nu n_\ell H(\theta_\nu^*)\, sign \cos(\alpha - \beta^*),$$

$$\Omega^{(\ell)}(\theta^*) = -\frac{n_\ell}{r^* \cos(\alpha - \beta^*)},$$

$$H(x) = \frac{\pi}{2} e^{ix} + \frac{1}{x} + \cos x\, six - \sin x\, cix.$$

Using functions (8.70) and material (8.56) we may determine the field quantities at any point of the area $z \notin \Gamma$. In the case where $z \to \zeta_0 \in \Gamma$, when calculating $\partial_\ell W_j(z)$ it is required to take into account the integrated terms appearing due to the singular character of kernel $G_{kj}^{(\ell)}(\zeta, z)$ at the point $z = \zeta$. Here we may show that the singular summands appearing in the functions $G_{kj}^{(\ell)}(\zeta, z)$ conform to the static case ($\omega = 0$). In reality, due to (8.60), (8.62), (8.63) and relations $\sum\limits_{\nu=1}^{2} \lambda_{\nu j}^{(k)} = b_j^{(k)} - \delta_j^3 \delta_k^3 \lambda_0$ we have

$$V_j^{(k)} - V_{j0}^{(k)} = \frac{\varepsilon_k R_k}{2\pi} \left\{ \frac{\Delta_j^{(k)}(\xi, \omega)}{\Delta(\xi, \omega)} - \frac{\Delta_j^{(k)}(\xi, 0)}{\Delta(\xi, 0)} \right\} = \tag{8.71}$$

$$= \frac{\varepsilon_k R_k}{2\pi c_{44}^3 A(\alpha)} \sum_{\nu=1}^{2} \lambda_{\nu j}^{(k)} \left(\frac{1}{\rho^2 - \gamma^2 \rho_\nu^2} - \frac{1}{\rho^2} \right).$$

We may transform (8.71) with the help of (8.59) and calculating the corresponding derivatives, we will obtain

$$\partial_\ell \left(W_j^{(k)} - W_{j0}^{(k)} \right) = \frac{\varepsilon_k R_k}{2\pi^2 c_{44}^3} \sum_{\nu=1}^{2} \int_0^\pi \lambda_{\nu j}^{(k)} \Psi_*^{(\ell)}(\theta_\nu) \frac{d\alpha}{A(\alpha)}, \tag{8.72}$$

$$\Psi_*^{(\ell)}(\theta_\nu) = -\gamma \rho_\nu n_\ell h(\theta_\nu)\, sign \cos(\alpha - \beta), \qquad h(x) = H(x) - \frac{1}{x}.$$

Hence, function (8.72) at $z = \zeta$ is reduced to zero, i.e. is regular.

The bulky procedure of the analytical calculation of the integrals corresponding to the statistic parts of thederivatives from fundamental solution (8.65) gives:

$$\partial_\ell W_{j0}^{(k)} = -\frac{\varepsilon_k R_k}{2\pi^2 c_{44}^3 r} \int_0^\pi \frac{b_j^{(k)} n_\ell d\alpha}{A(\alpha)\cos(\alpha-\beta)} = R_k \mathrm{Re} \sum_{\nu=1}^{3} \frac{\omega_{k\nu} m_{\ell\nu} A_{\nu j}^{(k)}}{z_\nu - z_{\nu 0}}, \quad (8.73)$$

$$\begin{aligned}
A_{\nu 1}^{(1)} &= -\left(c_{44}+c_{33}\mu_\nu^2\right)\left(\ni_{11}+\ni_{33}\,\mu_\nu^2\right)-\left(e_{15}+e_{33}\mu_\nu^2\right)^2, \\
A_{\nu 2}^{(1)} &= \mu_\nu\left[\left(c_{44}+c_{13}\right)\left(\ni_{11}+\ni_{33}\,\mu_\nu^2\right)+\left(e_{15}+e_{31}\right)\left(e_{15}+e_{33}\mu_\nu^2\right)\right], \\
A_{\nu 3}^{(1)} &= \mu_\nu\left[\left(c_{44}+c_{13}\right)\left(e_{15}+e_{33}\mu_\nu^2\right)-\left(e_{15}+e_{31}\right)\left(c_{44}+c_{33}\mu_\nu^2\right)\right], \\
A_{\nu 2}^{(2)} &= -\left(c_{11}+c_{44}\mu_\nu^2\right)\left(\ni_{11}+\ni_{33}\,\mu_\nu^2\right)-\mu_\nu^2\left(e_{15}+e_{31}\right)^2, \\
A_{\nu 3}^{(2)} &= \mu_\nu^2\left(e_{15}+e_{31}\right)\left(c_{44}+c_{13}\right)-\left(e_{15}+e_{33}\mu_\nu^2\right)\left(c_{11}+c_{44}\mu_\nu^2\right), \\
A_{\nu 3}^{(3)} &= \left(c_{11}+c_{44}\mu_\nu^2\right)\left(c_{44}+c_{33}\mu_\nu^2\right)-\mu_\nu^2\left(c_{31}+c_{44}\right)^2, \quad A_{\nu i}^{(k)} = A_{\nu k}^{(i)},
\end{aligned}$$

$$m_{1\nu} = 1, \quad m_{3\nu} = \mu_\nu, \quad z_\nu = x_1 + \mu_\nu x_3, \quad z_{\nu 0} = x_{10} + \mu_\nu x_{30},$$

where μ_ν ($\mathrm{Im}\ \mu_\nu > 0,\ \nu = 1,2,3$) are the roots of algebraic equation

$$\begin{aligned}
&a\mu^6 + b\mu^4 + c\mu^2 + d = 0, \qquad (8.74)\\
&a = -c_{44}c_{33}\ni_{33}\left(1+k_{33}^2\right), \quad k_{33}^2 = \frac{e_{33}^2}{c_{33}\ni_{33}}, \quad k_{15}^2 = \frac{e_{15}^2}{c_{44}\ni_{11}}, \\
&b = c_{44}\left(2c_{13}\ni_{33} - c_{33_{11}} + 2e_{31}e_{33}\right) - c_{11}c_{33_{33}}\left(1+k_{33}^2\right) - \\
&\quad - c_{33}\left(e_{15}+e_{31}\right)^2 + 2e_{33}c_{13}\left(e_{15}+e_{31}\right) + c_{13}^2\ni_{33}, \\
&c = c_{44}\left(2c_{13}\ni_{11} - e_{31}^2\right) + 2c_{13}e_{15}\left(e_{15}+e_{31}\right) + c_{13}^2\ni_{11} - \\
&\quad - c_{11}\left(c_{44}\ni_{33} + c_{33}\ni_{11} + 2e_{15}e_{33}\right), \\
&d = -c_{11}c_{44}\ni_{11}\left(1+k_{15}^2\right),
\end{aligned}$$

and the constants $\omega_{k\nu}$ are determined from three material systems of linear algebraic equations

$$\begin{aligned}
&\mathrm{Im}\sum_{\nu=1}^{3} d_{\nu j}^{(k)}\omega_{k\nu} = f_j^{(k)} \qquad \left(j=\overline{1,6}; k=\overline{1,3}\right), \\
&d_{\nu 1}^{(k)} = -\gamma_\nu^{(k)}\mu_\nu, \quad d_{\nu 2}^{(k)} = \gamma_\nu^{(k)}, \quad d_{\nu 3}^{(k)} = -r_\nu^{(k)}, \quad d_{\nu,i+3}^{(k)} = -A_{\nu i}^{(k)}, \\
&f_i^{(k)} = \delta_i^k \big/ 2\pi, \quad f_{i+3}^{(k)} = 0 \quad \left(i=\overline{1,3}\right), \\
&\gamma_\nu^{(k)} = c_{13}A_{\nu 1}^{(k)} + \left(c_{33}A_{\nu 2}^{(k)} + e_{33}A_{\nu 3}^{(k)}\right)\mu_\nu, \\
&r_\nu^{(k)} = \left(\ni_{33}\,A_{\nu 3}^{(k)} - e_{33}A_{\nu 2}^{(k)}\right)\mu_\nu - e_{31}A_{\nu 1}^{(k)}. \qquad (8.75)
\end{aligned}$$

Thus, integral representations (8.70) taking into account (8.72), (8.73) may be written in the form

$$\partial_\ell W_j(z) = \sum_{k=1}^{3} \int_\Gamma p_k(\zeta) F_{kj}^{(\ell)}(\zeta, z)\, ds \qquad (\ell = 1, 3),$$

$$F_{kj}^{(\ell)}(\zeta, z) = \mathrm{Re} \sum_{\nu=1}^{3} \frac{\omega_{k\nu} m_{\ell\nu} A_{\nu j}^{(k)}}{z_\nu - \zeta_\nu} + \frac{\varepsilon_k}{2\pi^2 c_{44}^3} \sum_{\nu=1}^{2} \int_0^\pi \lambda_{\nu j}^{(k)} \Psi_*^{(\ell)}(\theta_\nu^*) \frac{d\alpha}{A(\alpha)}. \tag{8.76}$$

Substituting the limiting values of derivatives (8.76) at $z \to \zeta_0 \in \Gamma$ into mechanical and electric boundary conditions, we come to a system of three singular integral equations of the second kind with respect to functions $p_k(\zeta)$

$$\begin{aligned}
&\frac{1}{2} p_\ell(\zeta_0) + \sum_{k=1}^{3} \int_\Gamma p_k(\zeta) M_{\ell k}(\zeta, \zeta_0)\, ds = N_\ell(\zeta_0) \quad (\ell = \overline{1,3}),\\
&M_{1k} = T_{11}^{(k)} n_1^* + T_{13}^{(k)} n_3^*, \quad M_{2k} = T_{13}^{(k)} n_1^* + T_{33}^{(k)} n_3^*,\\
&M_{3k} = \left[e_{15} \left(F_{0k2}^{(1)} + F_{0k1}^{(3)} \right) - \ni_{11} F_{0k3}^{(1)} \right] n_1^* + \left(e_{31} F_{0k1}^{(1)} + e_{33} F_{0k2}^{(3)} - \ni_{33} F_{0k3}^{(3)} \right) n_3^*,\\
&T_{11}^{(k)} = c_{11} F_{0k1}^{(1)} + c_{13} F_{0k2}^{(3)} + e_{31} F_{0k3}^{(3)},\\
&T_{13}^{(k)} = c_{44} \left(F_{0k2}^{(1)} + F_{0k1}^{(3)} \right) + e_{15} F_{0k3}^{(1)},\\
&T_{33}^{(k)} = c_{13} F_{0k1}^{(1)} + c_{33} F_{0k2}^{(3)} + e_{33} F_{0k3}^{(3)},\\
&F_{0kj}^{(\ell)} = F_{kj}^{(\ell)}(\zeta, \zeta_0),\\
&N_\ell(\zeta_0) = N_\ell^{(0)}(\zeta_0) + \sum_{j=1}^{2} N_\ell^{(j)}(\zeta_0),\\
&N_1^{(0)}(\zeta_0) = P_1, \quad N_2^{(0)}(\zeta_0) = P_3, \quad N_3^{(0)}(\zeta_0) = 0,\\
&N_1^{(1)}(\zeta_0) = c_{13} \left(1 + \frac{e_{31} e_{33}}{\ni_{33}\, c_{13}} \right) \tau_2 i \gamma_1 \left(\frac{\pi}{2} \right) n_1^* e^{-i\gamma_1\left(\frac{\pi}{2}\right) \mathrm{Im}\zeta_0},\\
&N_2^{(1)}(\zeta_0) = c_{33} \left(1 + k_{33}^2 \right) \tau_2 i \gamma_1 \left(\frac{\pi}{2} \right) n_3^* e^{-i\gamma_1\left(\frac{\pi}{2}\right) \mathrm{Im}\zeta_0},\\
&N_1^{(2)}(\zeta_0) = c_{44} \left(1 + k_{15}^2 \right) \tau_2 i \gamma_2(0)\, n_3^* e^{-i\gamma_2(0) \mathrm{Re}\zeta_0},\\
&N_2^{(2)}(\zeta_0) = c_{44} \left(1 + k_{15}^2 \right) \tau_2 i \gamma_2(0)\, n_1^* e^{-i\gamma_2(0) \mathrm{Re}\zeta_0},\\
&N_3^{(j)}(\zeta_0) = 0, \quad n_1^* = \cos\psi_0, \quad n_3^* = \sin\psi_0,\\
&\psi_0 = \psi(\zeta_0); \quad \zeta, \zeta_0 \in \Gamma = \cup \Gamma_m \qquad \left(m = \overline{1,n}; j = 1, 2 \right).
\end{aligned} \tag{8.77}$$

The right parts of (8.77) $N_\ell^{(j)}(\zeta_0)$ were found by taking into account expressions (8.66), (8.67); they correspond to two types of loadings: a plane quasilongitudinal wave, propagating in the negative direction of $x_3\,(j=1)$ -axis, and a plane quasitransversal wave radiating in the negative direction of $x_1\,(j=2)$ -axis. The third integral equation ($\ell=3$) in the system (8.77) refers to the electric boundary condition (8.68).

To determine the dynamic stress concentration in the medium with openings we should calculate the normal stress σ_ϑ on the contour Γ. Due to the integral representations (8.76) we have

$$\begin{aligned}
&\sigma_\vartheta = \mathrm{Re}\left(T_\vartheta e^{-i\omega t}\right),\\
&T_\vartheta(\zeta_0) = T_0 + \sum_{k=1}^{3}\left\{p_k(\zeta_0)\,t_k(\zeta_0) + \int_\Gamma p_k(\zeta)\,T_\vartheta^{(k)}(\zeta,\zeta_0)\,ds\right\},\\
&t_k(\zeta_0) = -\pi \mathrm{Im}\sum_{\nu=1}^{3}\frac{\left(\mu_\nu n_3^* + n_1^*\right)^2}{\mu_\nu n_1^* - n_3^*}\omega_{k\nu}\gamma_\nu^{(k)}, \qquad (8.78)\\
&T_\vartheta^{(k)} = T_{11}^{(k)}\left(n_3^*\right)^2 + T_{33}^{(k)}\left(n_1^*\right)^2 - 2T_{13}^{(k)}n_1^*n_3^*,\\
&T_0 = -i\gamma_1\left(\frac{\pi}{2}\right)\tau_2 e^{-i\gamma_1\left(\frac{\pi}{2}\right)\mathrm{Im}\zeta_0}\left[\left(c_{13} + \frac{e_{31}e_{33}}{\ni_{33}}\right)\left(n_3^*\right)^2 + c_{33}\left(1+k_{33}^2\right)\left(n_1^*\right)^2\right] +\\
&\qquad + 2i\gamma_2(0)\,\tau_2 c_{44}\left(1+k_{15}^2\right)n_1^*n_3^* e^{-i\gamma_2(0)\mathrm{Re}\zeta_0}.
\end{aligned}$$

The quantities $T_{ij}^{(k)}$ are defined in (8.77).

Let us investigating the stress concentration in a space (ceramics $PZT-4$) weakened by tunnel openings of "*elliptic*" or "*square*" type with parametric equations, respectively,

$$\begin{aligned}
&\mathrm{Re}\zeta = R_1\cos\eta, \qquad \mathrm{Im}\zeta = R_2\sin\eta, \qquad \eta\in[0,2\pi], \qquad (8.79)\\
&\mathrm{Re}\zeta = a\left(\cos\eta + C\cos 3\eta\right),\\
&\mathrm{Im}\zeta = a\left(\sin\eta - C\sin 3\eta\right), \qquad C = 0.14036.
\end{aligned}$$

Taking into account (8.79) the functions $p_k(\zeta)\ \left(k=\overline{1,3}\right)$ were calculated from (8.77) by the method of quadratures (See: Appendix 12.8, Par. B.2). By formula (8.78) we calculate stress $T_\vartheta(\zeta_0)$. The complex roots of (8.74) corresponding to ceramics $PZT-4$, are: $\mu_1 = 1.203802i$, $\mu_{2,3} = \pm 0.200427 + 1.069137i$.

Figure 8.16 illustrates the behavior of the modulus of the amplitude of relative stress $\chi = |T_\vartheta/\Lambda|$ at point $\eta=\pi$ of the contour of an elliptic cavity in functions of normalized wave number γR under the influence of quasilongitudinal wave ($R = 0.5\,(R_1+R_2)$). Curves 1 and 2 are constructed for values $R_1/R_2 = 2$ and 1, respectively. The dashed lines correspond to the values of piezoelectric constants $e_{33}=e_{13}=0,\ e_{15}=0.1cal/m^2$, which corresponds to the "*piezopassive*" material (at all $e_{ij}=0$ system (8.75) becomes confluent).

The changes of quantity χ on the contour of the square opening during the radiation of quasilongitudinal and quasitransversal waves is shown in Figs. 8.17 and 8.18, respectively. The curve with number m corresponds to the

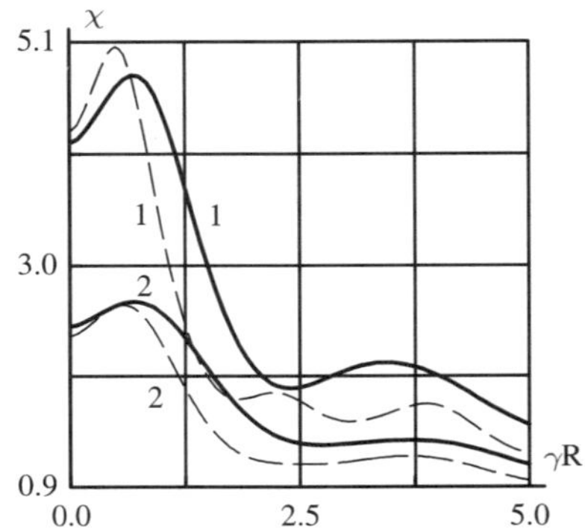

Fig. 8.16. Changes of quantity L at point $\eta = \pi$ of the contour of an elliptic cavity in function of γR under the influence of a quasilongitudinal wave

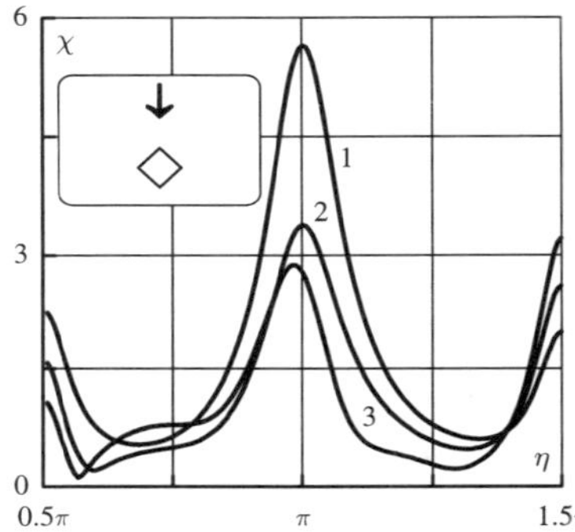

Fig. 8.17. Changes of quantity χ on the contour of a square opening at radiation of a quasilongitudinal wave

value of the normalized wave number $\gamma a = m$. The graphs in Fig. 8.19 (for the same values γa) refer to the case of loading of the cavity by uniform pulsating pressure of amplitude $\Lambda = P\,(P_1 = Pn_1^*,\ P_3 = Pn_3^*)$. The quantity $\Lambda = \left|T_{33}^0\right| = \tau_2\gamma\sqrt{c_{44}c_{33}\,(1+k_{33}^2)}$ represents the modulus of the amplitude of stress σ_{33} in the radiating quasilongitudinal wave, and $\Lambda = \left|T_{13}^0\right| = \tau_2\gamma c_{44}\sqrt{1+k_{15}^2}$ is the modulus of the amplitude of stress σ_{13} in the radiating quasitransversal wave.

From calculations it follows that under the dynamic excitation a redistribution of stress σ_ϑ takes place on the surface of the cavity. The influence of the inertial effect is manifested by the increase of the stress concentration at the cavity (for a range of changes of frequency ω) in comparison with the corresponding static case ($\omega = 0$). The effect of connection of the coupled fields, as it seen in Fig. 8.16, may influence the stress concentration considerably, what is also observed in the case of the deformation of a piezoceramic

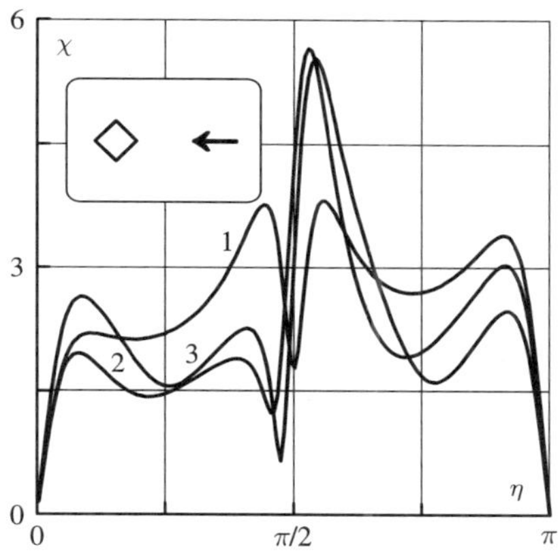

Fig. 8.18. Changes of quantity χ on the contour of a square opening at radiation of a quasitransverse wave

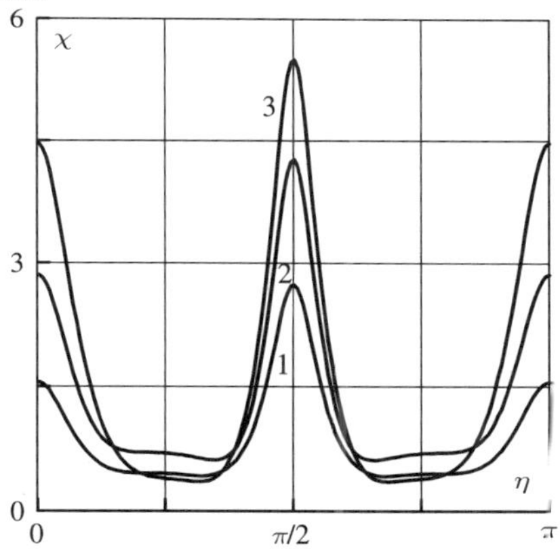

Fig. 8.19. Changes of χ on the contour of a square opening under the influence of normal pressure

medium with defects in the cavity perpendicular to the direction of the preliminary polarization of ceramics [145].

8.10 Oscillation of a Cylinder Under the Influence of Pulsating Pressure

The integral (8.77) may be used for investigating two-dimensional electroelastic fields in a continuous along x_2 -axis cylinder, excited by pulsating pressure of amplitude P (the outer medium is air). For this in system (8.77) before the integrated summands we should switched the signs, and its right part should be set as $N_1(\zeta_0) = -Pn_1^*,\ N_2(\zeta_0) = -Pn_3^*,\ N_3(\zeta_0) = 0$.

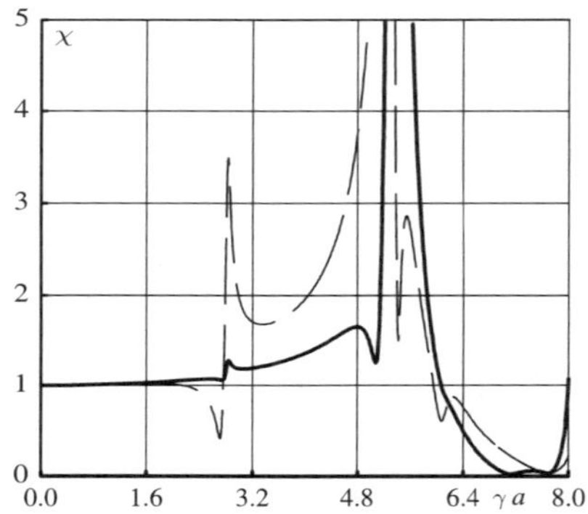

Fig. 8.20. Changes of $\chi = |T_\theta/N^*|$ at points $\eta = \pi$ and $\eta = 9\pi/19$ of the contour of a cylinder as a function of normalized wave number $\gamma\alpha$

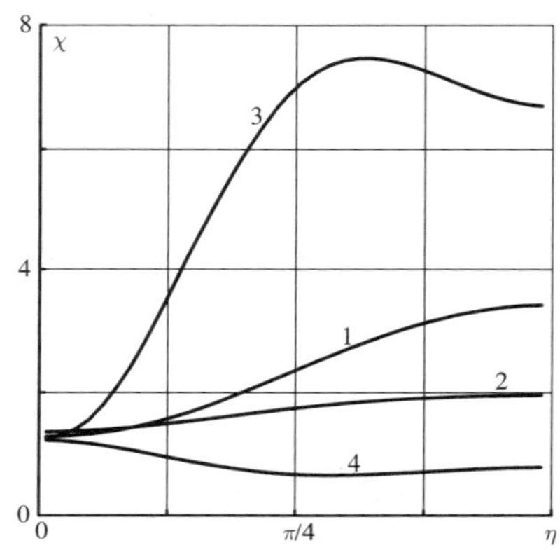

Fig. 8.21. Distribution of quantity χ on the contour of a cylinder for various parameter values of $\gamma\alpha$

To determine stresses σ_ϑ on the boundary of the cylinder we use (8.78), where we should set

$$t_k(\zeta_0) = \pi \operatorname{Im} \sum_{\nu=1}^{3} \frac{(\mu_\nu n_3^* + n_1^*)^2}{\mu_\nu n_1^* - n_3^*} \omega_{k\nu} \gamma_\nu^{(k)}.$$

As an example consider a circular cylinder (material - $PZT - 4$). Figure 8.20 illustrates the changes of $\chi = |T_\vartheta/N^*|$, at points $\eta = \pi$ and $\eta = 9\pi/19$ (dashed line) of the contour of the cylinder as a function of the normalized wave number γa (η is the polar angle, a is the cylinder radius). The distribution of quantity χ on the cylinder contour is shown in Fig. 8.21. Curves 1–4, corresponding to the distribution of stress σ_ϑ on the surface of the cylinder, are constructed for values $\gamma a = 2.84,\ 4,\ 5.07$ and 6, respectively.

9

Fundamentals of Magnetoelasticity

9.1 Magnetic Field and its Properties

Any current (moving charged particles) is characterized by its ability to interact with other currents by means of forces, which are called magnetic forces. In this case we have to do with field interaction: the moving electric charge, besides the electric field is characterized by a magnetic field that acts on the moving charged particles or the bodies.

The so-called induction of magnetic field is the fundamental power feature of the magnetic field. The role of the electrical point charge in the study of magnetic field plays a small, with respect to size, loop with the current and the direction of the magnetic force on the given point of field coincides with the direction of the vector of the magnetic moment of loop $\overrightarrow{p}_m$. However, it is necessary still to know additionally the module of this force. Here we have at our disposal the experimental law of Ampere, which connects the module of the force, which acts on a small, with respect to length, rectilinear portion with current, placed into the magnetic field with module of the induction vector of the magnetic field

$$\Delta F = kI\Delta lB \sin\alpha,$$

where Δl us the length of the small rectilinear portion of the conductor, along which the current I flows; k is the coefficient of proportionality, that depends on the selection of system of units; B is the module of the vector of the magnetic field in the place of the conductor; α is the angle between the direction of conductor and the direction of vector $\overrightarrow{B}$.

If the magnitudes $\Delta F, \Delta l, I, B$ are measured by the units of a system, then $k = 1$ and, further, if $\alpha = \frac{\pi}{2}$, then $\sin\alpha = 1$ and

$$B = \frac{\Delta F}{I\Delta l} \tag{9.1}$$

In this way, the induction of the magnetic field is a physical magnitude, the module of which is determined by the ratio of the module ΔF of force, which acts on the given place of the magnetic field, by the conductor of length Δl, the force of the current on which equals I, to the magnitudes Δl and I when the orientation of the conductor it perpendicular to the direction of vector $\overrightarrow{B}$.

In accordance with (9.1), the unit of measurement of the magnetic induction in the system SI is established. When $\Delta F = 1\mathrm{N}$, $\Delta l = 1m$, $I = 1\mathrm{A}$ we get:

$$B = \frac{1\mathrm{N}}{1A \cdot 1m} = 1\frac{\mathrm{N}}{A \cdot m} = 1\mathbf{T}. \quad (\mathrm{T} - \mathrm{tesla}).$$

Consequently, 1 Tesla is the induction of this uniform (identical in all points) magnetic field, in which to every meter of length of the conductor with current in 1A, located perpendicularly to the direction of the field $\overrightarrow{B}$, acts a force in 1N (this is a high value). In CGS magnetic induction is measured in Gausses.

An auxiliary vector $\overrightarrow{H}$, called intensity of the magnetic field, is usually introduced together with the vector of induction $\overrightarrow{B}$. The intensity of the magnetic field is defined as the ratio of the induction of field in the vacuum $\overrightarrow{B}$ to the magnetic constant μ_0:

$$\overrightarrow{H}_0 = \frac{\overrightarrow{B}_0}{\mu_0},$$

where $\mu_0 = \dfrac{1}{\ni_0 c^2} = 4\pi \cdot 10^{-7}\dfrac{\mathrm{N}}{\mathrm{A}^2} \left(or \ \dfrac{Wb}{A \cdot m}\right)$;

$$\ni_0 \text{ is the electric constant } \ni_0 = \frac{1}{4\pi k} = 8,85 \cdot 10^{-12}\frac{C^2}{Nm^2};$$

the coefficient of proportionality k is numerically equal to the force, with which the two unit charges interact on the unit distance; $c = 3 \cdot 10^8 m/s$ is the velocity of light in the vacuum.

The unit of measurement of the intensity of the magnetic field in the system SI is $1A/m$ (the intensity of the magnetic field in the center of the circular frame with radius 1m, along which current 2A flows). In the CGS system Oersted is used: $1Oe \approx 80\frac{\mathbf{A}}{\mathbf{m}}$.

Note: By analogy with the study of the electrical field, where vector $\overrightarrow{E}$ is called intensity of the electrical field, and auxiliary vector $\overrightarrow{D}$, which is called electrical displacement $\left(\overrightarrow{D} = \varepsilon_0 \overrightarrow{E}\right)$, in the theory of magnetism we had to call vector $\overrightarrow{B}$ intensity of the magnetic field and auxiliary vector, magnetic induction or magnetic displacement. The confusion in terminology appeared when the physical sense of the vectors of the field $\overrightarrow{B}$ and $\overrightarrow{H}$ was still not sufficiently clear.

9.1.1 Action of the Magnetic Field on the Moving Electric Charges

The moving electric charges are characterized not only by the fact that in their surrounding space magnetic fields are revealed, but also by the fact that the charges themselves are subject to the influence of external magnetic fields. The force, which acts on the moving charged particle, can be determined by means of the Ampere law; this force is called Lorentz force and its modulus is given by:

$$F_L = q\nu B \sin\alpha$$

where q is the charge of the particle; ν is the module of the average speed of the ordered movement of the particle; α is the angle between the vector $\overrightarrow{\nu}$ of the speed of the charged particle and the induction vector of the magnetic field $\overrightarrow{\boldsymbol{B}}$.

9.2 The Magnetic Properties of the Substance

It is shown in physics that the motion of electrons around the atomic nucleus is equivalent to micro-currents, which take place on closed loops. The vector of magnetic moment $\overrightarrow{p}_m$ connects the loop with the current. Thus, inside every substance, which consists of atoms, molecules and ions, there are many microscopic loops that act as currents, each of which is characterized by its magnetic field. The hypothesis that was once stated by Ampere about the presence of molecular currents inside the substance was turned out very fruitful, except that here we have not molecular, but electronic currents.

A question arises: why many substances under normal conditions do not possess magnetic properties manifesting themselves in their surrounding space? This is explained by the fact that even in a very small volume of substance the number of electron orbits is enormous. For example, in $1\,\text{cm}^3$ copperplate $\approx 2,43 \cdot 10^{24}$ electron orbits are counted. All these orbits are oriented in the space in different ways, and their magnetic moments can be compensated. Therefore, a piece of copper, under normal conditions, does not manifest itself magnetic properties, but they are inherent to every atom of it.

9.2.1 Action of External Magnetic Field on the Substance

The external magnetic field, into which a substance is introduced, affects the electrons of the substance and in this way changes its internal magnetic field. Indeed, every electron orbit is similar to a loop with current, and this loop in an external magnetic field is oriented in a specific manner. In this way, the action of an external magnetic field on the substance reorients the electron orbits (or the vectors of the magnetic moments). This reorientation ensures a magnetic field inside the substance, different from that, which was in it before the introduction into the external field.The resulting magnetic field within the limits of the capacity of the substance would be the result of the imposition of two magnetic fields – one external and one internal, induced in the substance by the action of the external field. The result of this imposition, as we will see below, can be different: the induction of the resulting magnetic field can be both less and more than the induction of the external field.

The dimensionless magnitude μ, which characterizes the degree of participation of the substance in the change of the resulting magnetic field, is called relative magnetic permeability of the substance:

$$\overrightarrow{B} = \mu \overrightarrow{B}_0 \tag{9.2}$$

where B, B_0 are the values of the induction of the magnetic field of one and the same sources in the substance and the vacuum, respectively. It is possible to show that $\overrightarrow{B} = \mu\mu_0 \overrightarrow{H}$.

The magnitude, which characterizes the degree of magnetization of the substance in the magnetic field is also introduced, that is the magnetization vector or magnetization. Magnetization vector is called the magnetic moment of the volume unit of the magnetized substance, that is the sum vector of the magnetic moments of all atoms included in the volume unit of the magnetized substance:

$$\overrightarrow{M} = \sum_{k=1}^{n} (\overrightarrow{p}_m)_k.$$

Dimension: $[\mathbf{M}] = \mathbf{A}/\text{м} = [\mathbf{H}]$.

9.2.2 Classification of Substances According to Magnetic Properties

All substances in nature can be considered as magnetic materials. The following kinds of magnetic materials distinguished together with their corresponding magnetic phenomena: diamagnetic (diamagnetism), paramagnetic (paramagnetism), ferromagnetic (ferro-magnetism), antiferromagnetic (antiferromagnetism), ferrimagnetic (ferrimagnetism); diamagnetic and paramagnetic materials are classified under weak magnetic materials and all the rest, under the strong ones.

Diamagnetism

Under diamagnetic are classified all inert gases, hydrogen, nitrogen, chlorine, ammonia, metals (zinc, gold, mercury), nonmetals (phosphorus, silicon, sulfur), tree, marble, glass, wax, water, petroleum and others.

Outwardly, diamagnetism manifests itself by the fact that the diamagnetic materials are ejected out of the magnetic field.

The diamagnetic materials are characterized by the fact that the sum orbital magnetic moment of all electrons of every atom is equal to zero. This means that the atom of a diamagnetic material in absence of an external magnetic field does not manifest itself any magnetic properties. The action of an external magnetic field on the atom of a diamagnetic material leads to magnetic moment in it, directed always against the induction of the external magnetic field. This is connected with the fact that the electrons in the external magnetic field experience the action of Lorentz forces. In this case, the frequency of revolution of the electrons around the orbits changes, and the orbits themselves can change their orientation. This means that a magnetic field appears inside the diamagnetic material, the induction vector of which is directed in the opposite way to the induction vector of the of external field. Within the limits of the diamagnetic material, the external field is weakened. Therefore, from (9.2) for the diamagnetic material $\mu < 1$.

For a diamagnetic material, the magnetization vector is proportional to the intensity of the magnetizing field:

$$\overrightarrow{M} = \chi \overrightarrow{H} \tag{9.3}$$

where the coefficient of proportionality χ is called the magnetic susceptibility of the substance; this is a dimensionless magnitude.

It is obvious that for a diamagnetic material $\chi < 0$. The magnetic susceptibilities of different substances in one and the same state of aggregation have little difference: for the gases $\chi \approx 10^{-9}$, for the liquids $\chi \approx 10^{-6}$, for the solid bodies $\chi \approx 10^{-5}$.

It can be shown that $\mu = 1 + \chi$. For a diamagnetic material, μ and χ are constants.

Paramagnetism

The considered mechanism of magnetization of diamagnetic materials is reduced to the induction of supplementary magnetic moment in the atom under the action of an external magnetic field. This phenomenon is characteristic for atoms of any substance and, therefore, all substances must possess diamagnetic properties. Question: hence these paramagnetic properties?

The atoms of paramagnetic materials even in absence of external magnetic field possess some magnetic moment, different from zero; however, because of their chaotic directivity in the macroscopic volume of the para-magnetic material magnetic properties do not appear.

When a paramagnetic material is introduced into an external magnetic field an internal magnetic field, the induction vector of which coincides in direction with the induction vector of the external field. There is also a diamagnetic effect, but it is weaker than the paramagnetic one.

For paramagnetic materials, $\mu > 1$. For gases (oxygen, nitrogen oxide), $\chi \approx 10^{-7}$, for solid bodies (platinum, sodium, lithium), $\chi \approx 10^{-4}$, for other materials, it is considerably less, i.e. $\mu \approx 1$.

Thus, the magnetic susceptibility of paramagnetic materials is approximately 100 times more than in diamagnetic ones. Therefore, the strong paramagnetic effect overlaps the weaker diamagnetic one.

Formulas (9.2), (9.3), where χ and μ are constants, also hold.

Ferromagnetism

Under the class of ferromagnetic materials fall iron, nickel, cobalt, their compounds and alloys, as well as some alloys of manganese, silver, aluminum, and others. Under relatively low temperatures, ferromagnetic materials are some rare-earth elements.

All ferromagnetic materials are characterized by:

1. Crystalline structure;
2. High positive value $\mu >> 1$;
3. Essential and nonlinear dependence of μ and χ on the intensity of field and temperature;
4. Ability to be magnetized to saturation under usual temperatures even in weak fields;
5. Hysteresis, i.e. dependence of magnetic properties on the previous magnetic state ("magnetic prehistory");
6. The Curie point, i.e., by a temperature, beyond which the material looses its ferromagnetic properties.

In contrast to diamagnetic and paramagnetic mate-rials, for which the magnetization vector is proportional to the intensity of the magnetizing field, in ferromagnetic materials its magnitude is fairly complicated nonlinear function of intensity. For nickel Fig. 9.1 is valid:

For convenience in the representation, the scale is disrupted: the domain of the field intensity in the section $0,1 \cdot 10^3 \ldots 2 \cdot 10^3$ A$/m$ is extended. It is

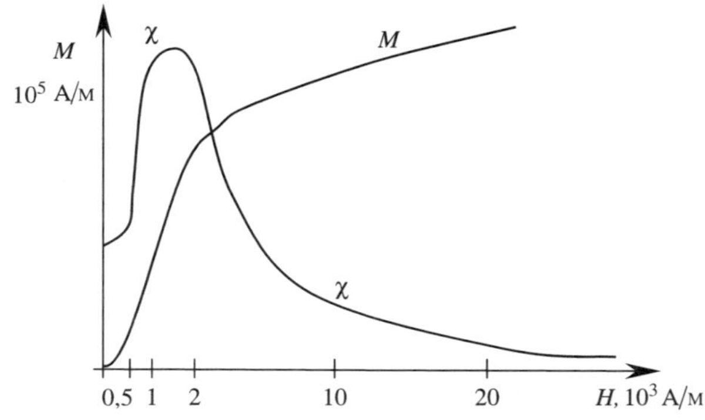

Fig. 9.1. Diagram of the dependence of magnetization and magnetic susceptibility on the intensity of the magnetic field

clear from the diagram that magnetization of nickel increases rapidly in the beginning with growth rate H. Then it reaches magnetic saturation, when magnetization practically ceases to depend on the magnitude of the magnetizing field. The expressions $\overrightarrow{M} = \chi\overrightarrow{H}$ and $\overrightarrow{B} = \mu\mu_0\overrightarrow{H}$ loose here their sense. They can be used only formally, having in mind that χ and μ are not constants, but complex functions of the intensity.

In contrast to diamagnetic and paramagnetic materials, which can be found in any state of aggregation, ferromagnetic properties are observed only in crystals. In liquid or gaseous state, ferromagnetic substances behave as usual paramagnetic materials, i.e. as it is confirmed by experience, ferromagnetic properties are determined not by special properties of the atoms of these substances, but by the special structure of their crystal lattice. A change in the structure of the lattice brings a corresponding change in the magnetic properties of ferromagnetic materials.

For every ferromagnetic material there is definite temperature – the Curie point, beyond which the ferromagnetic properties of the substance disappear: for **Fe** the Curie point is 770 °C, for Ni 360 °C, for permalloys (alloy 70% **Fe** and 30% Ni) is 70 °C.

Since ferromagnetism is present only in crystals, which are anisotropic, then a single crystal of ferromagnetic substance must be anisotropic in terms of magnetic properites. Experience actually confirms that the magnetic properties of single crystals are different in different directions.

The graphs below (Fig. 9.2) represent of magnetization of a single crystal **Fe**, which has a lattice of the kind of body-centered cube (bcc).

The edge of cube is the direction of soft magnetization; the diagonal of cube is the direction of hard magnetization.

In polycrystals the anisotropy of magnetization does not manifest itself, because the various microcrystal particles are oriented disorderly.

Ferromagnetic materials have hysteresis, consisting in the fact that the magnetization of a ferromagnetic material depends not only on the intensity

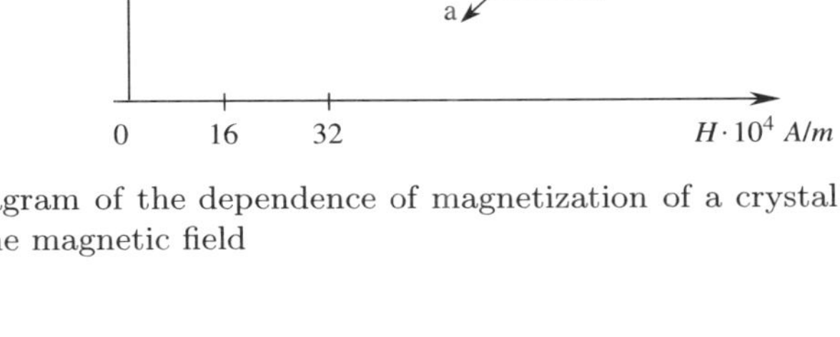

Fig. 9.2. Diagram of the dependence of magnetization of a crystal of iron on the intensity of the magnetic field

of the magnetic field in a given moment, but also on the pre-magnetization model, its "magnetic history."

Therefore, it cannot be generally indicated, what magnetization of a ferromagnetic material corresponds to a given value of the intensity of the magnetizing field, if it is not known, in what state it was found before. The same holds, naturally, for the values μ and χ.

In Fig. 9.3 point 0 refers to absence of magnetization. The section $0a$ is the dependence of $M\,(H)$ when magnetization is produced for the first time.

When a value $H > H_s$ is reached (to these values of field corresponds the magnetization saturation M_s) M stops increasing. In case when saturation is achieved, the intensity of the field H begins to decrease, then magnetization would drop not according to the curve $0a$, but according to ab, i.e. to the same values H, assumed in the reverse order, correspond larger values of M. When $H = 0$, magnetization does not disappear, but residual magnetization M_r is preserved (segment $0b$).

When $\overrightarrow{H}$ is applied in the opposite direction and a field with intensity $(-H_c)$ appears, magnetization equals to zero. The intensity of the magnetizing field H_c is called *coercive force.* With larger increase of the inverse in direction field H magnetizability of opposite sign appears. In this case, saturation can also be reached (point a').

If the magnetic field increases further, then magnetization increases according to the curve $a'b'c'$; moreover, $H = H_c$ $M = 0$.

Different ferromagnetic materials give very diverse hysteresis curves. With low values of H_c we get the so-called soft ferromagnetic materials (Fe, silicic steel, alloys Fe with Ni); with larger values of H_c, we get solid ferromagnetic materials (carbon and special steels).

The peculiarities of ferromagnetic materials can be explained, following the classical theory of ferromagnetism, proposed by P. Weiss. According to his theory, in temperatures lower than the Curie point ferromagnetic materials consist of microscopic regions, called *domains*, in each of which the magnetic

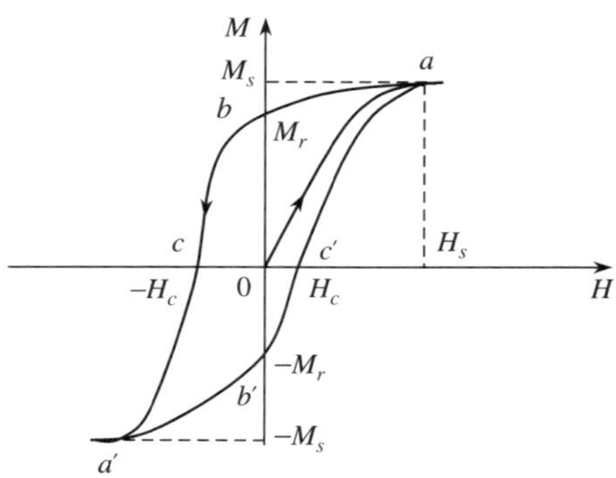

Fig. 9.3. Hysteresis loop

moments of the atoms are strictly located in one direction, which corresponds to the direction of the weak magnetization. Thus every domain is turned out magnetized to saturation independently from the presence of any external magnetic field and its magnitude. The size of the domains is 10^{-2}–10^{-3} cm, i.e., they can be possibly observed by a microscope. In absence of external magnetic field the domains in a single crystal are located in such a way, that their magnetic fields close each other and the resulting external magnetic field is proved equal to zero.

In the diagram, pointers show the direction of magnetization vectors inside the domains.

As D.D. Landau and E.M. Lifschitz have shown, this system of domains in a single crystal characterize the state of minimum energy of the magnetic field, which, from the standpoint of the laws of thermodynamics, ensures the stable equilibrium of entire system.

Polycrystals consist of single grains, in which the directions of the lightest magnetization are oriented disorderly. Each grain is split into several domains, directed along the line of the lightest magnetization.

Switching on a weak external magnetic field, we will always find many domains, in which the direction of magnetization vector coincides with the direction of the external field. These domains would possess minimum of energy and therefore it would result to a state of stable equilibrium. Adjacent domains possess maximum energy. Therefore, an energetically profitable situation is when the magnetic moments of some atoms change their direction and join to those domains, which energy is minimum. This process is called *shift of the domain boundaries.* When the intensity of the external magnetic field increases, the boundaries of the domains are shifted more strongly and, in this case, the walls of the domains begin to meet on their path the defects of the crystal (dislocation, voids, inclusions), which prevent a change in the magnetic moment in the atoms located close to the defects. When the field increases the energy of the atoms grows and the atoms change leap-wise their direction - the wall of the domain is stripped from this place and moved to the following defect. In this way, the process of the magnetization of polycrystals continues not smoothly, but leap-wise, which is confirmed experimentally.

In sufficiently strong fields, all walls move to the boundaries of the crystalline grains, and every grain turns out magnetized along the direction of lightest magnetization, which forms the smallest angle with the direction of the external field. If the intensity of the magnetizing field is strengthened further, the magnetic moments of the grains begin to orient along the field. When the magnetic moments of all grains would be oriented along the field, magnetic saturation begins.

In this way, using the idea of the domain structure of ferromagnetic materials, it is possible to explain all peculiarities of the process of their magnetization. The Curie point is proved to be the temperature, beyond which the domain structure destructs.

Two questions remain open: 1) what forces lead to the situation, when inside the domain the magnetic moments of all atoms spontaneously are oriented along the direction of weak magnetization. 2) Why some substances possess ferromagnetic properties, while others do not. Answers to these questions are provided by the quantum theory of magnetic phenomena.

Antiferromagneticism

The study of the properties of ferromagnetic materials led D.D. Landau to the conclusion that there must exist substances, in which, under low temperatures, the magnetic moments are oriented in antiparallel manner, i.e. the magnetic moments of two adjacent atoms are located in opposite direction to each other. Later, compounds like MnO, MnS, CrT3, NiCr and others were discovered experimentally.

Antiferromagnetic materials are characterized by a temperature (called Neel's point), beyond which the described magnetic order is destroyed and the substance behaves in a way like a paramagnetic material.

Ferrimagnetism

Let us think about a substance with antiferromagnetic order, in which half of the atoms Mn are substituted with other paramagnetic atoms, which have large magnetic moment, for example, Fe. In a temperature lower than the Neel point, these substances are turned out spontaneously magnetized, since the difference of the magnetic moments between two adjacent paramagnetic atoms is not equal to zero.

Such magnetic materials are called *ferrites*; they possess low electrical conductivity and are obtained by sintering the powders, which consist of thoroughly mixed oxides of ironwith oxides Li, Ni, Mn, and others.

9.3 General Relations of the Magneto-Elasticity of Electro-Conductive Bodies

Let us consider the uniform isotropic linear elastic body, which possesses large electrical conductivity and is placed into a magnetic field. Under mechanical action on this body, in the latter appear coupled electromagnetic and mechanical fields. In the solution of the problems of coupled magneto-elasticity for the above-described body Maxwell's equations are used [167]

$$\mathbf{rot}\,\overrightarrow{E} + \frac{\partial \overrightarrow{B}}{\partial t} = 0, \ \mathbf{div}\,\overrightarrow{D} = \rho_e, \tag{9.4}$$

$$\mathbf{rot}\,\overrightarrow{H} - \frac{\partial \overrightarrow{D}}{\partial t} = \overrightarrow{j}, \ \mathbf{div}\,\overrightarrow{B} = 0.$$

The equations of motion of the elastic and electrically conducting medium, must be augmented by mass Lorentz force and by the force, caused by the existence of free charges in the electrical field [135]

$$\partial_j \sigma_{ij} + f_i = \rho \frac{\partial^2 U_i}{\partial t^2} \left(i, j = \overline{1,3}\right), \qquad (9.5)$$
$$\vec{f} = \rho_e \vec{E} + \vec{j} \times \vec{B}.$$

Hooke law for linearly elastic bodies [104]

$$\sigma_{ij} = 2\mu e_{ij} + \lambda \delta_{ij} e_{kk} \qquad (9.6)$$

Cauchy relations [104]

$$e_{ij} = \left(\partial_i u_j + \partial_j u_i\right)/2 \qquad (9.7)$$

The state equations for a moving isotropic medium, which does not possess properties of polarization and magnetization are given by [135, 140]

$$\vec{D} = \varepsilon \cdot \vec{E} + \alpha \vec{\nu} \times \vec{H} \quad \vec{B} = \mu_e \vec{H} - \alpha \vec{\nu} \times \vec{E}; \qquad (9.8)$$

generalized Ohm's law [135]

$$\vec{j} = \rho_e \vec{\nu} + \sigma \left(\vec{E} + \vec{\nu} \times \vec{B}\right) \qquad (9.9)$$

where $\vec{E}, \vec{H}$ are the vectors of intensity of electrical and magnetic fields; $\vec{D}, \vec{B}$ are the vectors of electric flux density and magnetic induction; $\vec{j}$ is the density of the conductivity current; ρ_e is the density of free electrical charges; σ_{ij} is the tensor of elastic stresses; ρ is the density of the material; $\vec{u} = (u_1, u_2, u_3)$ is the displacement vector; μ, λ are the Lame constants; δ_{ij} is the Kronecker symbol; ε_{ij} is the tensor of elastic deformations; ε, μ_e are the electrical and magnetic permeability of substance respectively; $\vec{\nu} = \frac{\partial \vec{u}}{\partial t}$ is the velocity vector; $\alpha = \varepsilon\mu - \varepsilon_0\mu_0$; σ is the conductivity of the environment.

Equations (9.4)–(9.9) form a closed system of nonlinear equations of magneto-elasticity for the homogeneous isotropic medium. These equations should be solved with given initial conditions and taking into account the boundary conditions on the surface S of the section of the environments (in the general case with different electromagnetic properties).

The boundary conditions for the electromagnetic field have the form:

$$\left[\vec{E} + \vec{\nu} \times \vec{B}\right]_\tau = 0, \left[\vec{H} - \vec{\nu} \times \vec{D}\right]_\tau = j_m^* - \rho_e^* \nu_m, \qquad (9.10)$$
$$\left[\vec{B}\right]_n = 0, \left[\vec{D}\right]_n = \rho_e^*, \left[\sigma\left(\vec{E} + \vec{\nu} \times \vec{B}\right)\right]_n = -\frac{\partial \rho_e^*}{\partial t},$$

where $\vec{\tau}, \vec{m}$ are the mutually perpendicular directions, parallel to the tangent to the plane S; $\vec{n}$ is the direction, normal to S (so that vectors $\vec{\tau}, \vec{n}, \vec{m}$ form the right-handed triad); the symbol $[*]$ indicates the jump of the corresponding magnitude on the surface S; j_m^*, ρ_e^* is the density of surface current and the surface electric charge density.

In the formulation of boundary conditions for mechanical stresses and displacements it is necessary to consider that under the action of the electromagnetic field, stretching and compressive forces appear in the environment, which are determined by the Maxwell electrodynamic stress tensor. The indicated boundary conditions have the form:

$$[\sigma_{ij} + \tau_{ij}]\, n_j = 0, \ [u_i] = 0, \qquad (9.11)$$
$$\tau_{ij} = E_i D_j + H_i B_j - \frac{1}{2}\delta_{ij}\left(E_k D_k + H_k B_k\right),$$

where n_j are the unit vectors normal to the surface S; summation is performed according to the repetitive index.

If a body is bounded by vacuum, then in the region, occupied by the vacuum, the following relations hold:

- Maxwell's equations:

$$\mathbf{rot}\,\vec{E} + \frac{\partial \vec{B}}{\partial t} = 0, \ \mathbf{rot}\,\vec{H} - \frac{\partial \vec{D}}{\partial t} = 0, \qquad (9.12)$$
$$\mathbf{div}\,\vec{D} = 0, \ \mathbf{div}\,\vec{\mathbf{B}} = 0;$$

- Equations of state:

$$\vec{D} = \varepsilon_0 \vec{E}, \quad \vec{B} = \mu_0 \vec{H} \qquad (9.13)$$

In the domain of vacuum, $\sigma_{ij} = 0$.

In this way, the system of relations (9.4)–(9.13) describes the mechanical and electromagnetic fields in the conjugated media. As a result of the nonlinearity of this system the solution of concrete problems is connected with serious mathematical difficulties; for this reason, one is compelled to address to its linearization.

9.4 Linear Magneto-Elasticity of Diamagnetic (Para-Magnetic) Materials

Assume that into a stationary electromagnetic field a linear elastic and electroconductive body is placed. The body is found in a state of rest and nor surface currents or electric charges exist in it $\left(\vec{j}_m^{\,*} = 0, \ \rho_e^* = 0\right)$. Then for the determination of the electromagnetic field in the domain, occupied by the body, we get the following equations of electro- and magneto-statics.

$$\mathbf{rot}\,\vec{E}^0 = 0, \ \mathbf{div}\,\vec{D}^0 = 0, \ \mathbf{rot}\,\vec{H}^0 = \vec{j}^{\,0}, \ \mathbf{div}\,\vec{B}^0 = 0,$$
$$\mathbf{div}\,\vec{j}^{\,0} = 0, \ \vec{D}^0 = \varepsilon \vec{E}^0, \ \vec{B}^0 = \mu_e \vec{H}^0, \ \vec{j}^{\,0} = \sigma \vec{E}^0, \qquad (9.14)$$
$$\left[\vec{E}^0\right]_\tau = 0, \ \left[\vec{H}^0\right]_\tau = 0, \ \left[\vec{D}^0\right]_n = 0.$$

After the determination of the electromagnetic field, the corresponding stress-strained state of the body is determined from the solution of the problem of the static theory of the elasticity:

$$\partial_j \sigma_{ij}^0 + \left(\vec{j}^{\,0} \times \vec{B}^0\right)_i = 0 \quad \left(i, j = \overline{1,3}\right) \tag{9.15}$$

and the conjugation conditions (9.11), where

$$\tau_{ij}^0 = E_i^0 D_j^0 + H_i^0 B_j^0 - \frac{1}{2}\delta_{ij}\left(E_k^0 D_k^0 + H_k^0 B_k^0\right), \sigma_{ij} = \sigma_{ij}^0. \tag{9.16}$$

Let us carry out the linearization of the system of equations. At time $t = 0$, the motion of the above-described body is provoked by the application of external loads. The displacements of the points of the body and the connected electromagnetic fields caused in this way are described by small fluctuations u_j, e_j, h_j, such that

$$u_j = u_j^0 + u_j, \ E_j = e_j, \ H_j = H_j^0 + h_j \tag{9.17}$$

Substituting (9.17) into the equation of magneto-elasticity and disregarding squares and the products of small disturbances, we obtain the equations of state (9.8) in the form

$$\vec{D} = \varepsilon \vec{e} + \alpha \vec{\nu} \times \vec{H}^0, \quad \vec{B} = \mu_e \left(\vec{H}^0 + \vec{h}\right) = \vec{B}^0 + \vec{b} \tag{9.18}$$

the generalized Ohm's law (9.9):

$$\vec{j} = \sigma\left(\vec{e} + \vec{\nu} \times \vec{B}^0\right) \tag{9.19}$$

Maxwell's (9.4):

$$\begin{aligned} &\mathbf{rot}\,\vec{e} + \frac{\partial \vec{b}}{\partial t} = 0 \\ &\mathbf{rot}\,\vec{h} = \sigma\left(\vec{e} + \vec{\nu} \times \vec{B}^0\right) + \frac{\partial}{\partial t}\left(\varepsilon \vec{e} + \alpha \vec{\nu} \times \vec{H}^0\right) \\ &\varepsilon \mathbf{div}\,\vec{e} + \alpha \mathbf{div}\left(\vec{\nu} \times \vec{H}^0\right) = 0, \ \mathbf{div}\,\vec{h} = 0; \end{aligned} \tag{9.20}$$

the equations of motion (9.5) will become

$$\begin{aligned} &\left(\rho_e = 0 \Rightarrow \vec{f} = \vec{j} \times \vec{B}^0\right) \\ &\partial_j \sigma_{ij} + \left(\vec{j} \times \vec{B}^0\right)_i = \rho \frac{\partial^2 u_i}{\partial t^2} \quad \left(i, j = \overline{1,3}\right) \end{aligned} \tag{9.21}$$

the electromagnetic conditions of media conjugation given by (9.10):

$$\left[\vec{e} + \vec{\nu} \times \vec{B}^0\right]_\tau = 0, \ \left[\vec{h}\right]_\tau = 0, \ \left[\vec{b}\right]_n = 0, \qquad (9.22)$$
$$\left[\vec{D}\right]_n = 0, \ \left[\sigma\left(\vec{e} + \vec{\nu} \times \vec{B}^0\right)\right]_n = 0;$$

the mechanical conditions of conjugation (9.11):

$$\left[\sigma_{ij} + h_i B_j^0 + b_j H_i^0 - \mu_e \delta_{ij} H_k^0 h_k\right] n_j = 0 \ \left(i, j = \overline{1,3}\right). \qquad (9.23)$$

Usually, in the study of mechanical disturbances a quasi-static approximation is used:

$$\vec{D} = 0, \partial\vec{D}/\partial t = 0.$$

Then Maxwell's (9.20) can be written as follows:

$$\mathbf{rot}\,\vec{e} + \frac{\partial\vec{b}}{\partial t} = 0, \mathbf{rot}\,\vec{h} = \vec{j}, \mathbf{div}\,\vec{h} = 0 \qquad (9.24)$$

and, under the electromagnetic conditions of conjugation (9.22), the condition with respect to $\vec{D}$ disappears.

The equationss of motion (9.21), taking into account (9.6), (9.7) and (9.24), take the form

$$\mu\nabla^2\vec{u} + (\lambda+\mu)\,\mathbf{graddiv}\,\vec{u} + \mathbf{rot}\,\vec{h} \times \vec{B}^0 = \rho\frac{\partial^2\vec{u}}{\partial t^2} \qquad (9.25)$$

where $\nabla^2 = \partial_1^2 + \partial_2^2 + \partial_3^2$ is the Laplacian operator.

Eliminating $\vec{e}$ from (9.24) and taking into account the equality

$$\mathbf{rotrot}\,\vec{h} + \nabla^2\vec{h} - \mathbf{graddiv}\,\vec{h} = 0,$$

we obtain

$$\left(\nabla^2 - \sigma\mu_e\frac{\partial}{\partial t}\right)\vec{h} + \sigma\mu_e\mathbf{rot}\left(\frac{\partial\vec{u}}{\partial t} \times \vec{H}^0\right) = 0 \qquad (9.26)$$

The model of coupled magneto-elasticity described by (9.25), (9.26) can be simplified, if the medium in question is ideally electro-conductive ($\sigma = \infty$). We have

$$\vec{h} = \mathbf{rot}\left(\vec{u} \times \vec{H}^0\right),$$
$$\mu\nabla^2\vec{u} + (\lambda+\mu)\,\mathbf{graddiv}\,\vec{u} + \mu_e\mathbf{rotrot}\left(\vec{u} \times \vec{H}^0\right) \times \vec{H}^0 = \rho\frac{\partial^2\vec{u}}{\partial t^2}, \qquad (9.27)$$
$$\vec{e} = \vec{B}^0 \times \frac{\partial\vec{u}}{\partial t}, \ \left[\vec{h}\right]_\tau = 0, \ \left[\vec{b}\right]_n = 0.$$

9.5 Equations of Magneto-Elasticity Taking into Consideration the Effect of Magnetization

The action of an external magnetic field on ferromagnetic materials manifests itself in magnetization, which consists of the fact that the body, placed into an external magnetic field, acquires magnetic moment and undergoes magnetical action from the external field.

For soft-iron ferromagnetic materials (narrow hysteresis loop) the influence of induced currents is small in comparison with the effect of magnetization. Therefore, it is possible to use the quasi-static approximation, i.e. equations of magneto-statics and the fact that μ_e does not depend on the magnitude of the intensity of the field in case when magnetization is far from saturation,.

In the deduction of the equations of motion of magnetized environments it is necessary to consider that, in such environments, on the unit of mass, enclosed inside an arbitrary elementary volume, act body forces $f_k = \mu_0 M_j H_{k,j}$ and $I_k = \mu_0 \varepsilon_{ijk} M_j H_i$ (ε_{ijk} is the skew-symmetric symbol of Kronecker).

Taking into account these expressions, from the principles of momentum and angular momentum conservation, follow the equations of motion:

$$\partial_i t_{ij} + \mu_0 M_i \partial_i H_j = \rho \frac{\partial^2 u_j}{\partial t^2} \left(i, j = \overline{1,3}\right) \tag{9.28}$$

and the relation

$$t_{jk} - t_{kj} + \mu_0 \left(M_j H_k - M_k H_j\right) = 0 \tag{9.29}$$

where t_{ij} is the asymmetrical stress tensor (to its asymmetry leads the calculation of magnetization):

$$t_{ij} = \sigma_{ij} - \mu_0 (M_i H_j - M_j H_i)/2 \tag{9.30}$$

where σ_{ij} is the symmetrical stress tensor; M is magnetization.

The equations of motion (9.28) can be written in the form:

$$\partial_i \left(t_{ij} + T_{ij}\right) = \rho \frac{\partial^2 u_j}{\partial t^2} \tag{9.31}$$

where T_{ij} is the Maxwell asymmetrical stress tensor:

$$T_{ij} = \mu_0 \left\{H_i H_j - \left(H_k H_k\right) \delta_{ij}/2\right\} + \mu_0 M_i H_j$$

$\overrightarrow{u} = (u_1, u_2, u_3)$ is the displacement vector.

Equation (9.29) should necessarily be supplemented by the equations of magneto-statics:

$$\mathbf{rot}\, \overrightarrow{H} = 0, \ \mathbf{div}\, \overrightarrow{B} = 0 \tag{9.32}$$

and the conjugation conditions on the surface of the portion of the media:

$$\left[\vec{B}\right]_n = 0,\ \left[\vec{H}\right]_\tau = 0,\ [t_{ij} + T_{ij}]_n = 0 \tag{9.33}$$

If a body is bounded by vacuum, then for the magnetic field in the vacuum $\vec{H}^e$ we get the equation:

$$\mathbf{rot}\vec{H}^e = 0,\ \mathbf{div}\vec{B}^e = 0,\ \vec{B}^e = \mu_0\vec{H}^e \tag{9.34}$$

Relations (9.32)–(9.34) should be augmented with the equations of state for the magnetized media, connecting stresses and magnetic field with deformations and magnetization.

In the quasi-static linear theory of magneto-elastic interaction in soft ferromagnetic materials without consideration of hysteresis losses and effect of magnetostriction, magnetic magnitudes are represented in the form:

$$\vec{B} = \vec{B}_0 + \vec{b},\ \vec{M} = \vec{M}_0 + \vec{m},\ \vec{H} = \vec{H}_0 + \vec{h} \tag{9.35}$$

where $\vec{H}_0$, $\vec{B}_0$, $\vec{M}_0$ correspond to the non-deformed state of body; $\vec{h}$, $\vec{b}$, $\vec{m}$ are the fluctuations of the corresponding magnitudes, caused by deformations and connected with deformations and stresses by linear differential equations and boundary conditions.

The non-deformed state of body is determined by the equations of magnetostatics:

$$\begin{aligned} &\mathbf{rot}\vec{H}_0 = 0,\ \mathbf{div}\vec{B}_0 = 0,\ \vec{H}_0 = \chi^{-1}\vec{M}_0, \\ &\vec{B}_0 = \mu_0\left(\vec{H}_0 + \vec{M}_0\right) = \mu_0\mu_e\vec{H}_0,\ \left[\vec{H}_0\right]_\tau = 0, \left[\vec{B}_0\right]_n = 0. \end{aligned} \tag{9.36}$$

For the case $|M_{oj}\partial_j u_j| << |m_i|$ we get the following linearized relations [285]:

- field equations

$$\begin{aligned} &\partial_j t_{ij} + \mu_0 \left(M_{oj}\partial_j H_{oi} + M_{0j}\partial_j h_i + m_j\partial_j H_{0i}\right) = \rho\frac{\partial^2 u_i}{\partial t^2}, \\ &\mathbf{rot}\,\vec{h} = 0,\ \mathbf{div}\,\vec{b} = 0; \end{aligned} \tag{9.37}$$

- material equations

$$\begin{aligned} &\vec{m} = \chi\vec{h},\ \vec{b} = \mu_0\left(\vec{h} + \vec{m}\right) = \mu_0\mu_e\vec{h}, \\ &t_{ij} = \sigma_{ij} + \mu_0\chi\left(H_{0i}H_{oj} + H_{oi}h_j + H_{oj}h_i\right), \\ &T_{ij} = H_{oj}B_{oi} + H_{oj}b_i + B_{oi}h_j - 0,5\mu_0\delta_{ij}\left(H_{ok}H_{ok}\right) - \mu_0\delta_{ij}\left(H_{ok}h_k\right), \\ &\sigma_{ij} = \lambda\delta_{ij}\partial_k u_k + \mu\left(\partial_j u_i + \partial_i u_j\right); \end{aligned}$$

- Boundary conditions on the interfaces of the media

$$\begin{aligned} &[t_{ij} + T_{ij}]\,n_j = X_{in},\ \varepsilon_{ijk}\left\{n_j\,[h_k] - n_m\partial_j u_m\,[H_{0k}]\right\} = 0, \\ &n_i\,[b_i] - n_m\partial_i u_m\,[B_{0i}] = 0, \end{aligned}$$

where T_{ij} is the linearized Maxwell vector of stresses; X_{1n} is the stresses, which act on the body from the side of the external normal to its surface.

10

Influence of the Induced Currents on the Dynamic Intensity of Piecewise-Uniform Electro-Conductive Bodies in Magnetic Fields

We examine below some stationary dynamic problems of magneto-elasticity for a piecewise-uniform and electro-conductive medium which is placed in a magnetic field $\overrightarrow{H}^0$. Under the combined influence of the mechanical and magnetic field induced (vortex) currents appear, which lead to the appearance of Lorenz body forces. The consideration of these forces gives a supplementary tensor – the Maxwell stresses t_{ij}, which contribute to make important corrections into the intensity of the body.

We examine below the plane and anti-plane deformations of the medium, which contains stress concentrators such as cracks, openings or inclusions. Boundary-value problems are reduced to integral equations, which are solved numerically with the application of mechanical quadratures method.

The aim of our investigation is the evaluation of the influence of the initial magnetic field, the orientation of the incident magneto-elastic wave with respect to the vector $\overrightarrow{H}^0$, the configuration of the defect, the proximity of the border, the type and frequency of excitation to the stress intensity factors and concentration in the body. We actually investigate in pure form only the effects, connected with the currents appearing in the body.

10.1 Initial Relations

Let us suppose that a static magnetic field $\overrightarrow{H}^0$ acts in a magneto-elastic medium, found in a state of rest. The external excitation causes deformation of the body and, respectively an electromagnetic field, which can be described by the small fluctuations $\overrightarrow{e}$, $\overrightarrow{h}$. Further, it is assumed that this is quasi-static.

Assuming that $\overrightarrow{H}=\overrightarrow{H}^0+\overrightarrow{h}$, $\overrightarrow{E}=\overrightarrow{e}$, $\rho_e=0$, we obtain, on the grounds of Chapter 9, the following system of equations:

– field equations

$$\operatorname{rot} \vec{h} = \vec{j}, \quad \operatorname{rot} \vec{e} = -\mu_e \frac{\partial \vec{h}}{\partial t}, \quad \operatorname{div} \vec{h} = 0, \tag{10.1}$$

$$\mu \nabla^2 \vec{u} + (\lambda + \mu) \operatorname{graddiv} \vec{u} + \mu_e \vec{j} \times \vec{H}^0 = \rho \frac{\partial^2 \vec{u}}{\partial t^2};$$

– material equations

$$\vec{b} = \mu_e \vec{h}, \quad \vec{j} = \sigma \left(\vec{e} + \mu_e \vec{V} \times \vec{H}^0 \right); \tag{10.2}$$

– coupling conditions

$$\left[\vec{e} + \mu_e \vec{V} \times \vec{H}^0 \right]_\tau = 0, \left[\vec{h} \right]_\tau = 0, \quad \left[\mu_e \vec{h} \right] \cdot \vec{n} = 0, \tag{10.3}$$

$$\left[\sigma \left(\vec{e} + \mu_e \vec{V} \times \vec{H}^0 \right) \right] \cdot \vec{n} = 0,$$

$$\left[\sigma_{ij} + \mu_e \left(H_i^0 h_j + H_j^0 h_i - \delta_{ij} H_k^0 h_k \right) \right] n_j = 0.$$

The considered model of magneto-elasticity is valid for diamagnetic (paramagnetic) materials, for which the effect of magnetizability of the material is insignificant. Further, for materials, such as *Al*, *Cu* and others, it is useful to simplify the model further, by assigning ideal conductivity to the medium ($\sigma \to \infty$). In this case we finally obtain [135]

$$\vec{h} = \operatorname{rot} \left(\vec{u} \times \vec{H}^0 \right), \vec{e} = -\mu_e \left(\frac{\partial \vec{u}}{\partial t} \times \vec{H}^0 \right), \tag{10.4}$$

$$\mu \nabla^2 \vec{u} + (\lambda + \mu) \operatorname{graddiv} \vec{u} + \mu_e \operatorname{rot} \vec{h} \times \vec{H}^0 = \rho \frac{\partial^2 \vec{u}}{\partial t^2}.$$

It is necessary to add coupling conditions to the system (10.4) on the line of interface of the media

$$\left[\vec{h} \right]_\tau = 0, \left[\mu_e \vec{h} \right]_n = 0, \tag{10.5}$$

$$\left[\sigma_{ij} + \mu_e \left(H_i^0 h_j + H_j^0 h_i - \delta_{ij} H_k^0 h_k \right) \right] n_j = 0.$$

10.2 The Current-Conducting Medium with Tunnel Cracks in a Magnetic Field

Let us examine now the situation, when the magneto-elastic medium is not homogeneous, but along the x_3-axis tunnel cracks L_j $(j = 1, \dots, M)$ exist, and the vector of the initial magnetic field takes the form $\vec{H}^0 = (0, H^0, 0)$, $H^0 = const$.

The corresponding static field (which does not depend on coordinate x_3) is described by the system of equations

$$\mu\nabla^2 u_i^0 + (\lambda+\mu)\,\partial_i\theta^0 = 0 \quad (i=1,2)\,, \qquad (10.6)$$
$$\theta^0 = \partial_1 u_1^0 + \partial_2 u_2^0,$$
$$\mu\nabla^2 u_3 = 0, \quad \nabla^2 = \partial_1^2 + \partial_2^2$$

and the boundary conditions are

$$\sigma_{11}^0 n_1 + \sigma_{12}^0 n_2 = X_{1n} + \frac{\kappa\mu_0}{2} H_0^2\left(1+\kappa n_2^2\right) n_1, \qquad (10.7)$$
$$\sigma_{12}^0 n_1 + \sigma_{22}^0 n_2 = X_{2n} + \frac{\kappa\mu_0}{2} H_0^2\left(1+\kappa n_2^2\right) n_2,$$
$$\sigma_{31}^0 n_1 + \sigma_{32}^0 n_2 = X_{3n}, H_1^* = \kappa H_0 n_1 n_2, H_2^* = H_0\left(1+\kappa n_2^2\right),$$
$$\kappa = \frac{\mu_e}{\mu_0} - 1, \quad n_1 = \cos\psi, \quad n_2 = \sin\psi.$$

Here, μ_e, μ_0 are the magnetic permeability of the material and the substance, respectively, which fills the cavity of the crack (the vacuum); Ψ is the angle between the normal to the left edge L_j and the axis Ox_1 (Fig. 10.1). The zero on top indicates the components of the static field, and the star is related to the cavity of the crack.

Thus, static fields are divided into the states of plane and anti-plane deformations.

It should be noted that the magnetic permeability of many diamagnetic (paramagnetic) materials practically coincide with the magnetic permeability of the vacuum μ_0. Therefore it is possible to assume that $\kappa = 0$, which substantially simplifies the formulation of boundary conditions.

We pass to the determination of the field of fluctuations. We find, according to (10.2),

$$\vec{h} = \operatorname{rot}\left(\vec{u}\times\vec{H^0}\right) = \left\{H_0\partial_2 u_1, -H_0\partial_1 u_1, H_0\partial_2 u_3,\right\}, \qquad (10.8)$$

$$\operatorname{rot}\vec{h}\times\vec{H^0} = \left\{H_0^2\nabla^2 u_1,\ 0,\ H_0^2\partial_2^2 u_3\right\}.$$

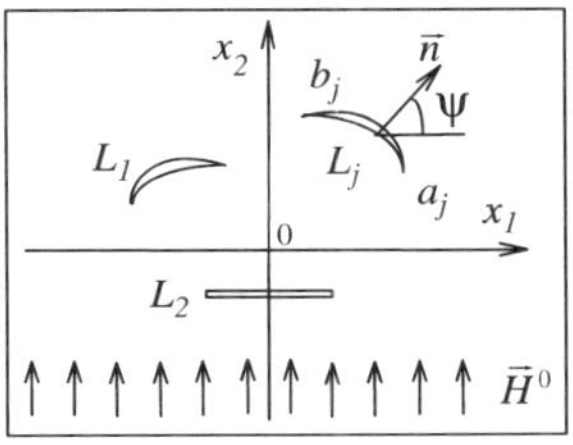

Fig. 10.1. Schema of piecewise-uniform conductor, placed into a magnetic field $\vec{H^0}$

Taking into account these equalities we get the complete system of linear relations, which determine the field of fluctuations.

Plane deformation:

- equation of motion

$$\left(\mu+\mu_e H_0^2\right)\nabla^2 u_1+(\lambda+\mu)\,\partial_1\theta=\rho\frac{\partial^2 u_1}{\partial t^2}, \tag{10.9}$$
$$\mu\nabla^2 u_2+(\lambda+\mu)\,\partial_2\theta=\rho\frac{\partial^2 u_2}{\partial t^2},$$

- components of the electromagnetic field

$$h_1=H_0\partial_2 u_1,\quad h_2=-H_0\partial_1 u_1,\quad h_3=0, \tag{10.10}$$
$$e_3=-\mu_e H_0\frac{\partial u_1}{\partial t},\quad e_2=e_1=0;$$

- boundary conditions on L_j $(j=1,\ldots,M)$

$$\sigma_{11}n_1+\sigma_{12}n_2=\mu_0\kappa H_0\left\{h_2+\kappa n_2\left(h_1n_1+h_2n_2\right)\right\}n_1+X_{1n}, \tag{10.11}$$
$$\sigma_{21}n_1+\sigma_{22}n_2=\mu_0\kappa H_0\left(1+\kappa n_2^2\right)\left(h_1n_1+h_2n_2\right)+X_{2n},$$
$$h_1^*=h_1\left(1+\kappa n_1^2\right)+\kappa h_2n_1n_2,$$
$$h_2^*=h_2\left(1+\kappa n_2^2\right)+\kappa h_1n_1n_2,$$
$$e_3^*+\mu_e H_0\frac{\partial u_1}{\partial t}=0.$$

Anti-plane deformation:

- equation of motion

$$\mu\nabla^2 u_3+\mu_e H_0^2\partial_2^2 u_3=\rho\frac{\partial^2 u_3}{\partial t^2} \tag{10.12}$$

- components of electromagnetic field

$$h_3=H_0\partial_2 u_3,\quad h_1=h_2=0, \tag{10.13}$$
$$e_1=\mu_e H_0\frac{\partial u_3}{\partial t},\quad e_2=e_3=0;$$

- boundary conditions on L_j

$$\sigma_{31}n_1+\sigma_{32}n_2=X_{3n}, \tag{10.14}$$
$$h_3^*=h_3=H_0\partial_2 u_3.$$

Here the stars are assigned to the field components in the cavity of the crack.

10.2.1 Antiplane Deformation of the Infinite Ideal Conductor with Tunnel Cracks

Under mechanical disturbances, which set the particles of the medium to motion, connected wave mechanical and electromagnetic fields u_i, σ_{ij}, h_i, e_i appear in it, which are described in Sect. 10.1. We assume here that plane magneto-elastic shear wave

$$u_3 = \mathrm{Re}\left(e^{-i\omega t}U_0\right), \qquad (10.15)$$
$$U_0 = \tau \exp\left\{-i\gamma_h\left(x_1\cos\beta + x_2\sin\beta\right)\right\}, \quad \tau = \mathrm{const},$$
$$\gamma_h = \gamma_2 \Big/ \sqrt{1 + h_c^2\sin^2\beta}, \quad h_c^2 = \mu_e H_0^2/\mu, \quad \gamma_2 = \omega/c_2.$$

is radiated from infinity.

In this case, the action of harmonically changing with time load $X_{3n} = \mathrm{Re}\left(e^{-i\omega t}X_3\right)$, $X_3 = X_3\left(x_1, x_2\right)$ is possible on the surfaces L_j.

Under such conditions in the magneto-elastic medium appears a stationary wave field that corresponds to anti-plane deformation. The mechanical wave field of displacements W is formed from the incident wave U_0 and the scattered by defects wave U. Let us represent the scattered wave in the following form:

$$U\left(x_1, x_2\right) = \frac{1}{2}\int_L p\left(\zeta\right)\left\{\frac{\partial}{\partial\zeta_1}H_0^{(1)}\left(\gamma_2 r_1\right)d\zeta_1 - \right. \qquad (10.16)$$
$$\left. -\frac{\partial}{\partial\bar{\zeta}_1}H_0^{(1)}\left(\gamma_2 r_1\right)d\bar{\zeta}_1\right\} + \int_L q\left(\zeta\right)H_0^{(1)}\left(\gamma_2 r_1\right)ds,$$

where

$$\zeta = \xi_1 + i\xi_2 \in L = \cup L_j, \quad \zeta_1 = \xi_1 + \frac{i\xi_2}{\sqrt{1 + h_c^2}},$$
$$r_1 = \left|\zeta_1 - z_1\right|, \quad z_1 = x_1 + \frac{ix_2}{\sqrt{1 + h_c^2}}.$$

The functions

$$p\left(\zeta\right) = \left\{p_j\left(\zeta\right), \zeta \in L_j\right\}, \; q\left(\zeta\right) = \left\{q_j\left(\zeta\right), \zeta \in L_j\right\}$$

are the unknown "densities"; $H_n^{(1)}\left(x\right)$ is the Hankel function of the first order of n order; ds is element of the contour arc L. The density $p\left(\zeta\right)$ has a simple mechanical sense: $p\left(\zeta\right) = -\frac{1}{2}\left[U\left(\zeta\right)\right]$, where $\left[U\left(\zeta\right)\right] = f$ is the jump of the displacement amplitude on L.

Function $u_3 = \mathrm{Re}\left(e^{-i\omega t}U\right)$, where U is determined by the formula (10.16), is the solution of (10.12); it also satisfies the emission conditions.

Consequently, let us represent the mechanical boundary conditions (10.14) on L in complex form

$$c(\psi)\left\{\frac{\partial}{\partial z_1}(U+U_0)\right\}^{\pm} + \overline{c(\psi)}\left\{\frac{\partial}{\partial \bar{z}_1}(U+U_0)\right\}^{\pm} = \pm X_3^{\pm}, \tag{10.17}$$
$$X_3^{+} = -X_3^{-} = X_3, \quad c(\psi) = \cos\psi + i\frac{\sin\psi}{\sqrt{1+h_c^2}}.$$

Here, the upper sign corresponds to the left lip of L_j (in the movement from its beginning a_j to the end b_j), ψ is the angle between the positive normal to the left edge and the axis Ox_1 (Fig. 10.1).

Let us require, in accordance with (10.17), that the vector of mechanical stress is continuously extended through the sections. Realizing the operations specified there, we find the relation between the densities

$$q(\zeta) = \frac{1}{4i}d(\psi)\frac{dp}{ds}, \tag{10.18}$$
$$d(\psi) = \frac{h_c^2}{\sqrt{1+h_c^2}}\sin 2\psi.$$

In view of equality (10.18), it is sufficient that the boundary conditions are satisfied only on one of the edges L_j $(j = 1,\ldots,M)$. Substituting in (10.17) the limiting values of the functions occurring there, we get to the following singular integro-differential equation

$$\int_L f'(\zeta)\,g(\zeta,\zeta_0)\,ds + \int_L f(\zeta)\,G(\zeta,\zeta_0)\,ds = N(\zeta_0), \quad \zeta_0 \in L_j, \tag{10.19}$$

$$g(\zeta,\zeta_0) = -\frac{\gamma_2}{4i}\frac{h_c^2\sin 2\psi}{\sqrt{1+h_c^2}}H_1^{(1)}(\gamma_2 r_{10}) \times \mathrm{Re}\left[c(\psi_0)\,e^{-i\alpha_{10}}\right] + \frac{1}{\pi}\mathrm{Im}\frac{c(\psi_0)}{\zeta_1-\zeta_{10}},$$

$$G(\zeta,\zeta_0) = \frac{\gamma_2^2}{4i}\left\{H_2(\gamma_2 r_{10})\,\mathrm{Im}\left[e^{2i\alpha_{10}}\overline{c(\psi_0)a(\psi)}\right]\right.$$
$$\left. + H_0^{(1)}(\gamma_2 r_{10})\,\mathrm{Im}\left[a(\psi)\,\overline{c(\psi_0)}\right]\right\},$$

$$N(\zeta_0) = 2i\tau\gamma_h\cos(\psi_0-\beta) \times \exp\{-i\gamma_h(\xi_{10}\cos\beta + \xi_{20}\sin\beta)\} + \frac{2}{\mu}X_3(\zeta_0),$$

$$H_2(x) = H_2^{(1)}(x) + 4i/\left(\pi x^2\right),$$

$$\zeta_{10} = \xi_{10} + \frac{i}{\sqrt{1+h_c^2}}\xi_{20}, \quad a(\psi) = \frac{i}{\sqrt{1+h_c^2}}\cos\psi - \sin\psi,$$

$$r_{10} = |\zeta_1-\zeta_{10}|, \quad \alpha_{10} = \arg(\zeta_1-\zeta_{10}), \quad f'(\zeta) = df/ds.$$

It is necessary that (10.19) would be examined in combination with the additional conditions

$$\int_{L_j} df = 0 \quad (j = 1, \dots, M). \tag{10.20}$$

The system of (10.19), (10.20) has a unique solution in the class of functions with unlimited derivatives in the ends L_j (on the assumption that the curvatures of the contours L_j and the function $X_3(\zeta)$ satisfy the Holder condition [103, 168]).

We obtain the formula for the determination of the stress intensity factor for the mechanical stresses in the tips of the defect. Let us parameterize contour L_j:

$$\zeta = \zeta(\delta), \quad \zeta_0 = \zeta(\delta_0), \quad -1 \le \delta,\ \delta_0 \le 1.$$

In accordance with this, we pose

$$f'(\zeta) = \frac{\Omega_0(\delta)}{\sqrt{1-\delta^2}\, s'(\delta)}, \quad \Omega_0(\delta) \in H[-1,1], \quad s'(\delta) = \frac{ds}{d\delta} > 0. \tag{10.21}$$

The singular part of "elastic" stress $\sigma_{n3} = \sigma_{13}n_1 + \sigma_{23}n_2$ on the extension beyond the edge of the defect is determined by relations (10.16), (1.6) and takes the following form

$$\begin{aligned} \sigma_{n3} &= \operatorname{Re}\left(S_{n3}e^{-i\omega t}\right), \\ S_{n3} &= S_{13}\cos\psi_c + S_{23}\sin\psi_c = \frac{\mu}{2\pi}\int_L f'(\zeta)\left[\operatorname{Im}\left(\frac{c(\psi_c)}{a(\psi)}\cdot\frac{d\zeta_1}{\zeta_1 - z_1}\right)\right. \\ &\left.+ \frac{d(\psi)}{2}\operatorname{Re}\left(\frac{c(\psi_c)}{a(\psi)}\cdot\frac{d\zeta_1}{\zeta_1 - z_1}\right)\right], \end{aligned} \tag{10.22}$$

where ψ_c is the angle between the normal to the left edge L_j in the tip c ($c = a$ or b) and the axis $0x_1$.

The asymptotic analysis of this expression, in consideration with the formula (10.21), gives

$$\begin{aligned} K_{\text{III}} &= \lim_{r\to 0}\sqrt{2\pi r}\operatorname{Re}\left(e^{-i\omega t}S_{n3}\right), \\ S_{n3} &= -\frac{\mu}{2}\frac{\sqrt{\pi}\,\Omega_0(\pm 1)}{\sqrt{s'(\pm 1)}}\left(\sqrt{1+h_c^2} + \frac{h_c^4\cos^2\psi_c}{\sqrt{1+h_c^2}}\sin^2\psi_c\right) \times \left(1 + h_c^2\sin^2\psi_c\right)^{-1}. \end{aligned} \tag{10.23}$$

Thus, the mechanical stress intensity factor K_{III} is expressed in terms of the solution of equations (10.19), (10.20). When $h_c = 0$, we obtain an algorithm, which corresponds to isotropic medium with the tunnel cracks [270]. It is evident from formula (10.23) that in the neighborhood of the tip of crack redistribution of that part of the stress tensor takes place, which has mechanical origin.

The summarizing stress intensity factor, which considers both the mechanical and Maxwell part of the stress tensor, is determined through the singular part of the expression

$$Q = (S_{13} + T_{13}) \cos \psi + (S_{23} + T_{23}) \sin \psi, \tag{10.24}$$
$$t_{j3} = \mathrm{Re} \left(T_{j3} e^{-i\omega t} \right).$$

Using relations (10.8) and the formulas

$$t_{31} = 0, \quad t_{32} = \mu_e H_0 h_3,$$

we find

$$Q = \mu \left[\cos \psi \partial_1 W + \left(1 + h_c^2\right) \sin \psi \partial_2 W \right].$$

The asymptotic analysis of this expression gives

$$Q = \pm \frac{\mu \sqrt{1 + h_c^2} \Omega_0 (\pm 1)}{2\sqrt{2rs' (\pm 1)}}.$$

The summarizing stress intensity factor has the form

$$K_{\mathrm{III}}^s = \lim_{r \to 0} \sqrt{2\pi r} \mathrm{Re} \left(Q e^{-i\omega t} \right) = \frac{\mu \sqrt{\pi \left(1 + h_c^2\right)}}{2\sqrt{s' (\pm 1)}} \left| \Omega_0 (\pm 1) \right| \cos \left(\omega t - \delta_s^{\pm} \right). \tag{10.25}$$

It should be noted here that the parameter of magnetic pressure h_c^2 does not implicitly appears in the value $\Omega (\pm 1)$, since the kernel and the right side of integral equation (10.19) depends on h_c^2.

10.3 Half – Space

Let us examine now a half-infinite ($x_2 \geq 0$) ideally conducting elastic medium, bordering to vacuum and weakened by tunnel cracks L_j along x_3-axis (Fig. 10.2). A static magnetic field $\overrightarrow{H}_*^0 = (0, H_0^*, 0)$ and a field $(0, H_0, 0)$ appear in the vacuum and the medium respectively, where $H_0 = H_0^* \mu_0 / \mu_e$. Let us assume that a magneto-elastic wave (10.15) is radiated from infinity, and on the contours L_j changing load $X_{3n} = \mathrm{Re} \left(e^{-i\omega t} X_3 \right)$ acts.

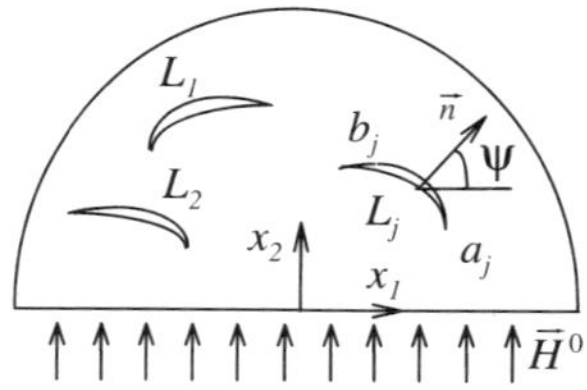

Fig. 10.2. Half-space with tunnel defects

The mechanical boundary conditions on the sections contours L_j take the form (10.14), and on the free from forces boundary of the half-space $x_2 = 0$ we have $\sigma_{23} = 0$.

The wave field of displacements u_3 is formed from the field of the incident wave (10.15), the reflected u_3^* and scattered by defects waves, according to relationship [275]

$$\begin{aligned} u_3 &= \mathrm{Re}\left(e^{-i\omega t}W\right), \quad W = U_0 + U_* + U, \qquad (10.26)\\ U_* &= \tau \exp\left\{-i\gamma_h\left(x_1\cos\beta - x_2\sin\beta\right)\right\},\\ U(x_1,x_2) &= \frac{1}{2}\int_L p(\zeta)\left\{\frac{\partial}{\partial\zeta_1}E(\zeta_1,z_1)\,d\zeta_1 - \frac{\partial}{\partial\overline{\zeta}_1}E(\zeta_1,z_1)\,d\overline{\zeta}_1\right\}\\ &\quad + \int_L q(\zeta)\,E(\zeta_1,z_1)\,ds,\\ E(\zeta_1,z_1) &= H_0^{(1)}(\gamma_2 r_1) + H_0^{(1)}(\gamma_2 r_1^*),\; r_1 = |\zeta_1 - z_1|,\; r_1^* = \left|\overline{\zeta}_1 - z_1\right|. \end{aligned}$$

The function W, determined by formula (10.26), is the solution of the corresponding Helmholtz equation, satisfies the conditions for emission and the above-indicated boundary condition on the line $x_2 = 0$ (Fig. 10.3).

The stated boundary problem, taking into account the formulas (10.26), is reduced to the singular integro-differential equation of the form (10.19), where the kernel and the right side are determined as follows:

$$\begin{aligned} g(\zeta,\zeta_0) &= \frac{1}{\pi}\mathrm{Im}\frac{c(\psi_0)}{\zeta_1 - \zeta_{10}} - \frac{\gamma_2}{4i}\frac{h_c^2\sin 2\psi}{\sqrt{1+h_c^2}}\\ &\quad\times\left\{H_1^{(1)}(\gamma_2 r_{10})\,\mathrm{Re}\left(c(\psi_0)\,e^{-i\alpha_{10}}\right) + \right.\\ &\quad\left. + H_1^{(1)}(\gamma_2 r_{10}^*)\,\mathrm{Re}\left(c(\psi_0)\,e^{-i\alpha_{10}^*}\right)\right\},\\ G(\zeta,\zeta_0) &= \frac{\gamma_2^2}{4i}\left\{H_2(\gamma_2 r_{10})\,\mathrm{Im}\left(e^{2i\alpha_{10}}\overline{c(\psi_0)}\overline{a(\psi)}\right) + \right. \qquad (10.27)\\ &\quad\left.\times\,\mathrm{Im}\left(e^{-2i\alpha_{10}^*}c(\psi_0)\,\overline{a(\psi)}\right) + H_0^{(1)}(\gamma_2 r_{10}^*)\,\mathrm{Im}\left(a(\psi)\,c(\psi_0)\right)\right\}, \end{aligned}$$

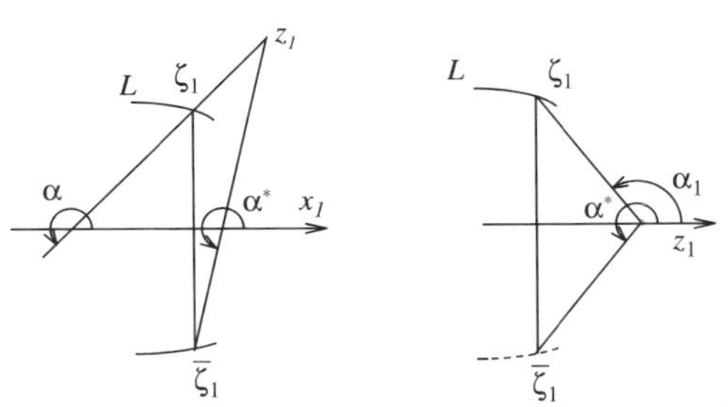

Fig. 10.3. To the satisfaction of the boundary condition on the plane $x_2 = 0$

$$N(\zeta) = 2i\tau\gamma_h \{\cos(\psi_0 - \beta) \exp[-i\gamma_h (\xi_{10} \cos\beta + \xi_{20} \sin\beta)] +$$
$$+ \cos(\psi_0 + \beta) \exp[-i\gamma_h (\xi_{10} \cos\beta - \xi_{20} \sin\beta)]\} + \frac{2}{\mu} X_3(\zeta_0),$$
$$r_{10}^* = |\overline{\zeta}_1 - \zeta_{10}|, \quad r_{10} = |\zeta_1 - \zeta_{10}|, \quad \alpha_{10}^* = \arg(\overline{\zeta}_1 - \zeta_{10}),$$
$$\alpha_{10} = \arg(\zeta_1 - \zeta_{10}).$$

Equations (10.19), (10.20), combined with formulas (10.27) determine uniquely the solution in the class of functions with unlimited derivatives at the ends of the arcs L_j. The summarizing stress intensity factor is computed from formula (10.25).

For the representation of the results of calculations, let us introduce the parameter T_h [286, 287]

$$T_h = \frac{i\mu\tau\gamma_2 \sin(\varphi - \beta)}{\sqrt{1 + h_c^2 \sin^2(\varphi - \beta)}}, \tag{10.28}$$

where $\varphi = 0$; $\pi/2$ and $\pi/4$ for the horizontal, vertical and inclined cracks, respectively.

Taking into account (10.28) we can write

$$K_{\mathrm{III}} = T_h \sqrt{\pi l} \alpha^{\pm} \cos(\omega t - \delta^{\pm}), \tag{10.29}$$
$$K_{\mathrm{III}}^s = T_h \sqrt{\pi l} \alpha_s^{\pm} \cos(\omega t - \delta_s^{\pm}),$$

where K_{III}, K_{III}^s are the elastic and complete stress intensity factors, respectively.

The results of calculations are presented in Figs. 10.4–10.10. The first figure is related to the unbounded medium with rectilinear elevation of

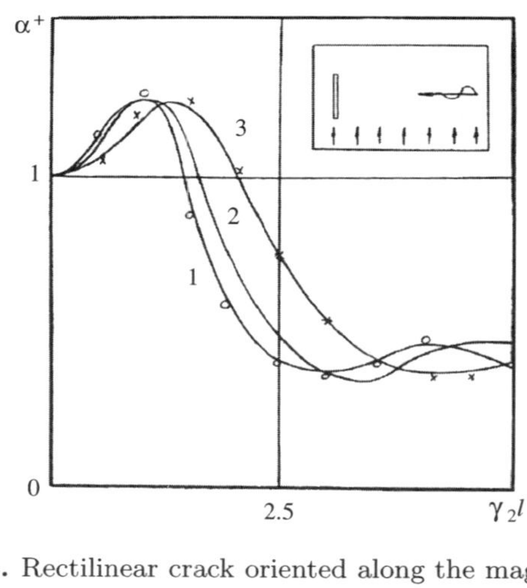

Fig. 10.4. Rectilinear crack oriented along the magnetic field

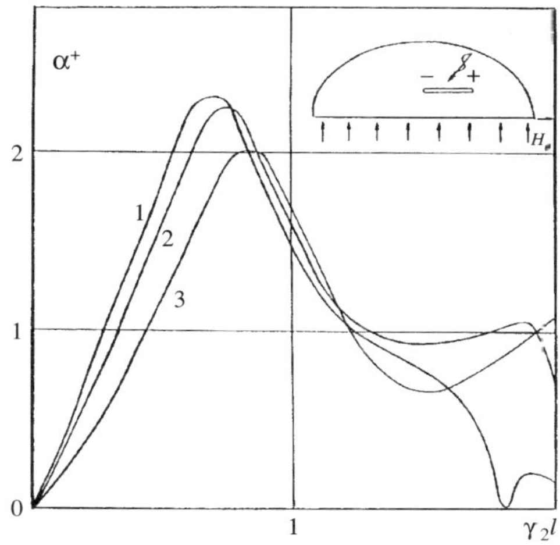

Fig. 10.5. Crack, perpendicular to the field at angle to an incident wave

length $2l$. A magneto-elastic wave (10.15) falls from infinity along the x_1-axis. Curves 1, 2 and 3 are built for the values of parameter $h_c = 0$; 0,5 and 1, respectively. Here, by points and small crosses are substituted results from the paper of Shindo [287].

The situation, when a magneto-elastic wave falls to a horizontal crack in the half-space at angles 60^0, 90^0 and 120^0, is presented in Figs. 10.5, 10.6 and 10.7, respectively. Here the distance from the border of the half-plane to the crack is equal to its length.

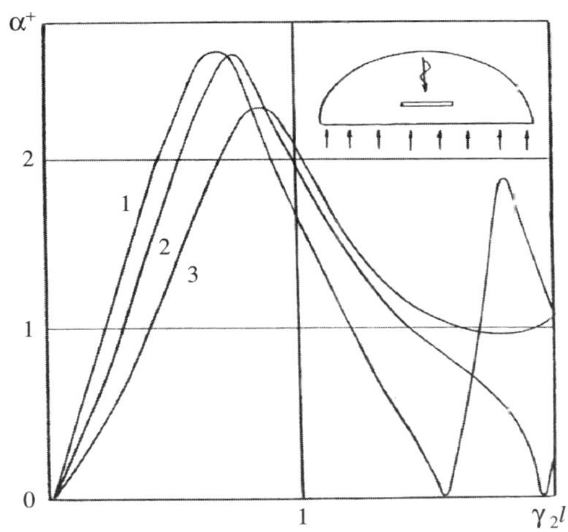

Fig. 10.6. Dependence of the relative coefficient of the intensity of stresses on the normalized wave number

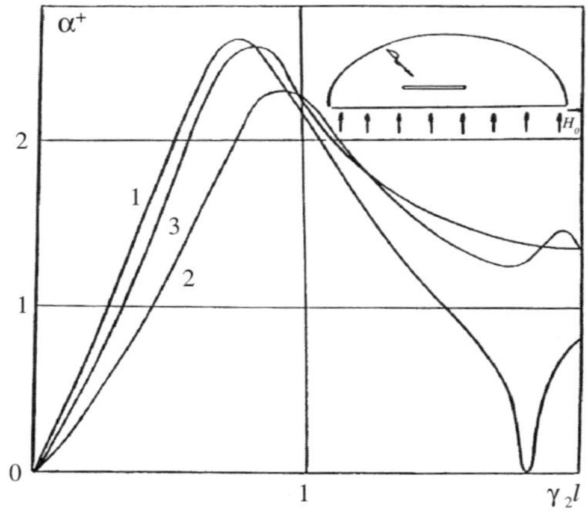

Fig. 10.7. Rectilinear crack, perpendicular to magnetic field, wave falls at angle of 120^0

In Fig. 10.8, the graphs are constructed for the crack, sloped toward the x_1-axis at angle 45^0; in this case, the wave falls to the crack along the x_2-axis (the distance from the center of the crack to the border $x_2 = 0$ equals to the length of the crack). The case, when there is no incident wave, and excitation is produced by harmonic in time shift load on the edges of crack, is examined in Fig. 10.9.

Figure 10.10 gives the result of calculation for the parabolic crack described by $\xi_1 = R_1\delta$, $\xi_2 = R\delta^2$, $(-1 \le \delta \le 1)$ when $\alpha = \alpha^{\pm}$ and $\alpha_s = \alpha_s^{\pm}$, in an infinite

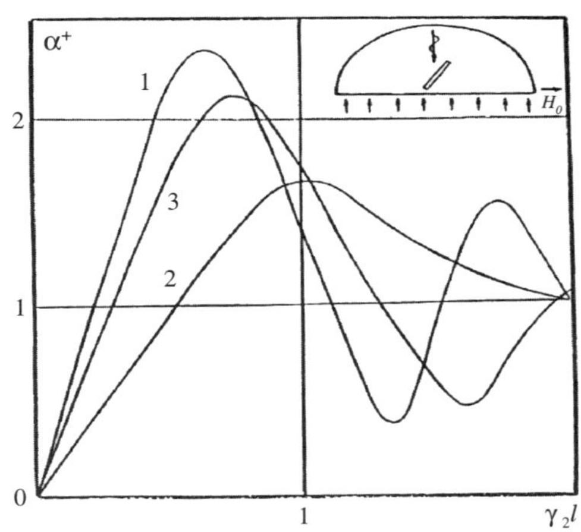

Fig. 10.8. Rectilinear crack, oriented at angle of 45^0 to the direction of the magnetic field and incident wave

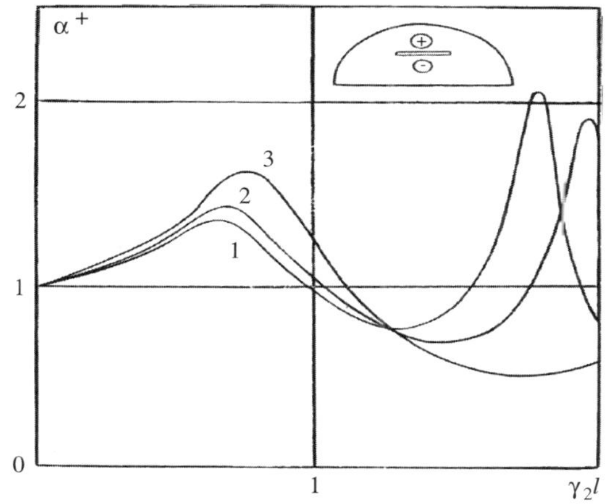

Fig. 10.9. Change in the relative stress intensity factor for the case of harmonic in time shift load on the edges of the crack

space. Curves 1, 3 are constructed for the straight crack ($R_1 = 1,\ R = 0$), curves 2, 4, 5, for the parabolic crack ($R_1 = 1,\ R = 1$). Curve 5 corresponds to $h_c = 0$, while the other ones to $h_c = 1$. Graphs 1, 2 illustrate the change of α_s, and the rest the change of α.

It should be noted that the influence of magnetic field is essentially shown (especially in the half-space with defect) close to the peak values of the normalized wave number. The initial magnetic field smoothes out the curves of the intensity coefficient, as if it was introduced a supplemental "inertness" of the medium.

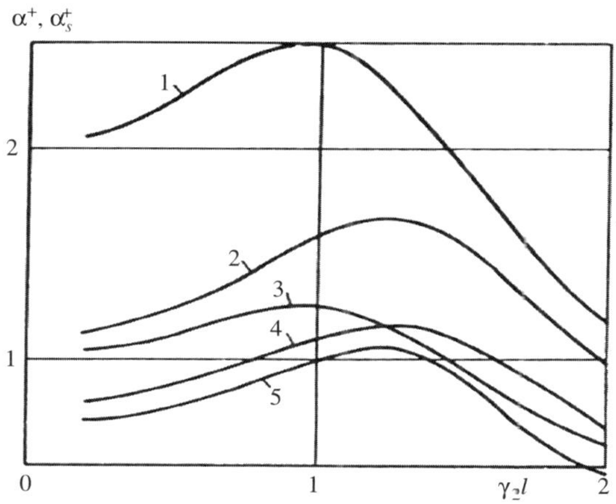

Fig. 10.10. Change of the magnitudes α^+ and α_s^+ for parabolic crack

10.4 Layer and Semi – Layer

Let us examine an ideally conducting elastic semi-layer $0 \leq x_1 \leq a$, $0 \leq x_2 < \infty$, $-\infty < x_3 < \infty$, weakened by a tunnel crack along the axis x_3 and found in a static magnetic field $\overrightarrow{H}^0 = (0, H_0, 0)$. Let a shear wave of displacement (10.15), when $\beta = \pi/2$, is radiated from infinity, and the surface crack is either free from forces or subjected to harmonic with time force which is independent from x_3 $(0, 0, X_{3n})$. In this case a stationary wave process appears in the body, which corresponds to the anti-plane deformation of the semi-layer $\overrightarrow{u} = \text{Re}\left(0, 0, We^{-i\omega t}\right)$.

We assume that the bases of the semi-layer are free from forces. In this case, $\partial_1 U_3 = 0$, when $x_1 = 0$ and $x_1 = a$. The conditions on the end border $x_2 = 0$ can be given in the form

$$A\,(A-1)\,U_3 + A\,(A+1)\,\partial_2 U_3 = 0. \tag{10.30}$$

In this case, if $A = 1$ or $A = -1$, we get the free or the fixed edge of the semi-layer, respectively; if $A = 0$ – we examine the layer $0 \leq x_1 \leq a$, $-\infty < x_2 < \infty$, $-\infty < x_3 < \infty$.

The boundary condition on the crack contour L can be written in the form

$$\left(\frac{\partial U_3}{\partial n}\right)^{\pm} = \frac{1}{\mu} X_3^{\pm}\,(\zeta)\,, \quad \zeta \in L, \tag{10.31}$$

where the upper sign corresponds to left edge of the section (see Fig. 10.11), moreover, $[X_3] = 0$ on L.

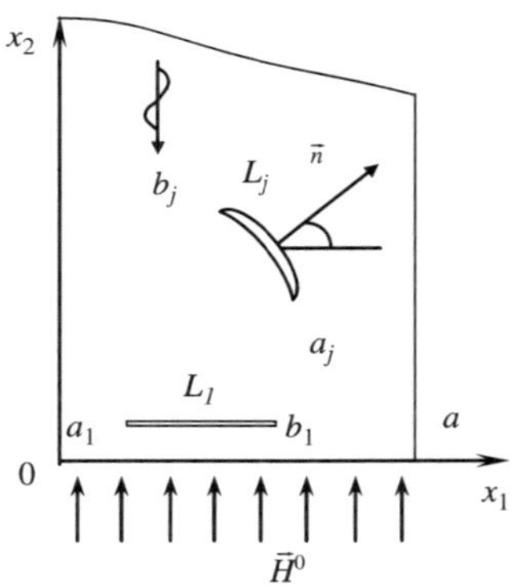

Fig. 10.11. Semi-layer with defects

The solution of the posed boundary problem can be represented in the form of superposition of the falling, reflected and scattered waves. We get

$$W = U_0 + AU_* + U, \tag{10.32}$$

$$U_0 = \tau \exp\left\{-i\gamma_2 x_2 \Big/ \sqrt{1+h_c^2}\right\}, U_* = \tau \exp\left\{i\gamma_2 x_2 \Big/ \sqrt{1+h_c^2}\right\},$$

$$U(x_1, y_1) = -i \int\limits_L p(\zeta) \left(\frac{\partial G}{\partial \zeta_1} d\zeta_1 - \frac{\partial G}{\partial \overline{\zeta_1}} d\overline{\zeta_1}\right) - \int\limits_L \frac{h_c^2 \sin 2\psi}{2\sqrt{1+h_c^2}} G p'(\zeta)\, ds,$$

$$G = G(\xi_1, x_1; \xi_2^*, x_2^*) = g(\xi_1, x_1; \xi_2^* - x_2^*) + Ag(\xi_1, x_1; -\xi_2^* - x_2^*),$$

$$g(\xi_1, x_1; \xi_2^* - x_2^*) = \frac{e^{i\gamma_2 |\xi_2^* - x_2^*|}}{2ia\gamma_2} - \frac{1}{a} \sum_{k=1}^{\infty} \frac{e^{-\lambda_k |\xi_2^* - x_2^*|}}{\lambda_k} \cos \alpha_k \xi_1 \cos \alpha_k x_1,$$

$$x_2^* = \frac{x_2}{\sqrt{1+h_c^2}}, \xi_2^* = \frac{\xi_2}{\sqrt{1+h_c^2}}, \alpha_k = \frac{\pi k}{a}, \zeta = \xi_1 + i\xi_2 \in L,$$

$$\zeta_1 = \xi_1 + i\xi_2^*, \quad \lambda_k = \sqrt{\alpha_k^2 - \gamma_2^2}\,(\gamma_2 < \alpha_k), \lambda_k = -i\sqrt{\gamma_2^2 - \alpha_k^2}\,(\gamma_2 > \alpha_k).$$

Here $p(\zeta)$ is the unknown density; U_* is the wave reflected from the end border of the semi-layer; G is the Green function of boundary-value problem for the half-strips $0 \le x_1 \le a,\ 0 \le x_2 < \infty$; g is the Green function for the strip $0 \le x_1 \le a,\ -\infty < x_2 < \infty$.

Integral representation (10.32) satisfies the differential equation $\nabla^2 U_3 + h_c^2 \partial_2^2 U_3 + \gamma_2^2 U_3 = 0$, which follows from the wave (10.12), the emission condition, and the boundary conditions on the faces of the semi-layer. It ensures the existence of the displacement jump, as well as the continuity of the stress vector in the section.

Subsequently, we will strengthen the convergence of the series, by introducing function g, and summing up its principal part. After the appropriate transformations, we get

$$\begin{aligned} g(\xi_1, x_1, \xi_2^* - x_2^*) = &\frac{1}{2\pi} \ln\left|4 \sin\frac{\pi}{2a}(\zeta_1 - z_1) \times \sin\frac{\pi}{2a}(\overline{\zeta_1} + z_1)\right| - \\ &- \frac{1}{2a}|\xi_2^* - x_2^*| + \frac{1}{2ia\gamma_2} e^{i\gamma_2 |\xi_2^* - x_2|} \\ &- \frac{1}{a}\sum_{k=1}^{\infty}\left(\frac{1}{\lambda_k} e^{-\lambda_k |\xi_2^* - x_2^*|} - \frac{1}{\alpha_k} e^{-\alpha_k |\xi_2^* - x_2^*|}\right) \\ &\cos \alpha_k \xi_1 \cos \alpha_k x_1, \end{aligned} \tag{10.33}$$

$$z_1 = x_1 + \frac{ix_2}{\sqrt{1+h_c^2}}, \quad \zeta_1 = \xi_1 + \frac{i\xi_2}{\sqrt{1+h_c^2}}, \quad \overline{\zeta} = \xi_1 - \frac{i\xi_2}{\sqrt{1+h_c^2}}.$$

Thus, function g has logarithmic singularity at the point of application of the concentrated functional. The general term of the series in (10.33) damps as k^{-3}, when $\xi_2^* = x_2^*$ and is exponential, when $\xi_2^* \ne x_2^*$.

For the satisfaction of the boundary conditions (10.31) in the sections it is necessary to compute the normal derivative of function (10.32). Further, regularizing the diverging integrals and repeatedly summing up the principal parts of the series for the second derivatives of Green function, we reduce the boundary equality (10.31) to the singular integro-differential equation

$$\int_L \{p'(\zeta)\, h(\zeta,\zeta_0) + p(\zeta)\, H(\zeta,\zeta_0)\} ds = N(\zeta_0), \quad \zeta_0 \in L, \tag{10.34}$$

$$\begin{aligned}
h(\zeta,\zeta_0) = {} & \operatorname{Im} \frac{c(\psi_0)}{\zeta_1 - \zeta_{10}} + \frac{1}{8a} \frac{h_c^2}{\sqrt{1+h_c^2}} \sin 2\psi \{ \operatorname{Re} [c(\psi_0) \\
& \times (\operatorname{ctg} \zeta_2 - \operatorname{ctg} \zeta_3 + 2i \operatorname{sign} r_2)] - \\
& \frac{2}{\sqrt{1+h_c^2}} \sin \psi_0 \times \left(\operatorname{sign} \eta_2 e^{-i\gamma_2 |\eta_2|} - A e^{i\gamma_2 \eta_3} \right) \\
& - 4 \sum_{k=1}^{\infty} c_{1k} \left(\alpha_k \varphi_{1k} s_{2k} \cos \psi_0 - \frac{\varphi_{2k}}{\sqrt{1+h_c^2}} c_{2k} \sin \psi_0 \right) \Bigg\},
\end{aligned}$$

$$\begin{aligned}
\mathrm{H}(\zeta,\zeta_0) = {} & -\frac{\pi}{8\alpha^2} \operatorname{Im}\{ c'(\psi) [c(\psi_0) (\zeta_2^{-2} - \cos ec^2 \zeta_2) + \\
& + \overline{c(\psi_0)} \cos ec^2 \zeta_3] \} + \frac{\gamma_2^2}{4\pi \sqrt{1+h_c^2}} [\cos(\psi - \psi_0) \ln |\sin \zeta_2| - \\
& - \cos(\psi + \psi_0) \ln |\sin \zeta_3| + \left(2 \ln 2 - \frac{\pi}{a} |\eta_2| \right) \sin \psi \sin \psi_0 \Big] - \\
& - \frac{i\gamma_2}{2a\sqrt{1+h_c^2}} \sin \psi \sin \psi_0 \left(e^{-i\gamma_2 |\eta_2|} + A e^{i\gamma_2 \eta_3} \right) + \frac{1}{a\sqrt{1+h_c^2}} \times \\
& \times \sum_{k=1}^{3} \Big[\varphi_{3k} c_{1k} c_{2k} \sin \psi \sin \psi_0 - \alpha_k^2 \varphi_{1k} s_{1k} s_{2k} \cos \psi \cos \psi_0 + \\
& + \alpha_k \varphi_{2k} \left(\frac{s_{1k} c_{2k}}{\sqrt{1+h_c^2}} \cos \psi \sin \psi_0 - \sqrt{1+h_c^2} c_{1k} s_{2k} \sin \psi \cos \psi_0 \right) + \\
& + \frac{\gamma_2^2}{\alpha_k} \left(s_{1k} s_{2k} \cos \psi \cos \psi_0 + c_{1k} c_{2k} \sin \psi \sin \psi_0 \right) e^{-\alpha_k |\eta_2|} \Big],
\end{aligned}$$

$$\begin{aligned}
N(\zeta_0) = {} & \frac{1}{\mu} X_3 + \frac{\tau}{2a\sqrt{1+h_c^2}} \sin \psi_0 \left[\exp\left(-i\gamma_2 x_2 \big/ \sqrt{1+h_c^2} \right) \right. \\
& \left. - A \exp\left(i\gamma_2 x_2 \big/ \sqrt{1+h_c^2} \right) \right],
\end{aligned}$$

$$c(\psi) = \cos \psi + \frac{i \sin \psi}{\sqrt{1+h_c^2}}, \quad c'(\psi) = \frac{dc(\psi)}{d\psi},$$

$$\begin{aligned}
\varphi_{1k} &= \frac{1}{\lambda_k} [\exp(-\lambda_k |\eta_2|) + A \exp(-\lambda_k \eta_3)] - \frac{1}{\alpha_k} \exp(-\alpha_k |\eta_2|), \\
\varphi_{2k} &= sign\ \eta_2 [\exp(-\lambda_k |\eta_2|) - \exp(-\alpha_k |\eta_2|)] - A \exp(-\lambda_k \eta_3), \\
\varphi_{3k} &= \lambda_k [\exp(-\lambda_k |\eta_2|) + A \exp(-\lambda_k \eta_3)] - \alpha_k \exp(-\alpha_3 |\eta_2|),
\end{aligned}$$

$$\zeta_2 = \frac{\pi}{2a}\left(\zeta_1 - \zeta_{10}\right), \zeta_3 = \frac{\pi}{2a}\left(\overline{\zeta_1} + \zeta_{10}\right),$$
$$\eta_2 = \frac{\xi_2 - \xi_{20}}{\sqrt{1+h_c^2}},$$
$$\eta_3 = \frac{\xi_2 + \xi_{20}}{\sqrt{1+h_c^2}},$$
$$\zeta_{10} = \xi_{10} + i\xi_{20}, \quad c_{1k} = \cos\alpha_k\xi_1,$$
$$c_{2k} = \cos\alpha_k\xi_{10}, \quad s_{1k} = \sin\alpha_k\xi_1,$$
$$s_{2k} = \sin\alpha_k\xi_{10}.$$

Kernel $h(\zeta,\zeta_0)$ is singular; $H(\zeta,\zeta_0)$ has logarithmic singularity. Equation (10.34) must be examined together with the condition of closure of the crack

$$\int\limits_L p'(\zeta)ds = 0. \qquad (10.35)$$

The defined algorithm uniquely determines the density $p(\zeta)$ in the class of functions with unlimited derivative.

Further, it is convenient to parameterize the contour of the section $\zeta = \zeta(\beta)$, $-1 \le \beta \le 1$. Accordingly, let us assume $p'(\zeta) = \Omega(\beta) / \left(s'(\beta)\sqrt{1-\beta^2}\right)$, where $\Omega(\beta) \in H[-1,1]$. Proceeding in the same way, as in Sect. 10.2, we find the summarizing stress intensity factor

$$K_{\mathrm{III}}^{s} = \pm\frac{\mu}{2}\sqrt{\frac{\pi\left(1+h_c^2\right)}{s'(\pm1)}}\,\mathrm{Re}\left\{e^{-i\omega t}\Omega(\pm1)\right\}. \qquad (10.36)$$

The numerical solution of (10.34), (10.35) was made by the method of mechanical quadratures with the use of Gauss-Chebyshev quadrature formulas for the regular and singular integrals [24]. Let us write formula (10.36) in the form

$$K_{\mathrm{III}}^{s} = \sqrt{a}T_{23}\alpha_s^{\pm}\,\cos\left\{\omega t - \arg\Omega(\pm1)\right\}, \qquad (10.37)$$

where $T_{23} = \mu\gamma_2\tau \Big/ \sqrt{1+h_c^2}$ is the module of mechanical stress in the incident wave; the upper sign corresponds to the tip of crack $b = \zeta(1)$; the lower one, to $a = \zeta(-1)$.

The results of calculations for the parabolic crack $\xi_1/a = 0,5{+}0,2\beta$, $\xi_2/a = 1 + R\beta^2$ $(-1 \le \beta \le 1)$, are represented in Figs. 10.12–10.14. In the first two figures, the dependence of value α_s^- from the normalized wave number $\gamma_2 a$ is illustrated. The value of curvature parameter $R = 0$ corresponds to the linear crack (Fig. 10.12); $R = 0,1$, to the parabolic crack (Fig. 10.13). Curve 1 is constructed for fixed $(A = -1)$; curve 2, for free $(A = 1)$ end edge of the semi-layer. Continual curves correspond to the value $h_c = 1$, dotted curves to $h_c = 0$.

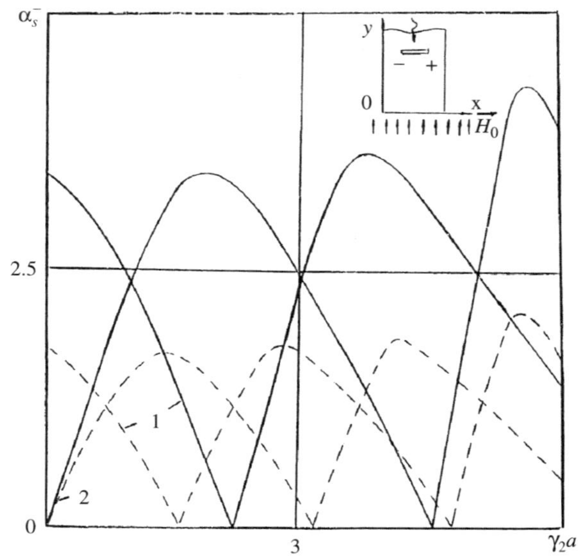

Fig. 10.12. Coefficient of the intensity of stresses for the parabolic crack in the semi-layer

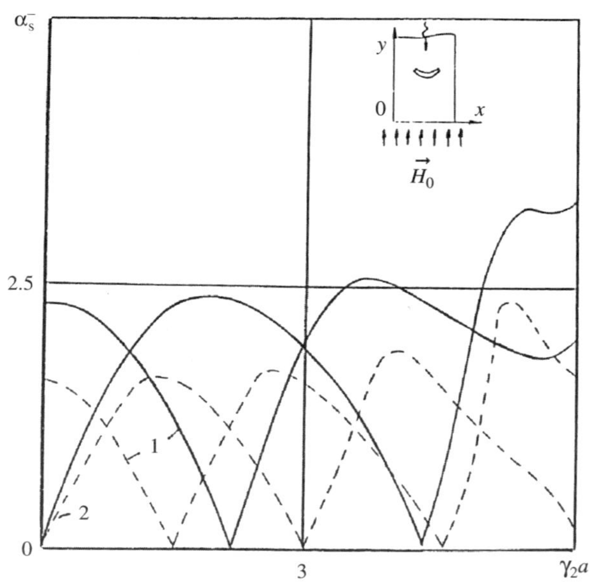

Fig. 10.13. Parabolic crack in the semi-layer

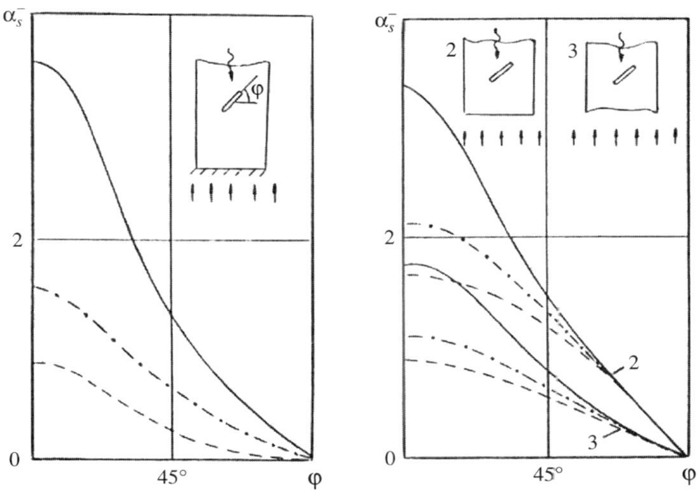

Fig. 10.14. Dependences of the magnitude α_s^- on the orientation angle of the crack

It is evident that the applied magnetic field shifts the points of extremes towards greater values $\gamma_2 a$. For a specific value of the distance between the linear crack and the end of the half space, magnitude $K_{\rm III}^s$ becomes zero, when $\gamma_2 a = (2k+1)\,\pi\sqrt{1+h_c^2}/2$ $(A=-1)$ and $\gamma_2 a = k\pi\sqrt{1+h_c^2}$ $(A=1)$. With increase $\gamma_2 a$ for the layer $(A=0)$, the value α_s^- increases insignificantly.

Figure 10.14 shows the change of α_s^- with the orientation angle φ of the rectilinear crack that is $0,4a$ long, for the semi-layer with fixed $(A=-1,\ \gamma_2 a=4,$ for curve 1) and free $(A=1,\ \gamma_2 a=\pi/2,$ for curve 2) end, as well as, for the layer $(A=0,\ \gamma_2 a=4,$ for curve 3). The equations of contour L take the form $\xi_1/a = 0,5+0,2\beta\cos\varphi$, $\xi_2/a = 1+0,2\beta\sin\varphi$ $(-1\le\beta\le 1)$. Continuous curves correspond to the value $h_c=1$, dotted-and-dashed ones, to $h_c=0,5$, dotted, to $h_c=0$.

From the results of computations follows that the nature of the influence of magnetic field on the stress intensity factor dependents on both the orientation angle of the crack and the form of the boundary condition on the end $x_2=0$. Thus, the applied static magnetic field, that corresponds to the value $h_c=1$, increases considerably the magnitude α_s^-, when $0\le\varphi\le\frac{\pi}{4}$, for all the considered values of the parameter A. At the same time, when $\pi/4\le\varphi\le\pi/2$, the magnitude α_s^- increases sharply, in the case of the fixed end, and increases insignificantly, in the case of layer and semi-layer with free end.

10.5 Stress Concentration in the Opening in the Conducting Half-Space in Presence of Magnetic Field

Let us examine below the influence of induced currents on the stress concentration in an opening in the state of anti-plane deformation. Let us assume that the diamagnetic (paramagnetic) half-space $x_2 \geq 0,\ -\infty < x_1, x_3 < \infty$ is weakened by tunnel cylindrical cavities with cross sections $\Gamma_j\ (j = 1, \dots, M)$ along the x_3–axis and bounded by vacuum along the forces-free plane $x_2 = 0$ (Fig. 10.15). Let us take the static magnetic field in vacuum, in the form $(0,\ H_0^*, 0)$, and in the body, in the form $\overrightarrow{H}^0 = (0,\ H_0,\ 0)$. Moreover, $H_0 = \mu_0 H_0^* / \mu_e$. As mechanical excitation let us take the independent from the coordinate x_3 shift load $X_{3n} = \mathrm{Re}\left(X_3 e^{-i\omega t}\right)$, acting on the surfaces of the cavities or the shift wave of displacement (10.15), falling from the infinity.

A mechanical field in half-space is formed from the fields of the incident (10.15), the reflected (10.26) and the scattered by cavities waves. The later can be presented in the form [276]

$$U(x_1, x_2) = \int_\Gamma p(\zeta) \left\{ H_0^{(1)}(\gamma_2 r_1) + H_0^{(1)}(\gamma_2 r_1^*) \right\} ds, \tag{10.38}$$

$$\zeta_1 = \xi_1 + i \frac{\xi_2}{\sqrt{1 + h_c^2}}, \quad z_1 = x_1 + i \frac{x_2}{\sqrt{1 + h_c^2}},$$

$$r_1 = |\zeta_1 - z_1|, \quad r_1^* = \left|\overline{\zeta}_1 - z_1\right|, \qquad \zeta = \xi_1 + i\xi_2 \in \Gamma = \cup \Gamma_j,$$

where $p(\zeta) = \{p_j(\zeta), \zeta \in \Gamma_j\}$ is the unknown density; integration along Γ_j is done counterclockwise.

Formula (10.38) satisfies the boundary condition $\sigma_{32} = 0$ on the axis $x_2 = 0$, no matter which is the choice of the function $p(\zeta)$, and the function $Re\left(e^{-i\omega t} U\right)$ is the solution of (10.12).

The boundary condition on Γ can be represented in amplitudes

$$\left(S_{13} + S_{13}^0 + S_{13}^1\right) \cos\psi + \left(S_{23} + S_{23}^0 + S_{23}^1\right) \sin\psi = X_3, \tag{10.39}$$

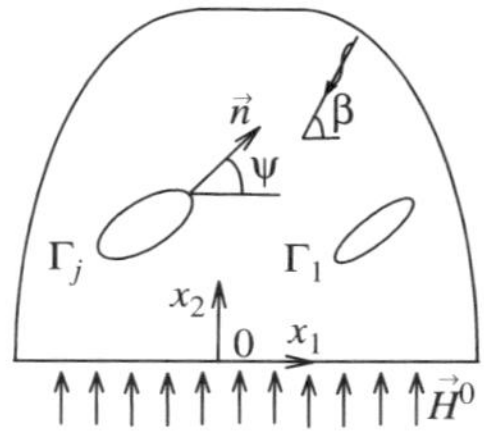

Fig. 10.15. Half-space with tunnel cavities in magnetic field

where S_{i3}, S_{i3}^0, S_{i3}^1 are the amplitudes of stress in the scattered, the incident and the reflected waves, respectively.

The procedure of satisfaction of this condition leads to the following integral equation with respect to the function $p(\zeta)$:

$$p(\zeta_0) + \int_\Gamma p(\zeta) G(\zeta, \zeta_0)\, ds = N(\zeta_0), \tag{10.40}$$

$$\eta(\psi_0) G(\zeta, \zeta_0) = \frac{2}{i\pi} \mathrm{Re} \left(\frac{c(\psi_0)}{\zeta_1 - \zeta_{10}} \right) + \gamma_2 \left\{ H_1(\gamma_2 r_{10}) \mathrm{Re} \left(c(\psi_0) e^{-i\alpha_{10}} \right) \right.$$
$$\left. + H_1^{(1)}(\gamma_2 r_{10}^*) \mathrm{Re} \left(c(\psi_0) e^{-i\alpha_{10}^*} \right) \right\},$$

$$\eta(\psi) N(\zeta) = \frac{X_3(\zeta)}{\mu} + i\gamma_h \left\{ U_0 \cos(\beta - \psi) + U_* \cos(\beta + \psi) \right\},$$

$$\gamma_h = \frac{\gamma_2}{\sqrt{1 + h_c^2 \sin^2 \beta}}, \eta(\psi) = -2i \mathrm{Im} \frac{c(\psi)}{a(\psi)},$$

$$a(\psi) = \frac{d}{d\psi} c(\psi), \quad c(\psi) = \cos\psi + \frac{i \sin\psi}{\sqrt{1 + h_c^2}},$$

$$r_{10} = |\zeta_1 - \zeta_{10}|, \quad r_{10}^* = \left| \overline{\zeta}_1 - \zeta_{10} \right|,$$

$$\alpha_{10} = \arg(\zeta_1 - \zeta_{10}), \quad \alpha_{10}^* = \arg\left(\overline{\zeta}_1 - \zeta_{10} \right),$$

$$\zeta_{10} = \xi_{10} + \frac{i\xi_{20}}{\sqrt{1 + h_c^2}}, \quad \zeta_0 = \xi_{10} + i\xi_{20} \in \Gamma.$$

It should be noted that in absence of an initial magnetic field ($h_c = 0$), this is a Fredholm integral equation of second order. If $h_c > 0$, then we get to singular equation of second order.

When determining the wave fields of stresses in the body, it is necessary to take into account the Maxwell additives. Therefore, the total shift stresses on the area, perpendicular to the contour of the opening, can be represented by the formula

$$\begin{aligned} \sigma_{3S} &= \left(\sigma_{23}^0 + \sigma_{23}^1 + \sigma_{23} + t_{23} \right) \cos\psi - \left(\sigma_{13}^0 + \sigma_{13}^1 + \sigma_{13} + t_{13} \right) \\ sin\psi &= \mathrm{Re}\left(T e^{-iwt} \right). \end{aligned} \tag{10.41}$$

The Maxwell stresses t_{i3}, in our case, have the form

$$t_{13} = 0, \quad t_{23} = \mu_e H_0^2\, \partial_2 u_3. \tag{10.42}$$

Appealing to the expressions for the stresses, taking into account formulas (10.15), (10.26) and theintegral representation (10.38), we obtain at point $\zeta_0 \in \Gamma$

$$T(\zeta_0) = -2i\mu\sqrt{1+h_c^2}p(\zeta_0)\,\mathrm{Re}\frac{c(\psi_0)}{a(\psi_0)} + \tag{10.43}$$

$$+\int_\Gamma p(\zeta)\,K(\zeta,\zeta_0)ds + i\mu\gamma_h\left\{U_0\sin(\psi_0-\beta) + U_*\sin(\psi_0+\beta)\right\},$$

$$K(\zeta,\zeta_0) = \frac{2i\mu}{\pi}\sqrt{1+h_c^2}\,\mathrm{Im}\frac{c(\psi_0)}{\zeta_1-\zeta_{10}} - \mu\gamma_2\left(bH_1(\gamma_2 r_{10}) + b_* H_1^{(1)}(\gamma_2 r_{10}^*)\right),$$

$$b = \cos\alpha_{10}\sin\psi_0 - \sqrt{1+h_c^2}\sin\alpha_{10}\cos\psi_0,$$

$$b_* = \cos\alpha_{10}^*\sin\psi_0 - \sqrt{1+h_c^2}\sin\alpha_{10}^*\cos\psi_0,$$

$$H_1(x) = \frac{2i}{\pi x} + H_1^{(1)}(x), \quad \zeta_0 = \xi_{10} + i\xi_{20},$$

the functions U_0, U_* are determined by (10.15) and (10.26), respectively, when $x_1 = \xi_{10}$, $x_2 = \xi_{20}$.

Let us consider an example. Let a half-space weakened by the cavity of elliptical cross section $\xi_1 = a_1\cos\varphi$, $\xi_2 = h + b_1\sin\varphi$. Let us assume that the surface of cavity is strainless and a magneto-elastic shear wave (10.15) falls from infinity along the axis x_2. The results of calculations of the magnitude $< T >= |T/T_0|$, depending on the normalized wave number $\gamma_2 R$, where $T_0 = \mu\tau\gamma_h$, $R = (a_1+b_1)/2$, are represented in Figs. 10.16 and 10.17 for the following values of the parameters: $a_1 = 1$, $b_1 = 0,75$, $h = 1,75$. The curves in Fig. 10.16 correspond to the point of the contour $\varphi = \pi/2$, in Fig. 10.17, to the point $\varphi = 0$. The graphs 1–3 are built for $h_c = 0$; 0,5 and 1, respectively. Curve 4 corresponds to the circular opening: $a_1 = b_1 = 1$; $h = 1,5$; $h_c = 0$; $\beta = 7\pi/8$; $\varphi = -\pi/2$. The points in Fig. 2.17 are data, obtained by a completely different method [263].

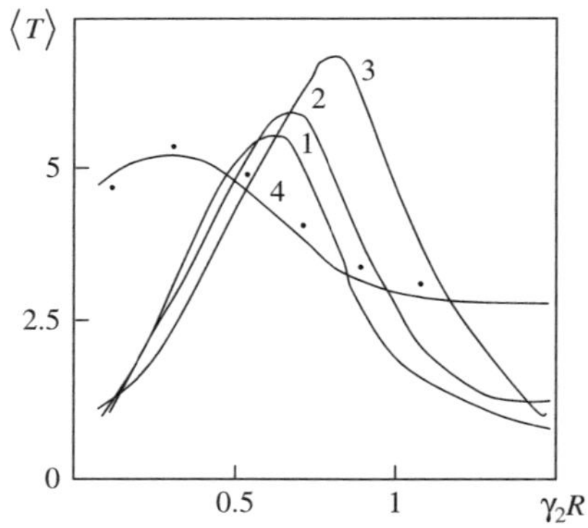

Fig. 10.16. Diffraction of magneto-elastic wave on the elliptical opening in presence of magnetic field

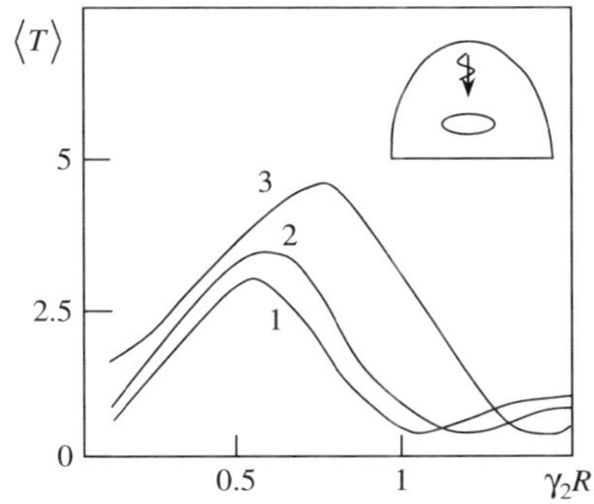

Fig. 10.17. Results of comparison with the data of the work [263]

The results of calculations show that the influence of the initial magnetic field when evaluating dynamic intensity of bodies with the stress concentrators is important and can't be ignored.

10.5.1 Note

The formula (10.43) is deduced without taking into account the straight and reverse traveling waves of Maxwell stresses. The consideration of these waves leads to the appearance of the following additional term in the right side of formula (10.43):

$$\Delta = i\mu h_c^2 \gamma_h \sin\beta \cos\psi_0 \left(U_* - U_0\right).$$

10.6 Interaction of Crack and Opening in the Current-Conducting Medium in Presence of Magnetic Field

Let us examine an unbounded current-conducting elastic medium, found in static magnetic field $\overrightarrow{H}^0 = (0, H_0, 0)$ and weakened along the axis x_3 (Fig. 10.18) by tunnel stress concentrators of the type of cracks L_j $(j = 1, 2, \ldots, M)$ and openings Γ_k $(k = 1, 2, \ldots, N)$. Let us assume that on the surfaces of the concentrators acts a harmonic in time shift load $X_{3n} = Re\left(e^{-i\omega t} X_3\right)$, $X_3 = X_3\left(x_1, x_2\right)$, and that emission of the shear magneto-elastic wave of displacement (10.15) is possible from the infinity. Under these conditions, a stationary wave process takes place in the system, which corresponds to the anti-plane deformation of the body.

The scattered field in space with defects can be represented in the form [278]

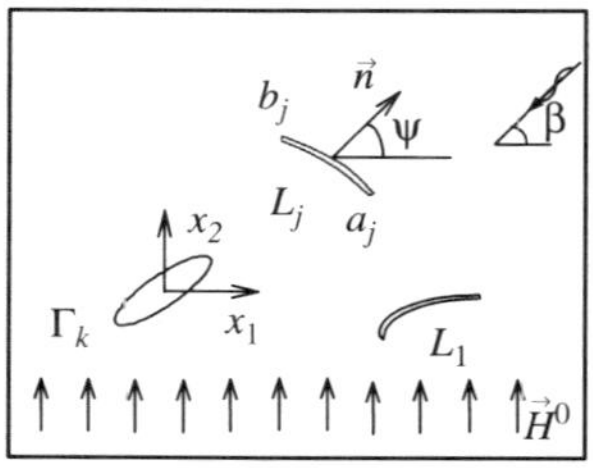

Fig. 10.18. Diagram of current-conducting space with tunnel heterogeneities

$$U(x_1, x_2) = \frac{1}{2} \int_L p(\zeta) \left\{ \frac{\partial}{\partial \zeta_1} H_0^{(1)} (\gamma_2 r_1)\, d\zeta_1 - \frac{\partial}{\partial \bar{\zeta}_1} H_0^{(1)} (\gamma_2 r_1)\, d\bar{\zeta}_1 \right\}$$
$$+ \int_L q(\zeta)\, H_0^{(1)}(\gamma_2 r_1) ds + \int_\Gamma w(\zeta)\, H_0^{(1)}(\gamma_2 r_1) ds,$$

where $p(\zeta) = \{p_j(\zeta), \zeta \in L_j\}$ and $w(\zeta) = \{w_k(\zeta), \zeta \in \Gamma_k\}$ are the unknown densities $L = \cup\, L_j$, $\Gamma = \cup\, \Gamma_k$; contour integration of openings is done counterclockwise, along the contours of the cracks – from the beginning – point a_j to the end b_j; all the remaining magnitudes are determined by (10.16), (10.18).

With the aid of the standard procedure the boundary-value problem (10.14) on the contours Γ_k and the arcs L_j is reduced to the system of integral equations, which is not exposed here.

Let us address ourselves directly to the results of calculations. The summarizing stress intensity factor is determined here by formula

$$K_{III}^s = P_h \sqrt{\pi l} \alpha_s^\pm \cos\left(\omega t - \delta^\pm\right),$$

where $P_h = -i\ \mu\ \tau\ \gamma$; the upper sign corresponds to the tip of crack b, the lower one, to the tip a.

The complete shift stress on the contour of the opening is determined by formula (10.41); moreover, its amplitude T can be represented in the form $T = P_h \langle T \rangle$.

Figures 10.19–10.21 give the results of calculations of the magnitude $\alpha_s^\pm$ for an unbounded medium, weakened by a cavity of elliptical cross section $\xi_1 = R_1 \cos\varphi$, $\xi_2 = R_2 \sin\varphi$ $(0 \le \varphi \le 2\pi)$, and a tunnel crack, whose cross section is the segment $\xi_1 = c_1\delta + c_2$, $\xi_2 = c_3\delta + c_4$ $(-1 \le \delta \le 1)$. A shear magneto-elastic wave (10.15) is radiated from the infinity along the axis x_2.

The graphs in Fig. 10.19 are constructed for the elliptical opening ($R_2/R_1 = 0,5$; $R_1 = 1$) and the horizontal crack ($c_1 = 1$, $c_2 = 2,2$, $c_3 = c_4 = 0$). The curves in Fig. 10.20 illustrate the situation, when a circular opening ($R_1 = R_2 = 1$) is located strictly in the shadow of a horizontal crack ($c_1 = 1$, $c_2 = c_3 = 0$, $c_4 = 2$). Finally, Fig. 10.21 corresponds to the case, when crack is oriented to

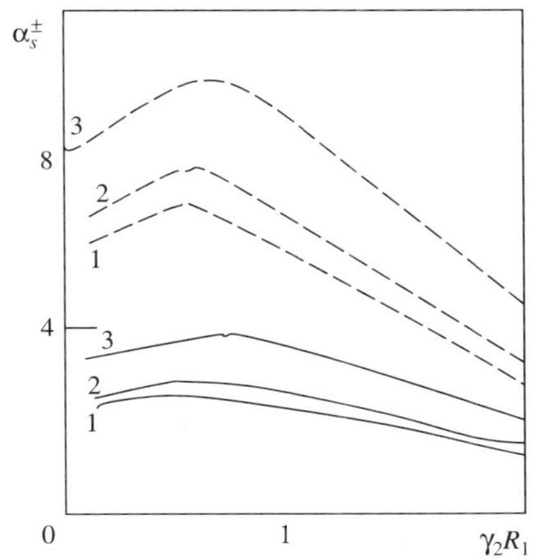

Fig. 10.19. Dependence of the magnitude $\alpha_s^{\pm}$ for the configuration of elliptical cavity and rectilinear crack

the axis $0x_1$ at angle 45^0 ($c_1 = 0,5$; $c_2 = 0$; $c_3 = -0,5$; $c_4 = 2$) and a circular opening ($R_1 = R_2 = 1$) is partially located in the shadow of the crack. The continual lines correspond to the magnitude α_s^-, the stroked ones, to the magnitude $\beta_S = |\langle T \rangle|$ at the point of ellipsis $\varphi = 0$. The curves 1, 2, 3 are constructed for the values $h_c = 0$; 0,5 and 1, respectively.

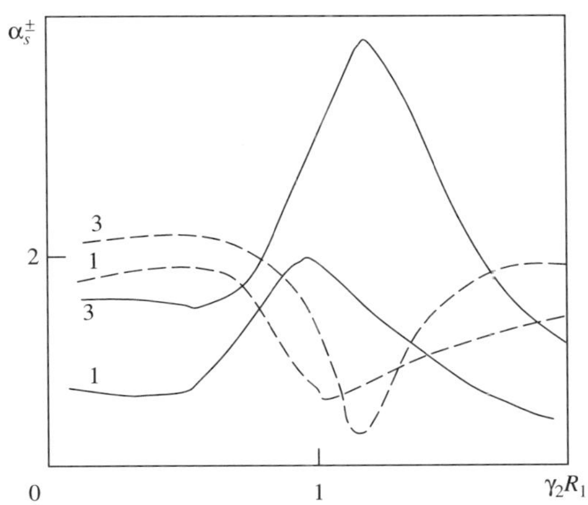

Fig. 10.20. Diffraction of magneto-elastic wave on the crack and the elliptical opening

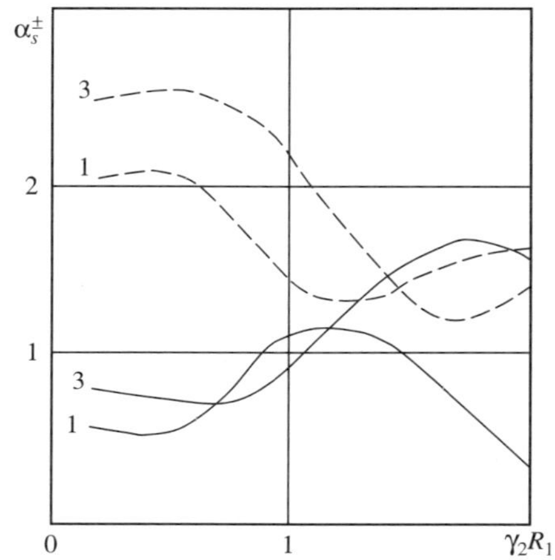

Fig. 10.21. Diffraction of magneto-elastic wave on the inclined crack and the circular opening

10.7 Diffraction of Shear Magneto-Elastic Wave in the Inclusions

Let us examine the unbounded ideally electro-conductive elastic medium, in reference to the Cartesian axes $0x_1x_2x_3$, which contains tunnel inclusions along the axis x_3 of another kind, the cross sections of which are limited by sufficiently smooth closed contours Γ_j $(j = 1, 2, \ldots, M)$.

A static magnetic field $\overrightarrow{H}^0 = (0,\ H_0,\ 0)$ acts on the medium, and mechanical excitation is caused by a plane magneto-elastic wave of type (10.15). We assume below that magnetic permeabilities of matrix and inclusions are identical and their magnetizability may be disregarded.

In this case, we get the anti-plane deformation of the medium with inclusions, which, according to (10.12)–(10.14), is described by the following system of equations:

– differential equations in matrix and inclusions

$$\begin{aligned} &\nabla^2 u_3 + h_c^2 \partial_2^2 u_3 = c_2^{-2} \partial_2 u_3 / \partial t^2, \\ &\nabla^2 u_3^{(j)} + h_{cj}^2 \partial_2^2 u_3^{(j)} = c_{2j}^{-2} \partial_2 u_3^{(j)} \Big/ \partial t^2, \\ &h_{cj}^2 = \mu_e H_0^2 / \mu_j; \end{aligned} \tag{10.44}$$

– electromagnetic components

$$\begin{aligned} &h_1 = h_2 = 0,\ h_3 = H_0 \partial_2 u_3,\ e_1 = -\mu_e H_0 \partial u_3 / \partial t, \\ &h_1^{(j)} = h_2^{(j)} = 0,\ h_3^{(j)} = H_0 \partial_2 u_3^{(j)},\ e_1^{(j)} = -\mu_e H_0 \partial u_3^{(j)} \Big/ \partial t, \\ &e_2 = e_3 = 0,\ e_2^{(j)} = e_3^{(j)} = 0; \end{aligned} \tag{10.45}$$

– conditions on the interface of media

$$\left[\sigma_{3i} + \sigma_{3i}^0 + t_{3i}\right] n_i = 0, \qquad (10.46)$$
$$\left[u_3 + u_3^0\right] = 0 \quad (i = 1, 2)\,.$$

In (10.44)–(10.46) the index j corresponds to inclusion with border Γ_j; h_i, e_i are the fluctuations of the vectors of magnetic and electrical intensities; σ_{3i}^0, σ_{3i} are mechanical stresses, which correspond to the incident wave and the scattered by inclusions field; t_{3i} is the Maxwell stresses; n_i are the components of the unit vector of the external normal to the border of the media interface; the symbol [*] indicates the jump of the corresponding function on Γ_j (Fig. 10.22).

Mechanical and Maxwell stresses in the matrix and the inclusions are expressed through displacements with the aid of equalities

$$\sigma_{n3} = \mathrm{Re}\left(e^{-i\omega t} S_{n3}\right), \quad \sigma_{n3}^{(j)} = \mathrm{Re}\left(e^{-i\omega t} S_{n3}^{(j)}\right), \qquad (10.47)$$
$$t_{n3} = \mathrm{Re}\left(e^{-i\omega t} T_{n3}\right), \quad t_{n3}^{(j)} = \mathrm{Re}\left(e^{-i\omega t} T_{n3}^{(j)}\right),$$
$$u_3 = \mathrm{Re}\left(e^{-i\omega t} W\right), \quad u_3^{(j)} = \mathrm{Re}\left(e^{-i\omega t} W_j\right),$$

$$S_{n3} = \mu \partial_n W, \quad S_{n3}^{(j)} = \mu_j \partial_n W_j,$$
$$T_{13} = 0, \quad T_{13}^{(j)} = 0,$$
$$T_{23} = \mu_e H_0^2 \partial_2 W, \quad T_{23}^{(j)} = \mu_e H_0^2 \partial_2 W_j.$$

For the formulation of the integral representation of the solution of the posed problem, let us introduce complex variables

$$z_1 = x_1 + \lambda x_2, \quad z_1^{(j)} = x_1 + \lambda_j x_2, \qquad (10.48)$$
$$\lambda = \frac{i}{\sqrt{1 + h_c^2}}, \quad \lambda_j = \frac{i}{\sqrt{1 + h_{cj}^2}} \quad (j = 1, 2, \ldots, M)\,.$$

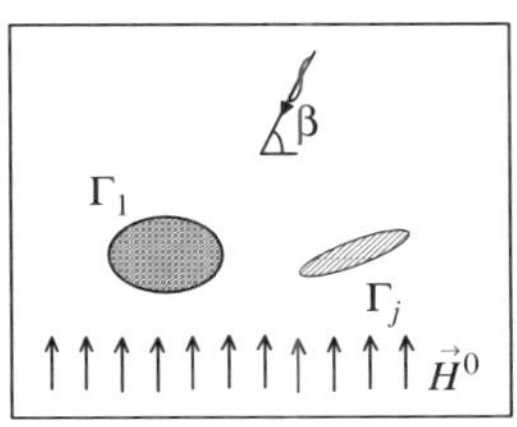

Fig. 10.22. Current-conducting elastic medium with inclusions in magnetic field

Let us assume [277]

$$
\begin{aligned}
& W = U_0 + U, \quad W_j = U_0^{(j)} + U_j, \qquad (10.49)\\
& U(x_1, x_2) = \int_\Gamma p(\zeta) H_0^{(1)}(\gamma_2 r_1)\, ds,\\
& U_j(x_1, x_2) = \int_\Gamma q(\zeta) H_0^{(1)}\left(\gamma_2^{(j)} r_1^{(j)}\right) ds,\\
& U_0^{(j)} = \tau \exp\left\{-i\gamma^{(j)}(x_1 \cos\beta + x_2 \sin\beta)\right\},\\
& \gamma^{(j)} = \frac{\gamma_2^{(j)}}{\sqrt{1 + h_{cj}^2 \sin^2\beta}}, \quad \gamma_2^{(j)} = \frac{\omega}{c_{2j}} \ (j = 1, 2, \ldots, M),\\
& r_1 = |\zeta_1 - z_1|, \quad r_1^{(j)} = \left|\zeta_1^{(j)} - z_1^{(j)}\right|, \quad \zeta_1 = \xi_1 + \lambda \xi_2,\\
& \zeta_1^{(j)} = \xi_1 + \lambda_j \xi_2, \ \zeta = \xi_1 + i\xi_2 \in \Gamma = \cup \Gamma_j.
\end{aligned}
$$

The densities $p(\zeta) = \{p_j(\zeta), \zeta \in \Gamma_j\}, q(\zeta) = \{q_j(\zeta), \zeta \in \Gamma_j\}$ are the unknown functions that we will calculate; ds is the element of the arc of contour Γ_j; $H_n^{(1)}(x)$ is the Hankel function of first type and n-th order.

Taking into account definitions (10.48), it is possible to show that the displacements $u_3 = \text{Re}\left(e^{-i\omega t} W\right)$, $u_3^{(j)} = \text{Re}\left(e^{-i\omega t} W_j\right)$, where functions W and W_j are given in (10.49), satisfy the generalized wave (10.44) and the emission conditions [43, 161]. In this way, the problem is reduced to the determination of the unknown "densities," $p(\zeta)$, $q(\zeta)$ from the conjugation conditions of fields on the interface of the media (10.46). It should be noted that the field $U_0^{(j)}$ is added to the wave field U_j, for convenience. So, in the degenerate case, when the material of any inclusion coincides with the material of matrix, the wave field in this region is automatically included in the wave field of the matrix.

The integral representations of mechanical and Maxwell stresses are obtained, by implementation of the specified in formulas (10.47) operations over functions (10.49). The substitution of the limiting values of the obtained expressions in the coupling conditions (10.46) leads to the system of integral equations

$$
\begin{aligned}
& p(\zeta_0) - \int_\Gamma p(\zeta) G(\zeta, \zeta_0)\, ds + \varepsilon_j \left\{ q(\zeta_0) + \int_\Gamma q(\zeta) G_j(\zeta, \zeta_0)\, ds \right\} = N_1(\zeta_0),\\
& \zeta_0 \in \Gamma = \cup \Gamma_j,\\
& \int_\Gamma p(\zeta) H_0^{(1)}(\gamma_2 r_{10})\, ds - \int_\Gamma q(\zeta) H_0^{(1)}\left(\gamma_2^{(j)} r_{10}^{(j)}\right) ds = N_2(\zeta_0), \qquad (10.50)
\end{aligned}
$$

$$G(\zeta,\zeta) = \frac{1}{\pi}\mathrm{Im}\frac{a(\psi_0)}{\zeta_1-\zeta_{10}} + \frac{i\gamma_2}{2}H_1(\gamma_2 r_{10})\,\mathrm{Im}\left\{a(\psi_0)e^{-i\alpha_{10}}\right\},$$

$$G_j(\zeta,\zeta_0) = \frac{1}{\pi}\mathrm{Im}\frac{a_j(\psi_0)}{\zeta_1^{(j)}-\zeta_{10}^{(j)}} + \frac{i\gamma_2^{(j)}}{2}H_1\left(\gamma_2^{(j)} r_{10}^{(j)}\right)\mathrm{Im}\left\{a_j(\psi_0)e^{-i\alpha_{10}^{(j)}}\right\},$$

$$N_1(\zeta_0) = \frac{S_n^0\cos(\psi_0-\beta)}{2\mu\sqrt{1+h_c^2}}\left\{e^{-i\gamma\left(\vec{r}_0\cdot\vec{k}\right)} - \frac{\gamma^{(j)}\mu_j}{\gamma\mu}e^{-i\gamma^{(j)}\left(\vec{r}_0\cdot\vec{k}\right)}\right\},$$

$$N_2(\zeta_0) = \frac{S_n^0}{\mu\gamma}\left\{e^{-i\gamma^{(j)}\left(\vec{r}_0\cdot\vec{k}\right)} - e^{-i\gamma\left(\vec{r}_0\cdot\vec{k}\right)}\right\}, \quad \psi_0 = \psi(\zeta_0),$$

$$\varepsilon_j = \frac{\mu_j\sqrt{1+h_{cj}^2}}{\mu\sqrt{1+h_c^2}}, \; a(\psi_0) = \lambda\cos\psi_0 - \sin\psi_0, \; S_n^0 = \mu\gamma\tau,$$

$$\zeta_1^{(j)} = \xi_1 + \lambda_j\xi_2, \quad \zeta_{10}^{(j)} = \xi_{10} + \lambda_j\xi_{20},$$

$$r_{10} = |\zeta_1 - \zeta_{10}|, \quad \xi_{10} + i\xi_{20} \in \Gamma,$$

$$r_{10}^{(j)} = \left|\zeta_1^{(j)} - \zeta_{10}^{(j)}\right|, \quad \alpha_{10} = \arg(\zeta_1 - \zeta_{10}),$$

$$\alpha_{10}^{(j)} = \arg\left(\zeta_1^{(j)} - \zeta_{10}^{(j)}\right),$$

$$H_1(x) = \frac{2i}{\pi x} + H_1^{(1)}(x), \; \vec{r}_0\cdot\vec{k} = \xi_{10}\cos\beta + \xi_{20}\sin\beta,$$
$$\zeta_1 = \xi_1 + \lambda\xi_2, \zeta_{10} = \xi_{10} + \lambda\xi_{20}.$$

The magnitude S_n^0 is the amplitude of stress in the incident wave; the kernels $G(\zeta,\zeta_0), G_j(\zeta,\zeta_0)$ are singular (Cauchy kernel), the remaining kernels have logarithmic singularity. When $h_c = 0$ (absence of initial magnetic field), the first (10.50) is converted into Fredholm integral equation of second order.

After the solution of the system (10.50) and the determination of the functions $p(\zeta)$ and $q(\zeta)$ by the formulas (10.49), the amplitudes of the distrubuted field of displacements U, U_j are restored and then the components of electrical, magnetic and mechanical fields are calculated by the relations (10.45), (10.47).

The total shift stresses on the interface of the media (from the side of the matrix) can be determined by formulas

$$\sigma_{3s}^- = \mathrm{Re}\left(e^{-i\omega t}T_s^-\right), \quad \sigma_{3n}^- = \mathrm{Re}\left(e^{-i\omega t}T_n^-\right),$$
$$T_s^-(\zeta_0) = \left(S_{23}^0 + S_{23} + T_{23}\right)^-\cos\psi_0 - \left(S_{13}^0 + S_{13} + T_{13}\right)^-\sin\psi_0,$$
$$T_n^-(\zeta_0) = \left(S_{13}^0 + S_{13} + T_{13}\right)^-\cos\psi_0 + \left(S_{23}^0 + S_{23} + T_{23}\right)^-\sin\psi_0,$$
$$\zeta_0 \in \Gamma.$$

Substituting here the limiting values of the amplitudes of mechanical and Maxwell stresses, we find

$$T_s^-(\zeta_0) = iS_n^0 \sin(\psi_0 - \beta)\, e^{-i\gamma\left(\vec{r_0}\cdot\vec{k}\right)} + \mu\left\{\frac{ih_c^2\sqrt{1+h_c^2}\sin 2\psi_0}{1+h_c^2\sin^2\psi_0}p(\zeta_0)\right.$$

$$\left. + \int_\Gamma p(\zeta)\, k(\zeta,\zeta_0) ds\right\},\ T_n^-(\zeta_0) = -iS_n^0\cos(\psi_0-\beta)\, e^{-i\gamma\left(\vec{r_0}\cdot\vec{k}\right)}$$

$$+ i\mu\sqrt{1+h_c^2}\left\{2p(\zeta_0) - \int_\Gamma p(\zeta) K_*(\zeta,\zeta_0)\, ds\right\},$$

$$K(\zeta,\zeta_0) = \frac{2}{\pi i}\left\{\mathrm{Re}\frac{a_*(\psi_0)}{\zeta_1-\zeta_{10}} + \gamma_2 H_1(\gamma_2 r_{10})\,\mathrm{Re}\left(a_*(\psi_0)\, e^{-i\alpha_{10}}\right)\right\},$$

$$K_*(\zeta,\zeta_0) = \frac{2}{\pi}\left\{\mathrm{Im}\frac{a(\psi_0)}{\zeta_1-\zeta_{10}} + i\gamma_2 H_1(\gamma_2 r_{10})\,\mathrm{Im}\left(a(\psi_0)\, e^{-i\alpha_{10}}\right)\right\},$$

$$a_*(\psi_0) = i\sqrt{1+h_c^2}\cos\psi_0 - \sin\psi_0.$$

In analogous manner, the total shift stresses on the interface from the side of the inclusion are calculated.

$$T_n^+(\zeta_0) = T_n^-(\zeta_0),\quad \zeta_0\in\Gamma_j\ (j=1,2,\ldots,M),$$

$$T_s^+(\zeta_0) = i\mu_j\gamma^{(j)}\tau\sin(\psi_0-\beta)\, e^{-i\gamma^{(j)}\left(\vec{\underset{0}{r}}\cdot\vec{k}\right)} + \mu_j\left\{\int_\Gamma q(\zeta) K_j(\zeta,\zeta_0)\, ds\right.$$

$$\left. - \frac{ih_{cj}^2\sqrt{1+h_{cj}^2}\sin 2\psi_0}{1+h_{cj}^2\sin^2\psi_0} q(\zeta_0)\right\},$$

$$K_j(\zeta,\zeta_0) = \frac{2}{\pi i}\left\{\mathrm{Re}\frac{a_*^{(j)}(\psi_0)}{\zeta_1^{(j)}-\zeta_{10}^{(j)}} + \gamma_2^{(j)} H_1\left(\gamma_2^{(j)} r_{10}^{(j)}\right) \times \mathrm{Re}\left(a_*^{(j)}(\psi_0) e^{-i\alpha_{10}^{(j)}}\right)\right\},$$

$$a_*^{(j)}(\psi_0) = i\sqrt{1+h_{cj}^2}\cos\psi_0 - \sin\psi_0.$$

As an example let us examine the unbounded medium with one elliptical inclusion $\xi_1 = R_1\cos\varphi,\ \xi_2 = R_2\sin\varphi,\ R_2/R_1 = 0,70;\ 0\le\varphi\le 2\pi$. A magneto-elastic wave (10.15) falls along the vertical axis ($\beta=\pi/2$) from infinity. Figs. 10.23–10.25 show the graphs of the magnitudes $\langle T_n\rangle = |T_n| / |S_n^0|$ and $\langle T_S^-\rangle = |T_S^-| / |S_n^0|$, along the inclusion contour, when the values of the parameters (ρ, ρ_1 are the densities of the material of the matrix and the inclusion): $\mu_1/\mu = 5,71;\ \rho_1/\rho = 0,83;\ \gamma_2(R_1+R_2)/2 = 0,9$ and $h_c = 0;\ 0,5$ and 1, respectively.

The representation of the orders of jumps of magnitude T_s on the section border of materials can be obtained from the Table 10.1, in which for the

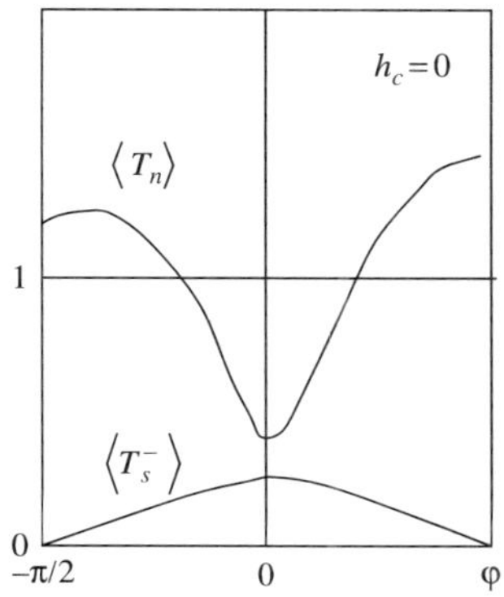

Fig. 10.23. Change in the contact stresses, when $h_s = 0$

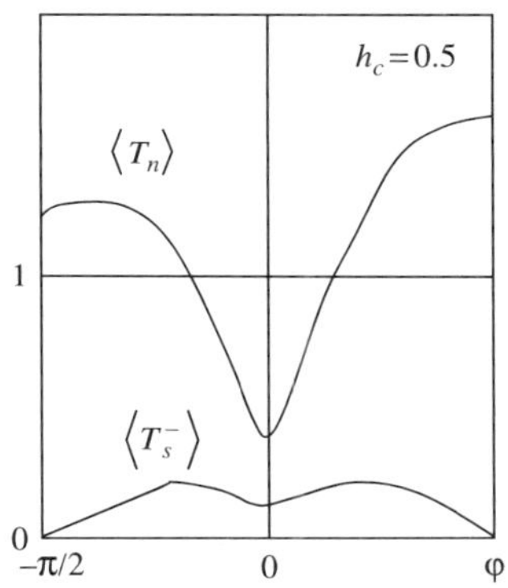

Fig. 10.24. Change in the contact stresses along the inclusion contour, when $h_c = 0.5$

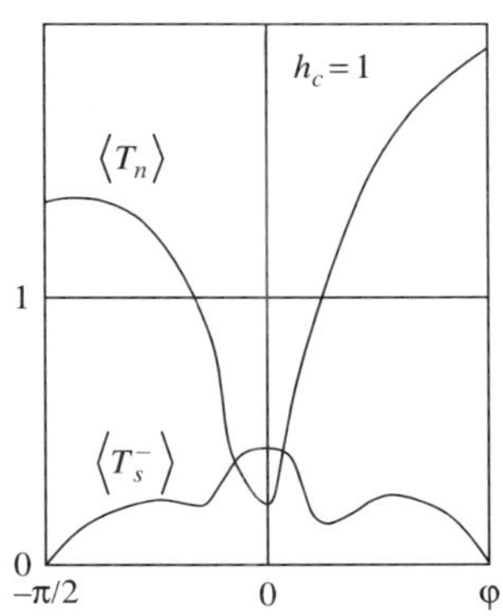

Fig. 10.25. Change in the contact stresses along the inclusion contour, when $h_c = 1$

Table 10.1. Values of quantities $|T_s^-|/S_n^0$ and $|T_S^+|/S_{n1}^0$ at the points of the contour of the circular inclusion

φ	–1,48	–1,29	–1,09	–0,904	–0,714	–0,524	–0,143
$\lvert T_s^-\rvert/S_n^0$	0,028	0,083	0,127	0,155	0,165	0,156	0,114
$\lvert T_s^+\rvert/S_{n1}^0$	0,020	0,051	0,101	0,141	0,179	0,215	0,269
φ	0,048	0,238	0,428	0,809	1,190	1,380	1,57
$\lvert T_s^-\rvert/S_n^0$	0,101	0,111	0,145	0,206	0,155	0,085	0,000
$\lvert T_s^+\rvert/S_{n1}^0$	0,284	0,287	0,277	0,220	0,122	0,062	0,000

circular inclusion, when $h_c = 0,5$ (remaining parameters are the same as above) contains the values $|T_s^-|/|S_n^0|$ and $|T_s^+|/|S_{n1}^0|$ $(S_{n1}^0 = \mu_1\gamma^{(1)}\tau)$, from the side of the matrix and the inclusion, respectively.

When $h_c = 0$, the initial magnetic field is absent, so that we get the ordinary problem about the scattering of the SH-wave on the inclusion.

10.8 Fundamental Solution of Two – Dimensional Equations of Magneto-Elasticity for Current-Conducting Medium

Let us examine an ideally conducting elastic medium in reference to rectangular Cartesian coordinate system $0x_1x_2x_3$, in which acts a harmonically changing with time concentrated force with amplitude $\vec{P} = (P_1, P_2)$. A constant magnetic field $\vec{H}^0 = (0,\ H_{02},\ 0)$ occupies the entire space.

Under mechanical excitation of the medium, as a result of elastic displacements of particles, vortical currents appear in it, which gives rise to the appearance of Lorenz body forces. The consideration of these forces gives the additional tensor of Maxwell stresses, which can exert essential influence on the wave fields in the medium. The question of their determination rises, taking into account the influence of induction currents. We proceed below from the relations for plane deformation (10.9)–(10.11).

The complete system of equations for our case takes the form

- equation of motion (in amplitudes)

$$\left(1 + \varepsilon_H \delta_m^1\right) \nabla^2 U_m^{(n)} + \sigma_* \partial_m \theta^{(n)} + \gamma_2^2 U_m^{(n)} = -P\delta_m^n \delta\left(x_1\right)\delta\left(x_2\right)/\mu, \tag{10.51}$$

$$\theta^{(n)} = \partial_m U_m^{(n)}, \quad \gamma_2 = \omega/c_2, \quad \partial_m = \partial/\partial x_m \ \ (m, n = 1, 2),$$
$$\varepsilon_H = h_c^2 = \mu_e H_{02}^2/\mu = c_A^2/c_2^2, \sigma_* = (\lambda + \mu)/\mu, \quad P_1 = P_2 = P;$$

– material equations

$$S_{ij} = \sigma_{ij} + t_{ij},\ \sigma_{ij} = \lambda\delta_i^j\theta + \mu\left(\partial_i U_j + \partial_j U_i\right), \tag{10.52}$$

$$t_{ij} = h_i B_{0j} + b_j H_{0i} - \mu_e \delta_i^j\ H_{0k}h_k + H_{0i}B_{oj},$$

$$\theta = \partial_k U_k,\ \vec{H} = \vec{H}_0 + \vec{h} = \left(h_1, H_{02} + h_2, 0\right),$$

$$\vec{B} = \mu_e\,\vec{H} = \left(b_1, B_{02} + b_2, 0\right);$$

– expression for the fluctuations of electromagnetic field

$$h_1 = H_{02}\partial_2 U_1, \quad h_2 = -H_{02}\partial_1 U_1, \quad h_3 = 0, \tag{10.53}$$

$$\vec{e} = \left(0, 0, e_3\right), \quad e_3 = i\mu_e\omega H_{02}U_1.$$

In formulas (10.51)–(10.53) the magnitude U_m^n is the amplitude of elastic displacement along the axis x_m from the force, which acts in the direction x_n; σ_{ij}, t_{ij} are the tensors of elastic and Maxwell stresses in the medium; c_A and c_2 are the alfven velocity [120] and the velocity of propagation of the transverse wave; h_i, b_i are the small disturbances of magnetic field strength and induction; λ, μ are the Lame constants; μ_e is the magnetic permeability of substance; ω is the angular frequency; δ_i^j is the Kronecker symbol; $\delta\left(x\right)$ is Dirac's delta function.

The system (10.51) differs from the classical equations of linear elasticity of isotropic body in the factor called the "parameter of magnetic pressure," ε_H, which introduces constructive anisotropy in the initial medium. For its integration let us use the Fourier transform on three-dimensional variables [43]

$$V\left(\xi\right) = \frac{1}{2\pi}\int U\left(x\right)e^{i(\xi,x)}dx, \quad x = \left(x_1, x_2\right), \tag{10.54}$$

$$U\left(x\right) = \frac{1}{2\pi}\int V\left(\xi\right)e^{-i(\xi,x)}d\xi, \quad \xi = \left(\xi_1, \xi_2\right).$$

As a result of standard transformations we obtain

$$V_m^{(n)}(\xi) = \frac{P\Delta_m^{(n)}(\xi)}{2\pi\mu\Delta(\xi)} \quad (m,\ n = 1,\ 2) \tag{10.55}$$

$$\Delta\left(\xi\right) = \left(1 + \sigma_*\right)\left(1 + \varepsilon_H\right)\Delta_0\left(\xi\right),$$

$$\Delta_0\left(\xi\right) = \left(\xi_1^2 + \xi_2^2\right)^2 - \left(\xi_1^2 + \xi_2^2\right)\left(\gamma_1^2 + \gamma^2 + \varepsilon^2\xi_1^2\right) + \gamma_1^2\gamma^2,$$

$$\Delta_m^{(m)}\left(\xi\right) = a_m\xi_1^2 + b_m\xi_2^2 - \gamma_2^2, \quad \Delta_1^{(2)} = \Delta_2^{(1)} = -\sigma_*\xi_1\xi_2,$$

$$a_1 = 1, \quad b_1 = 1 + \sigma_*, \quad a_2 = 1 + \varepsilon_H + \sigma_*,$$

$$b_2 = 1 + \varepsilon_H, \quad \gamma_1^2 = \frac{\gamma_2^2}{1 + \sigma_*}, \quad \gamma^2 = \frac{\gamma_2^2}{1 + \varepsilon_H}, \quad \varepsilon^2 = \frac{\sigma_*\varepsilon_H}{\left(1 + \sigma_*\right)\left(1 + \varepsilon_H\right)}.$$

By the introduction of polar coordinates $x_1 = r\cos\alpha$, $x_2 = r\sin\alpha$ and $\xi_1 = \rho\cos\varphi$, $\xi_2 = \rho\sin\varphi$, in (10.55), we can represent the characteristic polynomial Δ_0 in the form [280]

$$\begin{aligned}
&\Delta_0 = \left(1-\varepsilon^2\cos^2\varphi\right)\left(\rho^2-\gamma_2^2\rho_1^2\right)\left(\rho^2-\gamma_2^2\rho_2^2\right) \\
&\rho_1^2 = \frac{K_\sigma+K_\varepsilon-\sqrt{d(\varphi)}}{2\left(1-\varepsilon^2\cos^2\varphi\right)}, \quad \rho_2^2 = \frac{K_\sigma+K_\varepsilon+\sqrt{d(\varphi)}}{2\left(1-\varepsilon^2\cos^2\varphi\right)}, \\
&d(\varphi) = \left(K_\sigma-K_\varepsilon\right)^2+4\varepsilon^2K_\sigma K_\varepsilon\cos^2\varphi, \\
&K_\sigma = \frac{1}{\sigma_*+1}, \quad K_\varepsilon = \frac{1}{\varepsilon_H+1}.
\end{aligned} \tag{10.56}$$

Obviously, $\rho_m^2(\varphi) > 0$. The meaning of these parameters is that they show, in what sense the initial magnetic field changes the velocity of the running longitudinal and transverse monochromatic waves. Thus, if the normal to the wave front forms an angle β with the axis $0x_1$, then the velocity of propagation of these waves is determined by the relations $c_1^* = c_2\big/\rho_1(\beta)$, $c_2^* = c_2\big/\rho_2(\beta)$.

Analyzing the expressions for the transforms (10.55) into elementary fractions and conversing the Fourier transforms, we find (by summing over m)

$$\begin{aligned}
&U_n^{(n)}(x_1,x_2) = \frac{PK_\sigma K_\varepsilon}{2\pi^2\mu}\int\limits_0^\pi \frac{A_m^{(n)}(\varphi)\,d\varphi}{1-\varepsilon^2\cos^2\varphi} \times \int\limits_0^\infty \frac{\rho\cos\left[\rho r\cos(\alpha-\varphi)\right]}{\rho^2-\gamma_2^2\rho_m^2(\varphi)}d\rho, \\
&U_1^{(2)}(x_1,x_2) = \frac{PK_\sigma K_\varepsilon}{2\pi^2\mu}\int\limits_0^\pi \frac{A_m(\varphi)\,d\varphi}{1-\varepsilon^2\cos^2\varphi} \times \int\limits_0^\infty \frac{\rho\cos\left[\rho r\cos(\alpha-\varphi)\right]}{\rho^2-\gamma_2^2\rho_m^2(\varphi)}d\rho, \\
&A_m^{(n)}(\varphi) = (-1)^m\frac{\rho_m^2(\varphi)\left(a_n\cos^2\varphi+b_n\sin^2\varphi\right)-1}{\rho_2^2(\varphi)-\rho_1^2(\varphi)}, \\
&A_m(\varphi) = (-1)^{m-1}\frac{\sigma_*\rho_m^2(\varphi)\sin\varphi\cos\varphi}{\rho_2^2(\varphi)-\rho_1^2(\varphi)} \qquad (m,n=1,2).
\end{aligned} \tag{10.57}$$

In order the wave field to satisfy the emission conditions, the integrals, which appear in (10.57), must be understood in the following sense [43] appear:

$$\int\limits_0^\infty \frac{x\cos ax dx}{x^2-\beta^2-i0} = \lim_{\delta\to+0}\int\limits_0^\infty \frac{x\cos axdx}{x^2+\beta_*^2} = \frac{1}{2}\left\{e^{-ia\beta}E_1(-ia\beta)-e^{ia\beta}Ei(-ia\beta)\right\} \tag{10.58}$$

where $\beta_* = -i\sqrt{\beta^2+i\delta}$, $\delta > 0$, $a > 0$, $\mathrm{Re}\beta_* > 0$; $E_1(z)$, $Ei(z)$ are the integral exponential functions [22, 109].

Taking into account this remark we obtain formulas for the determination of the wave fields of displacements [280]

$$U_k^{(n)}(r,\alpha) = -\frac{PK_\sigma K_\varepsilon}{2\mu\pi^2}\int_0^\pi B_{m,k}^{(n)}(\varphi)\Phi(\theta_m)\,d\varphi \tag{10.59}$$

$$\Phi(x) = \cos x \text{ ci } x + \sin x \text{ si } x - \frac{\pi i}{2}e^{ix},$$

$$B_{m,n}^{(n)} = \frac{A_m^{(n)}(\varphi)}{1-\varepsilon^2\cos^2\varphi}, \quad B_{m,1}^{(2)} = B_{m,2}^{(1)} = \frac{A_m(\varphi)}{1-\varepsilon^2\cos^2\varphi},$$

$$\theta_m = r\gamma_2\rho_m(\varphi)\left|\cos(\alpha-\varphi)\right| \quad (m,n,k=1,2),$$

where ci x, si x are the integral cosine and sine, respectively [22, 109].

In the classical case, when there is no initial magnetic field ($\varepsilon_H = 0$), the inverse Fourier transform gives

$$U_n^{(n)}(r,\alpha) = \frac{i}{8\mu(1+\sigma_*)}\left\{2H_0^{(1)}(\gamma_1 r) + \sum_{k=1}^{2}(-1)^{k-1}\frac{\gamma_k^2}{\gamma_2^2}\right.$$
$$\left.\times\left[H_0^{(1)}(\gamma_k r) - (-1)^m H_2(\gamma_k r)\cos 2\alpha\right]\right\}$$
$$(n=1,2), \tag{10.60}$$

$$U_1^{(2)} = U_2^{(1)}(r,a) = \frac{i\sin 2\alpha}{8\mu(1+\sigma_*)}\sum_{k=1}^{2}(-1)^{k-1}\frac{\gamma_k^2}{\gamma_2^2}H_2(\gamma_k r),$$

$$H_2(x) = \frac{4i}{\pi x^2} + H_2^{(1)}(x).$$

Here $H_m^{(1)}(x)$ is the Hankel function of the first kind with order m [22].

The stresses in the medium can be found by the use of the relations (10.52), (10.53). We get

$$\sigma_{11}+\sigma_{22} = 2\mu\sigma_*\partial_k U_k, \quad \sigma_{22}-\sigma_{11}\pm 2i\sigma_{12} = -2\mu(\partial_1 \mp i\partial_2)(U_1 \mp iU_2). \tag{10.61}$$

The corresponding combinations of stresses of electromagnetic origin have form

$$t_{11}+t_{22} = 0, \quad t_{22}-t_{11}\pm 2it_{12} = -2\mu\varepsilon_H(\partial_1 \mp i\partial_2)U_1. \tag{10.62}$$

On the basis of these relations and definitions for the matrix of fundamental solutions (10.59), we find the componentsof the tensors of complete voltages.

We get

$$S_{11}^{(n)} + S_{22}^{(n)} = \frac{P\sigma_* K_\sigma K_\varepsilon}{\pi^2} \sum_{m=1}^{2}\sum_{k=1}^{2} \int_0^\pi c_{m,k}^{(n)}(\varphi, r, \alpha) n_k d\varphi \tag{10.63}$$

$$S_{22}^{(n)} - S_{11}^{(n)} \pm 2iS_{12}^{(n)} = -\frac{PK_\sigma K_\varepsilon}{\pi^2} \sum_{m=1}^{2}\sum_{k=1}^{2} \beta_k \int_0^\pi c_{m,k}^{(n)}(\varphi, r, \alpha) e^{\mp i\varphi} d\varphi,$$

$$(n = 1, 2),$$

$$C_{m,k}^{(n)}(\varphi, r, \alpha) = \gamma_2 B_{m,k}^{(n)}(\varphi) \rho_m(\varphi) \times \left(\Psi_*(\theta_m) + \frac{1}{\theta_m}\right) \text{sign } \cos(\alpha - \varphi),$$

$$\Psi_*(x) = \cos x \text{ si } x - \sin x \text{ ci } x + \frac{\pi}{2} e^{ix},$$

$$n_1 = \cos\varphi, \quad n_2 = \sin\varphi, \quad \beta_1 = 1 + \varepsilon_H, \quad \beta_2 = \mp i.$$

The right sides in (10.63) contain the singular terms, which determine the singularity at the point of application of the concentrated functional. These terms represent the corresponding static problem of magneto-elasticity, the solution of which can be obtained in closed form. Transforming the coefficients, when θ_m^{-1} on m and k, and then carrying out the necessary quadratures, we find

$$S_{11}^{(1)} + S_{22}^{(1)} = \frac{P\sigma_* \sqrt{1 - \varepsilon^2}}{\pi (1 + \sigma_* + \varepsilon_H)} \cdot \frac{\cos\alpha}{r(1 - \varepsilon^2 \sin^2\alpha)} + P\gamma_2 R_1(r, \alpha), \tag{10.64}$$

$$S_{11}^{(2)} + S_{22}^{(2)} = \frac{P\sigma_* \sqrt{1 - \varepsilon^2}}{\pi (1 + \sigma_*)} \cdot \frac{\sin\alpha}{r(1 - \varepsilon^2 \sin^2\alpha)} + P\gamma_2 R_2(r, \alpha),$$

$$S_{22}^{(1)} - S_{11}^{(1)} \pm 2iS_{12}^{(1)} = \frac{P}{\pi r} \{\lambda_{11}(\alpha, \varepsilon_H) \cos\alpha \pm i\lambda_{11}(\alpha, \varepsilon_H) \sin\alpha\}$$

$$- \gamma_2 P R_1^{\pm}(r, \alpha),$$

$$S_{22}^{(2)} - S_{11}^{(2)} \pm 2iS_{12}^{(2)} = \frac{P}{\pi r} \{\pm i\lambda_{21}(\alpha, \varepsilon_H) \cos\alpha + \lambda_{22}(\alpha, \varepsilon_H) \sin\alpha\}$$

$$- \gamma_2 P R_2^{\pm}(r, \alpha),$$

$$R_n(r, \alpha) = \frac{\sigma_* K_\sigma K_\varepsilon}{\pi^2} \sum_{m=1}^{2}\sum_{k=1}^{2} \int_0^\pi B_{m,k}^{(n)}(\varphi) \Psi_*(\theta_m) \rho_m(\varphi)$$

$$\times n_k \text{ sign } \cos(\alpha - \varphi) d\varphi,$$

$$R_n^{\pm}(r, \alpha) = \frac{K_\sigma K_\varepsilon}{\pi^2} \sum_{m=1}^{2}\sum_{k=1}^{2} \beta_k \int_0^\pi B_{m,k}^{(n)}(\varphi) \Psi_*(\theta_m) \rho_m(\varphi)$$

$$\times e^{\mp i\varphi} \text{ sign } \cos(\alpha - \varphi) d\varphi,$$

$$\lambda_{11}(\alpha, \varepsilon_H) = \frac{2 + \varepsilon_H}{\varepsilon_H} \lambda_0 - \frac{1 + \varepsilon_H}{1 + \sigma_* + \varepsilon_H} (\lambda_0 + 1),$$

$$\lambda_{22}(\alpha, \varepsilon_H) = \frac{\sqrt{1-\varepsilon^2}}{(1+\sigma_*)\left(1-\varepsilon^2 \sin^2\ \alpha\right)} + \frac{2+\varepsilon_H}{\varepsilon_H}\lambda_0,$$

$$\lambda_{12}(\alpha, \varepsilon_H) = \frac{\varepsilon_H - 2\lambda_0}{\varepsilon_H}, \quad \lambda_{21} = \frac{2\lambda_0}{\varepsilon_H} + 2\lambda_0 + 1,$$

$$\lambda_0 = \frac{\sqrt{1-\varepsilon^2}}{1-\varepsilon^2 \sin^2\ \alpha} - 1.$$

When $\gamma_2 = 0$, we hence obtain the solution of the static magneto-elasticity problem for the action of concentrated force in elastic conducting medium under an initial magnetic field. Obviously, for the determination of mechanical stresses in the static case it is sufficient to take the second combination S_{ij}, but with the upper sign.

For the explanation of the influence of induction currents on the dynamic intensity of the medium, we calculate the complete voltages $S_{11}^{(n)}$, $S_{22}^{(n)}$, under the action of concentrated force $P = 1$ along or across the initial magnetic field. The results are represented in the Table 10.2 (for value of wave number $\gamma_2 = 0,5$). It is evident that the initial magnetic field exerts influence on the amplitudes of stresses close to the harmonic source. Moreover, $\left|S_{11}^{(1)}\right|$, $\left|S_{11}^{(2)}\right|$ decrease, whereas, $\left|S_{22}^{(1)}\right|$, $\left|S_{22}^{(2)}\right|$ grow with increase of the parameter of magnetic pressure ε_H. The strongest influence of the initial magnetic field is manifested in the immediate proximity of the source. A more distant field of displacements is investigated in [262].

Table 10.2. Amplitude values of complete voltages in a medium with harmonic source

	$\lvert S_{11}^{(1)}\rvert/\lvert S_{11}^{(2)}\rvert$			$\lvert S_{22}^{(1)}\rvert/\lvert S_{22}^{(2)}\rvert$		
ε_H	$r=0,1$	$r=0,5$	$r=1$	$r=0,1$	$r=0,5$	$r=1$
0	$\frac{2,73}{0,46}$	$\frac{0,56}{0,10}$	$\frac{0,30}{0,06}$	$\frac{2,73}{0,46}$	$\frac{0,56}{0,10}$	$\frac{0,30}{0,06}$
0.5	$\frac{2,40}{0,38}$	$\frac{0,49}{0,08}$	$\frac{0,25}{0,05}$	$\frac{2,99}{0,67}$	$\frac{0,61}{0,14}$	$\frac{0,32}{0,08}$
1	$\frac{2,23}{0,33}$	$\frac{0,45}{0,07}$	$\frac{0,23}{0,04}$	$\frac{3,17}{0,81}$	$\frac{0,65}{0,17}$	$\frac{0,34}{0,09}$

10.9 Diffraction of Magneto-Elastic Waves on the Opening in a Magnetic Field

10.9.1 Statement of Boundary-Value Problem. Integral Representations of the Solutions

The matrix of the fundamental solutions $U_m^{(n)}$ obtained above can be used for the study of the dynamic intensity of conductors with stress concentrators in magnetic fields.

Let us consider an unbounded ideally conducting elastic medium in reference to the rectilinear Cartesian axes $0x_1x_2x_3$, weakened by a tunnel cavity along the x_3 – axis, whose cross section is limited by a simple, closed and sufficiently smooth contour Γ. The medium is placed in a uniform magnetic field described by vector $H^0 = \{0, H_0, 0\}$. On the surface of cavity normal $N \exp(-i\omega t)$ and tangential $T \exp(-i\omega t)$ forces act. It is evident that the forces are harmonically changing with time and independent from the coordinate x_3. There is also a possibility that a plane monochromatic magnetoelastic wave is radiated from infinity. Under these conditions the stationary wave process, which corresponds to plane deformation of the medium, takes place.

The appearance of fluctuations of electrical and magnetic fields $\vec{e} = \{0, 0, e_3\}$ and $\vec{h} = \{h_1, h_2, 0\}$, in the conductor, as well as components of the vector of elastic displacements u_i and the tensors of mechanical and Maxwell stresses σ_{ij}, t_{ij} are described, in linear approximation, by relations (10.51)–(10.53), when $P = 0$.

Since the fluctuations of the electromagnetic field are determined through the elastic displacements, the fundamental boundary-value problem is reduced to integration of the equation of motions (in amplitudes)

$$\left(1 + \varepsilon_H \delta_m^1\right) \nabla^2 U_m + \sigma_* \partial_m \theta + \gamma_2^2 U_m = 0 \quad (m = 1, 2). \tag{10.65}$$

In this case, it is necessary to satisfy the boundary conditions on Γ:

$$\sigma_{kj} n_j = X_k \quad (k, j = 1, 2). \tag{10.66}$$

Here, it is considered that the magnetic permeabilities of many magnetic materials coincide with the magnetic permeability of the vacuum; therefore the magnetic field before the deformation and in the process of deformation can be continuously extended into the cavity of the opening.

For the solution of the boundary problem (10.65), (10.66), we write the sought-for wave field of displacements in the ordinary form

$$U_m(x_1, x_2) = U_m^0 + U_m^* \tag{10.67}$$

where u_m^0, u_m^* are the displacements in the falling and scattered heterogeneity waves, respectively.

The scattered wave field can be represented in the form

$$U_m^*(x_1, x_2) = \sum_{n=1}^{2} \int_{\Gamma} \omega_n(\zeta) u_m^{(n)}(r, \alpha) ds \;\; (m = 1, 2), \tag{10.68}$$

where

$$r = |\zeta - z|, \quad \alpha = \arg(\zeta - z), \;\; z = x_1 + ix_2, \;\; \zeta = \xi_1 + i\xi_2 \in \Gamma,$$

$\omega_n(\zeta)$ are the searched-for functions on the contour Γ, ds is the element of length of arc Γ.

The representations (10.68) are correct in the sense that the displacements, determined by them, satisfy the equation of motion (10.65) and the conditions for emission, independently from the choice of densities $\omega_n(\zeta)$.

The combinations of the complete voltages amplitudes due to the scattered by heterogeneity field are determined by formulas (10 52). Performing the operations specified there and taking into account the relations (10.59), (10.61) and (10.62), we find

$$S_{11} + S_{22} = \frac{\sigma_* K_\sigma K_\varepsilon}{\pi^2}\gamma_2 \sum_{n=1}^{2} \int\limits_L \omega_n(\zeta) h_n(\zeta, z) ds, \tag{10.69}$$

$$S_{22} - S_{11} \pm 2iS_{12} = -\frac{K_\sigma K_\varepsilon}{\pi^2}\gamma_2 \sum_{n=1}^{2} \int\limits_L \omega_n(\zeta) h_n^{\pm}(\zeta, z) ds,$$

$$h_n(\zeta, z) = \sum_{m=1}^{2}\sum_{p=1}^{2} \int\limits_0^{\pi} n_m F_{p,m}^{(n)}(\zeta, z; \varphi)\, d\varphi, \quad \beta_1^{\pm} = 1 + \varepsilon_H, \quad \beta_2^{\pm} = \pm i,$$

$$h_n^{\pm}(\zeta, z) = \sum_{m=1}^{2}\sum_{p=1}^{2} \beta_m^{\pm} \int\limits_0^{\pi} e^{\pm i\varphi} F_{p,m}^{(n)}(\zeta, z; \varphi)\, d\varphi, \quad n_1 = \cos\varphi, \quad n_2 = \sin\varphi,$$

$$F_{p,m}^{(n)}(\zeta, z; \varphi) = B_{p,m}^{(n)} \Psi(\theta_p)\, \rho_p(\varphi) \operatorname{sign}\left[\cos(\alpha - \varphi)\right],$$

$$S_{ij} = \sigma_{ij} + t_{ij}, \quad \Psi(x) = \frac{1}{x} + \Psi_*(x),$$

$$\Psi_*(x) = \cos\, x \text{ si } x - \sin\, x \text{ ci } x - \frac{\pi}{2} e^{ix},$$

$$B_{p,n}^{(n)} = C(\varphi) A_p^{(n)}(\varphi), \quad B_{p,1}^{(2)} = B_{p,2}^{(1)} = C(\varphi) A_p(\varphi), \quad C(\varphi) = \left(1 - \varepsilon^2 n_1^2\right)^{-1}.$$

The combinations of stresses of mechanical origin σ_{ij} are calculated by the same formulas; it is necessary only to put $\varepsilon_H = 0$ in the expression for the coefficient $\beta_1^{\pm}$. Obviously, the knowledge of three combinations of stresses in (10.69) makes possible to determine all the components of stress tensors S_{ij} and σ_{ij}.

An important point for our future analysis is the explicit distinction of the singular parts of kernels h_n and $h_n^{\pm}$, connected with the occurrence of the term x^{-1} in the expression for the function $\Psi(x)$. Transforming m and p the coefficients when r^{-1} in these kernels, we calculate

$$\sum_{m=1}^{2}\sum_{p=1}^{2} B_{p,m}^{(1)} n_m = n_1 C(\varphi), \tag{10.70}$$

$$\sum_{m=1}^{2}\sum_{p=1}^{2} B_{p,m}^{(2)} n_m = (1+\varepsilon_H) n_2 C(\varphi),$$

$$\sum_{m=1}^{2}\sum_{p=1}^{2} \beta_m^{\pm} B_{p,m}^{(1)} = \left[(1+\varepsilon_H)\left(1+\sigma_* n_2^2\right) \pm i\sigma_* n_1 n_2\right] e^{\pm i\varphi} C(\varphi),$$

$$\sum_{m=1}^{2}\sum_{p=1}^{2} \beta_m^{\pm} B_{p,m}^{(2)} = -\left[(1+\varepsilon_H)\,\sigma_* n_1 n_2 \pm i\left(1+\varepsilon_H+\sigma_* n_1^2\right)\right] \times e^{\pm i\varphi} C(\varphi).$$

Now the combinations of complete voltages (10.69) can be presented in the form (summation over n)

$$S_{11}+S_{22} = \frac{\sigma_* K_\sigma K_\varepsilon}{\pi^2} \int\limits_L \omega_n(\zeta) \left\{ \frac{I_n(\alpha)}{r} + \gamma_2 H_n(\zeta, z) \right\} ds, \tag{10.71}$$

$$S_{22}-S_{11} \pm 2iS_{12} = -\frac{K_\sigma K_\varepsilon}{\pi^2} \int\limits_L \omega_n(\zeta) \left\{ \frac{I_n^{\pm}(\alpha)}{r} + \gamma_2 H_n^{\pm}(\zeta, z) \right\} ds,$$

where

$$I_1(\alpha) = \int\limits_0^{\pi} \frac{n_1 d\varphi}{\left(1-\varepsilon^2 n_1^2\right)\cos(\alpha-\varphi)}, \qquad I_2(\alpha) = (1+\varepsilon_H)$$

$$\int\limits_0^{\pi} \frac{n_2 d\varphi}{\left(1-\varepsilon^2 n_1^2\right)\cos(\alpha-\varphi)},$$

$$I_3^{\pm}(\alpha) = \int\limits_0^{\pi} \frac{(1+\varepsilon_H)\left(1+\sigma_* n_2^2\right) \pm i\sigma_* n_1 n_2}{\left(1-\varepsilon^2 n_1^2\right)\cos(\alpha-\varphi)} e^{\pm i\varphi} d\varphi,$$

$$I_4^{\pm}(\alpha) = -\int\limits_0^{\pi} \frac{(1+\varepsilon_H)\,\sigma_* n_1 n_2 \pm i\left(1+\varepsilon_H+\sigma_* n_1^2\right)}{\left(1-\varepsilon^2 n_1^2\right)\cos(\alpha-\varphi)} e^{\pm i\varphi} d\varphi.$$

Kernels H_n and $H_n^{\pm}$, that appear in (10.71), are determined by formulas (10.71) for h_n and $h_n^{\pm}$, respectively, in which it is necessary to replace functions $\Psi(\theta_p)$ with $\Psi_*(\theta_p)$. The integrals I_n and $I_n^{\pm}$ should be understood in the sense of principal value according to Cauchy. After the cumbersome procedure of their calculation, we come to the following result.

Let $z = x_1 + ix_2$, $\zeta = \xi_1 + i\xi_2 \in \Gamma$; we introduce the mapping

$$z_1 = \frac{x_1}{\sqrt{1-\varepsilon^2}} + ix_2, \quad \zeta_1 = \frac{\xi_1}{\sqrt{1-\varepsilon^2}} + i\xi_2, \quad \zeta - z = re^{i\alpha}, \quad \zeta_1 - z_1 = r_1 e^{i\alpha_1}. \tag{10.72}$$

Then the following representations hold

$$S_{11}+S_{22}=\frac{\sigma_*}{\pi}\int\limits_\Gamma \omega_n(\zeta)g_n(\zeta,z)ds+\frac{K_\sigma K_\varepsilon}{\pi^2}\sigma_*\gamma_2\int\limits_\Gamma \omega_n(\zeta)H_n(\zeta,z)ds, \tag{10.73}$$

$$S_{22}-S_{11}\pm 2iS_{12}=\frac{1}{\pi}\int\limits_\Gamma \omega_n(\zeta)g_n^{\pm}(\zeta,z)ds-\frac{K_\sigma K_\varepsilon}{\pi^2}\gamma_2\int\limits_\Gamma \omega_n(\zeta)H_n^{\pm}(\zeta,z)ds,$$

$$g_1(\zeta,z)=\frac{1}{1+\sigma_*+\varepsilon_H}\frac{\cos\alpha_1}{r_1},\quad g_2(\zeta,z)=\frac{1}{(1+\sigma_*)\sqrt{1-\varepsilon^2}}\frac{\sin\alpha_1}{r_1},$$

$$g_n^{\pm}(\zeta,z)=e_{1n}\frac{e^{\pm i\alpha}}{r}+e_{2n}\frac{\cos\alpha_1}{r_1}+e_{3n}\frac{\sin\alpha_1}{r_1}\quad (n=1,2),$$

$$e_{11}=-\frac{2+\varepsilon_H}{\varepsilon_H},\quad e_{21}=\frac{2+\varepsilon_H}{\varepsilon_H}-\frac{1+\varepsilon_H}{1+\varepsilon_H+\sigma_*},$$

$$e_{31}=\pm\frac{2i}{\varepsilon_H\sqrt{1-\varepsilon^2}},\quad e_{12}=\pm i\frac{2+\varepsilon_H}{\varepsilon_H},$$

$$e_{22}=\pm 2i\frac{1+\varepsilon_H}{\varepsilon_H},\quad e_{32}=\frac{1}{\sqrt{1-\varepsilon^2}}\left(\frac{2+\varepsilon_H}{\varepsilon_H}+\frac{1}{1+\sigma_*}\right),$$

For the stresses σ_{ij}^*, the formula (10.73) remain also valid, if it is assumed (all the remaining magnitudes do not change):

$$e_{11}=-\frac{2}{\varepsilon_H},\quad e_{21}=\frac{2}{\varepsilon_H}-\frac{1}{1+\varepsilon_H+\sigma_*},\ e_{12}=\pm\frac{2i}{\varepsilon_H},\ e_{22}=\pm i\frac{2+\varepsilon_H}{\varepsilon_H},$$
$$e_{31}=\pm i\frac{2+\varepsilon_H}{\varepsilon_H(1+\varepsilon_H)\sqrt{1-\varepsilon^2}},\quad e_{32}=\left(\frac{2}{\varepsilon_H}+\frac{1}{1+\sigma_*}\right)\frac{1}{\sqrt{1-\varepsilon^2}},\quad \beta_1^{\pm}=1. \tag{10.74}$$

Kernels g_n, $g_n^{\pm}$ $(n=1,2)$ in (10.73) are linear combinations of Cauchy kernels and complex-conjugated kernels of them in the physical and transformed planes, the kernels H_n and $H_n^{\pm}$ are regular. Thus, the limiting values of tensors S_{ij} and σ_{ij} on Γ are calculated in an elementary way by applying the formulas for the computation of the limiting values of Cauchy-type integrals [50, 103].

It is noteworthy that the "unlimited increase" of the functions $g_n^{\pm}$ whenever $\varepsilon_H\to 0$ (absence of initial magnetic field) is only apparent. In fact, they have finite limits. One can be easily convinced, if he considers the connections between the corresponding values in the physical and transformed planes

$$\frac{\cos\alpha}{r}=\frac{\left(1-\varepsilon^2\sin^2\alpha\right)\cos\alpha_1}{r_1\sqrt{1-\varepsilon^2}},\quad \frac{\sin\alpha}{r}=-\frac{\left(1-\varepsilon^2\sin^2\alpha\right)\sin\alpha_1}{(1-\varepsilon^2)\,r_1}.$$

Let us examine now the structure of plane magneto-elastic waves in isotropic ideally conductingmedium in a magnetic field. Seeking for the solutions of the (10.65) in the form $U_m^0 = A_m \exp\{i\gamma(x_1\cos\beta + x_2\sin\beta)\}$, we come to the homogeneous system of equations

$$\begin{aligned} &A_1\left[\gamma_2^2 - \gamma^2\left(1+\varepsilon_H+\sigma_*\cos^2\beta\right)\right] - A_2\gamma^2\sigma_*\sin\beta\cos\beta = 0,\\ &-A_1\gamma^2\sigma_*\sin\beta\cos\beta + A_2\left[\gamma_2^2 - \gamma^2\left(1+\sigma_*\sin^2\beta\right)\right] = 0, \end{aligned} \tag{10.75}$$

the determinant of which is

$$\Delta = (1+\varepsilon_H)(1+\sigma_*)\left(1-\varepsilon^2\cos^2\beta\right)\times\left(\gamma^2-\gamma_2^2\rho_1^2(\beta)\right)\left(\gamma^2-\gamma_2^2\rho_2^2(\beta)\right).$$

Hence we deduce that only waves with wave numbers $\gamma^{(n)} = \gamma_2\rho_n(\beta)$ and velocities $C_n^* = C_2/\rho_n(\beta)$ $(n=1,2)$ may propagate within the medium. Along the coordinate axes these waves can be propagated in the form of purely longitudinal and purely transverse waves.

Along the axis x_1:

$$U_1^0 = U_0\exp\left(\pm i\gamma^{(1)}x_1\right),\quad U_2^0 = 0,\quad \gamma^{(1)} = \frac{\gamma_2}{\sqrt{1+\varepsilon_H+\sigma_*}} = \frac{\gamma_1}{\sqrt{1+\varepsilon_H K_\sigma}},$$
$$U_2^0 = U_0\exp\left(\pm i\gamma^{(2)}x_1\right),\quad U_1^0 = 0,\quad \gamma^{(2)} = \gamma_2.$$

Along the axis x_2:

$$U_2^0 = U_0\exp\left(\pm i\gamma^{(1)}x_2\right),\quad U_1^0 = 0,\quad \gamma^{(1)} = \gamma_1,$$
$$U_1^0 = U_0\exp\left(\pm i\gamma^{(2)}x_2\right),\quad U_2^0 = 0,\quad \gamma^{(2)} = \frac{\gamma_2}{\sqrt{1+\varepsilon_H}}.$$

Moreover, in both cases when polarizing along the lines of the initial magnetic field the velocities of longitudinal and transverse waves coincide with the classical ones. When polarizing normal to the magnetic field

$$C_1^*/C_1 = \sqrt{1+\varepsilon_H K_\sigma} \geq 1,\quad C_2^*/C_2 = \sqrt{1+\varepsilon_H} \geq 1.$$

In the general case, the orientation of the wave vector for a wave of the first type (quasi-longitudinal wave), generates transverse polarization, and the wave of second type (quasi-transverse) generates longitudinal polarization in the plane $0x_1x_2$. We have, respectively

$$U_1^0 = U_0W_1,\quad U_2^0 = U_0C^{(1)}W_1,\quad U_1^0 = U_0W_2,\quad U_2^0 = U_0C^{(2)}W_2,$$

where

$$W_m = \exp\left\{i\gamma^{(m)}\left(x_1\cos\beta + x_2\sin\beta\right)\right\},$$
$$C^{(m)} = \frac{\sigma_*\rho_m^2(\beta)\sin\beta\cos\beta}{1-\rho_m^2(\beta)\left(1+\sigma_*\sin^2\beta\right)},\quad \sin\beta\cos\beta \neq 0.$$

In view of what is discussed above, the waves from infinity, as well as the stresses connected with them can be represented in the form

$$U_m^0 = \sum_{n=1}^{2} \lambda_{mn} W_n \quad (m = 1, 2), \tag{10.76}$$

$$S_{11}^0 + S_{22}^0 = 2\mu i \sigma_* \sum_{n=1}^{2} \sum_{m=1}^{2} \lambda_{mn} \gamma^{(n)} W_n n_m^*,$$

$$S_{22}^0 - S_{11}^0 \pm 2iS_{12}^0 = -2i\mu e^{\pm i\beta} \sum_{n=1}^{2} \sum_{m=1}^{2} \lambda_{mn} \gamma^{(n)} W_n \beta_m^{\pm},$$

where λ_{mn} are the amplitudes of displacements along the coordinate x_m in the wave of n-th type $n_1^* = \cos\beta$, $n_2^* = \sin\beta$.

The amplitudes of stresses σ_{ij}^0 are determined by the same formulas; it is only sufficient to assume that $\beta_1^{\pm} = 1$.

10.9.2 Integral Equations of the Boundary Value Problem

The boundary conditions on the contour of the opening can be represented in the form of two complex boundary equalities

$$\sigma_{11} + \sigma_{22} - e^{\pm 2i\psi} (\sigma_{22} - \sigma_{11} \pm 2i\sigma_{12}) = 2(N \pm iT), \tag{10.77}$$

$$\sigma_{ij} = \sigma_{ij}^0 + \sigma_{ij}^*,$$

where ψ is the angle between the normal to the contour of the opening and the axis $0x_1$; σ_{ij}^0 and σ_{ij}^* are stresses in the incident and scattered waves, respectively.

Substituting the limiting values of the corresponding combinations (10.73) in (10.77), and taking into account relations (10.76), we reduce the boundary-value problem to a system of singular integral equations of second order, relatively to the functions $\omega_1(\zeta)$, $\omega_2(\zeta)$, (summation over the repetitive indices):

$$a_m^{\pm}(\psi_0)\omega_m(\zeta_0) + \frac{1}{\pi} \int_{\Gamma} \omega_n(\zeta) D_n^{\pm}(\zeta, \zeta_0) ds = F^{\pm}(\zeta_0), \quad \zeta_0 \in \Gamma, \tag{10.78}$$

where

$$a_1^{\pm}(\psi_0) = -\frac{e^{\pm i\psi_0}}{(1+\varepsilon_H)(1-\varepsilon^2\cos^2\psi_0)}, \quad a_2^{\pm}(\psi_0) = \frac{\pm i(1-\varepsilon^2)\cos\psi_0 - \sin\psi_0}{1-\varepsilon^2\cos^2\psi_0},$$

$$D_n^{\pm}(\zeta, \zeta_0) = \Lambda_n^{\pm}(\zeta, \zeta_0) + \frac{K_\sigma K_\varepsilon}{\pi} \theta_n^{\pm}(\zeta, \zeta_0), \ b_1^{\pm} = 1,$$

$$\Lambda_n^{\pm} = \sigma_* g_n(\zeta, \zeta_0) - e^{\pm 2i\psi_0} g_n^{\pm}(\zeta, \zeta_0),$$

$$b_2^{\pm} = \mp i, \quad \theta_n^{\pm} = \sigma_* H_n(\zeta, \zeta_0) + e^{\pm 2i\psi_0} H_n^{\pm}(\zeta, \zeta_0),$$

$$F^{\pm}(\zeta_0) = 2(N \mp iT) - 2i\mu \sum_{m=1}^{2} \sum_{n=1}^{2} \lambda_{mn} \gamma^{(n)} \times \left\{ \sigma_* n_m^* + b_m^{\pm} e^{\pm i(2\psi_0 - \beta)} \right\} W_n(\zeta_0).$$

While kernels $\Lambda_n^{\pm}$ are singular, the kernels $\theta_n^{\pm}$ can possess not more than a weak singularity (on assumption that the curvature of contour Γ satisfies Holder condition); when determining functions g_n and $g_n^{\pm}$, one should use the formulas (10.73) with coefficients e_{mn} from (10.74).

10.9.3 Dynamic Intensity of Conductor in the Opening

The tangential normal stress $\sigma_{\psi\psi}$ on the contour of the opening is determined by relation

$$\sigma_{\psi\psi} = \sigma_{11} + \sigma_{22} - N. \tag{10.79}$$

Making use of the limiting values of the first combination of stresses from (10.71) and taking into account the formulas (10.76), we find

$$\begin{aligned}\sigma_{\psi\psi}(\zeta_0) = &-\frac{\sigma_* K_\sigma}{1-\varepsilon^2\cos^2\psi_0}\left\{K_\varepsilon\omega_1(\zeta_0)\cos\psi_0 + \omega_2(\zeta_0)\sin\psi_0\right\}\\ &+\frac{\sigma_*}{\pi}\sum_{m=1}^{n}\int_\Gamma \omega_m(\zeta)\times\\ &\times\left\{g_m(\zeta,\zeta_0)+\frac{K_\sigma K_\varepsilon}{\pi}\gamma_2 H_m(\zeta,\zeta_0)\right\}ds + 2i\mu\sigma_*\\ &\times\sum_{n=1}^{2}\sum_{m=1}^{2}\lambda_{mn}\gamma^{(n)}n_m^* W_n(\zeta_0) - N, \quad \zeta_0\in\Gamma.\end{aligned}$$

The complete voltage on Γ is determined by the equality

$$\begin{aligned}S_{\psi\psi} =& \frac{1}{2}(S_{11}+S_{22}) + \frac{1}{4}e^{2i\psi_0}(S_{22}-S_{11}+2iS_{12})\\ &+\frac{1}{4}e^{-2i\psi_0}(S_{22}-S_{11}-2iS_{12}),\end{aligned}$$

where the corresponding combinations of stresses are determined in (10.73) and (10.76).

For the numerical solution of the constructed algorithm we make parameterization of the contour of the opening $\Gamma: \ \zeta=\zeta(\beta), \ \zeta_0=\zeta(\beta_0), \ 0\le \beta, \ \beta_0\le 2\pi$. Applying the method of mechanical quadratures [24] to (10.78) and taking into account the interpolation formula (for the odd N)

$$f(\beta_{0l}) = \frac{1}{n}\sum_{i=1}^{n}(-1)^{i+l} f(\beta_i)\,\mathrm{cosec}\frac{\beta_i-\beta_{0l}}{2},$$

we reduce them to linear algebraic system of the form

$$\sum_{i=1}^{n} \omega_m (\beta_i) \left\{ a_m^{\pm} (\beta_{0l}) (-1)^{i+l} \cos ec \frac{\beta_i - \beta_{0l}}{2} + 2D_m^{\pm} (\beta_i, \beta_{0l}) \right\} s' (\beta_i)$$
$$= nF^{\pm} (\beta_{0l}) ,$$
$$\beta_i = \frac{\pi (2i-1)}{n}, \quad \beta_{0l} = \frac{2\pi (l-1)}{n} \quad (i, l = 1, 2, \dots, n) .$$

As an example, let us examine the elastic current-conducting medium, weakened by a circular opening of radius R, on the surface of which harmonically changing with time normal pressure is given, $N \exp(-i\omega t)$. Figs. 10.26 and 10.27 show a change of the relative contour stress $K = |S_{\psi\psi}/N|$ at the points of contour A and B as a function of the normalized wave number $\gamma_1 R$. The curves 1, 2, 3 are constructed for the values of the parameter of magnetic pressure $\varepsilon_H = 0$; $0,5$ and 1, respectively. The continuous lines correspond to the value of Poisson coefficient $\nu = 0,15$, the stroked ones, to the value $\nu = 0,45$.

Let us now consider a longitudinal wave $u_1^0 = u_0 \exp\left(-i\gamma^{(1)} x_1\right)$, $u_2^0 = 0$, $\gamma^{(1)} = \gamma_2/\sqrt{1 + \sigma_* + \varepsilon_H}$ that falls from infinity, along x_1-axis, on the elliptical opening $x_1 = R_1 \cos\beta$, $x_2 = R_2 \sin\beta$, $(0 \le \beta \le 2\pi)$. The results of the calculations of the magnitude $K = |S_{\psi\psi}/S|$, where $S = u\gamma_2 u_0 \sqrt{1 + \sigma_* + \varepsilon_H}$, as a function of relative wave number $\gamma_1 r_1$ $(2r_1 = R_1 + R_2)$ are presented in Figs. 10.28 and 10.29 for the circular $(R_1 = R_2 = R)$ and in Figs. 10.30 and 10.31 for the elliptical openings $(R_2/R_1 = 0,67)$. Here and below, the continual curves correspond to the value $\nu = 0,15$, and the stroked ones, to the value $\nu = 0,4$.

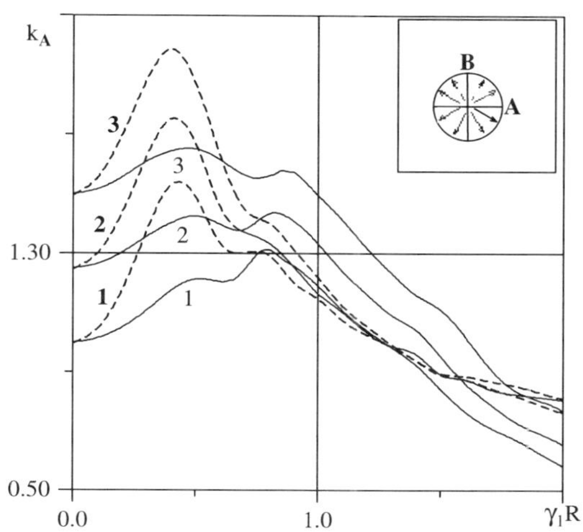

Fig. 10.26. Dependences of relative contour stress on the normalized wave number for the case, when pulsatory pressure acts on the contour of circular opening

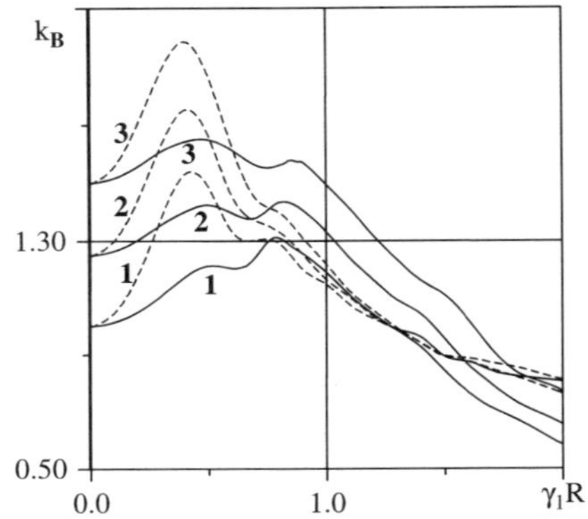

Fig. 10.27. Graphs of contour normal stress at the point B

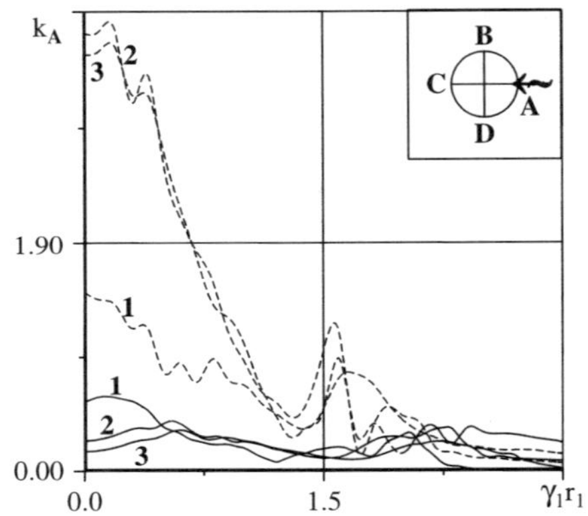

Fig. 10.28. Diffraction of longitudinal magneto-elastic wave on circular opening. Curves of relative normal stress at point A

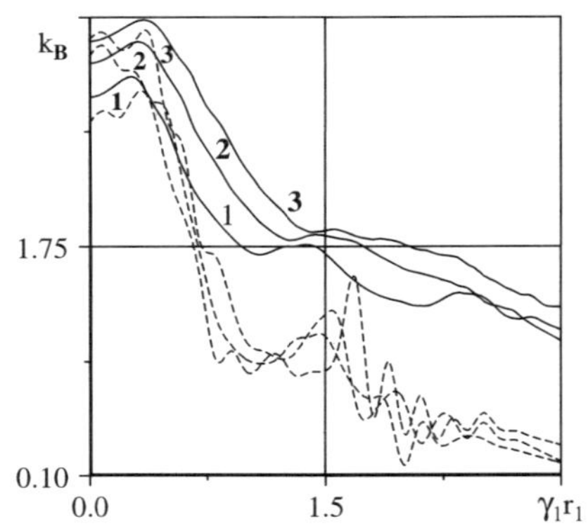

Fig. 10.29. Curves of relative normal stress at the point B on circular opening

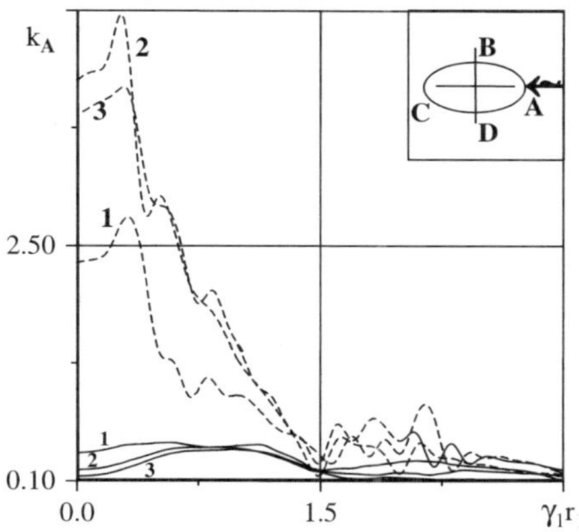

Fig. 10.30. Diffraction of longitudinal magneto-elastic wave on elliptical opening

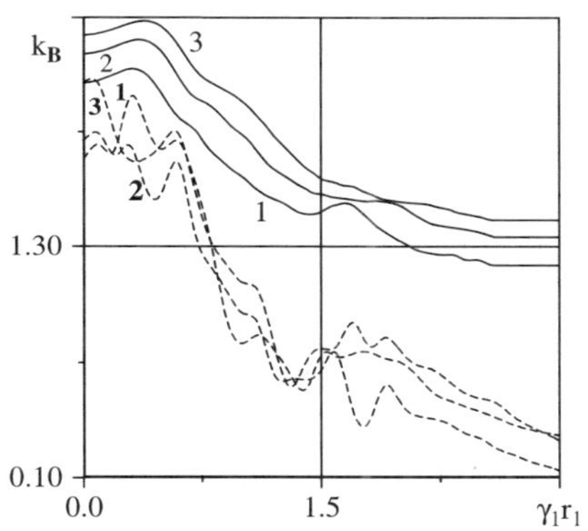

Fig. 10.31. Diffraction of longitudinal wave on elliptical opening. Curves correspond to the point B of the contour of elliptical opening

The situation, when a transverse wave $u_2^0 = u_0 \exp\left(-i\gamma^{(2)}x_1\right)$, $u_1^0 = 0$, $\gamma^{(2)} = \gamma_2$ falls on the circular or elliptical opening along the axis x_1 is presented in Figs. 10.32 and 10.33. Here, the graphs of the magnitude $K = |S_{\psi\psi}/S|$, where $S = \mu\gamma_2 u_0$ are plotted.

Analogous curves for the case, when longitudinal $u_2^0 = u_0 \exp\left(-i\gamma^{(1)}x_2\right)$, $u_1^0 = 0$, $\gamma^{(1)} = \gamma_1$ or transverse wave $u_1^0 = u_0 \exp\left(-i\gamma^{(2)}x_2\right)$, $u_2^0 = 0$, $\gamma^{(2)} = \gamma_2/\sqrt{1+\varepsilon_H}$ falls on the opening along the axis x_2, are presented in Figs. 10.34, 10.35, 10.36 and 10.37, respectively. When exciting with the aid of longitudinal wave the magnitude S has the next value $S = \mu\gamma_2 u_0\sqrt{1+\sigma_*}$, while for the transverse wave it is $S = \mu\gamma_2 u_0\sqrt{1+\varepsilon_H}$.

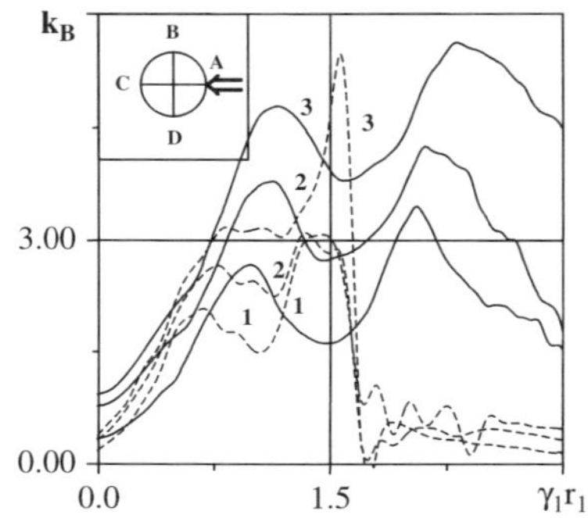

Fig. 10.32. Diffraction of transverse magneto-elastic wave on circular opening

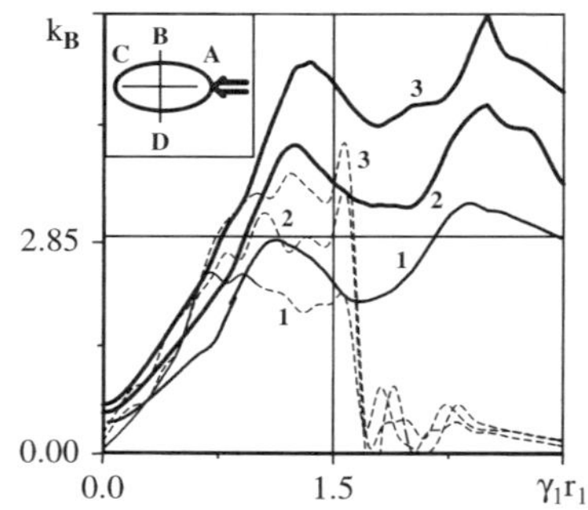

Fig. 10.33. Diffraction of transverse wave on elliptical opening

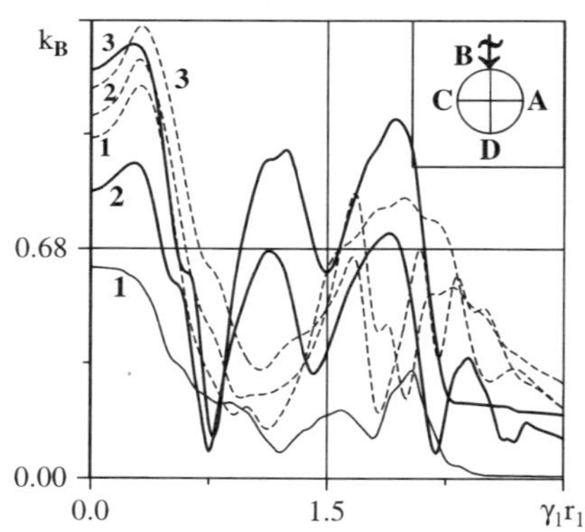

Fig. 10.34. Diffraction of longitudinal magnetoelastic wave on circular opening

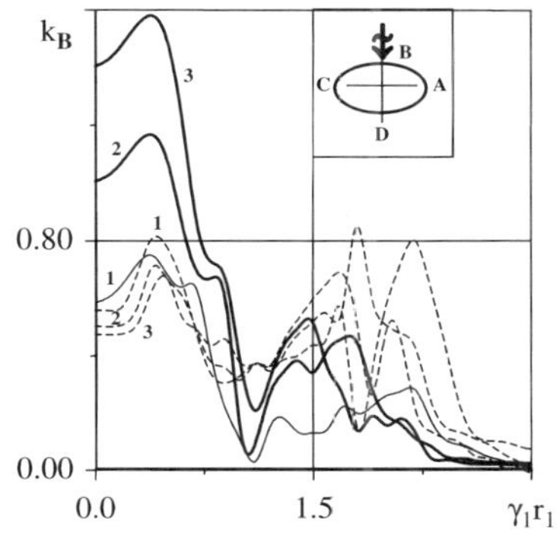

Fig. 10.35. Diffraction of longitudinal wave on elliptical opening

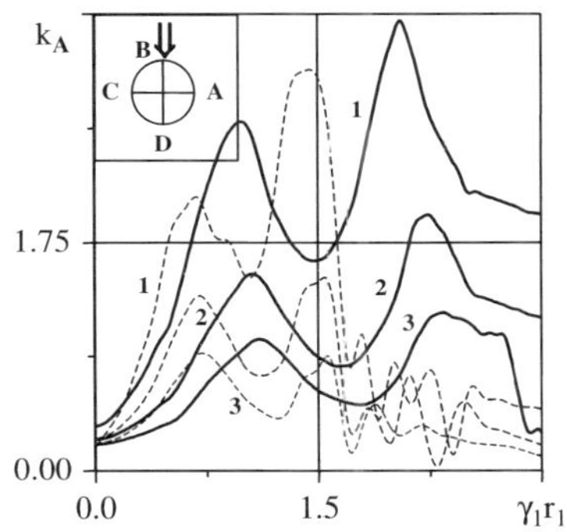

Fig. 10.36. Diffraction of transverse magneto-elastic wave on circular opening

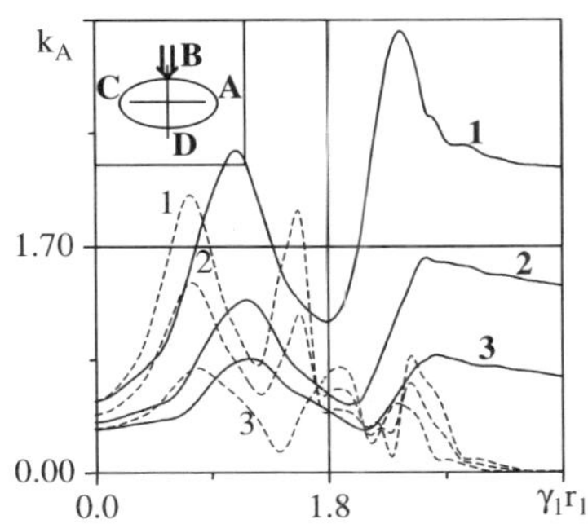

Fig. 10.37. Diffraction of transverse magneto-elastic wave on elliptical opening

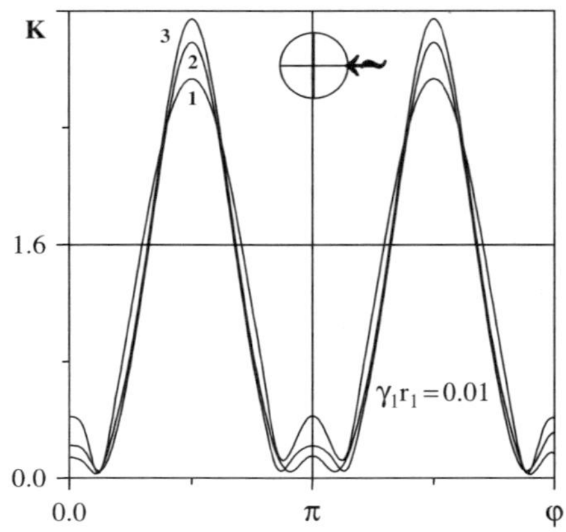

Fig. 10.38. Change in relative normal stress along the contour of circular opening for small normalized wave number

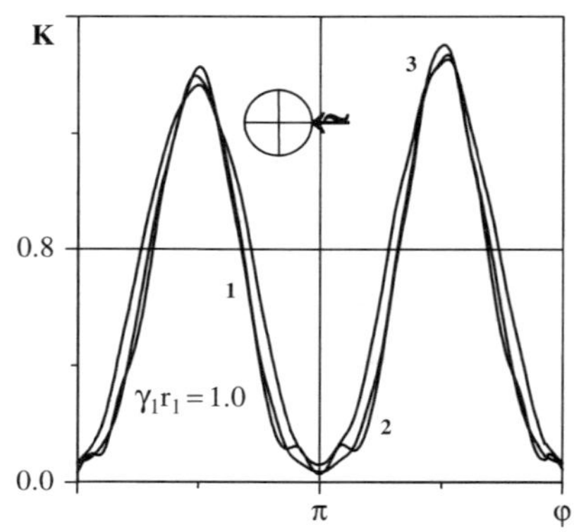

Fig. 10.39. Nature of redistribution of normal stress on the contour of circular opening with relatively great value of normalized wave number

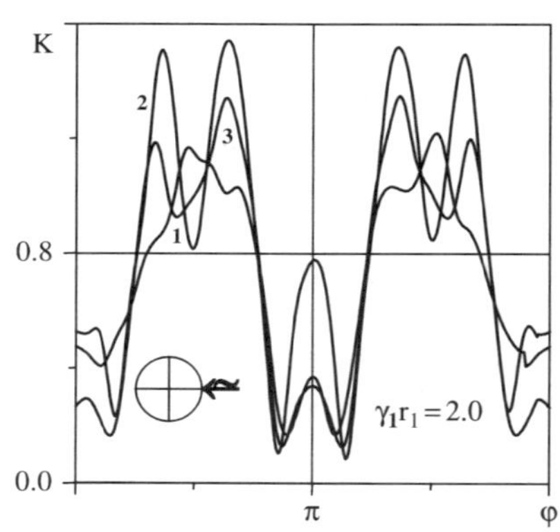

Fig. 10.40. Change in normal stress along the contour of circular opening

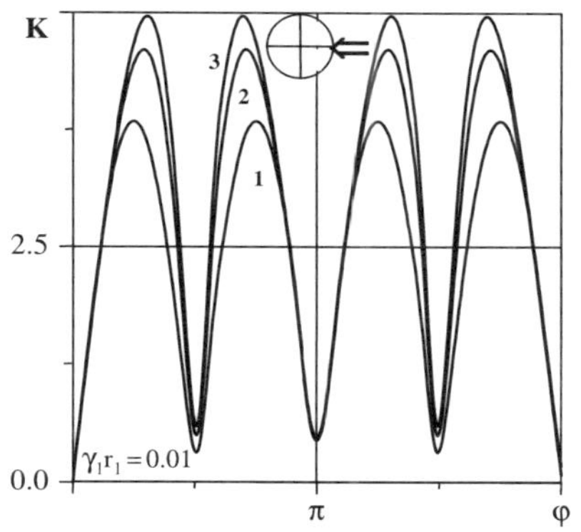

Fig. 10.41. Change in relative normal stress on the contour of circular opening for the case of transverse wave

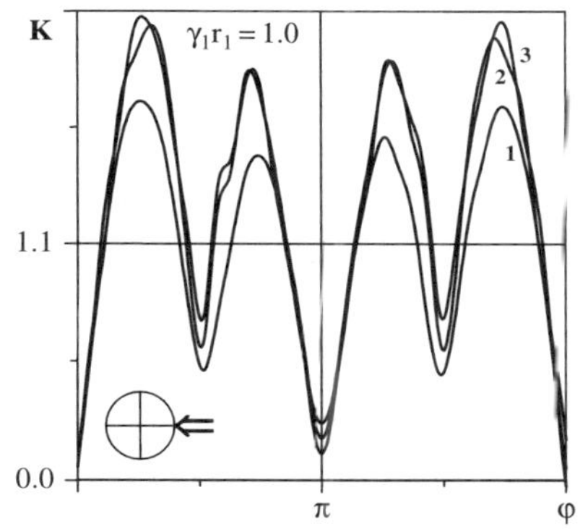

Fig. 10.42. Nature of redistribution of relative normal stress with great values of normalized wave number

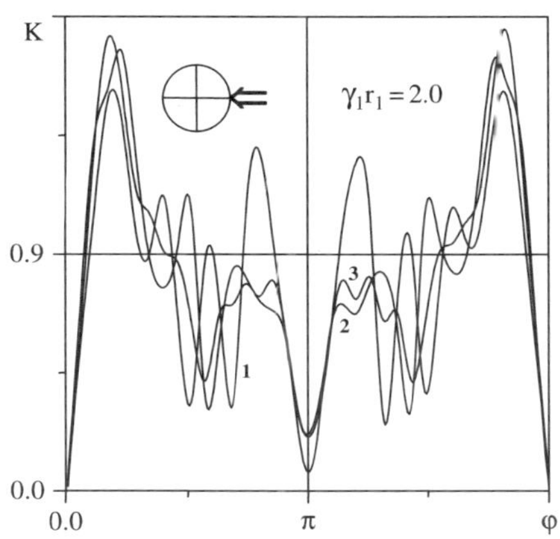

Fig. 10.43. Nature of redistribution of relative normal stress on the contour of circular opening when a transverse wave drops from the infinity

If the normalized wave number $\gamma_1 r_1$ is small, then the initial magnetic field weakly redistributes the magnitude K on the contour of the opening. Only its extreme values change (Figs. 10.38). When increasing the wave number, essential redistribution of the contour stress occurs (Figs. 10.39 and 10.40). This observation is especially related to the situation, when on the opening falls a transverse wave (Figs. 10.41–10.43).

11

Influence of Magnetizability of Material on the Stress State of a Ferromagnetic Medium with Heterogeneities

In this chapter some static problems of magneto-elasticity are examined for soft-iron unbounded medium, weakened by cracks or openings. For such materials, like iron-nickel alloys, transformer iron, etc., which possess small remnant magnetism (narrow loop of hysteresis), the influence of induced currents is small in comparison with the effect of magnetization. The linear theory [285, 286] is used below, which disregard hysteresis losses and the effect of magneto-elasticity.

The influence of the material magnetizability on the intensity of an unbounded soft-iron ferromagnetic medium with rectilinear cracks under strong static magnetic field is studied in the papers [140, 286]. We elaborate below new procedures, which enable examination of bodies with defects of sufficiently arbitrary configuration.

11.1 Initial Relations of Linear Magneto-Elasticity of Ferromagnetic Materials. Complex Representations of the Solutions

Let us examine a body of soft-iron ferromagnetic material in reference to a rectangular Cartesian coordinate system $0x_1x_2x_3$, weakened by tunnel heterogeneities along the x_3 - axis. Let us assume that the initial magnetic field (into which the ferromagnetic body is placed) is uniform throughout the space and oriented along the axis x_2. As a result of the magnetization of the material, the body acquires magnetic moment and undergoes mechanical action from the side of the external field. Furthermore, we consider that on the surfaces of heterogeneities the action of mechanical load $(X_{1n},\ X_{2n},\ 0)$, where $X_{in} = X_{in}\,(x_1, x_2)$ is possible.

Under the influence of all these forces in the body appear deformations, which cause an additional (induced) magnetic field. A question arises about the determination of the interacting mechanical and magnetic fields in the piecewise-uniform ferromagnetic body.

As a result of the deformation of the body, the initial magnetic field undergoes small changes. We will assume that the cause of deformation is a field of infinitesimal displacement vector $\vec{u} = (u_1, u_2, u_3)$

$$\vec{B} = \vec{B}_0 + \vec{b}, \quad \vec{H} = \vec{H}_0 + \vec{h}, \quad \vec{M} = \vec{M}_0 + \vec{m}, \tag{11.1}$$

where $\vec{B}_0 = (B_{01}, B_{02}, B_{03})$; $\vec{H}_0 = (H_{01}, H_{02}, H_{03})$, $\vec{M}_0 = (M_{01}, M_{02}, M_{03})$ are, respectively, the induction, the intensity of magnetic field and magnetization in the state of rigid (non-deformated) body, and $\vec{b}$, $\vec{h}$, $\vec{m}$ are the fluctuations of the indicated magnitudes, which have by hypothesis the same order of magnitude as the vector of elastic displacement.

Then, on the condition that $(M_{0j}\partial_j u_i (<< (m_i ($, the following system of linear relations [122, 285, 286] holds:

field equations (summation over the repetitive index)

$$\begin{aligned} &\partial_j t_{ji} + \mu_0 \left(M_{0j}\partial_j H_{0i} + M_{0j}\partial_j h_i + m_j \partial_j H_{0i}\right) = 0, \\ &\mathbf{rot}\, \vec{h} = 0, \quad \mathbf{div}\, \vec{b} = 0 \quad (i, j = 1, 2, 3)\,, \end{aligned} \tag{11.2}$$

material equations

$$\begin{aligned} &\vec{m} = \chi \vec{h}, \vec{b} = \mu_0(\vec{h} + \vec{m}) = \mu_0\mu_r \vec{h}, \quad \mu_r = \chi + 1, \\ &t_{ij} = \sigma_{ij} + \chi\mu_0 \left(H_{0i}H_{0j} + H_{0i}h_j + H_{0j}h_i\right), \\ &T_{ij} = H_{0j}B_{0i} + H_{0j}b_i + B_{0i}h_j - \\ &\quad - \frac{1}{2}\,\mu_0 \left(H_{0k}H_{0k}\right)\delta_{ij} - \mu_0 \left(H_{0k}h_k\right)\delta_{ij}, \\ &\sigma_{ij} = \lambda\delta_{ij}\partial_k u_k + \mu \left(\partial_j u_i + \partial_i u_j\right); \end{aligned} \tag{11.3}$$

boundary conditions on the surface of section of the media:

$$\begin{aligned} &\left(S_{ij} - S^e_{ij}\right) n_j = 0, \quad S_{ij} = t_{ij} + T_{ij}\,(i, j = 1, 2, 3)\,, \\ &\varepsilon_{ijk} \left\{n_j \left(h_k - h^e_k\right) - \left(H_{0k} - H^e_{0k}\right) n_m \partial_j u_m\right\} = 0, \\ &n_i \left(b_i - b^e_i\right) - \left(B_{0i} - B^e_{0i}\right) n_m \partial_i u_m = 0. \end{aligned} \tag{11.4}$$

In the relations (11.2)–(11.4), σ_{ij}, t_{ij} and T_{ij} are the tensors of elastic, magneto-elastic and Maxwell stresses, respectively; ε_{ijk} is the Levi-Civita tensor; index "e" refers to the medium (here, to the cavity of crack or opening); n_i is the projection of the unit vector of the normal to the surface of section to the axis x_i; $\mu_0 = 4\pi \cdot 10^{-7}\mathrm{N/A}^2$ is the absolute magnetic permeability; μ_r is the relative magnetic permeability; χ is the magnetic susceptibility of substance; λ, μ are the Lame constants.

In these relations, there are components of undisturbed magnetic field, which are determined from the solution of the problem of magneto-statics for coupled media (one of which is usually the vacuum):

in the body

$$\mathbf{rot}\overrightarrow{H}_0 = 0, \quad \mathbf{div}\overrightarrow{B}_0 = 0, \tag{11.5}$$
$$\overrightarrow{B}_0 = \mu_0\left(\overrightarrow{H}_0 + \overrightarrow{M}_0\right) = \mu_0\mu_r\overrightarrow{H}_0;$$

in the cavity

$$\mathbf{rot}\overrightarrow{H}_0^e = 0, \quad \mathbf{div}\overrightarrow{B}_0^e = 0, \tag{11.6}$$
$$\overrightarrow{B}_0^e = \mu_0\overrightarrow{H}_0^e.$$

Coupling conditions of fields in the body and the cavity of concentrator take the ordinary form [161, 167]

$$\varepsilon_{ij}n_j\left(H_{0k} - H_{0k}^e\right) = 0, \quad n_j\left(B_{0j} - B_{0j}^e\right) = 0 \tag{11.7}$$
$$(i, j, k = 1, 2, 3).$$

The system of (11.2)–(11.7) is complete, in the sense that it provides the possibility, in principle, to determine all mechanical and magnetic magnitudes in the body, as well as the magnetic field in the medium.

For the determination of fields of elastic displacements and the fluctuations of the magnetic field, we use the relations (11.2), (11.3) and the fact that the magnitudes t_{ij}, T_{ij}, h_i do not depend on the coordinate x_3. We have [288, 135, 140]:

differential equations of equilibrium of soft-iron elastic medium

$$\nabla^2 u_i + \sigma_* \partial_i \theta + 2\chi\mu_0\mu^{-1}H_{02}\partial_2 h_i = 0 \quad (i = 1, 2) \tag{11.8}$$

the equations of magneto-statics for the fluctuations of magnetic field

$$\overrightarrow{h} = grad\Psi, \quad \nabla^2\Psi = 0. \tag{11.9}$$

In the formulas (11.8), the magnitude $\theta = \partial_m u_m$; $\sigma_* = (\lambda + \mu)/\mu = = (1-2\nu)^{-1}$; ν; is the Poisson ratio.

From (11.8), (11.9) immediately follows, that the volume expansion $\boldsymbol{\theta}$ is harmonic function. Accordingly, let us assume that

$$\mu\theta = (\kappa - 1)\,\mathrm{Re}\Phi(z), \quad \Psi = \mathrm{Re}\{if(z)\}, \tag{11.10}$$
$$\kappa = 3 - 4\nu,$$

where $\Phi(z)$, $f(z)$ are arbitrary analytical in the domain, occupied by the body, functions of complex variable $z = x_1 + ix_2$.

If we write (11.8) in the form of one complex equation in relation to $u_1 + iu_2$ and we integrate it, in accordance to the determination of volume expansion (11.10), we find

$$2\mu\left(u_1+iu_2\right)=\kappa\varphi\left(z\right)-z\overline{\Phi\left(z\right)}-\overline{\varphi_1\left(z\right)}+\mu_0M_{02}\left(z\overline{F\left(z\right)}-f\left(z\right)\right), \quad (11.11)$$
$$\Phi\left(z\right)=d\varphi\left(z\right)/dz=\varphi'\left(z\right), \quad F\left(z\right)=f'\left(z\right).$$

Hence, taking into account formulas (11.10), we obtain

$$\begin{aligned} &t_{11}+t_{22}=4Re\Phi\left(z\right)-2\mu_0M_{02}\mathrm{Re}F\left(z\right)+\chi\mu_0H_{02}^2, \quad (11.12)\\ &t_{22}-t_{11}+2it_{12}=2\left\{\bar{z}W'\left(z\right)+W_1\left(z\right)\right\}+\chi\mu_0H_{02}^2,\\ &S_{11}+S_{22}=4\mathrm{Re}W\left(z\right)+2\chi\mu_0H_{02}^2,\\ &S_{22}-S_{11}+2iS_{12}=2\left\{\bar{z}W'\left(z\right)+W_1\left(z\right)\right\}-\\ &-2B_{02}F\left(z\right)+\left(2\chi+1\right)\mu_0H_{02}^2,\\ &h_1-ih_2=iF\left(z\right),\\ &W\left(z\right)=\Phi\left(z\right)-\mu_0M_{02}F\left(z\right),\\ &W_1\left(z\right)=\Phi_1\left(z\right)-\mu_0M_{02}F\left(z\right),\Phi_1\left(z\right)=\varphi_1'\left(z\right).\end{aligned}$$

Thus, the components of the tensor of magneto-elastic and Maxwell stresses, as well as the vectors of mechanical displacement and the fluctuations of intensity of magnetic field in the body are expressed in terms of three arbitrary analytic functions. When $B_{02}=0$, formulas (11.12) and (11.11) coincide with the classical representations of the two-dimensional problem of the theory of elasticity.

11.2 Ferromagnetic Medium, Weakened by the Tunnel Cracks

Let us consider here the situation, when the unbounded soft-iron ferromagnetic elastic medium, placed into constant magnetic field $\overrightarrow{B}_0=(0,B_{02},0)$, is weakened by tunnel cracks L_j $(j=1,2,\ldots,M)$. along the axis x_3. We consider that the magnetic field $\overrightarrow{B}_0$ is not sensitive to cracks, since the later in the non-deformed state are treated as mathematical cuts. Accordingly, on the axis of the cuts we have (Fig. 11.1)

$$H_{01}^e=\chi H_{02}n_1n_2, \quad H_{02}^e=H_{02}\left(1+\chi n_2^2\right), \quad (11.13)$$

where $n_1=\cos\psi$; $n_2=\sin\psi$; ψ is the angle between the normal to L_j and the axis $0x_1$.

The boundary conditions follow from the conditions for coupling (11.4). We get, taking into account (11.13) (11.14)

$$\begin{aligned} &n_1\left(b_1-b_1^e\right)+n_2\left(b_2-b_2^e\right)=\mu_0\chi H_{02}n_1U_1^*, \quad (11.14)\\ &-n_2\left(h_1-h_1^e\right)+n_1\left(h_2-h_2^e\right)=\chi H_{02}n_2U_1^*,\\ &U_1^*=n_m\frac{\partial u_m}{\partial s}, \quad \frac{\partial}{\partial s}=n_1\partial_2-n_2\partial_1,\end{aligned}$$

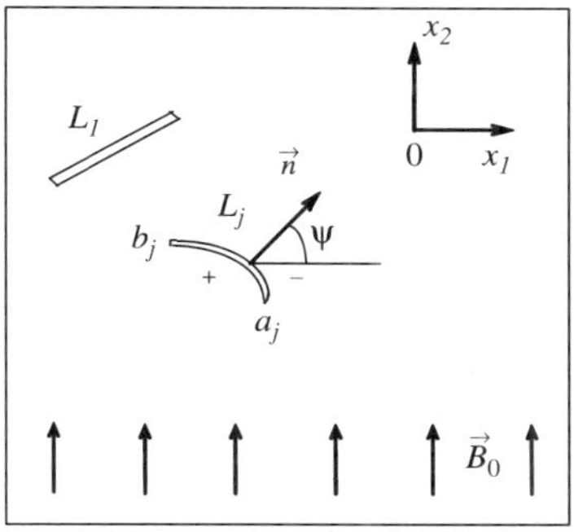

Fig. 11.1. The problem under study

Excluding the components related to the cavity of crack, we find

$$[h_1 - ih_2] = i\chi H_{02}\delta(\psi) U_1, \quad U_1 = [U_1^*], \tag{11.15}$$
$$\delta(\psi) = \left(\frac{n_1}{i\mu_r} - n_2\right) e^{-i\psi},$$

where the symbol $[*]$ denotes the jump of the corresponding magnitude upon transfer through the section.

In the cavity of crack (on the axis of crack L_j) we have

$$h_1^e = \left(\chi n_1^2 + 1\right) h_1 + \chi n_1 n_2 h_2 - \chi H_{02} U_1^* \cos 2\psi, \tag{11.16}$$
$$h_2^e = \chi n_1 n_2 h_1 + \left(\chi n_2^2 + 1\right) h_2 - \chi H_{02} U_1^* \sin 2\psi.$$

The mechanical boundary conditions on the tips of L_j are obtained taking into account the relations (11.3), (11.13), (11.9). Let us represent them in the form of complex equalities

$$(t_{11} + t_{22})^{\pm} - e^{2i\psi} (t_{22} - t_{11} + 2it_{12})^{\pm} + 2\chi\mu_0 H_{02}$$
$$\times \left(H_{02}\alpha(\psi) U_1^{*\pm} - \chi n_2 n_m h_m^{\pm}\right) =$$
$$= 2\left(N^{\pm} - iT^{\pm}\right) + \mu_0 \chi^2 H_{02}^2 n_2^2, \ \alpha(\psi) = \chi n_2 e^{-i\psi} - i. \tag{11.17}$$

Here $N^{\pm}$ and $T^{\pm}$ are the effective on the tips L_j normal and tangential forces; the upper sign concerns the left tips.

In the sequel, we consider that $[N - iT] = 0$. Boundary equality (11.17) should be satisfied only on one of the edges of each crack, if, in this case, the condition of continuous extendability of its left side upon the transfer through L_j is taken into account:

$$[t_{11} + t_{22}] - e^{2i\psi} [t_{22} - t_{11} + 2it_{12}] + 2\mu_0 \chi H_{02}^2 \beta(\psi) U_1 = 0, \tag{11.18}$$
$$\beta(\psi) = \frac{\chi}{\mu_r} n_1 n_2 - i\left(1 + \chi n_2^2\right)$$

We complete the statement of the boundary-value problem, by paraphrasing conditions (11.15), (11.17) and (11.18) in complex variables. We have

$$[F(\zeta)] = \chi H_{02}\delta(\psi) U_1(\zeta), \quad \zeta \in L = \bigcup_{j=1}^{k} L_j, \tag{11.19}$$

$$[\mathrm{Re}(\Phi(\zeta) + W(\zeta))] - e^{2i\psi}\left[\overline{\zeta}W'(\zeta) + W_1(\zeta)\right] + \chi\mu_0 H_{02}^2 \beta(\psi) U_1(\zeta) = 0,$$
$$2\mathrm{Re}W(\zeta) - e^{2i\psi}\{\bar{z}W'(\zeta) + W_1(\zeta)\} + \chi\mu_0 H_{02}\Lambda = R(\zeta),$$

where

$$\Lambda = \mathrm{Im}\left(F(\zeta)\overline{\alpha(\psi)}\right) + H_{02}\alpha(\psi) U_1^*(\zeta),$$
$$R(\zeta) = N - iT + \frac{1}{2}\chi\mu_0 H_{02}^2\left(\chi n_2^2 + e^{2i\psi} - 1\right),$$
$$2\mu U_1^*(\zeta) = \mathrm{Im}\left\{\chi\mu_0 H_{02} F(\zeta) - W(\zeta) - \kappa\Phi(\zeta) + e^{2i\psi}\left(\overline{\zeta}W'(\zeta) + \Phi_1(\zeta)\right)\right\}.$$

The representations of the sought-for analytic functions, are correct, in the sense that the conditions for the jumps on the cracks are satisfied independently from the densities occurring in them, and can be written, following the scheme in the paper [272]. We get

$$F(z) = \frac{\chi\mu_0 H_{02}}{2\pi i}\int_L \frac{\omega(\zeta)}{\zeta - z}d\zeta, W(z) = -\frac{1}{2\pi i}\int_L \frac{p_1(\zeta)}{\zeta - z}d\zeta,$$
$$W_1(z) = \frac{1}{2\pi i}\int_L \frac{\overline{\zeta}p_1(\zeta)}{(\zeta - z)^2}d\zeta + \frac{1}{2\pi i}\int_L \frac{\overline{p_2(\zeta)}}{\zeta - z}d\overline{\zeta}, \tag{11.20}$$
$$\Phi_1(z) = W_1(z) + \chi\mu_0 H_{02} F(z),$$
$$\Phi(z) = W(z) + \chi\mu_0 H_{02} F(z),$$
$$p_j(\zeta) = \frac{2\mu}{\kappa + 1}\left\{iU_1(\zeta) - U_2(\zeta) + e_0^2\Lambda_j(\psi) U_1(\zeta)\right\},$$
$$\Lambda_1(\psi) = \frac{\chi}{2}\left\{i + \chi e^{-i\psi}\left(\frac{in_1}{\mu_r}(2 - \kappa) - (\kappa + 1) n_2\right)\right\},$$
$$\Lambda_2(\psi) = \frac{i\chi}{2}\left\{-\kappa + \chi(2 - \kappa) e^{-i\psi}\frac{n_1}{\mu_r}\right\},$$
$$e_0^2 = H_{02}^2\mu_0/\mu\,(j = 1, 2), \quad \mu_0\omega(\zeta) = \delta(\psi) U_1.$$

In formulas (11.20) functions U_j are expressed as displacements according to the relation (ds is the element of the arc of contour L)

$$(U_1 + iU_2) e^{i\psi} = \frac{d}{ds}[u_1 + iu_2], \tag{11.21}$$

which is coherent with the definition of function U_1 in (11.14), (11.15).

Functions (11.20) do not still fully correspond to the physical content of the stated boundary problem. It is necessary to ensure that the conditions of uniqueness of displacements in the domain occupied by the medium are satisfied, as well as that the integral equalities

$$\int_{C_j} (n_1 b_1 + n_2 b_2)\, ds = 0, \quad \int_{C_j} (-n_2 h_1 + n_1 h_2)\, ds = 0 \tag{11.22}$$
$$(j = 1, 2, \ldots, M),$$

where C_j is arbitrary closed contour, which includes L_j.

The conditions of uniqueness of displacements are reduced, in virtue of (11.20), to the form

$$\int_{L_j} (U_1 + iU_2)\, d\zeta = 0 \quad (j = 1, 2, \ldots, M). \tag{11.23}$$

Equalities (11.22) can be represented, taking into account (11.12), in the equivalent form

$$\int_{L_j} \omega(\zeta)\, d\zeta = 0 \quad (j = 1, 2, \ldots, M). \tag{11.24}$$

Thus, integral representations (11.20) make sense only when the conditions (11.23), (11.24) are fulfilled. In fact, equalities (11.24) follow from (11.23). This can be easily shown, by appealing to the first relation (11.19) and definition (11.21).

The boundary-value problem (11.19), using representations (11.20), is easily reduced to the following system of singular integral equations of first order [272, 273]

$$\int_L \sum_{n=1}^{2} U_n(\zeta)\, dG_{mn}(\zeta, \zeta_0) = N_m(\zeta_0), \zeta_0 \in L\, (m = 1, 2), \tag{11.25}$$

where

$$\begin{aligned}
dG_{11} &= \operatorname{Im}\left\{ \left[H_2 H_3^0 g/2 - H_1 \left(H_4^0 (\kappa + 1) + H_3^0 h/2 \right) - \right.\right.\\
&\qquad \left.\left. i\chi^2 e_0^2 (\kappa + 1)\, \delta(\psi)\, H_5^0/4 \right] d\tau \right\},\\
d\tau &= \frac{d\zeta}{\zeta - \zeta_0},\\
dG_{12} &= \operatorname{Im}\left\{ \left[1 + i\chi^2 (\kappa + 1) e_0^2 n_1^0 n_2^0/4 + H_3^0 (h - g)/2 \right] d\tau \right\},\\
dG_{21} &= \operatorname{Re}\left\{ \left[H_6^0 (H_1 h - H_2 g)/22 + \chi \left(1 + \chi \sin^2 \psi_0 \right) \times e_0^2 (\kappa + 1) H/4 \right] d\tau \right\},\\
dG_{22} &= \operatorname{Re}\left\{ \left[H_6^0 (g - h)/2 - \chi \left(1 + \chi \sin^2 \psi_0 \right) \times (\kappa + 1)\, e_0^2/4 \right] d\tau \right\},
\end{aligned}$$

$$
\begin{aligned}
&g = e^{2i(\psi_0 - \psi)}, \\
&2N_1(\zeta) = \pi(\kappa + 1)\left\{N/\mu + \chi e_0^2\left(\chi n_2^2 + \cos 2\psi - 1\right)/2\right\}, \\
&2N_2(\zeta) = \pi(\kappa + 1)\left\{-T/\mu + \chi e_0^2 n_1 n_2\right\}, \quad h = e^{2i\psi_0}\frac{\overline{\zeta} - \overline{\zeta}_0}{\zeta - \zeta_0}, \\
&H_1 = i + e_0^2\Lambda_1(\psi), \quad H_2 = e_0^2\overline{\Lambda_2(\psi)} - i, \\
&H_3 = \frac{\kappa - 1}{\kappa + 1} + 2H_4, \quad H_4 = (\kappa + 1)^{-1} + i\chi^2 e_0^2 n_1 n_2/4, \\
&H_6 = 1 + \chi\left(1 + \chi n_2^2\right)e_0^2/2, \\
&H_5 = \overline{\alpha(\psi)} + \chi^2 e_0^2 n_1 n_2\left(1 - \kappa + e^{2i\psi}\right)/2, \quad H_i = H_i(\psi), \\
&H = H(\psi, \psi_0) = H_1 + \chi^2 e_0^2 \delta(\psi)\left(1 - \kappa + e^{2i\psi_0}\right)/2, \\
&H_i^0 = H_i(\psi_0).
\end{aligned}
$$

This system must be considered in conjunction with the additional conditions (11.23). On assumption that the contours L_j do not have common points and their curvature, as well as their function $N_i(\zeta)$ satisfy Holder's condition, we obtain a unique solution in the class of functions, unbounded on the ends of the arcs L_j.

For the determination of the stress intensity factor, we introduce a parametric representation of the contour L_j (in the sequel, the index j is omited): $\zeta = \zeta(\beta)$, $-1 \le \beta \le 1$. In accordance with the above, we assume

$$
U_i(\zeta) = \frac{\Omega_i(\beta)}{\sqrt{1 - \beta^2}}, \quad \Omega_i(\beta) \in H[-1, 1] \quad (i = 1, 2). \tag{11.26}
$$

Substituting functions (11.26) into the integral representations (11.20), we obtain, by the standard procedure of distinction of singularities of the solution at the ends L_j of the formula for the stress intensity factors in the tips of the crack

$$
\begin{aligned}
K_I^s - iK_{II}^s = \lim\sqrt{2r}\left(S_{nn} - iS_{ns}\right) = \pm\mu\left(s'(\pm 1)\right)^{1/2}\Bigg\{2\frac{i\Omega_2(\pm 1) - \Omega_1(\pm 1)}{\kappa + 1} + \\
+\frac{1}{2}\chi e_0^2\Omega_1(\pm 1)\left(-\frac{2}{\kappa + 1} + \sum_{\alpha,\delta=1}^{2} a_{\alpha\delta} n_\alpha^{\pm} n_\delta^{\pm}\right)\Bigg\},
\end{aligned} \tag{11.27}
$$

where

$$
\begin{aligned}
&a_{11} = \frac{2\chi(\kappa - 2)}{\mu_r(\kappa + 1)}, a_{22} = -2\chi, \\
&a_{12} = a_{21} = i\chi\left(1 + \frac{\kappa - 2}{\mu_r(\kappa + 1)}\right), \\
&n_1^{\pm} = \cos\psi(\pm 1), \quad n_2^{\pm} = \sin\psi(\pm 1).
\end{aligned}
$$

The upper sign concerns the tip $c = b$ $(\beta = 1)$, and lower one, to the tip $c = a\,(\beta = -1)$, S_{nn}, S_{ns} are the normal and tangent total stresses on the area with the normal $\overrightarrow{n} = (\cos\psi_c,\ \sin\psi_c)$ on the extension beyond the tip c.

As a first example we take a medium with rectilinear crack of length 2ℓ, oriented, with respect to the initial magnetic field, in the manner shown in Fig. 11.2.

We take the parametric equation L in the form: $\zeta = il\beta e^{i\psi}$, $\zeta_0 = il\beta_0 e^{i\psi}$, $-1 \le \beta,\ \beta_0 \le 1$. In this case, the integral (11.25) and the additional condition (11.23) take the form

$$\sum_{n=1}^{2} p_{mn} \int_{-1}^{1} \frac{\Omega_n(\beta)d\beta}{\sqrt{1-\beta^2}\,(\beta-\beta_0)} = N_m^*(\beta_0) \quad (m = 1, 2)\,, \tag{11.28}$$

$$\int_{-1}^{1} \frac{\Omega_1(\beta) + i\Omega_2(\beta)}{\sqrt{1-\beta^2}} d\beta = 0,$$

$$p_{11} = \sum_{j=0}^{4} d_j \chi^j, \quad p_{21} = \sum_{j=2}^{3} d_j^* \chi^j,$$

$$N_m^*(\beta) = N_m(\zeta), \quad p_{12} = \frac{\chi^2}{4} e_0^2 (\kappa - 1)\, n_1 n_2,$$

$$p_{22} = 1 + \frac{\chi}{4} e_0^2 \left(1 + \chi n_2^2\right)(1 - \kappa)\,, \quad d_0 = -1,$$

$$d_1 = \frac{e_0^2}{4}(\kappa - 1)\,, \quad d_2 = -\frac{e_0^2}{4}(5 - \kappa)\,\frac{n_1^2}{\mu_r},$$

$$d_3 = \frac{e_0^2}{4}(\kappa + 1)\, n_2^2, \quad d_4 = \frac{e_0^4}{4} n_1^2 n_2^2 \left(\frac{1 - 2\kappa}{\mu_r} + \kappa + 1\right),$$

$$d_2^* = \frac{e_0^2}{4} n_1 n_2 \left(\kappa + 1 + \frac{2\kappa - 4}{\mu_r}\right),$$

$$d_3^* = \frac{e_0^4}{4} n_1 n_2 \left(1 + \chi n_2^2\right)\left(\frac{2\kappa - 1}{\mu_r} - \kappa - 1\right).$$

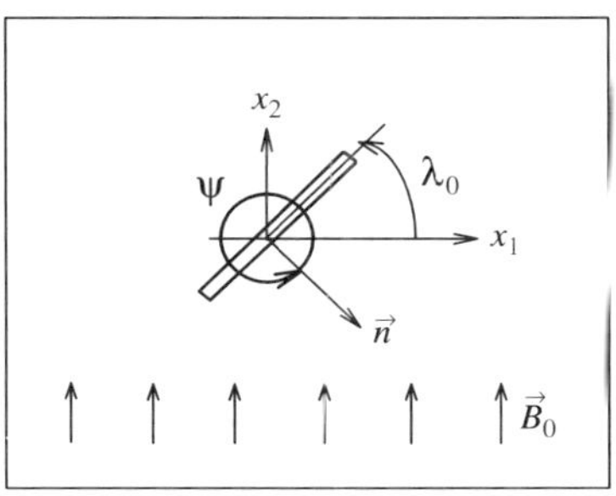

Fig. 11.2. Ferromagnetic medium with rectilinear crack in strong magnetic field

The solution of the characteristic system (11.28) is the following:

$$\Omega_n(\beta) = A_n\beta \quad (n = 1, 2)\,, \tag{11.29}$$

where

$$A_1 = \frac{1}{\pi\Delta}\left(p_{22}N_1^* - p_{12}N_2^*\right), \quad A_2 = \frac{1}{\pi\Delta}\left(p_{11}N_2^* - p_{21}N_1^*\right),$$
$$\Delta = p_{11}p_{22} - p_{21}p_{12}.$$

The sought-for analytic functions and their combination are determined accordingly (Fig. 11.3):

$$F(z) = \frac{\chi H_{02}}{2i}A_1\delta\left(\psi\right)\gamma_1(z), W(z) = \frac{\mu\left(A_2 - A_1H_1\right)}{i\left(\kappa+1\right)}\gamma_1(z), \tag{11.30}$$

$$W_1(z) = \frac{\mu e^{-2i\psi}}{i\left(\kappa+1\right)}\left\{\left(A_2 - A_1H_1\right)\gamma_2(z) + \left(A_2 - A_1H_2\right)\gamma_1(z)\right\},$$

$$\overline{z}W'(z) + W_1(z) = \frac{\mu}{\left(\kappa+1\right)} \times \Big\{\left(A_1H_1 - A_2\right)le^{i[\lambda_0 - 3(\theta_1+\theta_2)/2]}\left(\rho_1\rho_2\right)^{-12} \times$$
$$\times \sin\left(\theta_2 - \theta_1\right) - i\left[A_1\left(H_1 + H_2\right) - 2A_2\right]$$
$$\times \left(\rho e^{i[\theta - (\theta_1+\theta_2)/2]}\left(\rho_1\rho_2\right)^{-1/2} - 1\right)e^{-2i\lambda_0}\Big\},$$

$$\rho_1 e^{i\theta_1} = z - c, \quad \rho_2 e^{i\theta_2} = z + c, \quad z = \rho e^{i\theta},$$

$$\lambda_0 = \psi - 3\pi/2, \quad \gamma_1(z) = \frac{z}{\left(z^2 - c^2\right)^{1/2}} - 1,$$

$$\gamma_2(z) = \frac{z^3 - 2zc^2}{(z^2 - c^2)^{3/2}} - 1, \quad c = le^{i\lambda_0}.$$

With the aid of functions (11.30) we find the mechanical and magnetic magnitudes at any point of the body. In particular, on the straight line, parallel to axis of the crack we get

$$\left(S_{nn} - iS_{ns}\right)/\mu = \frac{2\varepsilon_1}{i}\left\{\frac{\rho}{\left(\rho_1\rho_2\right)^{1/2}}\cos\left(\theta - \frac{\theta_1+\theta_2}{2}\right) - 1\right\} + \tag{11.31}$$

$$-\bar{\varepsilon}_1 le^{3i\left(\frac{\lambda - \theta_1 + \theta_2}{2}\right)}\frac{\sin\left(\theta_2 - \theta_1\right)}{\left(\rho_1\rho_2\right)^{1/2}} + \frac{e_0^2}{2}\left(2\chi + \left(2\chi+1\right)e^{2i\lambda}\right),$$

$$\varepsilon_1 = \frac{A_1\overline{H}_1 - A_2}{\kappa+1} = \frac{i\chi}{4}e_0^2A_1\left(-\frac{2}{\kappa+1} + \sum_{\alpha,\delta=1}^{2}a_{\alpha,\delta}n_\alpha n_\delta\right) - \frac{A_2 + iA_1}{\kappa+1}.$$

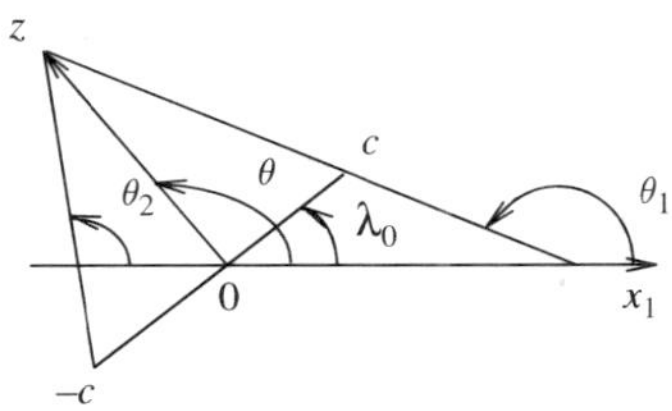

Fig. 11.3. On the determination of closed solution for rectilinear crack

However, if the straight line coincides with the axis of crack, then we find: when $\rho \leq l$,

$$S_{nn} - iS_{ns} = \mu \left\{ \frac{e_0^2}{2} \left(2\chi + (2\chi + 1)\, e^{2i\lambda}\right) + 2i\varepsilon_1 \right\}, \tag{11.32}$$

when $\rho > l$

$$S_{nn} - iS_{ns} = \mu \left\{ \frac{e_0^2}{2} \left(2\chi + (2\chi + 1)\, e^{2i\lambda}\right) + 2i\varepsilon_1 \left(1 - \frac{\rho}{\sqrt{\rho^2 - l^2}}\right) \right\}.$$

The effect of the orientation of crack with respect to the initial magnetic field $\overrightarrow{B}_0$ on the mechanical and magnetic magnitudes can be easily is found with the aid of these dependences. The simplest results are obtained for the horizontal or vertical crack. In this case, $n_1 n_2 = 0$, $p_{12} = p_{21} = 0$ and the system (11.28) splits into two independent equations; moreover the magnitudes A_n, which occur in (11.29), are such that:

$$A_1 = \frac{N_1^*}{\pi p_{11}}, \quad A_2 = \frac{N_2^*}{\pi p_{22}}, \tag{11.33}$$

where for the horizontal crack ($n_1 = 0,\ n_2 = -1$)

$$p_{11} = \frac{\chi e_0^2}{4} \left(\kappa - 1 - \chi \frac{5 - \kappa}{\mu_r}\right) - 1, \quad p_{22} = 1 - \frac{\chi e_0^2}{4} (\kappa - 1),$$

$$N_1^* = (\kappa + 1) \frac{\pi N}{2\mu}, \quad N_2^* = -(\kappa + 1) \frac{\pi T}{2\mu}.$$

for the vertical crack ($n_1 = 1,\ n_2 = 0$)

$$p_{11} = \frac{\chi e_0^2}{4} \left(\kappa - 1 - \chi \frac{5 - \kappa}{\mu_r}\right) - 1, \quad p_{22} = 1 - \frac{\chi e_0^2}{4} (\kappa - 1),$$

$$N_1^* = (\kappa + 1) \frac{\pi N}{2\mu}, \quad N_2^* = -(\kappa + 1) \frac{\pi T}{2\mu}.$$

From formulas (11.31)–(11.33), it is immediately evident that there are such values of the parameter $b_c = \mu_r e_0 = B_{02} / \sqrt{\mu_0 \mu}$, which are called critical, under which the components of the tensors of magneto-elastic and Maxwell stresses, as well as the vectors of intensity and induction of magnetic field in the ferromagnetic space with crack tend to infinity. This phenomenon was discovered earlier [286, 287], where the case of horizontal crack was considered. At the same time, the matter is different for the vertical crack (parallel to the initial magnetic field). As it follows from (11.33), when $N \neq 0,\ T = 0$, such a critical value does not exist at all, but when $N = 0,\ T \neq 0$, exists theoretically $\left(b_c^* = 6,3 \cdot 10^2,\ \text{when } \mu_r = 10^5,\ \nu = 0,25\right)$, but this value is not realized physically. If one takes into account the fact that in the homogeneous ferromagnetic medium under initial magnetic field $(0, B_{02}, 0)$ there are total stresses, namely $S_{11} = -\mu_0 H_{02}^2/2\, S_{12} = 0,\ S_{22} = (2\chi + 1/2)\, \mu_0 H_{02}^2$, then we can come to the conclusion that when $b_c = b_c^*$, the plane form of equilibrium of ferromagnetic medium with crack becomes impossible.

For a rectilinear crack, oriented to the direction of the initial magnetic field at a certain angle $\pi/2 - \lambda_0$, the critical value b_c^* is determined according to (11.29), as the smallest positive root of the equation $\Delta = 0$. This is an algebraic equation of fourth degree with respect to the parameter b_c. The magnitude b_c^* assumes its smallest value when $\lambda_0 = 0$ $\left(b_c^* = 3,6 \cdot 10^{-3} \text{when} \mu_r = 10^5,\right.$ $\nu = 0,25)$. Further, when λ_0 increases from 0 to $\pi/2$, it monotonically increases up to the above-indicated value for the vertical crack.

From the before said follows that the consideration of the overall situation shows that there is a whole spectrum of critical values of the parameter b_c, depending on the configuration of cracks, their mutual arrangements, orientations with respect to the initial magnetic field, and the type of load. Apparently, it makes sense to study the interaction of mechanical and magnetic fields in a ferromagnetic medium with cracks only when the values of the level of magnetic field are such that $0 \leq b_c < b_c^*$.

Another interesting case is when the medium is weakened by a singular parabolic crack $\xi_1 = p_1 \beta \cos\gamma - p_2 \beta^2 \sin\gamma,\ \xi_2 = p_1 \beta^2 \sin\gamma + p_2 \beta^2 \cos\gamma,\ -1 \leq \beta \leq 1$ (γ is the angle of revolution of the axes of the parabola with respect to the coordinate system $0x_1x_2$). The calculation of the stress intensity factors $K_{\mathrm{I,II}}$ was done making use of the system of integral (11.25), (11.23) and formula (11.27).

Figure 11.4 gives the graphs for the critical values of parameter b_c^2 depending on the orientation of the crack (angle γ) for different values of the parameter of curvature p_2. The curves, which characterize the dependence of the stress intensity factor K_{I} from the parameter b_c^2, for different orientations of the straight crack are presented in Fig. 11.5.

Let us consider the situation, when the medium is weakened by two rectilinear cracks with centers on the axis $0x_2$. The lengths of both cracks $2l = 2l_1 = 2$, one of which is located on the axis x_1, the second is turned with respect to it to the angle γ. The distance between the centers of cracks is equal

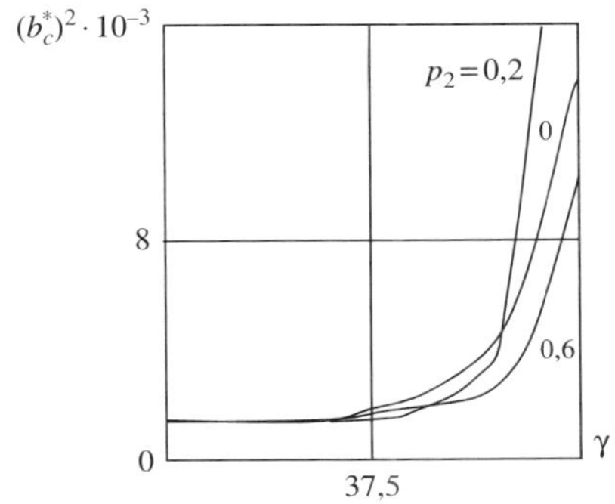

Fig. 11.4. Graphs of critical values $(b_c^*)^2$ depending on the orientation of the parabolic crack

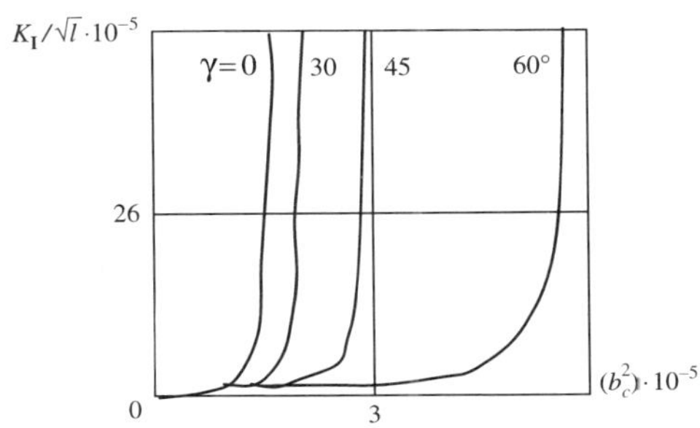

Fig. 11.5. Dependence of the coefficient of intensity on the parameter b_c^2

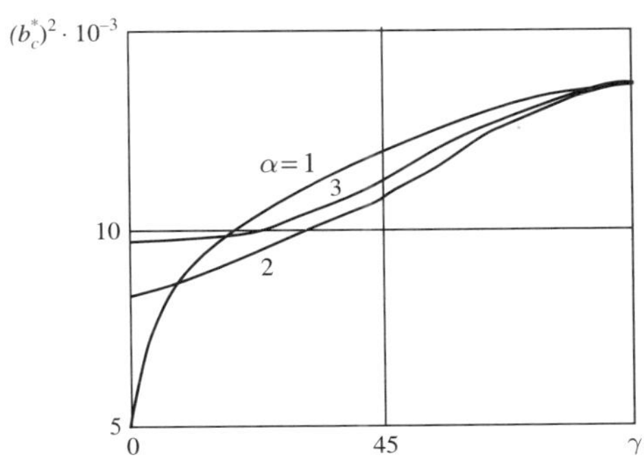

Fig. 11.6. Graphs of the critical values of the parameter b_c^2 for ferromagnetic medium with two cracks

to d. The results of the calculations for the critical values of the parameter b_c^2, depending on the angle γ for different d are shown in Fig. 11.6.

The analysis of numerical results shows that the initial magnetic field exerts essential influence on the stress intensity factors only close to its critical values. For example, for the horizontal crack a change of the magnitude K_{I} is observed close to the critical value $B_0^* \approx 1,2$ **T**. In the general case, the critical values of the level of the initial magnetic field depend on the mutual arrangement of defects and their configurations.

11.3 Generalized Kirsh Problem for Ferromagnetic Medium with Cavity in Strong Magnetic Field

11.3.1 Statement of the Problem. Complex Representation of the Solutions

Let us consider an unbounded medium made of soft-iron ferromagnetic material, in reference to a rectangular Cartesian coordinate system $0x_1x_2x_3$, weakened by a tunnel cylindrical cavity along the axis x_3. Let us assume that the initial magnetic field, into which a ferromagnetic body is placed, is uniform (infinitely) and given by a vector of magnetic displacement $\overrightarrow{B}_0 = (0, B_{02}, 0)$, and normal ($N$) and tangential ($T$) mechanical stresses act on the border of the opening, and the cavity of the opening is filled by vacuum (Fig. 11.7).

The initial undisturbed magnetic field that corresponds to the rigid (non-deformated) state of ferromagnetic body is determined by the equations of magneto-statics (11.5)–(11.7).

Let us introduce complex representations of the intensity of the initial magnetic field in the body and the cavity

$$
\begin{aligned}
H_1 - iH_2 &= \frac{iB_{02}}{\mu_0\mu_r} F(z), \\
H_1^e - iH_2^e &= \frac{iB_{02}}{\mu_0} F^e(z),
\end{aligned}
\tag{11.34}
$$

where $F(z)$ and $F^e(z)$ are functions, analytical on the external part of the opening (domain D), and in the domain D^e, occupied by the opening, respectively.

Then the problem of magneto-statics (11.5)–(11.7) is reduced to the classical boundary-value problem of the theory of functions (Γ is the contour of the cross section of cavity):

$$
\begin{aligned}
&\mathrm{Re}\left\{e^{i\psi}\left(F(\zeta) - \mu_r F^e(\zeta)\right)\right\} = 0, \\
&\mathrm{Im}\left\{e^{i\psi}\left(F(\zeta) - F^e(\zeta)\right)\right\} = 0, \quad \zeta \in \Gamma,
\end{aligned}
\tag{11.35}
$$

which is equivalent to the regular integral equation of second order

$$
p(\zeta_0) + \frac{\mu_r - 1}{\pi(\mu_r + 1)} \int\limits_{\Gamma} \frac{\cos(\alpha_0 - \psi_0)}{r_0} p(\zeta)\, ds = 2\frac{\mu_r - 1}{\mu_r + 1} \cos\psi_0. \tag{11.36}
$$

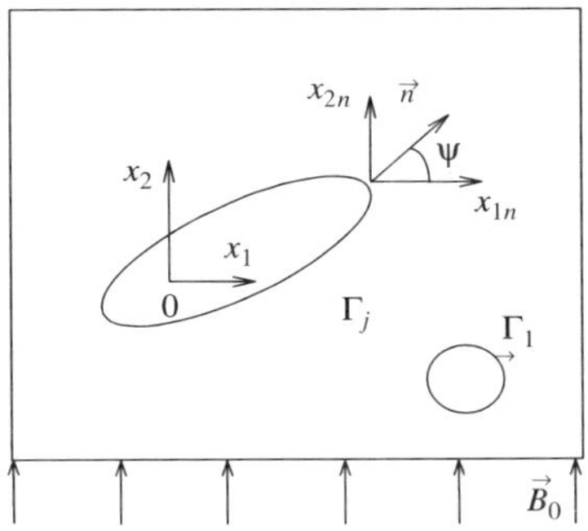

Fig. 11.7. Ferromagnetic elastic medium with the openings in the strong magnetic field

Here ψ, ψ_0 are the angles between the external normal to the contour Γ and the axis $0x_1$ at the points ζ and $\zeta_0 \in \Gamma$, respectively, $r_0 = |\zeta - \zeta_0|$, $\alpha_0 = \arg(\zeta - \zeta_0)$; the direction of integration is counterclockwise.

In this case, the functions $F(z)$, $F^e(z)$ are determined by the following formulas

$$\frac{1}{2\pi} \int_{\Gamma} \frac{p(\zeta)ds}{\zeta - z} - 1 = \begin{cases} F(z), & z \in D, \\ F^e(z), & z \in D^e, \end{cases} \tag{11.37}$$

where $\mathrm{Im} p(\zeta) = 0$.

Thus, the initial magnetic field can be considered known. For example, if cavity has circular cross section of radius a, it would be

$$F(z) = -1 - \frac{(\mu_r - 1)\, a^2}{(\mu_r + 1)\, z^2}, \quad F^e(z) = -\frac{2}{\mu_r + 1}. \tag{11.38}$$

As a result of the deformation of the body, small fluctuations $\overrightarrow{h} = (h_1, h_2)$, $\overrightarrow{b} = \mu_0 \mu_r \overrightarrow{h} = (b_1, b_2)$ and $\overrightarrow{h}^e = (h_1^e, h_2^e)$, $\overrightarrow{b}^e = \mu_0 h^e = (b_1^e, b_2^e)$ are added to the initial magnetic field. Accordingly, the summarizing magnetic fields in the body and the cavity are determined by the formulas

$$W_j = H_j + h_j, \; W_j^e = H_j^e + h_j^e \quad (j = 1, 2)\,. \tag{11.39}$$

The equations of equilibrium of soft-iron ferromagnetic body is written in the form (before linearization)

$$\partial_j t_{ji} + \chi \mu_0 \partial_j (W_i W_j) = 0, \tag{11.40}$$

where

$$t_{ij} = \sigma_{ij} + \chi \mu_0 W_i W_j,$$
$$\partial_j = \partial / \partial x_j \quad (i, j = 1, 2)\,.$$

From (11.40) follow the relations

$$\sigma_{ij} = \sigma_{ij}^0 - 2\chi\mu_0 W_i W_j, \quad \partial_j \sigma_{ji}^0 = 0\,(i,j=1,2)\,. \tag{11.41}$$

Introducing the function of stresses $\left(\sigma_{11}^0 = \partial_2^2 U,\ \sigma_{12}^0 = -\partial_1\partial_2 U,\ \sigma_{22}^0 = \partial_1^2 U\right)$, and applying the condition of compatibility of deformation, we come, taking into account the equalities $\partial_j W_j = 0,\ \partial_1 W_2 = \partial_2 W_1$, to the differential equation relative to the function of stresses

$$\nabla^2\nabla^2 U = 4\chi\mu_0\nu_* \left((\partial_1 W_1)^2 + (\partial_2 W_1)^2\right), \tag{11.42}$$
$$\nu_* = (1-2\nu)\,(1-\nu)^{-1}\,.$$

The fluctuations of the magnetic field can be presented in the form

$$h_1 - ih_2 = \frac{iB_{02}}{\mu_0\mu_r} R\,(z)\,, \quad h_1^e - ih_2^e = \frac{iB_{02}}{\mu_0} R^e\,(z)\,, \tag{11.43}$$

where $R\,(z)$ and $R^e\,(z)$ are functions analytical in the domains D and D^e, respectively.

Then (11.42) can be represented in the form (μ is the shear modulus)

$$\nabla^2\nabla^2 U = 8\nu_* m\chi\mu_r^{-1}\left|\Omega'\,(z)\right|^2, \tag{11.44}$$

where

$$m = \frac{\mu b_c^2}{2\mu_r}, \quad b_c^2 = \frac{B_{02}^2}{\mu\mu_0},$$
$$\Omega\,(z) = F\,(z) + R\,(z)\,, \quad \Omega'\,(z) = d\Omega/dz.$$

The integration of this equation gives

$$U = U_0 + U_*, \tag{11.45}$$
$$U_0 = \mathrm{Re}\left(\bar{z}\varphi\,(z) + \chi_*\,(z)\right), \quad U_* = \frac{\chi\nu_*}{2\mu_r} m\left|\omega\,(z)\right|^2,$$
$$\omega'\,(z) = \Omega\,(z)\,;$$

where $\varphi\,(z)\,,\ \chi_*\,(z)$ are arbitrary functions, analytical in the domain D.

The obtained expression for the function of stresses in combination with (11.40), (11.41) makes possible to write the elementary complex representations of the field magnitudes under consideration. We get

$$t_{11} + t_{22} = 4\mathrm{Re}\Phi\,(z) - \frac{2m\nu\chi}{(1-\nu)\,\mu_r}\left|\Omega\,(z)\right|^2,$$
$$t_{22} - t_{11} + 2it_{12} = 2\left(\bar{z}\Phi'\,(z) + \Psi\,(z)\right)$$
$$+\, 2\nu_* m\chi\mu_r^{-1}\Omega'\,(z)\,\overline{\omega\,(z)} - 2m\chi\mu_r^{-1}\Omega^2\,(z)\,, \tag{11.46}$$
$$2\mu\,(u_1 - iu_2) = \kappa\overline{\varphi\,(z)} - \bar{z}\Phi\,(z) - \psi\,(z) + 2m\chi\mu_r^{-1}$$
$$\left(\int \Omega^2\,(z)\,dz - \frac{1}{2}\nu_*\Omega\,(z)\,\overline{\omega\,(z)}\right),$$

where u_1, u_2 are the corresponding components of the vector of small elastic displacement in the ferromagnetic medium; $\kappa = 3--4\nu$ for the flat deformation; $\varphi'(z) = \Phi(z)$, $\chi'_* = \psi(z)$, $\psi'(z) = \Psi(z)$.

The tensor of Maxwell stresses T_{ij} is added to the tensor of magneto-elastic stresses t_{ij}, which is manifested in both the body and in the cavity of the opening. The components of this tensor take the form

$$T_{ij} = \mu_0 \mu_r W_i W_j - \frac{1}{2}\mu_0 \delta_{ij} (W_k W_k), \tag{11.47}$$

where δ_{ij} is the Kronecker symbol; summing is done on index k.

The corresponding analytical representations of the Maxwell stresses in the body and the cavity are such that:

$$\begin{aligned} &T_{11} + T_{22} = 2m\chi\mu_r^{-1} |\Omega(z)|^2, \quad T_{11}^e + T_{22}^e = 0, \\ &T_{22} - T_{11} + 2iT_{12} = 2m\Omega^2(z), \\ &T_{22}^e - T_{11}^e + 2iT_{12}^e = 2m\mu_r (\Omega^e)^2, \end{aligned} \tag{11.48}$$

where $\Omega^e = F^e(z) + R^e(z)$.

Thus, all field magnitudes in the body and the cavity are expressed as four analytic functions $\Phi(z)$, $\Psi(z)$, $R(z)$ and $R^e(z)$. For their determination we have (linearized) boundary conditions of the opening (11.4). The power boundary condition occurring there can be conveniently represented in the form

$$[S_{11} + S_{22}] - e^{2i\psi} [S_{22} - S_{11} + 2iS_{12}] = 2(N - iT), \tag{11.49}$$

where N, T are the given mechanical, normal and tangential forces on the contour of the openings; the angular parentheses indicate the jump of the corresponding magnitude on the interface of the media.

Taking into account complex representations (11.43), (11.45), (11.46) and (11.48), the boundary condition (11.49) and the electromagnetic conditions of conjugation in (11.4) can be expressed (after the appropriate linearization) in terms of the boundary values of the sought-for analytic functions. We have

$$\begin{aligned} &2\mathrm{Re}\Phi - e^{2i\psi}(\bar{z}\Phi' + \Psi) + m\Lambda_1 = N - iT + mA_1, \\ &\mathrm{Im}\left\{2\mu(R - R^e)e^{i\psi} + (F - F^e)Q_*\right\} = N_1 + mA_2, \\ &\mathrm{Re}\left\{2\mu\left(R\mu_r^{-1} - R^e\right)e^{i\psi} + \left(F\mu_r^{-1} - F^e\right)Q_*\right\} = N_2 + mA_3, \end{aligned} \tag{11.50}$$

where

$$\begin{aligned} &Q = 2\Phi - (\kappa - 1)\overline{\Phi}, \quad N_1 = \mathrm{Im}\left\{e^{i\psi}(F - F^e)(N + iT)\right\}, \\ &N_2 = \mathrm{Re}\left\{e^{i\psi}\left(F\mu_r^{-1} - F^e\right)(N + iT)\right\}, \quad \nu_1 = \nu_*\chi\mu_r^{-1}, \end{aligned}$$

$$A_1 = \left(\nu_1 F'\overline{f} + \mu_r^{-1} F^2 - \mu_r (F^e)^2\right) e^{2i\psi} - \nu_1 |F|^2,$$
$$A_2 = \mathrm{Im}\left\{(F^e - F)\Lambda\right\}, \quad A_3 = \mathrm{Re}\left\{\left(F^e - \mu_r^{-1}F\right)\Lambda\right\},$$
$$\Lambda_1 = 2\nu_1 \mathrm{Re}\left(\overline{F}R\right) - \left\{\nu_1\left(F'\overline{r} + R'\overline{f}\right) + 2FR\mu_r^{-1} - 2\mu_r F^e R^e\right\} e^{2i\psi},$$
$$\Lambda_2 = 4\nu_1 e^{i\psi} \mathrm{Re}\left(\overline{F}R\right) - \left\{(4\chi + 2)\mu_r^{-1}\overline{F}\,\overline{R} - 2\mu_r \overline{F}^e\overline{R}^e\right\} e^{-i\psi},$$
$$\Lambda = 2\nu_1 |F|^2 e^{i\psi} + \left\{\mu_r \left(\overline{F^e}\right)^2 - (2\chi + 1)\mu_r^{-1}\overline{F}^2\right\} e^{-i\psi},$$
$$Q_* = Qe^{i\psi} + m\Lambda_2.$$

In the boundary conditions (11.50), the coefficients of the sought-for functions are the limiting values of the known analytic functions, which characterize the initial (undisturbed) magnetic field in the body and the cavity. Since for many ferromagnetic materials $\chi\mu_r^{-1} \approx 1$ and $\mu_r >> 1$, the effect of magnetizability of material on the stressed state is determined by the parameter m.

11.3.2 Stress Concentration in Circular Opening

We assume below that the contour of the opening is free from load and that a uniform field of mechanical stresses $\langle\sigma_{ij}\rangle$, occurs at infinity, moreover $\langle\sigma_{12}\rangle = 0$. Let us represent the sought-for analytic functions in the form (a is the radius of the opening)

$$\Phi = \sum_{k=0}^{\infty} \gamma_k \left(\frac{a}{z}\right)^{2k}, \Psi = \sum_{k=0}^{\infty} (2k\gamma_k + \lambda_k)\left(\frac{a}{z}\right)^{2k} + \lambda_{-1}, \tag{11.51}$$
$$R = \sum_{k=0}^{\infty} R_k \left(\frac{a}{z}\right)^{2k}, R^e = \sum_{k=0}^{\infty} R_k^* \left(\frac{z}{a}\right)^{2k}, z = x_1 + ix_2.$$

In view of the power and geometric symmetry, all coefficients in series (11.51) would be real. The functions, which determine the initial magnetic field, are given by (11.38).

Let us require that the fluctuations of the magnetic field disappear at infinity and the mechanical stresses σ_{ij} are bounded ($\sigma_{11}^{\infty} = \langle\sigma_{11}\rangle$, $\sigma_{22}^{\infty} = \langle\sigma_{22}\rangle$, $\sigma_{12}^{\infty} = 0$). Then, taking into account the relations (11.46), (11.41) and (11.43) we obtain

$$4\gamma_0 = \langle\sigma_{11}\rangle + \langle\sigma_{22}\rangle + \frac{2m}{1-\nu}, \quad R_0 = 0, \tag{11.52}$$
$$2\lambda_{-1} = \langle\sigma_{22}\rangle - \langle\sigma_{11}\rangle + 4m.$$

The rest of the coefficients are determined from the boundary equalities (11.50). Using the appropriate standard procedure, we come to the next system of equalities:

$$
\begin{aligned}
\lambda_0 &= \frac{1}{2}\left(\langle\sigma_{11}\rangle + \langle\sigma_{22}\rangle\right) + \frac{m}{1-\nu} + 2m\nu_* R_1, \qquad (11.53)\\
\lambda_1 &= \frac{1}{2}\left(\langle\sigma_{22}\rangle - \langle\sigma_{11}\rangle\right) + \frac{4-6\nu}{1-\nu}m + 2m\left(2R_0^* - \nu_* R_1 + \frac{5}{3}\nu_* R_2\right),\\
\lambda_k &= 2m\left\{2R_{k-1}^* + k\nu_*\left(\frac{2k+3}{2k+1}R_{k+1} - R_k\right)\right\}(k=2,3,\ldots),\\
\gamma_1 &= \frac{1}{2}\left(\langle\sigma_{22}\rangle - \langle\sigma_{11}\rangle\right) + \frac{m}{1-\nu} + m\left(4R_0^* + \nu_* R_1 + \frac{1}{3}\nu_* R_2\right),\\
\gamma_k &= m\left\{4R_{k-1}^* + \nu_*\left(R_k + \frac{2k-1}{2k+1}R_{k+1}\right)\right\}(k=2,3,\ldots),\\
&2\mu\left(R_1 + R_0^*\right) + 4m\left(R_1 + R_2 + R_0^*\right) = (\kappa+1)\gamma_1,\\
&2\mu\left(\mu_r R_0^* - R_1\right) + 4m\left(R_1 + R_2 + R_0^*\right) = (\kappa+1)\gamma_1,\\
&2\mu\left(R_{k+1} + R_k^*\right) + 4m\left(R_k + 2R_{k+1} + R_{k+2} + R_k^* + R_{k-1}^*\right)\\
&= (\kappa+1)\left(\gamma_k + \gamma_{k+1}\right),\\
&2\mu\left(R_{k+1} - \mu_r R_k^*\right) + 4m\left(R_k - R_{k+2} - R_k^* + R_{k-1}^*\right) = (\kappa+1)\\
&\left(\gamma_k - \gamma_{k+1}\right) \quad (k=1,2,\ldots).
\end{aligned}
$$

The analysis of this system leads to the relations of connection ($\mu_r \pm 1 \approx \mu_r$)

$$
R_{k+1} = \frac{1}{2}\mu_r\left(R_{k-1}^* + R_k^*\right), \quad R_{-1}^* = 0 \ \ (k=0,1,\ldots) \qquad (11.54)
$$

and the equation relative to the coefficients R_k^*:

$$
\begin{aligned}
&\mu R_k^* + 4m\nu\left(R_{k-1}^* + \frac{2k+1+\nu^{-1}}{2k+3}R_{k+1}^*\right) = 0 \qquad (11.55)\\
&(k=1,2,\ldots),\\
&\mu R_0^* + \frac{4}{3}(1+\nu)\, m R_1^* = \frac{\kappa+1}{\mu_r}\left(\frac{\langle\sigma_{22}\rangle - \langle\sigma_{11}\rangle}{2} + \frac{m}{1-\nu}\right).
\end{aligned}
$$

When $\nu = 0.5$ we get an equation with constant coefficients, the solution of which, in consideration of the initial condition in (11.55), takes form

$$
\begin{aligned}
R_k^* &= C Z_1^k, \quad Z_1 = \sqrt{\left(\frac{\mu}{4m}\right)^2 - 1} - \frac{\mu}{4m}, \qquad (11.56)\\
C &= \frac{\langle\sigma_{22}\rangle - \langle\sigma_{11}\rangle + 4m}{\mu_r\left(\mu + 2mZ_1\right)}.
\end{aligned}
$$

With accuracy of $m/\mu = b_c^2/(2\mu_r)$, we can put

$$
\begin{aligned}
R_0^* &= \frac{1}{\mu\mu_r}\left(\langle\sigma_{22}\rangle - \langle\sigma_{11}\rangle + 4m\right), \qquad (11.57)\\
R_k^* &= 0 \quad (k=1,2\ldots).
\end{aligned}
$$

When $\nu < 0.5$, the difference equation has constant coefficients whenever k is sufficiently large. The detailed analysis of the system (11.55) gives (with accuracy of m/μ)

$$R_0^* = \frac{\kappa+1}{2\mu\mu_r}\left(\langle\sigma_{22}\rangle - \langle\sigma_{11}\rangle + \frac{2m}{1-\nu}\right), \tag{11.58}$$
$$R_k^* = 0 \quad (k = 1, 2 \ldots).$$

According to (11.58), we find the fluctuations of magnetic induction in the body and the cavities:

$$b_1 - ib_2 = \frac{iB_{02}}{\mu}\left\{(1-\nu)\left(\langle\sigma_{22}\rangle - \langle\sigma_{11}\rangle\right) + 2m\right\}\frac{a^2}{z^2}, \tag{11.59}$$
$$b_1^e - ib_2^e = \frac{2iB_{02}}{\mu\mu_r}\left\{(1-\nu)\left(\langle\sigma_{22}\rangle - \langle\sigma_{11}\rangle\right) + 2m\right\}.$$

We determine now the total normal stress on the contour $S_{\psi\psi}$ of the free from mechanical stresses opening. Taking into account the equalities (11.48) and the power boundary condition (11.49), we find

$$S_{\psi\psi} = S_{11} + S_{22} + \frac{1}{2}\mathrm{Re}\left\{e^{2i\psi}\left(T_{22}^e - T_{11}^e + 2iT_{12}^e\right)\right\} = 4\mathrm{Re}\Phi + 2m\nu_*\left|\Omega\right|^2$$
$$+ m\mu_r\mathrm{Re}\left\{e^{2i\psi}\left(\Omega^e\right)^2\right\}. \tag{11.60}$$

After we linearize expression (11.60) and substitute function (11.51) in it, we obtain the following relations for the circular opening, if we also consider relations (11.54), (11.58)

$$S_{\psi\psi} = S_{\theta\theta} = \langle\sigma_{11}\rangle + \langle\sigma_{22}\rangle + \frac{2(3-4\nu)}{1-\nu}m + 2\left(\langle\sigma_{22}\rangle - \langle\sigma_{11}\rangle + 4m\right)\cos 2\theta,$$

where θ is the polar angle of the point of contour.

Analogous calculations for the magneto-elastic and elastic stresses on the contour of the opening give

$$t_{\theta\theta} = \langle\sigma_{11}\rangle + \langle\sigma_{22}\rangle + \frac{2m(1-2\nu)}{1-\nu} + 2\left(\langle\sigma_{22}\rangle - \langle\sigma_{11}\rangle + 2m\right)\cos 2\theta,$$
$$\sigma_{\theta\theta} = \langle\sigma_{11}\rangle + \langle\sigma_{22}\rangle - \frac{2m}{1-\nu} + 2\left(\langle\sigma_{22}\rangle - \langle\sigma_{11}\rangle\right)\cos 2\theta.$$

Let us consider, for example, the case, when they is no stress at the infinite, $b_c^2 = 0,8 \cdot 10^{-4}$; $\mu = 0,7 \cdot 10^{11} N/m^2$, $\mu_r = 10^3$, $\nu = 0,3$. Then, $m = 2800\ N/m^2$, and the calculations give.

$$\max S_{\theta\theta} = \left(16 - \frac{2}{1-\nu}\right)m \approx 36,7 \cdot 10^3 \quad N/m^2.$$

Thus, the effect of magnetizability on the stress concentration in the opening in ferromagnetic elastic medium can be proved to be essential. From the solution follows that in the problem considered above, the palne form of equilibrium is always possible; the choise of the value for the parameter b_c^2 is irrelevant.

11.3.3 Opening of Arbitrary Configuration in Ferromagnetic Medium

We assume now that the cross sections of the tunnel cavities are sufficiently smooth, closed contours Γ_j $(j = 1, 2, \ldots, M)$, moreover, $\bigcap \Gamma_j = \emptyset$. The conditions of loading of body with cavities remain the same as before. Following [273], let us introduce, for the solution of the boundary problem (11.50), the integral representations of the sought-for analytic functions:

$$\Phi(z) = \frac{1}{2\pi i} \int\limits_{\Gamma} \frac{\omega(\zeta)\, d\zeta}{\zeta - z} + A, R(z) = \frac{1}{2\pi i} \int\limits_{\Gamma} \frac{q(\zeta)\, d\zeta}{\zeta - z}, \tag{11.61}$$

$$\Psi(z) = \frac{\kappa}{2\pi i} \int\limits_{\Gamma} \frac{\overline{\omega(\zeta)} d\bar{\zeta}}{\zeta - z} - \frac{1}{2\pi i} \int\limits_{\Gamma} \frac{\bar{\zeta}\omega(\zeta)\, d\zeta}{(\zeta - z)^2} + B, z \in D,$$

$$R^e(z) = \frac{1}{2\pi i} \int\limits_{\Gamma} \frac{q(\zeta)\, d\zeta}{\zeta - z}, \quad z \in D_j^e,$$

where $\omega(\zeta) = \{\omega_j(\zeta), \zeta \in \Gamma_j\}$, $q(\zeta) = \{q_j(\zeta),\ \zeta \in \Gamma_j\}$ are subjected to the definition of densities; D_j^e is the final simply connected domain, bounded by the contour Γ_j; D is the domain, occupied by the body.

The constants A and B must ensure the existence of the given uniform field $\langle \sigma_{ij} \rangle$ at the infinite.

From the relations (11.40), (11.50) we find

$$\sigma_{11} + \sigma_{22} = 4\mathrm{Re}\Phi(z) - \frac{2m}{1-\nu} \left\{ |F|^2 + 2\mathrm{Re}\left(F\overline{R}\right) \right\}.$$

$$\sigma_{22} - \sigma_{11} + 2i\sigma_{12} = 2\left\{\bar{z}\Phi'(z) + \Psi(z)\right\} + 2m\nu_* \times \left(F'\overline{f} + R'\overline{f} + F'\overline{r}\right) - 4m\left(F^2 + 2FR\right).$$

Hence, taking into account representations (11.61), we determine the constants A and B:

$$A = \frac{\langle \sigma_{11} \rangle + \langle \sigma_{22} \rangle}{4} + \frac{m}{2(1-\nu)},$$

$$B = \frac{\langle \sigma_{22} \rangle - \langle \sigma_{11} \rangle + 2i \langle \sigma_{12} \rangle}{2} + 2m.$$

For the computation of functions $\omega(\zeta)$, $q(\zeta)$, we apply the standard procedure of reduction of the boundary-value problem to a system of singular integral equations. In this case, we can split it into two independent systems, disregarding small (in comparison with the unity) terms.

After the necessary transformations, we obtain

$$(-1)^m(\kappa+1)X_m(\zeta_0)+\sum_{n=1}^{2}\int_\Gamma X_n(\zeta)\,dH_{mn}(\zeta,\zeta_0)ds$$
$$=N_m(\zeta_0)\,(m=1,2), \tag{11.62}$$

$$\sum_{n=1}^{2}\left\{B_{nm}(\zeta_0)Y_n(\zeta_0)+\int_\Gamma Y_n(\zeta)\,dK_{mn}(\zeta,\zeta_0)ds\right\}$$
$$=Q_m(\zeta_0)\quad(m=1,2), \tag{11.63}$$

where

$$X_1=\mathrm{Re}\,\omega(\zeta),X_2=\mathrm{Im}\,\omega(\zeta),Y_1=\mathrm{Re}q(\zeta),Y_2=\mathrm{Im}q(\zeta),$$
$$dH_{11}(\zeta,\zeta_0)=\frac{1}{\pi}\mathrm{Im}\left\{\left[2+e^{2i\psi_0}\frac{\overline{\zeta}-\overline{\zeta}_0}{\zeta-\zeta_0}+\kappa e^{2i(\psi_0-\psi)}\right]\frac{d\zeta}{\zeta-\zeta_0}\right\},$$
$$dH_{12}(\zeta,\zeta_0)=\frac{1}{\pi}\mathrm{Re}\left\{\left[2+e^{2i\psi_0}\frac{\overline{\zeta}-\overline{\zeta}_0}{\zeta-\zeta_0}-\kappa e^{2i(\psi_0-\psi)}\right]\frac{d\zeta}{\zeta-\zeta_0}\right\},$$
$$dH_{21}(\zeta,\zeta_0)=-\frac{1}{\pi}\mathrm{Re}\left\{\left[e^{2i\psi_0}\frac{\overline{\zeta}-\overline{\zeta}_0}{\zeta-\zeta_0}+\kappa e^{2i(\psi_0-\psi)}\right]\frac{d\zeta}{\zeta-\zeta_0}\right\},$$
$$dH_{22}(\zeta,\zeta_0)=\frac{1}{\pi}\mathrm{Im}\left\{\left[e^{2i\psi_0}\frac{\overline{\zeta}-\overline{\zeta}_0}{\zeta-\zeta_0}-\kappa e^{2i(\psi_0-\psi)}\right]\frac{d\zeta}{\zeta-\zeta_0}\right\},$$
$$B_{11}(\zeta_0)=-2G\sin\psi_0+mq_0\left(2+\frac{1}{\mu_r}\right)\mathrm{Im}\left(F^-e^{2i\psi_0}\right)+mq_0\mu_r\,Im\left(F^+e^{2i\psi_0}\right),$$
$$B_{12}(\zeta_0)=-2G\cos\psi_0+mq_0\left(2+\frac{1}{\mu_r}\right)\mathrm{Re}\left(F^-e^{2i\psi_0}\right)+mq_0\mu_r\,Re\left(F^+e^{2i\psi_0}\right),$$
$$B_{21}(\zeta_0)=G\left(1+\frac{1}{\mu_r}\right)\cos\psi_0-2m\nu_*\,Re\left\{e^{i\psi_0}\left(F^+(\zeta_0)-\frac{1}{\mu_r}F^-(\zeta_0)\right)\right\}$$
$$\mathrm{Re}F^-(\zeta_0)+m\left(2+\frac{1}{\mu_r}\right)\mathrm{Re}\left\{e^{i\psi_0}\left(\overline{F^+}(\zeta_0)--\frac{1}{\mu_r}\overline{F^-}(\zeta_0)\right)F^-(\zeta_0)\right\}$$
$$+m\mu_r\mathrm{Re}\left\{e^{i\psi_0}\left(\overline{F^+}(\zeta_0)--\frac{1}{\mu_r}\overline{F^-}(\zeta_0)\right)F^+(\zeta_0)\right\},$$
$$B_{22}(\zeta_0)=-G\left(1+\frac{1}{\mu_r}\right)\sin\psi_0-2m\nu_*\mathrm{Re}\left\{e^{i\psi_0}\left(F^+(\zeta_0)-\frac{1}{\mu_r}F^-(\zeta_0)\right)\right\}$$
$$\mathrm{Im}F^-(\zeta_0)-m\left(2+\frac{1}{\mu_r}\right)\mathrm{Im}\left\{e^{i\psi_0}\left(\overline{F^+}(\zeta_0)--\frac{1}{\mu_r}\overline{F^-}(\zeta_0)\right)F^-(\zeta_0)\right\}$$
$$-m\mu_r\mathrm{Im}\left\{e^{i\psi_0}\left(\overline{F^+}(\zeta_0)--\frac{1}{\mu_r}\overline{F^-}(\zeta_0)\right)F^+(\zeta_0)\right\},$$

$$dK_{11}(\zeta,\zeta_0) = \frac{1}{\pi}\mathrm{Re}\left\{mq_0e^{2i\psi_0}\left[\left(2+\frac{1}{\mu_r}\right)F^- - \mu_r F^+\right]\times\frac{d\zeta}{\zeta-\zeta_0}\right\},$$

$$dK_{12}(\zeta,\zeta_0) = \frac{1}{\pi}\mathrm{Im}\left\{mq_0e^{2i\psi_0}\left[-\left(2+\frac{1}{\mu_r}\right)F^- + \mu_r F^+\right]\times\frac{d\zeta}{\zeta-\zeta_0}\right\},$$

$$dK_{21}(\zeta,\zeta_0) = \frac{1}{\pi}\mathrm{Im}\left\{\left[G\left(1-\frac{1}{\mu_r}\right)e^{i\psi_0} + 2m\nu_*\overline{F}^-(\zeta_0)\times\right.\right.$$
$$\times\mathrm{Re}\left(e^{i\psi_0}\left(F^+(\zeta_0)-\frac{1}{\mu_r}F^-(\zeta_0)\right)\right) - m\left(2+\frac{1}{\mu_r}\right)e^{-i\psi_0}\left(\overline{F^+}(\zeta_0)-\right.$$
$$\left.-\frac{1}{\mu_r}\overline{F^-}(\zeta_0)\right)F^-(\zeta_0) + m\mu_r e^{i\psi_0}\left(\overline{F^+}(\zeta_0)-\frac{1}{\mu_r}\overline{F^-}(\zeta_0)\right)$$
$$\left.\left.\times F^+(\zeta_0)\right]\frac{d\zeta}{\zeta-\zeta_0}\right\},$$

$$dK_{22}(\zeta,\zeta_0) = \frac{1}{\pi}\mathrm{Re}\left\{\left[G\left(1-\frac{1}{\mu_r}\right)e^{i\psi_0} + 2m\nu_*\mathrm{Re}\left(e^{i\psi_0}\times\right.\right.\right.$$
$$\left.\left(F^+(\zeta_0)-\frac{1}{\mu_r}F^-(\zeta_0)\right)\right)\overline{F^-}(\zeta_0) - m\left(2+\frac{1}{\mu_r}\right)e^{i\psi_0}$$
$$\left.\left.\left(\overline{F^+}(\zeta_0)-\frac{1}{\mu_r}\overline{F^-}(\zeta_0)\right)F^-(\zeta_0)\right]\frac{d\zeta}{\zeta-\zeta_0}\right\},$$

$$N_1(\zeta_0) = 2\left\{N + m\mathrm{Re}A_1 - 2A + \mathrm{Re}\left(Be^{2i\psi_0}\right)\right\},$$
$$N_2(\zeta_0) = 2\left\{-T + m\mathrm{Im}A_1 + \mathrm{Im}\left(Be^{2i\psi_0}\right)\right\},$$
$$Q_1(\zeta_0) = -qT + mA_2 + q_0(\kappa+1)\mathrm{Im}\Phi(\zeta_0),$$
$$Q_2(\zeta_0) = \mathrm{Re}\left\{e^{i\psi_0}\left(F^+(\zeta_0)-\frac{1}{\mu_r}F^-(\zeta_0)\right)(N+iT)\right\}+$$
$$+mA_3 - \mathrm{Re}\left\{\left(F^+(\zeta_0)-\frac{1}{\mu_r}F^-(\zeta_0)\right)e^{i\psi_0}\times\left(2\Phi(\zeta_0)-(\kappa-1)\overline{\Phi}(\zeta_0)\right)\right\},$$
$$q_0 = \left(F^+(\zeta_0)-F^-(\zeta_0)\right)e^{i\psi_0}.$$

$\Phi(\zeta_0)$, $F^{\pm}(\zeta_0)$ are the corresponding limiting values of the indicated functions on the contour Γ.

The succession of calculations is the following. We first determine the undisturbed magnetic field. We apply the integral (11.36) and formulas (11.34), (11.37) to this effect. Then, we find the solution $\omega(\zeta)$ from system (11.62); after that the functions $\Phi(z)$ and $\Psi(z)$ are reconstucted, according to the formulas (11.61). Further, we solve system (11 63) and determine the functions $R(z)$ and $R^e(z)$, which makes possible the calculation of the fluctuations of the magnetic field in the body and the cavities. In the last stage, we find the complete voltages on the forces-free contour of the opening, according to formula (11.60).

As an example, let us consider a soft-iron ferromagnetic medium, weakened by a cavity with elliptical cross section described by $\xi_1 = r_1\cos\beta$, $\xi_2 = r_2\sin\beta$ $(0\le\beta\le 2\pi)$. Let that the stresses $\langle\sigma_{ij}\rangle = 0$ $(i,j=1,2)$, and that

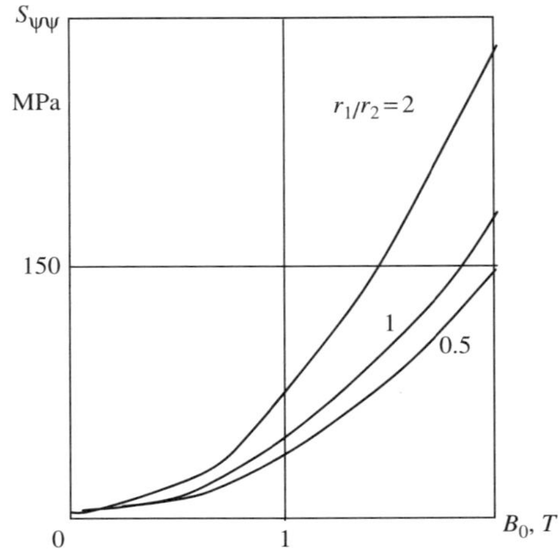

Fig. 11.8. Dependence of stress concentration on the contour of elliptical cavity for technical iron

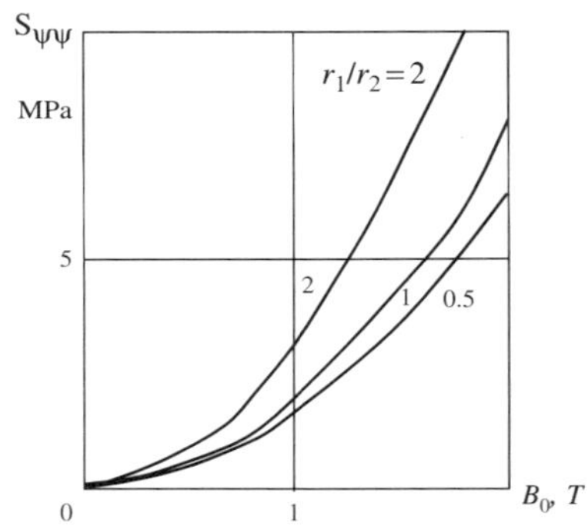

Fig. 11.9. Dependence of stress concentration on the contour of elliptical cavity for ferrite F-107

an initial magnetic field $\overrightarrow{B}_0 = (0, B_{02}, 0)$ occurs at infinity. The greatest stresses appear at the points $(\pm r_1, 0)$. The dependences of stresses $S_{\psi\psi}(r_1, 0)$ on the level of the initial magnetic field, for different ratios of the semi-axes of the ellipsis for technical iron ($\mu = 110$ GPa; $\mu_r = 2500$; $\nu = 0,28$) and ferrite F-107 ($\mu = 68$ GPa; $\mu_r = 110$; $\nu = 0,3$) are represented in Figs. 11.8 and 11.9, respectively.

12

Optimal Control of Physical Fields in Piezoelectric Bodies with Defects

In this chapter we study some optimal control inverse problems of variables such as stress intensity factor which govern the body behaviour. We derive the optimal boundary actions in concrete problems for the control of the characteristics of an electric field in piecewise-homogeneous piezoceramic bodies by minimizing the energy.

12.1 Optimization of Fracture Characteristics of Anisotropic Semi-Infinite Plate with Cracks

12.1.1 Direct Problem

Let us consider the anisotropic half-plane $y \geq 0$, weakened by crack-sections L_j $(j = 1, 2, \ldots, n)$. Let that on portions of the border of half-plane U and V, normal $P(x)$ and tangential $Q(x)$ distributed forces are applied respectively and the edges of the sections are free from forces. The direct problem of the theory of elasticity consists in the definition of the characteristics of the stress-strained state and parameters, which regulate the destruction of the body in the tip of the crack: the stress intensity factors, energy flow on the crack tip, etc.

The problem is similar to that set above and was considered [270]. Its solution can be presented in the form

$$\begin{gathered} r_1(t) = \Omega_0(\beta)\left(1-\beta^2\right)^{-1/2} \quad (t = t(\beta), -1 \leq \beta \leq 1), \\ \Omega_0(\beta) = \int\limits_U P(x)\,\Omega_0^{(2)}(\beta, x)\,dx + \int\limits_V Q(x)\,\Omega_0^{(1)}(\beta, x)\,dx, \end{gathered} \tag{12.1}$$

where functions $\Omega_0^{(k)}(\beta,\ x)$ are determined by the system of equations, indicated in [270].

The stress intensity factors in the tips L_j on the basis of (12.1) are defined as linear functionals of the form

$$K_i^{(j)\pm} = \int_U P(x) f_i^{(j)}(\pm 1, x)\, dx + \int_V Q(x) g_i^{(j)}(\pm 1, x)\, dx, \tag{12.2}$$

$$f_i^{(j)}(\pm 1, x) = \mathrm{Im}\left\{ R_i \overline{\Omega_0^{(2)}(\pm 1, x)} \right\}, \quad g_i^{(j)}(\pm 1, x) = \mathrm{Im}\left\{ R_i \overline{\Omega_0^{(1)}(\pm 1, x)} \right\},$$

where

$$R_i = \pm 2\sqrt{\pi s'(\pm 1)}\,(\bar{\mu}_1 - \bar{\mu}_2)(\mu_2 - \bar{\mu}_2)^{-1}\, \overline{a_1(\psi_c)} a_2\left(\psi_c + 1/2\pi\delta_2^i\right),$$
$$\delta_2^1 = 0,\ \delta_2^2 = 1 \quad (i = 1, 2; j = 1, 2, \ldots, n).$$

The upper sign corresponds to the crack tip $c = t(1) = b$, the lower one to $c = t(-1) = a$; the index j in the magnitudes $\Omega_0^{(i)}$ and R_i is omitted.

The energy flow on the tip of crack [196] is determined taking into account the asymptotic behavior of the fields of stresses and displacements in the neighborhood of the apex by the following quadratic functional (we henceforth use the repetitive indices summation convection) [269]:

$$G = A d_{km} x_k x_m \quad (k, m = 1, 2), \tag{12.3}$$
$$\alpha = \mu_1 - \bar{\mu}_2, \quad \beta = \mu_2 - \mu_1,$$
$$d_{11} = |\alpha|^2 + \mathrm{Re}(\alpha\beta),\, d_{22} = |\alpha|^2 - \mathrm{Re}(\alpha\beta),\ d_{12} = d_{21} = -\mathrm{Im}(\alpha\beta),$$
$$A = 1/4\pi a_{11} |\mu_2 - \mu_1|^2 (\mathrm{Im}\mu_2)^{-1},\ \mathrm{Im}\mu_1 > 0,\ \mathrm{Im}\mu_2 > 0,$$
$$x_m = \int_U P(x) r_m(\pm 1, x)\, dx + \int_V Q(x) h_m(\pm 1, x)\, dx,$$
$$r_1 + i r_2 = a_1(\psi_c)\, \Omega_0^{(2)}(\pm 1, x),\, h_1 + i h_2 = a_1(\psi_c)\, \Omega_0^{(1)}(\pm 1, x).$$

The quadratic form (12.3) is strictly positively defined. This follows from the positive sign of the principal minors $d_{11} = 2\mathrm{Im}\alpha\mathrm{Im}\mu_1 > 0,\ d_{11}d_{22} - d_{12}^2 = 4|\alpha|^2 \mathrm{Im}\mu_1\mathrm{Im}\mu_2 > 0$. It should be also noted that, as follows from (12.2), for a fixed crack can exist distributions that are different from zero (named $P(x)$ and $Q(x)$), for which the parameters of destruction in all tips of L_j cracks are equal to zero. In the sequel, such loads are excluded from our consideration.

12.1.2 Optimization

Under the influence on the construction of mechanical loads or generally of fields of different physical nature, all characteristics of the stress-strained state are functionals, defined on a permissible set of external interactions. This enables us to put and solve a number of elementary optimization problems, which illustrate the possibility of control by destruction due to external fields.

Let $\alpha_{ij}^{\pm} \geq 0$ are given. Let us consider the functional

$$J(P,Q) = \sum_{j=1}^{n} \left(\alpha_{ij}^{+} K_i^{(j)+} + \alpha_{ij}^{-} K_i^{(j)-} \right) \quad (i = 1, 2), \tag{12.4}$$

where $K_i^{(j)\pm}$ are determined in (12.2), $J : M \to R$, M is the space of piecewise-continuous functions.

Let us introduce the subspaces (A, B, C, D are given functions):
$M_Q = \{Q(x) \in M : 2C(x) \leq Q(x) \leq 2D(x),\ x \in V\}$.

Problem A_1

Find a point $(P^0, Q^0) \in M_P \times M_Q$, such that $|J(P,Q)| \leq |J(P^0, Q^0)|$, $\forall (P,Q) \in M_P \times M_Q$.

Solution. Taking into account (12.2), we write the functional $J(P,Q)$ in the form

$$J(P,Q) = \int_U f(x) P(x)\, dx + \int_V g(x) Q(x)\, dx, \tag{12.5}$$

where

$$f(x) = \sum_{j=1}^{n} \left(\alpha_{ij}^{+} f_i^{(j)}(1,x) + \alpha_{ij}^{-} f_i^{(j)}(-1,x) \right) \quad (i = 1, 2),$$

$$g(x) = \sum_{j=1}^{n} \left(\alpha_{ij}^{+} g_i^{(j)}(1,x) + \alpha_{ij}^{-} g_i^{(j)}(-1,x) \right) \quad (i = 1, 2).$$

Let us introduce the functions $P_1(x) = P(x) - (A+B)$, $|P_1(x)| \leq B - A = B_1(x)$,

$$Q_1(x) = Q(x) - (C+D), \quad |Q_1(x)| \leq D - C = D_1(x).$$

Evaluation of the left side of (12.5) gives

$$|J| \leq |\gamma| + \int_U B_1(x) |f(x)|\, dx + \int_V D_1(x) |g(x)|\, dx, \tag{12.6}$$

where

$$\gamma = \int_U (A+B) f(x)\, dx + \int_V (C+D) g(x)\, dx.$$

The exact equality in (12.6) is attained on the functions $P_1^0(x) = B_1(x)$ $\operatorname{sign}(\gamma f(x))$, $Q_1^0(x) = D_1(x) \operatorname{sign}(\gamma g(x))$. Consequently, the solution has the following form

$$P^0(x) = A(x) + B(x) + [B(x) - A(x)] \times \operatorname{sign}(\gamma f(x)) \quad (x \in U), \tag{12.7}$$
$$Q^0(x) = C(x) + D(x) + [D(x) - C(x)] \times \operatorname{sign}(\gamma g(x)) \quad (x \in V).$$

In case of symmetrical restrictions ($A = -B,\ C = -D$) in Chebyshev metric $|P(x)| \le 2B,\ |Q(x)| \le 2D$, we have $P^0(x) = 2B(x)\,\mathrm{sign} f(x)$, $Q^0(x) = 2D(x)\,\mathrm{sign}\, g(x)$. Analogous considerations also hold in other functional spaces.

Problem A_2

Maximize $|J(P, Q)|$ under the restrictions

$$\|P\|_{L^p(U)} \le \boldsymbol{I}_1, \quad \|Q\|_{L^{p'}(V)} \le \boldsymbol{I}_2, \quad (p, p' > 1)\,. \tag{12.8}$$

Solution. We apply the Holder inequality to the right side of the inequality $|J(P, Q)| \le \int\limits_U |f(x)|\,|P(x)|\,dx + \int\limits_V |g(x)|\,|Q(x)|\,dx$ [264]. We obtain

$$\begin{gathered} |J(P, Q)| \le \|f\|_{L^q(U)} I_1 + \|g\|_{L^{q'}(V)} I_2 \\ \left(p^{-1} + q^{-1} = 1, p'^{-1} + q'^{-1} = 1\right). \end{gathered} \tag{12.9}$$

We can easily confirm that the exact equality is attained for the functions

$$\begin{aligned} P^0(x) &= I_1 \|f\|_{L^q(U)}^{-q/p} |f|^{q-1} \,\mathrm{sign}\, f(x)\,, \\ Q^0(x) &= I_2 \|g\|_{L^{q'}(U)}^{-q'/p'} |g|^{q'-1} \,\mathrm{sign}\, g(x)\,. \end{aligned} \tag{12.10}$$

In this case restrictions (12.8) are satisfied. Consequently, functions (12.10) are extremal, and $\max |J(P, Q)| = \left|J\left(P^0, Q^0\right)\right|$ is determined by the right side of the inequality (12.9).

In the considered problems, the linear combination of the stress intensity factors on all cracks is maximized under definite restrictions on the control functions $P(x)$, $Q(x)$. Let us consider below two versions of dual problem, in which the "energetic" expenditures on control are minimized when the linear combination (12.4) is given.

Problem A_3

Find (the constants $\gamma,\ \delta,\ c$ are given):

$$\min_{(P,Q)\in(I_2)} \left[\gamma \int\limits_U P^2(x)\,dx + \delta \int\limits_V Q^2(x)\,dx\right] \quad (\gamma \ge 0, \delta \ge 0)\,, \tag{12.11}$$

where

$$\begin{gathered} (I_2) = \left\{(P, Q) : P \in L^2(U)\,, Q \in L^2(V)\,, \alpha + \beta = c\right\}, \\ \alpha = \int\limits_U P(x)\, f(x) dx, \\ \beta = \int\limits_V Q(x)\, g(x) dx. \end{gathered}$$

Solution. Let us analyze $L^2(U)$ ($L^2(V)$, respectively) into the orthogonal sum of one-dimensional subspace with basis $\{f\}$ ($\{g\}$, respectively) and its orthogonal complement. Then [124]

$$P(x) = af(x) + h, \quad Q(x) = bg(x) + h^*, \tag{12.12}$$

where $a\,||f|\,|^2 + b\,||g|\,|^2 = c, \quad (f, h) = (g, h^*) = 0,$

$$\mu(P, Q) = \gamma\,\|P\|^2 + \delta\,\|Q\|^2 = \gamma\left(a^2\,\|f\|^2 + \|h\|^2\right) +$$
$$+\delta\left(b^2\,\|g\|^2 + \|h^*\|^2\right) \geq \gamma a^2\,\|f\|^2 + \delta b^2\,\|g\|^2 \quad (\|\cdot\| = \|\cdot\|_{L^2}).$$

Hence follows that $\min\limits_{(P,Q)\in(I_2)} \mu(P, Q) = \min \Phi(a, b)$, $\Phi(a, b) = \mu(af, bg)$, $a\,||f|\,|^2 + b\,||g|\,|^2 = c$.

Thus, it is necessary to find the minimum of the function with two variables $\Phi(a, b)$ under the restriction (12.12). The solution of this elementary problem takes the form

$$\min_{(P,Q)\in(I_2)} \mu(P, Q) = \mu\left(P^0, Q^0\right) = \gamma\delta c^2 D, \tag{12.13}$$
$$P^0(x) = c\delta Df(x) \quad (x \in U),$$
$$Q^0(x) = c\gamma Dg(x) \quad (x \in V),$$

where $D = \left(\delta\,||f|\,|^2 + \gamma\,||g|\,|^2\right)^{-1}$, $||f|\,| = ||f|\,|_{L^2(U)}$, $||g|\,| = ||g|\,|_{L^2(V)}$.

Problem A_4

Find $\min\limits_{(P,Q)\in(I_\infty)} \{\max\left[\gamma \sup_U |P(x)|, \delta \sup_V |Q(x)|\right]\}$ $(\gamma, \delta \geq 0)$,

$$(I_\infty) = \{(P, Q) : P \in L^\infty(U),\, Q \in L^\infty(V),\, \alpha + \beta = c\},$$

where γ, δ, c are given constants.

Solution. The solution is done for the case $c > 0$. The magnitudes α and β are defined in (12.11); Holder inequality in this case gives

$|\alpha| \leq ||P|\,|_\infty\,||f|\,|_1$, $|\beta| \leq ||Q|\,|_\infty\,||g|\,|_1$ $(||\cdot|\,|_\infty = ||\cdot|\,|_{L^\infty}, ||\cdot|\,|_1 = ||\cdot|\,|_{L^1})$.

Consequently,

$$\nu(P, Q) = \max\{\gamma\,\|P\|_\infty, \delta\,\|Q\|_\infty\} \geq \max\{\gamma_1\,|\alpha|, \delta_1\,|c - \alpha|\} = \nu_1(\alpha), \tag{12.14}$$

where $\gamma = \gamma_1\,||f|\,|_1$, $\delta = \delta_1\,||g|\,|_1$.

Function $\nu_1(\alpha) \geq 0$ is piecewise-linear; therefore its minimum value must be attained in the point of inflection α^0, determined by the equality $\gamma_1\left|\alpha^0\right| = \delta_1\left|c - \alpha^0\right|$. Here, two cases must be considered: $\gamma_1 = \delta_1$ and $\gamma_1 \neq \delta_1$. After the appropriate analysis we come to the next result

$$P^0(x) = \alpha^0 \text{sign } f(x)/\|f\|_1, \tag{12.15}$$
$$Q^0(x) = \left(c - \alpha^0\right) \text{sign } g(x)/\|g\|_1,$$
$$\min_{(P,Q)\in(I_\infty)} \nu(P, Q) = \nu\left(P^0, Q^0\right) = c\gamma\delta\left(\delta\,\|f\|_1 + \gamma\,\|g\|_1\right)^{-1},$$

where $\alpha^0 = c\delta_1 (\delta_1 + \gamma_1)^{-1}$.

We considered above the linear combination of the stress intensity factors, that did not give the possibility to control the destruction conditions in the tip of each crack. Meanwhile, special interest shows the creation, for example, of such conditions, when crack is "locked" in the tip.

We denote $K_1^{(j)+} = \lambda_j$, $K_2^{(j)+} = \lambda_{n+j}$, $K_1^{(j)-} = \lambda_{2n+j}$, $K_2^{(j)-} = \lambda_{3n+j}$, $f_1^{(j)+} = f_j$, $f_2^{(j)+} = f_{n+j}$, $f_1^{(j)-} = f_{2n+j}$, $f_2^{(j)-} = f_{3n+j}$ $(j = 1, 2, \ldots, n)$ and assume $Q(x) = 0$. Then formulas (12.2) can be represented in the form of the equations of connection $\lambda_m = (P, f_m)$ $(m = 1, 2, \ldots, 4n)$. We assume that the system $\{f_m\}$ is linearly independent.

Problem B_1

Find $\min ||P||$ on the set $(I) = \{P : P \in L^2(U),\ (P, f_m) = \lambda_m,$ $m = 1, 2, \ldots, 4n\}$.

Its physical meaning is that we must find such a control $P(x)$, under which the stress intensity factors have given values λ_m. In this case the norm of control, which characterizes the energetic expenditures, must be minimum.

Solution. Let us analyze $L^2(U)$ into the orthogonal sum of finite-dimensional subspace with basis $\{f_m\}$ and its orthogonal complement. The following representation holds

$$P(x) = \sum_{n=1}^{4n} c_m f_m(x) + h, \quad (h, f_m) = 0. \tag{12.16}$$

Substituting (12.16) in the equations of connection, we come to the system of linear equations with respect to the coefficients c_m:

$$\sum_{m=1}^{4n} c_m (f_m, f_k) = \lambda_k \quad (k = 1, 2, \ldots, 4n). \tag{12.17}$$

This system has unique solution, since the Gram determinant is different from zero. Further, we have $\|P\|^2 = \|c_m f_m\|^2 + \|h\|^2 \to \min$ $(m = 1, 2, \ldots, 4n)$. Hence, we conclude that $h = 0$. Thus, the extremal element $P^0(x)$ is computed according to formula (12.16) with the coefficients c_m, defined from the system (12.7).

Under conditions of complex loading it is needed to optimize not the stress intesity factors, but the energy flow into the tip of crack. Let us assume that $P(x) \equiv 0$ and that there is only one crack.

Problem C_1

Find $\max G(Q)$ on the set $(I) == \{Q : Q \in L^2(V),\ ||Q|| \le \Lambda\}$. Its physical meaning is that the maximization of energy flow into one of the tips of the crack under the indicated restrictions on control.

Solution. According to Kuhn -Takker theorem [289] at the saddle point (Q^0, λ^0) of Lagrange functional the following relations hold

$$F'(Q^0, \varphi) = 0, \quad \forall \varphi \in L^2(V), \quad \lambda^0 (\|Q^0\| - \Lambda) = 0, \quad \lambda^0 \ge 0,$$
$$F(Q) = -G(Q) + \lambda (\|Q\| - \Lambda).$$

Here $F'(Q,\varphi)$ is the derivative according to Gateaux.

Differentiating the functionals occurring in $F(Q)$, using (12.3) and the arbitrariness of φ, we come to the relation (summation over the repetitive indices)

$$d_{km}h_m(1,x)\,x_k - 1/2\lambda Q(x)\left(A\,\|Q\|\right)^{-1} = 0\,(k,m=1,2)\,. \tag{12.18}$$

Hence, follows the "energetic" equality

$$\max G = Ad_{km}x_k^0x_m^0 = 1/2\lambda^0\left\|Q^0\right\| \tag{12.19}$$

In virtue of the strict positive definiteness of the quadratic form (12.3), we conclude that $\lambda^0 > 0$ and, may be $\left|\left|Q^0\right|\right| = \Lambda$.

Equality (12.18) is equivalent to the integral equation with separated kernel

$$\begin{gathered}\int Q(\xi)g(x,\xi)\,d\xi - 1/2\lambda Q(x)/(A\Lambda) = 0,\\ g(x,\xi) = d_{km}h_m(1,x)\,h_k(1,\xi)\quad(k,m=1,2)\,,\end{gathered} \tag{12.20}$$

the solution of which is elementary.

Eigenfunction $Q^0(x)$, which corresponds to the positive characteristic number λ^0, is fixed by restriction $\left\|Q^0\right\| = \Lambda$.

The problems of the families A_i, B_i, C_i can be generalized in different directions, bringing the statements closer to the actual conditions regulating the elements of the constructions.

Let us examine the semi-infinite plate from the boron-epoxy composite (BKM) with parameters $\mu_1 = 0,62i$, $\mu_2 = 5,12i$, weakened by arched crack $\xi_1 = R\cos\left[0,5\,(1+\beta)\,\varphi\right]$, $\xi_2 = \delta + R\sin\left[0,5\,(1+\beta)\,\varphi\right]$, $-1 \le \beta \le 1$. In the first case the coefficient of intensity K_1^- is maximized, in the second K_2^- (on the lower tip of crack). In this case, control is accomplished only by a function $P(x)$, $(|P(x)| \le I_1,\ Q(x) \equiv 0)$, or a function $Q(x)$, $(|Q(x)| \le I_2,\ P(x) \equiv 0)$.

If $R = 1$, $\delta = 0,1$, $\varphi = \pi/3$ and $U,V = [0,2]$, we obtain, in the first case, $P(x) = I_1\ (0 \le x \le 0,9)$, $P(x) = -I_1\ (0,9 < x \le 2)$, $Q(x) = -I_2\ (0 \le x \le 0,3)$, $Q(x) = I_2\ (0,3 < x \le 2)$. In the second case, $P(x)$ has the same form, and $Q(x)$ is determined by formula $Q(x) = -I_2\ (0 \le x \le 0,1;\ 1,6 \le x \le 2)$, $Q(x) = I_2\ (0,1 < x < 1,6)$. Here the solution of the problem A_1 was used, when $n = 1$ and for the corresponding values of the coefficients $\alpha_{ij}^{\pm}$.

The following example is related to problem B_1. Let that in a half-plane from BKM a horizontal crack of length $2l = 2$ exists at distance $\delta = 2$ from the border. It is required, making use of the control $P(x)$ with minimum norm, in the first case to "lock" the crack at the left end $\left(K_1^- = K_2^- = 0\right)$, and on the right end to obtain $K_1^+ = 0,5$, $K_2^+ = 0$; in the second case, $K_1^+ = K_1^- = 1$, $K_2^{\pm} = 0$. The results are shown in Fig. 12.1. The upper diagram corresponds to the first case, while the lower one to the second case.

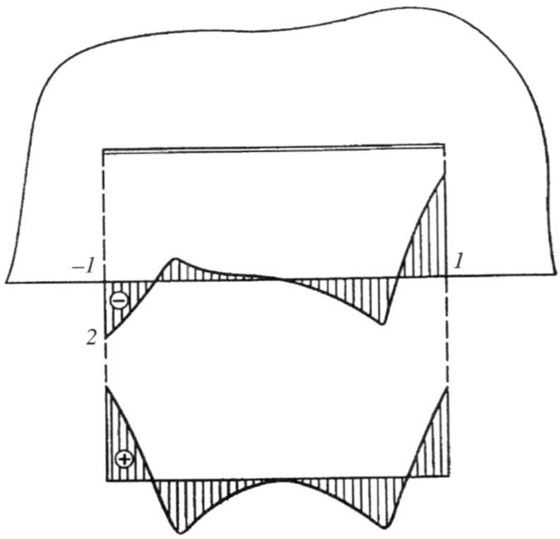

Fig. 12.1. The examples under study

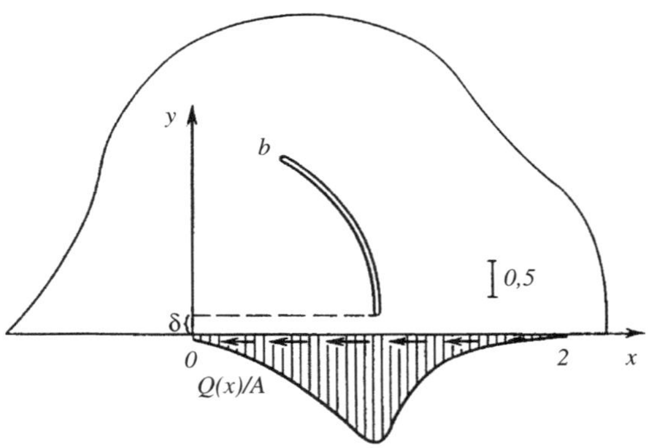

Fig. 12.2. The results for the third example

Finally, the third example is related to problem C_1. In the half-plane from BKM with arched crack, to maximize the energy flow on the tip of crack b making use of $Q(x)$, moreover, $\|Q\| \leq \Lambda$. The results are given in Fig. 12.2.

The obtained solutions illustrate the effectiveness of control by the parameters of destruction due to the external fields.

12.2 Statement of Certain Optimization Problems

In the previous chapters we considered a primal problem in electroelasticity, namely the determination of a critical parameter of fracture: the stress intensity factor. The solutions of extremal (inverse) problems in electroelasticity can be considerably simplified due to the fact that the parameters of fracture can be written as functionals, which are determined by the solutions of integral equations of the corresponding primal problems.

Problem 1

Let in a piezoelectric halfspace a single crack, and on some portion Λ of the boundary of the halfspace a distributed loading $q(x_1,t) = \mathrm{Re}\left\{Q(x_1)e^{-i\omega t}\right\}$ acting (either shear forces, or electric charges). Then, using the results of Sect. 4.3, the stress intensity factor taking into account (4.51) may be represented in the form

$$K_{III}^{\pm} = \pm c_{44}^{E}\sqrt{\frac{\pi}{s'(\pm 1)}}\left|\Omega_0(\pm 1)\right|\cos\left(\omega t - \arg\Omega_0(\pm 1)\right),$$
$$\Omega_0(\pm 1) = \int\limits_{\Lambda} Q(x_1)\Omega_1(\pm 1, x_1)\,dx_1. \tag{12.21}$$

Here, $\Omega_1(\delta, x_1)$ is the solution of integrodifferential (4.49) using as right part (4.61) at $P_0 = Q_0 = 1$ (i.e. the "*standard*" solution corresponding to the concentrated force or to the charge of a unit amplitude).

We consider the following problem: find a control law bounded by Chebyshev metrics $Q(x_1)$, $|Q(x_1)| \leq B(x_1)$, such that the quantity $|\Omega_0|$ is maximized, with respect to any tip of the crack (for example in tip b), i.e.

$$|\Omega_0(1)| \to \max, \qquad |Q(x_1)| \leq B(x_1), \qquad B(x_1) > 0. \tag{12.22}$$

Solution of the simplest problem under symmetrical restrictions on the control norm

$$Q(x_1) = B(x_1)e^{-i\arg\Omega_1(1,x_1)}. \tag{12.23}$$

Where

$$\max|\Omega_0(1)| = \int\limits_{\Lambda} B(x_1)\left|\Omega_1(1,x_1)\right|dx_1. \tag{12.24}$$

Problem 2

In the previous problem, as control we took a function characterizing tension distribution on boundary $x_2 = 0$ -. Here we consider dynamic loading, and the control law is written as a function of the change of loading with time (evolutional control).

Let a concentrated loading of form $q(x_1,t) = Q(t)\delta(x_1 - x_{10})$ acting on the free from forces boundary of the piezoelectric halfspace with an inner crack. Let a time instance $t = T$ and a function $c(t) > 0$. We seek to maximize $|K_{III}|$ with respect to one of the tips of the crack at $t = T$ at the expense of function $Q(t)$ under the restriction $|Q(t)| \leq c(t)$.

According to (4.71) appearing in Sect. 4.6, the above can be written as

$$|K_{III}(T)| = \pm A\frac{1}{\sqrt{s'(\pm 1)}}\left|\int\limits_{-\infty}^{\infty}\widehat{Q}(\omega)\Omega_0^*(1,\omega)e^{-i\omega T}d\omega\right| \to \max, \tag{12.25}$$
$$|Q(t)| \leq c(t), \qquad c(t) > 0.$$

Here $A = 1$, if by $q(x_1, t)$ we imply concentrated force, and $A = c_{44}^E / e_{15}$, if $q(x_1, t)$ are interpreted as a concentrated charge.

To solve the above problem we use $\hat{Q}(\omega)$ (the Fourier transform of control $Q(t)$) and substitute it into formula (12.25) to get

$$|K_{III}(T)| = \frac{A}{2\sqrt{\pi s'(1)}} \left| \int_{-\infty}^{\infty} Q(t) R(T-t) \, dt \right|, \tag{12.26}$$

$$R(T-t) = \int_{-\infty}^{\infty} e^{-i\omega(T-t)} \Omega_0^* (1, \omega) \, d\omega.$$

According to the results presented in Sect. 4.6 we get $\operatorname{Im} R(T-t) = 0$. Therefore, the maximum value of (12.26) takes place on the element

$$Q(t) = c(t) \, sign R(T-t) \tag{12.27}$$

and,

$$\max |\mathrm{K_{III}(T)}| = \frac{A}{\sqrt{\pi s'(1)}} \left| \int_{-\infty}^{\infty} c(t) R(T-t) dt \right| \tag{12.28}$$

The indicated optimization problem of the stress intensity factor was considered under symmetric restrictions on the control loading. The above results may be somewhat generalized if we remove the restriction of the symmetry.

Problem 3

$$|K_{III}(T)| \to \max, \qquad A(t) \le Q(t) \le B(t). \tag{12.29}$$

Here $A(t)$, $B(t)$ are prescribed piecewise-continuous functions, the meaning of $K_{III}(T)$ and $Q(t)$ follows from the preliminary problem.

Introducing the change $Q^*(t) = Q(t) - [A(t) + B(t)]/2$, we get a problem with symmetric restrictions

$$|Q^*(t)| \le \frac{B(t) - A(t)}{2}.$$

Hence, the final solution reads

$$Q(t) = \frac{1}{2} \{A(t) + B(t) - [A(t) - B(t)] \, sign\gamma(T) \, sign R(T-t)\},$$

$$\gamma(T) = \frac{1}{2} \int_{-\infty}^{\infty} [A(t) + B(t)] R(T-t) \, dt. \tag{12.30}$$

and

$$\max |K_{III}(T)| = \frac{A}{\sqrt{\pi s'(1)}} \left\{ |\gamma| + \frac{1}{2} \int_{-\infty}^{\infty} [B(t) - A(t)] |R(T-t)| \, dt \right\}. \tag{12.31}$$

To fulfill the initial conditions the functions $A(t)$, $B(t)$ should be replaced by $A(t)\eta(t)$ and $B(t)\eta(t)$, where $\eta(t)$ is the unit Heavyside function.

Up to now we considered the problem of the maximization of the stress intensity factor on any tip of one of the cracks. In this case quantity K_{III} cannot be controlled on other tips. Do deal with that we consider the following optimal control problem:

Problem 4

Let k-th cracks appearing in a half-space piezoelectric, which is bounded by vacuum and is free from forces everywhere except on some part of the boundary Λ. A distributed loading $q(x_1,t) = \operatorname{Re}\{Q(x_1)\exp(-i\omega t)\}$ acts on Λ.

Here we assume that on the tips of all cracks the stress intensity factors K_{III} and the amplitudes of mechanical and electric quantities at characteristic points of a body have prescribed values at minimum "*energetic*" expenses on control: $\|Q(x_1)\|_{L^2(\Lambda)} \to \min$ (L^2 is the complex Hilbert space [61] on set Λ).

Due to (12.21) the factor K_{III} for the j-th crack reads

$$
\begin{aligned}
K_{III}^{(j)} &= \pm c_{44}^{E}\sqrt{\frac{\pi}{s'(\pm 1)}}\left|\Omega_0^{(j)}(\pm 1)\right|\cos\left(\omega t - \Omega_0^{(j)}(\pm 1)\right), \\
\Omega_0^{(j)}(\pm 1) &= \int_{\Lambda} Q(x_1)\Omega_1^{(j)}(\pm 1, x_1)\,dx_1 \qquad \left(j = \overline{1,k}\right).
\end{aligned} \tag{12.32}
$$

The amplitudes of stress and of the components of the electric field at the characteristic points z_j $\left(j = \overline{2k+1,n}\right)$ are calculated with the help of formulas (4.29) and (4.48) and can be represented in the following way

$$
\Psi(z_j) = \int_{\Lambda} Q(x_1)\boldsymbol{G}(x_1,z_j)\,dx_1, \tag{12.33}
$$

where $\Psi = \{S_{13}, S_{23}, E_{1^*}, E_{2^*}, D_{1^*}, D_{2^*}\}$, $G = \left\{G_m(x_1,z_j),\ m = \overline{1,6}\right\}$ is the vector, the components of which are determined from the solution of a direct problem.

The above can be written as

$$
\|Q(x_1)\|_{L^2(\Lambda)} = \left\{\int_{\Lambda} |Q(x_1)|^2\,dx_1\right\}^{1/2} \to \min, \tag{12.34}
$$

under the restriction of the equality types

$$
\int_{\Lambda} Q(x_1)H_j(x_1)\,dx_1 = \alpha_j \quad \left(j = \overline{1,n}\right). \tag{12.35}
$$

Here

$$
H_j(x_1) = \begin{cases}
\Omega_1^{(j)}(-1,x_1), & j = \overline{1,k} \\
\Omega_1^{(j)}(1,x_1), & j = \overline{k+1,2k} \\
G_m(x_1,z_j), & j = \overline{2k+1,6n}
\end{cases}
$$

The values of the quantities α_j appearing in (12.35) are pre-assigned.

To solve the above problem we introduce the subspace M with the basis $\{\bar{H}_j(x_1)\}_{j=1}^{n}$ into L^2. Each element $Q \in L^2$ is represented in the form

$$p = \sum_{m=1}^{n} c_m \bar{H}_m + h^{\perp}, \tag{12.36}$$

where $h^{\perp}$ belongs to the orthogonal addition M in relation to L^2. The constants c_m are uniquely determined from equations of connection (12.35). We have

$$\sum_{m=1}^{n} (H_j, H_m) c_m = \alpha_j \qquad (j = \overline{1,n}) . \tag{12.37}$$

The determinant of system (12.37) (Gramm's determinant) differs from zero, therefore is has a unique solution.

Due to the equalities

$$\|Q\|^2 = \left\| \sum_{j=1}^{n} c_j \bar{H}_j \right\|^2 + \|h^{\perp}\|^2$$

and to the arbitrary selection of the element $h^{\perp} = 0$, $h^{\perp} = 0$ (from condition (12.34)).

Thus, the extremal element

$$Q = \sum_{j=1}^{n} c_j \bar{H}_j, \tag{12.38}$$

where c_j are constants, determined from (12.37).

12.3 Control of the Parameters of Fracture in Piezoceramic Half-Plane with Cracks

Let us consider a piezoceramic half-plane Ox_1x_3 $(x_3 \geq 0)$, polarized along x_3 and weakened by a crack L. We assume that the half-plane borders with the vacuum and that in a certain portion of the boundary a distributed load $\{P(x), Q(x)\}$ or a distributed electric charge $\rho(x)$ is given. The remaining part of the boundary is free from forces.

The stress intensity factors are represented in the form [51, 282]

$$K_i^{\pm} = \int_U P(x) f_i(\pm 1, x)\, dx \quad (i = 1, 2, 3), \tag{12.39}$$

$$f_i(\pm 1, x) = \operatorname{Im} \sum_{k=1}^{3} R_i \Omega_k^0 (\pm 1, x),$$

where

$$R_1 = \mu\sqrt{\pi s'(\pm 1)}a_k(\psi)\gamma_k,$$
$$R_2 = \pm\sqrt{\pi s'(\pm 1)}a_k(\psi)(\mu_k \sin\psi + \cos\psi)\gamma_k,$$
$$R_3 = \mu\sqrt{\pi s'(\pm 1)}a_k(\psi)r_k.\mu$$

The magnitudes γ_k, r_k are given, and functions $\Omega_k^0(\pm 1, x)$ are found as a result of the solution of the corresponding direct problem for a concentrated force (or charge), applied at point $x \in U$.

The schemes for the solution of the problems considered below are indicated in Sect. 12.1. Let us point out here only the final results.

Figures 12.3 and 12.4 provide the distributions of the controlling load $P(x)$, whenever $|K_I^-| \to \max$ or $|K_{II}^-| \to \max$, respectively, for straight crack ($x_1 = \beta,\ x_3 = 1/2,\ -1 \le \beta \le 1$). Figure 12.5 shows the distribution of the controlling load $Q(x)$ in $[-1, 1]$ for vertical crack, whenever $|K_{II}^-| \to \max$.

For second example let us take the solution of the problem A_1 for the case, when on the forces-free and non-electrodized boundary of the half-plane controlling function is sought in the form of distributed electric charge.

Let $A(x) = 0$, $B(x)$ take the form indicated in Fig. 12.6, $I(p) = K_I^-$, and the crack is a horizontal segment $x_1 = \beta$, $x_3 = 1/2$. Figures 12.6 and 12.7 show the charge distribution for the case $|K_I^-| \to \max$.

Finally, Fig. 12.8 give the distributions of the controlling loads $P(x)$ and $Q(x)$, respectively, for the problem B_1. Figure 12.8 corresponds to the case $\gamma_3^+ = \gamma_3^- = 1$, $\gamma_1^\pm = \gamma_2^\pm = 0$; Fig. 12.9 to the case $\gamma_2^\pm = \gamma_3^\pm = 0$, $\gamma_1^+ = \gamma_1^- = 1$. In this case,

$$\gamma_3^\pm = \lambda_3^\pm\sqrt{\pi l}Pd_{33}, \quad \gamma_1^\pm = \lambda_1^\pm\sqrt{\pi l}P. \tag{12.40}$$

The broken lines correspond to vertical, while the continual to horizontal crack.

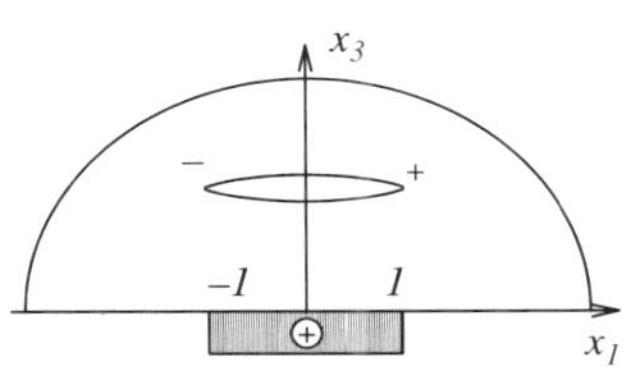

Fig. 12.3. The distribution of the control load

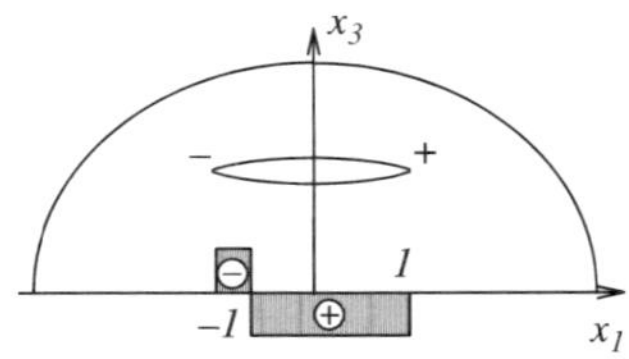

Fig. 12.4. The distribution of the control load

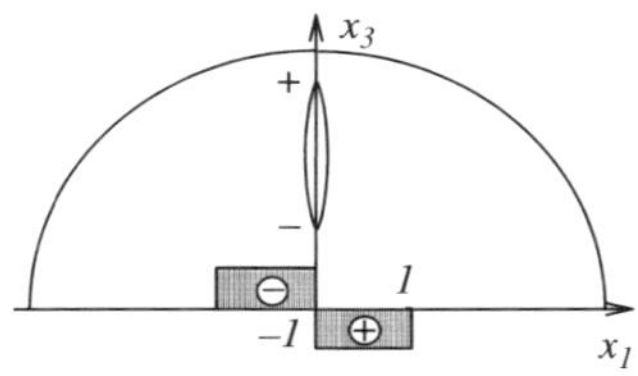

Fig. 12.5. The control load for a vertical crack

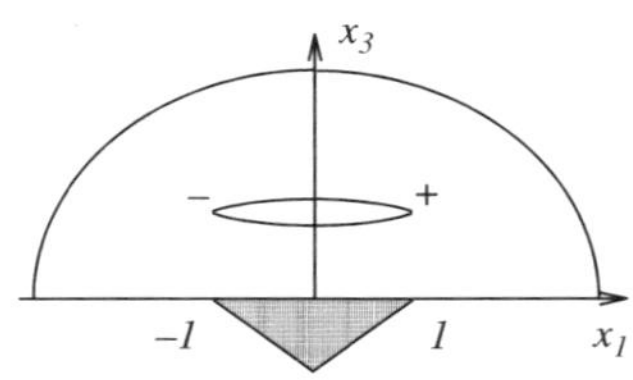

Fig. 12.6. The charge distribution for the problem

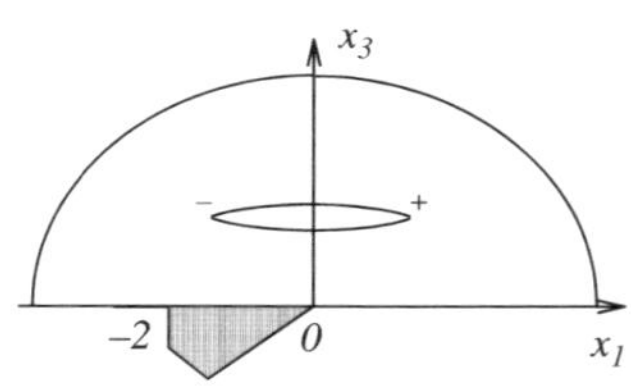

Fig. 12.7. The charge distribution for the problem

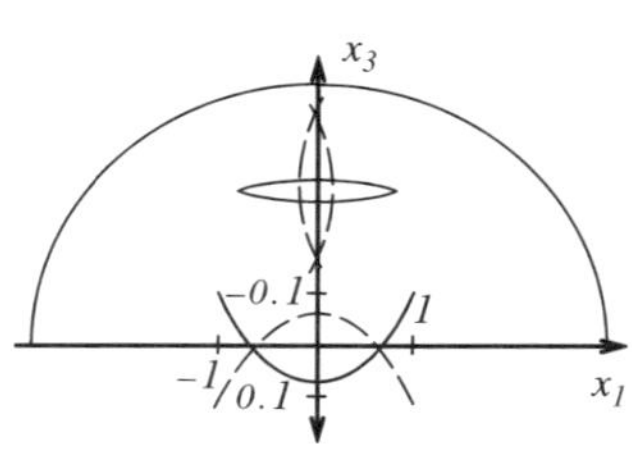

Fig. 12.8. The control loads for problem B_1

12.4 Control of the Stress Intensity Factor in a Bimorph with an Interphase Crack

Consider a piezoceramic bimorph with an interphase crack (Fig. 12.9). A static force $P_m(\delta)$ $(m = 1, 3)$ or electric charges $\rho(\delta)$ act along the parabolic arc in the upper halfspace. The parametric equations of the parabola have the form $x_1 = p_1\delta,\ x_3 = d + p\delta^2,\ \delta \in [-1, 1]$.

According to the results of Sect. 3.6, the stress intensity factors on the tips of an interphase crack are given by

$$K_j^{\pm} = \frac{\Lambda}{ah}\left\langle K_j^{\pm}\right\rangle \qquad (j = 1, 2) \tag{12.41}$$

where $\left\langle K_j^{\pm}\right\rangle$ are determined from (3.76), (3.77), and the quantity Λ is the amplitude of acting static forces or charges.

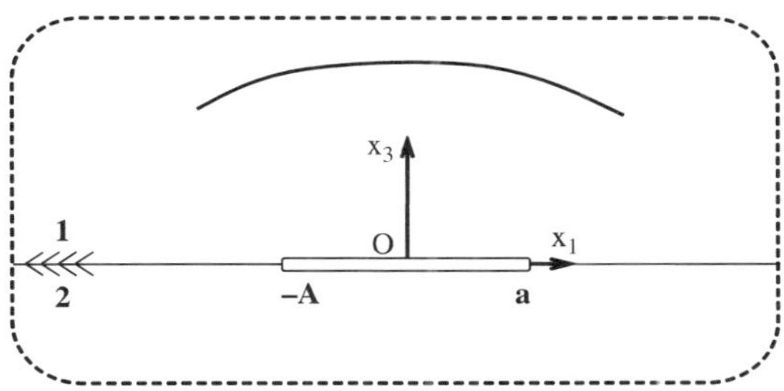

Fig. 12.9. The scheme of a bimorph with an interphase crack

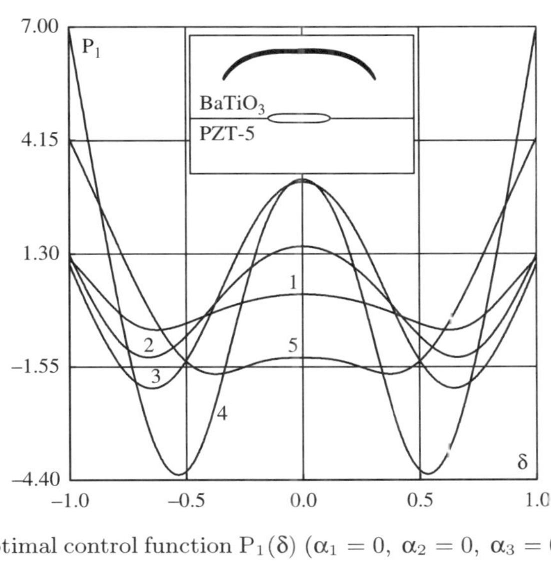

Fig. 12.10. Optimal control function $P_1(\delta)$ $(\alpha_1 = 0,\ \alpha_2 = 0,\ \alpha_3 = 0{,}01,\ \alpha_4 = 0{,}01)$

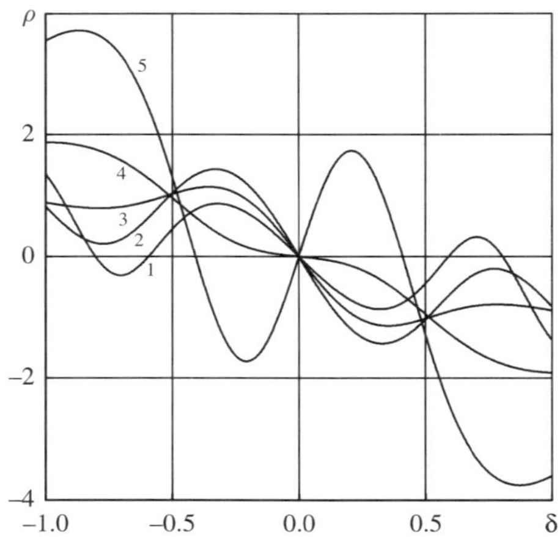

Fig. 12.11. Optimal control function p(δ) ($\alpha_1 = 0,\ \alpha_2 = 0,\ \alpha_3 = 0,01,\ \alpha_4 = 0,01$)

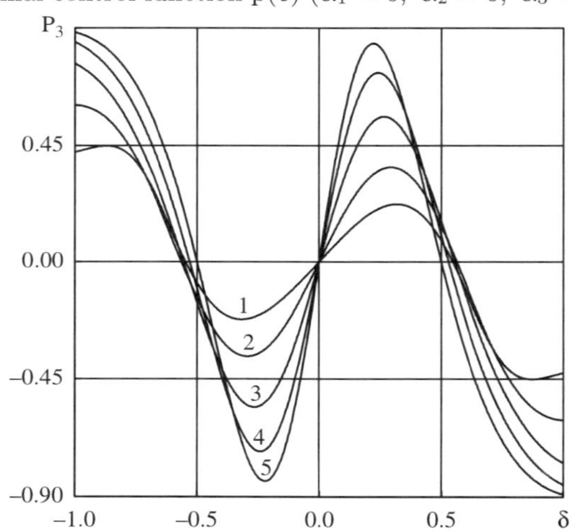

Fig. 12.12. Optimal control function $P_3(\delta)$ ($\alpha_1 = 0,\ \alpha_2 = 0,\ \alpha_3 = 0,01,\ \alpha_4 = 0,01$)

In the case where the loading is distributed along some arc L with intensity $\Lambda(\delta)$ (δ is the parameter),the factors $K_j^{\pm}$ (taking into account (12.41)) read

$$K_j^{\pm} = \frac{1}{ah} \int\limits_L \Lambda(\delta) \left\langle K_j^{\pm}(\delta) \right\rangle ds, \tag{12.42}$$

where ds is the element of the arc- length of contour L.

For the optimal control problem we assign the values of the quantities $K_j^{\pm}$ $(j = 1, 2)$, and we ask for the norm of the function $\Lambda(\delta)$ in Hilbert space L^2 to be minimal. According to the results presented in Sect. 12.2, the optimal control law reads

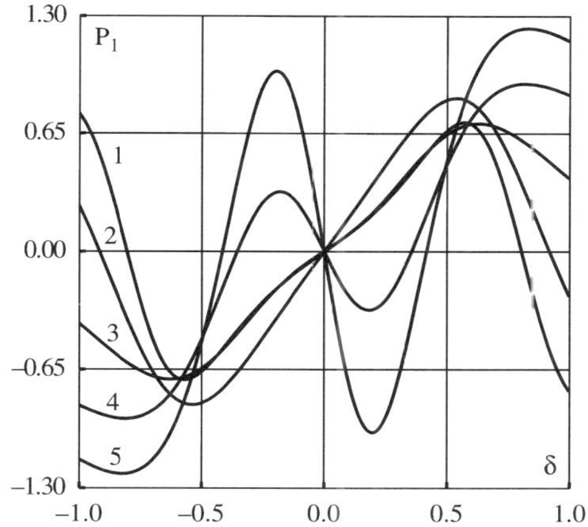

Fig. 12.13. Optimal control function $P_1(\delta)$ ($\alpha_1 = 0,01,\ \alpha_2 = 0,01,\ \alpha_3 = 0,\ \alpha_4 = 0$)

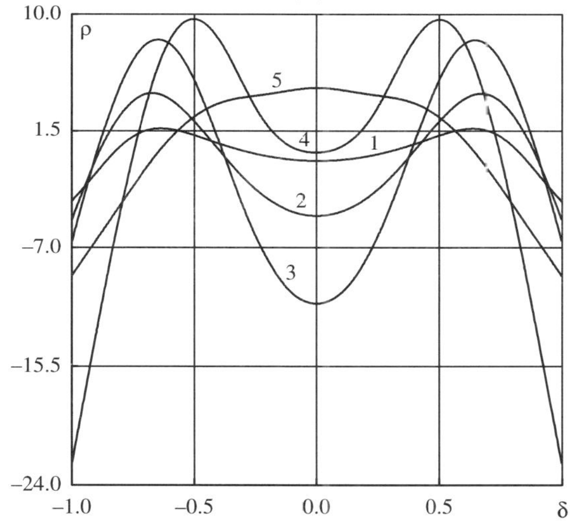

Fig. 12.14. Optimal control function $\rho(\delta)$ ($\alpha_1 = 0,01,\ \alpha_2 = 0,01,\ \alpha_3 = 0,\ \alpha_4 = 0$)

$$\Lambda(\delta) = \frac{s'(\delta)}{ah}\left(c_1\left\langle K_1^+\right\rangle + c_2\left\langle K_1^-\right\rangle + c_3\left\langle K_2^+\right\rangle + c_4\left\langle K_2^-\right\rangle\right) \tag{12.43}$$

The factors c_m $\left(m = \overline{1,4}\right)$ are determined from the system of linear algebraic equations, which is obtained as a result of substitution of expression (12.43) into the moment (12.42).

Figures 12.10–12.15 show the function of optimal control $\Lambda(\delta)$ for various types of loading; the curves with number m in Figs. 12.10–12.12 are constructed for $p_1/a = 3$, $d/a = 2$, $p/a = -1.8 + 0.9(m-1)$, and $a_1 = ahK_I^+ = 0$, $a_2 = ahK_I^- = 0$, $a_3 = ahK_{II}^- = 0.01$, $a_4 = ahK_{II}^- = 0.01$; in Figs. 12.13–12.15 for $a_1 = 0.01$, $a_2 = 0.01$, $a_3 = a_4 = 0$.

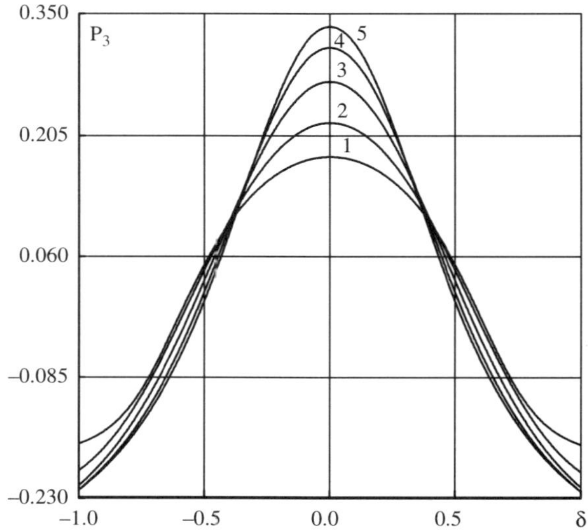

Fig. 12.15. Optimal control function $P_3(\delta)$ ($\alpha_1 = 0,01$, $\alpha_2 = 0,01$, $\alpha_3 = 0$, $\alpha_4 = 0$)

12.5 On the Application of the General Problem of Moments to Certain Optimization Problems of the Theory of Elasticity

12.5.1 Statement of the Problems

For subject of our study we take the elastic half-plane $y \geq 0$, weakened by stress concentrators (crack, openings, inclusions). On the border of the half-plane we distinguish the sets U and V, on which act normal $P(x)$ and tangential $Q(x)$ loads, respectively. Let us call these loads *controlling*. Under the term *objects of control* we would understand certain functionals, responsible for the stressed state and the strength of the body. This can be the stress concentration on the openings, the coefficients of intensity of stresses in the apexes of defects, the stressed state at some characteristic points or lines, the moments of stresses of different nature etc.

Depending on the nature of the optimization problem, the control (P, Q) must be selected from certain functional class.

We state the optimization problems for half-plane with stress concentrators [60]. We should note from the beginning that the problem of moments in arbitrary Banach space [264] is stated as problem of finding a linear continuous functional $\Phi \in B^*$, such that

$$\Phi(\varphi_k) = c_k \quad (k = 1, 2, \ldots, n), \quad \|\Phi\|_{B^*} \to \min, \tag{12.44}$$

where $\varphi_k \in B$ and the constants c_k are given.

It is known that this problem is always solvable (Hahn-Banach theorem [264]) for finite n; however, no effective procedure is known for finding the extremal functional in the general case.

If the functional is required to be non-negative (i.e. the functional is required to assume non-negative values on the intersection of certain cone in B and the linear closure of elements φ_k), then we come to the classical problem of moments [7]. In this case, obviously, the existence of functional $\Phi: \ \Phi(\varphi_k) = c_k \ (k = 1, 2, \ldots, n)$ cannot be guaranteed; so it is required for the concrete space B and the concrete set of elements $\{\varphi_k\}_{k=1}^{n}$ to find a criterion of solvability. Sometimes, the necessary assertions are deducible from the general theorems of solvability [7, 89], but usually they are independent part of the theory. The problem (12.44) is called *the l-problem of moments*, while the problem of seeking for a non-negative functional Φ, *the classical problem of moments.*

A more complicated version of the problem of moments is obtained by replacement of the exact equalities (12.44) by the following condition: if C is certain n-dimensional parallelepiped, then it is required that $\{\Phi(\varphi_k)\}_{k=1}^{n} \in C$. In particular, $|\Phi(\varphi_k) - c_k| \leq \varepsilon_k \ (k = 1, 2, \ldots, n)$.

Let us assume that $B = C(U)$ is the space of continuous functions, given on the compact set $U \subset R$. Riesz theorem [261] gives the general form of the linear continuous functional Φ on $C(U)$

$$\Phi(\varphi) = \int\limits_U \varphi(x)\, d\sigma(x), \quad \|\Phi\|_{C^*} = \underset{x \in U}{var}\, \sigma(x), \tag{12.45}$$

and for positive functional (the cone consists of all non-negative on U functions) the requirement is added that $\sigma(x)$ is a function not decreasing on U. In this case, since the number of moment functions ($n < +\infty$) is finite, the step-functions are optimal, in other words (in the sense of the generalized functions):

$$\sigma(x) = \sum_{j=1}^{m} \rho_j \delta(x - x_j) \quad (m < +\infty, x_j \in U). \tag{12.46}$$

In the classical problem of moments condition $\rho_j > 0$ is added and one can additionally require that the solution (certainly, in the case that the problem of moments is solvable) had minimum total variation $\sum \rho_j$, which has a physical sense.

Let us state the problem about concentrated boundary control, when $P(x) = \sigma'(x) \ (x \in U), \ Q(x) = \tau'(x) \ (x \in V)$. In the form of problems of moments, problem 1 (the classical problem of moments) is formulated as follows:

$$\int\limits_U \varphi_k(x)\, d\sigma(x) + \int\limits_V \psi_k(x)\, d\tau(x) = c_k \quad \left(k = \overline{1, n}\right), \tag{12.47}$$

where $d\sigma(x) \geq 0, \qquad d\tau(x) \geq 0, \ \text{var}\, \sigma(x) + \text{var}\, \tau(x) \to \min.$

Problem 1, consists, in fact, in the search for concentrated boundary stresses of one type ($\rho_j > 0$), which lead to given stressed state in the controlled domains. Problem 2 (the l-problem of moments) is formulated as follows:

$$d\sigma(x) \gtrless 0, \qquad d\tau(x) \gtrless 0, \quad \text{var}\, \sigma(x) + \text{var}\, \tau(x) \to \min. \tag{12.48}$$

In the sequel, we put $d\tau(x) \equiv 0$. Actually, it is not difficult to show that the general problem ($d\sigma(x) \equiv 0$, $d\tau(x) \equiv 0$) is equivalent to the problem of moments on the set $U \cup V$ relative to moment functions f_k : $f_k(x) := \varphi_k(x) \ (x \in U - V)$, $f_k(x) := \varphi_k(x) + \psi_k(x) \ (x \in U \cap V)$, $f_k(x) := \psi_k(x) \ (x \in V - U)$. Furthermore, in the case of solvability of this new problem of moments we have freedom in the choice of functions $\sigma(x)$, $\tau(x)$ on the set $U \cap V$ with fixed sum $\sigma(x) + \tau(x)$.

12.5.2 Approximational Approach

For the above-indicated class of problems of mechanics the moment functions $\varphi_k(x)$ are linear combinations of certain standard solutions of integral equations of the corresponding boundary-value problems [59] and can be found numerically on the system of points U' from U. In order to keep on the analytical approach to the problems of control, we can approximate the moment functions $\varphi_k(x)$ on U' in other one or another way, for example, by ordinary polynomials.

Let that the degree of the approximating function $\varphi_k(x)$ of the polynomial $R_k(x)$ is designated by n_k. The polynomials with minimal deviation from $\varphi_k(x)$ in Chebyshev metric, i.e. in the space $C(U')$ [261] are the best, from the view point of minimization of the absolute error in approximation. Further, we put

$$s_\nu = \int_U x^\nu d\sigma(x), \tag{12.49}$$

where s_ν is the ν-th degree moment of the distribution $\sigma(x)$.

The relations

$$\int_U \varphi_k(x)\, d\sigma(x) = c_k \quad \left(k = \overline{1,n}\right) \tag{12.50}$$

take the form of linear connections for exponential moments

$$\sum_{\nu=0}^{n_k} a_{\nu k} s_\nu = c_k \quad \left(k = \overline{1,n}\right). \tag{12.51}$$

Thus, disregarding the error in the approximation, the original problem is equivalent to the following one: to find the conditions for numbers c_k, under which a non-decreasing function bounded by the variation $\sigma(x)$ exists (the function of load distribution on the border of the half-plane) with moments s_ν, which satisfy the system of (12.51).

If we seek for approximating polynomials $R_k(x)$ such that $\max\{n_k\} = N$ would coincide with the number of equations in (12.51), then the exponential moments s_ν are uniquely determined and the problem is reduced to the determination of the necessary and sufficient conditions of solvability of the classical exponential problem of moments [7, 89] on the set U. Further investigation depends substantially on the form of the sets U, on which $\sigma(x)$ are given. If $U = (-\infty, \infty)$, then by the necessary and sufficient condition on the existence of measure $\sigma(x)$ with given moments

$$s_\nu = \int\limits_{-\infty}^{\infty} x^\nu d\sigma(x) \quad (\nu = 0, 1, 2, \ldots, N) \tag{12.52}$$

is the requirement that the quadratic form is not negative (Hankel form).

$$\sum_{\alpha,\beta=0}^{(N/2)} s_{\alpha+\beta}\xi_\alpha\xi_\beta. \tag{12.53}$$

In case of certain localization of the application of generalized forces on the border of the body we can use the results [89, 268]. Thus, if it $\sigma(x)$ is concentrated on certain $[a, b]$ and N is odd number, then necessary and sufficient condition of solvability of the problem is the two quadratic forms

$$\sum_{\alpha,\beta=0}^{(N-1)/2} (s_{\alpha+\beta+1} - a s_{\alpha+\beta})\,\xi_\alpha\xi_\beta, \qquad \sum_{\alpha,\beta=0}^{(N-1)/2} (b s_{\alpha+\beta} - s_{\alpha+\beta+1})\,\xi_\alpha\xi_\beta. \tag{12.54}$$

to be non-negative.

The requirement that the quadratic forms must be not negative appears as a result of application of the general M. Riesz criterion (the cone of the space B consists of non-negative continuous functions) and the representation of the non-negative on U polynomial. Thus, it is known that

$$\begin{gathered} P(x) \geq 0\,(x \in (-\infty, \infty)) \Leftrightarrow P(x) = [A(x)]^2 + [B(x)]^2, \\ 2\deg A(x) \leq \deg P(x), \quad 2\deg B(x) \leq \deg P(x). \end{gathered} \tag{12.55}$$

If $A(x) = \sum \xi_\alpha x^\alpha$, then $[A(x)]^2 = \sum \xi_\alpha\xi_\beta x^{\alpha+\beta}$ and $\Phi\left([A(x)]^2\right) = \sum s_{\alpha+\beta}\xi_\alpha\xi_\beta$.

If, for example, $U = [a, b]$, then for $\deg P(x) = N$ (where N is odd number) Lucas theorem holds [7]

$$P(x) \geq 0\,(x \in [a, b]) \Leftrightarrow P(x) = (x - a)[A(x)]^2 + (b - x)[B(x)]^2. \tag{12.56}$$

Hence follow the quadratic forms (12.54). The case for even N is analogous.

It is noteworthy that the requirement that the quadratic form must be non-negative is equivalent to the requirement that the principal minors of the matrix of the quadratic form must be non-negative. When the problem of moments is solvable a function $\sigma(x)$ exists; it has finite number of jumps ρ_k at the points $x_k \in U$:

$$P(x) = \sigma'(x) = \sum \rho_k \delta(x - x_k) \quad (\rho_k > 0) . \tag{12.57}$$

If U is the union of mutually disjoint segments $[a_\nu, b_\nu]$, then the situation is somewhat complicated, but the points of jumps x_k and the magnitudes of jumps ρ_k are found by a completely effective and sufficiently elementary method [268].

Thus, the program of solution of the stated problem consists of the following steps. Firstly, with the application of numerical methods we find the solution of the direct problem about the stressed state and the moment functions are formed. Secondly, the approximation of the moment functions is done by (in particular) polynomials.

On the grounds of the investigation of the exponential (in particular) problem of moments one can conclude about the solvability or non-solvability of the given stressed state in the controlled domain.

The approximational approach introduces a certain error into the values of the magnitudes of the controlled parameters $\{s_k\}_{k=1}^n$. Let us estimate this error. Let $\varphi_k(x) = R_k(x) + \varepsilon_k(x)$, $\varepsilon := \max\left\{\max|\varepsilon_k(x)| : x \in U, k = \overline{1,n}\right\}$, and $\sigma^*(x)$ a certain solution of the exponential problem of moments (12.49). Then, it is obvious that

$$\left| \int\limits_U \varphi_k(x)\, d\sigma^*(x) - s_k \right| \leq \varepsilon Var \sigma^*(x) = \varepsilon s_0 . \tag{12.58}$$

Consequently, the approximation of the moment functions and the introduced in this way error can change the original values of the parameters no more than on εs_0.

Let us consider in greater detail the optimization problem of the coefficients of intensity of stresses in anisotropic half-plane, weakened by internal crack L. Let that the border of the half-plane is forces-free, except the section U, on which acts the normal load $P(x)$, and the edges of the sections are free from forces. The function $P(x)$ is generalized function, particularly, $P(x)\,dx = d\sigma(x)$, where $\sigma(x)$ is the function of bounded variation.

In accordance with the aforesaid, let us consider two situations. In the first situation, $d\sigma(x) \geq 0$, whilein the second $d\sigma(x) \gtrless 0$. According to [269], the coefficients of intensity are linear functionals

$$K_I^{\pm} = \mu 2\sqrt{\pi s'(\pm 1)} \times \left\{ \int_U \operatorname{Im}\left[\bar{a}_1(\psi_c)\, a_2(\psi_c) \frac{\bar{\mu}_2 - \bar{\mu}_1}{\mu_2 - \bar{\mu}_2} \Omega_0(\pm 1, x) \right] d\sigma(x) \right\}, \tag{12.59}$$

$$a_k(\psi_c) = \mu_k \cos\psi_c - \sin\psi_c,\ \psi_c = \psi(\beta)|_{\beta=\pm 1}\ (k = 1, 2),$$

where U is the subset from $(-\infty, \infty)$, on which $\sigma(x)$ is defined; β is the parameter in the equation of the section $x = x(\beta)$, $y = y(\beta)$, $|\beta| \leq 1$. Values $\beta = \pm 1$ correspond to the ends of the crack, ψ is the angle between the positive normal to the left edge and the axis Ox_1. The stress intensity factor $K_{II}^{\pm}$ is given by formula (12.59), in which it was used $a_2(\psi_c + \pi/2)$ instead of $a_2(\psi_c)$.

Function $\Omega_0(\beta, x)$ is the solution of the integral equations system

$$\begin{aligned} &\int_{-1}^{1} \frac{\Omega_0(\beta, x) H(\beta, \beta_0) + \bar{\Omega}_0(\beta, x) H_*(\beta, \beta_0)}{\sqrt{1-\beta^2}} d\beta \\ &\quad + \int_{-1}^{1} \frac{\Omega_0(\beta, x)}{(\beta - \beta_0)\sqrt{1-\beta^2}} d\beta = N(\beta_0, x), \\ &\int_{-1}^{1} \frac{\Omega_0(\beta, x)}{\sqrt{1-\beta^2}} t_1'(\beta)\, d\beta = 0, \end{aligned} \tag{12.60}$$

where

$$\begin{aligned} H(\beta, \beta_0) &= \frac{1}{2} \frac{d}{d\beta} \left\{ \ln \frac{(\bar{t}_2 - \bar{t}_{20})(t_1 - t_{10})}{(\beta - \beta_0)^2 (t_1 - \bar{t}_{20})(\bar{t}_2 - t_{10})} \right. \\ &\quad \left. + \left| \frac{\mu_1 - \bar{\mu}_2}{\mu_1 - \mu_2} \right|^2 \ln \frac{(t_1 - \bar{t}_{20})(t_2 - \bar{t}_{10})}{(t_2 - \bar{t}_{20})(t_1 - \bar{t}_{10})} \right\} t_1'(\beta), \\ H_*(\beta, \beta_0) &= \frac{\bar{\mu}_1 - \mu_2}{2(\mu_1 - \mu_2)} \frac{d}{d\beta} \times \left\{ \ln \frac{(\bar{t}_2 - \bar{t}_{20})(\bar{t}_1 - t_{1C})(t_2 - \bar{t}_{10})}{(\bar{t}_1 - \bar{t}_{10})(\bar{t}_2 - t_{1C})(t_2 - \bar{t}_{20})} \right\} t_1'(\beta), \\ N(\beta_0, x) &= \frac{1}{4} \left\{ \frac{\mu_2}{\mu_1 - \mu_2} \frac{dt_{10}/d\beta_0}{t_{10} - x} - \frac{\bar{\mu}_1 (\mu_2 - \bar{\mu}_2)}{|\mu_1 - \mu_2|^2} \right. \\ &\quad \left. \times \frac{dt_{20}/d\beta_0}{\bar{t}_{20} - x} - \frac{\bar{\mu}_2 (\bar{\mu}_1 - \mu_2)}{|\mu_1 - \mu_2|^2} \frac{d\bar{t}_{10}/d\beta_0}{\bar{t}_{10} - x} \right\}. \end{aligned}$$

For the approximate-analytical solution of the system (12.39) we apply Gauss quadrature formulas with n nodes with respect to weight $\left(1 - \beta^2\right)^{-1/2}$.

We get

$$\frac{\pi}{n}\sum_{\nu=1}^{n}\frac{\Omega_0(\beta_\nu,x)}{\beta_\nu-\beta_0}+\frac{\pi}{n}\sum_{\nu=1}^{n}\left[H(\beta_\nu,\beta_0)\Omega_0(\beta_\nu,x)+H_*(\beta_\nu,\beta_0)\bar{\Omega}_0(\beta_\nu,x)\right]$$
$$=N(\beta_0,x), \tag{12.61}$$
$$\frac{\pi}{n}\sum_{\nu=1}^{n}\Omega_0(\beta_\nu,x)\,t_1'(\beta_\nu)=0,$$

where

$$\beta_\nu=\cos\theta_\nu,\quad \theta_\nu=\frac{2\nu-1}{2n}\pi.$$

We take the values $\beta_0=\beta_{0\mu}=\cos(\mu\pi/n)$ $(\mu=1,2,\ldots,n-1)$ for collocation points in (12.61). The obtained system of linear equations enables us to represent approximately the values $\Omega_0(\beta_\nu,x)$ in the form of linear combinations of functions $N(\beta_{0\mu},x)$.

The optimization problem can be formulated in the following way: let that the desired values of the stress intensity factors $K_{I,0}^{\pm}$, $K_{II,0}^{\pm}$ in the tips of crack L are given. In the general case the "domain" $\left|K_I^{\pm}-K_{I,0}^{\pm}\right|\le\varepsilon_1^{\pm}$, $\left|K_{II}^{\pm}-K_{II,0}^{\pm}\right|\le\varepsilon_2^{\pm}$ for the stress intensity factors $K_I^{\pm}$, $K_{II}^{\pm}$ are given. The external action $d\sigma(x)$, which enables to control the values of the coefficients of intensity, is found from the minimal condition of the total impulse $\mathrm{Var}\,\sigma(x)\to\min$.

Each of the problem of moments takes the form (we assume $\varepsilon_i^{\pm}=0$ for reasons of abbreviation)

$$\int_U\varphi_1^{\pm}(x)\,d\sigma(x)=K_I^{\pm},\quad \int_U\varphi_2^{\pm}(x)\,d\sigma(x)=K_{II}^{\pm}, \tag{12.62}$$

where $\varphi_1^{\pm}(x)=\mu 2\sqrt{\pi s'(\pm1)}\mathrm{Im}\left[\bar{a}_1(\psi_c)\,a_2(\psi_c)\,\frac{\mu_2-\bar{\mu}_1}{\mu_2-\bar{\mu}_2}\Omega_0(\pm1,x)\right]$,

$$\varphi_2^{\pm}(x)=\mu 2\sqrt{\pi s'(\pm1)}\mathrm{Im}\left[\bar{a}_1(\psi_c)\,a_2\left(\psi_c+\frac{\pi}{2}\right)\times\frac{\mu_2-\bar{\mu}_1}{\mu_2-\bar{\mu}_2}\Omega_0(\pm1,x)\right]$$

and $\Omega_0(\pm1,x)$ are calculated according to the Gauss method as the values of the interpolation polynomial

$$\Omega_0(\pm1,x)=\frac{1}{n}\sum_{\nu=1}^{n}(-1)^{\nu+r_\mp}\,\Omega_0(\beta_\nu,x)\,\mathrm{tg}\left(\frac{2\nu-1}{4n}\pi+\rho_\pm\frac{\pi}{2}\right), \tag{12.63}$$
$$r_+=1,\quad r_-=n,\quad \rho_+=\pi/2,\quad \rho_-=0.$$

The moment functions $\varphi_i^{\pm}(x)$ $(i=1,2)$ are approximated by polynomials of third degree

$$P_j^{\pm}(x)=\sum_{k=0}^{3}a_k(\pm1,j)\,x^k. \tag{12.64}$$

The more exact representation (12.64) is, the "less" set U is. Let us substitute for functions φ their approximate representation (12.64) and from the obtained system of linear equations we find the exponential moments $s_\nu = \int\limits_U x^\nu d\sigma(x)$ $(\nu = 0, 1, 2, 3)$.

If $U = [a, b]$, then we can decide about the solvability of problem 1 (version of the classical problem of moments) by using criterion (12.54), which is utterly constructive. The question about finding at least one extremal solution $\sigma(x)$ remains unanswered. We should note, first of all, that var $\sigma(x) = \int\limits_U d\sigma(x) = s_0$. This means that the approximation approach enables us to find immediately the optimal value of the total impulse Var $\sigma(x)$.

In order to find the values of pulses ρ_k and the points of their application $x_k \in U$ (see (12.57)), we can use the following algorithm [7, 89]. The sequence of moments $\{s_k\}_{k=0}^3$ is supplemented with the moment s_4, so that the sequence $\{s_k\}_{k=0}^4$ was singularly positive with respect to set U. After the determination of moment s_4 it is necessary to built in the ordinary way the system of orthogonal polynomials $P_j(x)$ of degree j $(j = 0, 1, 2)$ relative to the non-negative functional $I : I\left\{x^k\right\} = s_k$ $(k = 0, 1, \ldots, 4)$. In [7, 89] elementary evident formulas are given for the system of polynomials $P_j(x)$. Having chosen the number λ, so that among the roots of the polynomial $Q(x, \lambda) := P_2(x) + \lambda P_1(x)$ were the ends of the interval $U = [a, b]$ (in [89] in the terms of the index of representation (12.57) the overall situation is studied), and finding all the roots of the polynomial $Q(x, \lambda)$, we obtain the entire set of points $\{x_k\}$ from (12.57). The magnitudes of jumps are defined in that case as the values of the functional I on a specially constructed non-negative polynomial or as the solution of the system of linear equations $\sum \rho_i x_k^j = s_k$ $(k = 0, 1, 2, 3)$.

In problem 2, with the aid of the approximational approach we come to the exponential l–problem of moments, the complete algorithm of solution of which with the determination of the points x_k and the magnitudes ρ_k is exposed in [89].

Based on the example of the half-plane $y \geq 0$ we illustrate what we have exposed above [269]. Let that the internal crack is rectilinear, $|x| \leq 0,2,\ y = 1$, and the boron-epoxy half-plane has characteristics $\mu_1 = 5,12i,\ \mu_2 = 0,52i$. On a portion of the boundary $|x| \leq 0,1$, loads $d\sigma(x) \geq 0$ are allowed. We give the symmetrical values of the stress intensity factors K_{I}^+, $K_{\mathrm{I}}^- = -K_{\mathrm{I}}^+$, K_{II}^+, $K_{\mathrm{II}}^- = K_{\mathrm{II}}^+$.

Disregarding the very bulky values for the moment functions $\varphi_\nu^\pm(x)$, where

$$K_\nu^\pm = \int\limits_{-0,1}^{0,1} \varphi_\nu^\pm(x)\, d\sigma(x) \quad (\nu = 1, 2)\,; \varphi_1^-(x) = -\varphi_1^+(x)\,, \quad \varphi_2^-(x) = \varphi_2^+(x)\,, \tag{12.65}$$

we give their approximations: $R_1^+(x) = 0,3077 + 0,2470x - \ -0,8855x^2 + 0,5765x^3$; $R_2^+(x) = -0,1371 + 1,2887x + 1,1549x^2 + \ +0,5114x^3$ with the relative errors $\delta_1 = 0,15\%$, $\delta_2 = 0,4\%$, respectively. Applying the criterion of solvability of the exponential problem of moments on $[-0,1;\ 0,1]$, we obtain that for $K_1^{\pm} = \pm 0,3000$, $K_2^{\pm} = -0,1250$ there exists generalized load $d\sigma(x) = 0,5 \times [\delta(x+0,1) + \delta(x-0,1)]$. For $K_1^{\pm} = \pm 0,3034$, $K_2^{\pm} = -0,1313$ exists a generalized load $d\sigma(x) = 0,5 \times \times [(x - 0,0707) + \delta(x - 0,0707)]$, exists, whereas for $K_1^{\pm} = \pm 0,2192$, $K_2^{\pm} = -0,0216$ no loads $d\sigma(x) \geq 0$ exist in the portion $[-0,1;0,1]$

12.6 Pulse Boundary Control of the Stressed State of the Half-Plane

We study below the conditions for the existence of boundary control of the form $\sum \rho_\nu \delta(x - x_\nu)$ ($\delta(\cdot)$ – is the delta function, $\rho_\nu > 0$), which realize the possibility of accessibility of the given elastic mode in the half-plane $y > 0$. We find the necessary and sufficient conditions of solvability of the problem in the terms of nonnegativity of the quadratic forms. The solution is based on certain integral representations of special classes of analytic functions. We examine the examples [275].

12.6.1 Statement of the Problem

Let us consider the elastic anisotropic half-plane $x_2 > 0$, at the points of the set F of which are given the values of stresses σ_{11}, σ_{22}, σ_{12}. The problem about the choice of the normal and tangential forces $N(x)$, $T(x)$ $(x_1 \in E \subset R)$, on the border $x_2 = 0$ is posed, so that at the points of the set F the values $(\sigma_{11}, \sigma_{22}, \sigma_{12})$ would coincide with those given beforehand. Specifically, the necessary and sufficient conditions of solvability of this problem are needed.

R. Nevalinna functional classes [7, 89, 265]. The function $f(z)$ belongs to class R, if:

a) $f(z)$ is holomorphic in the upper half-plane;
b) $\mathrm{Im} f(z) \geq 0$, when $\mathrm{Im} z \geq 0$.

The following theorem gives the integral representation of the functions of class R.

Theorem A (R. Nevalinna)

A necessary and sufficient condition for the function $f(z)$ to belong to class $f(z)$ is to admit the following additive representation:

$$f(z) = \mu z + \nu + \frac{1}{\pi} \int_{-\infty}^{\infty} \left(\frac{1}{t - z} - \frac{t}{1 + t^2} \right) d\tau(t), \tag{12.66}$$

where μ, ν are real numbers; $\mu \geq 0$, $\tau(t)$ is non-decreasing function with convergent integral $\int_{-\infty}^{\infty} (1 + t^2)^{-1} d\tau(t)$, or that $f(z)$ admitsrepresentation

(with the same μ, ν)

$$f(z) = \mu z + \nu + \frac{1}{\pi} \int\limits_{-\infty}^{\infty} \frac{1+tz}{t-z} d\sigma(t), \tag{12.67}$$

where $\sigma(t)$ is non-decreasing function of bounded variation on $(-\infty, \infty)$.

If function $\tau(t)$ from (12.66) is normed by conditions $\tau(t) = 1/2\left[\tau(t+0) + \tau(t-0)\right]$, $\tau(0) = 0$, then it is uniquely determined by the R–function $f(z)$ and the conversion formula holds (Perron-Stieltjes)

$$\tau(t) = \lim_{\varepsilon \to 0+} \int\limits_0^t \operatorname{Im} f(\tau + i\varepsilon)\, d\tau \tag{12.68}$$

Furthermore, it is known that the function $y\operatorname{Im} f(iy)$ does not decrease on $(0, +\infty)$, but if $\mu = 0$ in (12.66), then

$$\int\limits_{-\infty}^{\infty} d\tau(t) = \lim_{y \to +\infty} \left[y \operatorname{Im} f(iy)\right]. \tag{12.69}$$

Both parts of this equality can be infinite.

The function $f(z)$ belongs to class $R[a, b]$ [7], if:

a) $f(z) \in R$;
b) $f(z)$ is holomorphic and positive in $(-\infty, a)$, holomorphic and negative in $(b, +\infty)$.

Theorem B

A necessary and sufficient condition for the function $f(z)$ to belong to class $R[a, b]$ is to admit the representation $f(z) = \int\limits_a^b d\sigma(t) / (t - z)$, where $\sigma(t)$ is bounded non-decreasing function.

Before proceeding to the concept of interpolation problem in the class R, let us elucidate the terminological conventions. The quadratic form

$$\sum_{\alpha,\beta=1}^{n} c_{\alpha\beta} \xi_\alpha \bar{\xi}_\beta \quad (c_{\alpha\beta} \in C, c_{-\alpha} = \bar{c}_\alpha) \tag{12.70}$$

is called *positive*, if the form assumes only positive values whenever $|\xi_1|^2 + \ldots + |\xi_n|^2 \neq 0$. If one can find simultaneously ξ_α that are not equal to zero, so that the quadratic form, in the general case, preserving sign, nevertheless tends to zero, then it is called *singular*. The terms "positive form," and "singular form" are brought to the common name "non-negative from."

Let that certain numerical sets F_z and F_w from the upper half-planes $\mathrm{Im}z > 0$, $\mathrm{Im}w > 0$ are given and a definite mapping $F_z \ni z_\alpha \to w_\alpha \in F_w$. The Nevanlinna-Pick interpolation problem in the class R consists in the finding of the conditions for the existence of the function $f \in R$ such, that

$$f(z_\alpha) = w_\alpha \quad \forall z_\alpha \in F_\alpha. \tag{12.71}$$

Theorem C

1. For the existence of the function $f(z) \in R$, which satisfies (12.71), it is necessary and sufficient all quadratic forms

$$\sum_{\alpha,\beta=1}^{n} \frac{w_{k\alpha} - \bar{w}_{k\beta}}{z_{k\alpha} - \bar{z}_{k\beta}} \xi_\alpha \bar{\xi}_\beta \tag{12.72}$$

not to be negative for any finite sets $\{z_{k\alpha}\} \subset F_z$.

2. If $F_z = \{z_0, z_1, \ldots, z_n, n < +\infty\}$, then it is possible to confine ourselves only to one quadratic form

$$\sum_{\alpha,\beta=1}^{n} \frac{w_\alpha - \bar{w}_\beta}{z_\alpha - \bar{z}_\beta} \xi_\alpha \bar{\xi}_\beta. \tag{12.73}$$

3. For the existence of a function $f(z) \in R[a,b]$ that is satisfying (12.71), it is necessary and sufficient that in R were solvable the two interpolation problems: $f_1(z_\alpha) = (z_\alpha - a) w_\alpha;\ f_2(z_\alpha) = (b - z_\alpha) w_\alpha,\ f_1, f_2 \in R$.
4. If any of the forms (12.72) is singular, then the function $f(z)$ is uniquely determined and the function $\sigma(t)$ corresponding to it is a function of jumps with finite number of jumps.

12.6.2 Control of Elastic State

Let us examine the elastic anisotropic half-plane (plate of thickness h) $y \geq 0$ and the set $F = \{z^{(\alpha)}\}_{\alpha \in A}$ of points of the open half-plane $y > 0$ (A is the set of indices). Let at the points of the set F the values of stresses $\left\{\sigma_{11}^{(\alpha)}, \sigma_{22}^{(\alpha)}, \sigma_{12}^{(\alpha)}\right\}$ are given. The problem consists of finding the conditions, under which normal $N(x)$ and tangent $T(x)$ loads exist on the border of the half-plane, which lead at the points of the set F to the given beforehand values of stresses $\left\{\sigma_{11}^{(\alpha)}, \sigma_{22}^{(\alpha)}, \sigma_{12}^{(\alpha)}\right\}$.

We would seek for $N(x),\ T(x)$ in the class of generalized functions of the form

$$N(x) = \sum \rho_\nu \delta(x - x_\nu), \rho_\nu > 0, \quad -\infty < x_\nu < \infty. \tag{12.74}$$

The relations, which express stresses in the half-plane through the analytic functions $\Phi_\nu(z_\nu)$ $(\nu = 1, 2)$ hold.

Let us assume that $z_j^{(\alpha)} = x^{(\alpha)} + \mu_j y^{(\alpha)}$, $\Phi_j\left(z_j^{(\alpha)}\right) = a_j^{(\alpha)} + ib_j^{(\alpha)}$, $\left(a_j^{(\alpha)}, b_j^{(\alpha)} \in R\right)$. We consider (for simplification of the computations) that $\mu_j = i\nu_j \ (\nu_j \in R)$; $a_j^{(\alpha)}$ are uniquely defined:

$$a_j^{(\alpha)} = (-1)^{j+1} \frac{\sigma_{11}^{(\alpha)} + \nu_{j^*}^2 \sigma_{22}^{(\alpha)}}{2\left(\nu_2^2 - \nu_1^2\right)}, \quad j^* = \begin{cases} 2\,(j=1), \\ 1\,(j=2), \end{cases} \tag{12.75}$$

and $b_j^{(\alpha)}$ is connected with the dependence

$$\nu_1 b_1^{(\alpha)} + \nu_2 b_2^{(\alpha)} = 1/2\sigma_{12}^{(\alpha)}. \tag{12.76}$$

In [266] the following relations are established

$$\Phi_1(z_1) = \frac{1}{2\pi i h(\mu_1 - \mu_2)} \int_{-\infty}^{\infty} \frac{\mu_2 N(\xi) + T(\xi)}{\xi - z_1} d\xi, \tag{12.77}$$

$$\Phi_2(z_2) = \frac{1}{2\pi i h(\mu_2 - \mu_1)} \int_{-\infty}^{\infty} \frac{\mu_1 N(\xi) + T(\xi)}{\xi - z_2} d\xi,$$

where $N(x)$, $T(x)$ are the normal and tangential loads on the border $y = 0$.

For given set F, let F_1, F_2 the corresponding sets of points $\{z_j = x + \mu_j y\colon x + iy \in F\}$. We assume for simplicity that $F_1 \cap F_2 = \emptyset$, and consider the case of absence of tangential load ($T(x) \equiv 0$). Further, let $G = F_1 \cup F_2$, $G_w = \left\{w_j^{(\alpha)}\right\}_{\alpha \in A}$, where

$$w_j^{(\alpha)} = (-1)^j\, 2\pi i h \frac{\nu_2 - \nu_1}{\nu_{j^*}} \left(a_j^{(\alpha)} + ib_j^{(\alpha)}\right). \tag{12.78}$$

The control problem stated above can be formulated as problem of moments [7, 89, 239]: to find the conditions for the existence of non-decreasing function $\sigma(x)$ ($d\sigma(x) = N(x)\,dx$) of bounded variation (measure), such that

$$w_j^{(\alpha)} = \int_{-\infty}^{\infty} \frac{d\sigma(\xi)}{\xi - z_j^{(\alpha)}}, \quad z_j^{(\alpha)} \in G \tag{12.79}$$

The conditions of solvability of the control problem with respect to the finite set F are indicated in Theorem 1.

Theorem 1.
We put $z_{2\alpha-1} := z_1^{(\alpha)}$, $1 \le j,k \le M$, $w_{2\alpha-1} := w_1^{(\alpha)}$, $w_{2\alpha} := w_2^{(\alpha)}$. For the solvability of the control problem of boundary generalized load $N(x)$ of the form (12.74) it is necessary and sufficient that existed such values $b_j^{(\alpha)}$, for which the quadratic form (12.73) would be non-negative, where $(N(x)\,dx = d\sigma(x))\,[a,b] \subset R$.

Proof.

1. If the problem is solvable, then a normal load $N(x)$ exists from the class of differentiated measures, such that at every point $-\infty < a \le x \le b < +\infty$ the stresses assumed the given values $\left\{\sigma_{11}^{(\alpha)}, \sigma_{22}^{(\alpha)}, \sigma_{12}^{(\alpha)}\right\}$. Consequently, the values $\Phi_j(z_j)$, found from (12.77), are connected with $\left\{\sigma_{11}^{(\alpha)}, \sigma_{22}^{(\alpha)}, \sigma_{12}^{(\alpha)}\right\}$ via the formulas (12.75). Thus, for the existing function $\sigma(x)$ the relations (12.79) hold.
2. Let the form (12.73) is non-negative. According to theorem C a function $\Phi(z) \in R$ exists

$$\Phi(z) = \mu z + \nu + \int_{-\infty}^{\infty} \frac{1+tz}{t-z} d\tau(t), \quad d\tau(t) \ge 0, \tag{12.80}$$

which gives the solution of the interpolation problem $\Phi(z_\alpha) = W_a$. Since the function $\tau(t)$ can be chosen concentrated on the bounded set of the real axis, then $\int_{-\infty}^{\infty} (1+t^2)\ d\tau(t) < \infty$.

Following the note [7, p. 118], let us assume

$$\sigma(t) = \int_{-\infty}^{t} (1+\nu^2)\, d\tau(\nu), \mu = 0, \quad \nu = \int_{-\infty}^{\infty} \frac{t}{1+t^2} d\sigma(t), \tag{12.81}$$

hence

$$\begin{aligned}\Phi(z) &= \int_{-\infty}^{\infty} \frac{d\sigma(u)}{u-z}, \quad d\sigma(u) \ge 0, \quad \operatorname{var} \sigma < \infty, \quad w_j^{(\alpha)} = \Phi\left(z_j^{(\alpha)}\right) \\ &= \int_{-\infty}^{\infty} \frac{d\sigma(t)}{t - z_j^{(\alpha)}},\end{aligned} \tag{12.82}$$

and the interpolation problem (12.79) is solvable, and *a fortiori* the stated problem.

12.6.3 Examples

1. Let $z_1 = x + i\nu_1 y$ the set F consists of one point $z = x + iy,\ y > 0$. Then, $z_2 = x + i\nu_2 y$. Let $\nu_2 > \nu_1$, and the stress tensor at the point z such that: $\sigma_{11}, \sigma_{22} = s\sigma_{11},\ \sigma_{12} = 0\ (s \in R)$.
According to formulas (12.75), (12.79) we find

$$\begin{aligned}&a_1 = \sigma_{11}\beta_2, \quad a_2 = -\sigma_{11}\beta_1, \quad b_2 = -\kappa b_1, \\ &w_1 = A(b_1 - i\sigma_{11}\beta_2), \quad w_2 = A\left(b_1 - iA\sigma_{11}\beta_1\kappa^{-1}\right), \\ &\beta_j = \frac{1+s\nu_j^2}{2(\nu_2^2 - \nu_1^2)}, \quad \kappa = \nu_1/\nu_2, \quad A = 2\pi h(1-\kappa).\end{aligned} \tag{12.83}$$

The quadratic form (12.73) has the form

$$\frac{w_1-\bar{w}_1}{z_1-\bar{z}_1}|\xi_1|^2+2\mathrm{Re}\left[\frac{w_1-\bar{w}_2}{z_1-\bar{z}_2}\xi_1\bar{\xi}_2\right]+\frac{w_2-\bar{w}_2}{z_2-\bar{z}_2}|\xi_2|^2$$
$$=-\frac{A\sigma_{11}}{\nu_1 y}\left[|\xi_1+\xi_1|^2+s\,|\nu_2\xi_1+\nu_1\xi_2|^2\right]. \tag{12.84}$$

If $s<0$, then the quadratic form cannot contain constant sign. If $s\geq 0$, $\sigma_{11}<0$, then the conditions of theorem 1 are satisfied and the control $N(x)$ from class (12.74) is possible. However, the elementary control $N(u)=\rho\delta(u-x)$, $x\in R$, $\rho>0$ is impossible. Indeed, in this case

$$\frac{\rho}{x_1-x-i\nu_1 y}=Ab_1-iA\sigma_{11}\beta_2,\qquad \frac{\rho}{x_1-x-i\nu_2 y}=Ab_1-iA\sigma_{11}\beta_1\kappa^{-1} \tag{12.85}$$

and then

$$\frac{\rho(x_1-x)}{(x_1-x)^2+\nu_1^2y^2}=\frac{\rho(x_1-x)}{(x_1-x)^2+\nu_2^2y^2}=b_1, \tag{12.86}$$

which is possible only when $x_1=x$, $b_1=0$. But in this case, $\rho=-A\sigma_{11}\beta_2\nu_1 y=-A\sigma_{11}\beta_1\nu_2 y\kappa^{-1}$, whence follows that $\nu_1=\nu_2$.

Let us consider now the function $N(x)=\rho_1\delta(x-x_1)+\rho_2\delta(x-x_2)$, $x_1<x_2$, $\rho_1>0$, $\rho_2>0$. Moment equalities (12.79) take the form

$$\frac{\rho_1}{x-x_1-i\nu_1 y}+\frac{\rho_2}{x_2-x-i\nu_1 y}=Ab_1-iA\sigma_{11}\beta_2, \tag{12.87}$$
$$\frac{\rho_1}{x_1-x-i\nu_2 y}+\frac{\rho_2}{x_2-x-i\nu_2 y}=Ab_1-iA\sigma_{11}\beta_1\kappa^{-1}.$$

We put $x_2-x=-(x_1-x)$ and separate the real and the imaginary parts in (12.87). Simple analysis shows that under the conditions $\nu_2>\nu_1$

$$x_1=x-y\big/\sqrt{5},\quad x_2=x+y\big/\sqrt{5}, \tag{12.88}$$
$$\rho_1=\rho_2=-\frac{A\sigma_{11}y\left(1+s\nu_1^2\right)\left(1+s\nu_2^2\right)}{4s\nu_1\left(\nu_2^2-\nu_1^2\right)}>0.$$

Note. If in the previous example $\sigma_{22}=0$, then direct computations show that the control problem is solvable if and only if $\sigma_{11}\leq 0$.

2. Let us consider M points $z_j=x_j+iy$, which lie on the horizontal line. Then $z_1^{(j)}=x_j+i\nu_1 y$, $z_2^{(j)}=x_j+i\nu_2 y$. Let $\sigma_{11}=\mathrm{const}$, $\sigma_{22}=0$, $\sigma_{12}=0$ the desired values of stresses at all M points. From (12.75), (12.76) follows

$$a_1^{(j)}=A\sigma_{11},\quad a_2^{(j)}=-a_1^{(j)},b_2^{(j)}=-\kappa b_1^{(j)},\quad A=\left[2\left(\nu_2^2-\nu_1^2\right)\right]^{-1}. \tag{12.89}$$

We put $\lambda:=2\pi h(1-\kappa)$, $b_j:=b_1^{(j)}$. Then $w_1^{(j)}=\lambda\left(b_1^{(j)}-iA\sigma_{11}\right)$, $w_2^{(j)}=\lambda\left(b_1^{(j)}-iA\sigma_{11}\kappa^{-1}\right)$.

It is convenient to represent the quadratic form (12.73) as sum of four expressions, which contain the differences $w_1^{(j)} - \bar{w}_1^{(k)}$, $w_2^{(j)} - \bar{w}_2^{(k)}$, $w_1^{(j)} - \bar{w}_2^{(k)}$, $w_2^{(j)} - \bar{w}_1^{(k)}$. Then

$$\sum_{\alpha,\beta=1}^{2M} \frac{w_\alpha - \bar{w}_\beta}{z_\alpha - \bar{z}_\beta} = \lambda \left\{ \sum_{j,k=1}^{M} \frac{b_j - b_k - 2iA\sigma_{11}}{x_j - x_k + 2i\nu_1 y} \xi_j^{(1)} \bar{\xi}_k^{(1)} \right.$$
$$+ \sum_{j,k=1}^{M} \frac{b_j - b_k - 2iA\sigma_{11}\kappa^{-1}}{x_j - x_k + 2i\nu_2 y} \xi_j^{(2)} \bar{\xi}_k^{(2)} +$$
$$+ \sum_{j,k=1}^{M} \frac{b_j - b_k - iA\left(1+\kappa^{-1}\right)\sigma_{11}}{x_j - x_k + i\left(\nu_1+\nu_2\right) y} \xi_j^{(1)} \bar{\xi}_k^{(2)} \tag{12.90}$$
$$\left. + \sum_{j,k=1}^{M} \frac{b_j - b_k - iA\left(1+\kappa^{-1}\right)\sigma_{11}}{x_j - x_k + i\left(\nu_1+\nu_2\right) y} \xi_j^{(2)} \bar{\xi}_k^{(1)} \right\}.$$

Recall that the values $b_1, \ldots, b_M$ can be chosen arbitrarily. Let us require that for all $1 \leq j, k \leq M$ the equalities $(b_j - b_k)/A\sigma_{11} = -(x_j - x_k)/\nu_1 y$ were held. It is not difficult to confirm that this system of equations is consistent. With this choice $b_1, \ldots, b_M$ the quadratic form coincides with the value

$$-\frac{\lambda A\sigma_{11}}{\nu_1 y} \left| \sum_{j=1}^{M} \left(\xi_j^{(1)} + \xi_j^{(2)} \right) \right|^2 \tag{12.91}$$

and is non-negative, when $\sigma_{11} < 0$.

Further, let us consider briefly the case of localized in $[a, b] \subset R$ loads.

Theorem 2.

In order to exist a generalized boundary load $N(x)$ $(N(x)\ dx = d\sigma(x))$, applied to the portion $-\infty < a \leq x \leq b < +\infty$ and such, that

a) $d\sigma(x) \geq 0$;

b) the relations $w_j^{(\alpha)} = \int_a^b d\sigma(t) / \left(t - t_j^\alpha\right)$ $(\alpha = 1, 2, \ldots, M)$ with given $\sigma_{11}^{(\alpha)}$, $\sigma_{22}^{(\alpha)}$, $\sigma_{12}^{(\alpha)}$ were fulfilled, it is necessary and sufficient that the quadratic forms

$$\sum_{j,k=1}^{2M} \frac{(z_j - a) w_j - (\bar{z}_k - a) \bar{w}_k}{z_j - \bar{z}_k} \xi_j \bar{\xi}_k, \quad \sum_{j,k=1}^{2M} \frac{(b - z_j) w_j - (b - \bar{z}_k) \bar{w}_k}{z_j - \bar{z}_k} \xi_j \bar{\xi}_k. \tag{12.92}$$

were non-negative with any choice $b_1^{(\alpha)}$ from (12.76)

The designation z_j, z_k was introduced above.

12.7 Boundary Control of the Stress Intensity Factors in a Halfspace with a Crack

Let us consider a crack in piezoceramic halfspace ($x_2 \geq 0$) which is bounded by vacuum and it is free from forces everywhere, except at a part of the boundary $x_1 \in [a, b]$, where shear force $p(x_1, t) = \mathrm{Re}\left(P(x_1)\, e^{-i\omega t}\right)$ or electric charges $q(x_1, t) = \mathrm{Re}\left(Q(x_1)\, e^{-i\omega t}\right)$ that are independent of coordinate x_3.

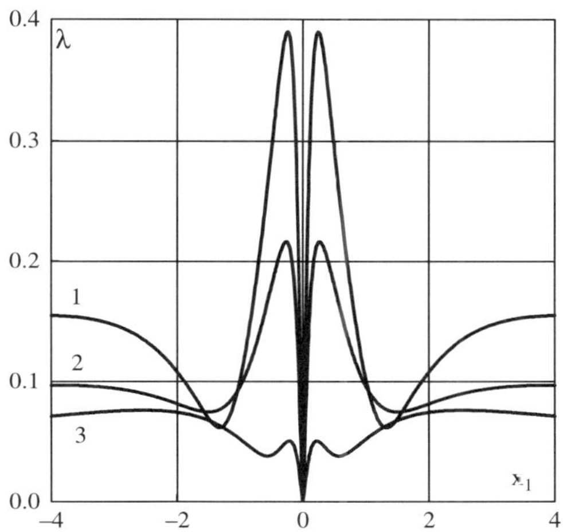

Fig. 12.16. Changes of the optimal control function λ on the boundary portion of a halfspace in case of a crack positioned perpendicular to the boundary of the halfspace

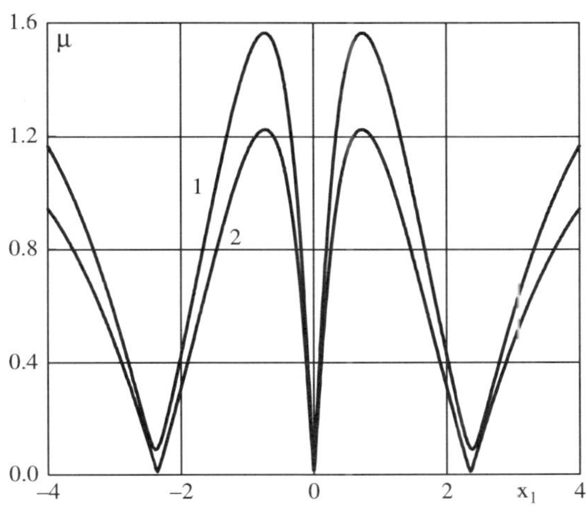

Fig. 12.17. Changes of the optimal control function μ on the boundary portion of a halfspace in case of a crack positioned perpendicular to the boundary of the halfspace

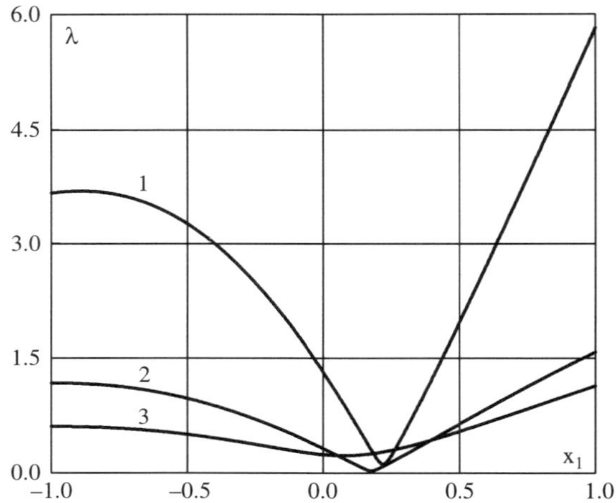

Fig. 12.18. Changes of the optimal control function λ in case of a crack oriented at angle $\pi/4$ to the boundary of the halfspace

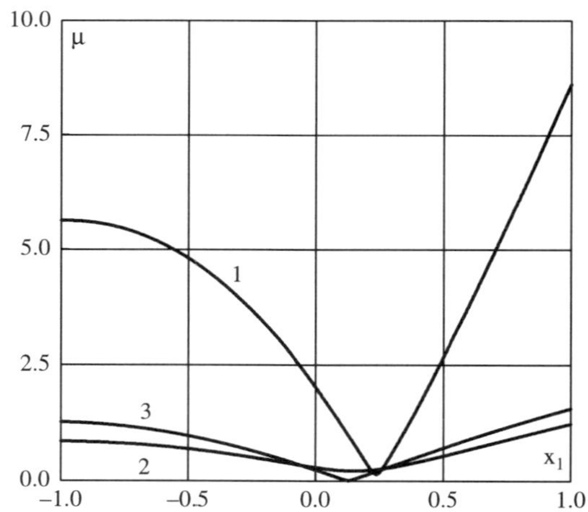

Fig. 12.19. Changes of the optimal control function μ in case of a crack oriented at angle $\pi/4$ to the boundary of the halfspace

We want to determine the control functions $P(x_1)$, $Q(x_1)$ from conditions (12.34), (12.35).

Example 1. A straightline crack of length 2ℓ is positioned perpendicular to the boundary and symmetric with the respect of portion $[a, b]$ (h is the distance from the centre of the crack to the boundary).

Figure 12.16 shows the changes $\lambda = |P(x_1)| / c_{44}^E$ for the values of parameters $a/\ell = -4, b/\ell = 4$, $\alpha_1 = \Omega_0^{(1)}(-1) = 0$, $\alpha_2 = \Omega_0^{(1)}(1) = 0.1$, $h/\ell = 1.2$

(the crack is closed in the vicinity of the vertex). Curves 1–3 have been constructed for the normalized values of wave number $\gamma^* \ell = 1$, 1.5 and 2.

The graphic of magnitude $\mu = |Q(x_1)| / e_{15}$ for the same values of the parameters and $\gamma^* \ell = 1$ is shown in Fig. 12.17, where curve 1 is for $\alpha_1 = 0$, $\alpha_2 = 0.1$ and curve 2 is for $\alpha_1 = 0.1$, $\alpha_2 = 0$.

Example 2. A straightline crack is positioned at an angle of $\pi/4$ to the boundary. For this case Figs. 12.18 and 12.19 show the graphs of quantity λ and μ at $h/\ell = 3$, $a/\ell = -1$, $b/\ell = 1$, $\alpha_1 = 0$, $\alpha_2 = 0.1$ (the crack is closed on the nearest to the boundary tip). Curves 1–3 conform to values $\gamma^* \ell = 1$, 2 and 3.

12.8 Control of Electric Charges on the Electrodes in a Layer with a Partially Electrodized Opening

Consider a piezoceramic layer $(0 \le x_1 \le a, -\infty < x_2 < \infty, -\infty < x_3 < \infty)$, in the Cartesian system of coordinates $Ox_1x_2x_3$, containing a tunnel opening along the x_3-axis, the cross-section of which is bounded by a smooth contour C. The bases of the layer are free from stress and coupled with vacuum everywhere except the portions $x_2 \in [b, d]$ $(x_1 = 0, a)$, where shear forces $p(x_2, t) = \text{Re}\left(P(x_2) e^{-i\omega t}\right)$ or electric charges $q(x_2, t) = \text{Re}\left(Q(x_2) e^{-i\omega t}\right)$ harmonically changing with time and independent of the coordinate x_3 act; (t is the time, ω is the circular frequency). On the free from forces surface of the opening there are two continuous in the direction of axis x_3 active electrodes symmetrically positioned with the prescribed difference of electric potential $2\phi^* = \text{Re}\left(\Phi^* e^{-i\omega t}\right)$ (unelectrodized portions of the opening are coupled with vacuum). It is assumed that the cross-section of the cavity has horizontal and vertical axes of symmetry and it is positioned symmetrically with respect to the bases of the layer (Fig. 12.20).

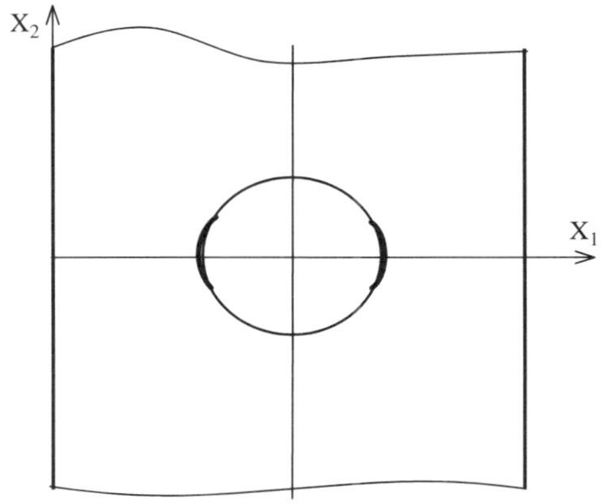

Fig. 12.20. The scheme of a layer with partially electrodized cavity

The optimization problem is formulated as follows: we seek to determine the intensity of distribution of shear forces $P(x_2)$ or electric charges $Q(x_2)$, assumed as the control function, along the portions $x_2 \in [b,d]$ $(x_1 = 0, a)$, in such a way, so that the total electric charges on the active electrodes reach the prescribed values. Besides, the norm of the control function in space $L^2_{[b,d]}$ is required to be minimal. In other words, we investigate the possibility of electric current control in the outer circuit of the voltage generator with minimal electric expenses due to the boundary mechanical and electric loading.

The solution of this inverse problem is carried out considering the corresponding direct boundary problem of electroelasticity.

Consider a piezoceramic layer with a tunnel cavity, in which the wave electroelastic field is excited by the difference of electric potentials $2\phi^*$ supplied on two continuous electrodes. The bases of the layer and the surface of the cavity are free from forces, the outer medium is vacuum. The additional factor of excitation of harmonic oscillations in the layer are the concentrated on its bases along lines $x_2 = \eta_2$, $-\infty < x_3 < \infty$, $x_1 = 0$ and $x_1 = a$, harmonically changing with time, not depending on coordinates x_3 shear forces $P_k = \mathrm{Re}\left(P_k^* \delta(x_2 - \eta_2) e^{-i\omega t}\right)$ or electric charges $Q_k = \mathrm{Re}\left(Q_k^* \delta(x_2 - \eta_2) e^{-i\omega t}\right)$ $(k = 1, 2)$. Here index $k = 1$ refers to boundary $x_1 = 0$, index $k = 2$ to $x_1 = a$, $\delta(x)$ to the Dirac δ-function.

Following the presentation of previous chapters we may show, that the given boundary problem may be reduced to the following system of integrodifferential equations

$$\nu(\zeta_0) + \int_C \nu(\zeta) g_1(\zeta,\zeta_0)\, ds + \int_C f'(\zeta)\, g_2(\zeta,\zeta_0)\, ds = N_1(\zeta_0),$$

$$-\frac{1}{2} f(\zeta_0) + \int_C \{\nu(\zeta)\, g_3(\zeta,\zeta_0) + f(\zeta)\, g_4(\zeta,\zeta_0)\}\, ds = N_2(\zeta_0), \qquad \zeta_0 \in C_\phi,$$

$$\int_C f'(\zeta) g_5(\zeta,\zeta_0)\, ds = N_3(\zeta_0), \qquad \zeta_0 \in C \backslash C_\phi,$$

$$g_1(\zeta,\zeta_0) = \frac{1}{2a} \mathrm{Re}\left\{ e^{i\psi_0} \left[ctg\frac{\pi(\zeta_0 - \zeta)}{2a} + ctg\frac{\pi(\zeta_0 + \bar{\zeta})}{2a} \right] \right\} + P_1 e^{i\psi_0} + P_2 e^{-i\psi_0},$$

$$g_2(\zeta,\zeta_0) = \frac{2e_{15}}{c_{44}^E (1 + k_{15}^2)} g_5(\zeta,\zeta_0),$$

$$g_3(\zeta,\zeta_0) = \frac{e_{15}}{\ni_{11}^\varepsilon} \left\{ -\frac{|\xi_{20} - \xi_2|}{2a} + \frac{1}{2\pi} \ell n \left| 4 \sin\frac{\pi(\zeta - \zeta_0)}{2a} \sin\frac{\pi(\zeta + \bar{\zeta}_0)}{2a} \right| + \right.$$
$$\left. + \frac{1}{2ia\gamma} e^{i\gamma|\xi_{20} - \xi_2|} - \frac{1}{a} \sum_{m=1}^{\infty} c_m (\xi_{20} - \xi_2) \cos\alpha_m \xi_1 \cos\alpha_m \xi_{10} \right\},$$

$$g_4(\zeta,\zeta_0) = \frac{1}{4a}\mathrm{Re}\left\{e^{i\psi}\left[ctg\frac{\pi(\zeta-\zeta_0)}{2a} + ctg\frac{\pi(\zeta+\bar{\zeta}_0)}{2a}\right]\right\}$$

$$+ \frac{\sin\psi}{2a} sign(\xi_{20}-\xi_2),$$

$$g_5(\zeta,\zeta_0) = \frac{1}{4a}\mathrm{Im}\left\{e^{i\psi_0}\left[ctg\frac{\pi(\zeta-\zeta_0)}{2a} + ctg\frac{\pi(\bar{\zeta}+\zeta_0)}{2a}\right]\right\},$$

$$P_1 = S - \frac{1}{a}(A_0 - iB_0), \qquad P_2 = -S - \frac{1}{a}(A_0 + iB_0),$$

$$S = \frac{1}{2ia} sign(\xi_2-\xi_{20})\left(1 - e^{i\gamma|\xi_2-\xi_{20}|}\right),$$

$$A_0 = \sum_{k=1}^{\infty}\beta_{1k}\alpha_k\cos\alpha_k\xi_1\sin\alpha_k\xi_{10},$$

$$B_0 = \sum_{k=1}^{\infty}\beta_{0k} sign(\xi_{20}-\xi_2)\cos\alpha_k\xi_1\cos\alpha_k\xi_{10},$$

$$\beta_{mk} = \frac{1}{\alpha_k^m}e^{-\alpha_k|\xi_2-\xi_{20}|} - \frac{1}{\lambda_k^m}e^{-\lambda_k|\xi_2-\xi_{20}|}, \qquad f'(\zeta) = \frac{df}{ds},$$

$$c_m(\xi_{20}-\xi_2) = \frac{1}{\lambda_m}e^{-\lambda_m|\xi_{20}-\xi_2|} - \frac{1}{\alpha_m}e^{-\alpha_m|\xi_{20}-\xi_2|},$$

$$N_1(\zeta_0) = \sum_{k=1}^{2}\frac{P_k^*}{ac_{44}^E(1+k_{15}^2)}\left\{sign(\xi_{20}-\eta_2)\sin\psi_0\left(e^{i\gamma|\xi_{20}-\eta_2|}-1\right)+\right.$$

$$+\frac{1}{2}\mathrm{Re}\left[e^{i\psi_0}\left(ctg\frac{\pi}{2a}(\overline{\mu_k}+\zeta_0) - ctg\frac{\pi}{2a}(\mu_k-\zeta_0)\right)\right]$$

$$\left.-2(A_k\cos\psi_0 + B_k\sin\psi_0)\right\} -$$

$$-\sum_{k=1}^{2}\frac{k_{15}^2Q_k^*}{ae_{15}(1+k_{15}^2)}\left[\sin\psi_0 sign(\xi_{20}-\eta_2)e^{i\gamma|\xi_{20}-\eta_2|}\right.$$

$$\left.-2(A_k\cos\psi_0 + B_k\sin\psi_0)\right]$$

$$N_2(\zeta_0) = \Phi^*(\zeta_0) + \sum_{k=1}^{2}\frac{P_k^*k_{15}^2}{e_{15}(1+k_{15}^2)}\left\{-\frac{|\xi_{20}-\eta_2|}{2a}\right.$$

$$+\frac{1}{2\pi}\ell n\left|4\sin\frac{\pi(\mu_k-\zeta_0)}{2a}\sin\frac{\pi(\mu_k+\bar{\zeta}_0)}{2a}\right| +$$

$$\left.+\frac{1}{2ia\gamma}e^{i\gamma|\xi_{20}-\eta_2|} - \frac{1}{a}\sum_{m=1}^{\infty}c_m(\xi_{20}-\eta_2)\cos\alpha_m\eta_1^{(k)}\cos\alpha_m\xi_{10}\right\} +$$

$$+\sum_{k=1}^{2}\frac{Q_k^*}{\ni_{11}^{\varepsilon}(1+k_{15}^2)}\left\{-\frac{|\xi_{20}-\eta_2|}{2a} + \frac{1}{2\pi}\ell n\left|4\sin\frac{\pi(\mu_k-\zeta_0)}{2a}\sin\frac{\pi(\mu_k+\bar{\zeta}_0)}{2a}\right| -\right.$$

$$-k_{15}^2\left[\frac{1}{2ia\gamma}e^{i\gamma|\xi_{20}-\eta_2|}-\frac{1}{a}\sum_{m=1}^{\infty}c_m\left(\xi_{20}-\eta_2\right)\cos\alpha_m\eta_1^{(k)}\cos\alpha_m\xi_{10}\right]\Bigg\}$$

$$N_3\left(\zeta_0\right)=\sum_{k=1}^{2}\frac{Q_k^*}{4a\ni_{11}^{\varepsilon}}\Big\{2sign\left(\eta_2-\xi_{20}\right)\sin\psi_0$$

$$+\mathrm{Re}\left[e^{i\psi_0}\left(ctg\frac{\pi}{2a}\left(\bar{\mu}_k+\zeta_0\right)-ctg\frac{\pi}{2a}\left(\mu_k-\zeta_0\right)\right)\right]\Big\}$$

$$A_k=-\sum_{m=1}^{\infty}\left[\frac{\alpha_m}{\lambda_m}e^{-\lambda_m|\xi_{20}-\eta_2|}-e^{-\alpha_m|\xi_{20}-\eta_2|}\right]\cos\alpha_m\eta_1^{(k)}\sin\alpha_m\xi_{10},$$

$$B_k=sign\left(\eta_2-\xi_{20}\right)\sum_{m=1}^{\infty}\left(e^{-\lambda_m|\xi_{20}-\eta_2|}-e^{-\alpha_m|\xi_{20}-\eta_2|}\right)\cos\alpha_m\eta_1^{(k)}\cos\alpha_m\xi_{10},$$

$$\psi=\psi\left(\zeta\right),\quad\psi_0=\psi\left(\zeta_0\right),\quad\zeta=\xi_1+i\xi_2,\quad\zeta_0=\xi_{10}+i\xi_{20},\quad\zeta,\zeta_0\in C. \tag{12.93}$$

here $\eta_1^{(1)}=0$, $\eta_1^{(2)}=a$, $\mu_1=i\eta_2$, $\mu_2=a+i\eta_2$, $\Phi^*\left(\zeta_0\right)$ is the piecewise-constant function, determining the value of the electric potentials on the electrodes, ψ is the angle between the normal to contour C and axis x_1 at point $\zeta\in C$.

It should be noted that the action of linear sources on the bases of the layer causes the appearance of electric charges of various signs on the active electrodes. Therefore in (12.93) we should set $P_1^*=-P_2^*$, $Q_1^*=-Q_2^*$. Violation of this requirement renders the system (12.93) unsolvable.

The expression for the density amplitude of distribution of electric charges $q_j\left(\beta\right)$ $\left(0\le\beta\le2\pi\right)$ on j-th electrode has the form

$$q_j\left(\beta_0\right)=q_j^{\phi}\left(\beta_0\right)+\sum_{k=1}^{2}Q_k^*R_k\left(\eta_2,\zeta_0\right)\qquad\left(\zeta_0\in C_{\varphi_j}\right),$$

$$q_j^{\phi}\left(\beta_0\right)=-\frac{\ni_{11}^{\varepsilon}}{4a}\int\limits_C f'\left(\zeta\right)\mathrm{Im}\left\{e^{i\psi_0}\left[ctg\frac{\pi\left(\zeta-\zeta_0\right)}{2a}+ctg\frac{\pi\left(\bar{\zeta}+\zeta_0\right)}{2a}\right]\right\}ds$$

$$R_k\left(\eta_2,\zeta_0\right)=\frac{1}{2a}sign\left(\eta_2-\xi_{20}\right)\sin\psi_0$$

$$+\frac{1}{4a}\mathrm{Re}\left[e^{i\psi_0}\left(ctg\frac{\pi}{2a}\left(\bar{\mu}_k+\zeta_0\right)-ctg\frac{\pi}{2a}\left(\mu_k-\zeta_0\right)\right)\right] \tag{12.94}$$

where C_{ϕ_j} is the part of contour C, where the j-th electrode is positioned.

Returning to the control problem we will write it in a formalized manner. The amplitude of the total charge of j-th electrode taking into account (12.94) may be represented as

$$Q_j=Q_j^{\phi}+Q_j^{(\Lambda)},$$

$$Q_j^{\phi}=\int\limits_{\beta_{2j-1}}^{\beta_{2j}}q_j^{\phi}\left(\beta_0\right)s'\left(\beta_0\right)d\beta_0\qquad\left(\beta_{2j-1}<\beta_0<\beta_{2j},j=1,2\right) \tag{12.95}$$

Here $Q_j^{(\Lambda)}$ is the additional charge appearing due to the action of the layers bases shear forces $p(x_2,t)$ or charges $q(x_2,t)$ on the prescribed portions $x_2 \in [b,d]$. The quantity $Q_j^{(\Lambda)}$ may be determined by

$$
\begin{aligned}
Q_j^{(\Lambda)} &= \int\limits_{\beta_{2j-1}}^{\beta_{2j}} q_j^*(\beta_0)\, s'(\beta_0)\, d\beta_0 \qquad (\beta_{2j-1} < \beta_0 < \beta_{2j}, j = 1,2) \\
q_j^*(\beta_0) &= \int\limits_C \left\{ \int\limits_b^d \Lambda(\eta_2)\, f'_{(\Lambda)}(\eta_2,\zeta)\, d\eta_2 \right\} G(\zeta,\zeta_0) ds + g \int\limits_b^d Q(\eta_2) L(\eta_2,\zeta_0)\, d\eta_2, \\
L(\eta_2,\zeta_0) &= R_1(\eta_2,\zeta_0) - R_2(\eta_2,\zeta_0), \qquad (12.96) \\
G(\zeta,\zeta_0) &= -\frac{\ni_{11}^{\varepsilon}}{4a} \mathrm{Im}\left\{ e^{i\psi_0} \left[ctg\frac{\pi(\zeta-\zeta_0)}{2a} + ctg\frac{\pi(\bar{\zeta}+\zeta_0)}{2a} \right] \right\},
\end{aligned}
$$

where $f'_{(\Lambda)}(\eta_2,\zeta)$ is the "*standard*" solution of system (12.93) corresponding to the action at the points $x_2 = \eta_2$, $x_1 = \eta_1^{(1)} = 0$ and $x_1 = \eta_1^{(2)} = a$ of linear sources (forces and charges) of unity intensity. $\Lambda(\eta_2)$ denotes either the intensity of forces $P(\eta_2)$ or the intensity charges $Q(\eta_2)$ interpreted as the control function. The quantity $g = 0$, if $\Lambda(\eta_2) = P(\eta_2)$ and $g = 1$ at $\Lambda(\eta_2) = Q(\eta_2)$.

The equality (12.96) can written in the following form

$$
\begin{aligned}
Q_j^{(\Lambda)} &= \int\limits_b^d \Lambda(\eta_2) m_j(\eta_2)\, d\eta_2 \qquad (j = 1,2) \\
m_j(\eta_2) &= \int\limits_{\beta_{2j-1}}^{\beta_{2j}} S_J^{(\Lambda)}(\eta_2,\zeta_0)\, s'(\beta_0)\, d\beta_0, \qquad (12.97) \\
S_J^{(\Lambda)}(\eta_2,\zeta_0) &= -\frac{\ni_{11}^{\varepsilon}}{4a} \int\limits_C f'_{(\Lambda)}(\eta_2,\zeta) \mathrm{Im}\left\{ e^{i\psi_0} \left[ctg\frac{\pi(\zeta-\zeta_0)}{2a} \right.\right. \\
&\left.\left. + ctg\frac{\pi(\bar{\zeta}+\zeta_0)}{2a} \right] \right\} ds + +gL(\eta_2,\zeta_0) \qquad (\zeta_0 \in C_{\phi_j})
\end{aligned}
$$

Here the total charges on the electrode must have the prescribed values and taking into account (12.95), (12.97) we come to relations

$$
Q_j = Q_j^{\phi} + \int\limits_b^d \Lambda(\eta_2) m_j(\eta_2)\, d\eta_2 = \kappa_j \qquad (j = 1,2), \qquad (12.98)
$$

where κ_j is the given complex number.

Thus, the inverse problem is reduced to ℓ-problem of moments: it is necessary to determine the control function $\Lambda(\eta_2)$ that has to satisfy the moments of equality

$$\int_b^d \Lambda(\eta_2) m_j(\eta_2)\, d\eta_2 = \kappa_j - Q_j^\phi \quad (j = 1, 2) \tag{12.99}$$

and observing the conditions of the minimum of the control norm

$$\|\Lambda(\eta_2)\|_{L^2_{[b,d]}} = \sqrt{\int_b^d |\Lambda(\eta_2)|^2\, d\eta_2} \rightarrow \min \tag{12.100}$$

From the results presented in Sect. 12.1 it follows that the extremal element is determined by

$$\Lambda(\eta_2) = \sum_{j=1}^{2} \chi_j \bar{m}_j(\eta_2), \tag{12.101}$$

where the constants χ_j are determined from system (12.99).

As an example consider a layer by piezoceramic with a circular tunnel cavity. Stress is imposed via a pair of electrodes, the amplitude potential of which differs by $2\Phi^*$. It is required to determine the intensity of distribution

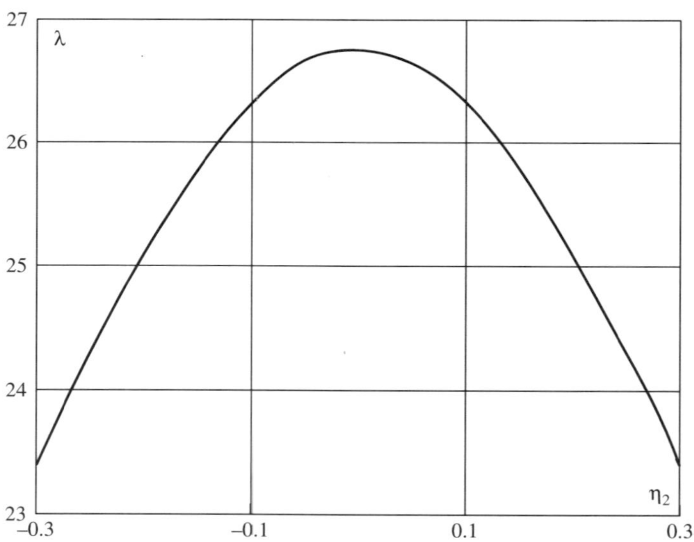

Fig. 12.21. Changes of the modulus of the optimal function amplitude $\lambda = |P(\eta_2)/\Phi^*|$ disconnecting the circuit of voltage generator at $R/\alpha = 0,1, \beta/\alpha = -0,3, d/\alpha = 0,3, \gamma\alpha = 3$

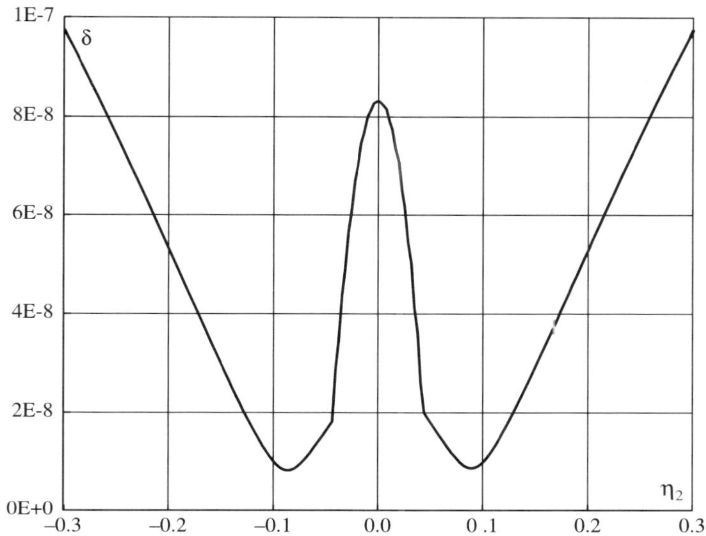

Fig. 12.22. Changes of the modulus of the optimal function amplitude $\delta = |Q(\eta_2)/\Phi^*|$ disconnecting the circuit of voltage generator at $R/\alpha = 0,1$, $\beta/\alpha = -0,3$, $d/\alpha = 0,3$, $\gamma\alpha = 3$

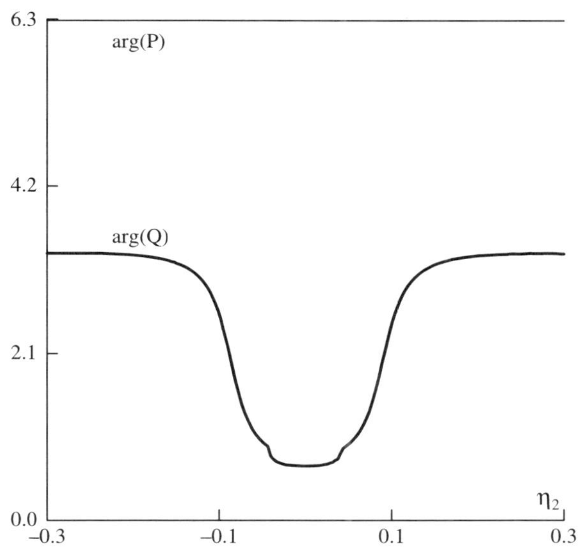

Fig. 12.23. Changes of the arguments of the amplitudes of the optimal shear forces and charges on the prescribed loading positions

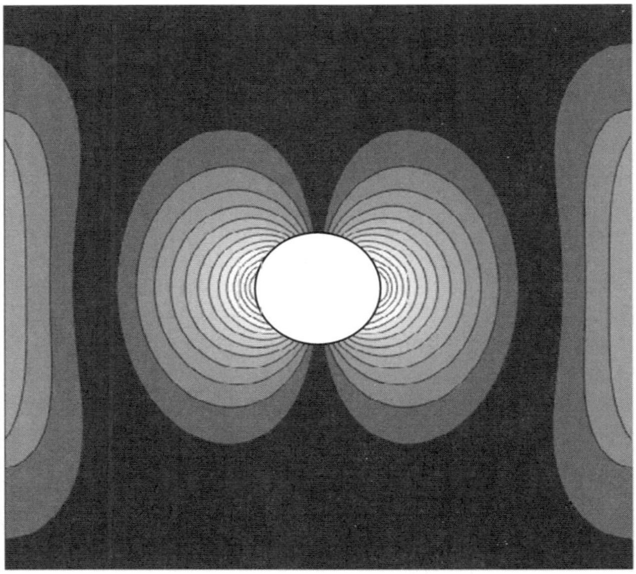

Fig. 12.24. The contour lines of the modulus of the displacement amplitude in a piece-wise homogenous layer under the influence of optimal shear forces $P(\eta_2)$

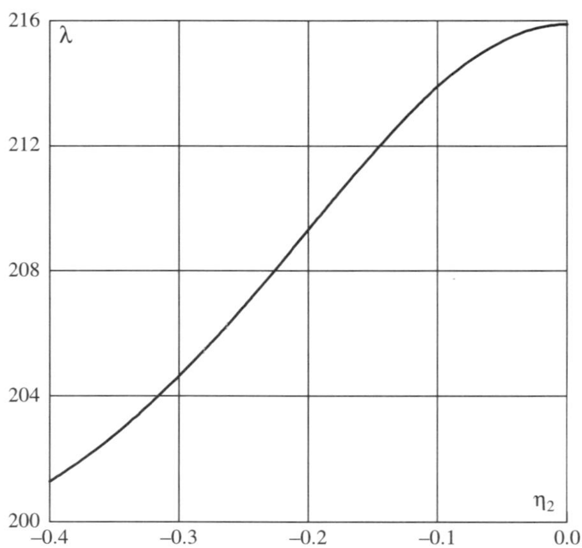

Fig. 12.25. Changes of the modulus of optimal control function amplitude $\lambda = |\mathrm{P}(\eta_2)/\Phi^*|$ at $R/\alpha = 0,1,\ \beta/\alpha = -0,4,\ d/\alpha = 0,\ \gamma\alpha = 4,\ \kappa_1 = 0\ (l = 1, 2)$

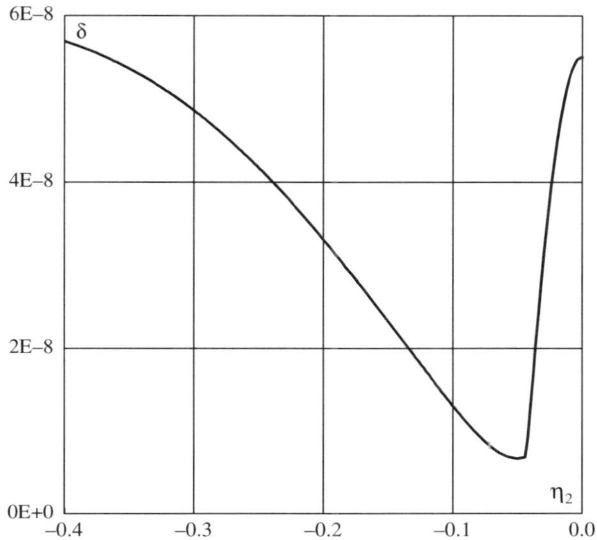

Fig. 12.26. Changes of the modulus of optimal control function amplitude $\delta = |Q(\eta_2)/\Phi^*|$

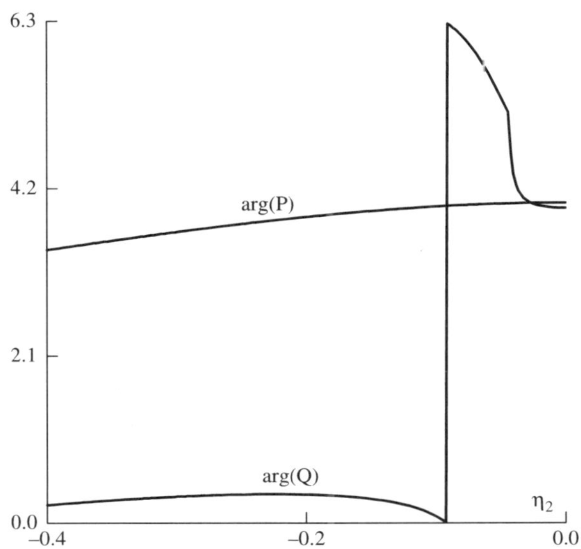

Fig. 12.27. Changes of the arguments of the amplitudes of the optimal shear forces and charges on the prescribed loading positions

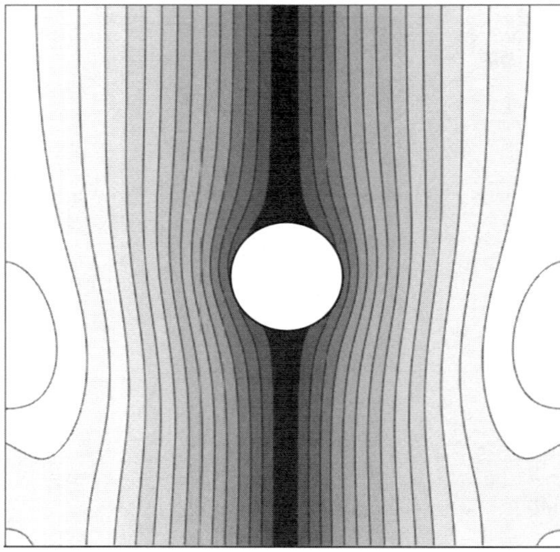

Fig. 12.28. The contour lines of the modulus of the displacement amplitude in a piece-wise homogeneous layer under the influence of optimal shear forces $P(\eta_2)$

of shear forces $P(\eta_2)$, either charges $Q(\eta_2)$ or conditions (12.99) at restrictions (12.100).

Figure 12.21 illustrates the change of the modulus of function of optimal control $\lambda = |P(\eta_2)/\Phi^*|$ for the values of parameters $R/a = 0.1$, $b/a = -0.3$, $d/a = 0.3$, $\gamma/a = 3$, $\kappa_\ell = 0$ $(\ell = 1, 2)$ (the outer circuit of the voltage generator is "disconnected". The graph of quantity $\delta = |Q(\eta_2)/\Phi^*|$ for the same values of parameters are represented in Fig. 12.22. The arguments of the functions of optimal control are shown in Fig. 12.23. The contour lines of the modulus of the displacement amplitude in a layer with an opening under the influence of optimal shear forces for this case are given in Fig. 12.24.

Figures 12.25–12.28 illustrate analogous results for asymmetrically positioned portions by the application of control actions for the values of parameters $R/a = 0.1$, $b/a = -0.4$, $d/a = 0$, $\gamma a = 4$, $\kappa_\ell = 0$ $(\ell = 1, 2)$. In the calculations we set $a = 1m$.

The numerical calculations show that the form of the control function is considerably determined by the frequency of harmonic excitation, the length and distribution of the loading portions, and also the configuration of heterogeneities.

Appendix A

Table A.1. Properties of some piezoceramics [25]

Quantity	Dimensions	Composition			
		$PZT-4$	$PXE-5$	$BaTiO_3$	$PZT-5$
c_{11}^E		13.9	10.3	15.0	12.1
c_{12}^E		7.78	5.80	6.6	7.54
c_{13}^E	$10^{10} N/m^2$	7.43	5.90	6.6	7.52
c_{33}^E		11.5	10.2	14.6	11.1
c_{44}^E		2.56	2.5	4.4	2.11
s_{11}^E		12.3	15.4	9.1	16.4
s_{12}^E		−4.05	−5.1	−2.7	−5.74
s_{13}^E	$10^{-12} m^2/N$	−5.31	−6.2	−2.9	−7.22
s_{33}^E		15.5	17.0	9.5	18.8
s_{44}^E		39.0	40.0	22.8	47.5
$\ni_{11}^s / \ni_0$		730	1008	840	916
$\ni_{33}^s / \ni_0$	$\ni_0 = 8,85 \cdot 10^{-12}$	635	893	820	830
$\ni_{11}^\sigma / \ni_0$	F/m	1475	1800	1450	1730
$\ni_{33}^\sigma / \ni_0$		1300	1750	1700	1700
e_{31}		−5.2	−7.78	−4.35	−5.4
e_{33}	cal/m^2	15.1	15.2	17.5	15.8
e_{15}		12.7	12.9	11.4	12.3
d_{31}		−123	−178	−78	−171
d_{33}	$10^{-12} cal/N$	289	356	190	374
d_{15}		496	515	260	584

Table A.2. The roots μ_i $(i = 1, 2, 3)$ of (3.8) for some piezoelectric media.

	Plane stress state		Plane deformation	
PZT-4	1.208661i,	$\pm 0.160878 + 1.010431i$	$1.229865i$,	$\pm 0.248901 + 1.054629i$
$PXE-5$	$1.018582i$,	$\pm 0.276178 + 1.023294i$	$1.106819i$,	$\pm 0.329118 + 1.067716i$
$BaTiO_3$	$0.923772i$,	$\pm 0.211637 + 1.000093i$	$0.950025i$,	$\pm 0.234453 + 1.003569i$
$PZT-5$	$1.023773i$,	$\pm 0.214681 + 1.038072i$	$1.102905i$,	$\pm 0.307261 + 1.065445i$

Appendix B

In order to solve the singular integral equations (SIE) we efficiently apply the direct method allowing to avoid the procedure of their regulation and is directly reduces them to the system of linear algebraic equations. One of the direct methods is the method of quadratures [24, 48, 70, 84, 137, 208, 210, 211, 216, 217, 218, 243, 244, 245, 246, 247, 248, 249].

B.1 The Approximate Solution of Singular Integrodifferential Equations Prescribed on Smooth Disconnected Contours

The procedure of the method of quadratures consider on the example of solution of singular integrodifferential equation similar to the equation considered in Par. 4.1.

$$\int_{-1}^{1} \frac{p'(\delta)\, d\delta}{\delta - \delta_0} + \int_{-1}^{1} p(\delta) G(\delta, \delta_0)\, d\delta = N(\delta_0), \quad \delta_0 \in [-1, 1]. \tag{B.1}$$

Appearing here function $G(\delta, \delta_0)$ may have a slight singularity.
Equation (B.1) is considered together with additional condition

$$\int_{-1}^{1} p'(\delta) d\delta = 0. \tag{B.2}$$

Seeking for the solution of SIE in the class of functions with derivatives not restricted at the ends of portion $[-1, 1]$ we assume

$$p'(\delta) = \frac{\Omega_0(\delta)}{\sqrt{1 - \delta^2}}, \quad \Omega_0(\delta) \in H[-1, 1]. \tag{B.3}$$

Construct the Lagrange interpolating polynomial for sought-for function $\Omega_0(\delta)$ in the form [126]

$$L[\Omega_0, \delta] = \frac{2}{n}\sum_{\nu=1}^{n}\Omega_\nu^0 \sum_{m=0}^{n-1}\cos m\Theta_\nu \cos m\Theta - \frac{1}{n}\sum_{\nu=1}^{n}\Omega_\nu^0,$$
$$\Theta_\nu = \frac{2\nu-1}{2n}\pi,\ \Omega_\nu^0 = \Omega_0(\delta_\nu),$$
$$\delta_\nu = \cos\Theta_\nu \quad (\nu = 1, 2, \ldots, n), \tag{B.4}$$

where Θ_ν are the zeros of Chebyshev polynomial of the first kind.

Applying Gauss formula [126] to the additional condition (B.2)

$$\int_{-1}^{1}\frac{\omega(\delta)\,d\delta}{\sqrt{1-\delta^2}} = \frac{\pi}{n}\sum_{\nu=1}^{n}\omega(\delta_\nu), \tag{B.5}$$

we obtain

$$\sum_{\nu=1}^{n}\Omega_\nu^0 = 0. \tag{B.6}$$

Therefore in (B.4) the last sum may be omitted. For the singular integral we have [208]

$$\int_{-1}^{1}\frac{\omega(\delta)\,d\delta}{\sqrt{1-\delta^2}(\delta-\eta)} = \frac{\pi}{n}\sum_{\nu=1}^{n}\frac{\omega(\delta_\nu)}{\delta_\nu-\eta} + \pi\omega(\eta)\frac{U_{n-1}(\eta)}{T_n(\eta)}, \tag{B.7}$$

where $T_n(\eta) = \cos(n\arccos\eta)$, $U_{n-1}(\eta) = \sin(n\arccos\eta)/\sqrt{1-\delta^2}$ are the Chebyshev polynomial of the first and second kind, respectively.

At points $\eta_m = \cos(\pi m/n)$ $(m = 1, 2, \ldots, n-1)$ which are zeros of the Chebyshev polynomial of the second kind, formula (B.7) coincides with the ordinary formula of quadrature of Gauss type (B.5)

$$\int_{-1}^{1}\frac{\omega(\delta)\,d\delta}{\sqrt{1-\delta^2}(\delta-\eta_m)} = \frac{\pi}{n}\sum_{\nu=1}^{n}\frac{\omega(\delta_\nu)}{\delta_\nu-\eta_m} \quad (m = 1, 2, \ldots, n-1). \tag{B.8}$$

Formulas (B.5) and (B.5) are coexact, if $\omega(\delta)$ is the power polynomial not exceeding $2n-1$ and $2n$ [126, 208], respectively.

Integrating expression (B.4) and taking into account (B.6) we find

$$p(\delta_\nu) = \int_{-1}^{\delta_\nu} dp(\delta) = -\frac{2}{n}\sum_{m=1}^{n} S_{m\nu}\Omega_m^0, \tag{B.9}$$
$$S_{m\nu} = \sum_{\ell=1}^{n-1}\frac{\cos\ell\Theta_m \sin\ell\Theta_\nu}{\ell}.$$

With the help of (B.5) and (B.9) we obtain the formula of quadrature

$$\int_{-1}^{1} p(\delta) G(\delta, \eta_m)\, d\delta = -\frac{2\pi}{n^2} \sum_{k=1}^{n} \sum_{\nu=1}^{n} S_{k\nu} \Omega_k^0 G(\delta_\nu, \eta_m) \sin\Theta_\nu. \tag{B.10}$$

Thus, using the quadrature formulas (B.5), (B.8) and (B.10) to the integrals in (B.1), (B.2) and taking into account (B.6) we come to the system of n linear algebraic equations with respect to the values of function $\Omega_0(\delta)$ in the nodes of interpolation δ_ν $(\nu = 1, 2, \ldots, n)$

$$\begin{cases} \left\{ \sum\limits_{\nu=1}^{n} c_{m\nu}\, \Omega_\nu^0 = N_m \right. \ (m = 1, 2, \ldots, n-1), \\ \sum\limits_{\nu=1}^{n} \Omega_\nu^0 = 0, \end{cases}$$

$$c_{m\nu} = \frac{\pi}{n} g(\delta_\nu, \eta_m) - \frac{2\pi}{n^2} \sum_{k=1}^{n} S_{k\nu} \sin\Theta_k G(\delta_k, \eta_m), \tag{B.11}$$

$$N_m = N(\eta_m), \ \Omega_\nu^0 = \Omega_0(\delta_\nu).$$

At points $\delta = \pm 1$ interpolation polynomial (B.4) has the form

$$L[\Omega_0, -1] = \frac{1}{n} \sum_{\nu=1}^{n} (-1)^{\nu+n} \Omega_\nu^0 tg \frac{2\nu - 1}{4n} \pi, \tag{B.12}$$

$$L[\Omega_0, 1] = -\frac{1}{n} \sum_{\nu=1}^{n} (-1)^{\nu} \Omega_\nu^0 ctg \frac{2\nu - 1}{4n} \pi.$$

Increasing n the sequence of the constructed according to the solution of system (B.11) of interpolating trigonometric polynomials is evenly converges

Table B.1.

n	$c_{44} = c_{55} = 1$, $c_{45} = 0$	$c_{44} = 1,\ c_{55} = 3$, $c_{45} = 0$	$c_{44} = 1,\ c_{55} = 6$, $c_{45} = 0.5$	
	$\langle K_{III}^- \rangle$	$\langle K_{III}^- \rangle$	$\langle K_{III}^- \rangle$	$\langle K_{III}^+ \rangle$
2	0.8830	1.0792	1.3663	1.3663
3	0.9673	1.0905	1.1717	1.2779
4	1.0396	1.1250	1.1455	1.2597
5	1.0365	1.1248	1.1548	1.2595
6	1.0334	1.1242	1.1595	1.2597
7	1.0337	1.1246	1.1624	1.2587
8	1.0340	1.1250	1.1630	1.2601
9	1.0339	1.1249	1.1625	1.2603
10		1.1248	1.1623	1.2601
11		1.1248		
12		1.1249		

to the solution of (B.1), (B.2) [210]. the uniform tendency of the approximate solution of the (B.1), (B.2) to accurate.

In order to illustrate the convergency of the given scheme in Table B.1 there are represented the value of relative stress intensity factors $\langle K_m \rangle$ on the tips of parabolic crack $\left(\xi_1 = \delta, \xi_2 = \delta^2/2, \delta \in [-1, 1]\right)$ in the anisotropic space under the influence on its edges of statistic shear forces $(x_3 = const)$ (see: Par. 5.1).

B.2 Solution of the Singular Integral Equations Given on Smooth Connected Contours

Consider the singular integral equation of the second kind which is similar to SIF obtained in Par. 5.1

$$f(\varphi_0) + \int_0^{2\pi} \frac{f(\varphi)\, d\varphi}{\varphi - \varphi_0} + \int_0^{2\pi} f(\varphi) F(\varphi, \varphi_0)\, d\varphi = M(\varphi_0), \tag{B.13}$$

$$\varphi_0 \in [0, 2\pi].$$

Here, kernel $F(\varphi, \varphi_0)$ may have not more than a slight singularity, $M(\varphi_0) - 2\pi$ is the periodical of the Hölder function.

Calculating the singular and regular integrals in (B.13) we use quadrature formula [137]

$$\int_0^{2\pi} y(\varphi) g(\varphi, \varphi_0)\, d\varphi = \frac{2\pi}{n} \sum_{j=1}^{n} y(\varphi_j) g(\varphi_j, \varphi_m^*), \tag{B.14}$$

where $\varphi_j = 2\pi(j-1)/n$ and $\varphi_j^* = \varphi_j + \pi/n\, (j = 1, 2, \ldots, n)$ are the nodes of interpolation and collocation, respectively.

The interpolation formula for function $f(\varphi)$ at odd n has the form [70]

$$L_m[f; \phi] = \frac{1}{n} \sum_{j=1}^{n} f(\varphi_j) \sin \frac{n(\varphi_j - \varphi)}{2} \cos ec \frac{\varphi_j - \varphi}{2}. \tag{B.15}$$

In the nodes of collocation $\varphi_m^* \left(m = \overline{1, n}\right)$

$$L[f; \phi_m^*] = \frac{1}{n} \sum_{j=1}^{n} (-1)^{m+j} f(\varphi_j) \cos ec \frac{\varphi_m^* - \varphi_j}{2}. \tag{B.16}$$

Taking into account the formulas (B.14), (B.16) integrals (B.13) are brought to the system of n algebraic equations with respect to the values of function $f(\varphi)$ in the nodes of interpolation φ_j $(j = 1, 2, \ldots, n)$

$$\frac{1}{n}\sum_{j=1}^{n} f_j^0 \left\{(-1)^{m+j} \cos ec\frac{\varphi_m^* - \varphi_j}{2} + 2\pi F^*(\varphi_j, \varphi_m^*)\right\} = M_m, \tag{B.17}$$

$$F^*(\varphi_j, \varphi_m^*) = F(\varphi_j, \varphi_m^*) + \frac{1}{\varphi_j - \varphi_m^*},$$

$$f_j^0 = f(\varphi_j), \ M_m = M(\varphi_m^*) \ (m = 1, 2, \ldots, n).$$

At $n \to \infty$ the sequence of the approximate solution evenly converges to the coexact solution of SIF (B.13).

B.3 Numerical Solution of the Singular Integrodifferential Equations Given on Connected Contours

Consider one of the methods of numerical solution of the system of the integrodifferential equations of type (5.18) and (6.42) based on the method of quadratures. Let us write down the Lagrange interpolation polynomial for sought-for functions $p(\zeta)$ and $f'(\zeta)$ in nodes $\beta_j = 2\pi(j-1)/N$ $(j = \overline{1, N})$ in the form

$$L_N[\{p_*(\beta), f_x'\}; \beta] = \frac{1}{N}\sum_{j=1}^{N}\{p_j^0, f_j^0\} \sin\frac{N(\beta_j - \beta)}{2} \cos ec\frac{\beta_j - \beta}{2}, \tag{B.18}$$

$$p(\zeta) = p_*(\beta), \ p_j^0 = p_*(\beta_j),$$
$$f(\zeta) = f_*(\beta), \ f_j^0 = f_*'(\beta_j)$$

Expressions (B.18) are valid for the odd number of the nodes of subdivision of contour C.

Integration of polynomial (B.18) for $f_*'(\beta)$ with the help of equality [57]

$$\int \frac{\sin(2m+1)x}{\sin x}dx = 2\sum_{k=1}^{m}\frac{\sin 2kx}{2k} + x,$$

brings to the following expression for function $f_*(\beta)$

$$M_N[f_*(\beta); \beta] = \frac{1}{N}\sum_{j=1}^{N} f_j^0 \Omega_j(\beta) + A, \tag{B.19}$$

$$\Omega_j(\beta) = -2\sum_{k=1}^{\frac{N-1}{2}}\frac{\sin k(\beta_j - \beta) - \sin k\beta_j}{k} + \beta.$$

Appearing here constant A should be determined from the condition of periodicity of function $f_*(\beta)$, which due to (B.19) have the form

$$\sum_{j=1}^{N} f_j^0 = 0. \tag{B.20}$$

Using (B.19) we also find the quadrature formula

$$\int_0^{2\pi} f_*(\beta)\, G(\beta, \beta^*)\, d\beta = \frac{2\pi}{N^2} \sum_{j=1}^{N} f_j^0 \sum_{m=1}^{N} \Omega_{jm} G(\beta_m, \beta^*) + \tag{B.21}$$
$$+ A \frac{2\pi}{N} \sum_{m=1}^{N} G(\beta_m, \beta^*),$$

where $\Omega_{jm} = \Omega_j(\beta_m)$. In the nodes of collocation $\beta_\ell^* = \pi(2\ell - 1)/N$ $\left(\ell = \overline{1, N}\right)$ at odd value N polynomial (B.18) takes value

$$L_N\left[p_*(\beta); \beta_\ell^*\right] = \frac{1}{N} \sum_{j=1}^{N} p_j^0 (-1)^{\ell+j} \cos ec \frac{\beta_\ell^* - \beta_j}{2} \quad \left(\ell = \overline{1, N}\right). \tag{B.22}$$

For the singular integral in (6.42) there is a formula similar to the calculation formula of regular integral

$$\int_0^{2\pi} f_*'(\beta)\, \mathrm{Im} \frac{e^{i\psi_0}}{\zeta(\beta) - \zeta_0(\beta_\ell^*)} s'(\beta)\, d\beta = \tag{B.23}$$
$$= \frac{2\pi}{N} \sum_{j=1}^{N} f_j^0 \mathrm{Im} \frac{e^{i\psi_0(\beta_\ell^*)}}{\zeta(\beta_j) - \zeta_0(\beta_\ell^*)} s'(\beta_j).$$

Substituting the integrals in (6.42) by the final seems over formulas (B.21), (B.23) and using equalities (B.19), (B.20) and (B.22) we will come to the system of $2N+1$ algebraic equations with respect to the values of functions $p(\zeta)$ and $f'(\zeta)$ in the nodes of interpolation β_j $\left(j = \overline{1, N}\right)$ and constant A.

Note. The approximation of type (B.18) of function $f'(\zeta)$ are not quite correct, because, as it follows from system (6.42), this function has root singularities on the edges of electrodes. However, at some distance from the edges, as the numerical investigation shows, we obtain the results appear to be satisfactory. Solving the system of SIF (5.18) over the described scheme, due to the continuity of its solution, the indicated incorrectness does not take place.

In conclusion we should note that the indicated here procedure were probed in the process of the numerical realization of the boundary problems of electroelasticity for bodies with defects and appeared to be rather efficient.

References

1. Avetisyan A.S. About the distribution of shear waves in a piezoelectric medium. Izv. Acad. Sci. ArmSSR. "Mekhanika". Issue 38, No. 1, pp. 12–19, 1987.
2. Agranovich V.M., Derevianko N.I.. An electroelastic field of direct piezoeffect at deformation of piezoceramic bodies. Applied Mekh 1974. T 10. No. 9. pp 3–8
3. Alexandrov V.M., Mkhitaryan S.M.. Contact problems of bodies with thin coatings and interlayers. M Nauka, 1983, 487p.
4. Ambrartsumyan S.A., Bagdasaryan G.E.. Electroconductive plates and shells in a magnetic field. M Phys. Math. Lit. 1996, 286p.
5. Ambrartsumyan S.A., Bagdasaryan G.E., Belubekyan M.V.. Magnetoelasticity of thin shells and plates. M.: Nauka, 1977, 272p.
6. Ambrartsumian S.A., Belubekian M.B.. Some problems of electromagnetic elastic plates. Yerevan State Univ. Pub. 1991, 143p.
7. Akhiezer N.I.. A classical problem of moments and some questions of analysis connected with it. M.: Fizmatgiz, 1961, 310p.
8. Babeshko V.A., Glushkov E.I., Zinehenko Z.F.. The dynamics of heterogenic linear elastic media. M.: Nauka, 1989, 344p.
9. Babich V.M., Buldirev V.S.. The asymptotic methods in the problems of diffraction of short waves. M.: Nauka, 1972, 456p.
10. Babich V.M., Molotkov I.A.. The mathematical methods in the theory of elastic waves. Summaries of science and technology. Ser. "Mechanics of deformed bodies", 1977, T 10, pp 5–62.
11. Bazenov V.M., Kutsenko G.V., Vlitko A.F., Distribution of plane electroelastic waves in a piezoceramic medium. Proc. Acad. Sci. USSR. Ser. A, 1977, No 2, pp 124–128.
12. Balakirev M.K., Gilinsky I.A.. Waves in piezocrystals. Novosibirsk: Nauka, 1982, 239p.
13. Bardzokas D.I., Kudryavtsev B.A., Senik N.A.. Rayleigh's waves in a half space with the finite system of surface electrodes. Izv. Russian Acad. Sci. Mechanics of solid bodies, 1996, No 1, pp 45–51.
14. Bardzokas D.I., Kudryavtsev B.A., Senik N.A.. About criteria of electromechanic fracture of piezoelectrics initiated by the edges of electrodes. Probl. Prochnosti, 1994, No. 7, pp 42–46.
15. Bardzokas D.I., Parton V.Z., Theocaris P.S.. Integral equations of the theory of elasticity for a multiconnected area with inclusions. Prikl. Math. and Mekh, 1989, V. 53, No 3, pp 485–495.

16. Bardzokas D.I., Rodriges P., Filstinsky M.L.. Tesion of piezoceramic bimorph with a crack crossing the interphase. Mekh. Komposit Materialov, 1997, V. 33, No 4, pp 482–488.
17. Bardzokas D.I., Senik N.A.. Excitation of symmetric and asymmetric Lamb waves in a piezostrip by surface electrodes. Pukl. Math. Mekh. 1993, V. 57, No 6, pp 91–99.
18. Bardzokas D.I., Filstinsky M.L., Concentration of electroelastic fields in a composite piezoceramic plate crossing the interphase materials. Mekh. Komposit. Materialov 1999, V. 35, No 3, pp 359–366.
19. Bardzokas D.I., Filstinsky M.L.. Impulse excitation of a piezoceramic halfspace with tunnel cavities. Izv. Rus. Acad. Sci. Mechanics of solid bodies 1996, No 3, pp 21–26.
20. Bardzokas D.I., Filstinsky M.L., Diffraction of a shear wave on cylindrical inclusions in a piezoelectric halfspace. Izv. Rus. Acad. Sci. Mechanics of solid bodies 1997, No 3, pp 77–84.
21. Barroute J. Introduction into the physics of ferroelectric phenomena. M.: Mir 1970, 343p.
22. Beitman Q., Erdein A., Heigher transcendental functions. M.: Nauka 1967, 300p.
23. Belokon A.V. Vorovich I.I., Some mathematical questions of the theory of electroelastic bodies. Aktualnie Probl. Mekhaniki Deformiruemikh Sred. 1979, pp 53–67.
24. Belotserkovsky S.M., Lifanov I.K.. The numerical methods in singular integral equations. M.: Nauka 1985, 254p.
25. Berlineourt D., Kerran D., Jaffe H.. Piezoelectric and piezomagnetic materials and their applications in converters. Fizicheskaia Akustika, V. 1, Methods and instruments of ultrasound researches. Part A., M.: Mir 1966, pp 204–326.
26. Birukov S.B.. Calculation of electrode transforms of surface waves in piezoelectrics. Z. Teckhn. Fisiks 1980, V. 50, No 8, pp 1655–1662.
27. Birukov S.B. Gorishnik L.L.. The theory interaction of surface waves in piezoelectrics with electrode structures. Z. Techn. Fisiki 1980, V. 50, No 8, pp 1647–1654.
28. Bitsadze A.B.. Boundary problems for elliptic equations of the second order. M.: Nauka 1966, 203p.
29. Bitsadze H.P.. Fundamental boundary conditions of the problem of electroelasticity for transversal-isotropic halfspace. Trans. Inst. Prikl. Matematiki, Tbilisi State Uni. 1985, No 16, pp 34–55.
30. Boriskovsky V.G., Parton V.Z.. The dynamic fracture mechanics. Itogi Nauki I Tekhn. VINITI. Ser. "Mechanics of deformable solid bodies" 1983, Vol. 16, 84 p.
31. Born M., Wolf E.. Principles of optics. M.:Nauka 1973, 720 p.
32. Borodachev N.M., Dynamic problem of a crack in case of a longitudinal shear deformation. Probl. Proehnosti. 1973, No 4, pp 23–25.
33. Brebbia C.A., Telles J.C., Wrobel L.C.. Boundary element techniques. M.: Mir 1987, 524 p.
34. Brekhovskich L.M.. Waves in laminated media. M.: Nauka 1973, 342 p.
35. Brichkov I.A., Prudnikov A.P.. Integral transformations of generalized functions. M.: Nauka 1977, 287 p.
36. Butkovsky A.Q.. The methods of control of the systems with disturbed parameters. M.: Nauka, 1975, 287 p.

37. Waganov R.B., Katsenelenbaum B.Z.. Bases of the theory of diffraction. M.: Nauka, 1975, 568 p.
38. Weinshtein L.A.. The theory of diffraction and the method of factorization. M. Sou Radio. 1966, 430 p.
39. Vatulian A.O., Kubikov V.L.. About boundary integral equations in electroelasticity. Prikl. Math. and Mech. 1989, V. 53, Issue 6.
40. Vekovischeva I.A.. A plane problem of the theory of electroelasticity for a piezoelectric plate. Prikl. Mekh. 1975, V. II, No 2, pp 85–89.
41. Vekua A.P.. The systems of singular integral equations and some boundary problems. M.: Nauka 1970, 380 p.
42. Verlan A.F., Sizikov V.S.. Integral equations; methods algorithms, programs. Directory. K. Nauka dumka 1986, 544 p.
43. Vladimirov V.S. Equations of the mathematical physics. M.: Nauka 1981, 512 p.
44. Voitovich N.V., Katsenelenbaum B.Z., Sivov A.N.. A generalized method of natural oscillations in the theory of diffraction. M.: Nauka 1974, 416 p.
45. Vorovich I.I., Alexandrov V.M., Babeshko V.A.. Non-classic mixed problems of the theory of elasticity. M.: Nauka 1974, 455 p.
46. Vorovich I.I., Babeshko V.A.. Dynamic mixed problems of the theory of elasticity for non-classic areas. M.: Nauka 1979, 319 p.
47. Methods of calculations in electrodynamics (Editor: R. Mittra). M.: Mir 1977, 485 p.
48. Gabdulhaev B.G.. Finite dimensional approximation of singular integrals and discrete methods of singular integral and integrodifferrential equations. In the book: Mathematical Analysis, Moscow 1980, pp 251–307 (Itogi Nauki I Tekhniki VINIITI, Issue 18)
49. Galishnikova T.N., Ilinsky A.S., Numerical methods in the problems of diffraction. M.: Pub. House of Moscow Uni., 1987, 208 p.
50. Gahov F.D.. Boundary Problems. M.: Nauka, 1977, 640 p.
51. Gahov F.D., Chersky U.I.. Equation of a convolution type. M.: Nauka, 1978, 295 p.
52. Gilinsky I.A., Popov V.V.. Excitation of electroacoustic waves in piezoelectrics by outer sources. Z. Tekhn. Fizizki, 1976.
53. Golovchan V.T.. Diffraction of an elastic transverse on cylinders in a halfspace. Proc. Acad. Sci. USSR, Ser. A, 1971, No 3, pp 246–249.
54. Gorishnik L.L., Kondratiev S.N.. Excitation of surface electroacoustic waves by electrode transforms. Radiotekhnika I Elektronika, 1974, V. 19, No 8, pp 1719–1728.
55. Gorishnik L.L., Kondratiev S.N.. The theory of electrode transforms of elastic surface waves in piezoelectrics. Radiotekhnika I Elektronika, 1978, V. 23, No 1, pp 151–159.
56. Gorishnik L.L., Kondratiev S.N.. Calculation of electrode transforms of surface waves. Radiotekhnika I Elektronika, 1978, V. 23, No 1, pp 160–166.
57. Gradshtein I.S., Rizik I.M.. Tables of integral sums, series and products. M.: Fizmatgiz, 1962, 1100p.
58. Grigoliuk E.I., Filshtinsky L.A.. Periodical piece-wise homogenous elastic structures. M.: Nauka, 1994, 336 p.
59. Grigoliuk E.I., Filshtinsky L.A.. Regular piece-wise homogenous structures with defects. M.: Nauka, 1994, 336p.

60. Grigoliuk E.I., Filshtinsky V.A., Filshtinsky L.A.. About application of the general problem of moments to some optimization problems of the theory of elasticity. Izv. Rus. Acad. Sci. Mechanics of solid bodies, 1992, No 2, p. 31–37.
61. Grilitsky D.V., Opanasovich V.K.. Distribution of stresses in a piece-wise homogenous plane with a slit under pressure. Izv. Acad. Sci. Armenian SSR. Ser. "Mechanics", 1974, V. 27, No 2, pp 24–39.
62. Grinchenko V.T.. Balance and steady-state oscillations of elastic bodies of finite dimensions. K.: Nauka Dumka, 1978, 264 p.
63. Grinchenko V.T., Meleshko V.V.. Harmonic oscillations and waves in elastic bodies. K.: Naukova Dumka, 1981, 253 p.
64. Guz A.N., Golovchan V.T.. Diffraction of elastic waves in multiply connected bodies. K.: Naukova Dumka, 1972, 253 p.
65. Guz A.N., Kubenko V.D., Cherevko M.A.. Diffraction of elastic waves. K. Naukova Dumka, 1978, 308 p.
66. Duduchava R.V.. Integral equations of convolution with discontinuous presymbols, singular integral equations with immobile singularities and their applications to the problems of mechanics. Tbilisi: Metsniereba, 1979, 135 p.
67. Dundurs I., Komninou M.. Revision and prospects of investigation of an anterphase crack. Fracture of composite materials. Proc. 1st Soviet-American Sym., Riga, 1979, pp 78–87.
68. Dieulesaintn E., Roger D.. Elastic waves in solid bodies. Application for processing of signals. M.: Nauka, 1982, 424 p.
69. Sommerfeld A.. Optics. Translation from German, edited by M.E. Elishevicha. M.: Inostr. Liter., 1953.
70. Ivanov V.V. The theory of approximate methods and its applications to the numerical solution of singular integral equations. K.: Nauka Dumka, 1968, 288 p.
71. Ivanov V.V. The methods of calculations on computers. Directory. K.: Naukova Dumka, 1986, 584 p.
72. Ivanov E.A. Diffraction of electromagnetic waves for two bodies. Minsk: Nauka I Tekhnika, 1968, 582 p.
73. Ilien V.P. The numerical methods of solutions of the problems of electrophysics. M.: Nauka, 1985, 336 p.
74. Iliensky A.S., Shestopalov U.V. Application of the spectral methods in the problems of distribution of waves. Moscow State Uni. Pub.
75. Iliushin A.A. Mechanics of a solid medium. 3rd Edition. Moscow State Uni. Pub., 1978, 288 p.
76. Inglind. A crack between two various media. Trans ASME J. Appl. Mech., 1965, Vol. 32, No 2, pp 165–168.
77. Zabreiko P.P., Koshelev A.I., Krasnoselky M.A. et al. Integral Equations. M.: Nauka, 1973, 496 p.
78. Isakovich M.A. General acoustics. M.: Nauka, 1973, 496 p.
79. Israilov M.S. The dynamic theory of elasticity and a wave diffraction. Moscow State Uni. Pub., 1992, 205 p.
80. Kino G.S.. Acoustic waves. M.: Nauka, 1990, 656 p.
81. Kalandia A.I. The mathematical methods of two dimensional elasticity. M.: Nauka, 1973, 306 p.
82. Kit G.S., Krivtsun M.G. Plane problems of thermoelasticity for bodies with cracks. K.: Naukova Dumka, 1983, 277 p.

83. Kokunov V.A., Kudryavtsev B.A., Senik N.A. A plane problem of electroelasticity for a piezoelectric layer with periodical system of electrodes on surfaces. Pukl. Math. I Mekh, 1985, T. 49, No 3, pp 489–491.
84. Korneichuk A.A. Quadrature formulas for singular integral. The numerical methods of solution of differential and integral equations and quadrature formulas. M.: Nauka, 1964, pp 64–74.
85. Kosmodamiansky A.S., Kravchenko A.P., Lozkin V.N. Influence of point electric charge on the boundary of an piezoelectric halfplane weakened by an elliptic opening. News of Acad. Sci. of Armenian SSR. Ser. "Mekhanica", 1977, V. 30, No 1, pp 13–20.
86. Kosmodamiansky A.S., Lozkin V.N. Generalized plane stress state of thin piezoelectric plates. Prikl. Mekh., 1975, Vol. 11, No 5, pp 45–53.
87. Kosmodamiansky A.S., Storozev V.I. Dynamic problems of the theory of elasticity for anisotropic media. K.: Naukova Dumka, 1985, 175 p.
88. Crouch S.L., Starfild A.M.. Boundary element methods in solid mechanics. London, 1983, 330p.
89. Krein M.G., Nudelman A.A. Problems of Markov moments and extremal problems. M.: Nauka, 1973, 551 p.
90. Krivoy A.F., Radiollo N.V. Singularities of a stress field in the vicinity of inclusions in a composite anisotropic plane. Izv. Acad. Sci. USSR. Meckanise of solid bodies, 1984, No 3, pp 84–92.
91. Kudryavtsev B.A. Mechanics of piezoelectric materials. Itogi nauki i techniki VINIITI. Ser. "Meckanics of solid bodies", 1978, V. 11, pp 5–66.
92. Kudryavtsev B.A. Electroelastic state of a halfspace from piezoceramics with two boundary electrodes. Probl. Prochnostic, 1982, No 7, pp 56–59.
93. Kudryavtsev B.A., Parton V.Z., Rakitin V.I. Mechanics of fracture of piezoelectric materials. A rectangular tunnel crack on a boundary with a conductor. Prikl. Mat. and Mech., 1975, V. 39, No 1, pp 149–159.
94. Kudryavtsev B.A., Raitin V.I. A Griffiths crack in a piezoelectric medium. Izv. Acad. Sci. USSR, Mech. of Solid bodies, 1979, N 1, pp 125–132.
95. Kudryavtsev B.A., Parton V.Z. About steady-state oscillations in electroelasticity. Sov. Rem Probl. Mekhaniki i Aviatsii. M.: 1982, pp 159–172.
96. Kudryavtsev B.A., Parton V.Z., Senik N.A. Mechanical models of piezoelectrics for electronic engineering. Itogi nauki i tekhniki. VINIIII. Ser. "Mechanics of solid bodies", 1984, V. 17, pp 3–62.
97. Kupradze V.D.. Boundary problems of the theory of oscillations and integral equations. M., L.:Gostekhizdat, 1950, 280 p.
98. Kupradze V.D.. The methods of potentials in the theory of elasticity. M.: Fizmatgiz, 1963, 472 p.
99. Kupradze V.D.. Fundamental problems of the mathematical theory of diffraction: Steady-state process. L;M.:ONII, 1935, 112 p.
100. Kupradze V.D., Burchuladze T.V.. Dynamic problems of the theory of elasticity. Itogi Nauki i tekhniki. Ser.: Modern problems of mathematics", VINIITI. M., 1975, V. 7, pp 105–187.
101. Kucherov I.I., Fedorchenko A.M.. Influence of a piezoeffect on elastic waves in bounded crystals. Ukr. Fiz. Z., 1971, V. 16, No 9, pp 1567–1569.
102. Cadu W.. Piezoelectricity and its practice application. M.: Izd. Inostr. Liter., 1949, 718 p.
103. Lavrentiev V.A., Shabat B.V.. The methods of the theory of functions of a complex variable. M.: Nauka, 1973, 736 p.

104. Landau L.D., Lifshitz E.M.. Electrodynamics of continuous media. M.: Fizmatgiz, 1959.
105. Levin L.. The theory of waveguides. M.: Radio and communication, 1981, 311 p.
106. Litvinchuk G.S.. Boundary problems and singular integral equations with a shear. M.: Nauka, 1977, 488 p.
107. Lions J.L.. Optimal control of the systems described by differential equations in partial products. M.: Mir, 1972, 414 p.
108. Loeber J.F., Si G.C.. Green's function for cracks in heterogeneous materials. Pukl. Mekh. Applied Mechanics, 1967, V. 34, No 1, pp 131–133.
109. Luke Y.L. Mathematical functions and their approximations. M.: Mir, 1980, 608 p.
110. Liamshev L.M., Sheviakhov N.S.. Diffusion of a plane axially shear wave by circular piezohalfconductor cylinder. Acoust. Zurhal, 1977, V. 23, No 1, pp 95–105.
111. Macklintoc F.A., Argon A.S.. Mechanical behaviour of materials. 1966, 443 p.
112. Malmeister A.K., Tamuzs V.P., Teters G.A.. Resistance of polymere and composite materials. Riga: Zinatne, 1980, 572 p.
113. Mandelshtam L.I.. Lectures on the theory of oscillations. M.: Nauka, 1972, 470 p.
114. Mechanics of fracture and strength of materials. Directory: 4 Volumes, edited by Panasiuk V.V., K.: Naukova Dumka, 1988–1989.
115. Mechanics of coupled fields in the elements of constructions. 5 volumes. Electroelasticity, Grinchenko V.T., Ulitko A.F., Shulga N.A., edited by Guz A.N., K.: Naukova Dumka, 1989, V. 5, 280 p.
116. Mittra R., Lee S.. Analytical techniques in the theory of guided waves. M.: Mir, 1974, 323 p.
117. Mikhlin S.G., Morozov N.F., Paukshto M.V.. Integral equations in the theory of elasticity. SP.: Pub. House of Saint-Petersburg, 1994, 271 p.
118. Morachkovsky O.K., Filshtinsky M.L.. Interaction of stress waves with cracks in an anisotropic medium under the conditions of antiplane deformation. Proc. Acad. Sci. Ukrainian SSR, Ser. A, 1985, No 6, pp 42–45.
119. Morozov N.F.. Mathematical questions of the theory of cracks. M.: Nauka, 1984, 256 p.
120. Morse P.M., Feshbach H.. Methods of the theoretical physics: 2 Vols., M.: Foreign literature, 1958, Vol.1. 930 p.
121. Muskhelishvili N.I.. Singular integral equations. M.: Nauka, 1968, 512 p.
122. Mason W.. Piezoelectric crystals and their application in ultrasounds. M.: Foreign literature, Pub. 1952, 447 p.
123. Nazarchuk Z.T.. Numerical investigation of diffraction of waves on cylindrical structures. K.: Naukova Dumka, 1988, 254 p.
124. Nye J.F.. Physical properties of crystals and their description with the help of tensors and matrixes. M.: Foreign literature press, 1960, 388 p.
125. Nayfeh A.H.. Perturbation methods. M.: Mir, 1976, 455 p.
126. Natanson I.P.. The constructive theory of functions. M.;L.: Gostekhizdat, 1949, 688 p.
127. Nefedov E.I., Diffraction of electromagnetic waves on dielectric structures. M.: Nauka, 1979, 722 p.
128. Nefedov E.I., Sivov A.N.. Electrodynamics of periodical structures. M.: Nauka, 1977, 208 p.

129. Nikolsky V.V.. The theory of an electromagnetic field. M.: Vishaia Shkola, 1961, 370 p.
130. Nikolsky V.V.. Variational methods for inner problems of electrodynamics. M.: Nauka, 1967, 460 p.
131. Nikolsky V.V., Nikolskaia T.I.. Decomposed approach to the problems of electrodynamics. M.: Nauka, 1967, 304 p.
132. Nikolsky S.M.. Quadrature formulas. M.: Nauka, 1974, 233 p.
133. Noble B.. Methods based on the Viener-Hopf technique. M.: Foreign literature Pub. 1962, 278 p.
134. Novatsky V.. The theory of elasticity. M.: Mir, 1975, 872 p.
135. Novatsky V.. Electromagnetic effects in solid bodies. M.: Mir, 1986, 160 p.
136. Panasiuk V.V., Savruk M.P., Datsishin A.P.. Distribution of stresses in the vincinity of cracks in plates and shells. K.: Naukova Dumka, 1976, 442 p.
137. Panasiuk V.V., Savruk M.P., Nazarchuk Z.T.. Method of singular integral equations in two dimensional problems of diffraction. K.: Naukova Dumka, 1984, 344 p.
138. Parton V.Z., Boriskovsky V.G.. Dynamics of brittle fracture. M.: Machinostroenie, 1988, 239 p.
139. Parton V.Z., Kudryavtsev B.A.. Dynamic antiplane problem of piezoelectric medium. Proc. Moscow Inst. Of Chem. Eng., 1974, No 56, pp 3–13.
140. Parton V.Z., Kudryavtsev B.A.. Electroelasticity of piezoelectric and electroconductive bodies. M.: Nauka, 1988, 472 p.
141. Parton V.Z., Kudryavtsev B.A., Senik N.A.. About one criterium of electric fracture of dielectrics in strongly heterogeneous fields. Proc. Acad. Sci. USSR, 1988, V. 298, No 3, pp 611–615.
142. Parton V.Z., Mustafaev J.M., Senik N.A.. Methods of the mechanics of fracture and the breakdown of dielectrics. Baku.: Inst. Of Phys. Of Acad. Sci. of Azerbaidgan SSR, preprint, 1991, No 7, 24 p.
143. Parton V.Z., Perlin P.I.. Integral equations of the theory of elasticity. M.: Nauka, 1977, 12 p.
144. Parton V.Z., Perlin P.I.. Methods of the mathematical theory of elasticity. M.: Nauka, 1981, 688 p.
145. Parton V.Z., Filstinsky M.L.. Dynamic problems of the theory of elasticity for a piezoelectric medium with tunnel cuts. News of Acad. Sci. Armenian SSR. Ser. "Mekhanika", 1989, V. 42, No 5, pp 17–24.
146. Parton V.Z., Filstinsky M.L.. General representations of coupled electromagnetic and mechanical fields in a piezoelectric medium. Proc. Acad. Sci. USSR, 1989, V. 308, No 1, pp 53–55.
147. Parton V.Z., Filstinsky M.L.. Stationary wave process in a piezoelectric layer and halfplanes weakened by tunnel cuts (antiplane deformation). Prikl. Math. i Mech., 1992, V. 56, No 3, pp 510–518.
148. Parton V.Z., Filstinsky M.L.. Dynamic problem of electroelasticity for a layer and halfplanes with tunnel cavities. Izv. Russian Acad. Sci. Mech. Of solid bodies, 1993, No 5, pp 82–88.
149. Pisarenko G.G.. Strength of piezoceramics. K.: Naukova Dumka, 1987, 231 p.
150. Podstrigach J.S., Bourak J.I., Gachkevich A.P., Cherniavskaia L.V.. thermoelasticity of electroconductive bodies. K.: Naukova Dumka, 1977, 247 p.
151. Podstrigach J.S., Bourak J.I., Kondrat V.F.. Magnetothermoelasticity of electroconductive bodies. K.: Naukova Dumka, 1982, 196 p.

152. Polovinkina I.B., Ulitko A.F.. On the theory of eqyilibrium of piezoceramic bodies with cracks. Teplovie Napriazenia v elementah kohstruktsi, 1978, Issue 18, pp 10–17.
153. Popov G.J.. concentration of elastic stresses in the vicinity of dies, cuts, thin inclusions and strengthens. M.: Nauka, 1982, 344 p.
154. Poruchikov V.B.. Methods of the dynamic theory of elasticity. M.: Nauka, 1986, 328 p.
155. Prossdorf S.. Some classes of singular equations. M.: Mir, 1979, 494 p.
156. Prusov I.A., Lunskia L.I.. Elastic state of a piece-wise homogenous orthotropic plane with cuts. Prikl. Mech., 1969, V. 32, No 8, pp 77–83.
157. Rise J.R., Sih G.C. Plane problems of cracks positioned on the interphase of two various media. Trans. ASME J. Appl. Mech.
158. Romalis N.B., Tamuzs V.P.. Fracture of structurally heterogenic bodies. Riga: Zinatie, 1989, 224 p.
159. Fracture. Translation from English. Edited by H. Liebowitz. M.: Mir, 1973–1976, V. 1–7.
160. Savruk M.P.. Two dimensional problems of elasticity for bodies with cracks. K.: Naukova Dumka, 1981, 324 p.
161. Sedov L.I.. Mechanics of a continuous media. M.: Nauka, 1983, V. 1, 2, 536 p, 584 p.
162. Seleznev I.T.. Simulation of wave and diffractional processes in continuous media. K.: Naukova Dumka, 1989, 204 p.
163. Senik N.A.. Rayleigh waves in a piezoelectric halfspace with a pair of surface electrodes. Izv. Russian Acad. Sci. Mech. of solid bodies, 1992, No 4, pp 106–115.
164. Slepiyan L.I.. Mechanics of cracks. L.: Shipbuilding, 1990, 296 p.
165. Sneddon I.. Fourier transform. M.: Foreign literature Pub. House, 1954, 296 p.
166. Directory of special functions with formulas and mathematical tables. Edited by M. Abramovitz, I Stigun. M.: Nauka, 1979, 830 p.
167. Tamm I.E.. The fundamental theory of electricity. M.: Nauka, 1976, 616 p.
168. Tikhonov A.N., Samarsky A.A.. Equations of the mathematical physics. M.: Nauka, 1977, 736 p.
169. Ulitko A.F.. On the theory of oscillations of piezoceramic bodies. Thermal stresses in the elements of constructions, Isssue 25, 1975, pp 90–99.
170. Ulitko A.F.. Method of natural vector functions in space problems of the theory of elasticity. K.: Naukova Dumka, 1979, 261 p.
171. Ulitko A.F.. About some singularities of the statement of boundary problems of the electroelasticity. Sovrem. Probl. Mekhaniki i aviatsii, 1982, pp 290–300.
172. Fedoruk M.V.. Asymptotics Integrals and Series. M.: Nauka, 1987, 544 p.
173. Felsen L., Markuvits N.. Radiation and diffusion of waves: 2 Vols., M.: Mir, 1978, 547 p, 555 p.
174. Filshtinsky V.A., Filshtinsky M.L.. About the choise of minimizing functional at optimal conversion of an elastic plate from initial state into prescribed final. Prikl. Mekhs, 1986, Vol. 22, No 2, pp 90–94.
175. Filshtinsky L.A.. Interaction of stress waves with curvilinear tunnel cracks of longitudinal shear in a halfspace. Math and mekh, 1982, V. 46, No 3, pp 482–487.
176. Filshtinsky L.A., Volkova L.V.. Dynamic problem of the theory of elasticity for an area with curvilinear cuts (plane deformation). Proc. Acad. Sci. USSR, 1983, Vol. 271, No 4, pp 831–834.

177. Filshtinsky L.A., Filshtinsky M.L.. Tension of a composite pieazoceramic plate weakened by crack cuts. Prikl. Mekh, 1993, V. 29, No 12, pp 66–71.
178. Filshtinsky L.A., Filshtinsky M.L.. Green's function for a composite piezoceramic plane with an interphase crack. Prikl. Math. and Mekh., 1994, Vol 58, Issue 2, pp 159–166.
179. Filshtinsky M.L.. Fracture control of a piezoceramic body with a crack. Dynamics and strength of machines: collection of articles. Kharkov, 1987, Issue 15, pp 93–96.
180. Filshtinsky M.L.. Dynamic reaction of piezoceramic halfspace with a crack of longitudinal shear on the action of concentrated forces. Teoret. i Prikl. Mekh., K.: Donetsk: Higher School, 1989, Issue 20, pp 50–55.
181. Filshtinsky M.L.. Interaction of waves of mechanical displacements with tunnel cracks of longitudinal shear in a piezoceramic halfspace. Izv. Acad. Sci. Armenian SSR, 1988, No 5, pp 50–55.
182. Filshtinsky M.L.. Harmonic oscillation of a piezoelectric halfspace with tunnel cavities (deformation of longitudinal shear). Acoust J., 1991, Vol 37, Issue 4, pp 777–781.
183. Filshtinsky M.L.. Dynamic loading of a piezoceramic halfspace with a crack. Acoust. J., 1993, V. 39, Issue 5, pp 921–928.
184. Filshtinsky M.L.. Unstationary dynamic problem of electroelasticity for an unbounded medium with curvilinear tunnel cracks. Prikl. Math. and Mekh., 1994, V. 58, No 4, pp 159–166.
185. Filshtinsky M.L.. About one approach to the investigation of electroelastic fields in cylinder excited by a system of surface electrodes. Vestnik Sumi Uni., Sumi 1998, No 1(9), pp 3–10.
186. Filshtinsky M.L.. Invastigation of oscillations of a piezoceramic cylinder with tunnel rigid inclusions. Vestnik Sumi State Uni., Sumi 1999, No 1(12), pp 12–20.
187. Filshtinsky M.L.. Solution of static and dynamic problems of electroelasticity for piecewise homogeneous bodies. Dr. Eng, Thesis, Donetsk, 1996, 284 p.
188. Filshtinsky M.L., Bardzokas D.I.. Concentration of stresses in a composite plane with openings. Prikl. Mekh., 1998, V. 34, No 1, pp 98–102.
189. Filshtinsky M.L., Bardzokas D.I.. Oscillations of a piezoceramic cylinder under the influence of pulsing pressure parallel to the symmetry of the material. Vestnik Sumi Uni., Sumi 1999, No 1(12), pp 20–28.
190. Filshtinsky M.L., Bardzokas D.I.. Method of boundary of integral equations in the problems of diffraction of electroelastic waves. Sumi Uni., 1999, 194 p.
191. Filshtinsky M.L., Mordik I.N.. Influence of heterogeneities of crack types on the amplitude frequency characteristics of piezoceramic cylinder. Vestnik Sumi Uni., Sumi 1997, No 1(7), pp 158–170.
192. Finkelshtein A.B. Waves in an electroelastic halfbounded medium with periodical system of electrodes. Izv. Math. Caucasic Sci. Center of Higher School of Natural Sciences, 1986, No 1, pp 59–62.
193. Fok V.A. Problems of diffraction and propagation of electromagnetic waves. M.: Sov. Radio, 1970, 518 p.
194. Henl X., Maue A., Vestpfal K.. The theory of diffraction. M.: Mir, 1964, 428 p.
195. Cherepanov G.P.. Diffraction of elastic waves on a cut. Mechanics of continuous media and relative problems of an analysis. M.: Nauka, 1972, pp 615–622.
196. Cherepanov G.P.. Mechanics of brittle fracture. M.: Nauka, 1974, 640 p.

197. Shestopalov V.P.. method of Riemann-Hilbert problems in the theory of diffraction and propagation of electromagnetic waves. Charkov: Pub. House of Kharkov Uni., 1971, 400 p.
198. Shestopalov V.P.. Adder equations in the modern theory of diffraction. K.: Naukova Dumka, 1983, 252 p.
199. Shulga N.A., Bolkisev A.M.. Oscillation of piezoceramic bodies. K.: Naukova Dumka, 1990, 228 p.
200. Erdogan F.E.. Propagation of stresses in connected various materials with cracks. Trans. ASME, J. Appl. Mech., 1965, Vol. 32, No 2, pp 169–177.
201. Achenbach J.D.. Wave propagation in elastic solids. Amsterdam, North Holland Publ. Co., 1973, 427p.
202. Achenbach J.D.. Elastic wave propagation problems in non destructive evaluations. Trends, Delft, 1989, pp 19–38.
203. Achenbach J.D., Keer L.M., Mendelson D.A.. Elastodynamic analysis of an edge crack. Trans. ASME J. Appl. Mech., 1980, Vol. 47, pp 551–556.
204. Adelman N.T., Stavski Y., Flexural-extensional behaviour of composite piezoelectric circular plates. J. Acoust. Soc. Amer., 1980, Vol. 67, No 3, pp 819–822.
205. Atkinson C.. On some integral equations arising in elastic biomaterial problems. Int. J. Eng. Sci., 1979, Vol. 17, No 5, pp 587–597.
206. Bardzokas D.I., Filshtinsky M.L., Rodriguez R.. Concentration of electroelastic fields in a composite piezoceramic plate with defects crossing the interphase. Mechanics of composite materials, 1998, Vol. 34, No 6, pp 549–556.
207. Bugdayci N., Body D.B.. A two dimensional theory for piezoelectric layers used in electromechanical transducers. Part 1. Derivation Int. J. Solids and Struct., 1981, Vol. 17, No 12, pp 1159–1178.
208. Chawla M.M., Ramakrishnan T.R.. Modified Gauss-Jacobi quadrature formulas for the numerical evaluation of Cauchy type singular integrals. BIT (Sver.), 1974, Vol. 14, No 1, pp 14–21.
209. Dunkin J.W., Eringen A.C.. On the propagation of waves in an electromagnetic elastic solid. J. Engn. Sci., 1963, Vol. 1, pp 461–495.
210. Erdogan F.E., Gupta G.D.. On the numerical solutions of singular integral equations. Quart. Appl. Math, 1972, Vol. 29, No 4, pp 525–534.
211. Erdogan F.E., Gupta G.D., Cook T.S.. The numerical solution of singular integral equations. Mechanics of Fract. Leyden: Noordhoff Int. Publ., 1973, Vol. 1, pp 368–425.
212. Filshtinsky M.L.. Diffraction of acoustoelectric waves by tunnel cavities in an unbounded piezoceramic medium. J. Appl. Mechs., 1998, Vol. 62, No 4, pp 649–655.
213. Filshtinsky M.L., Rodriguez-Ramos R.. Squeezing of composite piezoceramic plate under loading of regular normal pressure. Proc. 6th Annual International Conference on Composite Engineering (ICCE/6), Orlando (USA), 1999, p. 207.
214. Filshtinsky M.L., Rodriguez-Ramos R., Sanchez-Casals O. Fracture mechanics in piezoceramic composite plate. Technical Report of Concordia University, Montreal, 2000, No 4. 21p.
215. Filshtinsky M.L., Shaban K.. The steady-state wave process in a piezoceramic medium with cavities (plane deformation). Proc. 20th International Conference "The boundary element method" (BEM 20), Orlando (USA), 1998, Vol. 4, pp 167–176.

216. Gerasoulis A., Strivastav R.P.. A method for the numerical solution of singular integral equation with a principal value integral. Int. J. Eng. Science, 1981, Vol. 19, No 9, pp 1293–1298.
217. Ioakimidis N.I.. Genaral methods of solving crack problems in the theory of plane elasticity. (Dr. Eng. Thesis), Athens, 1976, 406 p. (in Greek).
218. Ioakimides N.I.. Application of the Gauss and Radou-Laguerre quadrature rules to the numerical solution of the Cauchy type singular equations. Comp. & Struct., 1981, Vol. 14, No 1–2, pp. 63–70.
219. Joshiu S.G., White R.M.. Excitation and detection of surface elastic waves in piezoelectric crystals. J. Acoust. Soc. Amer., 1969, Vol. 46, No 1, Pt. 1, pp 17–27.
220. Kudryavtsev B.A., Parton V.Z., Senik N.A.. Mechanics of piezoelectric materials. Applied Mechanics: Soviet Review. Electromagnetoelasticity. Vol. 2, Hemisphere Publ. Corp., 1990, pp 1–58.
221. Kurlandska Z.T.. A crack problem in elastic dielectric. I. The statement of the problem and diffraction of electro-magnetic wave. Bul. Acad. Polon., Ser. Sci. Techniqyes, 1977, Vol. 25, No 4, pp 109 [225]–117[263].
222. Kurlandska Z.T. Influence of electromagnetic field on crack propagation in elastic dielectric. Bul. Acad. Polon., Ser. Sci. Techniqyes, 1978, Vol. 28, pp 497 [971].
223. Lee E.B., Markus L.. Foundations of optimal control theory. John Wiley and Sons, Inc., New York, London, Sydney, 1967.
224. Loeber J.F., Sih G.C. . Diffraction of anti-plane shear waves by a finite crack. J. Acoust. Soc. Am., 1968, Vol. 40, pp 90–98.
225. Luong W.C., Keer L.M., Achenbach J.D.. Elastodynamic stress intensity factors of a crack near an interphase. Int. J. of Sol. And Struct., 1975, Vol. 11, pp 919–925.
226. Maue A.W. Die Beugung elasticher wellen an der halbebene. ZAMM, 1953, Bd. 33, No 1–2, pp 1–10.
227. Maue A.W. Die Entspannugswelle bei plotzlichem Einschitteines gespannten elastischen Korpers. ZAMM, 1954, Bd. 34., No 1–2, pp 1–12.
228. Maugin G.A.. Continuum mechanics of electromagnetic solids. Amsterdam-New York: North-Holland, 1988, 488p.
229. Maugin G.A., Epstein M., Trimarco C.. Theory of elastic inhomogeneities in electromagnetic materials. Int. J. Eng. Science, 1992, Vol. 30, pp 1441–1449.
230. Moon F.G.. Scattering of wave by a cylindrical piezoelectric inclusion. J. Acoust. Soc. Amer., 1970, Vol. 48, Part 2, pp 253–262.
231. Mindlin R.D.. High frequency vibration of piezoelectric crystal plates. Int. J. of Sol. And Struct., 1972, Vol. 8, No 7, pp 895–906.
232. Mindlin R.D.. Equations of high frequency vibration of thermopiezoelectric crystal plates. Int. J. of Sol. and Struct., Vol. 10, No 6, pp 625–637.
233. Pao Y.H., Mow C.C.. Diffraction of elastic waves and dynamic stress concentrations. 1973, 685 p.
234. Pak Y.E.. Crack extension force in a piezoelectric material. Trans. ASME J. Appl. Mech., 1990, Vol. 57, No 3, pp 647–653.
235. Parton V.Z.. Fracture mechanics for piezoelectric materials.Acta Astronautica, 1976, Vol. 3, No 9–10, pp 671–683.
236. Parton V.Z. Kudryavtsev B.A.. Electromagnetoelasticity. New York: Gordon & Breach, 1998.

237. Sih G.C. (ed.). Elastodynamic crack problems. Leyden: Noordhoff, 1977.
238. Sih G.C., Loeber J.F.. Wave propagation in an elastic solid with a line of discontinuity or finite crack. Quart. Appl. Math, 1969, Vol. 27, No 2, pp 193–213.
239. Shohat J., Tamarkin J.. The Problem of Moments. New York, 1943, 140 p.
240. Sommerfeld A. Elecrodynamics. New York: Academic Press, 1952.
241. Stone S.F., Ghosh M.L., Mal A.K.. Diffraction of anti-plane shear waves by an edge crack. Trans. ASME J. Appl. Mech., 1980, No 2, pp 359–362.
242. Suo Z., Kuo C.M., Barnett D.M., Willis J.R.. Fracture mechanics for piezoelectric ceramics. J. Mech. Phys. Solids, 1992, Vol. 40, No 4, pp 739–765.
243. Theocaris P.S.. Numerical solution of singular integral equations: Methods. J. Eng. Mech. Div. Pros. ASCE, 1981, Vol. 107, No 5, pp 733–752.
244. Theocaris P.S.. Numerical solution of singular integral equations: Applications. J. Eng. Mech. Div. Pros. ASCE, 1981, Vol. 107, No 5, pp 753–770.
245. Theocaris P.S., Ioakimidis N.I. Numerical solution of Cauchy type singular integral equation. Athens: Academy of Athens Publ., 1977, 65 p.
246. Theocaris P.S., Ioakimidis N.I. On the Gauss-Jacobi numerical integration method applied to the solution of singular integral equations. Bull. Calcutta Math. Soc., 1978, Vol. 71, pp 29–43.
247. Theocaris P.S., Ioakimidis N.I. The inclusion problem in plane elasticity. Quart. J. Mech. Appl. Math., 1977, Vol. 30, No 4, pp 437–448.
248. Theocaris P.S., Tsamashyros G.j. Numerical solution of systems of singular integral equations with variable coefficients.
249. Tsamashyros G.J., Theocaris P.S.. Equivalence and convergence of direct and indirect methods for the numerical solution of singular integral equations. Computing, 1981, Vol. 27, No 1, pp 71–80.
250. Voigt W. Lehrbuch der Kristall-Physik. Leipzig: Teubner, 1910.
251. Wolf P. Low frequency diffraction by a hard strip. SIAM J. Appl. Math., 1975, Vol. 29, No 2, pp 273–287.
252. Yeh C. The diffraction of waves by a penetrable ribbon. J. Math. Phys., 1963, Vol. 4, No 1, pp 65–71.
253. Curie J., Curie P. Constactions et dilatations produities par des tensions electriques dans les cristaux hemeedres a faces inclinnees. Comp. rend, 1881, Vol. 93, p. 1137–1142.
254. Bardzokas D.I., Filshtinsky M.L.. Electroelasticity of piecewise-uniform bodies. Sumy, University Book Publ., 2000, 308 p.
255. Bardzokas D.I., Zobnin A.I., Senik N.A., Filshtinsky M.L. Mathematical modeling in the problems of the mechanics of coupled fields. Moscow, Edition URSS, 2005, Vol. 1, 311 p.,Vol 2, 374 p.
256. Bardzokas D.I., Zobnin A.I., Senik N.A., Filshtinsky M.L.. Problems in thermopiezoelectricity with analytical solutions. Moscow, Editor URSS, 2005, 173 p.
257. Bardzokas D.I., Zobnin A.I.. Mathematical modeling of Physical processes in Composite Materials of periodical Structure. Moscow, Editor URSS, 2003, 334 p.
258. Bardzokas D.I., Kudryavtsev B.A., Senik N.A. Wave propagation in electromagnetoelastic media. Moscow, Editor URSS, 2003, 301 p.
259. Bardzokas D.I., Sfyris G.I.. Excitation of shear waves in a piezoceramic medium with partially electrodized tunnel openings strengthened by a rigid stringer. MPE 2005, Vol. 2. pp 231–244.

260. Bardzokas D.I., Sfyris G.I., Sfyris D.I.. Antiplane Mixed Boundary Value Problem of Electroelastric Media, for a Piezoelectric Medium with Defects in th Form of Holes. The 5th National NDT Conference of the Hellenic Society of Non Destructive Testing. Athens, 2005, 33 p.
261. Akhiezer N.I. "Lectures on the approximation theory" Moscow, Nauka, 1965, 407p.
262. Grigoryan E.H. Vibration of the magneto-elastic medium that is excited by a concentrated harmonic force. Izv. Armenian Academy of Sciences Mechanica 1978, Vol. 38, N 5, pp 48–62.
263. Guz A.N., Golouchan V.T. Diffraction of the elastic waves in multiply connected bodies. Kiev, Nauk. Dumka, 1972, 253 p.
264. Kantorovich L.V., Akilov G.P Functional Analysis in normed spaces. Moscow, Figmatgiz, 1959, 684 p.
265. Kats N.S., Krein M.G., R-functions, analytical functions mapped in the upper half-plane on themselves. From the book Atkinson F. "Discrete and continuous boundary value problems", Moscow, Mir, 1968, pp 629–648.
266. Lechnitskii S.G. Anisotropic plates, Moscow, Gostechizdat, 1967, 356 p.
267. Sedov L.I. Plane problems of hydrodynamics and aerodynamics. Moscow, Gostechizdat, 1950, 444p.
268. Filshtinsky V.A. Exponential problem of moments on the axis for a given finite number of intervals in the spectrum. Notes of the mechanical-mathematical School of the Charkov University and of the Mathematical Society of Charkov, 1964, Vol. 30, N 4, pp 186–200.
269. Filshtinsky V.A., Filshtinsky L.A. The control of fracture of cracked bodies. Izv. Acad. Sci. USSR Mechanics of solid bodies, 1988, N 1, pp 158–163.
270. Filshtinsky L.A. Two dimensional static and dynamic problems of elasticity for cracked bodies. Theroy and calculation of the elements of thin-shelled constructions. Izd. Moscow State Uni. 1986 pp 107–117.
271. Filshtinsky L.A. Difraction of elastic waves on cracks, openings and inclusions within an isotropic medium. Izv. Acad. of Sci USSR, Mech. of solid bodies, 1991, N 4, pp 119–127.
272. Filshtinsky L.A. Plane problem of magneto-elasticity for a ferromagnetic medium weakened by cracks. Prikl. Math. and Mech., 1993, Vol. 57, N 5, pp 143–152.
273. Filshtinsky L.A., Kravets E.M. Plane problem of magneto-elasticity for a ferromagnetic medium with holes. Fizika Chim. Mech. Mater., 1995, N 6, pp 38–46.
274. Fishtinsky L.A., Kravets E.M., Hvorost V.A. Stress state of ferromagnetic plates with cracks under the influence of strong magnetic field. Prikl. Mech. and Tech. Fizika, 1993, N 6, pp 147–151.
275. Filshtinsky L.A., Fomenko L.I. Dynamic intensity of a conductive space with cracks under the influence of a strong magnetic field. Prikl. Math. and Mech., 1992, Vol. 56, N 1, pp 143–149.
276. Filshtinsky L.A., Fomenko L.I., Dynamic intensity of a conductive half-space with cracks under the influence of a strong magnetic field. Prikl. Math. and Tech. Fizika, 1991, N 6, pp 16–20.
277. Filshtinsky L.A., Fomenko L.I., Hvorost V.A. Diffraction of a shear wave on inclusions in diamagnetic and paramagnetic media, under the influence of strong magnetic fields. Fiz. Chim. Mech. Mater., 1993, Vol. 29, N 6, pp 82–86.

278. Filshtinsky L.A., Chestalova V.V. Diffraction of a magnetoelastic wave on a hole and a crack (antiplane problem). Fiz. Chim. Mech. Mater., 1992, Vol. 29, N 3, pp 14–18.
279. Filshtinsky L.A., Kravets E.M., Hvorost V.A. Application of the singular integral equations technique in the study of the stationary stress of a ferromagnetic medium with cracks. Proc. of the VII Inter. Congress "Methods of discrete singularities in Mathematical Physics., Charkov, 1997, pp 42–45.
280. Filshtinsky L.A., Kravets E.M., Hvorost V.A. Stationary dynamic problem of magnetoelasticity for the partially-homogenous infinite and electrical conducting medium. Proc. of the VII Inter. Congress "Methods of discrete singularities in Mathematical Physics., Charkov, 1997, pp 191–194.
281. Filshtinsky L.A., Kravets E.M., Hvorost V.A. Stress state of the ferromagnetic plate with cracks under the influence of a strong magnetic field. Prikl. Math. and Tech. Fizika, 1993, N 6, pp 147–151.
282. Filshtinsky M.L., Oleinik V.M., Filshtinsky L.A. Optimal control of the static and dynamic S.I.F for a piezoelectric halfspace. VI Conf. USSR on control of mechanical systems. Lvov, 1988, p.156.
283. Brown W.F. Electric and magnetic forces. Am. J. of Phys., 1951, Vol. 19, pp 290–304, pp 333–350.
284. Erdogan F.E. Approximate solutions of systems of singular integral equations. SIAM J. Appl. Math., 1969, Vol. 17, N 6, pp 1041–1059.
285. Pao Y.H., Yeh C.S. A linear theory for soft feroomagnetic elastic solids. Int. J. Eng. Sci., 1973, N 11, pp 415–436.
286. Shindo Y. Mechanics of electroelastic solid bodies under the influence of a magntic field, 1983, Vol. A49, N 448, pp 1467–1474.
287. Sindo Y. diffraction of antiplane shear waves bya finite crack in the presence of a magnetic field. Z. angew Matg. und Mech., 1976, Vol. 56, N 1, pp 33–41.
288. Kudryavtsev B.A., Parton V.Z. Magnetoelasticity. Itogi Nauki I techniki VINITI, 1981, Vol. 14, 39p.
289. Sea G. Optimization, Theory and Algorithms, Moscow, Mir, 1973, 244p.